AF898961

APPLIED SUPERCONDUCTIVITY, Metallurgy, and Physics of Titanium Alloys

Volume 1
Fundamentals

THE INTERNATIONAL CRYOGENICS MONOGRAPH SERIES

General Editors

K. D. Timmerhaus, *Engineering Research Center University of Colorado, Boulder, Colorado*

Alan F. Clark, *National Bureau of Standards U.S. Department of Commerce, Boulder, Colorado*

Founding Editor

K. Mendelssohn, F.R.S. *(deceased)*

Recent Volumes in this series

M. G. Zabetakis
Safety with Cryogenic Fluids

D. H. Parkinson and B. E. Mulhall
The Generation of High Magnetic Fields

W. E. Keller
Helium-3 and Helium-4

A. J. Croft
Cryogenic Laboratory Equipment

A. U. Smith
Current Trends in Cryobiology

C. A. Bailey
Advanced Cryogenics

D. A. Wigley
Mechanical Properties of Materials at Low Temperatures

C. M. Hurd
The Hall Effect in Metals and Alloys

E. M. Savitskii, V. V. Baron, Yu. V. Efimov, M. I. Bychkova, and L. F. Myzenkova
Superconducting Materials

W. Frost
Heat Transfer at Low Temperature

I. Dietrich
Superconducting Electron-Optic Devices

V. A. Al'tov, V. B. Zenkevich, M. G. Kremlev, and V. V. Sychev
Stabilization of Superconducting Magnetic Systems

G. Walker
Cryocoolers, Part 1: Fundamentals
Cryocoolers, Part 2: Applications

E. W. Collings
Applied Superconductivity, Metallurgy, and Physics of Titanium Alloys
Volume 1: Fundamentals
Volume 2: Applications

APPLIED SUPERCONDUCTIVITY, Metallurgy, and Physics of Titanium Alloys

Volume 1
Fundamentals

Alloy Superconductors: Their Metallurgical, Physical, and Magnetic-Mixed-State Properties

E. W. Collings

Senior Research Scientist
Battelle Memorial Institute
Columbus, Ohio

PLENUM PRESS • NEW YORK AND LONDON

Library of Congress Cataloging in Publication Data

Collings, E. W.
Applied superconductivity, metallurgy, and physics of titanium alloys.

(The International cryogenics monograph series)
Includes bibliographies and index.
Contents: v. 1. Fundamentals.
1. Titanium alloys—Electric properties. 2. Titanium alloys—Metallurgy. 3. Superconductivity. I. Title. II. Series.
TN693.T5C625 1985 673′.7322 85-12063
ISBN-13: 978-1-4612-9241-8 e-ISBN-13: 978-1-4613-2095-1
DOI: 10.1007/978-1-4613-2095-1

© 1986 Plenum Press, New York
Softcover reprint of the hardcover 1st edition 1986
A Division of Plenum Publishing Corporation
233 Spring Street, New York, N.Y. 10013

All rights reserved

No part of this book may be reproduced, stored in a retrieval system, or transmitted in any form or by any means, electronic, mechanical, photocopying, microfilming, recording, or otherwise, without written permission from the Publisher

To

BETTY

JANE

AND

CHRISTOPHER

also to an international community of low-temperature physicists and cryogenic engineers whose work is represented on these pages

Type-2 Superconductor in the Critical Mixed State. →

Carbon-replicated electron micrograph of a ferromagnetic-particle-decorated representation of the arrangement of axially directed flux lines in an annealed-and-deformed Pb-In(6.3 at.%) 4-mm ø rod in the trapping critical state (see, e.g., Figure 16-7). The arrangement of the flux lines was found to change continuously from long-range-ordered in the relatively high field (~ 70 G) interior region of the sample (≥ 400 μm deep) [left-hand plate], to amorphous in the low-residual-field region just below the surface (≤ 100 μm deep) [right-hand plate]. In this material, for which $\kappa = 2$ at 1.2 K, the flux-lattice parameter at 70 G is about 0.6 μm. Original micrograph courtesy of U. Essmann, Max-Planck-Institute für Metallforschung, Institut für Physik, Stuttgart, West Germany. (*Reference:* H. Träuble and U. Essmann, *J. Appl. Phys.* **39**, 4052–4059 (1968)).

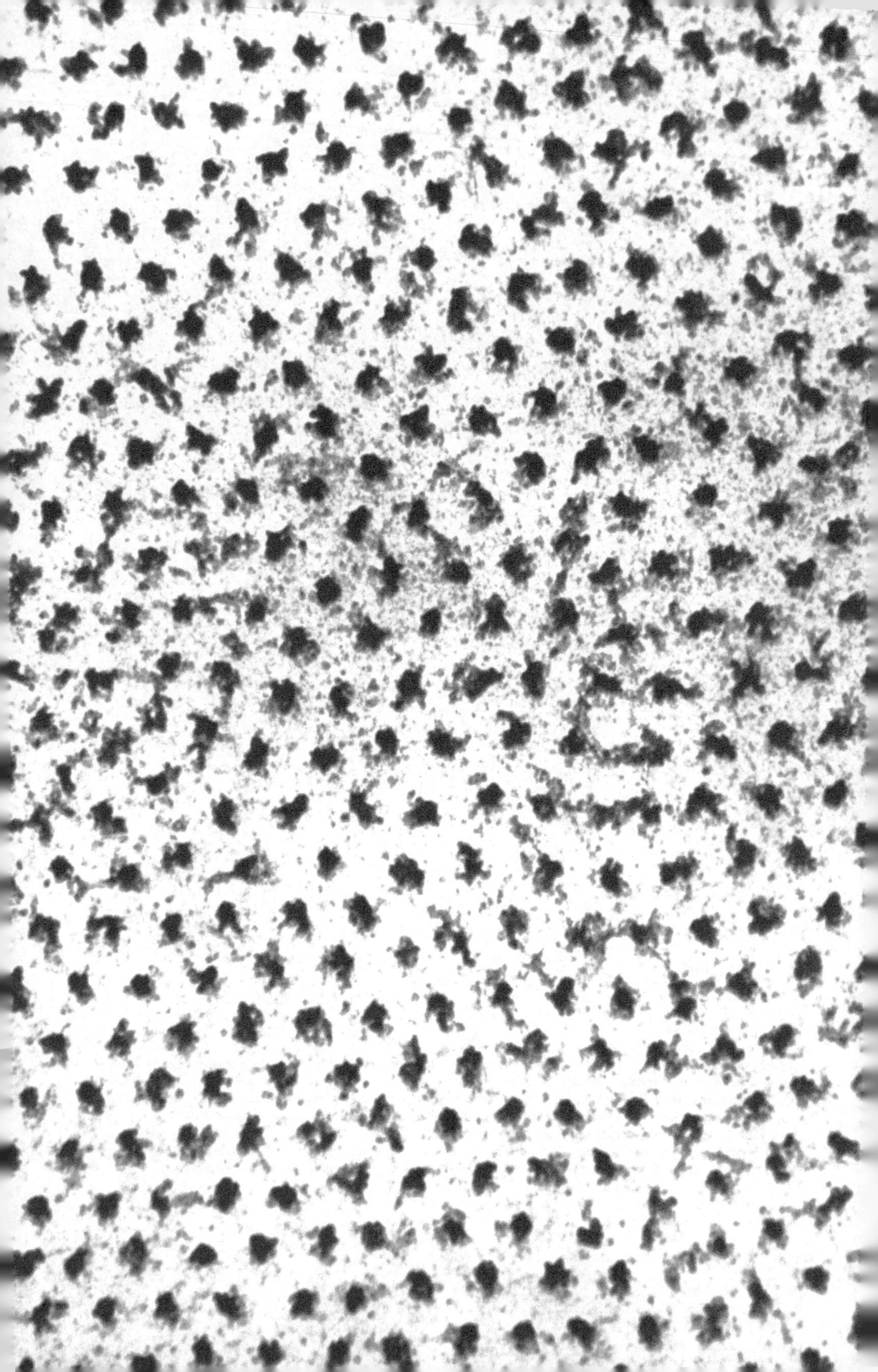

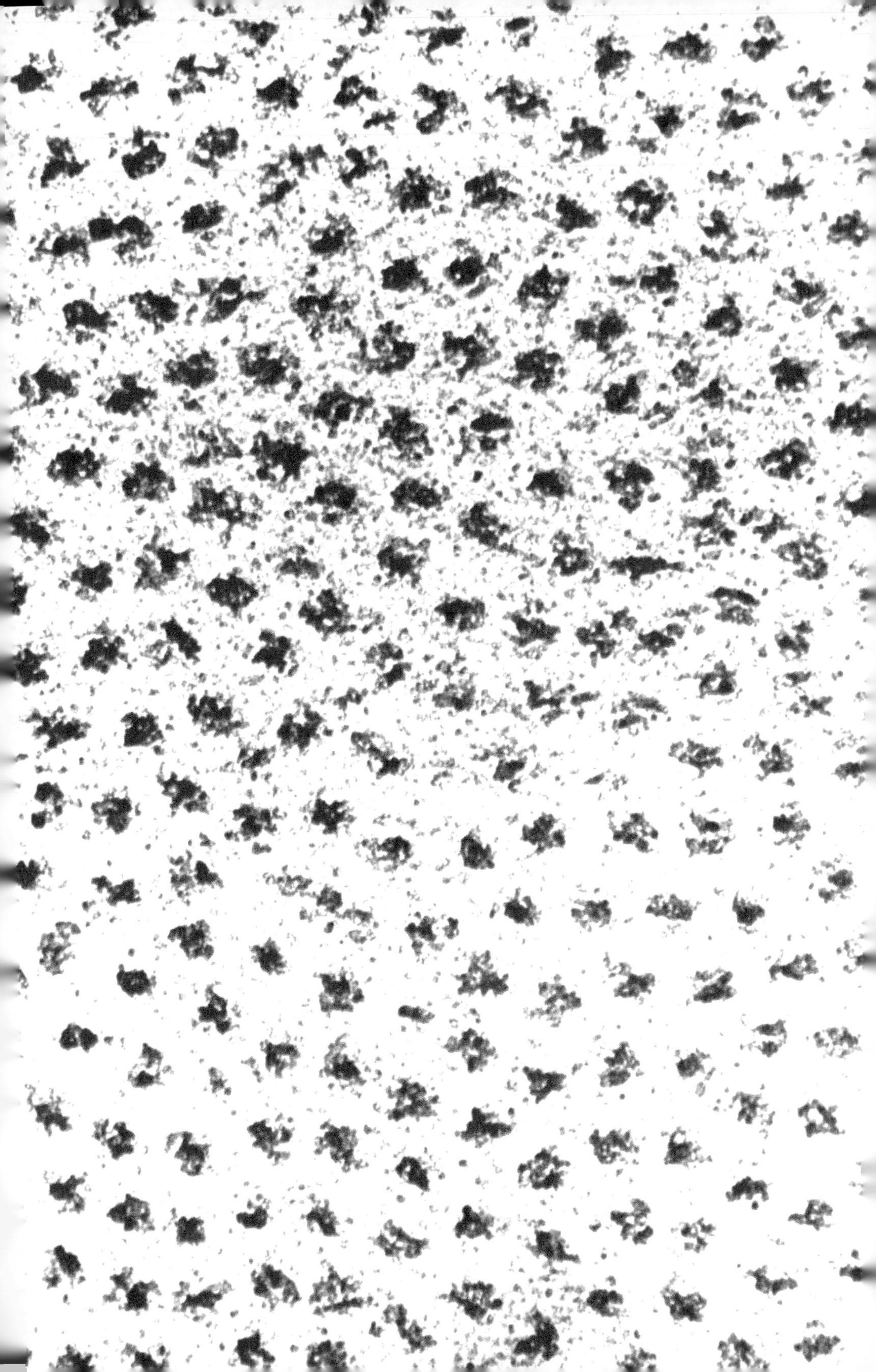

General Preface

Scope and Purpose

Although conductors based on the A15 intermetallic compound Nb_3Sn possess desirable high-field superconducting properties, manufacturing and handling difficulties, coupled with the tendency of their critical current densities to degrade rapidly under stress, have generally restricted their use to fairly straightforward, usually small-scale solenoidal-magnet applications. Likewise the A15 compound V_3Ga, which has a wider critical strain window than Nb_3Sn but a uniformly lower upper critical field, has not entered widespread service. Strain has been found to have no measurable influence on either the critical fields or the critical current densities of compound superconductors with B1 and C15 crystal structures, but as yet they are still in the research and development stages. On the other hand, conductors using the binary alloy Ti–Nb or multicomponent alloys based on it, because of their relative ease of manufacture, excellent mechanical properties, and relatively low strain sensitivities, are now being pressed into service in numerous large-scale devices. Such conductors are being wound into magnets for use in energy storage, energy conversion (i.e., generators and motors), and high-energy particle detectors and beam-handling magnets.

The use of cold-rolled or drawn Ti–Nb-alloy wire for superconducting magnet applications was first proposed in 1961. During the ensuing ten years, while progress was being made in the development of Cu-clad filamentary-Ti–Nb-alloy conductors, Ti–Nb and other Ti-base binary transition-metal (TM) alloys were being employed as model systems in the fundamental study of type-II superconductivity. The next decade saw the development of improved current-carrying capacity in the binary Ti–Nb alloy along with its incorporation into multifilamentary monoliths, cables, and braids capable of carrying tens of thousands of amperes. Also, there were improvements to both the economic and technical attractiveness of the binary alloy through the incorporation into it of ternary and quaternary TM additions. Superconductivity in Ti–TM alloys in general, and Ti–Nb base ones in particular, and the roles that those alloys played in the development of type-II superconductivity, are the subjects of a companion work: *A Sourcebook of Titanium Alloy Superconductivity*, E. W. Collings, Plenum Press, 1983, which is a materials-addressable text that refers to this work for fundamental support.

This two-volume text has several purposes: (i) It is intended to be a discussion of the metallurgical, physical, and superconductive properties of Ti-base alloys. (ii) It attempts to discuss the fundamental bases of applied superconductivity in general—transition temperature, proximity effect, the mixed state and its lower and upper bounds, and flux pinning—using Ti–TM alloys as systems for case study. (iii) It considers the design and fabrication of composite conductors, again with reference to alloy superconductors but in such a way that many of the results are generally applicable. In this respect it plays the role of a conductor-design handbook. (iv) Finally under the group heading "Conductor Properties," it deals with two very important topics—stress effects and radiation effects—with, however, specific reference to Cu/Ti–Nb composites.

The work is divided, for conceptual and practical reasons, into two volumes: Volume 1, *Fundamentals*, covers the metallurgy and physics of Ti–TM alloys, the superconducting transition, and the mixed state. The effects of transport current are not introduced in Volume 1. Its final chapter is transitional, however, in that it describes how appropriate measurements of magnetic-mixed-state properties can lead to determinations of the critical current density. Volume 2, *Applications*, focusses attention on composite conductors carrying electrical transport currents. Its first group of chapters describes the interaction of such currents with the Abrikosov flux lattice—strongly pinned at first, and then in motion when the dislodgement forces become sufficiently strong. The second group of chapters deals with all aspects of design and fabrication. Finally, as indicated above, the volume concludes with discussions of the responses of composite conductors to two important influences—stress and high-energy radiation.

It is hoped that *Applied Superconductivity* will appeal to: (i) Teachers and students of physics, physical metallurgy, and electrical engineering, to whom the following topics are addressed: aspects of the physical metallurgy of β-Ti alloys and their low-temperature properties; low-temperature metal physics of Ti-base and other technical alloys; the electrical and magnetic properties of superconductors; the historical development of theories of the mixed state in type-II superconductors and of practical hard type-II superconductivity, traced with reference to Ti-base alloys. (ii) Research and development groups assigned the tasks of alloy design and superconductor property optimization. (iii) Superconductor and superconducting magnet design engineers for whom the chapters on Stability, AC Loss, and Conductor Design are intended. (iv) Manufacturers of "NbTi" alloy superconductors who wish to become reapprised of the Ti-alloy metallurgy and metal physics that underlies the behavior and properties of their product.

Administrative Credits and Genesis

The idea of preparing a review of the literature of Ti-alloy superconductors arose in November 1976 during conversations with Professor U. Zwicker, Universität Erlangen-Nürnberg, who had earlier prepared a comprehensive treatise on the metallurgical properties of Ti and its alloys. In the following year a search for financial support took place, as a result of which joint funding was obtained from Battelle Memorial Institute's Corporate Technical Development (CTD) office and from the then Energy Research and Development Administration's (ERDA) Division of Magnetic Fusion Energy. The combined backing of F. G. Dawson (CTD, Battelle) and E. Ziurys (ERDA) gave the project its initial impetus. As the work proceeded, ERDA became the Department of Energy (DOE) and the host division became the Office of Fusion Energy; officers within this organization who continued to lend their support to the project were G. M. Haas, E. N. C. Dalder, and subsequently D. S. Beard. In Battelle continued work on the review was strongly advocated by L. A. Rancitelli during his association with CTD, while completion of this book and its companion volume have been made possible by F. J. Milford, Associate Director for Research, and by D. E. Niesz, Manager of the Materials Department, who has given generous and whole-hearted practical support to the project in all of its phases.

Although a preliminary literature collection had been on hand for several years prior to 1977, its updating and expansion began in earnest at the beginning of 1978. A comprehensive bibliography was prepared in January 1978 by the Cryogenic Data Center of the Institute for Basic Standards, National Bureau of Standards, Boulder, Colorado, under the direction of N. A. Olien. The result of this computer survey conducted by the Center, which is now defunct, unfortunately, provided a core of documents to which numerous others were added as writing proceeded. Most of the literature identified by the search was already available in Battelle's Main and Foreign-Science Libraries. In the latter, the Soviet literature was sought and copied by K. Cottrill. In the Main Library, many hundreds of articles were acquired and copied under the direction of L. S. Vahey, to whom another debt of gratitude is owed. Literature to be collected consisted of: (i) regular journal articles, (ii) review articles, (iii) symposia proceedings,

(iv) government reports, (v) Ph. D. and other theses, (vi) books, and (vii) patent documents. All the interlibrary arrangements were conducted by L. S. Vahey. Reports were obtained through our Report Library by D. L. E. Welch and L. A. Masters; patents were acquired from or through our Patent Library by S. L. Wakefield.

At the very beginning a scheme of chapters, sections, and subsections was intuitively developed. The task of collecting, scanning, and sorting the literature into these categories was completed by the end of January 1979, after which the writing commenced. One of the chapters was to deal with "materials," and all the rest with "properties" (i.e., superconductive, metallurgical, and physical ones). Writing commenced with the "materials" manuscript which was completed in first-draft form by the end of August 1979 and set to one side. Work then began on the "properties" manuscripts. It had already become apparent that "materials" was far too large a subject to be condensed into a single chapter. Accordingly it was decided, on the advice of A. F. Clark, National Bureau of Standards, Boulder, Colorado, that a pair of conjoint volumes should be prepared—the "materials" manuscript was developed into *A Sourcebook of Titanium Alloy Superconductivity*, which was published by Plenum Press in May, 1983; the "properties" manuscript was to join Plenum's Cryogenic Monograph Series as the present pair of volumes short-titled *Applied Superconductivity*. The now-published Sourcebook is thus a materials-oriented companion to the latter and refers to it frequently for fundamental support. By the beginning of 1982 the first complete and fully revised version of *Applied Superconductivity* was ready for critical review and began to be sent out chapter by chapter. The reviewers are acknowledged below in the prefaces to the individual volumes. Towards the latter part of that year, with the reviewers' comments in hand, final revision commenced and continued well into 1983; at the same time the results of all relevant literature that came to my notice during this period were blended into the text. Although the complete manuscripts were submitted for publication in December 1983, the results of important new work that came to fruition during 1984 were able to be introduced into the page proofs. Thus in areas that have undergone recent advances, this pair of volumes is current as of late 1984.

Technical and Scientific Acknowledgments

During the wriring of this review and its companion volume, numerous technologists and scientists have submitted new information, discussed problems, helped to solve difficulties, or contributed samples or photographs. They are, in alphabetical order (and with the general topic of discussion and/or other type of contribution given in parentheses): M. A.

Abdou, Argonne National Laboratory, Argonne, Illinois, (neutron-irradiation experimentation with particular reference to tokamak reactors; Figure 30-3); E. Adam, Airco Superconductors, Carteret, New Jersey, (conventional and compacted monolith and cable fabrication; Figures 28-23 and 28-42); A. C. Anderson, University of Illinois, Urbana-Champaign, Illinois, (SQUID *versus* capacitance low-temperature dilatometry); E. Arnold, Wardwell Braiding Machine Co., Central Falls, Rhode Island, (construction and operation of rotary braiders; Figure 28-48); C. R. Barnes, Battelle Memorial Institute, Columbus, Ohio, (SEM micrography, including Figure 28-42); M. R. Beasley, Stanford University, Stanford, California, (spin-orbit coupling in mixed-state theory); L. H. Belz, Kawecki-Berylco Industries, Reading, Pennsylvania, (special method of comelting Ti and Nb in Ti–Nb ingot manufacture); A Blackmore, Ceeco Manufacturing Ltd., Maple, Ontario, Canada, (construction and operation of cabling (stranding) machines; Figure 28-33(a)); R. L. Cappelletti, Ohio University, Athens, Ohio, (strong-coupling theory and the low temperature specific heat of superconductors); W. J. Carr, Jr., Westinghouse Research and Development Center, Pittsburgh, Pennsylvania, (AC loss in composite conductors; Figure 26-12); A. L. Clark, National Bureau of Standards, Boulder, Colorado, (thermal expansion of superconductors and magnet structural materials); R. R. Coltman, Jr., Oak Ridge National Laboratory, Oak Ridge, Tennessee, (neutron irradiation damage in tokamak materials; information on the BSR/LTIF); D. N. Cornish, Lawrence Livermore National Laboratory, Livermore, California, (the MFTF conductor and its stability testing); C. W. Curtis, Teledyne Wah Chang Albany, Albany, Oregon, (superconducting alloy manufacturing processes; Figure 28-2 and its caption, Figure 28-10); J. Dennehy, New England Butt Company, Providence, Rhode Island, (construction and operation of maypole braiders; Figure 28-50); T. A. de Winter, Magnetic Corporation of America, Waltham, Massachusetts, (composite superconductor manufacture and properties—Figures 28-13(a) and (b); sample of the Fermilab cable for Figure 27-11); L. Dresner, Oak Ridge National Laboratory, Oak Ridge, Tennessee, (stabilities of stable and metastable composite conductors); D. S. Easton, Oak Ridge National Laboratory, Oak Ridge, Tennessee, (composite conductors and the effects of stress on them; part of Figure 3-25, Figure 3-26, Figure 3-27(a), and Figure 28-22); J. W. Ekin, National Bureau of Standards, Boulder, Colorado, (stability and stress effects in composite conductors and technical guidance for the writing of Chapter 25; Figure 29-6, a previously unpublished $J_c - H_{c2} - \varepsilon_c$ diagram for Ti–Nb); U. Essmann, Max-Planck-Institüt für Metallforschung, Stuttgart, Germany, (micrograph of an iron-decorated fluxoid lattice); F. R. Fickett, National Bureau of Standards, Boulder, Colorado, (critical current density measurement standardization; magnetoresistances of Al and Cu); D. K. Finnemore, Iowa State University, Ames, Iowa, (superconductive proximity effect in Pb–Cd lamellar *in situ* composites); R. J. Fiorentino, Battelle Memorial Institute, Columbus, Ohio, (extrusion metalworking and, in particular, hydrostatic extrusion; Figures 28-12 and 28-15); E. S. Fisher, Materials Technology Corporation, Sarasota, Florida, late of Argonne National Laboratory, (elastic constants, particularly of Ti–TM alloys, and some stress effects in composite conductors); T. H. K. Frederking, University of California, Los Angeles, California, (thermal conductivity temperature dependence of a Ti–Nb alloy); M. Garber, Brookhaven National Laboratory, Upton, New York, (performance and development of the ISAbelle braid and a sample of it for Figure 27-14; J_c measurement of the CBA conductor and unpublished data for Figure 23-6); M. A. Green, Lawrence Berkeley Laboratory, Berkeley, California, (technical advice prior to the writing of Chapter 25); K. L. Hammond (dec.), Battelle Memorial Institute, Columbus, Ohio, (sectional optical micrography of the Fermilab cable—Fig. 27-11); K. T. Hartwig, University of Wisconsin, Madison, Wisconsin, (information and photographs of the Wisconsin-SMES and other Al-stabilized conductors—Figures 28-25 and 28-24, respectively); E. Helfand, Bell Telephone Laboratories, Murray Hill, New Jersey, (special reduced-field and spin-orbit-coupling formulations); K. Hemachalem, Intermagnetics General Corporation, Guilderland, New York, (optimization of commercial composite superconductors, including alloy composition and the use of multiple heat treatment); H. Hillmann, Vacuumschmelze GmbH, Hanau, Germany, (superconductor processing in general, drawing texture in superconducting filaments, flux pinning and its anisotropy under flattening; part of Figure 3-25, Figures 3-27(b), 27-6, 27-8, 28-18(a) and (b), 28-45, and 28-47); E. Hodge, Lawrence Livermore National Laboratory, Livermore, California, (stress effects in the MFTF magnet); P. C. Hohenberg, Bell Telephone Laboratories, Murray Hill, New Jersey, (special reduced field and spin-orbit-coupling formulations as used in the "WHH" series of papers); T. Horiuchi, Kobe Steel, Ltd., Kobe, Japan, (superconducting composites and processing machinery—Figure 28-19(c); samples of composite conductors); Y. Iwasa, Francis Bitter National Magnet Laboratory, Cambridge, Massachusetts, (technical advice prior to the writing of Chapter 25; minimum propagating zones, a special superconductor joining technique); M. A. Janocko, Westinghouse Research and Development Center, Pittsburgh, Pennsylvania, (photographs and micrographs related to experimental lattice braiding—Figures 27-15 and 28-51(a)); S. H. Kim, Argonne National Laboratory, Argonne, Illinois, (implications of the results of the stress-effect experiments of Fisher and Kim); T. A. Kozman, Lawrence Livermore National Laboratory, Livermore, California, (the MFTF-B, its superconductor and constructional details; Figure 30-4); C. C. Koch, Oak Ridge National Laboratory, Oak Ridge, Tennessee, (special J_c-related precipitation effects in Ti–Nb alloys); E. J. Kramer, Cornell University, Ithaca,

New York, (flux-pinning mechanisms and models); D. Larbalestier, University of Wisconsin, Madison, Wisconsin, (advances in J_c and flux pinning; Figure 21-25); M. A. R. LeBlanc, University of Ottawa, Ottawa, Ontario, Canada, (verbal and written discussions of "longitudinal," i.e., force-free, nearly force-free, and helical current flow); H. M. Ledbetter, National Bureau of Standards, Boulder, Colorado, (elastic constants in general; data for the elastic constants of Ti–Nb prior to publication); T. S. Luhman, Boeing Aerospace Company, Seattle, Washington, late of Brookhaven National Laboratory, (metallurgical improvements to the ISAbelle braid); W. K. McDonald, Teledyne Wah Chang Albany, Albany, Oregon, (superconducting alloy manufacturing processes; the ultimate consequences of coring in Ti–Nb ingots); V. D. McGinniss, Battelle Memorial Institute, Columbus, Ohio, (radiation damage to organic polymers); M. Nagata, Sumitomo Electric Industries, Ltd., Osaka, Japan, (structural and electrical details of numerous commercial superconducting composites—Figures 3-27(c), 27-3, 27-9, 28-13(c), (d), and (e), 28-21, 28-26(b), 28-27, 28-28, 28-29(b) and (c), and 28-46); H. Nakamoto, Japan Atomic Energy Research Institute, Ibaraki-ken, Japan, (permission to quote from JAERI Report M 8785 for use in Figures 21-43 and 21-44); D. F. Neal, IMI Titanium, Birmingham, UK, (TEM micrographs—Figures 2-11(a) and (b)); M. Nishi, Japan Atomic Energy Research Institute, Ibaraki-ken, Japan, (pictorial details of the cross section and surface of the 8-T conductor for the Japan Large Coil Task—Figures 27-2(a) and (b)); B. Obst, Kernforschungszentrum Karlsruhe, Karlsruhe, Germany, (temperature- and concentration-dependence of the electrical resistivity of Ti–Nb); E. K. Opperman, Hanford Engineering Development Laboratory, Richland, Washington, (information on the Fusion Materials Irradiation Test Facility, FMIT); T. P. Orlando, Francis Bitter National Magnet Laboratory, Cambridge, Massachusetts, late of Stanford University, (a corrected set of equations describing the properties of the mixed state, and permission to reproduce them—Table 14-1); D. H. Polonis, University of Washington, Seattle, Washington, (superconducting properties of numerous Ti-base alloys; the ω-phase reversion effect); S. J. Poon, University of Virginia, Charlottesville, Virginia, (the transition temperature of unalloyed β-Ti; the upper critical field and spin-orbit-scattering effects); R. Pott, Universität zu Köln, Köln, Germany, (part of Figure 9-12 and Figure 9-13); D. T. Read, National Bureau of Standards, Boulder, Colorado, (Figures 3-28 and 12-5); R. Remsbottom, Fermi National Accelerator Laboratory, Batavia, Illinois, (features and properties of the "Fermilab cable" and several use-permissions relating to it); R. W. Rollins, Ohio University, Athens, Ohio, (the relationship between induction and applied field strength in critical state theory; the roles of the "equilibrium" and "dynamic" permeability factors, especially in flux-profile measurement); S. L. Sass, Cornell University, Ithaca, New York, (sets of

(100) and (112) electron diffractograms for Ti–Nb alloys from which a selection was made for use in Figure 1-22(a)); R. M. Scanlan, Lawrence Livermore National Laboratory, Livermore, California, (the MFTF conductor—its strain during cycling and its shielding in the mirror-fusion machine); C. Schmidt, Kernforschungszentrum Karlsruhe, Karlsruhe, Germany, (acoustic emission from Ti–Nb during strain cycling—Figure 3-12); J. D. Scudiere, Intermagnetics General Corporation, Guilderland, New York, (information on commercial superconducting composites and their manufacture—Figures 27-10, 27-11, and 28-19(a)); H. R. Segal, IBM Instruments, Inc., Stanbury, Connecticut, late of Holec Draad, Netherlands, and Magnetic Corporation of America, (critical current measurement, in particular the transition criteria used in the measurement of various Ti–Nb-base alloy systems; comments on small-coil testing); M. G. Shields, Teledyne Wah Chang Albany, Albany, Oregon, (photographs of superconductor processing—Figures 28-2 and 28-19(b)); S. Shimamoto, Japan Atomic Energy Research Institute, Ibaraki-ken, Japan, (SEM micrographs of Ti–Nb filaments—Figures 28-18(c) and (d)); E. G. Smith, Battelle Memorial Institute, Columbus, Ohio, (extrusion metalworking and in particular, hydrostatic extrusion; Figures 28-12 and 28-15); R. D. Smith, Battelle Memorial Institute, Columbus, Ohio, (optical metallography of Ti–TM alloys and superconducting strands—Figures 1-13, 1-18, 1-19, 10-14, 11-15(b), (c), and (d), 27-1(a), and 27-3); E. Springer, Vacuumschmelze GmbH, Hanau, Germany, (information on the properties of commercial conductors and cables—in association with H. Hillmann); S. J. St. Lorant, Stanford Linear Accelerator, Stanford, California, (applications of superconductivity in high energy physics, high energy radiation damage in organic insulators); E. Tada, Japan Atomic Energy Research Institute, Ibaraki-ken, Japan, (determination of H_{c2} from J_c *versus* applied field results for some commercial superconductors); Y. Tanaka, Furukawa Electric Company, Ltd., Tokyo, Japan, (structural and electrical details of numerous commercial superconducting composites; Figures 28-29(a), 28-30, 28-31, 28-32, 28-39, 28-40, and 28-43); G. L. Taylor, Airco Superconductors, Carteret, New Jersey, (processing and properties of commercial superconducting composites, and in association with E. Adam, Figures 27-1(b), 27-14, and 28-41); B. I. Verkin, Institute for Low Temperature Physics and Engineering, Kharkov, USSR, (low temperature mechanical properties of Ti alloys—Figure 3-21; low temperature specific heat data for Ti–Nb—part of Figure 8-4); H. Wada, National Research Institute for Metals, Ibaraki, Japan, (the properties of various Ti-base alloy systems and information concerning the measurement of their H_{c2}s and J_cs); M. S. Walker, Intermagnetics General Corporation, Guilderland, New York, late of Westinghouse Research and Development Center, (AC loss in composite conductors); A. W. West, University of Wisconsin, Madison, Wisconsin, (the composition limit for

α-phase precipitation in Ti–Nb; TEM and STEM micrography and EDAX examination—Figures 1-8, 21-20, and 21-21); W. Westerholt, Ruhr-Universität Bochum, Bochum, Germany, (part of Figure 9-12); W. J. Weider, late of Battelle Memorial Institute, Columbus, Ohio, (macrophotography of cable and braid—Figures 27-11 and 27-14); G. K. White, CSIRO Division of Applied Physics, Lindfield, Australia, (thermal expansion and its measurement); J. J. White III, Battelle Memorial Institute, Columbus, Ohio, (influence of deformation induced microstructural changes on calorimetrically observed superconducting transitions in Ti–TM alloys; application of the NONLIN-4 code to superconductive transition rounding); J. Willbrand, Krupp Forschungsinstitut, Essen, Germany, (thermomechanical processing of composite conductors; TEM micrography of Ti–Nb alloys—Figure 21-17); J. C. Williams, Carnegie-Mellon University, Pittsburgh, Pennsylvania, (physical metallurgy of Ti alloys, especially electron microscopy—Figures 1-15, 1-16, 1-17, 1-22(b), 2-2, 2-3, and 2-5); S. L. Wipf, Los Alamos National Laboratory, Los Alamos, New Mexico, (technical guidance for the writing of Chapter 25, discussions of flux-jump instability and cryostability); J. J. Wollan, Los Alamos National Laboratory, Los Alamos, New Mexico, (basic strand and conductor design details—Figures 28-26(a) and 28-44); V. E. Wood, Battelle Memorial Institute, Columbus, Ohio, (development of a billet-packing analysis for hexagonal elements—Section 28.3.2; verification of many of the equations and numerical analyses presented in these volumes); U. Zwicker, Universität Erlangen-Nürnberg, Erlangen, Germany, (precipitation and superconducting properties of Ti-base alloys; optical metallography—Figure 2-10).

Manuscript Preparation

The initial MS was drafted from hand-written copy and recorded dictation by C. L. Owens who also typed and retyped most of the tables and the first draft of the figure captions. The final revised text and figure captions were typed by J. Bulford with the assistance of A. L. Maynard, who also made all the final corrections and gave the finishing touches to the tables. The 550-odd individual diagrams which constitute the line drawings of these volumes were exquisitely inked by J. S. Ward, technical illustrator, who went on to perform the corrections and final checking. C. L. Conrad composed the lettering for the photographs and the drawings. The latter were photographically reduced to size by J. S. Bucher. B. W. Gonser provided valuable assistance in organizing the references and subsequently in checking one of the final lists. The references for Volume 1 were arranged and typed by C. L. Roseberry; those for Volume 2 were arranged and typed by P. S. Staubs, C. L. Roseberry, and B. E. Caminiti. Of course, numerous other members of Battelle's Report and Photographic units also contributed their services in one way or another to the preparation of the final manuscript.

It has taken five years to complete *Sourcebook* and *Applied Superconductivity.* Throughout this long period of writing, which has voraciously encroached not only into evenings and weekends but also into vacation time, my wife Betty Collings has provided constant support and encouragement. As a practicing sculptor she has been able to appreciate the sacrifices demanded by intense creative labor of which book-writing is just one example.

Preface to Volume 1

Alloy superconductors based on Ti–Nb have now come of age. A complete understanding of their behavior as current-carrying superconductors requires, as basis, an understanding of their fundamental metallurgical, physical, and mixed-state properties. The purpose of the first volume of *Applied Superconductivity* is to satisfy that requirement.

Chapter 1 discusses the *Equilibrium and Nonequilibrium* microstructural states of Ti-base binary transition-metal alloys and the compositional ranges of the metastable-ω and equilibrium-α precipitating phases which have played important roles in flux pinning. *Deformation* metalworking and subsequent heat treatment, or *Aging*, play essential roles in superconductor processing; these topics are considered in Chapter 2. Of particular interest in Chapter 3, *Mechanical Properties* are the anomalous responses of Ti–Nb alloys to tensile and compressive stresses and the interrelationships between tensile strength and hardness. The tensile properties of both bare Ti–Nb and Cu/Ti–Nb composites are also considered in that chapter.

Chapter 4, which deals with the physics and dynamic measurement of the *Elastic Modulus*, a quantity considered phenomenologically and statically in Chapter 3, forms a bridge between the "metallurgical" and "physical" groups of chapters. The results of *Electrical Resistivity* and *Low-Temperature Specific Heat* measurements enter into expressions for the upper critical field; information on *Thermal Conductivity* and *Thermal Expansion* are needed in the design of most devices intended for operation at low temperatures; the *Magnetic Susceptibility* of alloy superconductors is positive (paramagnetic) in the normal state, a property that influences their attainable upper critical fields. Accordingly, all of these topics are considered in the Physical Properties subdivision of this volume—Chapters 4 to 9.

The four variables defining the state of a practical current-carrying superconductor are: temperature, applied field strength, transport current density, and longitudinal strain. Critical values of the first two of these are considered in this volume, the latter two in Volume 2. The Superconducting Transition is the general subject of Chapters 10, 11, and 12. Of all possible measurements of the superconducting transition temperature, calorimetric ones provide the most comprehensive body of information. Of special interest in Chapter 10, which deals with *Calorimetric Studies of the Transition*, is the analysis of deformation-induced superconductive-transition rounding. *Proximity Effect* is a very important phenomenon; Chapter 11 emphasizes this and shows how an otherwise normal material can be drawn into the superconducting state under the influence of an adjacent superconductor. This has particular significance for alloy superconductors when α-phase precipitates are expected to provide a large fraction of the flux pinning. Chapter 12 deals with the *Superconducting Transition Temperature* itself, both in theory and in practice.

The most interesting and important properties of superconductors concern their responses to applied magnetic fields—these are discussed in Chapters 13 through 18. When a sufficiently strong magnetic field is applied to a type-II superconductor it enters a "mixed state" characterized by the presence of a long-range-ordered lattice of microscopic normal zones. The fundamental *Magnetic Properties* are considered in Chapter 13, a tutorial chapter. In continuing further, it is important to distinguish between the mixed state of an idealized superconductor with zero normal-state magnetic susceptibility and that of a strongly paramagnetic one. Most treatments confuse these situations; accordingly, for heuristic and practical reasons these two subclasses are accorded separate chapters—the idealized *Mixed State* is Chapter 14, and the *Paramagnetic Mixed State* is Chapter 15. The flux lattice may be either unpinned (ideal or reversible) or pinned. The upper bound of the mixed state in a pinned (hard or irreversible) superconductor, corresponding to the threshold for breakaway of the flux lattice, is the *Critical State*—Chapter 16. It is, in part, defined by a *Critical Field*, a quantity that is considered in theory and in practice in Chapter 17. The strongly pinned critical mixed state breaks down by way of flux creep and flux jumping, the unpinned or weakly pinned state by flux flow. These topics are considered in Chapter 18, which emphasizes two important features of *Flux in Motion*: (a) its diffusional aspects, and (b) its transport-property aspects, with the fluxoids as the carriers. The connection between *Magnetization and Critical Current Density* in the pinned mixed state is the subject of Chapter 19. That chapter, an appropriate termination to Volume 1, also forms a link with Volume 2, which considers conventional transport-current or potentiometric methods of critical-current-density measurement and, indeed, interactions in general between transport current and the pinned mixed state.

Prior to the final revision of this volume, each of the chapters was subjected to critical review by one or more recognized authorities on its topic. I do hasten to add, however, that although the reviewers were invited to consider not only editorial aspects but, more importantly, the accuracy of the text, formulae, and equations, and their comments were in most cases scrupulously honored, the final responsibility for these matters rests solely on my shoulders. With this caveat, it is indeed a pleasure and a privilege to acknowledge the invaluable assistance of: T. S. Luhman, Boeing Aerospace Co., Seattle, Washington, late of Brookhaven National Laboratory, (Chapters 1 and 3); R. M. Rose, Massachusetts Institute of Technology, Cambridge, Massachusetts, (Chapter 2); E. S. Fisher, Materials Technology Corporation, Sarasota, Florida, late of Argonne National Laboratory, (Chapter 4); H. M. Ledbetter, National Bureau of Standards, Boulder, Colorado, (Chapter 4); K. Schroder, Syracuse University, Syracuse, New York, (Chapter 5); P. G. Klemens, University of Connecticut, Storrs, Connecticut, (Chapter 6); L. N. Mulay, Pennsylvania State University, University Park, Pennsylvania, (Chapter 7); R. L. Cappelletti, Ohio University, Athens, Ohio, (Chapters 8 and 10); J. D. Livingston, General Electric Research Laboratory, Schenectady, New York, (Chapter 9—first version); J. A. Rayne, Carnegie-Mellon University, Pittsburgh, Pennsylvania, (Chapter 9—first version); G. K. White and J. G. Collins, CSIRO Division of Applied Physics, Lindfield, Australia, (Chapter 9—present version); I. Bannerjee, Argonne National Laboratory, Argonne, Illinois, (Chapter 11); P. B. Allen, State University of New York, Stony Brook, New York, (Chapter 12); W. L. McLean, Rutgers University, Piscataway, New Jersey, (Chapter 13); M. R. Beasley, Stanford University, Stanford, California, (Chapters 14, 15, and 17); A. M. Campbell, University of Cambridge, Cambridge, UK, (Chapters 16 and 18); R. W. Rollins, Ohio University, Athens, Ohio, (Chapter 16); D. M. Kroeger, Oak Ridge National Laboratory, Oak Ridge, Tennessee, (Chapter 18); C. C. Koch, North Carolina State University, Raleigh, North Carolina, late of Oak Ridge National Laboratory, (Chapter 19).

E. W. Collings

Materials Department
Battelle Memorial Institute
Columbus, Ohio, USA
March 1985

Units and Conventions

Units

An attempt has been made to employ units which are convenient and recognizable. This has generally resulted in the abandonment of the SI system (which yields, for example, a very difficult unit for magnetic susceptibility) except in simple cases, such as when dealing with mechanical forces and pressures, in favor of the "practical" version of the unrationalized c.g.s.-e.m.u. system in which the units centimeters, seconds, amperes, volts, ohms, gauss, oersted, etc., are used. In presenting critical-current data, however, in deference to common practice, the magnetic field is often expressed in tesla (T), the SI unit of induction, since with *it* (in contrast to the use of $A\ m^{-1}$ for magnetic field strength) a simple conversion to and from kG (or kOe) is possible.

Whatever system is employed, the units being used are always clearly defined.

References

References for Volumes 1 and 2 are listed at the ends of those respective volumes. The reference format adopted is usually derived from the first three letters of the first author's name and the year of publication—thus for example, a paper by Smith published in 1979 would be referred to as [Smi79].

Equations

Equations are numbered thus: (P-Q), where P is the chapter number and Q is the serial number within that chapter. Occasionally it is convenient to repeat an equation; if so, its previously assigned number is also given, but in square brackets.

Compositions

Depending on the point of view adopted in the original publication, or other considerations, alloy compositions are expressed either (i) in weight percent in the formats A–B(n wt.%) or A–nB, or (ii) in atomic percent in the format A–B(n at.%) although in the latter case, $A–B_n$ is occasionally used for brevity. The atomic-percent compositions of multicomponent alloys are given in the format $A_l–B_m–C_n$. With regard to the binary Ti–Nb alloys, the research and technical literatures combine to present compositions in terms of either: (i) at.% Nb, (ii) wt.% Nb, or (iii) wt.% Ti (e.g., Nb–46.5Ti). Since these variations occur principally in Chapter 21, that chapter has been provided with a table, Table 21-4, to enable the reader, whatever his or her preference with regard to compositional format, to easily convert between representations.

The short-hand notation [M] is occasionally used to denote "concentration of M."

Symbols and Abbreviations

An alphabetical list of symbols and abbreviations is provided at the end of each volume.

Contents

I: METALLURGY

Chapter 1. Equilibrium and Nonequilibrium Phases

Chapter 3. Mechanical Properties

PART 1. HARDNESS

PART 2. ANOMALOUS MECHANICAL PROPERTIES

PART 3. NORMAL MECHANICAL PROPERTIES OF TITANIUM–NIOBIUM ALLOYS AND COMPOSITE CONDUCTORS

II: PHYSICS

Chapter 4. Dynamic Elastic Modulus

Chapter 5. Electrical Resistivity

Chapter 6. Thermal Conductivity

Chapter 7. Magnetic Susceptibility

Chapter 8. Low-Temperature Specific Heat

Chapter 9. Low-Temperature Thermal Expansion

III: THE SUPERCONDUCTING TRANSITION

Chapter 10. Calorimetric Studies of the Superconducting Transition and the Mixed State

Chapter 11. The Superconductive Proximity Effect

Chapter 12. The Superconducting Transition Temperature

IV: THE MIXED STATE

Chapter 13. Magnetic Properties of Superconductors

Chapter 14. The Mixed State

Chapter 15. The Paramagnetic Mixed State

Chapter 16. The Critical State

Chapter 17. The Upper Critical Field

Chapter 18. Flux in Motion under the Influence of a Field Gradient

Chapter 19. Magnetization and Critical Current Density

I

Metallurgy

Chapter 1

Equilibrium and Nonequilibrium Phases

1.1 Equilibrium Phases

1.1.1 Electron/Atom Ratio Systematics

The question of lattice stability plays an important role in any discussion of the physics of pure metal or alloy systems. This is particularly true of Ti alloys, whose lattice stability (i.e., structural phase stability) has technical as well as fundamental significance. The crystal structures of the three long periods of transition elements change more or less systematically from hcp through bcc to fcc as their group numbers increase from IV to VIII. Whether or not it is associated with an underlying physical significance, a useful correlation also exists between the crystal structures of binary transition-metal (TM) alloys and their *average* group numbers or electron/atom ratios. Frequently designated "e/a ratio," this quantity, defined as the average number of valence $s + d$ electrons, takes on the values 4 to 10 when applied to the members of the seven columns of the transition-metal block of the periodic table headed by the elements Ti to Ni. It will become apparent as this book develops that several important electronic properties (including superconductive ones) may also be indexed in terms of the above-defined "conventional electron/atom ratio," or quantities related to it such as the atomic-volume-corrected "electron concentration" of JENSEN, MATTHIAS, and ANDRES [JEN65] or the "effective electron/atom ratio," N_{eff}, of DESORBO *et al.* [DES 65]. Another quantity, advocated by LUKE, TAGGART, and POLONIS [LUK64] as being appropriate for the indexing of the compositional threshold for martensitic transformation in Ti–TM alloys, is an average Pauling valence which, although equal to the conventional e/a for the transition elements of groups IV to VI, never exceeds 6 for elements of later groups.

1.1.2 Electronic Structure and Phase Stability

The crystal structures, particularly of simple metals, have been justified from several fundamental standpoints. BREWER and ENGEL [BRE67] have related structure to the spectroscopic states of the individual participating atoms, as has PAULING [PAU67] in considering the nature of the metallic bond.

The orthogonal-plane-wave (OPW) method of calculating the electronic band structures of pure metals also utilized atomic spectroscopic states, but in a more satisfactory manner, by starting with an array of bare ions and then replacing the electrons in such a way that the wavefunctions represent tightly bound electrons near the cores and nearly free electrons in the spaces between. A breakthrough in the calculation of the electronic structures of *disordered alloys* followed the development of the coherent potential approximation (CPA). The first successful replacement of the old heuristic rigid-band model, it was applied by EHRENREICH and colleagues [KIR70] to the Cu–Ni system, again to yield an electronic density of states $n(E)$ *versus* energy (E) function which conformed to experimental results. Treating the d electrons from a tight-binding (TB) standpoint and the other conduction electrons as nearly free (NFE), hence designated NFE-TB-CPA, it was the forerunner of more sophisticated techniques for calculating the electronic properties of concentrated disordered alloys [FAU82]. The most obvious improvement to this method combined important features of the CPA approach with those of the earlier Korringa–Kohn–Rostoker (KKR) method which had been successfully applied to some nonsimple metals. Cu–Ni was again used as the test system, this time by STOCKS, TEMMERMAN, and GYORFFY [STO78]. The results of CPA–KKR calculations, and their testing by experiment (angular resolved photoemission is a suitable technique), are available for only a relatively few alloys, leaving the field open for extensive exploration.

To go from electronic structure to crystal structure (i.e., phase stability) is a large step. The MOTT and JONES interpretation of the Hume–Rothery rules for simple metals was the first; but although unsuccessful in detail [BLA67], it certainly stimulated much further work in the field. With regard to transition metals, the remarkable experimental observation that their crystal structures vary in a regular manner from hcp through bcc to fcc as e/a increases from 4 to 10, is still without a general theoretical interpretation. Attempts to deal electronically with phase stability in transition metals have been made by INGLESFIELD [ING79] and PETTIFOR [PET72]. The situation with regard to alloys is much more difficult; however, with the aid of a CPA approach, PETTIFOR [PET79] has at least been able to calculate the heats of formation of some binary systems.

The closest approach to an exact calculation of phase stability in a transition-metal alloy system, in particular Zr–Nb, has been made by

MYRON, FREEMAN, and MOSS [MYR75], who developed an electronic mechanism for the formation, not of the equilibrium phases, but of the metastable ω-phase. Adequately discussed in their paper (see also SINHA and HARMON [SIN75, SIN76]), their technique coupled a KKR band structure and Fermi-surface calculation for bcc Zr with the effect of "rigid-band" modifications brought about by the addition of Nb. The goal was to demonstrate that an electronically instigated enhancement of the natural dip in the bcc-lattice phonon spectrum at $\frac{2}{3}\langle 111\rangle$ could lead, in a manner to be discussed below, to the ω-phase transformation.

The purely electronic description of equilibrium phase stability has been strongly criticized, from two standpoints, by KAUFMAN and NESOR [KAU73]. They noted, first, that in many treatments competition between phases was completely ignored, and second, that electronic property data acquired at low temperatures were used to justify high-temperature electronic transformations, with no account being taken of entropy differences. KAUFMAN and NESOR recommended the use of a thermodynamic procedure, in which energetic competition between the candidate phases was fully taken into account when attempting to define the lattice stabilities of metallic elements and alloys. Full discussion of a quantitative thermodynamic approach leading to a computer-assisted calculation of binary and multicomponent phase diagrams is to be found in the work of KAUFMAN [KAU70].

In the following two sections we return to phenomenological justifications (not predictions), in electronic terms, for the existence of the two principal classes of Ti alloys—the bcc and hcp alloys—called β-Ti and α-Ti, respectively.

1.2 β-Titanium Alloys

The transition-metal block of the periodic table may be regarded as commencing with group III (Sc, Y, and La or perhaps more precisely, Lu) and ending with the group-VIII elements (Ni, Pd, and Pt), with the alkaline-earth metals (Ca, Sr, and Ba) as "pre-transition metals" and the noble metals (Cu, Ag, and Au) as "post-transition metals." As indicated in most periodic charts of the elements, the structures of all transition metals change from hcp to bcc as e/a increases from 4 to 6. It is possible that stabilization of the bcc structure can be justified within the framework of a screening model, in terms of which a high conduction-electron concentration which enhances the screening of the ion cores may favor a symmetrical, hence cubic structure. If so, the increased electron density of the elements of groups V and VI, as compared to those of group IV, could be responsible for their bcc structures. Symmetrization may also be accomplished by an increase in the amplitude of the lattice vibrations; thus, all six of the group-

III and group-IV elements transform to the bcc structure at sufficiently high temperatures. With regard to alloys, the addition of transition elements to Ti increases the electron density and, through an augmentation of the screening, stabilizes the bcc, or β, structure. Thus, as a general rule, the transition elements are "β-stabilizers."

According to ZENER [ZEN48], and subsequently FISHER [FIS70, FIS75], [STE81], who has considered the problem of bcc stability in considerable detail, the magnitude of the elastic shear modulus, $C' = (C_{11} - C_{12})/2$, is a useful parameter for ranking the stabilities of bcc transition metals and alloys. The variation of C' with conventional e/a ratio is plotted in Figure 1-1. With reference to the figure, two important features are in evidence: (i) with *decreasing* e/a, C' vanishes at $e/a = 4.1$, which then corresponds to the threshold of martensitic transformation, to be further discussed below; (ii) the alloying of group-IV elements with others to the right of them in the periodic table increases the bcc stability, which (as gauged by C') rises rapidly to a maximum near $e/a = 6$, corresponding to the elements Cr, Mo, and W. As for the phase diagram, such a continuous increase of bcc stability with higher-group-number solute content manifests itself as a *lowering* of the $(\alpha+\beta)/\beta$ transus temperature (Figure 1-2). Within the context of β-stabilization, two subclasses of phase diagram exist: the "β-eutectoid" and the "β-isomorphous," corresponding to whether or not a solid-solution/compound eutectoid exists at some elevated temperature. This is also illustrated in Figure 1-2.

Superconductivity in Ti alloys at temperatures within the liquid-He range usually requires the cubic structure. Numerous papers by COLLINGS *et al.* have emphasized the parallelism, or scaling, that exists between the T_c of Ti–TM alloys and their electronic Fermi density-of-states-related

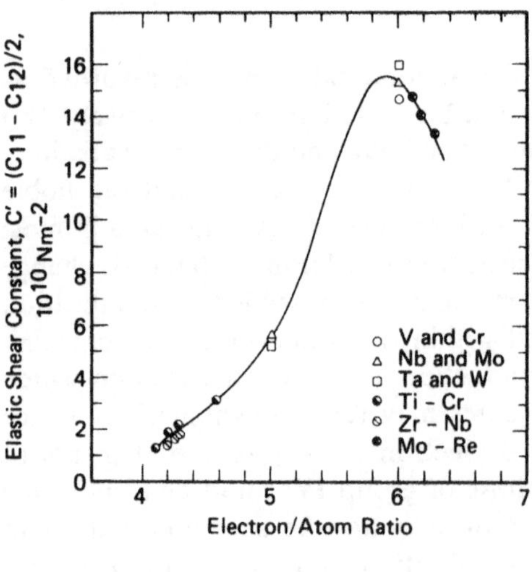

Figure 1-1. Elastic shear modulus $C' = (C_{11} - C_{12})/2$ for various early transition elements and their binary alloys as a function of e/a ratio—after COLLINGS *et al.* [COL73, COL75[d]].

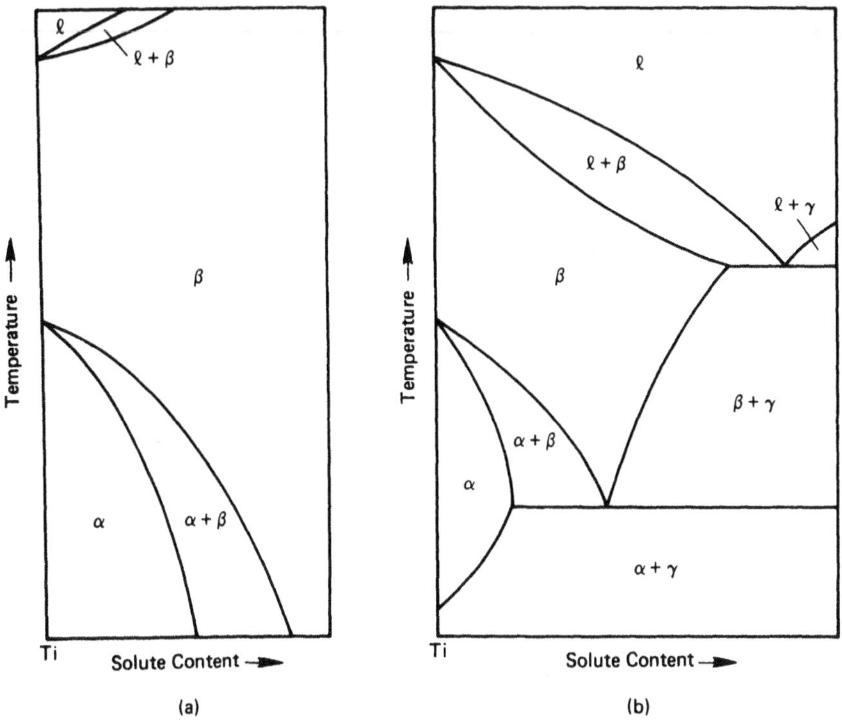

Figure 1-2. Principal features of (a) the β-isomorphous, and (b) the β-eutectoid classes of β-stabilized equilibrium Ti-base transition-metal alloy (Ti–TM) equilibrium phase diagram. An intermetallic compound is represented by γ.

parameters, such as γ (the electronic specific heat coefficient) and χ (the total magnetic susceptibility), as well as possible relationships between T_c and the stability of the bcc phase [CoL71, CoL72, CoL73[a]]. All the technically important Ti-alloy superconductors are concentrated β-phase solid solutions.

1.3 α-Titanium Alloys

The α-stabilizing solutes are those which, as a function of concentration, tend to *elevate* the temperature of the (α+β)/β transus (Figure 1-3). Such solutes are generally nontransition metals (i.e., "simple metals," SM). An explanation of α-stability based on the screening argument of the previous section proceeds as follows. When simple metals (Al is a good example) are dissolved in Ti, very few electrons appear at the Fermi level, most of them going to states within the lower part of the band. The Ti *d* electrons tend to avoid the SM cells, which thereby appear to "dilute" or

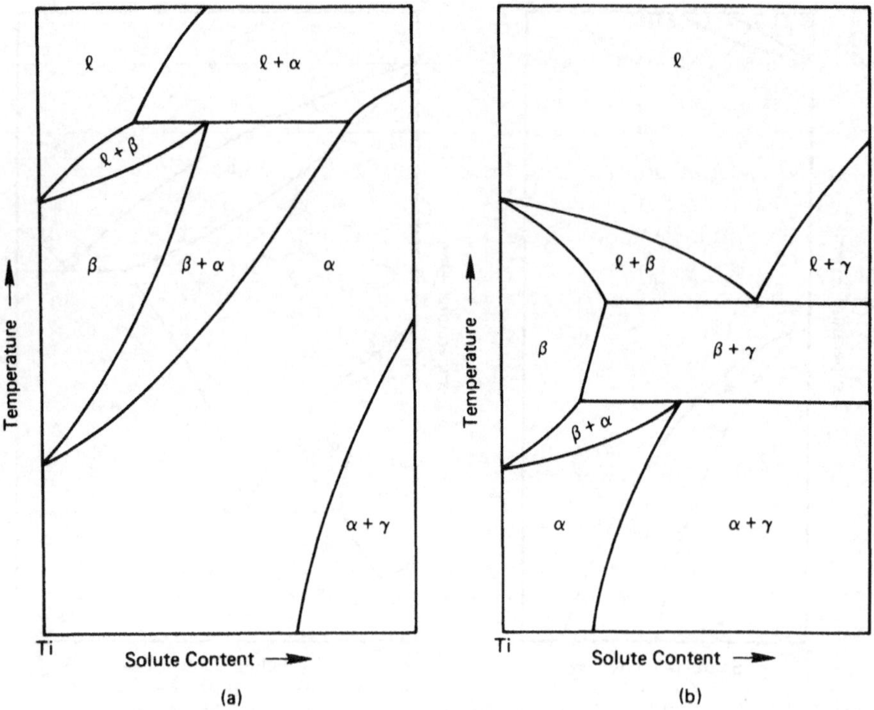

Figure 1-3. Principal features of (a) the simple-peritectic, and (b) the β-peritectoid classes of α-stabilized Ti-alloy equilibrium phase diagram. The symbol γ may represent a compound.

"expand" the Ti sublattice. The consequence of this is to emphasize any preexisting Ti-bond directionality and, thus, to preserve the hcp structure characteristic of the Ti crystal. In general, when simple metals are added to Ti, the fields of "Ti-like" α-stability are eventually terminated by intermetallic compounds, of composition Ti_3SM,[1] which are also hexagonal in structure. This bond argument is consistent with the observation that α-stabilizers are quite rapid solution strengtheners both of Ti and of β-Ti alloys [CoL75[d]].

Unlike the β-stabilized Ti alloys, concentrated α-phase alloys, with their generally low T_cs, are not important from a superconducting standpoint. On the other hand, it is important to emphasize the roles played by *small* additions of α-stabilizers (particularly the interstitial elements, O and N) to β-stabilized Ti–TM alloys. First of all, their presence enhances the speed of α-precipitation when there is already a tendency for this to occur, such as within the dislocation network of heavily deformed alloys; and second, the poor intrinsic superconducting properties of the resulting α-precipitates are responsible for their effectiveness as flux-pinning sites.

[1]An exception is supposed to be Ti_4Pb.

1.4 Binary Titanium–Transition-Metal Alloys

1.4.1 Further Classification Schemes for Titanium-Alloy Phases

MOLCHANOVA [MOL65] has presented a useful classification scheme for α- and β-stabilization in binary alloys of Ti (Figure 1-4). According to the figure, α-stabilized alloys are further subdivisible into those whose phase diagrams exhibit (i) "peritectic transformations" and (ii) "peritectoid transformations" between solid-solution phases. The β-stabilized alloys are also subdivisible into two classes, in this case according to whether they exhibit (i) "simple transformations" or (ii) "eutectoid transformations."

It is also possible to classify low-concentration *additions* to a Ti–TM base (e.g., Ti–Nb) according to this four-part scheme. Figure 1-5, which illustrates this, is from the work of RASSMANN and ILLGEN [RAS73], who in a series of useful papers discussed the effects of structure [RAS72], α-stabilizing additions [RAS72[a]], "neutral-element" additions [RAS72[b]], and β-stabilizing additions [RAS73] on the critical current density of Ti–Nb and Ti–Nb-base ternary alloys. Their opinions as to whether members of the four respective classes of solute increased (↑) or decreased (↓) the J_c of partially optimized Ti–Nb wire are indicated by arrows in the figure. Representatives of all the alloy classes referred to in Figure 1-5 are considered in detail in a companion volume [COL83].

The superconductive properties of all the binary alloys of Ti with other transition elements are also considered fully in the companion volume

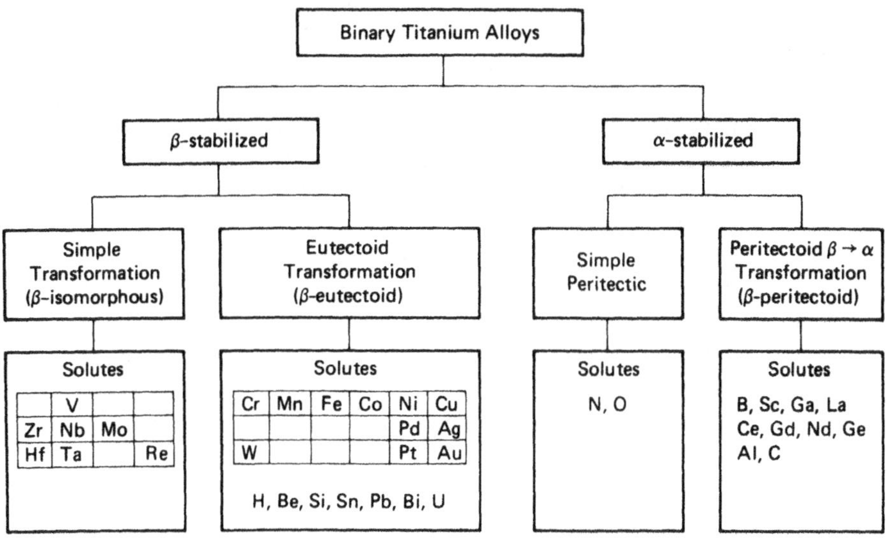

Figure 1-4. Classification scheme for binary alloys of titanium—after MOLCHANOVA [MOL65, p. 154].

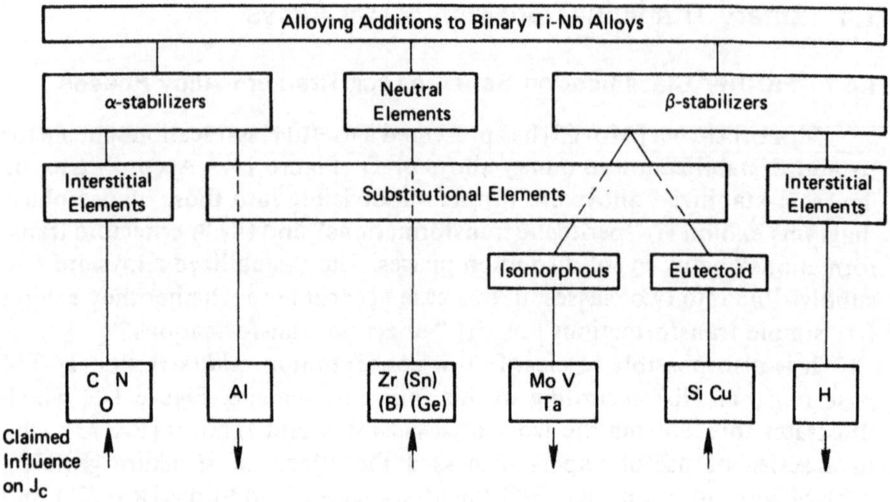

Figure 1-5. Classification scheme for ternary alloys based on Ti–Nb, with particular reference to the influence of ternary additions on the critical current density—after RASSMANN and ILLGEN [RAS73].

[COL83]. Although the equilibrium phase diagrams of most systems are adequately treated in the standard reference works of HANSEN [HAN58], SHUNK [SHU69], and ELLIOTT [ELL65], and especially MOLCHANOVA [MOL65], who has dealt exclusively with Ti-base alloys, it would be instructive in the present context to offer a group of simplified, compositionally truncated, equilibrium phase diagrams, arranged according to the positions that the solute elements occupy in the transition metal block of the periodic table. Some representative diagrams selected from such a postulated arrangement are given in Figure 1-6. In order to focus attention on the composition range of greatest interest, the limiting composition in each group has been selected in order to restrict the e/a ratio to 5.0. In so doing it has been assumed that the numbers of $s + d$ valence electrons belonging to the elements in the columns headed by Fe, Co, Ni, and Cu, are 8, 9, 10, and 11, respectively. Some interesting systematics to be noted in the figure are: (i) as the solute element "moves" to the right, the phase diagram changes from β-isomorphous to β-eutectoid; and (ii) along the 3d row the eutectoid temperature undergoes a more or less monotonic increase. Extrapolating this trend to the left suggests that Ti–V could also be thought of as eutectoid, but with an inaccessibly low eutectoid temperature.

1.4.2 The Ti–Cr System

An important representative of the β-eutectoid type of alloy is Ti–Cr, whose complete equilibrium phase diagram is similar to that shown in

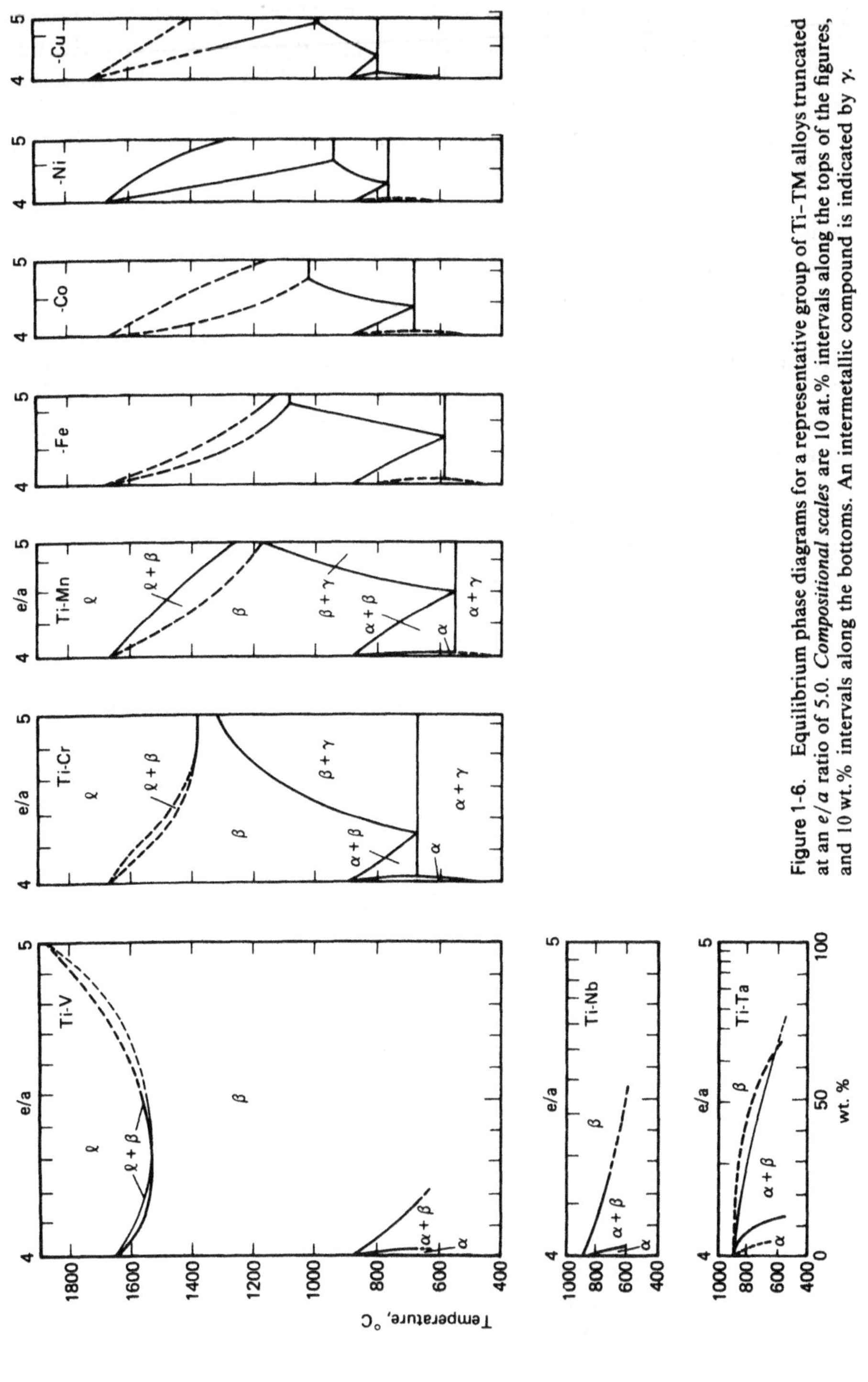

Figure 1-6. Equilibrium phase diagrams for a representative group of Ti–TM alloys truncated at an e/a ratio of 5.0. *Compositional scales* are 10 at.% intervals along the tops of the figures, and 10 wt.% intervals along the bottoms. An intermetallic compound is indicated by γ.

Figure 1-2(b). Of particular interest in systems of this type is the eutectoid decomposition, which takes place very sluggishly at temperatures below 660–685°C. Once formed, the products of eutectoid decomposition are readily redissolved during heating in the β-field. The line compound in this case is of nominal composition $TiCr_2$, but has a homogeneity range of some 2 percentage points. According to SHUNK [SHU69, p. 283], $TiCr_2$ is polymorphic, existing as the hcp ($MgZn_2$-structured) "β-$TiCr_2$" phase at high temperatures and as the fcc ($MgCu_2$-structured) "α-$TiCr_2$" phase at lower temperatures. The transformation temperature of $TiCr_2$ seems to be uncertain [SHU69, p. 283]. According to MOLCHANOVA [MOL65, p. 36] the hexagonal modification exists above 1300°C and the cubic below 1000°C, both phases coexisting in the intervening temperature range. POLONIS, TAGGART, and LUHMAN employed Ti-Cr as a model system in studying correlations between metallurgical and superconductive properties [LUH69, LUH70, LUH70[a], LUH71], [POL69, POL70, POL71]. They made numerous investigations into the influences of composition, and metallurgical treatments such as β-quenching, intermediate-temperature aging (at 196 and 300°C), and up-quenching (to 450°C), on the superconducting transition and the mixed-state magnetization.

1.4.3 The Ti–Nb System

The binary system of paramount importance in superconductivity is, of course, Ti–Nb, a typical β-isomorphous system, whose equilibrium and metaequilibrium phase diagrams are frequently referred to in metallurgical discussions of technically important superconducting alloys. A composite equilibrium phase diagram for Ti–Nb, based primarily on the work of HANSEN *et al.* [HAN51] and IMGRAM *et al.* [IMG61], is reproduced in Figure 1-7. Since thermodynamic $\alpha+\beta$-phase equilibrium is difficult to achieve at temperatures below about 600°C, the experimentally deduced line diagram is not continued below that temperature. At temperatures below the $\beta/(\alpha+\beta)$ transus, or its projection, a slow approach to equilibrium is made via the decomposition of the metastable $\omega+\beta$- or β-phases, the rate of which is accelerated in the presence of oxygen or the dislocation products of heavy deformation.

The recent renewal of interest in the occurrence of α-Ti–Nb precipitation in rather concentrated Ti–Nb alloys, of up to about 53.5 wt.% (37 at.%) Nb, and in its role in flux pinning has not only rekindled a corresponding level of interest in the equilibrium phase diagram for temperatures in the vicinity of 400°C but has also led to the ready availability of heavily-cold-worked and heat-treated samples upon which transmission electron microscopy (TEM) could be carried out. HILLMANN *et al.* [PFE68], [BES79] and WILLBRAND and SCHLUMP [WIL75] were among the first to examine the

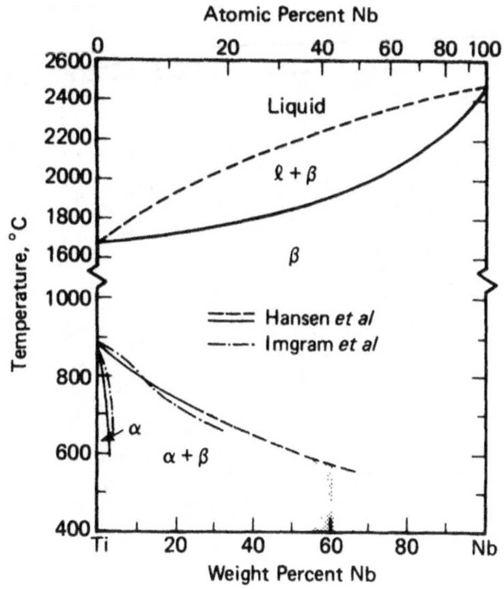

Figure 1-7. The Ti-Nb equilibrium phase diagram—after HANSEN *et al.* [HAN51] and IMGRAM *et al.* [IMG61] and including a practical limit for the "equilibrium" $\alpha+\beta$-field (shading) based on the arguments of Section 1.4.3; see also Section 2.13.2.

occurrence and morphology of α-phase precipitation in alloys of 50 wt.% Nb after moderate-time aging at temperatures near 380°C. Precipitation from cold-worked Ti–Nb(36 at.%) has been carefully studied by OSAMURA *et al.* [OSA80] using specimens in the form of (i) *foils*—solution treated 1 h/800°C/slow cooled, cold rolled to thickness reductions of up to 90%, and aged at 380°C; and (ii) *fine wires*—obtained from Cu-clad composites which had been reduced by 90 to 99.994% and aged at 380°C. The foils were the subjects of low-angle x-ray scattering experiments (to determine the average diameter and spacing of the precipitated particles, see Section 21.13.2(c)) and TEM observation (in order to obtain further information about the particles and their distribution). Wide-angle x-ray diffraction, the only technique that could be satisfactorily applied to the fine wires, served to confirm that α-phase was the dominant precipitate present in the sample of cold-worked-and-aged Ti–Nb(36 at.%). More recently WEST and LARBALESTIER [WES80, WES82] have demonstrated conclusively the presence of α-Ti–Nb precipitates in Ti–53.5Nb (taken from the Fermilab Cu/Ti–Nb multifilamentary composite) after two-stage heat treatments (associated with cold drawing) of 80 h/375°C and 40 h/375°C, respectively.

In discussing detection techniques, WEST has pointed out [WES82] that since α-phase precipitates are not easy to identify in cold-worked structures, particularly since dark-field imaging is complicated by the close positioning of matrix and precipitate reflections in selected area diffraction (SAD) patterns, best results are obtained through the use of the more recently perfected techniques of scanning transmission electron microscopy (STEM) and associated energy-dispersive x-ray analysis (EDAX). The results of

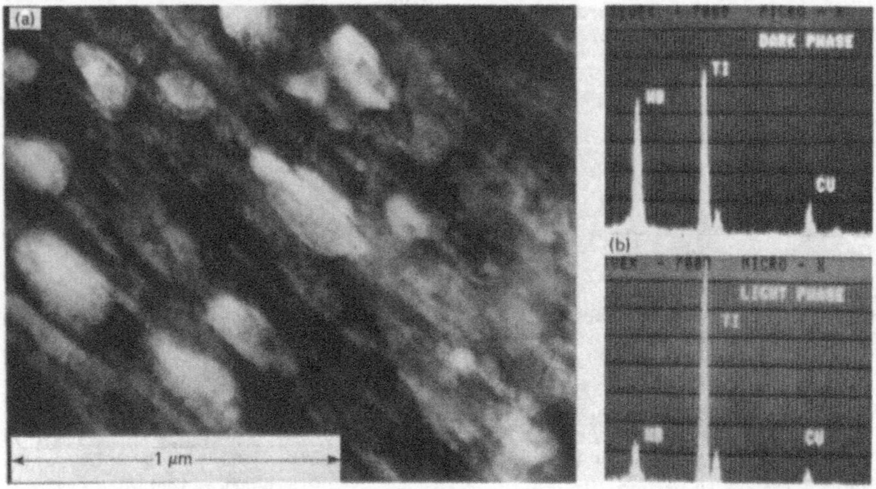

Figure 1-8. (a) STEM micrograph of Fermilab superconductor Ti–53.5Nb after 80 h/375°C at 3.66 mm$^{\phi}$ plus cold work plus 40 h/375°C at 1.5 mm$^{\phi}$ [WES82]. (b) EDAX pictures of the dark (Nb-rich matrix) and light (Nb-lean, hence α-phase) regions, respectively. Photographs courtesy of A. W. West.

such observational methods applied to a sample of Ti–53.5Nb are given in Figure 1-8.

On the other hand, Ti–Nb(42 at.%) (i.e., 58 wt.%, analyzed composition) after heavy cold deformation and aging, usually for relatively short periods of time at temperatures of 350–500°C, yielded only *traces* of α-phase. The failure of NEAL *et al.* [NEA71] to detect appreciable precipitation in this case may have been due partly to deficiencies in the detection techniques used and partly to the short aging times. With regard to the latter, it should be recognized that in the long-time aging of Ti alloys contamination by oxygen, an α-stabilizer, must be suspected whenever unanticipated α-phase precipitation takes place.

Based on the results presented above, and other results, it is concluded that 56 ± 1 wt.% (i.e., 42 at.%) is the practical limit of macroscopic α-phase precipitation at about 400°C. In placing this result in Figure 1-7, it is recognized that, when suitably stimulated by cold work and heat treatment, thermodynamically permissible submicroscopic precipitation may develop in these and in even more concentrated alloys. Thus, for example, LARBALESTIER and WEST [LAR84], [WES84] have produced evidence to suggest that a thin film-like precipitate (albeit not positively identifiable as α-phase) coats the cell walls (see below) of deformed-and-aged Ti–Nb alloys with concentrations as high as those studied by NEAL *et al.*

For further discussion of this subject, within the framework of experimental flux pinning, and with reference to some very recent results, see Chapter 21, Part 2.

1.5 Multicomponent Titanium–Transition-Metal Alloys

1.5.1 Titanium-Base Multicomponent Alloys in General

The five classes of multicomponent Ti-base alloys, considered in the companion to this volume [CoL83], consist of ternary and quaternary alloys conforming to the following titles and descriptions:

(a) Ti–TM–SM, Ti–TM_1–TM_2, Ti–TM–Intl (Nb excluded)
(b) Ti–Nb–SM
(c) Ti–Nb–Intl
(d) Ti–Nb–TM
(e) Ti–Nb-base quaternary.

where TM = transition metal; SM = simple metal, noble metal, or metalloid; Intl = interstitial element such as B, C, N, or O. The availability of alloy phase data[2] for these systems is indicated in Table 1-1. The scarcity of complete equilibrium-phase descriptions of Ti–Nb-base alloys is evident by a glance at the table, but understandable in view of the specialized and nonstructural nature of their applications.

Prominent in the history of applied superconductivity are Zr–Nb (now obsolete) and Ti–Nb alloys, whose equilibrium- and nonequilibrium-phase properties have been extensively treated. The product of intermixing this pair of alloys is Ti-Zr–Nb, a technical alloy once of considerable interest to both Soviet and Japanese manufacturers. Because of its potential importance as a superconductor, Ti–Zr–Nb has been subjected to considerable microstructural study by Soviet [ALE67[a]], [SAV73, p. 307], Japanese [DOI66[a]], and other [ZWI68] researchers. The essential features of their results are summarized below

1.5.2 The Ti–Zr–Nb System

The equilibrium phase diagrams of Ti-Zr–Nb alloys, after quenching from anneals at the temperatures listed in Table 1-2, have been investigated by ALEKSEEVSKII *et al.* [ALE67[a]] and DOI *et al.* [DOI66] using the standard techniques of optical metallography and x-ray diffraction. In general, the

[2]Where possible the emphasis is placed on the availability of such data within the context of superconductivity.

Table 1-1. Multicomponent Systems for Which Alloy Phase Data[a] Are Available (TM = transition metal; SM = simple metal, nontransition metal, or metalloid; Intl = interstitial element such as B, C, N and O)

(a) Ternary Ti-TM-SM, $Ti-TM_1-TM_2$ and Ti-TM-Intl Alloys (Nb Excluded)

Alloy Classification and Constitution	Literature Reference for Partial or Comprehensive Phase Diagram Information: Reference	Code[b]
Ti-TM-SM		
Ti-V-Al	Standard Sources	
Ti-V-Sn	Efimov [EFI70]	(EX)
Ti-Mo-Al	Molchanova [MOL65]	(P)
$Ti-TM_1-TM_2$		
Ti-V-Zr	Savitskii [SAV73]	(P)
	Molchanova [MOL65]	(P)
	Efimov [EFI70]	(EX)
Ti-V-Ta	Savitskii [SAV73]	(EX)
Ti-V-Cr	Savitskii [SAV73]	(EX)
Ti-V-Mo	Savitskii [SAV73]	(EX)
Ti-V-W	Efimov [EFI70]	(EX)
Ti-TM-Intl		
Ti-Ta-C	Molchanova [MOL65]	(P)

(b) Ternary Ti-Nb-SM Alloys

Constitution	Literature Reference for Partial or Comprehensive Phase Diagram Information: Reference	Code[b]
Ti-Nb-Al	Guertler [GUE69]	(P)
	Rassmann [RAS72[a]]	(P)
	Savitskii [SAV73]	(EX)
Ti-Nb-Si	Rassmann [RAS73]	(S)
Ti-Nb-Cu	Rassmann [RAS73]	(S)
	Löhberg [LOH71]	(P)
	Savitskii [SAV73]	(P)
Ti-Nb-Ge	Rassmann [RAS72[b]]	(S)
	Heller [HEL71]	(P)
Ti-Nb-Sn	Rassmann [RAS72[b]]	(S)

Table 1-1 (*cont.*)

(c) Ternary Ti-Nb-Intl Alloys

Constitution	Literature Reference for Partial or Comprehensive Phase Diagram Information: Reference	Code[b]
Ti-Nb-B	Rassmann [RAS72[b]]	(S)
Ti-Nb-C	Kunakov [KUN70]	(S)
	Rudy [RUD69]	(P)
Ti-Nb-N	Baker [BAK70]	(S)
Ti-Nb-O	Rodriguez-Gonzales [ROD70]	(S)

(d) Ternary Ti-Nb-TM Alloys

Alloy Classification and Constitution	Literature Reference for Partial or Comprehensive Phase Diagram Information: Reference	Code[b]
Group IV TM Additions		
Ti-Nb-Zr	Alekseevskii [ALE67[a]]	(P)
	Savitskii [SAV73]	(P)
	Doi [DOI66[a]]	(P)
	Zwicker [ZWI68]	(P)
	Rassmann [RAS72[b]]	(P)
Ti-Nb-Hf	Savitskii [SAV73]	(S, EX)
Group V TM Additions		
Ti-Nb-V	Molchanova [MOL65]	(P)
	Savitskii [SAV73]	(P)
	Rassmann [RAS73]	(P)
Ti-Nb-Ta	Savitskii [SAV73]	(S, EX)
Group VI TM Additions		
Ti-Nb-Mo	Molchanova [MOL65]	(P)
	Rassmann [RAS73]	(P)
	Savitskii [SAV73]	(S, EX)
Ti-Nb-W	Savitskii [SAV73]	(P, EX)
Group VII TM Additions		
Ti-Nb-Re	Bychkova [BYC70[a]]	(S)
	Savitskii [SAV73]	(S)

Continued

Table 1-1 (*cont.*)

(d). Ternary Ti–Nb–TM Alloys (*cont.*)

Alloy Classification and Constitution	Literature Reference for Partial or Comprehensive Phase Diagram Information: Reference	Code[b]
Group VIII TM Additions		
Ti-Nb-Fe	Nishimura [NIS68]	(S)
Ti-Nb-Ru, Ti-Nb-Os, Ti-Nb-Rh, Ti-Nb-Ir, Ti-Nb-Pd, Ti-Nb-Pt	Zwicker [ZWI65]	(S)

(e). Quaternary Ti-Nb-Base Alloys

Constitution—Combinations of Four Elements Selected From: Group IV: Ti	Zr	Hf	Group V: V	Ta	Nb	Other	Literature Reference for Partial or Comprehensive Phase Diagram Information: Reference	Code[b]
Ti	Zr	Hf			Nb		Rayevskii [RAV69]	(P)
Ti	Zr		V		Nb		Kitada [KIT70[d]]	(S)
Ti	Zr			Ta	Nb		Kitada [KIT70[d]]; Horiuchi [HOR73[a]]	(S); (S)
Ti	Zr				Nb	Mo	Kitada [KIT70[g]]	(S)
Ti			V	Ta	Nb		Savitskii [SAV73]	(S)

[a] Where possible the emphasis is placed on the availability of such data within the context of superconductivity.

[b] (P) indicates complete or partial phase diagrams available. (EX) indicates that alloy phase data are claimed to be available in references cited within. (S) indicates that some solubility data are available.

results obtained by the two research groups were in good agreement in spite of the large differences in the annealing times to which the two respective sets of alloys had been subjected. Evidence for this is: (i) the fact that the diagram for 570°C due to DOI *et al.*[3] fitted logically between those for 550

[3] In analyzing the x-ray data, DOI *et al.* [DOI66] chose not to distinguish between the two β-phases when they coexisted with the α-phase, and labeled regions as $\alpha+\beta$ that ALEKSEEVSKII *et al.* [ALE67[A]] designated (using present terminology) $\alpha+\beta'+\beta''$. In Figure 1-9 some reasonable liberties have been taken with the data of DOI *et al.* for 570°C, after which their diagram falls nicely into position with respect to the surrounding Soviet results.

Table 1-2. Prequench Annealing Conditions during the Alloy Phase Analysis of Ti–Zr–Nb, the Results of Which Are Depicted in Figure 1-9

Alekseevskii *et al.* [ALE67[a]]		Doi *et al.* [DOI66[a]]	
Temperature °C	Time h	Temperature °C	Time h
—	—	1100	24
1050	0.5–1	—	—
900	30	900	620
800	100	800	1010
700	300	700	1010
600	800	—	—
550	1–3	570	2400
500	1500	—	—

and 600°C due to ALEKSEEVSKII *et al.*; (ii) the remarkable similarity of the two 800°C equilibrium phase diagrams.

A summary of the combined results is presented in Fig. 1-9. At 1050°C the ternary systems show unlimited mutual β solid solubility. Below 975°C a region of two-β-phase immiscibility, $\beta'+\beta''$, begins to develop from the Zr–Nb edge and expands as the temperature falls. Between 700 and 600°C an $\alpha+\beta$-region develops along the Zr–Nb edge separated from the $\beta'+\beta''$ lobe first by a line and then at lower temperatures by a region $\alpha+\beta'+\beta''$. By this time one $\alpha+\beta$-region is developing from the Ti corner and another from the Zr corner, severely restricting the remaining region of bcc stability. These changes in phase relationships take place very rapidly with temperature between 570 and 500°C, at which point an $\alpha+\beta$-phase occupies almost the entire diagram. The locations of the compositions of the Soviet and Japanese (Hitachi Ltd.) ternary technical alloys on the Ti–Zr–Nb composition triangle are shown in Figure 1-10. Considering these positions in the light of the equilibrium phases present at various temperatures according to Figure 1-9, we note that the "Ti-Nb-rich" Hitachi alloy and the Soviet alloys (35 BT, 50 BT, and 65 BT) would be expected to respond to heat treatment in ways similar to those of the corresponding binary Ti–Nb, and consequently that ω-phase precipitation would be anticipated upon quenching and α-phase during moderate-temperature aging. On the other hand, for the Zr–Nb-rich class of alloy, the development of $\beta'+\beta''$ compositional modulation is possible as a result of elevated temperature annealing. For this reason a 700°C heat treatment was preferred during the optimization of wire produced from Ti_{10}–Zr_{40}–Nb_{50}.

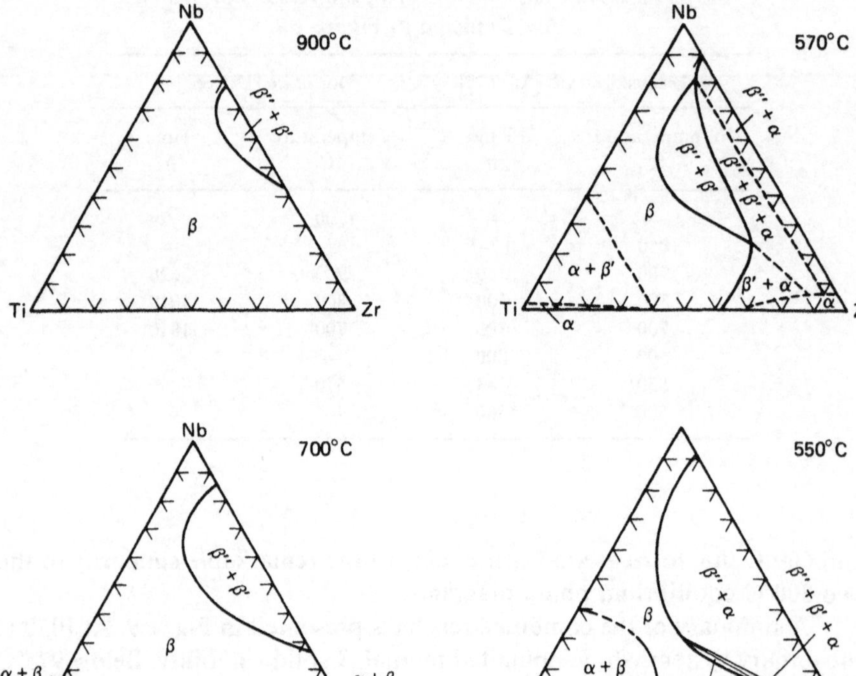

Figure 1-9. Ti–Zr–Nb equilibrium phase diagrams (at.% linear) for temperatures between 500°C and 900°C. Of particular interest is the development of the $\beta'+\beta''$-phase field—after DOI *et al.* [DOI66[a]] and ALEKSEEVSKII *et al.* [ALE67[a]].

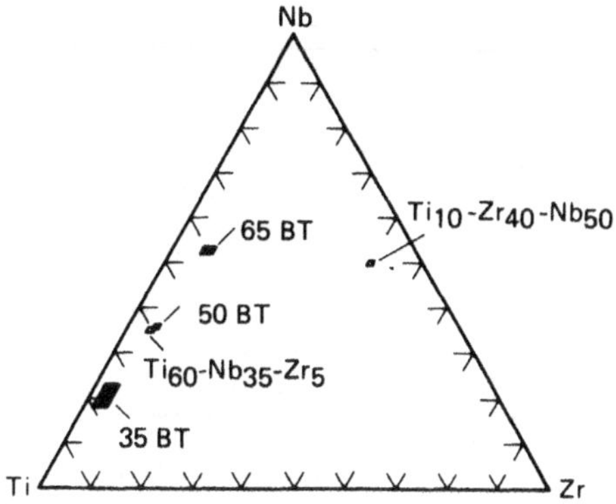

Figure 1-10. Composition triangle (at.% linear) indicating the locations of the Soviet technical alloys 35 BT, 50 BT, and 65 BT, and the Ti-Nb-rich and Nb-Zr-rich alloys Ti_{60}-Nb_{35}-Zr_5 and Nb_{50}-Zr_{40}-Ti_{10}, respectively, initially recommended by Hitachi Ltd.

1.6 Nonequilibrium Phases

1.6.1 The Martensitic and Athermal ω-Phases in Quenched Titanium–Transition-Metal Alloys

Equilibrium phase diagrams of the type discussed in the previous section are generally developed by deducing the *initial* states of alloys that have been quenched to room temperature. The nonequilibrium phases to be considered below represent the *final* states of such quenching processes. The structures assumed by rapidly β-quenched[4] Ti–TM alloys are illustrated in Figure 1-11.

Below a start temperature, M_s, the bcc structure begins a spontaneous allotropic transformation via a complicated shearing process to a structure known as martensite and designated α' or α'' depending upon whether the transformation product is hcp or orthorhombic. When the structural distinction between α' and α'' is unimportant for our purposes the martensites will be represented collectively by the symbol α^m. Being of second order, the martensitic transformation is anticipated by a regime of structural fluctuations called ω-phase. As represented in Figure 1-11(a), ω-phase, when formed as a result of very rapid quenching, exists as a crystalline precipitate plus a fluctuating component over a narrow composition range overlapping the boundary of the martensitic zone. In practice, however, the range over which it occurs during the brine quenching of macroscopic samples is quite broad, and an attempt has been made in Figure 1-11(b) to depict it as a region in which precipitate abundance decreases gradually with increasing solute content. The free energy of α^m is usually lower than that of ω-phase, and consequently during the partial martensitic transformation of an $\omega+\beta$-structure the martensite needles generally consume any ω-phase precipitates that lie in their paths.

[4]That is, quenched from the elevated temperature single-phase-bcc field.

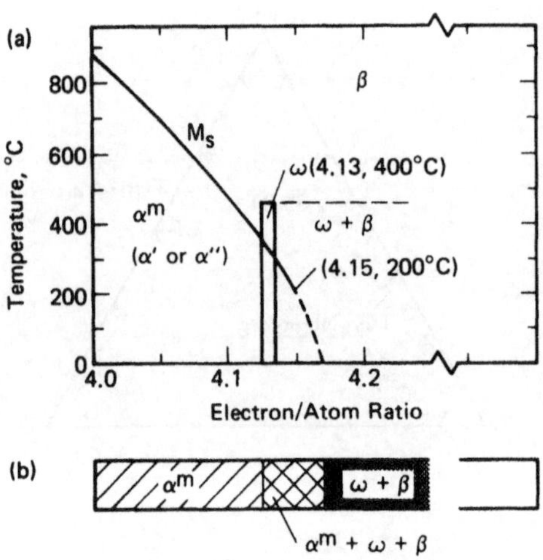

Figure 1-11. Schematic representation of the occurrences of the martensitic α'- and α''-phases (i.e., α^m, collectively) and the ω-phase in Ti-TM alloys. Both "quenched data" and "aged data" are included—see also Table 1-3 (α^m-phases) and Table 1-4 (ω-phase).

1.6.2 The Quenching Process

In studies of quenched microstructures an important if not always accomplished goal is the control and quantification of the quench rate. Should the quench be too slow, diffusional processes intervene to obscure the result. When the primary aim is to study microstructure (rather than to produce material for physical or mechanical property testing) and the fastest possible quench rates are mandatory, it is usual to take thin foils or wires, heat them in a controlled environment, and subject them to *in situ*[5] gas or liquid quenching. In HICKMAN's [HIC68, HIC69[a]] method, for example, rolled strips self-heated under high vacuum to 1250°C by the passage of direct current were quenched by admitting He gas to a pressure of 0.1 atm, restoring sample temperature, and then switching off the current. Such techniques are generally capable of quench rates of 50°C s^{-1} to 2×10^4°C s^{-1}, and in HICKMAN's case about 10^3°C s^{-1} was claimed. Gaseous He is about three times as effective a quench medium as Ar under comparable conditions. In the method selected by BROWN, JEPSON, and HEAVENS [BRO65], specimens varying in thickness from 0.002 to 0.20 in. were first heated indirectly under vacuum; Ar at a pressure of 150 to 700 torr was then allowed to enter the specimen chamber in order to suppress the boiling of the iced water or refrigerated calcium chloride solution subsequently admitted to quench the sample. In this way, quench rates of 2.5×10^4 to 2×10^5°C s^{-1} were achieved [JEP70]. BALCERZAK and SASS [BAL72], whose results are discussed below, used an Inconel specimen holder to transfer rolled strips (0.002–0.003 in.′) from the hot zone of a vacuum resistance furnace to a waiting pool of water-cooled silicone oil. Although it was known to be rapid, the characteris-

[5]In which the quenching medium is introduced to the sample rather than *vice versa*.

tic quench rate of this method was not specified. The quench rates achievable by all of the above methods are of course much faster than those obtained during the ice-brine quenching of the massive samples (up to 40 g) needed for mechanical and physical property study (especially of low-temperature specific heat). Accordingly, some discrepancies must be expected between the microstructural results obtained from thin foils and those derived from the quenching of bulk specimens.

The measured M_s temperature for any given alloy composition is itself a function of quench rate. In Ti-Nb(5 at.%) for example, JEPSON *et al.* [JEP70] noted that the M_s temperature decreased from 760°C to 710°C as the cooling rate increased from 10^{-3} to 10°C s^{-1}, but that once a critical cooling rate of 30°C s^{-1} was exceeded M_s became independent of it. The critical threshold itself was a function of alloy composition and decreased from 200°C s^{-1} to ~0.4°C s^{-1} as the Nb content increased from 0 to 15 at.%.

In Ti–TM alloys, as with other systems, the quench rates necessary to transform the structure, from high-temperature-equilibrium to room-temperature-metastable, while preserving compositional homogeneity are strongly constitution dependent. Thus, whereas bulk dilute alloys of Ti with early transition elements can be water quenched without evidencing serious decomposition, the same is not true of alloys such as Ti–Fe, -Ni, and -Co, whose anomalous superconducting properties (to be discussed below) could be partially interpreted in terms of compositional and structural segregation. The pronounced differences between the properties of the quenched dilute Ti-V, -Nb, etc., alloys and those of Ti–Fe, -Co, and -Ni can be fairly simply explained in terms of differences among the solute tracer diffusion coefficients in β-Ti at 1000°C. As shown in Figure 1-12, the diffusion coefficients for V, Nb, and Mo are less than 1.3×10^{-9} cm^2 s^{-1}, while those

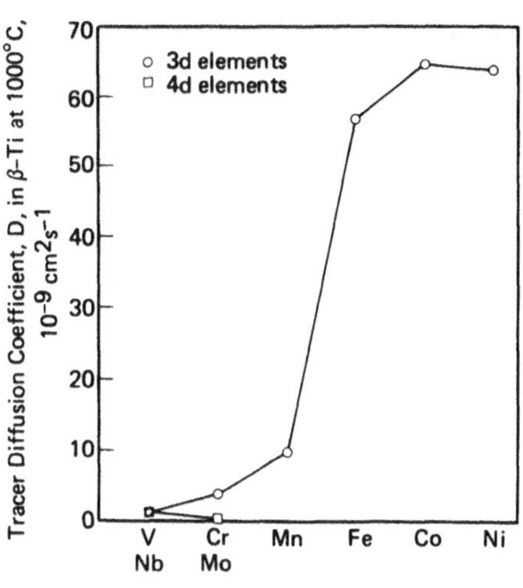

Figure 1-12. Tracer diffusion coefficients for the 3*d* solutes V, Cr, Mn, Fe, Co, and Ni, and the 4*d* solutes Nb and Mo in β-Ti at 1000°C—computed from frequency-factor and activation-energy data of ZWICKER [ZWI74, p. 174].

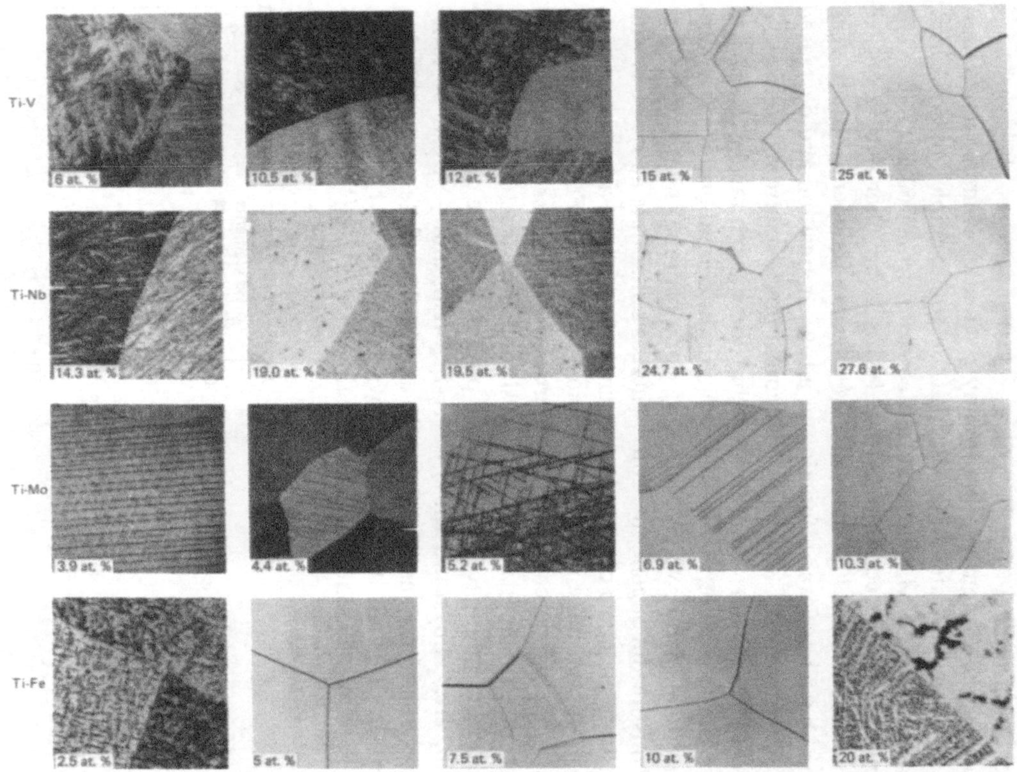

Figure 1-13. Optical micrographs from massive tokens of Ti–TM alloys, quenched from the β-phase into iced brine, showing the transition between the α^m- and $\omega+\beta$-regimes. Compositions of Ti–V and Ti–Fe are nominal; those of Ti–Nb and Ti–Mo are analyzed. Magnifications of the original $9 \times 9\ cm^2$ micrographs, $50\times$; a $300\times$ micrograph of Ti–Fe(20 at.%) is also shown.

for the Fe-group elements are $\sim 60 \times 10^{-9}$ cm^2 s^{-1}. The extreme examples are the Fe-group elements on one hand and Mo on the other, whose diffusion coefficients in β-Ti at 1000°C are in the ratio 200:1. Such large differences in diffusivity must be taken into consideration when comparing the properties of the two classes of Ti-base alloy, and in selecting a quenching technique.

1.6.3 Stability Limit of the β-Phase in Titanium–Transition-Metal Alloys

The optical microstructures and composition ranges of the martensites characteristic of ice-brine-quenched massive samples of Ti–V, -Nb, -Mo, and -Fe are exemplified by Fig. 1-13. Compositionally, Ti–Fe and Ti–Nb are extreme examples, their M_s curves delimiting those of all other measured Ti–TM alloys to intersect a 200°C (for example) isothermal at 3.3 and 20.5 at.% respectively, [Zwi74, p. 174]. In Table 1-3 are listed the $M_{s,\,200°C}$ compositions of ten Ti–TM alloys together with their conventional e/a ratios. Quite remarkable is the fact that all the e/a ratios, aside from those of Ti–Co and Ti–Ni, which seem to be exceptional cases, lie within ±0.03

Table 1-3. Compositions of the $M_{s,200°C}$ Intercepts Expressed in Terms of Conventional e/a Ratio

Element	Concentration Corresponding to M_s at 200°C, at.%[a]	Group Number	Conventional[b] e/a Ratio Based on Group Number
V	13.3	V	4.13
Nb	20.5	V	4.21
Ta	19.1[c]	V	4.19
Cr	6.0	VI	4.12
Mo	6.7	VI	4.13
W	8.2	VI	4.16
Mn	5.0	VII	4.15
Fe	3.3	VIII	4.13
Co	6.0	VIII	4.24 4.30[d]
Ni	7.6	VIII	4.30 4.46[d]
Mean[e]			4.15 ± 0.03

[a] After Zwicker [Zwi74, p. 174].
[b] Not atomic size corrected.
[c] Assuming Ta/Nb ratio following Molchanova [Mol65, p. 158].
[d] Based on number of valence $s + d$ electrons.
[e] Excluding Co and Ni.

of a common value, 4.15, suggesting that the martensitic transformation in Ti–TM alloys is of common origin and related to electronic factors. Electronic and lattice properties can, at least in principle, be coupled through calculations of the elastic constants. An empirical connection between martensitic transformation and the vanishing of the elastic shear modulus, $C' = (C_{11} - C_{12})/2$, in Ti–TM alloys is implied in Figure 1-1. It has also been considered elsewhere by COLLINGS *et al.* [COL72, COL73] with particular reference to Ti–Mo and Ti–Cr alloys.

In discussing the occurrence of ω-phase in Ti alloys, Figure 1-11 makes a useful starting point. As represented there, a narrow region exists in which ω-phase appears *athermally* during very rapid quenching from the β-phase, while over a broader composition range ω-phase will occur as a precipitation product of β-decomposition during moderate temperature ($\lesssim$400°C) *isothermal* aging. Alloys that have been *rapidly quenched* into this broader region are host to a "diffuse ω-phase," so called because of the existence of straight or curvilinear lines of diffuse intensity in the selected area electron diffractograms [SAS72]. The occurrence, composition, and structure of ω-phase, and its relationship to the competing α- and β-phases, have been discussed in review papers by HICKMAN [HIC69], SASS [SAS72], and WILLIAMS [WIL73, WIL82], some of which make reference to the earlier and often-quoted studies of SILCOCK [SIL58] and BAGARIATSKII *et al.* [BAG59].

During moderate-temperature ($\lesssim$400°C) aging for several days, a metastable-equilibrium $\omega+\beta$ two-phase state can be attained, x-ray analysis of which yields the compositions of the ω- (solute lean) and β- (solute rich) phase boundaries. The ω-phase regime does not, however, extend to zero solute concentration, but is tightly confined by the nearby lower-energy α^m-phase field. As a result of the limited composition range of the athermal ω-phase, good agreement is usually obtained between the "saturation composition of ω-phase after aging at 400°C" [HIC69[a]] and the "solute concentrations which yield ω-phase on quenching" [BAG59], as shown in Table 1-4. The latter is of course a difficult quantity to determine accurately, since away from the zone of the sought-for athermal ω-phase, rapidly quenched alloys support the diffuse ω-phase and more slowly quenched larger samples will contain isothermal-ω. Table 1-4 also lists the conventionally calculated e/a ratios for ω-phase, separated into two listings according to the source and method of data acquisition. It is notable that in each case the critical e/a ratios for the eight alloys listed are constant within 20% and that the two independent mean values (which differ by <0.01) agree within experimental scatter. Also, a comparison of Tables 1-3 and 1-4, which yield, respectively, $\langle e/a \rangle_{\alpha^m} = 4.15 \pm 0.03$ and (from an overall mean) $\langle e/a \rangle_{\omega} = 4.13 \pm 0.03$, emphasizes that athermal ω-phase occurs at the threshold of martensitic transformation and suggests that the ω- and α^m-transformations are related and electronic in origin. Taken together, the results expressed

Table 1-4. Compositions of the $\omega/(\omega+\beta)$ Data Points Expressed in Terms of Conventional e/a Ratio

Element	Group Number	Saturation Composition of ω-Phase (at.%) After Aging at ~400°C[a]	Solute Concentration (at.%) in Ti Alloys for Which ω-Phase is Formed on Quenching[b]	Conventional e/a Ratio Based on Group Number: "Aged Data"	Conventional e/a Ratio Based on Group Number: "Quenched Data"
V	V	13.8 ± 0.3	13	4.14	4.13
Nb	V	~9 ± 2	18	4.09	4.18
Cr	VI	6.5 ± 0.2	7	4.13	4.14
Mo	VI	4.3 ± 0.4	4.5	4.09	4.09
W	VI		7.5		4.15
Mn	VII	5.1 ± 0.2	5.5	4.15	4.16
Re	VII		4.5		4.14
Fe	VIII	4.3 ± 0.2	3	4.17	4.12
Means				4.13 ± 0.03	4.14 ± 0.03

[a] Hickman [Hic69[a]].
[b] Bagariatskii, Nosova, and Tagunova [Bag59].

Table 1-5. Solute Concentrations and e/a Ratios for 100% β-Stabilization in Ti–TM Alloys—After Luke *et al.* [LUK64]

Element	Critical Solute Concentration at.%	Group Number	Pauling Valency[a]	Conventional e/a Ratio Based on Group Number	Pauling e/a ratio	Phase With Respect to Which the β-phase Was Considered Stable[b]		References
V	15	V	5	4.15	4.148		ω	1, 2
Ta	15–21	V	5	4.15[c]	4.148	m		3
Cr	7.5	VI	6	4.15	4.147		ω	4
Mo	7.2	VI	6	4.14	4.143	m		5
	7.4	VI		4.15	4.148		ω	2
W	7	VI	6	4.14	4.14	m		6
Mn	5.5	VII	6	4.17	4.118		ω	7
	7.5			4.23	4.148	m or	ω	8
	8			4.24	4.158		ω	9
Fe	6–8[d]	VIII	6	4.28	4.139		ω	10
Co	6	VIII	6	4.24	4.12	m		11
	7.5			4.29	4.148		ω	12
	5–7[d]			4.24	4.12	m		13
Ni	7	VIII	6	4.28	4.14	m		14
	7-8[d]			4.29	4.15		ω	15
Mean conventional e/a ratio						4.20 ± 0.06	4.22 ± 0.06	
Mean Pauling e/a ratio						4.14 ± 0.01	4.14 ± 0.01	

[a] Reference [PAU56].
[b] i.e., whether an M_s composition (m) or absence of ω-phase (ω) was considered.
[c] The lower value is taken.
[d] Mean taken.

(1) F. R. Brotzen, E. L. Harmon and A. R. Troiano, Decomposition of Beta Titanium, *Trans AIME* **203**, 413 (1955).
(2) J. M. Silcock, An X-ray Examination of the Omega Phase in Ti–V, Ti–Mo, and Ti–Cr Alloys, *Acta Met.* **6**, 481 (1958).
(3) D. J. Maykuth, H. R. Ogden and R. I. Jaffee, Titanium-Tungsten and Titanium-Tantalum Systems, *Trans AIME* **197**, 231 (1953).
(4) S. A. Spachner, Comparison of Structure of Omega Transition Phase in Three Titanium Alloys, *Trans AIME* **212**, 57 (1958).
(5) D. J. DeLazaro, M. Hansen, R. E. Riley and W. Rostoker, Time, Temperature, Transformation Characteristics of Titanium–Molybdenum Alloys, *Trans AIME* **194**, 265 (1952).
(6) OTS PB 171424, DMIC Report 136 B, May 29 (1961), p. 106.
(7) R. I. Jaffee, Physical Metallurgy of Titanium Alloys, *Progress in Metal Physics* **7**, 65 (1958).
(8) P. D. Frost, W. M. Parris, L. L. Hirsch, J. R. Doig and C. M. Schwarz, Isothermal Transformation of Titanium–Manganese Alloys, *ASM Trans Quart.* **46**, 1056 (1954).
(9) R. M. Lerenman *et al.*, Electron Microscope Study of Structural Transformation in Titanium–Manganese and Titanium–Chromium, *Physics of Metals and Metallography* **9**, No. 3, 99 (1960).
(10) D. H. Polonis and J. G. Parr, Martensite Formation in Powders and Lump Specimens of Ti–Fe Alloys, *Trans AIME* **203**, 64 (1955).
(11) P. R. Swann and J. G. Parr, Phase Transformations in Titanium-Rich Alloys of Titanium and Cobalt, *Trans AIME* **212**, 276 (1958).
(12) F. W. Yakymyshyn, G. R. Purdy, R. Taggart and J. G. Parr, The Relationship Between the Constitution and Mechanical Properties of Titanium-Rich Alloys of Titanium and Cobalt, *ASM Trans Quart.* **53**, 283 (1961).
(13) F. R. Orrell and M. G. Fontana, The Titanium-Cobalt System, *ASM Trans Quart.* **47**, 554 (1955).
(14) H. Margolin, E. Ence and J. P. Neilsen, Titanium-Nickel Phase Diagram, *Trans AIME* **197**, 243 (1953).
(15) J. W. Barton, G. R. Purdy, R. Taggart and J. G. Parr, Structure and Properties of Titanium-Rich Titanium–Nickel Alloys, *Trans AIME* **218**, 844 (1960).

in those tables reconfirm that the regimes of α^m- and athermal ω-phase can, in fact, be juxtaposed in the manner depicted in Figure 1-11.

The results of a literature study of collected transformation data for nine Ti–TM alloys, conducted independently of that of BAGARIATSKII *et al.* [BAG59] and prior to the work of HICKMAN [HIC69[a]], and indexed in terms of an e/a ratio based on Pauling valence [PAU56] rather than group number, led to a similar conclusion. In conducting this survey of the limits of bcc stability LUKE, TAGGART, and POLONIS [LUK64] chose not to categorize the actual nature (i.e., whether α^m- or ω-phase) of the transformation product. But as a result of reviewing the original references cited, and separating the results into "α^m-limited" and "ω-phase-limited" β-phases as in Table 1-5, it is again possible to conclude—in agreement with Tables 1-3 and 1-4 considered jointly—that: (i) in terms of e/a, the threshold for the α^m-transformation (presumably $M_s \cong$ room temperature) is constant to within about ±0.03, (ii) the composition of athermal ω-phase, expressed as an e/a ratio, is also constant to within about ±0.03, and (iii) the martensitic and ω-phase "thresholds" agree to well within that scatter. Results such as these encouraged LUKE, TAGGART, and POLONIS [LUK64] to be among the first to explicitly cite electronic factors as being important determiners of bcc Ti–TM lattice stability.

The presence of athermal ω-phase particles, which with diameters of 20–40 Å are too small to be detected optically, can be unequivocally diagnosed using TEM and electron or x-ray diffractometry. The occurrence of ω-phase in ice-brine-quenched Ti–Mo(5 at.%), for example, which is just on the edge of the quenched-α^m regime (Figure 1-13), has been confirmed by the use of electron diffractometry followed by dark-field electron microscopy [COL71]. The results are shown in Figure 1-14.

The occurrence of athermal ω-phase in such close proximity to the α^m-phase boundary is a phenomenon of general validity and of fundamental importance in the theory of the ω-transformation (to be considered later in more detail). Although from the standpoint of the crystallographer the absence of a habit-plane description precludes the defining of the ω-phase transformation as martensitic [WIL73], both transformations, according to SUZUKI and WUTTIG [SUZ75] and CLAPP [CLA73], possess a common ingredient in the form of a soft-phonon instability. Arguments in support of this view are developed in subsequent sections. Within the framework of critical-phenomenon theory, it can be argued that after a material is quenched through an equilibrium-phase boundary it encounters the boundary of a metastable phase. Just outside of this, a region of statistical fluctuations is expected; this is referred to as "critical opalescence" since it was first observed optically in the study of liquid/vapor transformation. As a particular solid-state example, near a *structural* transformation *strain*

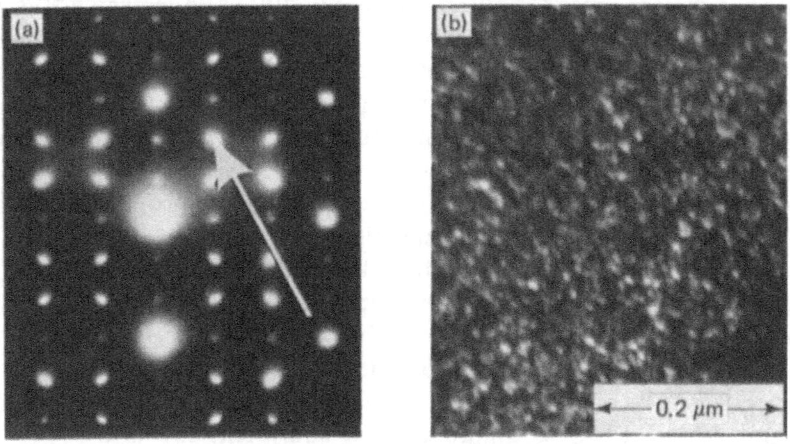

Figure 1-14. Transmission electron micrographs of Ti–Mo(5 at.%) quenched into iced brine from the β-field. (a) Electron diffractograph showing a superposition of two principal spot patterns: a rectangular arrangement of round spots originating from the β-matrix, and groups of elongated spots originating from the ω-phase. (b) Dark-field electron micrograph of the ω-phase precipitate originating from the diffraction spot indicated by the arrow in the diffractograph; using the dark-field technique, the bright patches of the photograph are specific to the ω-phase, however only one-quarter of the precipitate is visualized at one time—after COLLINGS and HO [COL71].

fluctuations would be appropriate. It is known from the neutron diffraction studies of MOSS *et al.* [MOS73] and the Mössbauer effect measurements of BATTERMAN *et al.* [BAT73] that ω-phase possesses a fluctuating component. It is, therefore, tempting to suggest that the dynamic (or "diffuse") ω-phase is the critical opalescence of the athermal precipitate and/or martensite.

1.7 Formation and Structures of the Martensitic Phases

In contrast to nucleation-and-growth types of phase change which rely on thermally activated atomic diffusion, martensitic transformation involves a spontaneous cooperative movement of atoms in a microscopically homogeneous transformation from one crystal lattice into another. In principle, the martensitic process is not thermally activated and should, therefore, take place at high temperature-independent speeds. But in real alloys a clear-cut separation of transformation processes into "nucleation-and-growth" and "martensitic" is not possible. Although in pure metals such as unalloyed Ti a simple athermal martensitic transformation is conceivable, in an *alloy* the situation becomes more complicated. While the long-range effects of alloying—as described, for example, in terms of changes of the elastic parameters—may simply change the conditions under which athermal transformation takes place, the local effects of alloying may inhibit the prescribed movement of atomic planes. This inhibition has the effect of reducing the distances over which atomic regions can cooperate, and consequently of perturbing the microstructure of the transformation product. Furthermore, it has the effect of reducing the speed of the transformation, thereby bringing it into competition with nucleation-and-growth processes—the Widmanstätten reaction, for example.

The influence of solute concentration on the martensitic transformation in Ti–Nb alloys has been examined in a carefully controlled series of experiments by JEPSON, BROWN, and GRAY [JEP70], who pointed out that the effect of increasing the Nb content was to make the transformation more sluggish. In particular, an examination of their data suggests that a cooling rate of 10^2°C s^{-1} in pure Ti is equivalent to 0.3°C s^{-1} in Ti-Nb(15 at.%). In considering what in Ti–TM alloys might be referred to as a "solute-controlled transformation at solute-inhibited transformation speed," in which athermal shear competes with diffusion, a given quench rate cannot be expected to yield similar martensitic structures for all concentrations and all types of Ti–TM alloys. The properties of the quenched product will depend on several fundamental variables, one of the most important of which is the diffusion coefficient of the particular solute species in β-Ti. Figure 1-12, which compares the 1000°C diffusion coefficients of eight transition elements in β-Ti, shows that a quench rate that will completely suppress diffusion in Ti–Nb ($D_{Nb} = 1.3 \times 10^{-9}$ $cm^2\,s^{-1}$) and Ti–Mo ($D_{Mo} = 0.3 \times 10^{-9}$ $cm^2\,s^{-1}$) may be quite inadequate to do so in Ti–Fe, -Ni, and -Co (for which $\langle D \rangle_{Fe,Ni,Co} = 61.6 \times 10^{-9}$ $cm^2\,s^{-1}$). Against this background it is possible to believe in the existence of a series of morphological variants of the martensitic and martensitic-like structures extending from "bulk martensite" through "acicular martensite" to the Widmanstätten structure defined below.

1.7.1 Morphology of Martensites

When conditions are particularly favorable, transformation from β to martensite takes place completely, on a large scale, and with considerable structural coherence. The result is the so-called "massive martensite" (otherwise known as packet, or lath, martensite) consisting of large irregular zones, on the scale of 50–100 μm, subdivided into parallel arrays of fine platelets less than 1 μm across (Figure 1-15). In massive martensite, the lack of retained β-phase prevents direct determination of the habit plane. With increasing solute concentration the coherence between the platelets

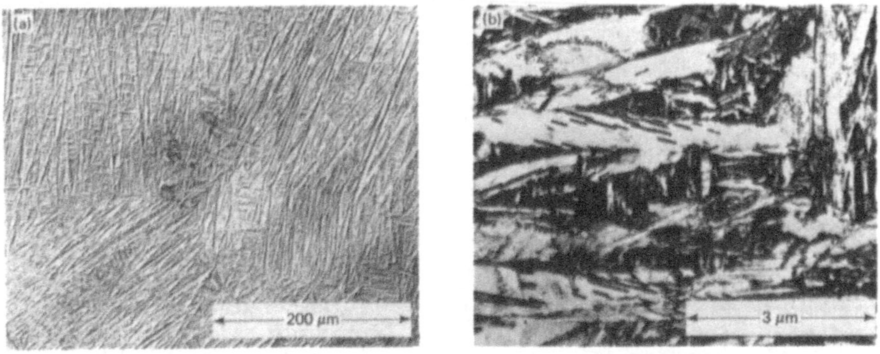

Figure 1-15. Massive martensite in Ti-1.78Cu quenched from 900°C. (a) Optical micrograph showing large colonies. (b) Electron micrograph showing individual plates within the colonies—after WILLIAMS [WIL73]. Micrographs courtesy of J. C. Williams.

(which would otherwise make up a massive colony) is lost, resulting in the partially disordered array of individual platelets referred to as "acicular martensite" (Figure 1-16). Further increase in solute concentration prevents a complete martensitic transformation from taking place; β-phase, trapped between the platelets of the acicular martensite, then makes it possible to accomplish direct habit-plane determination. Not far removed from the β-plus-acicular-martensitic quenched structure is the Widmanstätten arrangement consisting of groups of α-phase needles lying with their long axes parallel to the $\{110\}$ planes of the parent retained β (Figure 1-17).

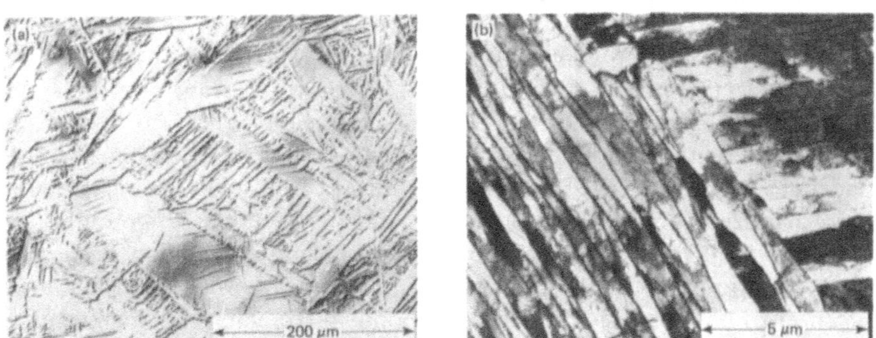

Figure 1-16. Acicular martensite in Ti-12V quenched from 900°C. (a) Optical micrograph. (b) Electron micrograph showing lenticular-shaped plates, some of which are internally twinned—after WILLIAMS [WIL73]. Micrographs courtesy of J. C. Williams.

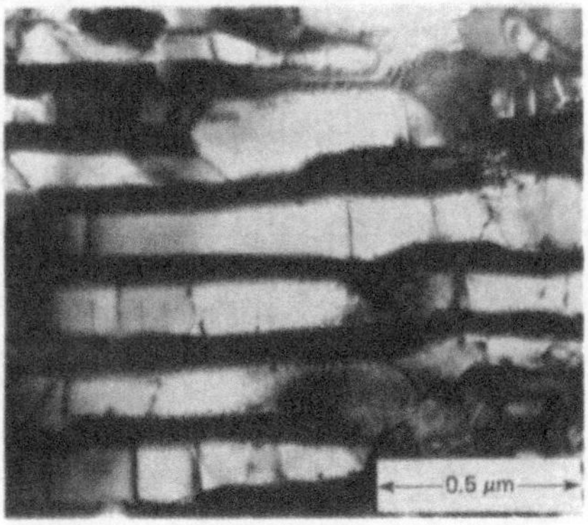

Figure 1-17. Widmanstätten α-phase. Bright-field micrograph showing Widmanstätten α-plates separated by the β-phase matrix—after WILLIAMS [WIL73]. Micrograph courtesy of J. C. Williams.

Although this structure is characteristic of solute-lean Ti–TM alloys its development during aging at temperatures too high for ω-phase precipitation leads to its being treated as a product of nucleation and growth. It is mentioned in the present context for two reasons: (i) the existence of nucleation in martensitic transformation should not be excluded [CHR65, pp. 527–531], [JEP70]; and (ii) its presence is anticipated in *quenched* alloys when the solute diffusion coefficients are sufficiently high.

The optical microstructures of a series of β-quenched Ti–TM alloys are shown in Figure 1-18. The five depicted alloys, Ti-Mo, -V, -Nb, -Mn, and -Fe, possess comparable e/a ratios, had similar masses, and were ice-brine quenched under similar conditions. Noticeable in the photomicrographs of these alloys, which are arranged in ascending order of solute diffusion coefficients in β-Ti (Figure 1-12), is what appears to be a gradual transition from acicular martensite to the Widmanstätten structure. These results indicate that the atomic diffusion coefficient must be introduced as a scaling factor when comparing quench rates in Ti–TM alloys.

Metastable bcc Ti–TM alloys are also known to transform under the application of mechanical stress. Some confusion has arisen in the past over the structure of the deformation product, but the situation has been adequately clarified by WILLIAMS [WIL73]. The solute-lean β-alloy is characterized by a low shear modulus; thus, close to its M_s boundary the application of stress will trigger a martensitic transformation designated "deformation martensite" or "stress-induced martensite." Further away from the boundary, the bcc lattice responds directly to the influence of an applied stress by twinning, which, as pointed out by SUZUKI and WUTTIG [SUZ75] can in this context be regarded as a special case of martensitic transformation to a crystallographically equivalent structure. Strain transformation in a heavily deformed Ti–Mo(5 at.%) alloy is illustrated in Figure 1-19. Optically indistinguishable from deformation martensite, mechanical

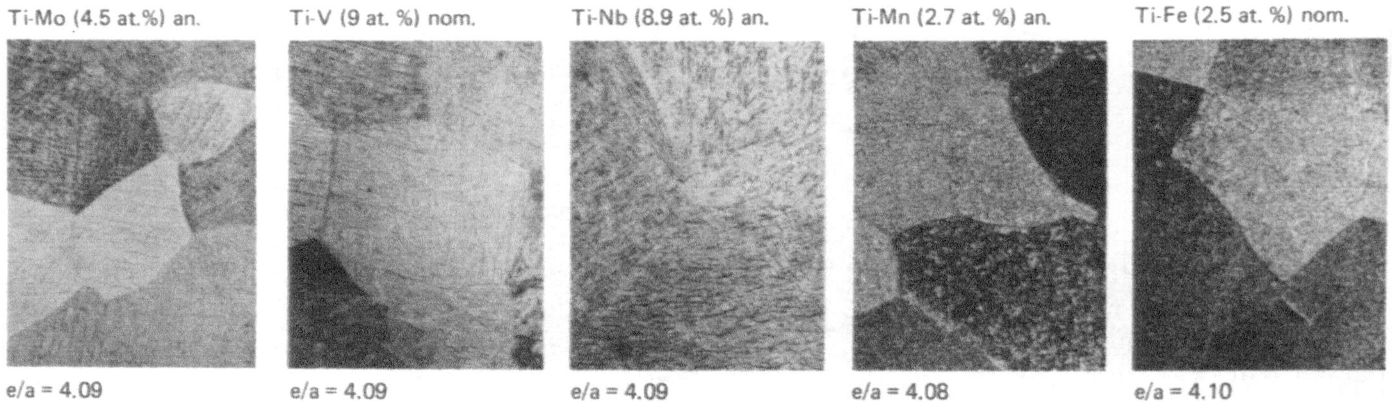

Figure 1-18. Optical micrographs of five low-concentration Ti–TM alloys after quenching into iced brine from the β-phase, arranged in ascending order of solute atomic diffusion coefficient in β-Ti (see Figure 1-12). Indicated are the analyzed solute concentrations (nominal in the cases of Ti–V and Ti–Fe) and the conventional e/a ratio ($\cong 4.09$ = constant). Magnification of the original $11.5 \times 9\ cm^2$ micrographs, 50×.

Figure 1-19. Ti–Mo(5 at.%) quenched from the β-phase and deformed by compression ($\varepsilon = 23\%$) showing deformation martensite or twins. Magnification of the original $9 \times 9\,cm^2$ micrograph, 50×.

twins are easily identified by diffractometry, their crystallographic structure of course being identical to that of the parent lattice.

1.7.2 Structure of the Martensites

The transformation structure assumed by the pure elements Ti, Zr, and Hf and the dilute Ti–TM alloys is hcp and has been assigned the symbol α' [BAG59], [WIL73]. Otherwise, Ti–TM alloys with compositions exceeding certain limits which differ from system to system transform on quenching to an orthorhombic structure designated α''. There is some disagreement as to whether Ti–V is to be included in this category; although WILLIAMS

Table 1-6. Compositions and e/a Ratios of the α'/α'' Boundary in Some Ti–TM Alloys—After Bagariatskii *et al.* [BAG59]

	α'/α'' Boundary		
Element	wt.%	at.%	Conventional e/a Ratio
V	9.4	8.9	4.09
Nb	10.5	5.7	4.06
Ta	26.5	8.7	4.09
Mo	4	2.0	4.04
W	8	2.2	4.04
Re	<10	<2.8	<4.08

[WIL73] did not succeed in detecting the α''-variant in quenched Ti-11.6V, FLOWER *et al.* [FLO82] obtained good indirect evidence for its presence in Ti-10.6V. The α'/α'' compositional interfaces for six Ti-TM systems, as determined by BAGARIATSKII *et al.* [BAG59], are given in Table 1-6.

Since stress-induced martensite can occur only in an alloy not already transformed by quenching, it is obviously confined to the higher concentration ranges; in the light of the foregoing observations on the structures of quenched martensites it is, therefore, not surprising to find that deformation martensite is invariably of the orthorhombic variety [WIL73]. As mentioned above, when the distinction between the martensite variants is unimportant, they are collectively represented here by the symbol α^m.

1.7.3 Crystallographic, Thermodynamic, and Acoustic Aspects of the Martensitic Transformation

(a) Crystallography. Details of the crystallographic $\beta \rightarrow \alpha^m$ transformation in Ti alloys have been reviewed and discussed by several authors, for example: HAMMOND and KELLY [HAM69, HAM70], BLACKBURN [BLA70], OTTE [OTT70], WILLIAMS [WIL73], SHIBATA and ONO [SHI77], and FLOWER *et al.* [DAV70], [FLO82]. The crystallographic approach focuses attention on the influence of various possible shear systems on the habit plane of the transformed plates in the hope of determining which are consistent with the results of habit-plane-orientation measurements and, consequently, which dislocations dominate the transformation. By a process of elimination, the justification for which forms the basis for discussion and is the essence of the exercise, a multitude of possible dislocation paths are reduced to just a few. According to OTTE [OTT70], the entire $\beta \rightarrow \alpha'$ transformation process was reducible to the shear system operations

$$\left.\begin{aligned} &[111]_\beta(11\bar{2})_\beta \rightarrow [2\bar{1}\bar{1}3]_{\alpha'}(\bar{2}112)_{\alpha'} \\ \text{and}\quad &[111]_\beta(\bar{1}01)_\beta \rightarrow [2\bar{1}\bar{1}3]_{\alpha'}(\bar{1}011)_{\alpha'} \end{aligned}\right\} \tag{1-1}$$

The crystallography of the α''-precipitate has been discussed by HATT and RIVLIN [HAT68].

(b) Thermodynamics and Acoustics. Underlying the phenomenological crystallographic theory of the martensitic transformation are several levels of semimechanical and, finally, mechanistic interpretations. Earlier in this section it was shown that similar e/a ratios characterized the limits of bcc stability with respect to either the ω-phase or the martensitic transformation. The underlying common ingredient is a localized soft-phonon mode, whose relationship to those two types of transformation can be traced through a series of papers by SUZUKI and WUTTIG [SUZ75], CLAPP [CLA73],

COOK [COO73, COO74], and DE FONTAINE and colleagues [DEF70, DEF71]. In the first pair of papers, a formalism has been developed for comparing martensitic transformation to spinodal decomposition, which in turn has been coupled by others with ω-phase precipitation. A formal identification is achieved by comparing—in macroscopic terms—composition fluctuation to displacement fluctuation (i.e., deformation). Thus, just as spinodal decomposition, with which the names of HILLERT [HIL61] and CAHN [CAH61] are associated, can be treated as a free energy (g) instability with respect to a composition (c) fluctuation, the threshold of which is defined by $\partial^2 g/\partial c^2 = 0$, the martensitic transformation, according to SUZUKI *et al.* [SUZ75] and CLAPP [CLA73], can be triggered by a strain (ε) fluctuation in the vicinity of a strain spinodal, $\partial^2 g/\partial \varepsilon^2 < 0$. The approach taken by CLAPP to describe the initial nucleation of the martensitic transformation commenced with an expression for the free energy density as a function of strain to third order in the elastic constants. For a cubic lattice the vanishing of one or more of the elements of 6×6 matrix representing the second derivatives of g with respect to strain, *viz*,

$$\partial^2 g/\partial \varepsilon_I \partial \varepsilon_J \equiv g_{IJ}(\varepsilon), \qquad I, J = 1, \ldots, 6 \tag{1-2}$$

defined a strain spinodal for the corresponding directions in strain space. In lattices which are prone to transform, g_{IJ}, already small, may be forced to zero by the application of a few percent strain (the deformation-martensite effect).

The condition for static instability may also be approached via the introduction of lattice imperfections such as point and line defects, and even the surface itself, which may place the crystal close to one or another of its strain spinodals. An appropriately applied dynamic strain, in the form of one half-cycle of a lattice vibration, will then be able to nucleate a burst of transformation.

1.8 Occurrence and Structure of the Quenched ω-Phase

A discussion in Section 1.6 led to the conclusion that the ω-phase and martensitic transformations are closely related. A connection between them in terms of soft-mode lattice dynamics is now apparent. As we shall see, the DE FONTAINE theory [DEF70, DEF71] of the ω-phase requires the bcc lattice to be unstable with respect to longitudinal phonons centered about $\frac{2}{3}\langle 111\rangle$.

For several compelling reasons, ω-phase is of considerable interest from the standpoint of irreversible type-II superconductivity: (i) the size (10–15 Å) and interparticle spacing (15–25 Å) [SAS72] of the athermal precipitate makes it an extremely effective pinner of the Abrikosov flux lattice; (ii) although absent at room temperature from some Ti–TM alloys, ω-phase has been noted to appear reversibly upon reduction of the temperature to 100 K; (iii) Ti–TM superconductors are operated at liquid-He temperatures; since these have until recently [OBS80] been inaccessible to TEM techniques, the existence of athermal ω-phase precipitation at liquid-He temperatures in the solute-rich intercellular regions of α-phase precipitated technical superconducting alloys has not yet been confirmed or denied. This section deals solely with the products of rapid quenching—i.e., the "athermal" and the "diffuse" ω-phases. The aging phenomenon is discussed in Chapter 2. Speculation as to the existence of ω-phase at low temperatures in technically interesting alloys is confined to this chapter. Subsequently, ω-phase is considered only insofar as its presence can be confirmed either directly (by electron or x-ray diffractometry) or indirectly (by hardness measurements, etc.) at room temperature.

The development of our current understanding of ω-phase precipitation can be traced—and the key literature retrieved—with the aid of Figure 1-20. During the 1950s, ω-phase precipitation was shown to occur athermally over a very narrow composition range during rapid quenching from the elevated-temperature β-field of numerous Ti–TM alloys. X-ray studies carried out at the time identified its structure as either hexagonal (SILCOCK [SIL58]) or trigonal (BAGARIATSKII *et al.*—referred to in [BAG59]) depending (as known from more recent work) on solute concentration. The compositions of athermal ω-phase in Ti–V, -Nb, -Cr, -Mo, -W, -Mn, and -Re, as determined by BAGARIATSKII *et al.* [BAG59] on quenched alloys using x-ray techniques, are presented in Table 1-4. They can be seen to be in excellent agreement with the results obtained a decade later by HICKMAN [HIC69a] in his studies of the compositions of aged alloys. In a series of investigations using single-crystal x-ray techniques, HICKMAN [HIC68, HIC69, HIC69a] studied the occurrence of ω-phase in the systems Ti–V, -Cr, -Fe, -Co, and -Ni, paying particular attention to the composition range over which it appeared during quenching and to the compositions of the products of isothermal aging to the quasi-equilibrium $\omega+\beta$-state.

Between about 1967 and 1972, a period of renewed interest in ω-phase, the use of TEM, x-ray diffraction, and electron diffraction led rapidly from a number of phenomenological studies, with which the names of BRAMMER and RHODES (Ti–Nb [BRA67]), BLACKBURN and WILLIAMS (Ti–V, -Mo [BLA68]), and KRAMER and RHODES (Ti–Nb [KRA67]) are associated, to an elegant series of quantitative investigations by HICKMAN (mentioned above) and by SASS and his students, who studied the systems Ti–Zr [SAS69], Zr–Nb [DAW70], Ti–V [MCC71], and Ti–Nb [BAL73]. Together with an important investigation by DE FONTAINE, PATON, and WILLIAMS [DEF71] centered on the Ti–Mo system, this work formed the basis of our current

Figure 1-20. Interrelationships between studies of ω-phase precipitation in Ti–TM alloys.

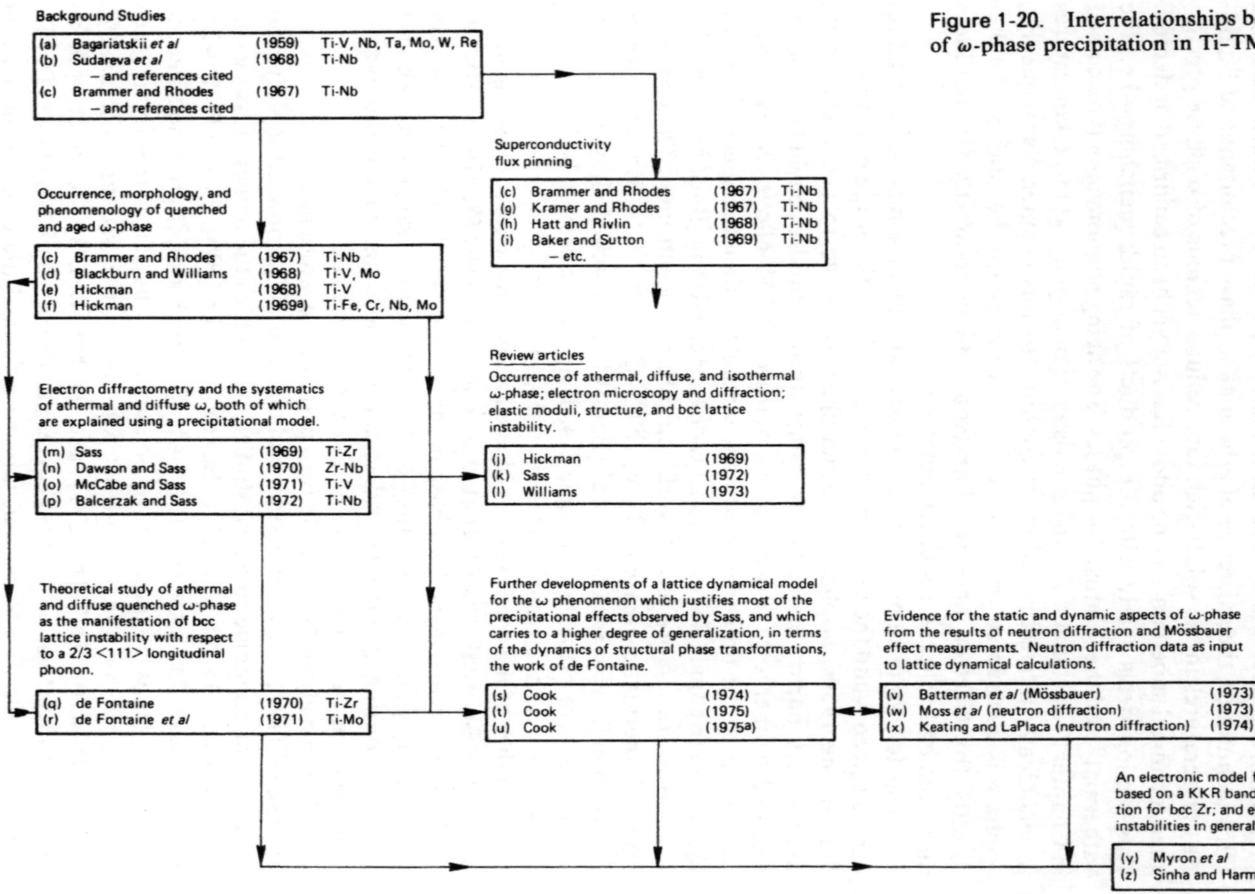

understanding of the athermal and "diffuse" quenched ω-phases, and indeed the entire ω-phase precipitation phenomenon.

It was early recognized [Bag58], [Sil58] that the relationship between the ω-precipitate and the parent lattice, defined by

$$(0001)_\omega \| (111)_\beta;\ [2\bar{1}\bar{1}0]_\omega \| [1\bar{1}0]_\beta$$

could be developed by applying to pairs of adjacent $(110)_\beta$ planes in the bcc lattice an equal and opposite shear in the $\langle 111 \rangle_\beta$ direction through a distance about equal to 1/6 of the separation of the $(111)_\beta$ planes. The arrows in Figure 1-21 indicate the planes, or rows, of atoms involved and the directions of the shear required. The coherent β/ω-interface at the boundary $(100)_\beta$ plane is an important feature. If d is the separation of the $(111)_\beta$ planes ($=a\sqrt{3}/2$, where a is the bcc lattice parameter), displacements of the A and B atoms by $\pm(1/6)d$ lead to the hexagonal structure of Figure 1-21, proposed originally by Silcock [Sil58] as a result of measurements on a low-solute-content ω-phase. Displacements of $\pm 0.15d$ yield the trigonal structure originally proposed by Bagariatskii (see [Bag59]) and now understood to be characteristic of higher-solute-content ω-phases. According to the constant-diameter (ϕ) hard-sphere model of Figure 1-21, the separation of the $(110)_\beta$ planes, originally $2\sqrt{2}\phi$, shrinks to $(1+\sqrt{3})\phi$ upon transformation to ω-phase, a 3% contraction that justifies the tendency for ω-phase to form in Ti and Zr under pressure [Jay63]. Adoption of this conventional crystallographic approach enabled most of the gross features of quenched ω-phase, as determined by TEM and electron diffraction [Sas69], [Daw70], [Mcc71], [Bal72], to be interpreted. Figure 1-21 suggests a $\langle 111 \rangle$ texture to the ω-precipitation. Indeed, the model proposed by Sass *et al.* [Daw70], [Mcc71], [Bal72] was based upon $\langle 111 \rangle$ rows of particles,

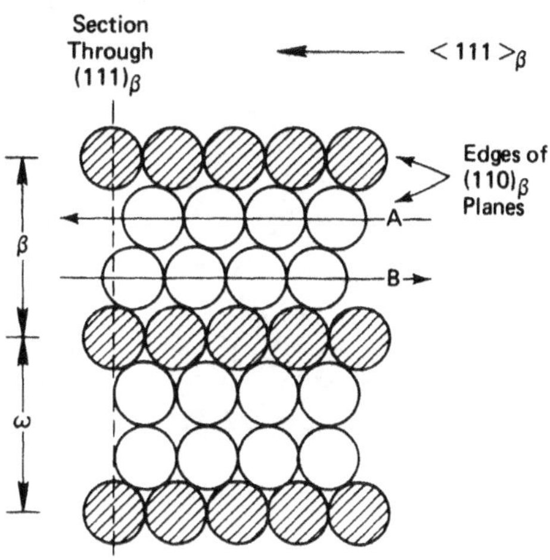

Figure 1-21. The "linear-fault" model for ω-phase transformation: $(101)_\beta$ section (in plane of page) through bcc crystal showing the transformation to ω-phase as a displacement of adjacent (here labeled A and B) $(110)_\beta$ planes—after Bagariatskii [Bag59].

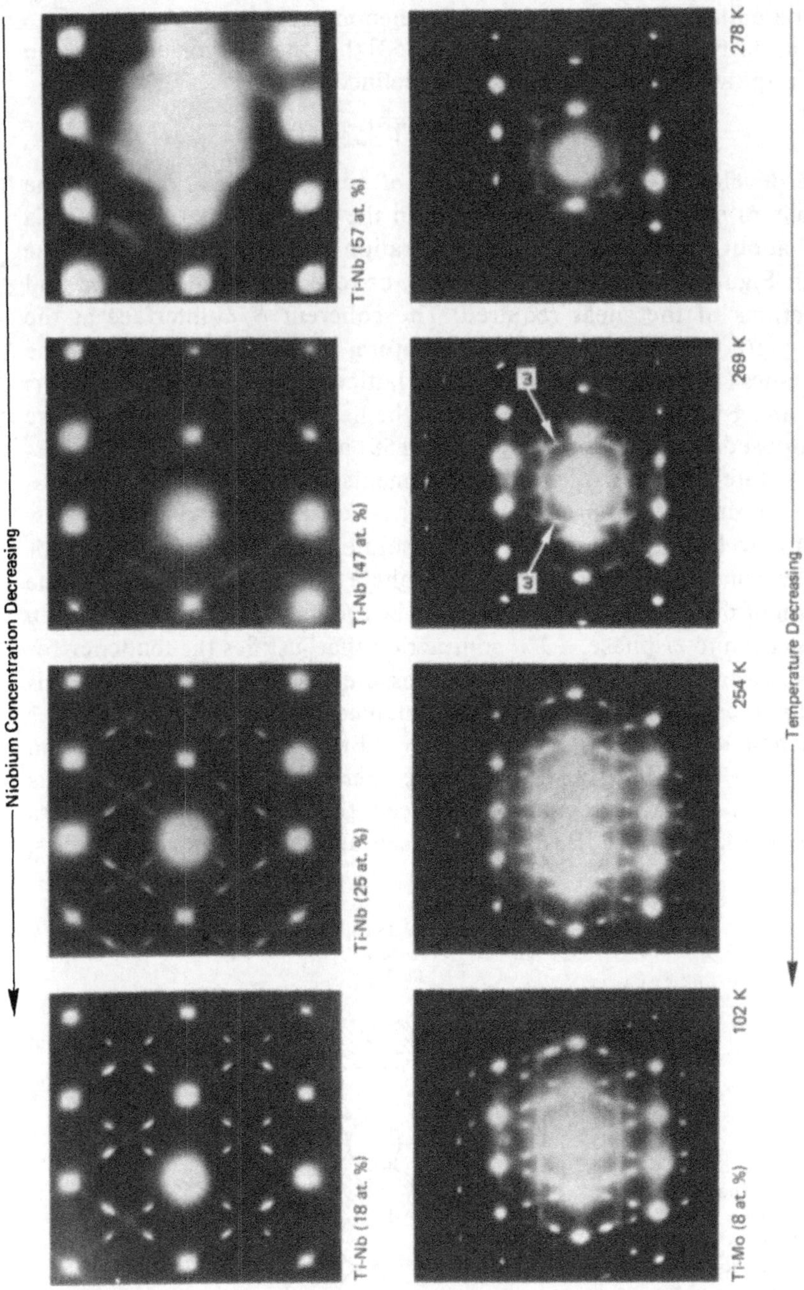

Figure 1-22. Changes from diffuse to sharp ω-reflections from quenched Ti–TM alloys in response to either decrease in the solute content or decrease in the temperature. (a) As-quenched Ti–Nb alloys showing the (110) reciprocal-lattice section—after BALCERZAK and SASS [BAL72], [SAS72]; photographs courtesy of S. L. Sass. (b) As-quenched Ti–Mo (8 at.%) showing the (131) reciprocal-lattice section—after DE FONTAINE *et al.* [DEF71] (see also [SAS72]); photographs courtesy of J. C. Williams.

10 to 15 Å in diameter and 15 to 25 Å apart. Athermal ω-phase, whose electron diffractograms (Figure 1-22) are characterized by sharp spots and straight "lines of intensity," was thought to be made up of clusters of such rows, while the broad reflections and the either straight or curved "lines of intensity" of the diffuse-ω were supposed to originate from either individual rows of particles or isolated particles, respectively.

But the reversible nature of ω-precipitation under temperature cycling between room temperature and 100 K calls for more than a static crystallographic interpretation. Figure 1-22 compares the composition dependence of the electron diffractograms of a Ti–TM alloy with its temperature dependence. Qualitatively the effect of lowering the temperature is similar to that of lowering the solute concentration; in particular, the curvilinear lines of diffuse intensity become, in both cases, straight and well defined. The soft-phonon mechanistic model of the ω-phase effect, originating with the work of DE FONTAINE [DEF70] and developed more fully in a paper by DE FONTAINE, PATON, and WILLIAMS [DEF71], provided a satisfactory rationalization in lattice-dynamical terms for both the temperature dependence and composition dependence of the quenched (both athermal and diffuse) ω-phases.

The *dynamical* equivalent of SASS's linear-fault crystallographic model centers around the proposed existence of a longitudinal phonon propagating in the ⟨111⟩ direction. The wavelength necessary to achieve (with the aid of anharmonicity) the necessary displacements of the A and B atoms of Figure 1-21 is illustrated in Figure 1-23, which depicts a unit cell of the

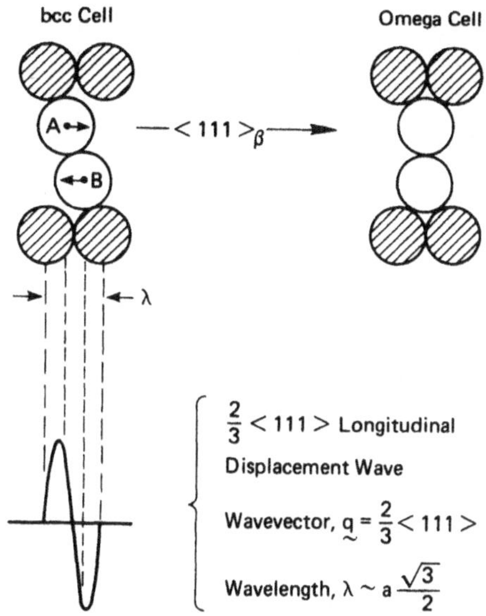

Figure 1-23. Development of the ω-structure through the application of $\frac{2}{3}$ ⟨111⟩ longitudinal displacement wave to the bcc lattice—after SASS [SAS72].

earlier figure. Clearly, if the shaded atoms are to remain unmoved, while A and B are to be shifted in opposite directions as shown, a longitudinal wave of wavelength equal to the separation of the $(111)_\beta$ planes is needed; *viz.*, a longitudinal phonon with wave vector $\frac{2}{3}\langle 111\rangle$. It is the instability of the bcc lattice to this disturbance that is responsible for the athermal transition. If the crystal is removed from a region of instability, by a change of either composition or temperature, the $\frac{2}{3}\langle 111\rangle$ phonon becomes responsible for the so-called reciprocal lattice "streaking" effect, also known as diffuse-ω (Figure 1-22). With particular reference to the Ti–V system, Table 1-7 summarizes and compares the lattice-dynamical and linear-fault approaches as they would apply to quenched ω-phase in Ti–TM alloys in general. Both philosophies agree in the athermal-precipitate regime in that they represent, respectively, "cause" and "effect." In the diffuse regime, the static approach is inadequate; since diffuse-ω domains cannot be imaged in the electron microscope, the linear-fault is incapable of explaining the curvilinear "lines of intensity." The phonon model achieves this by invoking the coupling of displacement modes—the combined process, for example, being represented by a hyperbola asymptotic to the straight lines of the independent processes [DEF70].

Further developments and generalizations of the lattice-dynamical model for phase stability, with particular reference to the quenched ω-phase phenomenon, have been made by COOK [COO75, COO75[a]] with help from the observations of SASS and DE FONTAINE and the results of the neutron diffraction studies, chiefly of Zr–Nb alloys, by AXE, KEATING, and MOSS [MOS73], [KEA74]. Lattice fluctuation phenomena in the region of the diffuse ω-phase have been probed using both neutron diffraction [MOS73], [KEA74] and Mössbauer effect measurements [BAT73]. There seems to be little doubt that the $\frac{2}{3}\langle 111\rangle$ soft mode, already present in the pure "solute" elements, interacting with a lattice of temperature-and-composition-dependent relative stability is responsible not only for the athermal precipitate but also for the diffuse-ω, which represents to varying degrees dynamical fluctuations between the crystalline β- and ω-phases.

It is, of course, important to remember that even the lattice-dynamical model is phenomenological in that (i) it relies on a virtual-crystal type of approximation (identical atoms), and (ii) the electronic properties are disguised as force constants. A more fundamental understanding of the transformation [COO73] requires an examination of what is generally referred to as the electronic structure of the alloy. It is not possible or appropriate to deal with the problem in this space. It is sufficient to mention that SINHA and HARMON [SIN75] constructed a dielectric screening model for the treatment of general so-called "electronically driven" instabilities, while MYRON, FREEMAN, and MOSS, with the aid of KKR calculations of the band structure and Fermi surface of bcc Zr and some assumptions regarding

Table 1-7. Electron Diffraction and Lattice-Dynamical Approaches to ω-Phase Formation in Group IV Base Transition-Metal Binary Alloys. (The Ti–V system has been selected as an example. The headings of each of the last two columns of the table indicate consecutively: Technique, Author, Class of Alloy, and Approach [Col74].)

Typical Solute Concentration Ranges in Ti–V	Diffraction Effects as Seen in Electron Microscopic Studies of Quenched Alloys[a]	• Electron Microscopy • S. L. Sass[a] • Real Alloy • Static[c]	• Lattice-Dynamical Calculations • D. de Fontaine[b] • Virtual Crystal • Dynamical[c]
13 at.% V	Sharp ω reflections and straight lines of intensity	Clusters of rows of particles plus individual rows	Propagating (long-range) $\frac{2}{3}\langle 111\rangle$ longitudinal mode of wavelength $(\sqrt{3}/2)$ times (bcc lattice parameter)
13, 15 and 19 at.% V	Broad ω reflections and straight lines of intensity	Rows of 20-Å particles; spaced 22 Å, typically 5 to a row	$\frac{2}{3}\langle 111\rangle$ vibrational modes tending to become more localized and less coherent
25 at.% V	Broad ω reflections, lines of intensity curved	Isolated 15-Å particles	
25–55 at.% V	Weak, broad lines of intensity, becoming diffuse and circular as solute concentration increases	Not interpretable in terms of precipitation	Curvilinear diffuse reflections due to pronounced coupling between local vibrational modes

[a] Reference [Sas72].
[b] Reference [Def70, Def71].
[c] These two columns are coupled by the following: (i) the dynamical effect yields a static (observable) atomic displacement through anharmonicity; (ii) the ω-phase becomes "frozen in" as a result of solute–solvent interdiffusion in a real alloy.

the perturbation of this structure during the addition of Nb, directly addressed the subject of the ω-instability [MYR75].

All aspects of the ω-phase phenomenon have been fully reviewed by SIKKA, VOHRA, and CHIDAMBARAM [SIK82], to whose work the reader is directed for further discussion of the subject.

1.9 Summary—The Occurrence of the Martensitic and ω-Phases in Quenched Titanium–Niobium Alloys

For a useful discussion of martensitic transformation kinetics in Ti–Nb alloys, the work of JEPSON, BROWN, and GRAY is recommended [JEP70]. For the α'-variant, an upper concentration limit of 6–8 at.% has been reported by BAGARIATSKII *et al.* [BAG59]. The upper limit of 3 at.% Nb referred to by HATT and RIVLIN [HAT68], and perpetuated by some subsequent authors, may have originated in what seems to have been an error in Figure 1(d) of [BAG59] but which was corrected by implication later on in the same paper.

According to BAGARIATSKII *et al.* [BAG59], the solute concentration in Ti–Nb alloys for which ω-phase is formed on quenching is 17–18 at.%. This should not be confused with the composition of the ω-phase that forms during the prolonged aging of higher concentration alloys at temperatures below about 450°C (the "saturation composition of the ω-phase"), which, according to HICKMAN [HIC69[a]], falls within the range 6–11 at.% Nb. During quenching, the precipitation of ω-phase from alloys situated within the composition range 6–11 at.% Nb to 17–18 at.% Nb is preempted by martensitic transformation. The picture with regard to alloys in the concentration range 18–22 at.% Nb is also complicated, although there are no significant points of disagreement among the several contributors to the subject. It has been claimed that the quenching microstructures depend not only on the quench rate (as is usually the case), but also on the temperature at which the sample is held immediately prior to the quench. For example, BALCERZAK and SASS [BAL72] showed that the oil-quenched microstructure of Ti–Nb(18.4 at.%) was essentially either α'' or $\omega+\beta$, depending on whether the prior annealing temperature had been 900 or 1000°C, respectively. In general, the transformation to α'' seems to preempt the formation of ω-phase, although a *trace* of α'' can coexist with $\omega+\beta$. For Ti–Nb(20.7 at.%) HATT and RIVLIN [HAT68] showed that water quenching from 900°C yielded α'' plus a trace of β (whereas slow cooling, of course, led to $\omega+\beta$), in agreement with BAKER and SUTTON [BAK69], who obtained the same result with a similar alloy. In Ti–Nb(22.6 at.%), BALCERZAK and SASS [BAL72] obtained essentially $\alpha''+\beta$ during a quench from 800°C, and $\omega+\beta$ in quenching from either 900 or 1000°C. From a heat-content standpoint, it is reasonable that a quench from the lower annealing temperature should be the more rapid; this is true to the extent that there may not, in fact, be a real distinction between "quench rate" and "prior annealing temperature." HICKMAN [HIC69[a]] achieved relatively rapid cooling rates (10^3°C s^{-1}) during the He-gas quenching of resistively heated strips of alloy and was thereby able to suppress ω-phase in all the alloys examined, *viz.*, Ti-Nb($\geq$22 at.%). In particular, Ti-Nb(22 at.%) yielded $\alpha'' + \beta$, in agreement with the results of the 800°C quench of BALCERZAK and SASS [BAL72]; gas-quenched Ti-Nb(25 at.%) was also $\alpha''+\beta$, whereas the structure of Ti-Nb(25.6 at.%) oil-quenched from 900°C by BALCERZAK and SASS [BAL72] was essentially $\omega+\beta$. HICKMAN'S Ti-Nb(27 at.%) was found to be "all β" [HIC69[a]], while for Ti-Nb(34 at.%) the quenched structure obtained by BALCERZAK and SASS [BAL72], although designated "β," yielded an SAD pattern exhibiting the so-called "lines of intensity." These lines, which are also present in association with ω-reflections in the lower concentration alloys, persisted in decreasing intensity through Ti-Nb(57 at.%) and have actually been noted in pure Nb [SAS72]. It is permissible to regard the "lines of intensity" as arising from "diffuse-ω" since they are believed to originate from the influence on the bcc lattice of that same vibrational state [SAS72] that gives rise, in the lower concentration alloys, to the athermal ω-phase itself. The occurrence of this effect as a function of composition in Ti-TM alloys has been summarized in tabular form (with particular reference to Ti-V) by COLLINGS [COL74] (Table 1-7). The lattice-instability phenomenon itself has been discussed from lattice-dynamical and thermodynamic standpoints by COOK [COO75, COO75[a]] and from an electronic point of view by SINHA and HARMON [SIN75] and MYRON *et al.* [MYR75].

Chapter 2

Aging and Deformation

2.1 The Aging of Quenched β-Titanium Alloys

Heat treatments are referred to as "annealing" or "aging" depending on whether they take place at elevated or moderate temperatures, respectively. The equilibrium structures considered in Chapter 1 were of course achieved by annealing or aging followed by quenching to room temperature. But of particular interest in this chapter are the nucleation-and-growth processes that occur and the metastable states that are achieved at intermediate stages during moderate-temperature isothermal aging. To be considered below, with reference to binary and multicomponent Ti–TM alloys, are:

(1) the aging of the $\omega+\beta$-structure;
(2) the precipitation from β-Ti–TM alloys of a second, solute-lean, bcc phase designated β' (the so-called "phase-separation" reaction);
(3) the decomposition of β into $\alpha+\beta$ and practical methods of distinguishing between the α-phase and the aged ω-phase precipitates;
(4) the effects of various ternary additions on the kinetics and products of the aging reaction;
(5) finally, consideration is given to β-phase decomposition into a pair of *equilibrium* bcc phases designated β' (solute lean, again) and β'' (solute rich). The purpose of considering this reaction is twofold: to draw attention to the distinction between the $\beta \rightarrow \beta'+\beta$ nonequilibrium phase-separation reaction and the reaction $\beta \rightarrow \beta'+\beta''$ where $\beta/(\beta'+\beta'')$ represents an equilibrium solid-state miscibility gap; and because this "β-phase immiscibility reaction" once played a role in the development of the ternary Ti–Zr–Nb technical superconductors.

2.2 The Athermal and Isothermal ω-Phases

2.2.1 Athermal ω-Phase

Athermal ω-phase has been shown to occur as a crystalline precipitate within a narrow composition range in quenched Ti–TM alloys. The sensitivities of the precipitation process to composition and experimental conditions have been explored in numerous Ti–TM alloy systems, of which Ti–Nb, considered below, is only one example. While BAGARIATSKII *et al.* [BAG59], on the basis of x-ray and hardness data, claimed a composition of 18 at.% for athermal-ω (a value which agrees with the results of an electron diffraction study by BALCERZAK and SASS [BAL72], HATT and RIVLIN [HAT68] found α″ to be the water quenched product in Ti–Nb(20.7 at.%) as did HICKMAN [HIC69[a]] in gas-quenched ribbons of Ti–Nb(22 and 25 at.%). On the other hand, in a pair of papers discussing the structure and superconducting properties of Ti–Nb(22 at.%), BRAMMER and RHODES [BRA67] and KRAMER and RHODES [KRA67], respectively, showed that the compositional limit for athermal ω-phase was already exceeded, and that the diffractographically defined "diffuse-ω," with its unresolvable real-space counterpart, was the quenched product.

2.2.2 Isothermal ω-Phase

During the moderate temperature aging ($\lesssim$ 450°C) of an alloy exhibiting the diffuse-ω reflections, decomposition into the isothermal ω- (solute lean) and β-phases take place. Numerous descriptions of this process as well as photomicrographs of the isothermal-ω precipitate are to be found in the literature. Reference to a few of the most important studies is made in Table 2-1. An early model [COU69] for the physics and kinetics of *isothermal* ω-phase development pictured an initial *structural* transformation of the lattice into ω and β (as for athermal ω-phase in *its* composition regime) followed by an exchange of solute and solvent across the ω/β-interface. DE FONTAINE, PATON, and WILLIAMS [DEF71], on the other hand, suggested that the first step was a *compositional* fluctuation. This was thought to be followed by structural β→ω transformations within the solute-lean zones instigated by $\frac{2}{3}\langle 111\rangle$ longitudinal lattice vibrations, which are already a property of the bcc lattice and have been shown (see Section 1.8) to be responsible for the *athermal* transformation at low solute levels. If such lattice vibrations indeed play a role in the isothermal process, it is not surprising that during the aging of Ti–V [MCC71] and Ti–Nb [BAL72] alloys the precipitates that form (cubic or ellipsoidal, respectively) appear to result from the growth of clusters of $\langle 111\rangle$ rows of particles.

Provided that the temperature is below about 400°C, prolonged aging results in a metastable ω+β-state characterized, at any given temperature,

Table 2-1. Important TEM Studies and Reviews of Isothermal Aging in Ti–TM Alloys

Alloy	Literature
Ti–V	Blackburn and Williams [BLA68] McCabe and Sass [MCC71] Williams[WIL73]
Ti–Cr	Luhman [LUH70]
Ti–Fe	Williams [WIL73]
Ti–Mo	Blackburn and Williams [BLA68] de Fontaine, Paton, and Williams [DEF71]
Ti–Zr–Mo–Sn	Williams [WIL78] Williams [WIL73]
Ti–Nb	Rhodes [ATO66] Kramer [ATO67] Kramer and Rhodes [KRA67] Brammer and Rhodes [BRA67] Sudareva [SUD70] Balcerzak and Sass [BAL72]

by fixed (in volume fraction and composition) ω-phase and β-phase end points [HIC69a]. Using x-ray techniques of the kind described by HICKMAN [HIC68, HIC69a], the volume fraction of the ω-phase may be obtained from the corrected relative intensities of the ω- and β-reflections. Then if da/dc is not too small, compositions can be calculated to ±0.1 at.% from calibrated lattice parameter measurements. It has been determined [WIL78] that the

Table 2-2. Time (h) at Temperature (°C) for the Appearance of α-Phase during the Aging of Quenched Ti–TM Alloys

	Time (h) at Temperature			
Alloy	400°C	450°C	500°C	Reference
Ti–V (15, 19, at.%)	20–30	<20		[HIC68]
Ti–V (19 at.%)			<4	[VET68]
Ti–V (25 at.%)	20–30	(No ω)		[HIC68]
Ti–Cr (9.3 at.%)	50			[HIC69a]
Ti–Mn (6.7 at.%)	68			[HIC69a]
Ti–Fe (6.0 at.%)	150	12	(No ω)	[HIC69a]
Ti–Nb (22 at.%)	72			[HAT68]
Ti–Mo (8 at.%)		320	50	[HIC69a]
Ti–Mo (10 at.%)		150	(No ω)	[HIC69a]

Table 2-3. End Point Data Relating to the Aging at 450°C to Metastable Equilibrium of Ti–Nb Alloys [HIC69[a]]

Nb Concentration at.%	Aging Time h	Volume Fraction of ω-Phase	Nb Concentration at.% ω-Phase	Nb Concentration at.% β-Phase
22	10	0.36	6–11	29 ± 1
	30	0.34	5–10	30 ± 1
	50	0.33	6–11	31 ± 1
25	10	0.25	7.5–12	30 ± 1
	24	0.26	7.5–12	30 ± 1

aged product bears the same crystallographic relationship to the parent lattice as does the athermal (quenched) ω-phase. After sufficiently long aging times at 450 and 500°C α-phase precipitation can be expected, as indicated in Table 2-2. The relative sluggishness of the decomposition process in Ti–Mo compared to that in Ti–Fe is a reflection of the great difference between the diffusivities in Ti of these two solute atoms (see Figure 1-12).

The compositions of the metaequilibrium isothermal ω-phases have already been given in Table 1-4 within the context of a discussion of the limits of β-phase stability. With the aid of corresponding data for the β-phase [HIC68, HIC69[a]], such as that presented in Table 2-3 for Ti–Nb(22 at.%), a semiquantitative metastable-equilibrium diagram for ω+β may be assembled, as in Figure 2-1. Represented in the figure are two

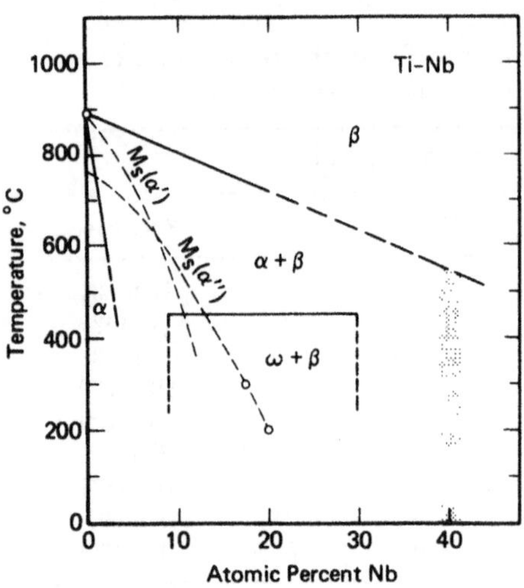

Figure 2-1. The locations of the α- and β-equilibrium transi, the $M_s(\alpha')$ and $M_s(\alpha'')$ transi and the regimes of occurrence of athermal and isothermal ω-phases in Ti–Nb—data sources: α- and β-transi, MOLCHANOVA [MOL65, p. 20]; M_s transi, JEPSON *et al.* [JEP70], and FLOWER *et al.* [FLO82]; ω+β-phase data, HICKMAN [HIC69[a]].

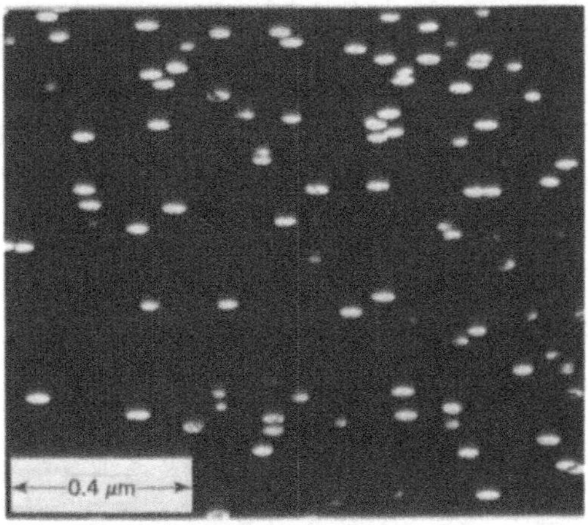

Figure 2-2. Dark-field electron micrograph of cubic ω-phase precipitation in Ti-10Fe—after WILLIAMS [WIL73] (see also [WIL78, Figure 17(a)]); micrograph courtesy of J. C. Williams.

important experimental results: (i) martensitic transformation tends to supersede athermal-ω transformation in quenched alloys (cf., for example, the results of BAGARIATSKII *et al.* [BAG59] and HATT and RIVLIN [HAT68] mentioned earlier) and (ii) at 450°C the maximal Nb concentration for isothermal ω-phase precipitation is 30 at.%. This is probably also true for lower aging temperatures, as judged from the results of the isothermal aging of the alloys Ti-V(15 at.%) [HIC68] and Ti-Cr(9.3 at.%) [HIC69[a]], whose $(\omega+\beta)/\beta$-transi, in the temperature zone 300–400°C, appeared to be practically vertical. The isothermally precipitated ω-phase particles assume one of two shapes—cubic or ellipsoidal—depending on the magnitude of the linear lattice misfit as represented by the function, $(V_\omega - V_\beta)/3V_\beta$, where V represents the unit cell volume divided by the number of atoms per unit cell [HIC69[a]]. If this is large (1–3%), as in Ti-V, -Cr, -Mn, and -Fe,

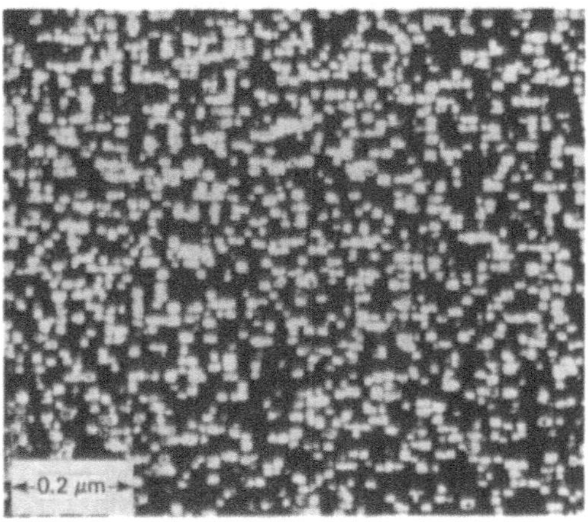

Figure 2-3. Dark-field electron micrograph of ellipsoidal ω-phase precipitation in Ti-11.5Mo-6Zr-4.5Sn—after WILLIAMS [WIL73] (see also [WIL78, Figure 17(b)]); micrograph courtesy of J. C. Williams.

minimization of elastic strains in the cubic matrix dictates a cubic morphology, as shown in Figure 2-2; if the misfit is small (<0.5%), as in Ti–Mo and Ti–Nb, the morphology is dominated by surface-energy considerations leading to the ellipsoidal particle shape depicted in Figure 2-3. The influence of misfit on ω-particle morphology has been graphically demonstrated by WILLIAMS *et al.* [WIL71] in experiments in which 5.5 at.% Zr was added to Ti–V(20 at.%); the resulting decrease in the misfit from 1.5–2.0% to ~0.25% was accompanied by a change in the precipitate shape from cubic to ellipsoidal.

2.3 β-Phase Separation

2.3.1 Occurrence of the Reaction

When the temperature is too high [LUH70], [WIL71] or the alloys too concentrated [WIL78] to support ω-phase precipitation, a solute-lean bcc precipitate, designated β', separates out. The relationship in temperature/composition space between the metastable $\omega+\beta$- and $\beta'+\beta$-phase fields is schematically shown in Figure 2-4, which is based on Figures 1-11 and 2-1 as well as some suggestions by WILLIAMS *et al.* [WIL71]. Like $\omega+\beta$, the $\beta' + \beta$-phase mixture is metastable. But unlike ω, which can be generated by a displacement wave in a virtual crystal if need be, the β'-precipitate stems from the chemical differences between the solute and solvent atoms. The $\beta \rightarrow \beta'+\beta$ phase-separation reaction is the result of the tendency towards like-atom clustering, which is exhibited by alloy systems that show positive heats of mixing [CHA72[a]] or equivalent manifestations of a tendency for the alloying constituents to unmix. Partly as a consequence of the similarity in time-temperature conditions required for both β'-formation and ω-phase precipitation, the former (or at least the precursor stages of it) was for a time confused with ω-phase. This is no longer the case, the

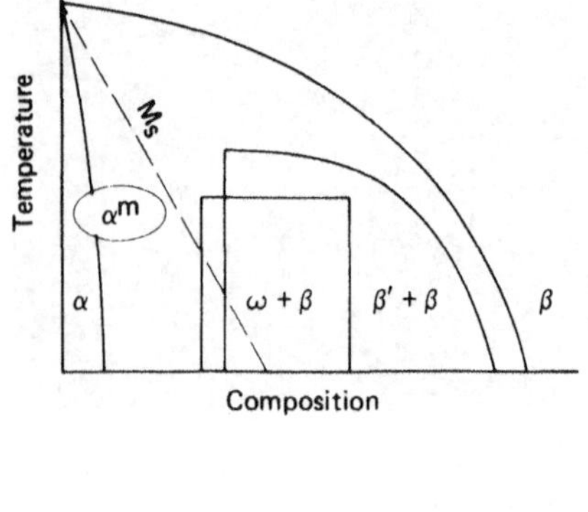

Figure 2-4. Schematic representation of the locations of the two metastable phases $\omega+\beta$ and $\beta'+\beta$, within the equilibrium $\alpha+\beta$-phase field in a typical Ti–TM alloy—after WILLIAMS *et al.* [WIL71].

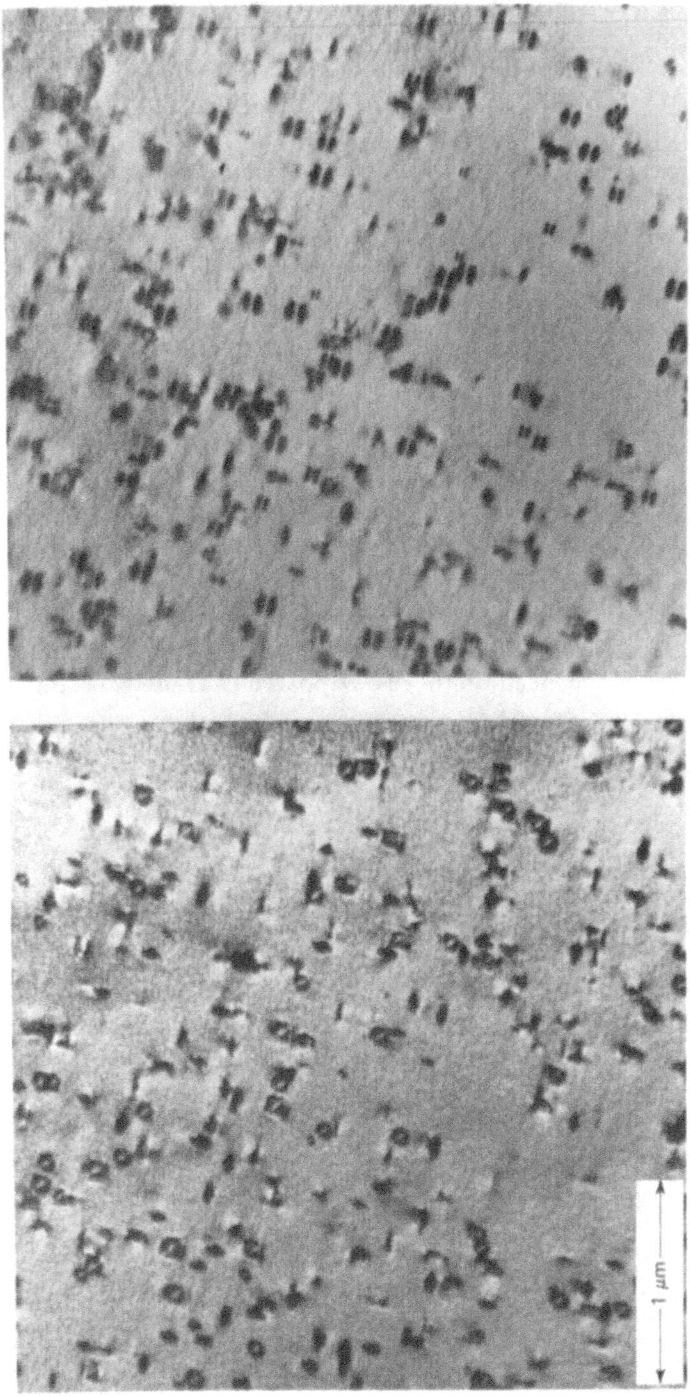

Figure 2-5. Thin foil micrograph showing β'-precipitates in bcc Ti_{72}–Zr_8–V_{20} in two orientations. Left-hand micrograph after WILLIAMS [WIL73]; both micrographs courtesy of J. C. Williams.

status of β' as a metastable bcc precipitate now being well established [WIL73, WIL78].

The $\beta \rightarrow \beta' + \beta$ reaction has been studied in considerable detail using Ti-Cr [NAR71], [LUH70], Ti-Mo [KOU70, KOU70[a]], [NAR70], [CHA72[a]], and Ti-Nb [RUD64], [MEN71] as model systems. Furthermore, its occurrence in multicomponent alloys, (e.g., Ti_{72}-Zr_8-V_{20}) is also well documented [WIL73, WIL78]. A typical example of β'-precipitate morphology is given in Figure 2-5. It is interesting to note that a structure very similar to the one depicted in the figure has been discovered in the Soviet alloy 65 BT (composition about Ti_{40}-Zr_7-Nb_{53}) after water quenching from 1250°C and aging 2 h/700°C [BUY70].

2.3.2 Ti–Cr

Ti-15Cr, when quenched, appeared to be single-phase bcc [NAR71], and after $5\frac{1}{2}$ min/300°C displayed the characteristics of diffuse ω-phase. But after holding the temperature at 450°C for 30 min, β'-precipitation in the form of thin discs, ~0.1 μm across and lying parallel to the {100} planes of the matrix [LUH70], was clearly discernible when photographed at a magnification factor of 52,000. For aging times longer than 90 min at 450°C α-precipitation commenced, in this case apparently at the β'/β-interfaces [LUH70].

2.3.3 Ti–Mo

Although early studies had indicated that Ti-Mo was a short-range-ordering system the converse is now known to be true [CHA72[a]]. HOCH and VISWANATHAN [HOC71], using a relative vapor pressure technique, have determined the activity of Mo in Ti and have shown it to be positive. At about the same time, GEHLEN, using the more direct technique of low-angle x-ray scattering, was able to show that Ti-Mo was a clustering system [HO69]. Thus, in a temperature-composition zone lying just outside the $\omega+\beta$-field the tendency for compositional modulation (which underlies the establishment of the equilibrium $\alpha+\beta$-field itself) stimulates in Ti-20Mo, for example at 450°C [CHA72[a]], a $\beta \rightarrow \beta' + \beta$ phase-separation reaction similar to that noted above in Ti-15Cr.

In Ti-Mo and also Ti-V alloys, WILLIAMS, HICKMAN, and LESLIE have considered the effects of composition, temperature, and third-element additions, on the relationship between the $\omega+\beta$- and $\beta'+\beta$-phases, noting that the electron-diffraction effect referred to as reciprocal-lattice streaking, which is characteristic of the existence of the $\frac{2}{3}\langle 111 \rangle$ phonon referred to earlier, was always present [WIL71].

2.3.4 Ti–Nb

RUDMAN [RUD64], in an x-ray diffuse scattering study, showed that Ti-Nb(25, 50, and 75 at.%) exhibited clustering-type local order. Subsequently, MENDIRATTA *et al.* [MEN71] found unmistakable evidence for β'-precipitation in aged Ti-Nb alloys, *viz.*, Ti-35Nb (aged 9 h/365°C), Ti-41Nb (aged 100 h/375°C), and Ti-45Nb (aged 100 h/400°C).

2.3.5 Thermodynamics of the Phase-Separation Reaction

The above examples serve to illustrate the general clustering tendencies of solid solutions of Ti with other transition metals. These tendencies may be gauged by, for example, the algebraic sign of the thermodynamic interaction parameter, Ω_{ij}. A list of interaction parameter values for some representative transition metals and simple metals against Ti are given in Table 2-4. As discussed by COLLINGS and GEGEL [COL75d], simple metals, which have negative interaction parameters, yield α-stabilized short-range-ordered systems when alloyed with Ti. Conversely, the transition elements, which have positive interaction parameters, are clustering-type β-stabilizers. KOUL and BREEDIS [KOU70, KOU70a], NARAYANAN and ARCHBOLD [NAR70], and GULLBERG *et al.* [GUL71] employed a bimodal pair of free-energy *versus* composition curves to describe phase separation as an equilibrium event. It must be remembered, however, that $\beta'+\beta$ is a *metastable* state of the β-phase, and as such must be described in terms of a single *inflected* free-energy (g) *versus* composition (c) curve (Figure 2-6), of the kind that gives rise to spinodal decomposition wherever $\partial^2 g/\partial c^2 < 0$.

Table 2-4. Regular Solution Thermodynamic Interaction Parameter, Ω_{ij} kJ mole^{-1}, between the Elements Listed and Titanium

Element	Interaction Parameter Ω_{ij} kJ mole^{-1}	Reference
Fe	+33.9	[COL75d]
Mo	+12.1 ± 2.9	[HOC71]
Cu	+8.0 ± 1.2	[HAC71]
V	+7.6 ± 0.5	[ROL71]
Cr	+7.5	[COL75d]
Ti	0.0	
Al	−37.7	[COL75d]
	−51.8 ± 3.4	[GEG73a]
Ga	−51.9 ± 2.9	[GEG73a]
Sn	−92.5 ± 3.3	[GEG73a]

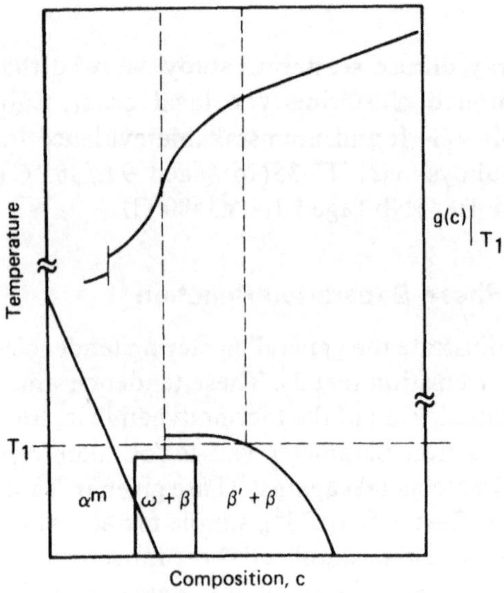

Figure 2-6. Schematic metastable equilibrium phase diagram for quenched-and-aged Ti–TM alloys accompanied by a postulated free-energy (g) curve for some temperature, T_1, showing a segment with negative curvature corresponding to the phase-separation region of the phase diagram.

The bimodal type of g *versus* c relationship, to be considered below, implies the existence of a *miscibility gap* in the β-field. Whereas this is not a feature of any of the Ti–TM phase diagrams, the Zr–Nb system does possess an elevated temperature field containing, in equilibrium, two immiscible β-phases designated β' (solute lean, as before) and β'' (Nb rich). The equilibrium $\beta'+\beta''$-phase is discussed below in connection with the Ti–Zr–Nb system.

2.4 α-Phase Precipitation from β-Titanium Alloys

2.4.1 Direct Precipitation

Outside the composition and temperature ranges of the ω- or β'-phases, represented schematically in Figure 2-4, but still within the equilibrium $\alpha+\beta$-field, α-phase precipitation must take place "directly" from the β-phase. Mechanisms of direct α-phase precipitation are outside the scope of this work but have been reviewed in detail by WILLIAMS [WIL73, WIL78]. A well-known morphology is the Widmanstätten α-phase, which, since it forms as platelets parallel to $\{110\}_\beta$, takes on a characteristic basket-weave-like microstructure under optical magnification. The interface between the α-plates and the matrix is then the source of further, apparently athermal [WIL78], α-phase. Other forms of direct α-phase precipitation, as discussed by WILLIAMS, have the appearance of large clumps $>1\ \mu$m across [WIL73] or of dispersions of fine particles $\sim$400 Å in diameter [WIL78].

Of greater interest in the present context is the manner in which α-phase nucleates *heterogeneously* from "defects" such as grain boundaries, dislocations, and the β'- and ω-phase precipitates.

2.4.2 Precipitation from the $\beta'+\beta$-Phase

LUHMAN [LUH70] showed that the prolonged aging of Ti-15Cr in the $\beta'+\beta$-field (e.g., >90 min/450°C) results in a nucleation of α-phase at the β'/β-interfaces. WILLIAMS *et al.* [WIL71], on the other hand, were careful to point out that for the Ti-Mo-Al alloy that they studied nucleation of the α-phase took place *within* the β'-region, indicating that in this case the composition difference between precipitate and matrix was the controlling factor. With only two examples, insufficient information is available to establish any systematics concerning the relative roles of lattice mismatch and composition gradient in the $\beta' \rightarrow \alpha$ nucleation reaction. Whatever the mechanism, an ability to nucleate α seems to be a general property of β'-precipitates. The subsequent growth of the α-phase may be into either acicular (a fine dispersion of star-like objects) or globular particles [WIL78].

2.4.3 Precipitation from the $\omega+\beta$-Phase

As indicated in Table 2-2, overaging of the $\omega+\beta$-phase generally results in α-phase precipitation. Several precipitation modes have been identified, none of which involve a direct conversion of ω-phase to α. As a matter of fact, in only one alloy system is the ω-phase precipitate itself involved in the conversion process; in Ti-V, BLACKBURN and WILLIAMS [BLA68] demonstrated the growth of α-plates at the ω/β-interfaces. The exceptional role played by Ti-V in this regard is not surprising in view of the fact that the ω-phase in Ti-V possesses one of the largest lattice misfits [HIC69]. In Ti-Mo and Ti-Nb, whose ω-phases have the lowest misfits, α-precipitation takes place elsewhere than near the ω-precipitate site. In Ti-Mo, large-scale lamellae of α- and β-phases have been observed growing out from a grain boundary [BLA68], and α-plates have been noted to nucleate at dislocations. Similarly, in Ti-Nb, α-phase nucleating at β-grain boundaries grew to consume the β- and ω-phases [BRA67]. With alloys such as Ti-Nb, lattice defects not only act as nucleation sites for α-precipitation, but also provide strain energy to accelerate the transformation process. The influence of prior deformation on the aging process is considered below.

In aging situations where ω- and α-precipitation might compete, it is important to be able both to distinguish between precipitate types when they are extremely fine and to measure their relative abundances when they are simultaneously present. With regard to the first problem, electron diffraction in crystallographic directions that include $\langle 111 \rangle$ provides an obvious means of distinguishing between ω- and α-precipitates, since in that direction all the allowed ω-reflections superimpose on the β-reflections [BAL72]. In experiments by RICKETTS [RIC69] the relative abundances of ω- and α-phases in response to aging were determined, using x-ray diffractometry, by comparing the intensities of the $(11\bar{2}0)_\alpha$ and $(2\bar{1}\bar{1}1)_\omega$ peaks. In addition, sample-to-sample normalization of the respective peak heights to that of the $(211)_\beta$ reflection enabled the aging responses of the individual phases to be determined.

2.5 Down-Quenching and Up-Quenching—ω-Reversion

By taking an alloy, previously aged isothermally in the $\omega+\beta$-field and, by raising the temperature, placing it in the $\beta'+\beta$-field, it is possible to exert some control over the β-phase separation reaction [LUH70]. The isothermal ω-phase "reverts" at the higher temperature to a β-phase leaner in solute content than the matrix. The reaction is not reversible—the β-phase precipitates remain in place as the alloy is returned to room temperature. The new phase could be thought of as β', since it is stabilized by the inherent tendency of phase separation (clustering) to occur in the β-phase [LUH70]. In studies of Ti–15Cr, LUHMAN [LUH70] observed that a larger number-density of finer precipitates resulted from up-quenching to 450°C for $2\frac{1}{2}$ min from isothermal $\omega+\beta$ aged at 300°C, than were found in the same alloy down-quenched to 450°C. Since, as before, β' provides sites for the nucleation of α-precipitates, these too can be produced as a fine dispersion as a result of the duplex aging treatment. Such heat treatments thereby afford a means of obtaining the dispersion strengthening characteristic of fine uniform precipitation, without the embrittlement effect of ω-phase [LUH70]. But even prior to α-phase precipitation the modulated $\beta'+\beta$-structure itself is associated with enhanced strength compared to that of the quenched homogeneous-β alloy and greater ductility than that of $\omega+\beta$ [POL71].

2.6 Effects of Third Element Additions on Precipitation in Quenched-and-Aged Titanium–Transition-Metal Alloys

The manner in which third-element additions influence the $\omega+\beta$- and $\beta'+\beta$-decomposition reactions can be extrapolated from the results of a 1970 investigation by WILLIAMS, HICKMAN, and LESLIE [WIL71] into the effects of Al, O, Zr, and Sn on precipitation in Ti–V(20 at.%) and Ti–Mo(6 at.%). Their conclusions, considered below, can be readily appreciated in the light of the preceding sections. As before, precipitation effects in the $\omega+\beta$- and $\beta'+\beta$-regimes (see Figure 2-4) will be treated separately.

2.6.1 The Ternary $\omega+\beta$-Phase Regime

Ternary additions of Al and O reduced the time of stability of the ω-phase during aging by promoting early precipitation of α-phase. In Ti-V(20 at.%) for example, the presence of 3 at.% Al reduced the time required at 400°C to produce a given volume fraction of α-precipitation from 100 to 24 h, and whereas in the binary alloy grain boundary nucleation was preferred, Al in solid solution promoted the formation of a uniform dispersion of precipitates.

The presence of 5.5 at.% Zr in Ti-V(20 at.%) places it, although within the $\omega+\beta$-field, close to its boundary with $\beta'+\beta$. As a consequence, $Ti_{74.5}$-$Zr_{5.5}$-V_{20} yields a relatively small volume-fraction of ω-phase upon aging. In contrast to the lattice misfit associated with ω-Ti-V, that associated with the ω-phase precipitated from the Zr-containing ternary alloy is sufficiently small that its particle shape is ellipsoidal rather than cubic (cf. Section 2.2).

Of particular interest from a superconductivity standpoint are the results of aging in the Ti-Zr-Nb system. A Zr-Nb-rich alloy will be dealt with later, but it is appropriate to consider in this section the erstwhile technically important Ti-Nb-rich superconductor, Ti_{60}-Zr_5-Nb_{35}, the location of which on the Ti-Zr-Nb equilibrium phase diagram has been given in Figure 1-10. Kitada and Doi [Kit69] conducted a detailed study of precipitation from previously homogenized (2 h/1000°C) and quenched material, in response to aging at between 350 and 500°C, temperatures which associate the alloy with the equilibrium $\alpha+\beta$-field (Figure 1-9). During aging, the β-quenched alloy decomposed according to the following scheme [Kit69]:

$$\beta \rightarrow \text{G.P.} + \beta \qquad (<1000\,\text{h}/350°\text{C})$$

$$\beta \rightarrow \text{G.P.}+\beta \rightarrow \kappa+\beta \quad \begin{bmatrix} <1000\,\text{h}/400°\text{C} \\ \ll 1000\,\text{h}/500°\text{C} \end{bmatrix}$$

$$\beta \rightarrow \text{above} \rightarrow \alpha+\beta \qquad (\gtrsim 1000\,\text{h}/\gtrsim 450°\text{C})$$

The "G.P. zone," some 50–200 Å in size, and the "intermediate precipitate," κ, are probably identifiable as the "early" and "late" stages, respectively, of isothermal ω-phase development. If so, the aging processes, which terminate with the development of a fine dispersion of α-precipitates, are in accord with those noted above in connection with the decompositions of Ti-V(20 at.%) and $Ti_{74.5}$-$Zr_{5.5}$-V_{20}.

2.6.2 The Ternary $\beta'+\beta$-Phase Regime

As was noted above in connection with binary-alloy decomposition, the aging of ternary alloys in the $\beta'+\beta''$-regime, which fringes the $\omega+\beta$-field

(Figure 2-4) eventually results in a β'-nucleated α-precipitation. Additions of solutes such as Zr and Sn to a binary Ti–TM alloy tend to stabilize the bcc structure when dissolved in it.[1] This is not to say they are "α-stabilizers" in the conventional sense, but rather stabilizers of the bcc lattice against the ω-instability. Zr and Sn thus impose constraints on the $\omega+\beta$-regime to such an extent that alloys like Ti_{72}-Zr_8-V_{20} and Ti_{84}-Mo_{10}-Sn_6 are left outside it and consequently lie in the $\beta'+\beta$-field. Aging then brings about the phase-separation reaction, which results in the precipitation of β', providing sites for the eventual precipitation of finely dispersed α-phase.

Finally, it is important to recall that, irrespective of whether the alloys were in the $\omega+\beta$- or $\beta'+\beta$-fields, the electron diffraction patterns were characterized by the reciprocal-lattice "streaking" effect discussed earlier in connection with the $\langle 111\rangle_\beta$ longitudinal displacement-driven ω-phase reaction. Such results confirm that, whether or not an ω- or a β'-transformation is eventually to take place, bcc transition-metal alloys (along with other classes of bcc metals (see, for example, SMITH *et al.* [SMI76])) tend towards $\frac{2}{3}\langle 111\rangle$ soft-phonon instability.

2.7 β-Phase Immiscibility

In contrast to the $\beta\to\beta'+\beta$ phase-separation reaction discussed above is a second class of β-decomposition process (this time an equilibrium one), with which, in one way or another, it has been frequently confused. Below a transus delineating the upper boundary of a region referred to as a "miscibility gap", a previously homogeneous single-phase-β solid solution decomposes into a macroscopic array of solution-lean (β') and solute-rich (β'') lamellae or cells. The reason for discussing it in this context is twofold: to draw attention to the nonequilibrium nature of the β'-precipitation process, and because the $\beta\to\beta'+\beta''$ reaction (in the terminology of DOI *et al.* [DOI66[a]]) has played an important role during the process development of Hitachi Ltd.'s Zr–Nb-rich superconducting alloy Ti_{10}-Zr_{40}-Nb_{50}.

A canonical example of $\beta\to\beta'+\beta''$ decomposition is provided by the system Zr–Nb whose equilibrium phase diagram is comparable to that of the general Ti–TM eutectoidal alloy system (Figure 1-2(b)), but with the Ti replaced by Zr and the intermetallic compound $TiTM_x$ (or γ) replaced by β''. Indeed, common to precipitation from both the aged $\beta'+\beta''$- (and in fact $\alpha+\beta''$) and the aged $\alpha+\gamma$-fields is the lamellar, cellular, or pearlitic morphology characteristic of classical discontinuous growth (see CHRISTIAN [CHR65, p. 472]). Precipitation from within the $\beta'+\beta''$-field does, however, possess an additional feature (to be discussed in detail below) not shared

[1]The solution strengthening of both α-stabilized and β-stabilized Ti–TM alloys has been discussed by COLLINGS and GEGEL [COL75[d]] [GEG73].

Table 2-5. Index of Literature to Alloy Phases in the Ti-Zr-Nb System and the Morphologies and Reaction Kinetics of Phases in Ti_{10}-Zr_{40}-Nb_{50}

Alloy	Properties Considered	Literature
Entire System	Ternary phase diagrams (500–800°C)	Alekseevskii *et al.* [ALE67[a]] Savitskii *et al.* [SAV73]
Entire System	Ternary phase diagrams (570–900°C)	Doi *et al.* [DOI66[a]]
Ti_{10}-Zr_{40}-Nb_{50}	Microstructure during aging (400–500°C) of 1100°C/quenched alloys	Kitada and Doi [KIT70[c]]
Ti_{10}-Zr_{40}-Nb_{50}	Microstructure during aging (620–800°C) of 1100°C/quenched alloys	Kitada and Doi [KIT70[a]]
Ti_{10}-Zr_{40}-Nb_{50}	Microstructure during aging (640–750°C) of cold-deformed alloys	Kitada and Doi [KIT70[b]]
Ti_{10}-Zr_{40}-Nb_{50}	Kinetics of $\alpha+\beta$-decomposition (400–550°C) and $\beta'+\beta''$-decomposition (650–800°C) in quenched and cold-drawn (25–99%) alloys	Soeno and Kuroda [SOE69]

by the $\alpha+\gamma$-precipitation process. An index to literature sources of Ti–Zr–Nb equilibrium alloy phase data and to studies of the morphology and reaction kinetics of Ti_{10}–Zr_{40}–Nb_{50} is provided in Table 2-5. In what follows, attention is focused on the $\beta\rightarrow\beta'+\beta''$ decomposition of the quenched alloy, which, according to the phase diagram of Figure 2-7, takes place within the temperature range 600–850°C. Passing reference will be made to $\alpha+\beta$-decomposition (temperature range 400–550°C as in Figure 2-7), which has

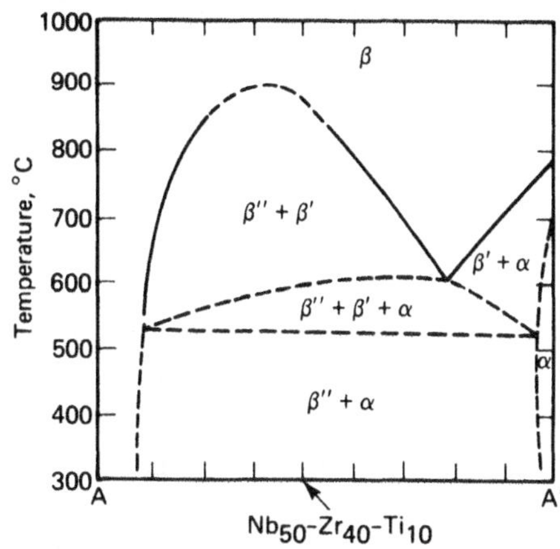

Figure 2-7. Equilibrium phase diagram (at.% linear) representing a slice through the Ti-Zr-Nb ternary prism at [Ti] = 10 at.%. Note that the eutectoid horizontal of the binary diagram (Figure 2-8) has developed into the three-phase $\alpha+\beta'+\beta''$ zone in the ternary slice—after KITADA and DOI [KIT70[c]].

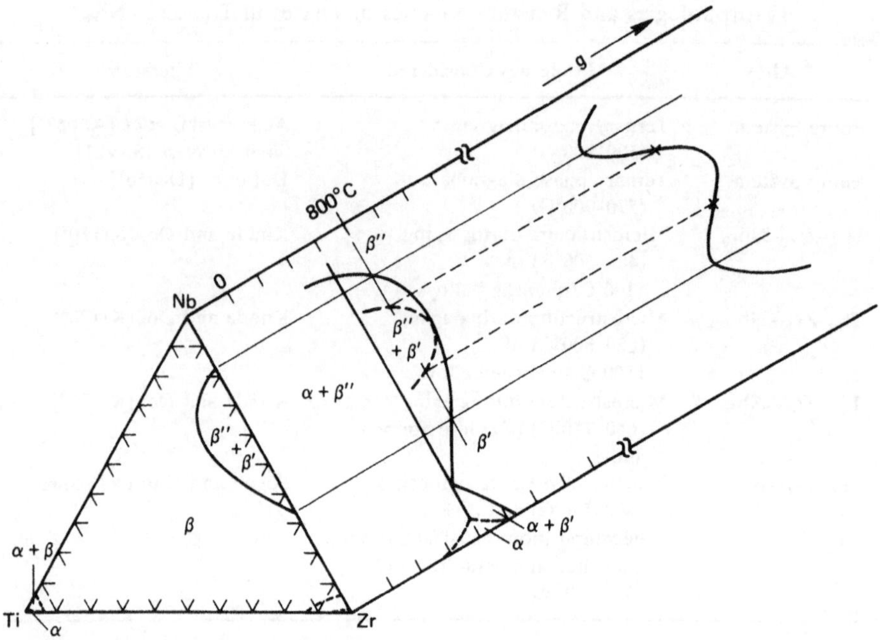

Figure 2-8. Binary equilibrium phase diagram for Zr–Nb (after LOVE and PICKLESHEIMER [LOV66]) projected from a side of the ternary (at.% linear) Ti–Zr–Nb 800°C-equilibrium phase diagram (after DOI *et al.* [DOI66[a]]). The two minima in the bimodal $g(c)$ curve define a region of equilibrium β-phase immiscibility, $\beta'+\beta''$, within which is a region of spinodal decomposition delimited by the points of inflection on $g(c)$.

a similar pearlitic morphology, but the influence of prior cold work on the aging process is postponed to a later section.

The development of the equilibrium $\beta'+\beta''$-region from the Zr–Nb edge of the composition triangle is shown in Figure 2-8, where it is juxtaposed against the appropriate isothermal section of the Zr–Nb binary phase diagram. Associated with the binary diagram is a bimodal free-energy curve representing the existence in *equilibrium* of the two β-phase solid solutions. Such a curve has been incorrectly applied to the $\beta' + \beta$ *metastable* situation [KOU70, KOU70[a]], [NAR70], [GUL71]. In contrast to $\alpha+\beta$-equilibrium, which is described by two independent parabolas, equilibrium phase-separation with its absence of structural change must be described by a *continuous* curve with two minima. An important corollary to this is the necessary presence of an intervening maximum with its associated description in terms of free energy: $\partial^2 g/\partial c^2 < 0$. In other words, away from the edges of the two-phase field, the $\beta \rightarrow \beta'+\beta''$ equilibrium process is preceded by spinodal decomposition. In Figure 2-8, two points on the spinodal

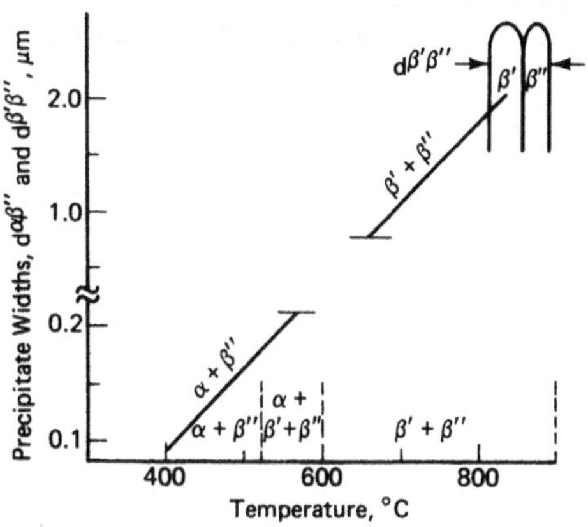

Figure 2-9. Interlamellar spacings of the $\alpha+\beta''$- and $\beta'+\beta''$-precipitates (cf. Figure 2-7) in Ti_{10}-Zr_{40}-Nb_{50} as a function of aging temperature—after KITADA and DOI [KIT70[a], KIT70[c]].

(indicated by the dashed curve on the phase diagram) are determined by the two points marked with an x on the *g versus c* curve for which $\partial^2 g/\partial c^2 = 0$. Precipitation phenomena interpretable in this manner have been seen during metallographic studies of quenched-and-aged Ti_{10}-Zr_{40}-Nb_{50} [KIT70[a]]. During the early stages of aging, according to KITADA and DOI [KIT70[a]], a very fine precipitate develops, presumably spinodally, throughout the grain interior such that after 10 h/700°C the hardness has increased from 280 to 338 kg mm^{-2}. In the meantime a lamellar or pearlitic arrangement of alternate β'- and β''-regions with a much lower hardness (~200 kg mm^{-2}) begins to develop outwards from the grain boundaries. The situation at 500°C is quite similar to this: a fine intragranular precipitate of β'' first appears, to be consumed by the growth of $\alpha+\beta''$-lamellae again originating at the grain boundaries [KIT70[c]]. Figure 2-9, a combined plot of precipitate width in both the $\alpha+\beta''$- and $\beta'+\beta''$-regions as a function of temperature, shows that the $\alpha+\beta''$-lamellae, with overall widths of typically 0.15 μm, are a factor of 10 finer than the $\beta'+\beta''$-lamellae.

The existence over certain temperature and composition ranges of β-phase immiscibility is an important characteristic of Ti–Zr–Nb alloys. At 700°C, the region of immiscibility has expanded to its maximum extent, which is consistent with the suppression of the $\alpha+\beta$-field along the Zr–Nb edge of the composition triangle (Figure 1-9). With its composition of Ti_{10}-Zr_{40}-Nb_{50}, Hitachi Ltd.'s Zr-Nb-rich alloy lay near the center of the 700°C $\beta'+\beta''$-field and was thereby in a position to take full advantage, for flux-pinning purposes, of the presence of the fine, evenly dispersed product of spinodal decomposition, augmented of course by the lamellate or pearlitic precipitate structure, which appeared during *aging* at that temperature.

2.8 Effects of Cold Deformation on the Microstructures of Quenched β-Titanium Alloys

2.8.1 Low- and High-Level Deformation Microstructures

The influence of stress on metastable quenched β-Ti–TM alloys, leading to the stress-induced martensitic transformation, has been discussed in Section 1.7.1. Considered below are the effects of extreme plastic deformation on the microstructures of higher concentration β-Ti–TM alloys. The subject is one of perennial interest. Studies of lightly strained material (e.g., after 0.2% deformation[2]) have enabled deformation processes to be identified with specific types of dislocation motion. For example in bcc Ti–V, electron microscopic observations have shown that at ordinary temperatures highly mobile edge dislocations move through the lattice trailing screw dislocations which can be said to exert control over the process [PAT74].

With particular reference to Ti–Nb and Ti–Nb-base alloys, studies have also been made of the effects of low levels of strain, such as: (i) that introduced by quenching and no additional cold work (Ti–Nb, [PRO68]), and (ii) the 2–3% and 12% strains, respectively, introduced by cold rolling (Ti–Nb(75 at.%), [NAR66]) in studies of the types of dislocation involved in the deformation processes and the finally attained dislocation structures. It has been noted for both Ti–50Nb [PRO68] and Ti_{10}–Zr_{40}–Nb_{50} [KIT70^a] that the act of quenching previously homogenized bcc alloys is itself sufficient to introduce a fine dislocation mesh. For example, the microstructure of quenched Ti–50Nb has been found to possess two interesting features: an elongated (~4 μm × 10 μm) dislocation network and subgrain boundaries formed by dense dislocation pileups [PRO68]. With regard to the effects of rolling, NARLIKAR and DEW-HUGHES [NAR66] identified the presence of screw dislocations during the early deformation stages (2–3% strain) prior to the formation of a well-developed cell structure at 12% strain.

Our present understanding of heavy deformation in metals can be traced back to an important series of papers by HIRSCH and coworkers (see references [53], [54], and [55] of [NAR66]), the essential results of which, as they apply to bcc metals, have been elegantly summarized by NARLIKAR and DEW-HUGHES [NAR66]. With the aid of a "microbeam" Laue back-reflection technique, in which the x-ray beam could be collimated through a capillary less than 20 μm in diameter, HIRSCH and coworkers were able to recognize the presence of crystallites of diameters as small as 1 μm. The importance of this work lay in its being able to show quantitatively that: (i) heavy deformation did not lead to a uniformly *disordered* structure but

[2]The stress required to produce 0.2% permanent strain is commonly accepted as the yield strength of a metal.

rather, in their words, to a "foam"-like structure consisting of particles of low dislocation density imbedded in a continuous three-dimensional net of highly dislocated material; and (ii) the resulting structure, instead of being the product of grain disintegration (and consequently true grain refinement), was in fact generated by dislocation motion within the grains such that the resulting dislocation network formed low-angle subgrain boundaries. In the English scientific literature these subgrains have also been referred to as "cells" or "subcells" and the dislocation network defining them as cell or subcell "walls" or "boundaries." The German[3] and more recent English literatures seem to prefer the use of "subband" and "subband wall (boundary)" to describe the same features. The presence of individual dislocation-free cells was disclosed by the appearance of spots on the Laue microdiffraction rings (see references [53] and [54] of [NAR66]). Later investigators, studying heavily cold-drawn [ARN74] or cold-swaged [LOH71] wire, using ordinary x-ray diffraction methods, have interpreted the spot patterns, which were noted in terms of ⟨110⟩ texture in the direction of the wire axis.

2.8.2 Further Studies of Cold Rolling

Relatively little attention has been given to the development of rolling texture in β-Ti alloys. Such information that is available has been reviewed by ZWICKER [ZWI74, p. 134]. The first study of the effects of cold rolling on the *superconductivity* of a Ti-Nb alloy was undertaken by HAKE, LESLIE, and RHODES [HAK63[a]]. They noted that the cold reduction of Ti-Nb(60 at.%) to a thickness ratio of 24:1 resulted in a pronounced laminar arrangement of wafer-like cells about 2 μm thick lying parallel to the rolling plane. As a result, the magnitude of the critical transport current, $I_c(\mathbf{H}_a)$, depended strongly upon whether the direction of the applied field, $\mathbf{H}_a$, was perpendicular or parallel to the plane of the laminations. For example, with $\mathbf{H}_a$ parallel to the rolling plane, I_c was about sixty times greater than when $\mathbf{H}_a$ was perpendicular to that plane. Of particular interest was the development of a minor I_c peak when $\mathbf{H}_a$ was rotated into a direction such that it was parallel to the plane of a group of laminae that had been distorted into a new direction during the shearing of the test specimen from a wider rolled strip. TEM observations of the in-plane structure of Ti-Nb(75 at.%), cold rolled to more than 90%, revealed a well-defined network of cells about 0.4 μm in size [NAR66]. In order to completely evaluate the cell structure of highly anisotropic cold rolled Ti-Nb, BAKER and TAYLOR [BAK67] carried out a TEM study of 97% cold rolled Ti-45Nb, taking observations both normal to the rolling plane and parallel to it along the rolling direction.

[3]According to HILLMANN [HIL73] "deformation bands" are 10^{-4} to 10^{-5} cm in width while "subbands," formed as a result of stronger deformations, are 10^{-5} to 2×10^{-6} cm wide.

The diameters of the in-plane cell cross sections, which were fairly equiaxed, were 0.25 μm or larger, in good agreement with the earlier result [NAR66]. But as expected, the end-view was one of severely flattened cells, about 0.1 μm thick and 2.5 μm or more in width.

KITADA and DOI conducted extensive investigations of the properties of ribbon conductors prepared from rolled sheets of Ti–Zr–Nb. In a series of papers, in which reference was made to (i) the laminar rolled structure (consisting of alternate "perfect" and "imperfect" layers[4] [KIT70]), (ii) the (110) and (112) pole figures of cold-rolled and cold-rolled-plus-heat-treated Ti_{60}–Zr_5–Nb_{35} [KIT72], and (iii) the microstructures and pole figures of heat-treated Ti_{10}–Zr_{40}–Nb_{50} [KIT70], attention was directed primarily towards the critical current, its anisotropy, and its response to variation of the thermomechanical processing variables. Further details of this work, which emphasized severe cold reduction (~99%) followed by a final low-temperature (~350°C) heat treatment for the development of flux-pinning sites in the Ti–Nb-rich alloy, but modest cold reduction (e.g., 80% and 50%) associated with intermediate or final heat treatments in the equilibrium $\beta'+\beta''$-field for the Zr–Nb-rich alloy, are given in Section 21.13.4.

2.8.3 Swaging

Swaging is employed commercially to reduce a Ti–Nb rod to a size suitable for drawing (Section 28.1) and has often been used on a laboratory scale for the same reason or to fabricate short lengths of bare wire [LOH73] or monofilamentary composite wire [MCI73] for experimental purposes. Many of the alloys studied by ZWICKER and his students have been cold reduced from 6-mm$^{\varnothing}$ cast rods to 1-mm$^{\varnothing}$ wire (corresponding to a cold reduction of 97.2%) using this technique (e.g., [LOH73]). After swaging, the longitudinal section exhibits a regular array of elongated cells not unlike those encountered following rod or wire drawing. Figure 2-10(a), a micrograph of such a structure developed in 97.2% cold-swaged Ti–50Nb, shows a cell diameter of 1–2 μm. The cross-sectional structure, on the other hand, is usually spiral in texture (Figure 2-10(b)) as a result of the rotation of the rod during its passage through the swaging machine. For this reason, swaging is quite unsuitable for the reduction of multifilamentary composite wire.

2.8.4 Flattening

Flattening is applied by light rolling to wire already deformed by drawing. The structure encountered is, therefore, a hybrid between those just discussed and the drawn structures to be considered below. BEST *et al.*

[4]Reference was being made to the dislocation density within the respective layers.

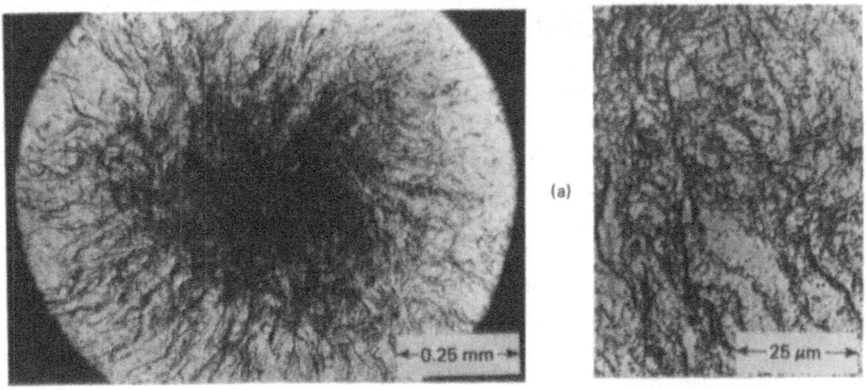

(b)

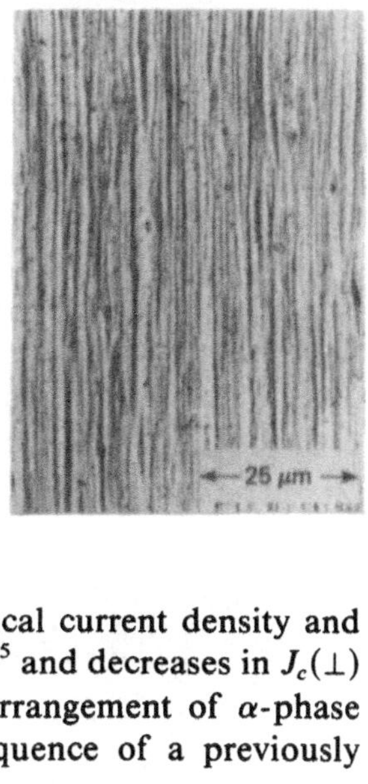

Figure 2-10. Ti-50Nb 1 mm$^{\varnothing}$ wire, cast plus cold-swaged 97.2%. (a) Transverse section, aged 16 h/400°C; optical micrographs, original magnifications, 100× and 1000×, respectively—after HELLER [HEL71]. (b) Longitudinal section, as cold worked; optical micrograph, original magnification, 1000×; etchant 30% HNO_3; 10% HF, 60% H_2O—after LÖHBERG [LOH71]; all photographs courtesy of U. Zwicker.

[BES79] considered the effect of flattening on critical current density and microstructure. The simultaneous increases in $J_c(\|)$[5] and decreases in $J_c(\perp)$ that took place were claimed to be due to a rearrangement of α-phase precipitate particles, already present as a consequence of a previously administered intermediate heat treatment.

BAKER and TAYLOR [BAK67], some of whose work has already been discussed above, investigated flattening under purely cold worked conditions. A wire, cold drawn 99.87% to 0.033 in.$^{\varnothing}$, was lightly cold rolled (4.7:1) to a thickness of 0.007 in. In consequence, an initial structure of parallel fibers some 0.1 μm in diameter underwent a coarsening when viewed normal to the rolling plane, and dislocation cells appeared within some of the fibers. It was expected that further cold reduction of this type would have transformed the structure to that observed in rolled material, which

[5] $J_c(\|)$ signifies J_c in the presence of a magnetic field applied parallel to the flattening (or rolling) plane; $J_c(\perp)$ has to do with a perpendicular orientation of $\mathbf{H}_a$.

had been shown to consist of equiaxed cells within the rolling plane and a fibrous structure at right angles to it. Indeed in a similar experiment, in which heavily cold drawn (99.994%) Ti–50Nb wire was reduced from a diameter of 0.25 mm to a thickness of 0.1 mm by cold rolling, PFEIFFER and HILLMANN [PFE68] noted that, although the in-plane structure was smeared out, especially along the sides, when viewed edge-on the initial fibrous structure of the wire appeared to have been scarcely disturbed.

The technical interest in flattening arises through its importance as a final metalworking operation in superconductor fabrication. By means of flattening, the filaments of a twisted monolithic composite can acquire partial transposition (see Section 28.10.2).

2.8.5 Wire Drawing

Wire drawing, a uniaxial deformation process, produces uniform deformation across a plane at right angles to the wire axis. Numerous photomicrographs are available in the literature showing an equiaxed cell structure in cross section and the usual elongated bands in the axial direction. One such pair of micrographs, for Ti–58Nb cold drawn to an area reduction ratio of $5 \times 10^4:1$ (99.998%) [NEA71] (Figure 2-11), depicts the microstructure of heavily cold-drawn wire as bundles of "pencil-shaped subcells" [NEA71] or "fibers" [BAK67] running parallel to the wire axis. The fibrous substructure of cold-drawn wire, the fundamentally important starting point from which the optimized microstructures of all technically important Ti–Nb and

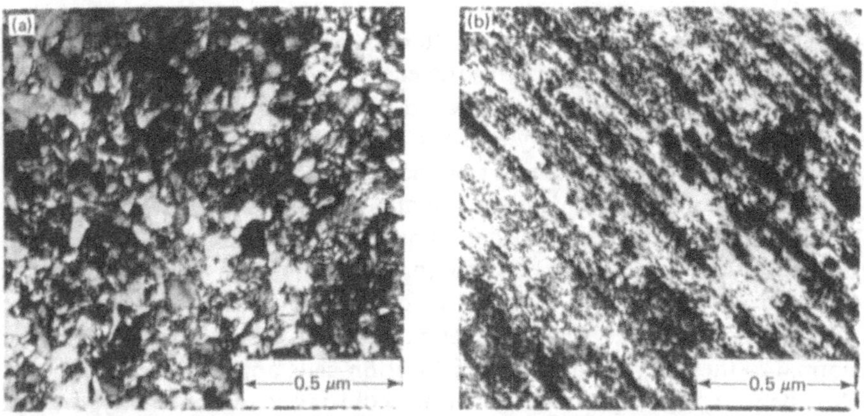

Figure 2-11. Ti–58Nb wire sample prepared for TEM from a single-filament Cu/Ti–Nb composite (core diam., 0.25 mm) in the as-cold-drawn ($5 \times 10^4:1$) condition: (a) transverse section, (b) longitudinal section. Original magnification, 110,000×—after NEAL *et al.* [NEA71]—micrograph courtesy of D. F. Neal. Copyright © 1971, Pergamon Press, reprinted with permission.

Ti-Nb-base alloy superconductors are developed, has been subjected to considerable scrutiny. In what follows, attention is drawn to the several key papers in the field.

In studies of cold-drawn texture, since sample smallness precluded the use of the usual methods of x-ray pole-figure determination, ARNDT and EBELING [ARN74], [WIL75[a]] employed an electron beam (20 $\mu m^{\varnothing}$) microdiffraction technique to view thinned cross sections of the wire. Although heat-treated wire, in which some cell growth had taken place, yielded single-crystal diffraction patterns, results interpretable in terms of texture could be obtained from wires with narrow, well-defined, straight subbands. In such cases a diffraction pattern consisting of sharp spots, representative of a superposition of many crystal orientations, suggested a strong ⟨110⟩ texture in the axial direction. The results of other drawing texture studies of Ti-Nb [HIL79[a]], as well as those of Ti-50Nb-1Cu and Ti-Mo [ZWI74, p. 136], were in agreement with this.

The use of the word *texture* suggests a microstructure consisting of grains separated by large-angle boundaries. Indeed, as pointed out by LARBALESTIER and WEST (see [LAR84] and the papers by LANGFORD *et al.* referred to therein) the drawn microstructure of Fe alloys changes from subgrain- (or subband-) dominated to textured-grain-dominated at strain levels corresponding to ARR = 4×10^2. In the case of Ti-Nb, this of course is much smaller than the ARRs encountered in superconductor processing.

Although the subbands in as-deformed wire are far from regular, it has been possible to quantify the rate at which the fiber diameter decreases in response to increasing amounts of cold work. For example PFEIFFER and HILLMANN [PFE68] showed that a subband density (i.e., number of cells per cm^2 visible in the transverse section) of $1.3 \times 10^{10}\ cm^{-2}$ in 99.78% cold reduced ($32 \rightarrow 1.5\ mm^{\varnothing}$) Ti-50Nb increased to $4.9 \times 10^{11}\ cm^{-2}$ after 99.993% ($1.42 \times 10^4 : 1$) reduction from 32 to 1.5 $mm^{\varnothing}$. The diameters of the cells decreased from 0.09 to 0.01 μm. Earlier results by BAKER and TAYLOR [BAK67] for 99.87% cold drawn Ti-55Nb with a cell diameter of about 0.1 μm were in good accord with this, as were later data by NEAL *et al.* [NEA71] for Ti-58Nb and by ARNDT and EBELING [ARN74], [WIL75[a]] for Ti-50Nb. A detailed study of deformation structures in Ti-54.3Nb after cold drawing to ARRs ranging from $3.0 \times 10^2 : 1$ to $8.8 \times 10^3 : 1$ (after extrusion) has been undertaken by WEST and LARBALESTIER [WES84] (see also [LAR84]). Some of these results are recorded in Table 2-6 and are depicted, along with the other data from that table, in Figure 2-12.

When plotted semilogarithmically as in the figure, the subband diameter, *d*, in the range 375–1000 Å is seen to decrease roughly linearly with $\log_{10}$ARR. The approximate formula

$$\log_{10}\text{ARR} \cong 5.58 - 3.61 \times 10^{-3} d(\text{Å}) \qquad (2\text{-}1)$$

represents a least-squares fit (correlation, −66%) to the tabulated data.

Table 2-6. Fiber Diameters as a Result of Cold Reduction by Drawing in Titanium-Niobium Alloys

Early Results[a]			Recent Data[b]		
Wt.% Nb	Area Reduction Ratio by Cold Drawing	Subband Diameter Å	Wt.% Nb	Area Reduction Ratio by Cold Drawing	Subband Diameter Å
50	4.6×10^2	880	54.3	2.96×10^2	690
	(1.6×10^4	140)[c]		4.59×10^2	800
				5.65×10^2	640
54	1.2×10^2	770		7.25×10^2	780
	4.4×10^3	375		1.11×10^3	660
				1.37×10^3	580
55	7.7×10^2	~1000		1.76×10^3	620
				2.11×10^3	570
58	5×10^4	443		2.67×10^3	650
	1.4×10^5	473[d]		3.40×10^3	620
	5×10^4	530[e]		4.49×10^3	560
	1×10^3	572[d]		5.49×10^3	470
	1×10^2	639[d]		7.06×10^3	540
				8.84×10^3	510

[a] Literature: Ti-50Nb [PFE68]; Ti-54Nb [ARN74]; Ti-55Nb [BAK67]; Ti-58Nb [NEA71]—represented by □ in Figure 2-12.
[b] Literature: Ti-54.3Nb [WES84]—represented by ○ in Figure 2-12.
[c] Not included in slope analysis.
[d] 1 h/385°C aging.
[e] Mean data for 1, 5, 25 h/385°C aging.

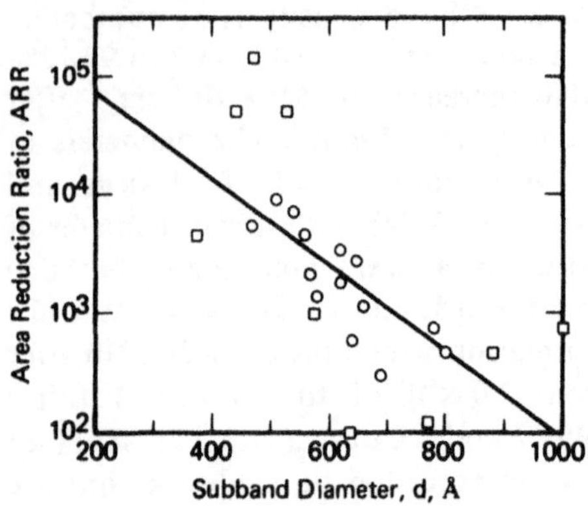

Figure 2-12. Subband diameter as a function of area-reduction ratio (ARR) by cold drawing. The straight line, which represents (2-1), is a least-squares fit (corr., −66%) to the plotted data (from Table 2-6).

2.8.6 Summary

In later sections, which deal with the influence of lattice defects on flux pinning, it will be shown that a subband structure aligned along the conductor axis, a few hundred Å wide when viewed in the direction of the applied magnetic field, and suitably decorated with precipitates is necessary for the transport of high current densities in strong magnetic fields. The fact that a high subband density, and not the simultaneous increase in dislocation density itself, is responsible for favorable flux pinning has been demonstrated by comparing the critical current density of wire with that of a ribbon flat rolled from the same wire [PFE68]. Since the dislocation density certainly did not diminish during the rolling, the considerable decrease in critical current density that took place was attributed to the deterioration of the previously existing favorable subband structure. Flux pinning by subbands and precipitates is considered in Section 21.15.

Flat rolling results in a fine laminar structure when viewed parallel to the rolling plane. It may lead to a useful conductor provided either that the anisotropy can be tolerated or, if not, that it can be reduced by the application of further processing.

Cold drawing to area reduction ratios of $10^3:1$ (99.9%), $10^4:1$ (99.99%), and $10^5:1$ (99.999%) results in usefully fine fiber diameters, especially if it is accompanied by suitable heat treatment.

2.9 Influence of Stress, Strain, and Interstitial-Element Additions on the Transformation Kinetics of Quenched β-Titanium Alloys

The equilibrium and metastable phases which appear during the aging of quenched β-Ti alloys were considered in Chapter 1. These *same phases* appear during the aging of deformed alloys but at *different rates* determined by the degree of strain and the level of retained stress. The following sections deal with factors that control the *kinetics* of the aging process. Since the results are intended to apply to Ti–Nb and Ti–Nb-base superconducting alloys, which inevitably contain as impurities the interstitial elements C, N, and O, mention is also made of the influences of stress and strain in the presence of interstitial elements on the transformation kinetics of quenched β-Ti alloys. The particular strain being considered is the deformation structure introduced by heavy cold work—usually wire drawing. Accordingly, the following sections are logical combinations and extensions of previous components of the chapter and, as such, also touch on the influence of heat treatment on the fiber morphology of wire-drawn alloys including of course those in which, by virtue of their composition, precipitation is not expected to take place.

The traditional descriptor of transformation kinetics is the time-temperature-transformation (T–T–T) diagram, in which phase boundaries are plotted semilogarithmically against temperature and time. Sets of the typically C-shaped T–T–T diagrams for the systems Ti–V, -Cr, -Mn, -Fe,

-Ni, and -Cu are given in a work by MAYKUTH *et al.* [MAY61] and the further references contained therein. Diagrams such as these are usually obtained by detecting isothermal transformation in a sample quenched from a single-phase elevated-temperature field to the temperature of observation. However, it must be borne in mind that, although of the same general form, such curves will not be coincident with those obtained by performing isothermal aging on specimens previously quenched to room temperature [MAY61]. The characteristic C-shape of the T–T–T diagram demands an explanation. Microstructural studies of samples quenched from various regions of the temperature–time diagram indicate that at high temperatures transformation begins at lattice defects such as grain boundaries and grows into the grains as a discontinuous precipitate, while at the low-temperature extreme the precipitation tends to take place uniformly throughout the grains (continuous precipitation). These effects result from the fact that the driving force for nucleation (the Δg between the pairs of phases involved) scales with some measure of the departure of the quenched alloy from its equilibrium state. Once nucleated, the precipitate growth rate increases with temperature; consequently, the transformation rate, which is proportional to the product of nucleation rate and growth rate, is maximum at some intermediate temperature [CHR65, p. 453]. The separation of the C-curve into a low-temperature nucleation-controlled region and a high-temperature growth-controlled region leads to an understanding of the manner in which stress, strain, and interstitial atoms influence its shape. Thus, if the density of nuclei for transformation is increased, by deformation or some other means, the effect will be to increase the transformation speed at high temperatures where nucleation was difficult, more so than at low temperatures where the driving force for nucleation was high and the alloy was already well supplied with nucleation sites. The result of combining a more nearly even distribution (with temperature) of nucleation rates with a growth rate that increases with temperature is to shift the C-curve towards the temperature axis by an amount that increases uniformly with temperature; in so doing this also has the effect of raising the temperature of the "nose," as illustrated in Figure 2-13. Three specific examples of this effect are presented below.

The time-temperature data-pairs representing some selected fraction of material transformed (e.g., 5% for the start of the transformation and 95% for its end) may be obtained either *at-temperature*, using a diagnostic technique such as dilatometry or resistivity measurement [HOR73[a]], or from interpretation of *quenched structures* using hardness or metallography, which in any case should always be employed as a semidirect observation and control. Two mechanical variables, stress level and degree of plastic deformation, each of which has a strong influence on the *decomposition kinetics* of β-Ti alloys, are considered below. Also to be discussed is the influence

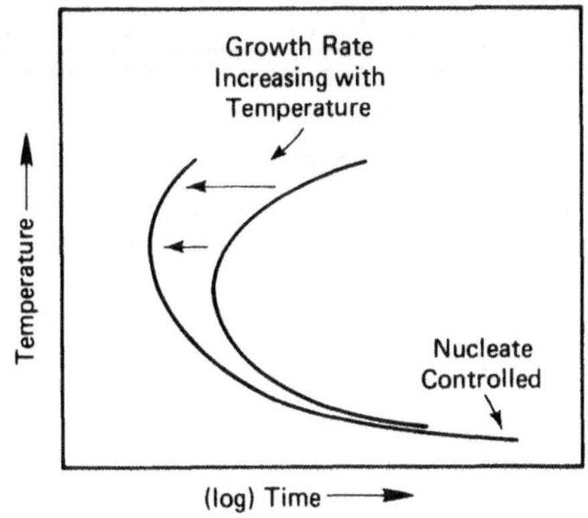

Figure 2-13. Schematic representation of the influence of a temperature-dependent increase in precipitate growth rate on the shape of the T-T-T curve. For a given nucleation rate the indicated increase in reaction rate (=product of nucleation and growth rates) has the effect of raising the "nose" of the T-T-T curve and giving it a "forward tilt."

of third-element addition. This has, of course, just been considered in connection with aging transformation in *quenched* β-alloys; but inasmuch as interstitial elements have a pronounced tendency to accumulate at dislocations and grain boundaries during aging, they also play an important role in the phase decomposition of plastically deformed material.

2.10 Influence of Stress on the Transformation

Studies by GOLDENSTEIN *et al.* [GOL59] on a series of Ti–Cr alloys, although dealing with the eutectoid transformation $\alpha+\beta\rightarrow\alpha+TiCr_2$, provide a useful example of stress-assisted[6] decomposition. Application of a relatively low level of stress (capable of producing 1% plastic strain in 1000 h) during isothermal aging reduced the time needed at 500°C to complete the eutectoid reaction in Ti–12Cr from 200 to 32 h. In a like manner the so-called "internal stress fields" (see TYSON [TYS75]) of lattice defects may also speed the transformation process.

2.11 Influence of Heavy Plastic Deformation

It is common practice in alloy preparation laboratories to heavily deform the as-cast billet in order to accelerate the equilibration reaction during a subsequent homogenization heat treatment. With the alloy superconductors, the primary purpose of deformation is simply to produce wire or ribbon from a massive starting billet. But it was soon found that: (i) the fibrous substructure resulting from the drawing of cold material through

[6]The term *stress* as used here implied elastic strain as distinct from the (heavy) plastic strain to be discussed later.

cold dies had a most favorable influence on the current-carrying capacity; (ii) although the fiber structure generally deteriorated under heat treatment, this disadvantage was outweighed many fold by the development of a suitable arrangement of flux-pinning precipitates. Superconductor process optimization couples these factors with one having more to do with precipitation kinetics; namely that flux-pinning precipitation is accelerated by the presence of the fibrous subband structure.

To be discussed below are the influence of cold deformation on the kinetics of precipitation and the influence of aging on the cold-worked fiber structure.

2.11.1 Influence of Heavy Deformation on the Kinetics of Precipitation

The literature is replete with descriptions of the microstructures and precipitates that occur during the aging of cold-deformed alloys. Furthermore, information concerning the heat-treatment times necessary to attain maximal critical current densities under specific conditions, which is certainly related to kinetics, is commonly available. But although β-Ti–Nb and related alloys can decompose in several possible modes, such as

$$\beta \rightarrow \begin{cases} \omega+\beta \rightarrow \alpha+\beta \\ \beta'+\beta \rightarrow \alpha+\beta \end{cases}$$

$$\beta \rightarrow \quad \beta'+\beta''$$

very few quantitative inquiries have been made into the combined influences of deformation and temperature on the speeds of these reactions. T–T–T diagrams as functions of deformation have, however, been developed by SOENO and KURODA [SOE69] for two compositions of a nominal Ti_{10}–Zr_{40}–Nb_{50} alloy (whose binary equivalent, Ti–Nb(50 at.%), is not technically useful) and by HORIUCHI *et al.* [HOR73[a]] for the technically interesting quaternary alloy Ti_{33}–Zr_{30}–Ta_7–Nb_{30}. For the purpose of discussion, simplified diagrams are presented in Figures 2-14 and 2-15, respectively.

(a) Decomposition of Ti_{10}–Zr_{40}–Nb_{50}. As indicated in Figure 1-9, and again in Figure 2-7, the Zr–Nb-base alloy Ti_{10}–Zr_{40}–Nb_{50} may exist in one of four distinct equilibrium-phase fields: (i) $\alpha+\beta$ up to 525°C; (ii) a narrow $\alpha+\beta'+\beta''$-region between 525 and 580°C; (iii) the miscibility gap region $\beta'+\beta''$ between 580 and about 900°C; (iv) the single-phase β-region above that temperature. Figure 2-14, constructed from electrical resistivity (ρ) data, describes the kinetics of transformation of a related alloy, after deformation, into: (i) $\alpha+\beta$ from a predeformation equilibration in the

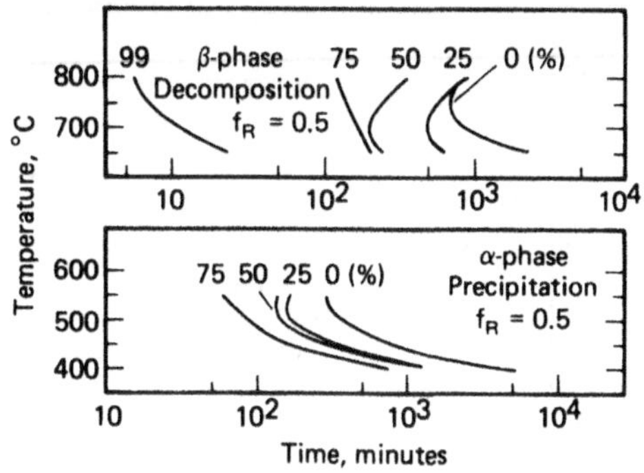

Figure 2-14. T-T-T diagrams representing the β-phase decomposition and α-phase precipitation reaction kinetics in cold-drawn $Ti_{4.4}$-$Zr_{40.7}$-$Nb_{54.9}$ (percent reductions indicated). The degree of completeness of the transformation, in resistometric terms, is $f_R = 0.5$ (where f_R is the fractional change in resistance)—after SOENO *et al.* [SOE69].

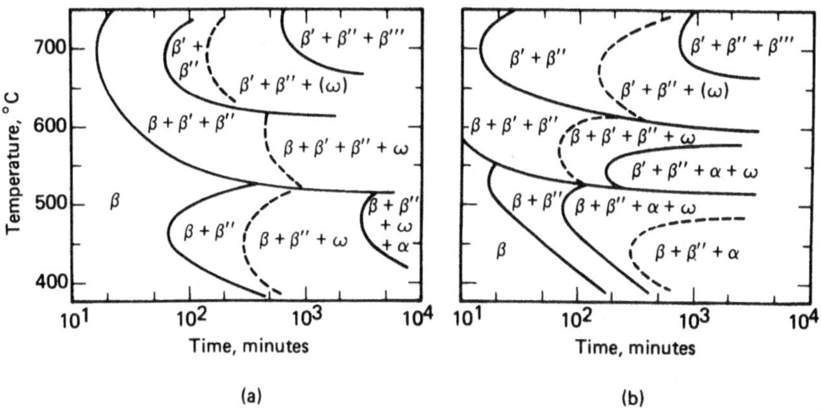

Figure 2-15. T-T-T diagrams representing β-phase decomposition and precipitation kinetics in Ti_{33}-Zr_{30}-Nb_{30}-Ta_7 cold reduced (a) 90%, and (b) 99%—after HORIUCHI *et al.* [HOR73[a]].

miscibility gap region at 700°C; (ii) discontinuous $\beta'+\beta''$ from predeformation equilibration in the β-regime at 1200°C. In order to quantify the effect of deformation on aging kinetics, the time of "50%" isothermal transformation (i.e., $f = \Delta\rho/\Delta\rho_\infty = 0.5$) at the temperatures 500°C ($\alpha+\beta$) and 700°C ($\beta'+\beta''$) may be plotted logarithmically *versus* area reduction ratio as in Figure 2-16.

(b) Decomposition of Ti_{33}–Zr_{30}–Ta_7–Nb_{30}. Figure 2-15, a transformation-start diagram for the quaternary alloy Ti_{33}-Zr_{30}-Ta_7-Nb_{30}, whose binary Ti-Nb "equivalent," Ti-Nb(37 at.%), would lie somewhere within the intermediate-concentration range to be defined below, shows the expected influences of heavy deformation on the reaction kinetics. Focusing attention on the $\beta\rightarrow\beta+\beta''$ or $\beta\rightarrow\beta'+\beta''$ transformations, for example, the effect of deformation is seen to be greatest towards the upper temperature limits of each range of stability. Thus, each C-curve is shifted, and tilted

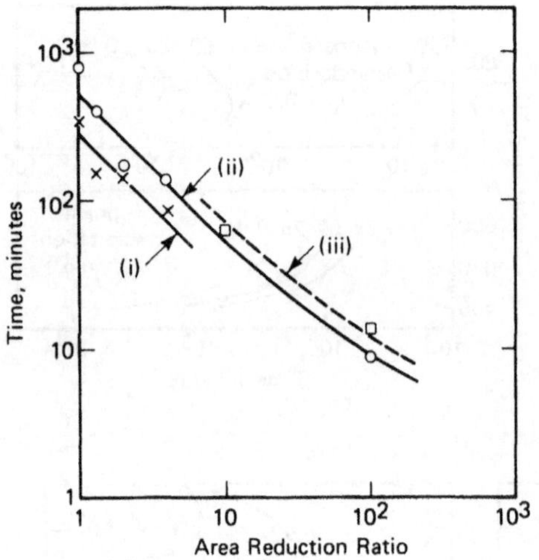

Figure 2-16. Influence of area reduction ratio on the time to 50% conversion in the reactions: (i) α-precipitation at 500°C; (ii) $\beta \rightarrow \beta' + \beta''$ decomposition at 700°C from the alloy $Ti_{4.4}$-$Zr_{40.7}$-$Nb_{54.9}$ (cf. Figure 2-14); (iii) $\beta \rightarrow \beta' + \beta''$ decomposition at 700°C from the alloy Ti_{33}-Zr_{30}-Nb_{30}-Ta_7 (cf. Figure 2-15).

towards the temperature axis, and its "nose" temperature is raised. The response of the 700°C $\beta \rightarrow \beta' + \beta''$ reaction-start time to a change of cold-deformation level, if plotted as in Figure 2-16, is seen to be in remarkably good agreement with the previously plotted results for Ti_{10}-Zr_{40}-Nb_{50}. Taken together these results lead to a useful rule of thumb: an increase by a factor of ten in the area reduction ratio decreases the precipitation time by a factor of ten; the effects of successive levels of cold reduction, although leading to further significant increases in the reaction kinetics, are not as strong as the first.

2.11.2 Influence of Aging on the Fibrous Cell Structure

During the early stages of moderate-temperature aging, the intracell dislocations migrate to the cell walls; then at higher aging temperatures, extremely rapid cell growth sets in. In what follows, the mechanism and kinetics of this process will be discussed.

(a) Dislocation Motion. A strongly cold-worked alloy with subband diameters in the range 100–700 Å may contain a dislocation density, within the cells, of about 10^{12} cm^{-2} [PFE68].[7] During moderate-temperature heat treatment ($\lesssim$300°C [NEA71]), migration of the dislocations to the cell walls begins to take place by some sort of thermally activated slip process. Evidence for the occurrence of this perfectly general phenomenon (discussed in standard works on dislocations, e.g., [FRI64]) is to be found in the sharpening of the x-ray diffraction pattern [CHA70]. After 1 h at 385°C,

[7] A cell of 500 Å diameter (area = 1.96×10^{-11} cm^2) would enclose 20 dislocations.

dislocation density is still high in the large cells that result from low levels of cold deformation, but much lower, according to NEAL *et al.* [NEA71], in the narrow cells of heavily reduced materials with their higher levels of strain energy and shorter path lengths. Viewed in cross section, the deformation cells possess curved boundaries and are irregular in size; then during the early stages of heat treatment, the cell walls become straight [NEA71] as in the early stages of classical polygonalization [FRI64], [CHA70]. In reference to the 1-h annealing of heavily cold-worked ($5 \times 10^4:1$) Ti–58Nb, the intracell dislocation density is lowest at 400°C, but the cell structure itself is on the threshold of growth. For this reason, annealing at 385°C is a safe compromise because at this temperature the cell structure seems to be fairly stable [NEA71].

(b) Cell Growth. Cell growth during the heat treatment of deformed Ti–Nb alloys has been considered by several authors [NAR66], [REU66], [PFE68], [BAK70], and investigated quantitatively by NEAL *et al.* [NEA71] and ARNDT and EBELING [ARN74], [WIL75[a]], whose results on this subject are combined in Figures 2-17(a) and (b). During annealing at 385°C, the

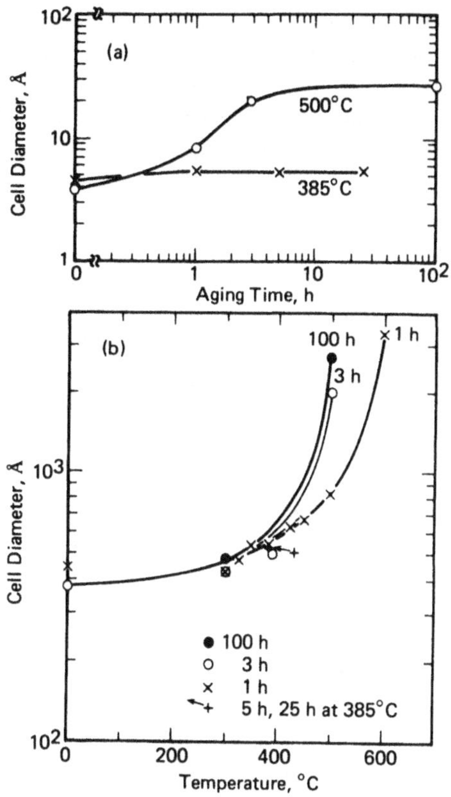

Figure 2-17. Deformation cell growth as a function of (a) time, and (b) temperature—after ARNDT *et al.* [ARN74] (prior deformation 99.98% (○)) and NEAL *et al.* [NEA71] (prior deformation 99.998% (×)).

Table 2-7. Influence of Heat Treatment at the Temperatures Specified on the Average Subband Diameter of Lightly Cold Worked Ti–53.5Nb—After West and Larbalestier [WES81]

Aging Time h	Subband Diameter, Å		
	300°C	375°C	400°C
0	690	690	690
1	980	—	770
3	840	870	940
30	860	1170	1160
100	920	1110	1290
150	—	1280	—
200	—	1040	—

cell diameter increased from 443 to 536 Å in the first hour, and then remained stable at 530 ± 6 Å for at least another 24 hours [NEA71]. At 500°C the cells grew to diameters of more than 2000 Å after just a few hours [ARN74] (Figure 2-17(a)). When cell diameter is plotted as function of 1, 3, and 100 h aging temperature as in Figure 2-17(b), it can be seen that at temperatures less than 385°C the cells grow (by less than 20%) to stable size within about one hour, but that at 400°C and above the cell size is strongly dependent on both aging time and temperature. WEST and LARBALESTIER [WES81] have studied the influence of heat treatment on subband sizes in commercial Ti–53.5Nb (the Fermilab alloy), first in a lightly cold-worked condition (ARR ≅ 255 : 1, i.e., 99.6%, after extrusion), and again after a further 96.7% cold reduction. The results of the first series of measurements are summarized in Table 2-7; although based on a relatively large initial average subband diameter (690 Å) the results do seem to confirm the earlier conclusions. Some of their more recent results are reported in [LAR84], [WES84].

2.12 The Influence of Interstitial-Element Additions

The influences of interstitial-element (B, C, N, and O) additions on critical current density in Ti–Nb alloys, a subject that has been considered by numerous writers, is reviewed in [COL83, Chap. 8]. Considered below are the *metallurgical* consequences of the presence of these elements in cold-worked and aged alloys. Precipitational effects associated with the presence of O in Ti–Nb(35 and 45 at.%) have been discussed by WITCOMB and DEW-HUGHES [WIT73] and REUTER *et al.* [REU66], and the influence of N on Ti–56Nb has been considered by BAKER [BAK70]. The kinetics of O diffusion has been discussed by CHARLESWORTH and MADSEN [CHA70] and the influence of O on the kinetics of β-phase decomposition by DELAZARO and ROSTOKER [DEL53] and WILLIAMS *et al.* [WIL71].

Interstitial elements are extremely potent strengtheners of Ti alloys [CON75]. As a consequence of its ability to pin dislocations, it has been claimed that the presence of N in Ti–56Nb necessitated increasing the 1-h aging temperature by 100°C in order to maximize the critical current density [BAK70]. But in the experiments of NEAL *et al.* [NEA71], it was probably the coarser starting subcell structure, rather than impaired dislocation mobility during aging, that retarded the kinetics of the optimization heat treatment.

Although it is unlikely, in view of the high mobility of interstitials, that dislocation pinning takes place *at temperature*, a study of O-doped Ti–39Nb [COM67] indicated that in some cases a coarser deformation substructure can be expected to result from the presence of strengthening centers. But the converse has also been observed: in some wire-drawing studies of steels (which are also bcc alloys), EMBURY *et al.* [EMB66] noted that, through their ability to inhibit dynamic recovery, the presence of interstitial elements can lead to the establishment of a finer subband structure than might otherwise be the case.

The diffusion of interstitial and substitutional elements in Ti has been thoroughly reviewed by ZWICKER [ZWI74, Chapter 5], some of whose collected data have already been used in the construction of Figure 1-12, which describes the relative rates of diffusion of the 3*d* and some 4*d* transition elements in β-Ti. Characteristic data (i.e., frequency factor and molar activation energies) describing the diffusion of C, N, O, and H in α- and β-Ti have been provided by ZWICKER [ZWI74, p. 104]; those for O in β-Ti are reproduced in Table 2-8. Unfortunately, the reliability of some of the published data is open to question; Table 2-8 shows that for O in β-Ti

Table 2-8. Diffusion Parameters for Oxygen in β-Ti—After Zwicker [ZWI74, p. 104][a]

Temperature Range °C	Frequency Factor, D_0 $cm^2\ s^{-1}$	Molar Activation Energy cal mole^{-1}	Diffusivity at 1000°C, D $cm^2\ s^{-1}$
930–1150	31,400	68,700	5.0×10^{-8}
932–1142	330	58,000	3.6×10^{-8}
950–1415	1.6	48,200	8.4×10^{-9}
907–1030	0.45	36,000	3.0×10^{-7}
1135–1355	8.3×10^{-2}	31,200	3.6×10^{-7}

[a] See Table 5.1 of [ZWI74] for literature sources.

Table 2-9. Diffusion Parameters for Interstitial Elements in Nb, β-Ti, and Ti–Nb(45 at.%)—After Charlesworth and Madsen [CHA70][a]

Interstitial Element	Solvent	Temperature Range °C	Frequency Factor, D_0 $cm^2\ s^{-1}$	Atomic Activation Energy eV	Diffusivity at 380°C, D $cm^2\ s^{-1}$
C	Nb	130–280	4×10^{-3}	1.43	3.7×10^{-14}
		150	1.5×10^{-2}	1.15	2.0×10^{-11}
	β-Ti	950–1150	108	2.1	
	Ti–Nb (45 at.%)			1.68–1.80	
N	Nb	150–300	8.6×10^{-3}	1.51	1.9×10^{-14}
		300	9.8×10^{-2}	1.65	1.8×10^{-14}
	β-Ti	900–1570	3.5×10^{-2}	1.46	
	Ti–Nb (45 at.%)			1.49–1.57	
O	Nb	40–150	2.0×10^{-2}	1.17	1.9×10^{-11}
		150	1.5×10^{-2}	1.2	8.2×10^{-12}
	β-Ti	950–1414	1.6	2.09	
	Ti–Nb (45 at.%)			1.68–1.69	

[a] See Table III of [CHA70] for literature sources.

at 1000°C, for example, values lying between 3×10^{-7} and 8×10^{-8} $cm^2\ s^{-1}$ are obtainable. CHARLESWORTH and MADSEN [CHA70] also assembled a set of diffusivity data (including *atomic* activation energies expressed in eV) for C, N, O, and H in β-Ti as well as in unalloyed Nb at moderate temperatures. These are reproduced in Table 2-9. In the absence of published diffusivity data for interstitials in Ti–Nb alloys, those authors assumed that an activation energy for diffusion in Ti–Nb(45 at.%) could be adequately extracted by linear interpolation between the published values for Nb and β-Ti; the values so obtained are included in Table 2-9. Bearing in mind the general inaccuracy of the available diffusivity data for interstitial elements in transition metals, it was suggested by CHARLESWORTH and MADSEN that an atomic activation energy of 1.55 eV for C, N, and O in Ti–Nb(45 at.%) would approximate the experimental situation reasonably well. Finally, a value of 2.0×10^{-2} $cm^2\ s^{-1}$ (cf. the O entry in Table 2-9) was suggested for the frequency factor of O, the most mobile of all three interstitials in Nb. At a temperature of 380°C, for example, these data yielded a diffusivity for O of 2.2×10^{-4} $cm^2\ s^{-1}$. In order to establish a benchmark for comparison, a set of diffusivity data for ^{95}Nb in a series of Ti–Nb alloys (with appropriate extrapolations where necessary) may be consulted (Figure 2-18). In so doing it will be found that O diffuses through Ti–Nb(45 at.%) at 380°C about as rapidly as Nb does at 800°C. This leads immediately to the interesting conclusion that as cell growth in Ti–Nb begins to accelerate with increase

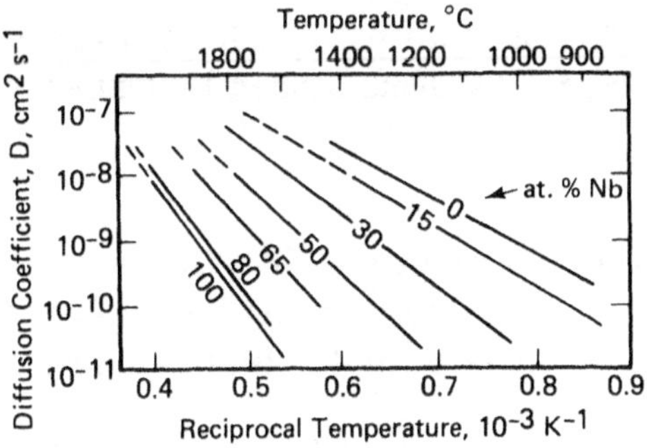

Figure 2-18. Temperature dependence of the diffusion coefficients of ^{95}Nb in Ti-Nb alloys—after ZWICKER [ZWI74, p. 112].

in temperature above 400°C (as in Figure 2-17(b)), the interstitial atoms will long since have diffused to, and become trapped by, the cell walls. Alternatively, an application of linear random walk theory according to which an atom moves a root-mean-square distance from its starting point of $\sqrt{2Dt}$ cm in t seconds, will show that an interstitial atom is able to migrate a distance of 1300 Å within the Ti-Nb alloy during a heat treatment of 1 h/380°C; by the same token, the time taken for it to move halfway across a typical subband at that temperature is about $1\frac{1}{2}$ min. During superconductor processing, therefore, any dissolved O will become associated with the cell walls during the first few minutes of moderate-temperature aging at the final wire diameter.

The implications of interstitial diffusion are in terms of cell-wall precipitation. The elements B, C, N, and O are all stabilizers of the α-phase in Ti (Figure 1-4) and as such are expected to encourage α-phase precipitation as they accumulate at the cell boundaries or residual dislocations. In addition, as a solution strengthener, dissolved O tends to stabilize the bcc lattice. Taken together these observations suggest that O retards ω-precipitation but accelerates the $\beta \rightarrow \alpha + \beta$ decomposition reaction. The accelerating

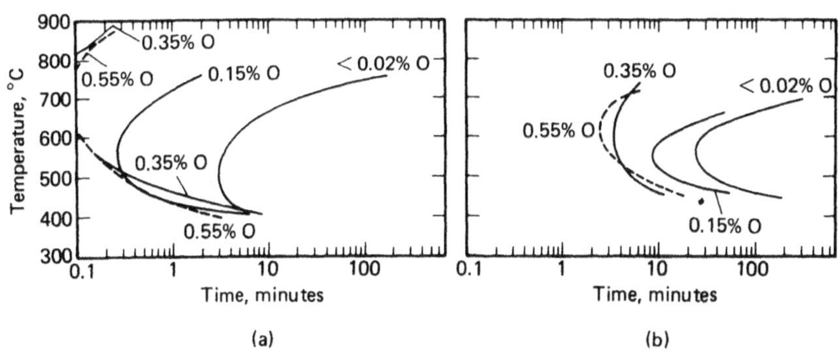

Figure 2-19. Influence of oxygen on the $\beta \rightarrow \alpha + \beta$ transformation kinetics of the β-isomorphous alloy Ti-11Mo; (a) transformation start, (b) transformation finish—after DELAZARO and ROSTOKER [DEL53].

effect of O on the kinetics of α-phase precipitation in Ti–Mo has been graphically demonstrated by DELAZARO and ROSTOKER [DEL53], and is shown here in Figure 2-19.

2.13 Summary—The Occurrence of Isothermal ω- and Equilibrium α-Phases in Deformed and/or Aged Titanium–Niobium Alloys

As a result of the work of KITADA, DOI and coworkers a complete picture is available of the metastable and equilibrium phases, and transformation kinetics in quenched-and-aged and quenched-deformed-and-aged Ti_{60}-Zr_5-Nb_{35} [KIT69], Ti_{10}-Zr_{40}-Nb_{50} [SOE69], [KIT70[a], KIT70[b], KIT70[c]], and related alloys [DOI66], [HOR73[a]]. But although the properties of numerous binary Ti–TM alloys were the subject of intense study during the 1950s [MAY61], Ti–Nb was not included among the long list of alloys which had, with *structural* applications in mind, been systematically investigated during that period. Subsequently, as a result of studies conducted during the 1960s, a substantial body of useful information having to do with microstructures, precipitation, and transformation kinetics in Ti–Nb found its way into the *superconductivity* literature. Finally, because of an increasing interest in alloy superconductivity and for other reasons, the 1970s saw numerous detailed studies of Ti–Nb microstructures and related properties under way in Germany [ARN74], [HIL74], [WIL75], [OBS80], [HOC81], [PAT81], in Japan [OSA80], and in the United States [MEN71], [BAL72], [WES80, WES81, WES82]. As a result of this work, a reliable picture of Ti–Nb metallurgy, at least in the intermediate-concentration range, has emerged.

2.13.1 The Isothermal ω-Phase

According to Section 2.2 (see also Table 2-3), the maximum Nb concentration for isothermal ω-phase precipitation from Ti–Nb at intermediate temperatures is 30 at.%. This estimate, due to HICKMAN [HIC69[a]], was acquired as the result of two independent observations: (i) ω-phase was unable to be detected in Ti–Nb(30 at.%) and Ti–Nb(35 at.%); and (ii) according to x-ray lattice parameter measurements, the Nb content of the ω-phase component of isothermally aged (for up to 30 h at 450°C) Ti–Nb(22 at.%) and Ti–Nb(25 at.%) was 30 ± 1 at.%. In partial agreement with this were the observations by OSAMURA *et al.* [OSA80] on ω-phase precipitation from cold-worked-and-aged Ti–Nb(35 at.%) foils and wires: *viz.*, that, during aging at 380°C, although no such precipitation was observed after times of up to about 160 h, a small amount of ω-phase was detected in alloys that had been aged for longer times. Apparently, ω-phase precipitation is generally quite sluggish in the higher-concentration Ti–TM alloys. Since, however, extremely long heat-treatment times are not contemplated for technical Ti–Nb superconductors, it is not unreasonable to adopt 30 at.% Nb as the practical limit of the isothermal $\omega+\beta$-phase field in aged Ti–Nb.

2.13.2 The Equilibrium α-Phase

As a consequence of the sluggishness of the $\beta \rightarrow \alpha+\beta$ decomposition reaction, it has not been possible to extend the equilibrium $(\alpha+\beta)/\beta$-transus to temperature much below about 700°C. Conventional information regarding phase equilibria in the temperature range around 400–500°C is, therefore, lacking. However, as discussed above, a significant increase in reaction kinetics will occur if the lattice has been severely strained as a consequence of heavy plastic deformation. This is, of course, the condition of as-drawn Ti–Nb alloy wire. It is to be expected, therefore, that the phases to be found after moderate-time aging of such material will reflect the conditions of thermodynamic equilibrium at the aging temperature.

In several pioneering studies HILLMANN *et al.* [PFE68], [HIL74] and WILLBRAND and SCHLUMP [WIL75] produced evidence for the presence, after moderate aging (e.g., several h at 380°C), of α-phase precipitation in Ti-50Nb wires. TEM experiments on heavily drawn wire are difficult to perform and interpret. Consequently, prior to about 1980, detailed information was scarce concerning the occurrence and morphology of α-phase precipitation in response to aging in the technically interesting temperature range near 400°C, especially in the case of alloys in the high Nb concentration range. The situation began to improve in later years after WEST and LARBALESTIER began to conduct a series of careful TEM/SAD [WES80] and STEM/EDAX [WES82] studies of several Ti–Nb samples (some of which had been extracted from commercial multifilamentary Cu/Ti–Nb composites) possessing Nb concentrations ranging from 47.3 wt.% to 53.5 wt.%. At the time of the measurements, the results for the latter were particularly interesting in that Ti-53.5Nb then became the highest concentration alloy known to give rise to α-phase precipitation. NEAL *et al.* [NEA71] in their study of Ti-58Nb (analyzed composition)—an alloy recognized as residing in the *thermodynamically permissible*, if not practically realizable, $\alpha+\beta$-phase regime—had failed to detect significant amounts of α-phase. At the time of the investigation they had, furthermore, concluded that its abundance was "far too small to provide a significant degree of fluxoid pinning." But by 1984, again thanks to the work of LARBALESTIER and his group, the flux-pinning and precipitation picture in Ti–Nb alloys over the entire relevant concentration range had undergone considerable clarification. It is now known that all technically interesting Ti–Nb alloys, certainly those of from 50 to 58 wt.% Nb, may yield α-phase precipitation during appropriate thermomechanical processing. The previously assigned division at 56 wt.% Nb (i.e., 40 at.% Nb) into "intermediate-concentration" and "high-concentration" alloys [CoL83, pp. 267–274] must now be regarded as having more to do with the kinetics of precipitation than with its thermodynamics.

Using Ti–54.3Nb (analyzed composition, nominally Ti–53.5Nb, from a sample of the Fermilab conductor) as a representative alloy for case study, LARBALESTIER and WEST [LAR84], [WES84] conducted a TEM/SAD investigation of precipitation in response to severe cold reduction (e.g., see Table 2-6, group *b*), followed by aging at 300 and 375°C, and that followed by further cold work and aging (see Chapter 21, Part 2 for further details). They were able to show that after heat treatment as mild as 3h/300°C a Ti-rich film some 30–60 Å thick coated the subband (or grain—see Section 2.8.5) boundaries. They were also able to demonstrate conclusively that such precipitates had also coated the deformation-cell walls of the Ti–58Nb alloy studied by NEAL *et al.* in response to similar thermomechanical processing. After the heat-treatment time was extended to 160 h, or after shorter times (e.g., 40 h) at higher temperatures (e.g., 375°C) precipitates, which eventually will grow into large (up to 500 Å) equiaxed α-phase particles, began to manifest themselves at the cell (grain) triple-junctions. The thickness of the subband film did not seem to be sensitive to the length of the heat treatment. In addition to these precipitates, some very fine Ti-rich plates, again about 30–60 Å thick, were discerned *within* the subbands. The influence of further cold work and heat treatment on these precipitates, especially as it relates to their flux-pinning potencies, is considered in Chapter 21, Part 2.

Chapter 3

Mechanical Properties

PART 1. HARDNESS

3.1 The Hardness Test

The measurement of hardness is a useful but simple technique for characterizing mechanical properties and investigating phases in quenched and aged alloys. Conventional techniques currently in use for measuring hardness, as well as the history of the test, are fully discussed by HANKE [HAN54]. In the Vickers method, a weighted square pyramid, usually of diamond, is allowed to rest for a specified length of time on the polished surface of the specimen. Since the cross-sectional area of the resulting indentation (mean diagonal, d, mm) is proportional to the load (L, kg), a load-independent hardness number can be formed from the quotient L/d^2. The Vickers hardness number is defined as $H_V = 1.8544\ L/d^2\ \mathrm{kg\ mm^{-2}}$. A 5-kg load, for example, resting on a diamond point bearing down on a surface of Vickers hardness $150\ \mathrm{kg\ mm^{-2}}$ would produce a 0.25-mm (diagonal) impression, as in the studies of Ti–V and Ti–Nb alloys to be considered below. If it is desired to investigate the individual grains of a fine-grain polycrystalline sample, miniaturized versions of the test are available. Using for example the Leitz Miniloader, loads in the range of 25–100 g produce measurable impressions less than 10 μm in diagonal. Since samples that have been mounted and polished for optical metallography are ideally prepared for hardness measurement, the two investigations are frequently associated with each other in studies of precipitation and aging. Although very useful in following the course of an aging reaction, the hardness measurement is obviously not capable of identifying the nature of the precipitating phases. This determination must be made by separate x-ray or TEM investigations, in the absence of which several alternative possibilities, particularly with regard to submicroscopic continuous precipitation processes (such as, for example, the reactions: $\beta \rightarrow \beta' + \beta \rightarrow \alpha + \beta$, $\beta \rightarrow \beta' + \beta''$, or $\beta \rightarrow \omega + \beta$), could be easily confused.

With the aid of formulae such as those due to HILL [HIL67] and MARSH [MAR64, MAR64[a]], to be discussed below Vickers hardness can be related to yield strength. This approach allows estimates of yield strength to be made on samples that are unable to be tested in the conventional way, or on very fine filaments, such as those of a composite superconductor, which, for hardness testing, do not need to be removed from their matrices. Likewise, with the aid of microhardness measurements on small test melts and the relationship between hardness and yield strength, the influences of interstitial and other third-element additions on the yield strengths of binary alloys can be conveniently and rapidly surveyed.

3.2 Hardness of Quenched Titanium–Transition-Metal Alloys

Vickers hardness measurements, augmented by x-ray diffraction studies, were employed by BAGARIATSKII *et al.* [BAG59] in a well-known investigation of the limits of the quenched α'-, α''-, and ω-phase regimes in the systems (i) Ti-V, -Ta, -Nb; (ii) Ti-Mo, -W; and (iii) Ti-Re. The occurrence of athermal ω-phase in each system was recognized by the appearance of a pronounced local hardness maximum with an amplitude of typically 80 kg mm^{-2}. GUZEI *et al.* [GUZ66] measured the hardnesses of a series of Ti-Nb alloys quenched from equilibration at various temperatures. RASSMANN and ILLGEN [RAS72] investigated hardness as function of composition (10–80 at.%) in a series of Ti-Nb alloy samples prepared in the form of sheet (cold deformed) or cast ingot from iodide Ti or Kroll-process Ti. Although the hardnesses, following either preparation process, increased rapidly with reduction of the Nb content below 30 at.%, those of the cold-deformed alloys were much higher throughout the concentration range, presumably on account of work hardening in general and deformation-martensitic transformation (to α'') at the lower Nb levels, in particular. The microstructural possibilities in response to 1-h annealing, which leads to a pronounced hardness increase at 300°C, are even more complicated. It should be noted that the several attempts which have been made to explore deformation, precipitation, and aging effects, using *only* hardness as a tool, exceeded the capabilities of the technique.

The results of some recent studies of hardness in Ti-V and Ti-Nb alloys are presented in Figure 3-1. The alloys of both systems are seen to become extremely hard as the solute concentration decreases below about 13 at.% ($e/a = 4.13$). But in contrast to earlier practice [BAG59], [GUZ66], no attempt is made here to generate smoothly rounded maxima from the low-concentration data. The "lambda-like" discontinuities shown, which also appear in the composition-dependences of physical properties such as magnetic susceptibility, electronic specific heat coefficient, and Debye tem-

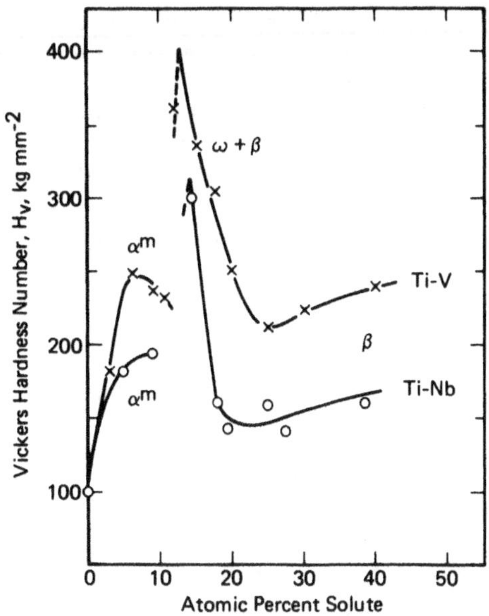

Figure 3-1. Vickers hardness (5-kg load) of β-quenched Ti-V and Ti-Nb alloys as function of composition and microstructure. The hardness discontinuity shown is a consequence of the presence of athermal ω-phase. Similar discontinuities at the boundary between the ω- and the α^m-regimes are also exhibited by many of the physical properties such as magnetic susceptibility and electronic specific heat coefficient—after COLLINGS [COL84].

perature, are associated with the presence in each alloy of a high volume fraction of athermal (as-quenched) ω-phase within a limited composition range. To the left of the discontinuity each of the alloys has transformed martensitically.

3.3 Hardness of Aged Titanium–Niobium Alloys

Table 3-1 lists several representative studies of age-hardening in Ti-Nb alloys, some of which were carried out isothermally for long time periods, while others were for short times (usually 1 h) but over wide ranges of temperature. In a typical such study, LÖHBERG [LOH71] noted that in both homogenized and as deformed samples of Ti-Nb(33 at.%) aged at 400°C an initial hardness decrease was followed by a monotonic increase. In the *homogenized* alloy, x-ray studies identified the presence of ω-phase after 15 h of aging, whereas in the *deformed* sample it was claimed that ω-phase appeared within the first two h of the aging process, giving way to α-phase after a few tens of hours. Granted that ω-phase is able to appear at all in such a Nb-rich alloy (possibly due to compositional undulations as a result of coring), these observations on the change of ω- and α-precipitation kinetics in response to cold work are in accordance with the discussions of Section 2.9. In a typical study of the influence of temperature on the 1-h aging of Ti-Nb alloys, RAUCH *et al.* [RAU68, RAU68[a]] demonstrated that in 99.9% cold-deformed Ti-50Nb the hardness rose to a sharp peak centered

Table 3-1. Index to Age-Hardening Studies of Ti-Nb Alloys

Nb Content at.%	Starting Conditions	Aging Conditions	Literature
20	15 min/1000°C; cw zero and 99.9%	1 h/(200-800°C)	Rassmann [RAS72]
22	4 h/800°C; cw 99.95%	various/(250-500°C)	McInturff [MCI73]
33	3 h/1000°C; cw zero and 97.2%	(0-150 h)/400°C	Löhberg [LOH71]
34, 44, 67	7-12 h/1500°C; cw 99.9%	1 h/(200-750°C)	Rauch [RAU68, RAU68[a]]
40, 52	1 h/800°C; cw zero, 9, and 90%	(0-167 h)/380°C	Katoh [KAT75]

at 375°C. In the absence of x-ray study, some uncertainty is associated with the origin of this peak. Its height and temperature range of occurrence suggested ω-phase, but the composition of the alloy placed it, according to the results of subsequent investigators, outside the range within which ω-phase can precipitate in such a short period of time from a homogeneous alloy (see Section 1.9). To provide a definitive answer to the precipitation question it is necessary to resort to direct metallographic observation (TEM, SAD, STEM, etc.), or x-ray intensity studies as in the investigations by RICKETTS [RIC69] of the aging of Ti-40Nb.

3.4 Influence of Third-Element Additions on the Hardnesses of Unalloyed Titanium and Titanium–Niobium Alloys

Low concentrations of simple-metal, semimetal, and interstitial-element additions have been made to Ti-Nb alloy superconductors in order to alter the precipitation kinetics of the host alloy and/or to provide additional flux-pinning centers. Systems for which data are available are: Ti-Nb-Ge [LOH71], Ti-Nb-Cu [HEL71], Ti-Nb-Fe [NIS68], and Ti-Nb-Cu, -Ag, -Al, -Ge, and -Zr [ZWI70].

It is important to be able to estimate the influences that such third-element additions might have on the mechanical and metalworking properties of the host alloys. To contribute to this understanding, a survey has been made of the hardening rates which accompany additions of: (i) the interstitial elements B, C, N and O; (ii) the valence-3 simple metals Al and Ga; (iii) the valence-4 metals and semimetals Si, Ge, and Sn to unalloyed Ti [COL84]. The results were then related to available data for the hardening of Ti–Nb(33 at.%) in response to additions of some of these elements [Zwi70]. Common elements selected were Al and Ge, and the hardening of Ti–Nb(33 at.%) by small additions of Zr was also added to the survey. Unfortunately, with regard to the Ti–Nb-base results, it was necessary to accept data corresponding to two different metallurgical conditions: *homogenized-and-quenched* for Ti–Nb_{33}–Cu, Ge and *cold-worked-and-aged* 34 h/400°C in the case of Ti–Nb_{33}–Al, Ge, Zr; consequently the comparison, although instructive, is semiquantitative at best. In least-squares fitting the H_V *versus* composition (c) data, it was noted that the interstitial-element hardening closely followed a $c^{1/2}$ relationship (correlation coefficient was, in most cases, >99%). With the other elements, the hardening rates were sufficiently low that they could be adequately represented over the range concerned by linear composition dependences. Table 3-2, in which the data and the results of the analyses are presented, shows that at the 0.1 at.% level the instantaneous hardening rates of unalloyed Ti by interstitials (*viz.*, 269–378 kg mm^{-2} per at.%) are up to 20 times greater than that produced by, say, Al or up to 9 times faster than that by Si. That this result (*viz.*, the exceptionally rapid hardening produced by interstitial elements in Ti) will also carry over to Ti–Nb as a host, can be judged by noting that some metalloids and simple metals produce similar hardening effects when added to either Ti or Ti–Nb. In particular:

(1) Ge, which appears as a solute in: (i) unalloyed Ti; (ii) Ti–Nb(33 at.%) homogenized-and-quenched [Hel71]; (iii) Ti–Nb(33 at.%) cold-worked-and-aged 34 h/400°C [Zwi70], produces linear hardening rates of, respectively, 33, 14, and 60 kg mm^{-2} per at.%;
(2) Al, when dissolved in unalloyed Ti and Ti–Nb(33 at.%), cold-worked-and-aged, yields hardening rates of, respectively, 16 and 19 kg mm^{-2} per at.%.

Numerical comparisons, such as those in Table 3-2, and graphical comparisons, such as Figure 3-2, serve to illustrate that the hardening of Ti or Ti–Nb alloys is subdivisible into three regimes according to whether the rate of hardening is "low" (neighboring transition elements such as Zr), "moderate" (substitutional simple metals or semimetals such as Al, Sn, Si, Ge) or "rapid" (the interstitial elements—B, C, N, and O). The strengthening of group-IV elements by interstitial elements has been considered in detail by Conrad *et al.* [Con75, Con81] and Tyson [Tys75], and the physical basis of strengthening by various classes of alloying addition to Ti has been discussed by Collings and Gegel [Col73, Col75[d]].

Anticipating the results of an analysis to be presented below concerning the proportionality of yield strength to hardness, it can be concluded that the addition of interstitial elements to Ti (hence to Ti–Nb alloys) can be

Table 3-2. Hardening Rates in Response to the Addition of Interstitial

Base Material	Alloying Addition	Concentration Range at.%	Condition	Law
Ti	B	0-0.2	120 h/800°C/IBQ	$c^{1/2}$
Ti	C	0-0.5	120 h/800°C/IBQ	$c^{1/2}$
Ti	N	0-5	120 h/800°C/IBQ	$c^{1/2}$
Ti	O	0-3	120 h/800°C/IBQ	$c^{1/2}$
Ti	Al	0-10	100 h/850°C/IBQ	c
Ti	Ga	0-5	as-cast	c
Ti	Si	0-2	1 h/1000°C/IBQ	c
Ti	Ge	0-5	1 h/1000°C/IBQ	c
Ti	Sn	0-7	as-cast	c
Ti-Nb(33 at.%)	Cu	0-5	3 h/1000°C/WQ	c
Ti-Nb(33 at.%)	Ge	0-4	3 h/1000 or 1080°C/WQ	c
Ti-Nb(33 at.%)	Al	0-2.5	cold deformed/34 h/400°C	c
Ti-Nb(33 at.%)	Ge	0-2	cold deformed/34 h/400°C	c
Ti-Nb(33 at.%)	Zr	0-6	cold deformed/34 h/400°C	c

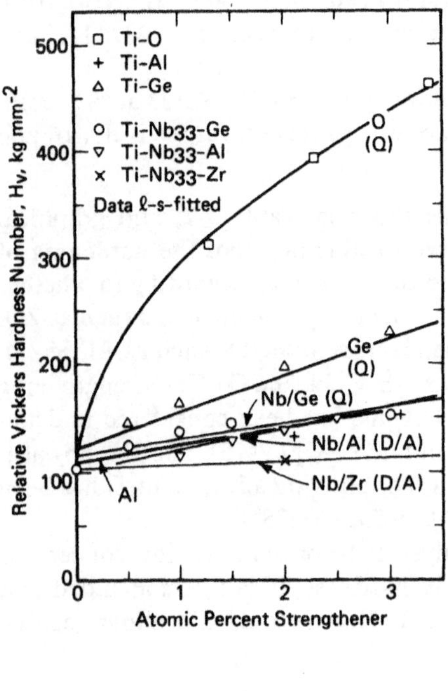

Figure 3-2. Influence of alloying additions on the rates of Vickers hardening of previously unalloyed Ti and Ti-Nb(33 at.%). Q indicates quenched, D/A indicates deformed-plus-aged. For the purpose of comparison the ternary data have been adjusted to $H_V = 103\ \text{kg mm}^{-2}$ at zero solute concentration, a point which is then included in the least-squares-fitting of all the data. Sources: *binary alloys*, COLLINGS [COL84]; Ti-Nb-Ge, HELLER [HEL71]; Ti-Nb-Al/Ge, ZWICKER *et al.* [ZWI70].

Elements, Simple Metals, Semimetals and Zr to Ti or Ti-Nb(33 at.%)

Slope kg mm^{-2} at.%$^{-1/2}$ or kg mm^{-2} at.%$^{-1}$	Intercept kg mm^{-2}	Correlation Coefficient %	Hardening Rate $\partial H_V/\partial c$ kg mm^{-2} at.%$^{-1}$ at 0.1 at.%	at 1.0 at.%	Literature
218	110	92	344	109	Collings [COL84]
170	104	99.9	269	85	Collings [COL84]
239	98	99.8	378	120	Collings [COL84]
194	100	99.9	307	97	Collings [COL84]
15	102	99.6	—	15	Collings [COL84]
24	108	99	—	24	Collings [COL84]
39	125	83	—	39	Collings [COL84]
33	120	98	—	33	Collings [COL84]
24	112	99	—	24	Collings [COL84]
14	193	99.5	—	14	Löhberg [LOH71]
14	165	93	—	14	Heller [HEL71]
19	195	99	—	19	Zwicker [ZWI70]
60	198	99	—	60	Zwicker [ZWI70]
3.7	198	99.9	—	3.7	Zwicker [ZWI70]

expected to increase yield strength (at the 0.1 at.% addition level) at the rate of about 30% per 0.1 at.%. On a wt.% basis the strengthening rate due to oxygen, at the 0.1 wt.% level, is a factor of $\sqrt{3}$ greater.

3.5 Hardness of Ternary and Quaternary Transition-Metal Alloys

3.5.1 Composition Dependence

Hardness data covering the full composition triangle are available for the alloys Ti-V-Nb (furnace cooled from 600°C) [MOL65, p. 210]. ZWICKER *et al.* [ZWI68] presented a full set of hardness data for members of the Ti-Zr-Nb system in both the as-cast state and after 91.5% deformation plus 30 min/900°C/water quench. Hardness data are also available for the related technically important superconducting systems Ti-Zr-Nb-Ta(5 at.%) and Ti-Zr-Nb-Ta(10 at.%), and HORIUCHI *et al.* [HOR73] (see also [HOR74]) presented results for alloys defined by: $Ti_{20\text{-}63}$-$Zr_{5\text{-}35}$-$Nb_{24\text{-}60}$-Ta_5 and $Ti_{20\text{-}60}$-$Zr_{5\text{-}30}$-$Nb_{24\text{-}60}$-Ta_{10} in the deformed-and-heat-treated (2 h/1000°C) condition.

Maximal hardness values encountered in the above systems were:

Ti-20V-3Nb (600°C/furnace cooled)	485 kg mm^{-2}
Ti_{18}-Zr_{18}-Nb_{64} (as-cast)	336 kg mm^{-2}
Ti_{25}-Zr_{31}-Nb_{44} (deformed + 30 min/900°C/WQ)	275 kg mm^{-2}
Ti_{30}-Zr_{35}-Nb_{30}-Ta_5 (deformed + 2 h/1000°C)	260 kg mm^{-2}
Ti_{30}-Zr_{30}-Nb_{30}-Ta_{10} (deformed + 2 h/1000°C)	273 kg mm^{-2}

3.5.2 Aging Studies

During the aging of Ti_{60}-Zr_5-Nb_{35}, Ti_{10}-Zr_{40}-Nb_{50}, and Ti_{33}-Zr_{30}-Nb_{30}-Ta_7 hardness measurements have been undertaken, diagnostically, in order to follow the courses of previously identified precipitation processes. In particular:

(1) Ti_{60}-Zr_5-Nb_{35}, quenched from 24 h/1100°C was aged (0–6 h)/500°C [ISH68] during which the simultaneous maximization of H_V and the critical supercurrent, I_c, after about 1 h of aging time, signalled the occurrence of optimal α-phase precipitation.

(2) Ti_{10}-Zr_{40}-Nb_{50} was investigated under the following conditions: (i) solution treated 5 h/1100°C, deformed 98.4% and aged (0–1 h)/640–750°C) [KIT70[b]]; (ii) solution treated 5 h/1100°C and aged (0–1000 h)/(400–500°C)[KIT70[c]], in investigations of both continuous and discontinuous (grain boundary) precipitation during the accompanying $\beta \rightarrow \alpha + \beta$ reaction; (iii) solution treated 3 h/1000°C and aged (0–200 h)/(620–725°C) [KIT70[a]], during which both the continuous-precipitational and discontinuous-precipitational aspects of the $\beta \rightarrow \beta' + \beta''$ reaction were studied using hardness measurements. Naturally, the *continuous* precipitate, a very fine dispersion, was a much more effective hardener than the lamellar heterogeneously nucleated *discontinuous* $\beta' + \beta''$-phase.

(3) Finally, data are available that are descriptive of the aging for (0–2000 min)/(450–650°C) of 99%-cold-worked Ti_{33}-Zr_{30}-Nb_{30}-Ta_7, during which the hardness responses to $\beta \rightarrow \alpha + \beta$ transformation (<500°C), $\beta \rightarrow \alpha + \beta' + \beta''$ transformation (500°C), and $\beta \rightarrow \beta' + \beta''$ precipitation (>500°C), were recorded [HOR73[a]].

3.6 Theoretical Relationships Between Hardness and Strength

Studies of plastic flow in metals and glass by HILL [HIL67] and MARSH [MAR64, MAR64[a]], respectively, have resulted in a pair of complementary relationships connecting the hardnesses and strengths of a wide range of materials. An excellent summary of the results of the HILL and MARSH theories has been offered by DAVIS [DAV75] in connection with a study of the hardness/yield-strength ratio in metallic glasses.

3.6.1 Hill's Theory

The model assumed by HILL [HIL67], which pictured an elastically rigid yet plastic body with the material displaced by the indentor squeezed up into a rim around the edges of the imprint, led to the commonly observed relationship $H_V \cong 3Y$ connecting the hardness, H_V, to the yield strength, Y, when expressed in the same units (usually kg mm^{-2}).

3.6.2 Marsh's Theory

The theory of MARSH [MAR67, MAR67[a]], which pictures a cavity being pushed into an elastic body, results in the semiempirical formula

$$H_V/Y = 0.28 + 0.6\,B \ln Z \tag{3-1}$$

where

$$\begin{cases} B \equiv 3/(3-l) \\ Z \equiv 3/(l + 3m - lm) \end{cases}$$

with

$$\begin{cases} l \equiv (1-2\nu)\dfrac{Y}{E} \\ m \equiv (1+\nu)\dfrac{Y}{E} \end{cases}$$

where ν is Poisson's ratio and E the Young's modulus.

3.6.3 Relationship Between the Models

These models represent extreme limits with regard to the relative magnitudes of E and Y. It turns out that if $133\,Y/E$ is "small" (<1) we have the plastic/elastically-rigid situation discussed by HILL, whereas if $133\,Y/E$ is "large" (>1) the situation is identifiable as the unyielding nonplastic-elastic condition required by MARSH. The crossover value of E/Y is 133.

The substitution of some realistic numerical values enables (3-1) to be reduced to a useful simplified practical form:

(1) For many polycrystalline pure metals and alloys, $\nu \cong 0.3$.

(2) The ratio Y/E is approximately 10^{-2}; e.g., for Ti–Nb(40 at.%), with $Y = 8.8 \times 10^8$ N m^{-2} [Koc77] and $E = 8.7 \times 10^{10}$ N m^{-2} [Rei73], $Y/E = 1.01 \times 10^{-2}$.

(3) It follows from the above conditions that

$$l \equiv (1-2\nu)\frac{Y}{E} \cong 0.4 \times 10^{-2}$$

thus

$$B \cong 1 \tag{3-2}$$

(4) It also follows from the above that

$$m = (1+\nu)\frac{Y}{E} \cong 1.3 \times 10^{-2}$$

(5) With $l \cong m \ll 1$, it also follows that

$$Z \equiv 3/(l + 3m - lm) \cong 3/(l + 3m) \tag{3-3}$$

$$= E/1.43Y$$

from the definitions of l and m.

(6) Finally, the substitution of (3-2) and (3-3) into (3-1) reduces it to

$$H_V/Y \cong 0.065 + 0.6 \ln(E/Y) \tag{3-4}$$

applicable, as mentioned above, for $E/Y < 133$, which, according to numerical solution of the equation, corresponds to $E/H_V < 44.3$.

3.7 Application of the Marsh Formula to the Determination of the Yield Strength of a Wire

The microhardness, H_V, is determined upon a metallographically mounted cross section of the wire. Young's modulus, E, is determined using either sonic or ultrasonic sound-velocity measurement techniques. If $E/H_V > 44.3$, the Hill relationship, $Y = H_V/3$, is immediately accepted. Otherwise the Marsh approach is adopted and (3-4) solved transcendentally for E/Y, hence Y. The Newton–Raphson method, which when applied to (3-4) converges very rapidly, is well suited for this purpose.

3.7.1 Newton–Raphson Solution of the Simplified Marsh Equation

If X_i approximately satisfies $F(X_i) = 0$, then a better root is X_{i+1} where

$$X_{i+1} = X_i - \frac{F(X_i)}{F'(X_i)} \tag{3-5}$$

By iterating (3-5) a few times excellent solutions can be obtained.

In the present case, having obtained values for H_V and E, the root, Y, of the function $F(Y) \equiv 0.065 - H_V/Y + 0.6 \ln(E/Y)$ is required. This is quickly obtained by substituting $y = Y^{-1}$ and iterating the equation

$$y_{i+1} = y_i - \frac{0.065 - H_V y_i + 0.6 \ln(E y_i)}{-H_V + 0.6/y_i} \tag{3-6}$$

3.7.2 Trial-and-Error Solution of the Simplified Marsh Equation

In

$$H_V/Y \cong 0.065 + 0.6 \ln(E/Y) \tag{3-4}$$

if

$$E/H_V \equiv R \tag{3-7}$$

then

$$H_V/Y = 0.065 + 0.6 \ln R + 0.6 \ln(H_V/Y)$$

Thus, the equation to be solved, by trial-and-error, for H_V/Y is

$$\left.\begin{aligned} H_V/Y - 0.6 \ln(H_V/Y) &= R' \\ \text{where} \qquad R' &\equiv 0.065 + 0.6 \ln R \end{aligned}\right\} \tag{3-8}$$

3.7.3 Graphical Solution of the Marsh Equation

Figure 3-3 has been constructed using (3-4), (3-7), and (3-8). Below the Hill lower bound, *viz.*, $E/H_V = 44.3$, values of H_V/Y or E/Y corresponding to measured values of E/H_V can be simply read from the graph.

3.7.4 Applications to Ti–Nb

An interesting case study of the applications of the Hill and Marsh methods to Ti–Nb superconductors is presented in Sections 3.14 and 3.15.

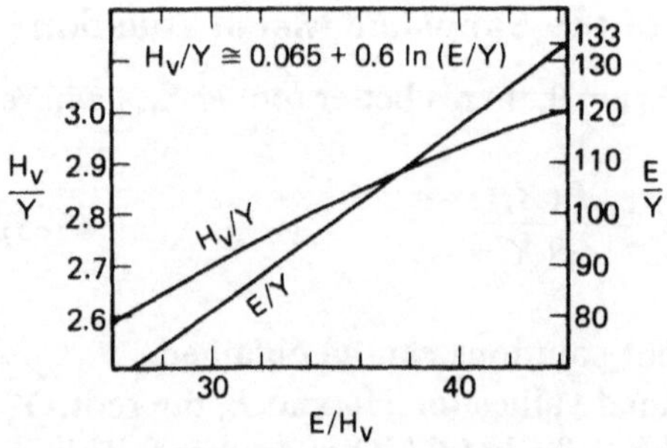

Figure 3-3. Parametric representation of the relationship $H_V/Y \cong 0.065 + 0.6 \ln(E/Y)$, (3-4).

3.8 Normal and Anomalous Tensile Properties of Superconductors

As magnets become larger both in physical dimensions and with respect to the magnetic fields and transport currents they are expected to maintain, the tensile properties of the conductor itself (initially ignored in the struggle to improve its electrical properties: critical field, critical current density, and stability) become important engineering quantities to be taken into serious consideration. This is particularly true with respect to large-scale high-field applications. Although important large coils are still being designed to operate at current densities lying well below the short-sample critical surface (see Tables 27-3 and 27-4), many present and future requirements call for operating currents as high as 90% of the short-sample critical values (see Table 27-7). Under such conditions intrinsic conductor mechanical instabilities (in addition to mechanical instabilities associated with the coil structure) represent potential sources of localized heat input to the conductor, thereby endangering its electrical stability. The superconducting magnet is a complicated assemblage of metal and insulation exposed to:

(1) static mechanical stresses related to the mass of the structure and the preloading of the wire during winding
(2) large variable, and in some cases cyclical, electromagnetic forces
(3) stresses imposed by thermal contraction, and contraction differentials, during-cool down from room temperature to liquid-He temperatures[1]

[1]According to Tables 9-1 and 9-6, Cu and Ti–50Nb undergo contractions of 0.33% and 0.17%, respectively, on cooling from room temperature to 4 K. When the Cu and the alloy are in intimate contact, as in a multifilamentary composite superconductor, the differential contraction results in a prestrain of 0.16% between the superconductor and the cladding material [PAS78]. The effect of this would be in addition to those of any stresses that may accumulate during magnet fabrication [HEI74].

In this light, it is clearly necessary for the magnet engineer to be fully cognizant of the tensile properties of composite superconductors, particularly those at low temperatures. In an attempt to address this need, the remainder of this chapter considers what might be termed the "normal" and "anomalous," respectively, static tensile properties.

3.8.1 The "Normal" Static Tensile Properties

The normal static tensile properties are parameterized in terms of the following features of the stress (σ) *versus* strain (ε) curve: (i) the slope of its linear segment, namely the Young's modulus, E; (ii) the limit of the linear segment—the elastic or proportional limit—defined as the stress at 0.1% plastic strain, $\sigma_{0.01}$; (iii) the yield strength, usually defined as the stress at 0.2% plastic strain, and represented by symbols such as Y or $\sigma_{0.2}$; (iv) the ultimate tensile strength, Y_{Ult} (i.e., the maximal engineering stress level achieved in the $\sigma(\varepsilon)$test); (v) the rupture strength, σ_B, which is another engineering quantity derived from the load-at-rupture divided by the unnecked cross-sectional area of the specimen. The $\sigma(\varepsilon)$ curve has many possible signatures. In brittle material, such as bcc alloys at low temperatures, the ultimate tensile strength and the fracture stress are coincident. On the other hand, at elevated temperatures, where fracture, if it takes place, does so in a ductile manner, Y_{Ult} and σ_B are separately recognizable.

3.8.2 The "Anomalous" Static Tensile Properties

Departures from normal stress-strain behavior manifest themselves in number of ways. The strain, instead of being a smooth function of stress, can undergo a kind of relaxation oscillation known as "serrated yielding." The sample may also undergo plastic deformation of a kind that recovers in response to stress reversal in such a way that the net strain vanishes with removal of the load; this is "pseudoelasticity." Other anomalous mechanical properties depend directly upon temperature change: examples are "thermoelasticity" and "shape-memory effect." Deformation mechanisms that contribute to these effects are (i) slip, (ii), twinning, (iii) martensitic transformation, and (iv) rearrangements of (or internal structural changes taking place within) a preexisting martensite.

In Part 2 of this chapter, the anomalous mechanical properties referred to above are all to be considered (with particular reference, whenever possible, to Ti–Nb alloys) together with two other important, more or less anomalous phenomena which have been observed to occur during the stress-strain cycling of superconducting composites, *viz.*, the Bauschinger effect and acoustic emission.

PART 2. ANOMALOUS MECHANICAL PROPERTIES

3.9 Anomalous Tensile and Related Properties

3.9.1 Serrated Yielding

(a) The General Phenomenon. The jagged oscillations in stress at constant strain-rate that characterize the $\sigma(\varepsilon)$ curves of many metals under certain conditions are referred to as serrated yielding. In the output trace of a screw-driven tensile machine, for example, sudden length dilatations occurring at rates faster than the screw-controlled deformation speed are accompanied by sharp unloadings of the specimen. The phenomenon has been well discussed in a description of some low-temperature (4.2 K) tensile studies by BASINSKI [BAS57], who, by using Al and some of its alloys as samples, was thereby able to disassociate the general effect from both twinning and martensitic transformation. Necessary conditions for serrated yielding in BASINKI's experiment were (i) a strong negative temperature dependence of the yield stress, and (ii) the usual small low-temperature specific heat. Since under such conditions local slip-nucleation is sufficient to trigger the thermomechanical instability, serrated yielding may be regarded as a quite general low-temperature mechanical effect [BAS57]. The local mean temperature increase associated with a serration was calculated by BASINSKI to be ~60 K. He also obtained the experimental lower limit for the magnitude of the temperature increase in an Al specimen through the use of an imbedded Nb wire whose transition to the normal state could be detected.

(b) Serrated Yielding in Ti–Nb. Serrations are noted whenever alloys of compositions near Ti-50Nb are strained at liquid-He temperatures beyond the linear elastic region [EVA73], [HIL73[a]], [EAS75, EAS77], [ALB76], [KOC77], [PAS78]. Figure 3-4 shows an example of serrated yielding in Ti-Nb(36 at.%) at 4.2 K. Since the effect does not take place in Ti-50Nb at room temperature, a transition from "normal" to "anomalous" tensile behavior must occur at some intermediate temperature. Figure 3-5, a series of typical $\sigma(\varepsilon)$ curves for bare Ti-Nb(36 at.%) wire, shows the onset of serrated yielding taking place at about 28 K.

EVANS [EVA73], in a search for a mechanism to explain the "training" of superconducting magnets, seems to have been the first to apply to Ti-Nb the general concepts of unstable low-temperature plastic flow described earlier by BASINSKI within the framework of some tensile studies of Al alloys. Those experiments had demonstrated that no exotic deformation mechanisms were needed, provided that a sufficiently steep negative yield-

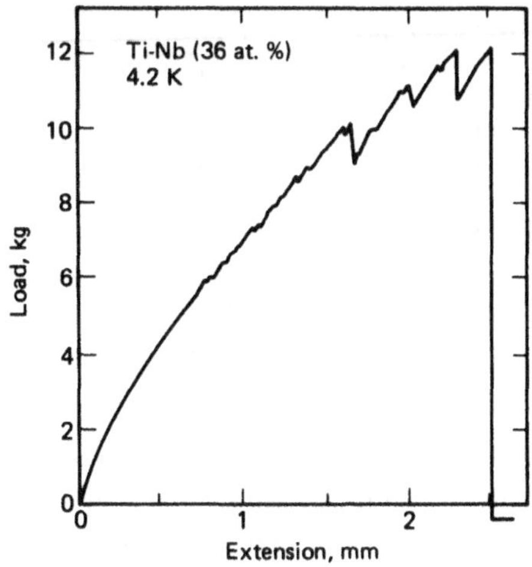

Figure 3-4. Stress-strain diagram showing serrated yielding of a bare Ti-Nb(35 at.%) tensile bar at 4.2 K—after KOCH and EASTON [KOC77].

stress temperature dependence was present. Thus, for an alloy such as Ti-Nb, with its inherent lattice instability manifesting itself as thermoelasticity, pseudoelasticity, or shape-memory effect (all to be considered below), low-temperature serrated yielding is ensured. As far as training is concerned, it is now recognized that effects more subtle than serrated yielding must be considered [PAS78], especially when all artifactual contributions (such as conductor motion, insulation cracking, etc.) have been eliminated.

In studies of the deformation and fracture of Ti-50Nb at 4.2 K, HILLMANN (following BASINSKI [BAS57]) attributed the serrations to spontaneous slip. It was concluded that slip took place on the {110}, or principal, bcc glide planes. ALBERT and PFEIFFER [ALB76], in tests of cold-worked (>90%) Ti-50Nb wires possessing the usual ⟨110⟩ texture parallel to the wire axis, after noting that the fracture surface was at 45° to the wire axis, came to the conclusion that slip in these samples took place on the {100} planes. A conventional slip-plasticity model was adequate to explain the effects of prior deformation or load alternation. With such a model, the

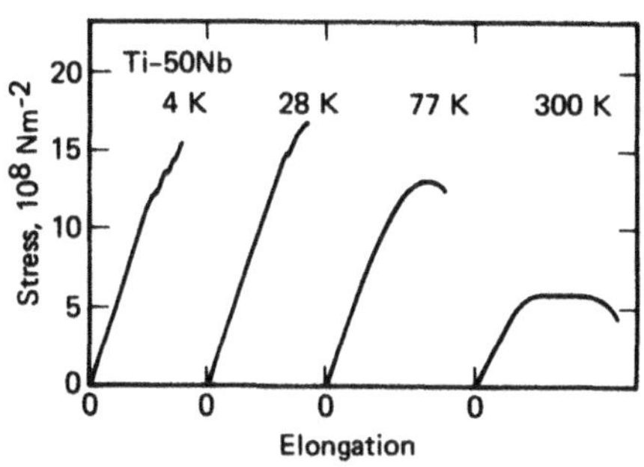

Figure 3-5. Stress-strain characteristics of bare Ti-50Nb tensile bars tested at 300, 77, 28, and 4 K, respectively, indicating the onset of serrated yielding at about 28 K—after ALBERT and PFEIFFER [ALB76].

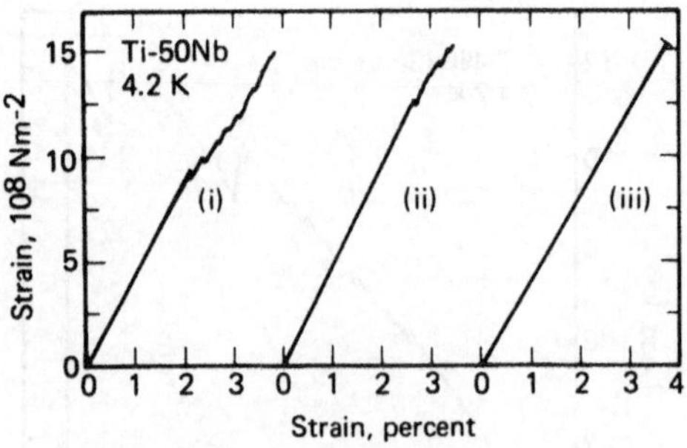

Figure 3-6. Stress–strain characteristics of bare Ti–50Nb tensile bars in the conditions: (i) 50% cold worked, (ii) 50% cold-worked-and-annealed 1 h/600°C, and (iii) 50% cold-worked plus twenty alternations of stress between 2.1×10^8 N m^{-2} and 12.6×10^8 N m^{-2}—after HILLMANN [HIL73[a]].

introduction of work-induced defects (i.e., work hardening) could be expected to inhibit slip and move the serrations up to higher and higher levels of strain as in Figure 3-6. Serrated yielding in Ti–50Nb at 4.2 K in the 1 h/600°C annealed and 50% cold-worked conditions has been observed by HILLMANN [HIL73[a]], and after 90% cold work, by ALBERT and PFEIFFER [ALB76]. In studies of the temperature dependence of the tensile properties (see Figure 3-5) the latter authors noted that, with decreasing temperature, instabilities appeared for the first time at 28 K. Although this value is in good accord with the extrapolated $\beta \rightarrow \alpha''$ transformation curve for Ti–Nb suggested by KOCH and EASTON [KOC77] and reproduced in Figure 3-7, it has been noted by HILLMANN [HIL76] that the transition from "normal" parabolic $\sigma(\varepsilon)$ behavior to "anomalous" behavior characterized by serrated yielding is gradual and not associated with a definite transition temperature. Detailed studies of the temperature dependence of the tensile properties of

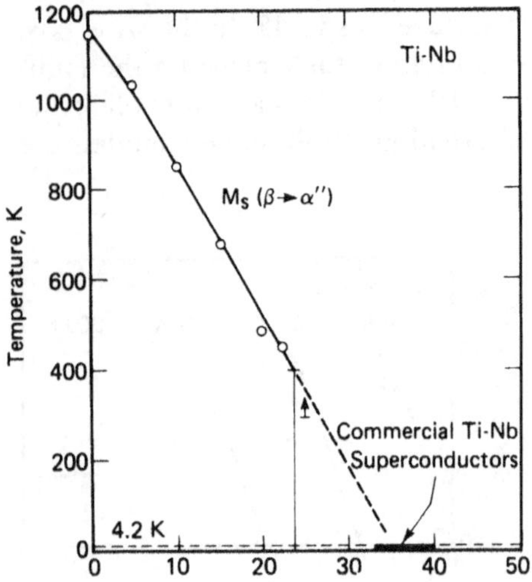

Figure 3-7. The martensitic $\beta \rightarrow \alpha''$ transformation curve for Ti–Nb based on the results of DUWEZ [DUW53], BROWN *et al.* [BRO64], and BAKER [BAK71], and its extrapolation by KOCH and EASTON [KOC77].

Ti–Nb by HILLMANN *et al.* [HIL76], one of the few groups to include 20 K in the measurement program, have shown that the entry into the serrated-yielding region is accompanied by several important departures from the behavior expected simply by extrapolating from higher temperature results; *viz.*,

(1) considerable scatter in the values of σ_B, $\sigma_{0.2}$, and $\sigma_{0.01}$ obtained from repeated experiments
(2) a σ_B which passes through a broad maximum in the vicinity of 20 K
(3) yield strengths, $\sigma_{0.2}$ and $\sigma_{0.01}$, which increase rapidly below about 30 K, but which for cold deformation $\gtrsim$90% also maximize at about 20 K
(4) an elongation which drops rapidly from 5–6% to below 1%

All these effects, which are clearly related to each other and to the serrated-yielding effect in Ti–Nb, warrant further study.

READ [REA78] detected serrated stress–strain behavior in Ti–45Nb at 4 K. The appearance of the fracture surface indicated that at least some of the deformation had taken place by irreversible twinning. EASTON and KOCH have noted serrated yielding in Ti–Nb(36 at.%) wire (0.0105 in.$^{\varnothing}$) at 4.2 K [EAS75],[KOC77], and in a filamentary sample (~280 $\mu m^{\varnothing}$) of comparable composition etched from a commercial multifilamentary composite conductor. In agreement with SCHMIDT [SCH76], they recorded that the serrations appeared in response to both increasing *and* decreasing loads, a reversibility that suggested the operation of a pseudoelastic effect (see below) such as reversible stress-induced martensitic transformation or stress-reversible twinning.

In any investigation of serrated yielding, the possibility that the results are being perturbed by the presence of sample artifacts and the characteristics of the testing machine must be considered. REED *et al.* [REE77] discussed the appearance of spurious load discontinuities associated with the cracking of a formvar coating on the gripped portion of the specimen. SCHMIDT [SCH77] considered in some detail the influence of sample-extension and load-train dynamics on the measured elongation and load drop during serrated yielding. With regard to the former, SCHMIDT pointed out that the rate of instantaneous elongation is determined in part by the inertia of the loading system and is, therefore, apparatus dependent; as for the load drop, he noted that during the instant of localized yield the rest of the sample relaxes elastically within a few microseconds, a time that is negligible when compared to the response time of a relatively massive load train, which, thus in effect, remains "rigid" during the process. Accordingly, it was concluded that when the specimen and equipment time-constants are related in such a way, the magnitude of the load drop should be independent of the apparatus.

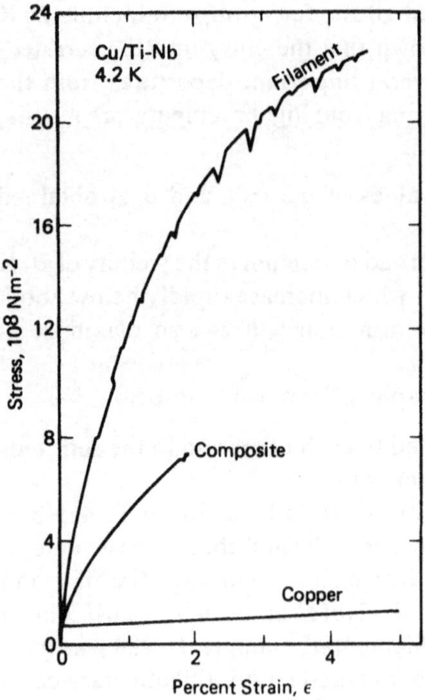

Figure 3-8. Stress-strain characteristic of a Cu/Ti-Nb(36 at.%) (2.8:1) 18-filament superconductive composite at 4.2 K compared with those of the individual components. As a consequence of pseudoelastic deformation, inhibited when Cu-clad, the filament shows greater elongation than the composite (cf. its 300 K behavior in Figure 3-31)—after EASTON and KOCH [EAS77] and KOCH and EASTON [KOC77].

Figure 3-8, which compares the $\sigma(\varepsilon)$ curve for a bare superconducting filament with that of a Cu-clad multifilamentary composite, suggests that the presence of the Cu inhibits serrated yielding; in addition, as pointed out by KOCH and EASTON [KOC77], the reduction in this manner of the possibility of pseudoelastic elongation leads to premature fracture. Although VERKIN *et al.* [VER76] reported seeing plastic instability in 6-strand Cu-clad T 60 (Ti-Nb(40 at.%)) cable and in 24-strand Cu-clad NT 50 (Ti-Nb(50 at.%)) cable just beyond the yield point at 4.2 K, it must be remembered that the Soviet experimental alloys are usually only thinly coated with Cu (see, e.g., [LAZ72, LAZ73[a]]). The influence of the matrix on the deformation of composites at ambient temperature has been investigated by PIEHLER [PIE65]. SCHMIDT, also, observed that serrated yielding, which was always found in bare Ti-Nb during loading (and occasionally during unloading), was not usually encountered in the Cu-clad wire [SCH76], [PAS78]. In addition, he noted that irreversible serrated yielding occurred (i) only for strains greater than about 1%, in agreement with an observation by EVANS [EVA73] that the effect began almost immediately upon exceeding the elastic limit (much greater than the 0.1-0.2% strain levels usually reached in superconducting magnets); (ii) in *repeated* loading only at stress levels higher than the maximal stress of the previous cycle [SCH76].

3.9.2 Thermoelasticity

Three interrelated thermomechanical properties—thermoelasticity, pseudoelasticity, and shape-memory effect—are described schematically in Figure 3-9. Thermoelasticity (Figure 3-9(a)) is a special kind of martensitic transformation that takes place continuously and reversibly with change of temperature, in contrast to the usual martensitic transformation which is characterized by sudden appearances, or bursts, of platelet groups.

Implicit in an extrapolated martensitic transformation diagram for Ti–Nb due to EASTON and KOCH [EAS77], [KOC77] (Figure 3-7) is the suggestion that Ti–Nb alloys of compositions close to those used commercially should be thermoelastic in the He-temperature regime. Although an intriguing possibility, there is at present no direct microscopic evidence for such a low-temperature reversible transformation. Indeed indirect evidence refutes the conjecture, favoring instead the reversible athermal ω-phase transformation considered in Section 1.6. Comparative studies of the room-temperature and He-temperature physical properties of Ti–TM alloys indicate that the composition of the martensitic-phase boundary remains fixed as the temperature decreases below room temperature. Thus, whereas a reasonable extrapolation of the M_s curve for Ti–V [ZWI74, p. 174] passes through, for example, the point (−200°C and 19 at.%), a comparison of the room-temperature magnetic susceptibility composition dependence with that of the superconducting transition temperature shows that the extent of the martensitic regime is no greater at He temperatures than it is at 298 K—certainly less than 12 at.% V in both cases [COL75, COL75[b], COL82]. On the other hand there is abundant, irrefutable, direct and indirect evidence for the reversibility of athermal ω phase in alloys of this type. The microstructural observations have been considered in Section 1.8 (see Figure 1-22)

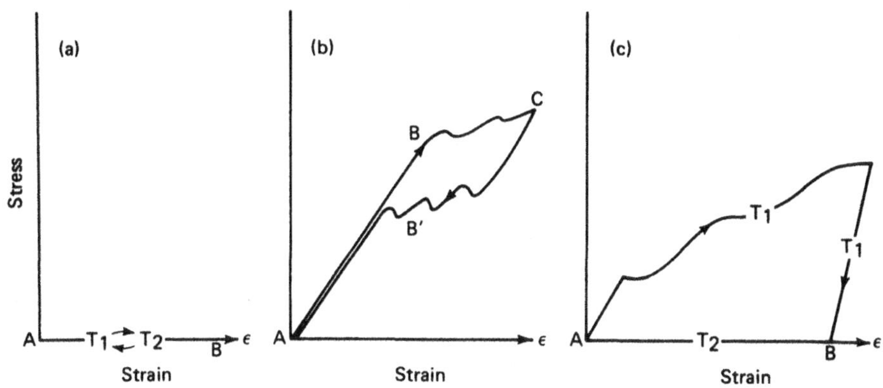

Figure 3-9. Schematic representations of (a) thermoelasticity, (b) pseudoelasticity, and (c) shape-memory effect in terms of their stress-strain-temperature signatures. In (a) and (c), T represents temperature, and $T_2 > T_1$—after DELAEY *et al.* [DEL74].

and the (indirect) physical property evidence is presented below. Having rejected, with respect to Ti–TM alloys, reversible low-temperature martensitic transformation in favor of ω-phase, one can only speculate at this stage as to *its* ability to lead to a measurable thermoelastic effect. A simple geometrical calculation leads to the prediction of a net 3% contraction perpendicular to the $(\bar{1}12)_\beta$ plane during the $\beta \to \omega$ transformation ([LUH70], see also Section 1.8). Since this can be accommodated by local coherency strains [HIC69, HIC69[a]], it must be concluded that a thermoelastic effect via this mechanism is unlikely.

3.9.3 Pseudoelasticity

(a) The General Phenomenon. The pseudoelastic effect is illustrated schematically in Figure 3-9(b). In it the sample, after a period of elastic extension to B, is shown deforming in an apparently plastic manner to C, and upon unloading returning spontaneously to B′ with net dissipation of energy, then elastically almost back to the initial state, A, [DEL74]. In this mechanical analog of the thermoelastic effect, martensitic transformation proceeds continuously with increasing applied stress and is reverted continuously when that stress is removed. At sufficiently low stress levels the sample exhibits normal elastic properties. The apparent plastic deformation is, thus, the result of a (reversible) stress-induced martensitic transformation. The *deformation* behavior of the sample suggests that the effect could be described by the title "pseudoplasticity;" but the term used, "pseudoelasticity," arises from the fact that the sample returns to its original state after removal of the applied stress, as in normal elastic behavior, in spite of its having experienced an appreciable amount of nonelastic deformation.

There is, of course, abundant evidence for stress-induced martensitic transformation in Ti–TM alloys. Section 1.7.3 has shown how the transformation can be triggered by the application of additional strain in the vicinity of a strain spinodal defined by $\partial^2 g/\partial\varepsilon^2 < 0$. That is to say, in a lattice that is prone to transform—a situation signalled for example by a low value of the elastic shear modulus $C' = (C_{11} - C_{12})/2$—the application of a few percent strain is sufficient to initiate the martensitic reaction. The occurrence at room temperature in Ti–Mo(5 at.%) (whose composition places it just above $M_{s,\,300\,\mathrm{K}}$) of deformation martensite and its effect on the superconducting transition is considered below in Section 10.7. It is important to bear in mind, however, that pseudoelasticity can also occur, in the absence of martensitic phase change, through some form of mutual accommodation between pairs of twinned structures [DEL74].

(b) Pseudoelasticity in Ti–Nb. The extrapolated M_s curve of Figure 3-7, although perhaps not to be taken literally, does suggest, without defining

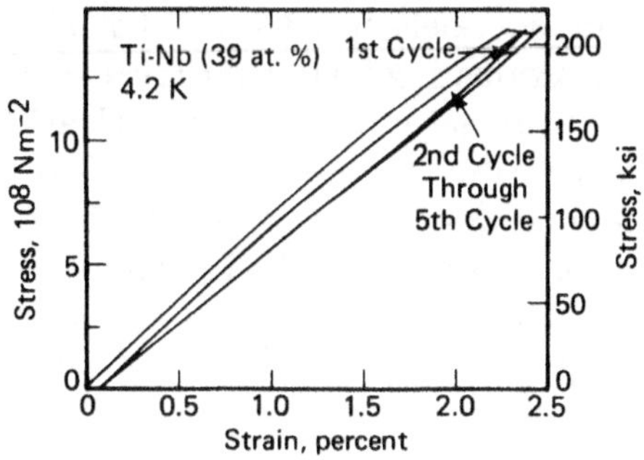

Figure 3-10. Evidence for pseudoelasticity during the repeated load-unload cycling of a bare Ti-Nb(39 at.%) tensile bar—after KOCH and EASTON [EAS75], [KOC77].

the mechanism, the possibility of some kind of low-temperature $\beta \rightarrow \alpha''$ transformation. This, and other evidence of lattice instability, such as the relatively low shear modulus for alloys of this class (Section 4.4) and the proven existence of diffuse ω-phase (Section 1.8), is evidence in support of a possible stress-induced martensitic transformability for alloys whose temperature/composition states lie just above the dashed line; this includes the commercial superconductive alloy Ti-50Nb at low temperatures. If the transformation is reversible (i.e., pseudoelastic) no trace of its ever having taken place will remain after the stress has been removed and the sample returned to room temperature [EAS75], [PAS78].

It is, however, possible to acquire *direct evidence* for pseudoelastic transformation, as in the *in situ* cold-stage TEM experiments of OBST *et al.* [OBS80], who noted what appeared to be a reversible transformation in reversibly strained Ti-45Nb. *Indirect evidence* for pseudoelastic transformation may be deduced from the form of the $\sigma(\varepsilon)$ curve. Thus, stress-strain behavior, which was thought to be characteristic of pseudoelasticity, has been detected in Ti-55Nb at 4.2 K by EASTON and KOCH [EAS75], [KOC77] (Figure 3-10). During the first loading cycle of the low-temperature tensile test, slip took place partially reversibly at a strain level of about 2.2%, following which the sample went on to execute closed hysteresis loops with no further accumulation of residual strain up to 100 cycles of deformation. The hysteretic deformation, which set in above the 0.4% strain level [EAS75], [PAS78] and which was accompanied by acoustic emission during both loading and unloading [KOC77], was interpreted as a manifestation of the pseudoelastic martensitic transformation.

As mentioned above, at higher strain levels ($\gtrsim$1%) bare Ti-Nb exhibited serrated yielding.

3.9.4 The Shape-Memory Effect

(a) The Shape-Memory Phenomenon in General. The shape-memory effect (Figure 3-9(c)) may be regarded as a combination of the thermoelastic and pseudoelastic effects in that after an initial stress-induced "permanent-transformation" phase the reverse transformation can be induced by raising the temperature of the sample [DEF71]. The effect is usually associated with martensitic-thermoelastic (so-called "marmem") alloys, whose behavior and properties have been considered in detail in reviews assembled by PERKINS *et al.* [PER75, PER75[a]]. The temperature range within which the effect is to be found is quite broad, and many alloys exhibit it. The classical Ti-base shape-memory material is, of course, the intermetallic compound TiNi (so-called "Nitinol").

(b) The Shape-Memory Effect in Ti–Nb. In experiments on β-quenched Ti-Nb(22 at.%), which fully transforms to α'' on quenching through ~175°C, BAKER [BAK71] showed that 1% "permanent" strain could be almost fully (~90%) erased by heating to 184°C. Similar effects were noted for the $\omega+\beta$-state of the same alloy (produced by aging for 2 min/400°C and 4 min/400°C) except that an almost 100% recovery from $1\frac{1}{2}$% strain was achieved. Experiments have indicated that a shape-memory effect can occur as a result of a martensitic $\beta\rightarrow\alpha''$ transformation, or, in a specimen whose initial structure is already fully martensitic, by way of a reorientation of the martensite plates [DEL74].

Since the $\beta\rightarrow\alpha''$ transformation is accompanied by considerable dilatation, its repeated occurrence during a tensile test either isothermally (pseudoelastic effect) or adiabatically (shape-memory effect) will give rise to load drops and corresponding $\sigma(\varepsilon)$ anomalies (serrated yield) of a form that is characteristic of both the sample properties and the testing machine response times [SCH77].

3.9.5 The Bauschinger Effect

(a) The Bauschinger Effect in General. It could be argued that since the Bauschinger effect is generally observed during the stress–strain cycling of metals and alloys, it is a *normal* mechanical property and as such should be considered in Part 3 of this chapter. On the other hand, the effect *is* anomalous[2] in the sense that it involves a stress–strain asymmetry explicable only in microscopic materials-property terms.

When a sample is subjected to repeated application and release of a unidirectional stress greater than its virgin yield stress, it eventually work

[2]There is, of course, an "anomalous Bauschinger effect." This is observed in pseudoelastic materials in which stress-induced transformation or twinning is responsible for premature third-quadrant yielding. Anomalous Bauschinger effects are also exhibited by alloys such as Ti-6Al-4V, which, after being strained in tension to a permanent 1% set, suffers a 40% loss of compressive yield strength as a result, it was thought [GIL76], of grain-boundary yielding.

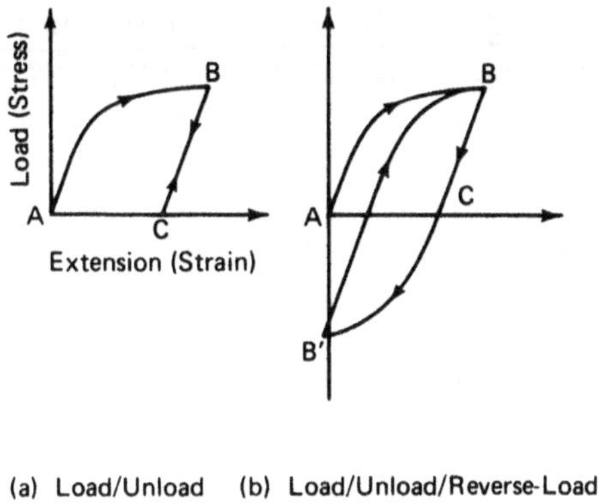

Figure 3-11. (a) Typical nonhysteretic stress–strain behavior of a sample successively loaded in tension and unloaded. (b) Hysteretic stress–strain behavior of a sample experiencing load–unload half-cycles alternately in tension and compression. Schematically represented in this case is the Bauschinger effect which consists of a premature yielding during a reverse-load quarter-cycle (C → B′) as compared to that which took place during initial loading (A → B)—see for example TWEEDDALE [TWE64] and FISHER *et al.* [FIS77].

hardens sufficiently to permit a simple linear elastic stress–strain behavior (Figure 3-11(a)). In contrast to this, the application of load-unload-reverse-load cycling leads to hysteretic stress–strain loops of the kind illustrated in Figure 3-11(b). Furthermore, it is generally found that the third quarter-cycle is characterized by yielding at a stress level that is low compared to that of the first. This is the Bauschinger effect, which may, therefore, be defined as *reversible-stress-softening that permits premature yielding of a material upon stress reversal after an initial half-cycle involving plastic deformation* (*in either the compressive or tensile mode*). It has been found that the effect is generally smaller the greater the strain is during the initial half cycle, and that it disappears completely with large amounts of strain hardening [TWE64, p. 74].

GITTUS [GIT75[a], p. 76] has discussed the Bauschinger effect in terms of a dislocation mechanism. TIMOSHENKO [TIM56, p. 413], in attempting an explanation, invoked the granularity of a polycrystalline metal and developed a mechanical model for the Bauschinger effect based on *intragranular slip and its anisotropy.* Expanding on this and related ideas, MARGOLIN *et al.* [MAR78] suggested that the Bauschinger effect could arise as a result of residual stresses that accumulate in the presence of grains with different yield stresses. Assuming a tension/compression sequence, grains with lower yield stresses then become placed in compression by the other grains in addition to the applied stress. The models of TIMOSHENKO and MARGOLIN both invoke an additional hidden component of stress to augment the applied stress and bring about the apparently "premature" third-quadrant yielding. Presumably the same mechanism is also operative, but at a more macroscopic level, in certain bimetallic composites that by their very natures generally consist of pairs of metals with widely differing mechanical properties. It is this mechanism that is particularly relevant in the present context.

(b) The Bauschinger Effect in Cu/Ti–Nb Composites. An excellent example of a bimetallic composite is the well-known Cu/Ti–Nb multi-filamentary superconductor, in which hundreds of fine filaments of Ti–50Nb, a material with a relatively high modulus, are imbedded in annealed OFHC Cu. Under tension, the Cu yields at a much lower stress than the Ti–50Nb, consequently it is possible for the latter still to be stretching elastically long after the Cu has passed its yield point. As a result, on the return stroke of the testing machine, the Cu is placed in compression by a combination of the applied stress and the elasticity of the Ti–Nb. The effect of cyclic stress on superconducting composite materials has been discussed by FISHER *et al.* [FIS77], and the possible influences of the Bauschinger effect on magnet performance and training has been considered by HEIM [HEI74]. Thermomechanical heating associated with the execution of Bauschinger loops is considered below in Section 3.12 (see Figure 3-13 therein).

3.10 Acoustic Emission from Copper and Titanium–Niobium

Acoustic emission is a general effect that occurs over a wide temperature range in association with both microscopic and macroscopic deformation processes. Acoustic emission can be a product of microplastic flow [SCH63], but has also been detected in association with the normal plastic deformation of Cu at strain levels above 0.3%, during which noise in the 100–300 kHz frequency range was emitted continuously during straining [PAS78]. The cooperative motion of glide dislocations along slip planes has been advanced as one possible mechanism for the effect [FIS67]. Martensitic transformation has been another possible mechanism, as during the testing of austenitic stainless steel at temperatures below 45 K when tensile load drops were accompanied by loud clicks [BAS57].

Audible clicking associated with load drops were occasionally noted by REED *et al.* during the tensile testing at 4.0 K of a Ti–Nb superconductor [REE77]. In a more extensive study of audio-frequency emission during the alternate loading and unloading of a 2-mm$^{\varnothing}$ tensile test sample of Ti–55Nb, EASTON and KOCH [EAS75] noted that

(1) although emission occurred *almost immediately* upon application of the stress at 4.2 K, none was detected at 77 or 300 K
(2) emission took place upon both loading *and* unloading and *bursts of noise(clicks) preceded any load-drop that occurred*
(3) during repeated loading the emission *increased in amplitude and frequency* when the immediately preceding stress level was exceeded

It is clear from a close examination of these results and intercomparison of the italicized statements, that pseudoelastic serrated yielding is a sufficient but not necessary condition for acoustic emission in Ti–Nb at low temperatures. These results, which are fundamental to our current understanding of the training phenomenon, were amplified and clarified in the reports of a detailed investigation by PASZTOR and SCHMIDT [SCH77a] [PAS78]. In studies of ultrasonic emission at 100-300 kHz during the straining of Ti-50Nb wire at 4.2 K, the above authors were able to describe the effect in terms of reversible and irreversible components depending on whether

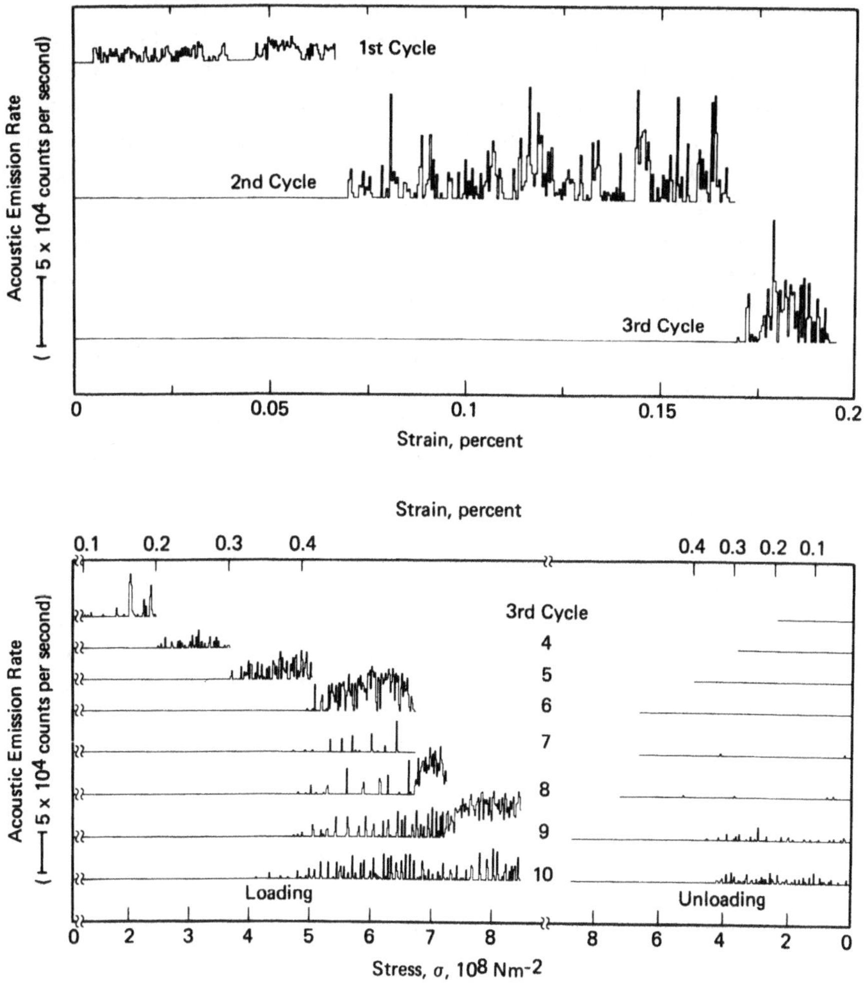

Figure 3-12. Acoustic emission from Ti–50Nb wire as function of strain, during two loading and eight subsequent loading-unloading cycles at 4.2 K. Amplifier characteristics: gain, 95 dB (upper) and 92 dB (lower); bandwidth, 100-300 kHz—after SCHMIDT and PASZTOR [SCH77a], [PAS78].

or not emission would take place during the unloading half-cycle. The threshold for the *reversible* component was a strain level of about 0.4%, which also marked the onset of hysteretic stress–strain behavior of the type referred to as pseudoelastic by EASTON and KOCH [EAS75], and illustrated in Figure 3-10. The *irreversible* component was present throughout the entire stress range and, after increasing the amplifier gain by 6 db, was detectable even at room temperature. The noise output took place in bursts of some 100 μs duration, commencing at extremely low levels of strain (~0.005%). In repeated loading to successively higher levels of strain, the irreversible noise signals (and the only noise generated at all at strains $\lesssim$0.3–0.4%) began only when the previous stress level was exceeded. These effects are well illustrated in Figure 3-12, which is assembled from the work of PASZTOR and SCHMIDT [SCH77[a]], [PAS78]. Finally, it should be noted that acoustic emission in a sample of Cu/Ti–Nb composite superconductor (10.8 × 0.19 mm^2; 200 filaments; twist pitch, 75 mm; Cu/SC ratio, 4.4:1) has been the subject of a detailed study by NOMURA *et al.* [NOM80].

As summarized in Table 3-3, three important conclusions can be drawn from the work of KOCH and EASTON [EAS75], [KOC77] and its extension by PASZTOR and SCHMIDT [SCH77[a]], [PAS78]:

(1) At 4.2 K, irreversible serrated yielding accompanied by its acoustic emission begins in Ti–Nb at about 1% strain [SCH77[a]] and pseudoelasticity, accompanied by $\sigma(\varepsilon)$ hysteresis and reversible acoustic emission [EAS75], begins at about 0.4% strain [EAS75], [SCH77[a]].

(2) Both of the above effects take place at stress levels of typically 50 ksi (~3.5×10^8 N m^{-2}), values which are higher than those expected during the operation of superconducting coils.

(3) Also detected has been an irreversible component of the acoustic emission. Taking place as a steady stream of bursts beginning at extremely small levels of strain ($\lesssim$0.01%) and releasing insignificant amounts of energy, this phenomenon was regarded as being symptomatic of movement at the atomic level.

3.11 Mechanical Fatigue of Composite Conductors

Some aspects of the cyclic yielding process in Cu/Ti–Nb composites have been referred to above, and "stress effects" in composite conductors is the subject of Chapter 29. Mechanical fatigue in composite conductors, the subject of this section, has been studied by EKIN [EKI78] using a 180-filament Cu/Ti–Nb (Cu/SC = 1.8:1) composite wire cycled to various stress amplitudes in the positive $\sigma(\varepsilon)$ quadrant at temperatures of 4, 77, and 295 K. The limiting component was found to be the Cu which suffered fatigue damage within the elastic range of the Ti–Nb. At 4 K the composite could endure up to 10^3 cycles [Eki78], [Sei79] at a stress amplitude as high as 10^3 MN m^{-2} with little change in property, but deteriorated rapidly beyond that point at a rate that depended on stress amplitude [Eki78]. For example, in the material cited above [Eki78], failure occurred at: 10^4 cycles/700 MN m^{-2}; 4×10^5 cycles/540 MN m^{-2} (52% ultimate); and 1.5×10^6 cycles/480 MN m^{-2} (46% ultimate), which represented essentially the "endurance-stress limit." The corresponding 4 K strain limit was about 0.35%. Cycling to stress levels lower than the endurance limit had what might be regarded as a beneficial strengthening effect [Sei79]: In the constant-stress cycling of a 1710-filament, 1.45-mm$^{\varnothing}$ Cu/Ti–Nb monolithic conductor, it was found that the strain persisted at its initial value (set at 0.2%) out to about 5×10^4 cycles; but beyond that point, as a consequence of slight work hardening, the strain dropped rather rapidly such that by 10^5 cycles a new constant value, some 12% lower than before, had set in [Sei79].

3.12 Thermomechanical Heating

Two aspects of stress-induced heating are to be considered here, neither of which are now regarded as threats to magnet stability. The first to be treated is the *heat pulse associated with serrated yielding*, which, as we have just seen, takes place in Ti–Nb beyond the plastic yield threshold ($\lesssim$1% strain), a point which is not reached during normal magnet operation. The second has to do with the steady heat output associated with *cyclical plastic deformation of the Cu cladding* in a composite Cu/Ti–Nb conductor, an effect initially thought to be a factor in magnet training [Hei74]. Although the ultimate training mechanism is now acknowledged to be a much more subtle effect [Eas75], [Eki77], [Sch77ᵃ], [Pas78] (see also Section 29.2), this type of thermomechanical heating must still be recognized, if only as a possible additional burden on conductor cooling in large magnets subject to a cyclical or repetitive pulse mode of operation [Kro77].

3.12.1 Strain Heating Due to Serrated Yielding

Some of the initial quantitative studies of load drops and temperature increases associated with yield instability were carried out by Basinski [Bas57], who estimated the local mean temperature rise accompanying the load-drops to be about 60 K. Subsequently Erdmann and Jahoda [Erd68], in tensile studies of vacuum-isolated annealed Cu–28Ni wires initially at 5 K, calculated that the local temperature increase associated with discontinuous slip taking place within a volume of some 1.4×10^{-3} cm^3 was about 35 K.

Table 3-3. Summary of Serrated Yielding, Pseudoelastic, and Acoustic Emission Effects in Ti–Nb at 4.2 K

Effect	Range of Strain	Observations	Literature
Serrated Yielding	$\gtrsim 1\%$	Experiments mostly proceeded to fracture, which took place at an angle of 45° to the sample axis.	Hillmann [HIL73[a]] Evans [EVA73] Albert [ALB76] Verkin [VER76]
		(Permanent) twinning was observed on the fracture surface.	Read [REA78]
		Appears mostly during the loading 1/4-cycle for strains $\gtrsim 1\%$; and in *rare cases* during unloading. Large temperature spikes ($\gtrsim 50$ K) and internal heat pulses ($\gtrsim 1$ mJ) are generated.	Schmidt [SCH76] Schmidt [SCH77[a]] Pasztor [PAS78]
		Serrations are noted during loading and unloading.	Easton [EAS75]
Pseudoelasticity	$\gtrsim 0.4\%$	$\sigma(\varepsilon)$ curves for bare Ti–Nb exhibit hysteresis. During the first cycle an almost fully reversible discontinuous yield was noted; subsequent load–unload loops, although hysteretic, are fully completed.	Easton [EAS75]
		The hysteresis appears above about 0.4% strain.	Pasztor [PAS78]

Acoustic Emission—Reversible	≳0.4%	Reversible ultrasonic acoustic emission commences at about 0.4% strain corresponding to the onset of pseudoelasticity and occurs during unloading down to low levels of strain. Serrated yielding, if it occurs, is also accompanied by audio-frequency emission (audible clicks) which are, therefore, expected to accompany any loading (≳1% strain) and (rare instances of) unloading serrations.	Easton [EAS75] Pasztor [PAS78]
Acoustic Emission—Irreversible or "Trainable"	≳0.005%	Audible clicking is also associated with "normal" or irreversible serrated yielding which sets in at ≳1% strain.	
		Audible emission has been detected almost immediately upon application of stress.	Easton [EAS75]
		Irreversible ultrasonic emission has been noted to commence at a strain level as low as 0.005%.	Schmidt [SCH77[a]]
		The energy associated with a burst of acoustic emission (calculated using reversible emission data) is ~1 μJ, a factor of 10^3 less than that for a discontinuous yield.	Pasztor [PAS78]

Detailed studies of strain heating in Ti–Nb were conducted by PASZTOR and SCHMIDT [PAS78]. Tensile tests were conducted on a 4-cm length of Ti–Nb wire immersed in liquid He and connected to an electrical circuit as if for four-terminal resistivity measurement. The appearance of a voltage pulse of width 0.2–0.3 ms during load drop indicated that a local temperature increase to $T > T_c$ (9.5 K) had occurred. Secondly, from the magnitude of the voltage pulse, the length ($l \cong 1$ mm), hence the volume or mass, of the instantaneously created normal zone was able to be calculated. The heat input to the normal zone was estimated from $Q = L\Delta\varepsilon$, the product of the load, L, and the instantaneous length increase, $\Delta\varepsilon$, of the zone in which the transformation took place. This yielded $Q = 3$ mJ, enabling the mean temperature rise to be calculated from

$$Q = V \int_{4.2\,\mathrm{K}}^{T} C^v(T)\, dT \tag{3-9}$$

where V is the volume of the normal zone and C^v represents the heat capacity per unit volume. Using an estimated value for the temperature-dependent specific heat of Ti–Nb, for which experimental data were not readily available at the time, the mean adiabatic temperature rise was calculated to be 95 K, a value in good accord with the early results of BASINSKI reported above.

To complete the analysis, PASZTOR and SCHMIDT [PAS78] inquired into the effect of the surrounding liquid He pool on the temperature rise. Because of its general interest and importance, the calculation is reproduced in full: As a consequence of the low thermal diffusivity of liquid He, the heat released during a load drop remains near the sample/He interface and evaporates a surface layer of He. Using the data of IWASA *et al.* on transient heat transfer [IWA77], the energy per unit surface area absorbed by the He before the onset of film boiling was obtained as a function of pulse duration. For a 0.3-ms pulse this was 0.6 mJ cm^{-2}, corresponding to an energy in the present case of 5 μJ. The rate of heat transfer during the subsequent film boiling is given by $\dot{Q} = h\Delta T$ with the transfer coefficient, h, equal to 0.05 W cm^{-2} K^{-1} [JAC69];[3] during the period of the pulse this amounted to about 8 μJ. Comparing this 13 μJ transferred across the surface of the wire with the ~1.5 mJ released internally as heat during the load drop, it is clear that the yield instability took place practically adiabatically even when bare Ti–Nb wires were immersed in liquid He [PAS78], a conclusion which had already been drawn by ERDMANN and JAHODA [ERD68]. The same was also found to be true for immersion in superfluid He at 1.5 K [PAS78].

[3]But see Section 25.40.2 for a discussion of steady-state and transient heat transfer to liquid He.

It is interesting to compare the heat outputs of serrated yielding and acoustic emission. Using some results from reversible acoustic emission, the integrated output of which was assumed to equal the area of a corresponding hysteresis loop, PASZTOR and SCHMIDT [PAS78] arrived at an energy output of about 1 μJ per burst, a factor of 10^3 less than the per-pulse output of discontinuous yield.

3.12.2 Strain Heating in a Cu/Ti–Nb Composite

The site of stress-induced heating in a Cu/Ti–Nb composite at low temperatures is the Cu matrix which is forced to execute *hysteretic* cycles of extension and compression while the Ti–Nb cores merely undergo *elastic* extension and relaxation in the positive $\sigma(\varepsilon)$ quadrant. The Cu undergoes plastic cycling since: (i) its relatively low yield strength, with respect to that of Ti–Nb, causes it to deform plastically while the latter is still extending elastically (Figure 3-13(a)); (ii) when the Ti–Nb relaxes on release of the tension, it eventually puts the Cu into compression such that, during subsequent load-unload cycles, it executes Bauschinger loops and dissipates energy instead of undergoing linear, work-hardened, $\sigma(\varepsilon)$ excursions.

Three classes of deformation are, of course, available to two-component composites:

(1) both components may deform elastically
(2) one component may deform plastically while the other remains elastic
(3) both components may deform plastically

In the Cu/Ti–Nb superconductor we are concerned with the "class-2" type of deformation process, since the elastic limit of Cu is attained at 0.05% strain [EAS76^a][4] while that of Ti–Nb (excluding pseudoelasticity) is at about 1% [PAS78]. In Figure 3-13(a) the projection of AC represents the permanent elongation of the conductor at the end of the first cycle, and indicates the amount to which the filaments are elastically strained in tension against the matrix.

The stress–strain behavior of the Cu alone is isolated in Figure 3-13(b), which can be interpreted with the help of Figure 3-11. Referring to Figure 5-11(a), during its first extension the Cu deforms plastically to point B and work hardens, then on release of the tension it returns almost linearly to C. After 10 or so cycles between B and C any last traces of hysteresis disappear; the Cu has by then work hardened sufficiently that its performance is perfectly elastic, as evidenced by the lack of any further dissipation of energy [EAS76^a]. On the other hand, if the Cu is put into compression

[4]The threshold of acoustic emission from Cu is 0.3% strain [PAS78].

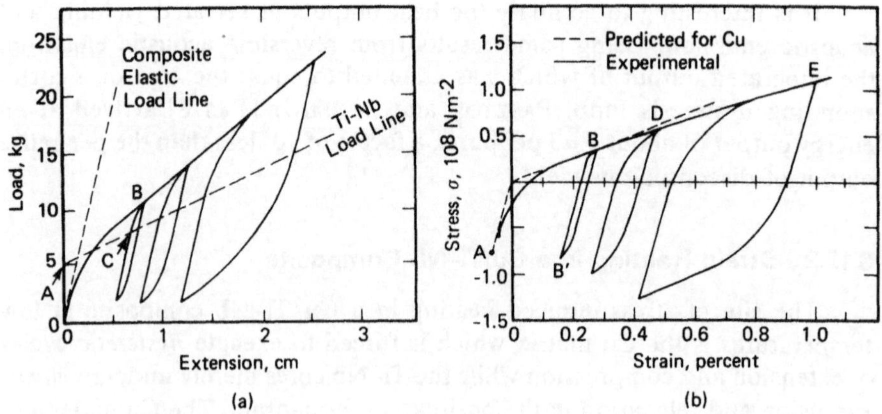

Figure 3-13. (a) Load-extension characteristics obtained during the tensile testing of a Cu/Ti-Nb composite (Cu/SC ratio, 4:1; strand diameter, 1.02 mm; filament number, 84; filament diameter, 50 μm). (b) Derived stress-strain characteristics of the Cu matrix. Initially under compressive loading as a result of residual fabrication strain (A) the Cu deforms as the Ti-Nb extends elastically to point B (where the magnet may quench due to "adiabatic heating"). During the second charging cycle the Cu can proceed to point D before the magnet quenches, and so on until it can be safely cycled to E at its design field—after HEIM [HEI74].

(Figure 3-11(b)) it yields long before the stress level at B (but in the opposite direction) has been reached. This premature yielding in compression more or less repeats itself for as long as the cycling between B′ and B is maintained [EAS77], [KRO77]. Early investigators [COR67] studying the effect of cyclic loading on composite superconductors neglected to consider the Bauschinger premature compressive yield effect, and supposed the Cu to work harden continuously and the hysteresis loops to become correspondingly smaller with time, as they would in the absence of the Ti-Nb cores.

In a fatigue test of a twisted 180-filament Cu/Ti-Nb composite (Cu/SC = 1.8:1) to a stress level of 8×10^8 N m^{-2}, EKIN [EKI78] noted that the width of the hysteresis loop did not remain strictly constant, but decreased noticeably in the first few cycles until the zero-stress closure reached completion. After stabilization of the plastic extension, the then-closed hysteresis loop continued to become narrower as work hardening of the copper accumulated; a minimum area was achieved after about 115 cycles. Of course at strains below 0.35% the area of the loop will settle down to a constant value.

Several authors have measured the temperature increases and have calculated the heat outputs associated with both single-extension and cyclical deformations of Cu and Cu/Ti-Nb composite superconductors [EAS75, EAS76[a]], [KRO77], [PAS78]. Assuming that only a small fraction of the total work done per deformation cycle is stored as strain energy within the specimen [KRO77], the heat generated (being approximately equal to the net work done on the specimen) can be obtained from the area of the hysteresis loop (an intrinsic quantity) and the diameter of the specimen. Then, by knowing the specific heat as a function of temperature, the temperature rise in the specimen can be estimated [EAS75] by solving (3-9) as before. Applying this approach on the assumption that the heat was confined to the Cu, the seat of the dissipation, EASTON and KOCH [EAS75] were able to construct a theoretical curve of temperature rise *versus* loop area. Their result is presented in Figure 3-14. The extreme value shown of $\Delta T = 33.2$ K corresponds to the stressing to 10^2 ksi (7×10^8 N m^{-2}) at 4.2 K of an 18-filament, 0.01-in.$^{\varnothing}$, Cu/Ti-52Nb composite of cross-sectional area 0.05×0.12 in.^{2} Because of experimental constraints it is not possible to measure such temperature rises directly.

In the serrated yielding experiment, the heat is generated locally in a relatively low conductivity Ti-Nb wire, whose local temperature rise can be measured directly by means of a miniature thermometer. On the other hand, in the fatigue experiment much of the heat, which is generated uniformly throughout the high-conductivity Cu matrix of the composite, is quickly conducted out to the grips. Consequently an indirect calorimetric approach is called for. Using a vacuum calorimeter attached to a cryogenic tensile probe, KROEGER, EASTON, and MOAZED have measured the thermomechanical heat outputs from samples of pure Cu, and from several Ti-Nb-strengthened composites with matrices of Cu, Cu/Cu-Ni, and an Al alloy [EAS76[a]], [KRO77]. Heat output was determined from the measured temperature rise ($\lesssim 6$ K) of a system whose electrically calibrated heat capacity was dominated by that of a pair of stainless steel grips or ballasts.

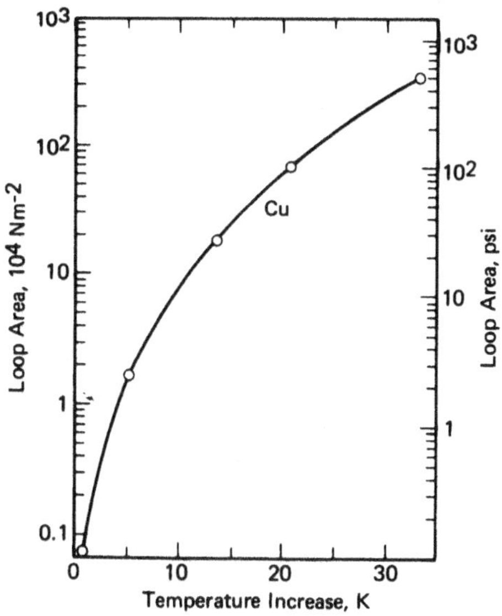

Figure 3-14. Temperature rise in Cu in response to hysteretic stress-strain cycling—after EASTON and KOCH [EAS75].

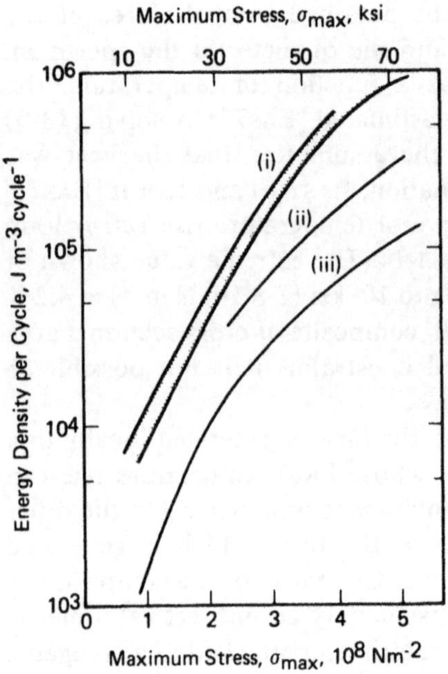

Figure 3-15. Cyclical heat outputs as functions of peak stress for various commercial composite superconductors: (i) four Cu/Ti–Nb strands (two with Cu/SC ratio 2.8:1, 18 filaments, and two with Cu/SC ratio 1.6:1, 1530 filaments); (ii) a Cu/Cu-Ni/Ti-Nb strand (matrix/SC ratio, 1.5:1; 1591 filaments); (iii) a 5056Al/Ti-Nb/Al strand (matrix/SC ratio, 1.63:1; 54 filaments)—based on data from KROEGER *et al.* [KRO77].

Several useful observations were made: (i) the agreement between the work done and the thermomechanical heat output was very good for such an experiment (*viz.*, ± ~25%); (ii) with the pure-Cu sample, temperature rises that did not occur in the elastic region ($\lesssim 75 \times 10^6$ N m^{-2}) appeared as soon as plastic deformation commenced; (iii) with pure Cu during repeated cycling to the same stress level, the heat output per cycle diminished to zero within 10–15 cycles as work hardening proceeded; (iv) the composite conductors generated no heat when stretched elastically, but under "class-2" deformation conditions (see above) a fixed quantity of heat was liberated at each cycle of deformation depending on the nature of the matrix and, of course, on the stress level. The cyclical heat outputs of various conductors are plotted as functions of maximum stress in Figure 3-15. It is interesting to note that the Al-alloy-"stabilized" conductor[5] evolved nearly an order of magnitude less heat than that stabilized with Cu [KRO77].

Using data from Figure 3-15 it is instructive to compare the thermomechanical heat generated during the cycling of a large magnet with another important internal source of cyclical heating, AC loss: a reference design for the poloidal field coil of the Oak Ridge Tokamak Experimental Power Reactor Study (Reference [8] of [KRO77]) suggested that, at a field of 70 kOe, an average hoop stress in the windings of 70×10^6 N m^{-2}

[5]Aluminum alloy type 5056, used in wire etc., manufacture is nominally Al-5.2Mg-0.1Mn-0.1Cr.

(~10 ksi) would be reasonable; accordingly a positive half-cycle of charging would liberate 7×10^3 J m^{-3} of thermomechanical heat. This may be compared with the $1.6_5 \times 10^3$ J m^{-3} of electromagnetic loss incurred in 2 seconds during a full (±70 kOe) charge/discharge cycle [KRO77]. AC loss itself is the subject of Chapter 26.

PART 3. NORMAL MECHANICAL PROPERTIES OF TITANIUM–NIOBIUM ALLOYS AND COMPOSITE CONDUCTORS

The Young's modulus, E, the slope of the initially linear portion of the $\sigma(\varepsilon)$ curve, may be obtained from the results of static or quasistatic (i.e., low frequency) tensile tests. Alternatively, the slope about the origin of $\sigma(\varepsilon)$ may be derived from the results of a sound velocity experiment taking advantage of the relationship $E = \rho_d v_{L,T}^2$ where ρ_d is the density, and $v_{L,T}$ are the velocities of long-wavelength longitudinal or transverse vibrations, respectively, (see Section 4.1). E and the shear modulus, G, can also be derived from the results of ultrasonic measurements on either polycrystals (Section 4.2) or single crystals (Section 4.3), in regard to which an intermediate step is the determination of the monocrystalline elastic constants, C_{ij}. It is generally agreed that elastic-constant studies lie within the purviews of both mechanics and physics. This point of view is preserved here in the decision to treat Young's modulus, when derived from tensile testing, in the present chapter, while that same modulus together with other bulk moduli, when derived from dynamic measurements, are to be treated along with the elastic constants themselves in Chapter 4—the first of a group of chapters making up the Physical Properties component of this book.

3.13 Young's Modulus of Titanium–Niobium Superconductors

During the 1970s, tensile Young's-modulus data were acquired on bare experimental superconductive wires, miniature tensile test bars of Ti–Nb, and commercial superconducting composites, both monofilamentary and twisted multifilamentary and from which the Cu had been chemically removed. A collection of numerical results is presented in Table 3-4. Figure 3-16 intercompares the data and also relates the implied Young's-modulus composition dependence to an "accepted composition dependence" for the composition range 20–40 at.% Nb as derived from the results of the

Table 3-4. Young's Moduli of Ti–Nb and Some Ti–Nb Base Superconducting Alloys

Code Number (Figure 3-16)	Alloy	Condition	Test Temperature K[a]	Young's Modulus, E^b 10^{10} N m^{-2}	Comments	Literature
1	Ti–Nb(34 at.%)	Optimized commercial wire Final cold worked	4.2 [300	10.0 8.3]	Single-core conductor from Vacuumschmelze, with Cu chemically removed. 300 K value estimated from 4 K value by scaling by 5.3/6.4	Schmidt [SCH76]
2	Ti–Nb(36 at.%)	—	300 4.2	8.2 8.3	"Sub-size" tensile sample of Ti–Nb alloy (as distinct from etched commercial composite)	Easton [EAS75]
3	Ti–Nb(36 at.%)	—	300	8.6	—	Koch [KOC77]
4	Ti–Nb(36 at.%)	—	300 77 4.2	8.2 8.24 8.28	—	Easton [EAS77]
5	Ti–Nb(38 at.%)	As commercially processed	298 76 4	5.3 5.8 6.4	180 twisted filaments (pitch, 1.27 cm) removed by etching from commercial composite, and formvar coated	Reed [REE77]

6	Ti–Nb(39 at.%)	—	300 4.2	8.2 8.3	"Sub-size" tensile sample of Ti–Nb alloy (as distinct from etched commercial composite)	Easton [EAS75]
7	Ti–Nb(40 at.%)	—	300	6.9	From Rutherford Lab Central Engineering Group Design Note EDN 0001 (1967)	Koch [KOC77]
8	Ti–Nb(20 at.%)	Cast and 97.2% cold worked	300	6.0_7	Stress-strain data acquired for wires 1 mm$^{\varnothing}$ × 100 mm^{l}. The influence of 400°C aging time on E was examined	Heller [HEL71]
9	Ti–Nb(26 at.%)	Cast and 97.2% cold worked	300	6.5_0		
10	Ti–Nb(33 at.%)	Cast and 97.2% cold worked	300	7.0_6		
11	Ti–Nb(44 at.%)	Cast and 97.2% cold worked	300	8.3_9		
	Ti_{65}–Nb_{33}–Ge_2	Cast and 97.2% cold worked	300	7.1_6		
12	Ti–Nb(33 at.%)	Cold worked plus 34 h/400°C	300	7.2_8	0	Zwicker [ZWI70]
	Ti_{65}–Nb_{33}–Zr_2	Cold worked plus 34 h/400°C	300	7.7_2	0.4_4	
	Ti_{65}–Nb_{33}–Al_2	Cold worked plus 34 h/400°C	300	7.9_9	0.7_1 Modulus increase ΔE due to solute addition	
	Ti_{65}–Nb_{33}–Cu_2	Cold worked plus 34 h/400°C	300	8.6_8	1.4_0	
	Ti_{65}–Nb_{33}–Ag_2	Cold worked plus 34 h/400°C	300	8.4_9	1.2_1	
	Ti_{65}–Nb_{33}–Ge_2	Cold worked plus 34 h/400°C	300	7.9_3	0.6_5	

[a] Room temperature, if not specified is quoted here as 300 K.
[b] In many cases, numerical data retrieved from curves.

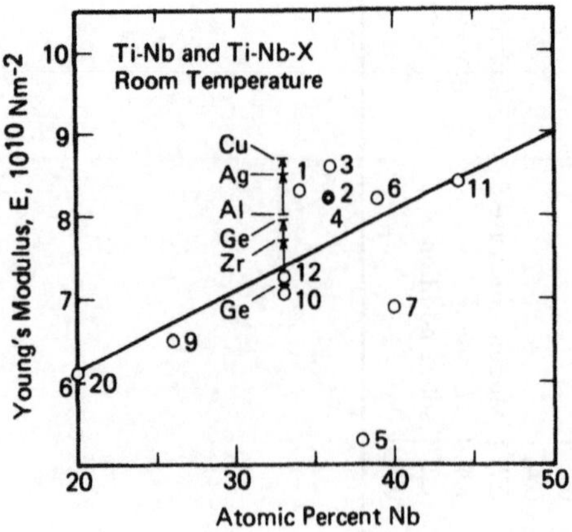

Figure 3-16. Young's moduli at room temperature for Ti–Nb binary and several Ti_{65}–Nb_{33}–SM_2 ternary alloys. The numbers against the points identify them with data listed in Table 3-4. The straight line, which is offered as an "acceptable composition dependence of Young's modulus for Ti–Nb," is consistent with the dynamic modulus results of Chapter 4 (see Figure 4-7).

dynamic-modulus testing discussed in Chapter 4. For the most part, the fit of the static Young's-modulus data to the dynamically derived consensus is satisfactory. Only two points in the figure deviate markedly from the line: Point No. 5, which was obtained from tensile tests on a twisted bundle of 180 bare 32-$\mu m^{\varnothing}$ filaments, and Point No. 7, whose history is not known. Also summarized in Table 3-4 and Figure 3-16 are the effects of 2 at.% additions of Zr, Al, Cu, Ag, and Ge to Ti–Nb(33 at.%) in the cold-worked plus 34 h/400°C-aged condition. Small increases in E accompany additions of all these "strengthening" elements to the base alloy. As a "neighboring" transition element, Zr in solid solution has only a small effect on the modulus of Ti–Nb. Cu gives the strongest increase in E, followed in turn by Ag, Al, and Ge. But these are not necessarily solid-solution effects—for example, the solubility of Ge in Ti–Nb(33 at.%) at temperatures below 600°C is only a few tenths of a percent, the remainder being present as Ti_5Ge_3.

3.14 Hardness and Modulus of Titanium–Niobium Superconductors

For numerous Ti–Nb binary alloys it is fortunate, thanks to the work of Albert and Pfeiffer [Alb72], Heller [Hel71], and Zwicker *et al.* [Zwi70], that hardness and modulus data pairs are available (Table 3-5). Remembering that two types of hardness imprint (the "Hill" type and the "Marsh" type—Section 3.6) occur depending on whether E/H_V is greater than or less than 44.3, we plot in Figure 3-17 E/H_V *versus* the Nb content, and draw attention to some remarkable results: (i) apart from some explicably outlying points, E/H_V for (a) quenched and (b) cold-worked-and-aged $\lesssim$400°C alloys are linearly increasing functions of Nb content for $20 \lesssim [Nb] \lesssim 50$ at.%; (ii) each group of alloys proceeds from the Marsh to the

Table 3-5. Comparison of Young's Modulus and Hardness in Some Ti–Nb Alloys

Code Number (Figure 3-17)	Nb Content at.%	Young's Modulus, E: 10^{10} N m^{-2}	Young's Modulus, E: 10^{6} g mm^{-2}	Vickers Hardness H_V, kg mm^{-2}	E/H_V	Comments	Literature
1	20	6.1	6.2_2	150	42	"Uncoupled" data pairs—E, from the continuous line in Figure 3-16 representing alloys in various conditions; H_V, from the smoothed curve of Figure 3-1 for annealed-and-quenched alloys	
2	30	7.1	7.2_4	155	47		
3	40	8.0_5	8.2_1	160	51		
4	22	9.2_5	9.4_3	338	28	95% cw + 10 h/350°C } E data from dynamic modulus measurements	Albert [ALB72]
5	22	8.2_2	8.3_8	237	35	95% cw + 10 h/450°C }	
6	34	7.5_0	7.6_5	205	37	95% cw + 10 h/400°C }	
7	44	8.2_8	8.4_4	181	47	95% cw + 10 h/400°C }	
8	33	7.0_6	7.2_0	180	40	cw 97.2%	Heller [HEL71]
9	33	7.2_8	7.4_2	198	37	cw + 34 h/400°C	Zwicker [ZWI70]
		7.3_6	7.5_0	—	38		Heller [HEL71]

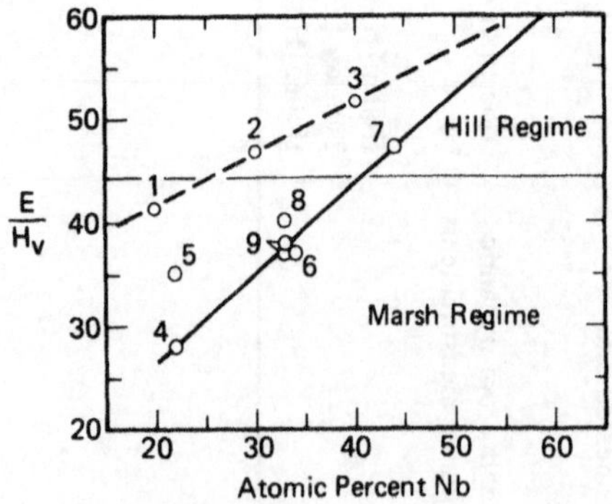

Figure 3-17. Young's-modulus/Vickers-hardness ratio, E/H_V, for Ti-Nb alloys. The numbers associated with the points identify them with data listed in Table 3-5.

Hill regime with increasing Nb content; (iii) finally, for cold-worked-and-aged $\lesssim$400°C alloys the $E/H_V > 44.3$ field is entered at 42 at.% Nb—precisely on the "practical" $(\alpha+\beta)/\beta$-phase boundary at 400°C (see Figure 1-7). The metallurgical implication of this observation is that as precipitation hardening dies out, E increases with Nb content more rapidly than H_V.

Table 3-6. Comparison of Young's Modulus and Hardness in Some Ti-Nb(33 at.%)-Base Ternary Alloys[a]

Alloy	Young's Modulus, E 10^{10} N m^{-2}	10^6 g mm^{-2}	Vickers Hardness H_V, kg mm^{-2}	E/H_V	Condition	References
Ti-Nb$_{33}$-Zr$_2$	7.7_2	7.8_7	205	38	b	[Zwi70]
Ti-Nb$_{33}$-Zr$_4$	7.4_0	7.5_5	212	36	b	[Zwi70]
Ti-Nb$_{33}$-Zr$_6$	7.6_7	7.8_2	220	36	b	[Zwi70]
Ti-Nb$_{33}$-Al$_1$	7.9_1	8.0_7	210	38	b	[Zwi70]
Ti-Nb$_{33}$-Al$_{1.5}$	7.9_1	8.0_7	224	36	b	[Zwi70]
Ti-Nb$_{33}$-Al$_2$	7.9_9	8.1_5	233	35	b	[Zwi70]
Ti-Nb$_{33}$-Al$_{2.5}$	8.4_5	8.6_2	244	35	b	[Zwi70]
Ti-Nb$_{33}$-Cu$_{0.5}$	7.7_1	7.8_6	206	38	b	[Zwi70]
Ti-Nb$_{33}$-Cu$_1$	8.5_6	8.7_3	266	33	b	[Zwi70]
Ti-Nb$_{33}$-Cu$_3$	8.8_3	9.0_0	285	32	b	[Zwi70]
Ti-Nb$_{33}$-Cu$_5$	8.8_8	9.0_5	260	35	b	[Zwi70]
Ti-Nb$_{33}$-Ag$_1$	8.2_5	8.4_1	261	32	b	[Zwi70]
Ti-Nb$_{33}$-Ag$_2$	8.4_9	8.6_6	274	32	b	[Zwi70]
Ti-Nb$_{33}$-Ge$_2$	7.1_6	7.3_0	260	28	c	[Hel71]
Ti-Nb$_{33}$-Ge$_4$	7.7_5	7.9_0	295	27	c	[Hel71]
Ti-Nb$_{33}$-Ge$_{0.5}$	7.5_3	7.6_8	223	34	b	[Zwi70]
Ti-Nb$_{33}$-Ge$_1$	7.8_5	8.0_0	269	30	b	[Zwi70]
Ti-Nb$_{33}$-Ge$_{1.5}$	7.7_2	7.8_7	284	28	b	[Zwi70]
Ti-Nb$_{33}$-Ge$_2$	7.9_3	8.0_9	318	25	b	[Zwi70]

[a] Numerical values retrieved from published curves.
[b] Cold worked plus 34 h/400°C.
[c] Cold worked 97.2%.

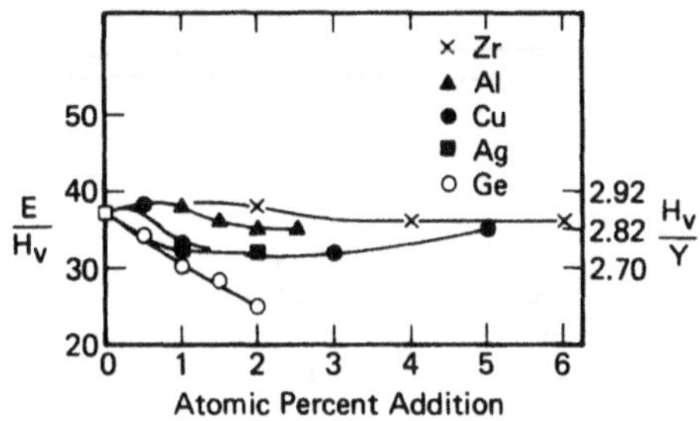

Figure 3-18. Young's-modulus/Vickers-hardness ratio, E/H_V, for ternary alloys based on Ti-Nb(33 at.%) plotted against third-element solute concentration in at.%—data from ZWICKER *et al.* [ZWI70], [HEL71] (see Table 3-6). Also shown are the corresponding Marsh-calculated hardness/yield-strength ratios, H_V/Y (see Figure 3-3).

Before moving on to a discussion of yield strength, a quantity which may be derived from H_V with the aid of either the Hill or the Marsh relationships, whichever is appropriate, it is interesting to consider the influence of third-element additions on the E/H_V ratios of Ti-Nb(33 at.%), an alloy for which coupled data pairs are available [ZWI70], [HEL71]. Figure 3-18, plotted from the data of Table 3-6, shows that E/H_V values for all the Ti-Nb(33 at.%)-base cold-worked-and-aged 34 h/400°C ternary alloys cluster within a band defined by $E/H_V = 35 \pm 5$, suggesting that E/H_V is approximately constant for a given *class* of alloy. Ti-Nb_{33}-$Ge_{0.5-2}$ are exceptions, possibly due to the effects of precipitation—the loss of Ge from solution would tend to decrease E, while the simultaneous formation of Ti_5Ge_3 would tend to maintain the hardness.

3.15 Hardness, Modulus, and Yield Strength in Titanium–Niobium Superconductors

3.15.1 The Hill and Marsh Relationships Applied to Ti–Nb Alloys

The Hill and Marsh equations (Sections 3.6 and 3.7), *viz.*,

$$H_V/Y = 3 \qquad \text{(Hill)} \qquad \begin{cases} E/Y > 133 \text{ or} \\ E/H_V \gtrsim 44.3 \end{cases} \qquad (3\text{-}10)$$

$$H_V/Y \cong 0.065 + 0.6\ln(E/Y) \qquad \text{(Marsh)} \qquad \begin{cases} E/Y < 133 \text{ or} \\ E/H_V \lesssim 44.3 \end{cases} \qquad [3\text{-}4](3\text{-}11)$$

have established relationships between coupled values of hardness, modulus, and yield strength. The Ti-Nb_{33}-SM class of alloys, whose E/H_V ratios are centered near 37, lie in the regime of Marsh, the solution of whose equation yields $H_V/Y = 2.87$. With $H_V \cong 200$, as for cold-worked-and-aged Ti-Nb(33 at.%) itself, the yield strengths would be expected to be about

Table 3-7. Case Study of the Interrelationships Between

	Experimental Input[a]				Model Regime Based on E/H_V: Hill (H) Marsh (M)
		Modulus, E			
Alloy	Hardness, H_V kg mm^{-2}	10^{10} N m^{-2}	10^6 g mm^{-2}	E/H_V	
Condition: "*Annealed*"—*800°C and water quenched*					
Ti-Nb(30 at.%)	145	6.6	6.7_3	46	H
Ti-Nb(39 at.%)	150	7.8	7.9_5	53	H
Ti-Nb(50 at.%)	147	8.5	8.6_7	59	H
Condition: "*Processed*"—*Ti-Nb*(30, 39 *at.*%): (800°*C*/ *WQ*)/ *extruded* 86%/(48 *h*/375°*C*)/ *drawn* 56%. *Ti-Nb*(50 *at.*%): *as-received cold-worked condition*					
Ti-Nb(30 at.%)	293	7.1_5	7.2_9	25	M
Ti-Nb(39 at.%)	174	7.8_8	8.0_4	46	H
Ti-Nb(50 at.%)	152	8.7_7	8.9_4	59	H

[a] All input data and measured yield strengths are due to Read [REA78] except the enclosed modulus data which have been derived from the dynamic-modulus data of Albert and Pfeiffer [ALB72] (see Chapter 4) for Ti-Nb(34, 44 at.%) in the condition 95% cw plus 40 h/400°C.

70 kg mm^{-2} (i.e., 6.8×10^8 N m^{-2})—a quite reasonable value in that it compares favorably with the $Y = 7.2 \times 10^8$ N m^{-2} obtained by READ [REA78] in a tensile test of cold-worked plus 48 h/375°C-aged Ti-Nb(39 at.%). In fact READ's results themselves form the basis of an interesting case study of the interrelationship between H_V, E, and Y in both of the above-defined E/H_V regimes.

3.15.2 Case Studies of Ti–Nb(30, 39, and 50 at.%)

READ [REA78] has measured the hardness and yield strengths of the alloys Ti-Nb(30, 39, and 50 at.%) in the "annealed" and "processed" conditions as defined in Table 3-7. The moduli of the "annealed" alloys were also reported but not those of the "processed." To replace the missing data, values of E were deduced by extrapolation and interpolation from the dynamic-modulus results of ALBERT and PFEIFFER [ALB72] for Ti-Nb(34 and 44 at.%). The first step in deriving Y from the measured values of H_V and E is to determine whether E/H_V (as listed in the 5th column of Table 3-7) is greater than or less than 44.3. In all cases but one $E/H_V > 44.3$, enabling Y to be obtained from the Hill relationship $Y = H_V/3$. The last three columns of Table 3-7 compare the calculated and measured values of Y. In most cases the agreement is remarkably good, the sole exception being Ti-Nb(39 at.%) in the "processed" condition, whose calculated yield

H_V, E, and Y at 300 K in Three Ti-Nb Alloys

Hardness/Strength Ratio from Application of Model H_V/Y	Yield Strength Calculated from Hardness, Y_{calc} kg mm^{-2}	10^8 N m^{-2}	Measured Yield Strength Y_{meas} 10^8 N m^{-2}	Deviation, $100\Delta Y/Y_{meas}$
3.0	48	4.7	4.5	4%
3.0	50	4.9	4.6	7%
3.0	49	4.8	4.1	17%
2.5_6	114	11.2	10.3	9%
3.0	58	5.7	7.0	19%
3.0	51	5.0	5.2	4%

strength is 19% too low, a deficiency possibly traceable to a low hardness number.

In confirmation of the correctness of Figure 3-17, and in particular that the alloy Ti-Nb(39 at.%) should be close to the line of demarcation between Hill- and Marsh-type behavior, we note that although from the standpoint of E/H_V (=46) the alloy lies just within the Hill regime, the value of the quotient $E/Y_{measured}$ (=116) places it just within the regime of Marsh.

3.16 Tensile Strengths of Titanium-Alloy Superconductors

The room-temperature mechanical properties of Ti-TM alloys have been reviewed by ZWICKER [ZWI74], who presented numerous data descriptive of the composition dependences of σ_B (the fracture strength), $\sigma_{0.2}$ (the stress at 0.2% plastic deformation), ψ (the percent area reduction at fracture), and δ (the percent elongation just prior to fracture). The data for Ti-Nb(0–40 wt.%) presented here are from some old work by BERGER *et al.* [BER58], who measured the tensile and yield strengths of alloys melted from commercial-grade starting materials and prepared for testing by the hot forging and rolling-in-air of unhomogenized ingots. Room-temperature tensile-strength measurements on the remainder of the series, *viz.*, Ti-Nb(30–95 wt.%), have been conducted on 0.25-mm$^{\varnothing}$ wire samples by BARON *et al.* [BAR65[a]]. The fact that two of the data points—those for Ti-94.5Nb and Ti-39Nb—were reproduced as common extrapolated points in studies of the effects of dilute third-element additions on the tensile strengths of those respective alloy bases is a testimonial to the consistency of their data. The results of the above-mentioned two studies [BER58], [BAR65[a]] of the composition dependence of room-temperature σ_B are assembled in Figure 3-19.

ZWICKER [ZWI74] has reviewed in detail the tensile properties of Ti–Mn and Ti–Mo alloys, and POLONIS and colleagues [POL71] have studied the deformation and fracture behavior of Ti–10Cr and Ti–15Cr as functions of heat treatment. Of particular interest in the latter work were the roles of athermal and aged ω-phase and "reverted ω-phase" (see Section 2.5) on the mechanical properties and the fracture characteristics.

Of considerable interest from the present standpoint is the relatively unknown work of the Physico-Technical Institute for Low Temperatures (Kharkov), which explores some relationships between the low-temperature tensile and the superconductive properties of metals and alloys. The relevant recent literature on the subject has been referred to by SOLDATOV *et al.* [SOL76], and some representative results of an extensive study of low-temperature plasticity in pure metals and binary alloys—in particular the α-Ti alloys Ti–Al(1.5, 3, 5.5 wt.%), Ti–Zr(1, 3, 8 wt.%), Ti–V(0.5, 1.5, 3 wt.%), and Ti–Nb(0.5, 2, 4 wt.%)—have been discussed by MOSKALENKO *et al.* [MOS70, MOS80]. Some of the results are summarized in Figure 3-20. In studying low-temperature mechanical properties it is very tempting simply

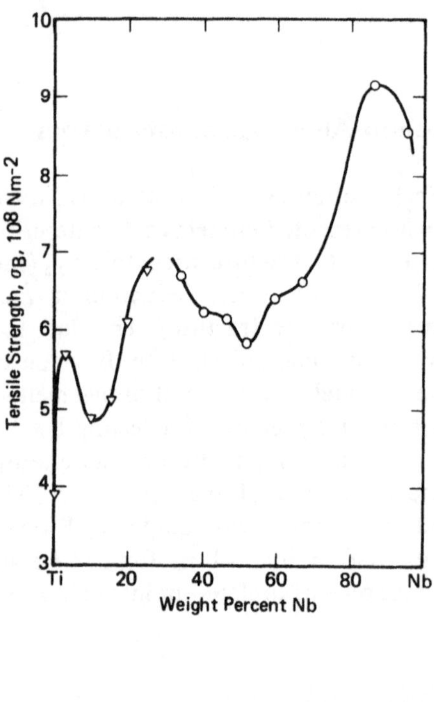

Figure 3-19. Tensile strengths of Ti–Nb alloys at room temperature. Data for the Ti-rich alloys (▽) are from measurements by BERGER *et al.* [BER58] on a hot-rolled (1500°F) sheet (data for Ti–(30, 35)Nb, omitted); data for the Nb-rich alloys (○) are from the work of BARON *et al.* [BAR65[a]] on 0.25 mm$^{\varnothing}$ wires.

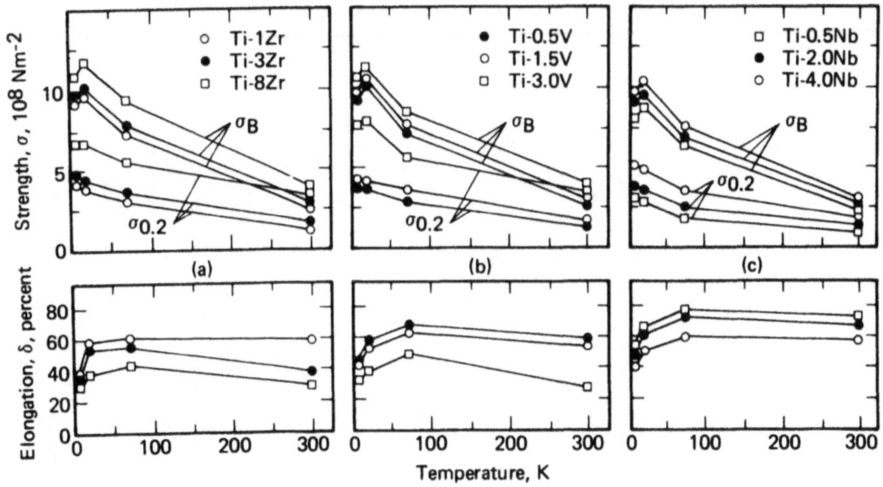

Figure 3-20. Yield strength (0.2%) and tensile strength, with associated elongations at fracture, of low-concentration Ti–Zr, Ti–V, and Ti–Nb alloys as functions of the test temperature—after MOSKALENKO *et al.* [Mos70, Mos80].

to make measurements at the temperatures of the usual liquid cryogens, N and He. But by omitting to take data at some intermediate temperature such as 20.5 K (liquid H) or 27 K (liquid Ne) it is possible for important anomalies in the low-temperature tensile behavior to be overlooked. This possibility was averted by MOSKALENKO *et al.* [Mos70], who conducted tensile tests within this intermediate-temperature range, and by so doing were able to identify the presence of pronounced maxima in both σ_B and $\sigma_{0.2}$ near 20 K (Figure 3-20). The drops in strength between 20 and 4.2 K are accompanied by decreases in δ.

The microstructures of Ti–0.5Nb alloys after fracture at various temperatures are depicted in Figure 3-21. It is interesting to note that the 77 K and 4.2 K fracture subsurfaces (the metallographically polished and etched fracture surfaces) are densely twinned. In a discussion to follow of the low-temperature deformation and fracture properties of the β-Ti–Nb "superconducting" alloys, attention is drawn to the twinning exhibited by the low-temperature fracture surface of Ti–Nb(30 at.%) [REA78] and a possible causal relationship between low-temperature tensile instability (serrated yielding) and a low-temperature strength anomaly similar to that reported by MOSKALENKO *et al.* Certainly the serrated-yield load-drop effect is sufficient to guarantee a tensile maximum in the β-Ti–Nb alloys, but on the basis of the work on low-concentration α-Ti–TM and α-Ti–SM alloys reported above it appears not to be the only mechanism.

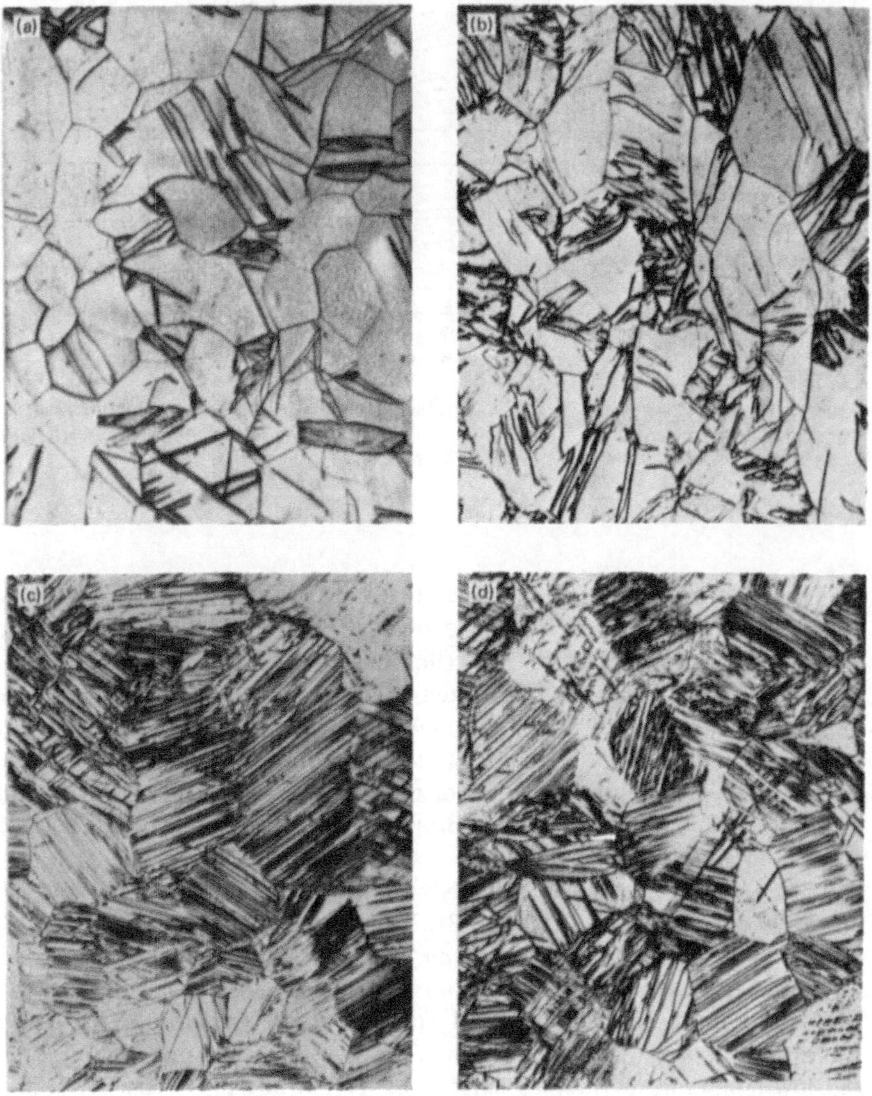

Figure 3-21. Microstructures of Ti-0.5Nb alloys in response to tensile deformation, ε, at various temperatures: (a) $\varepsilon = 12\%$, 293 K; (b) $\varepsilon = 25\%$, 293 K; (c) $\varepsilon = 12\%$, 77 K; and (d) $\varepsilon = 13\%$, 412 K. Original magnification, 340×—after MOSKALENKO and PUPCHOVA [Mos70]. Photographs courtesy of B. I. Verkin.

3.17 Tensile Properties of Titanium–Niobium Technical Superconducting Alloys

Since the mid 1960s numerous studies of the tensile properties, chiefly the ultimate tensile strength, of technically interesting superconducting Ti-Nb alloys have been undertaken. A recent review of the literature is available [KOC77] and the present status is summarized in Table 3-8 and the parts of Figure 3-22 that have been constructed from it. Many samples have been measured in the form of wire, typically 0.1-0.3 mm in diameter, and occasionally obtained from Cu-clad monolithic composites from which the Cu had been chemically removed. In this way both monofilamentary and multifilamentary (and twisted [REE77]) bare Ti-Nb samples have been acquired for measurement. BYCHKOVA [BYC70], using cold-worked wire samples, studied both the composition dependences of σ_B and the influence of 1-h aging temperature (300-900°C) on the strength. Alloys machined into the form of standard tensile specimens have been measured by EASTON and KOCH [EAS75], HILLMANN *et al.* [HIL73[a], HIL76], READ [REA78], and CURTIS and MCDONALD [CUR79]. When displayed as a function of Nb content, for example, the strength data appear at first glance to be completely scattered. However, by carefully isolating the variables of composition, heat treatment, and cold work as far as is possible, some useful information does emerge regarding:

(1) the composition dependence of σ_B
(2) the influence of 99.9% deformation on the σ_B of formerly recrystallized alloys
(3) the influence of area-reduction by cold work on σ_B
(4) the influence of temperature on the tensile properties of given classes of alloy

A picture of the manner in which strength responds to the variables—composition, degree of deformation, heat treatment, and measurement temperature in the 4.2-300 K range—can be assembled from a study of the papers of BYCHKOVA [BYC70], READ [REA78], HILLMANN *et al.* [HIL76], [ALB76], and CURTIS and MCDONALD [CUR79]. The data of READ [REA78], whose results have been discussed above in connection with the relationship between H_V, E, and Y, form a basis for investigating the composition dependence of σ_B. The low-lying band in Figure 3-22 represents recrystallized alloys. Also plotted in the figure are the results of BYCHKOVA [BYC70] and others for 99.9%-cold-worked alloys. These groups of results show: (i) that in the recrystallized alloys the composition dependence of tensile strength is rather weak, at least within the composition range 20-50 at.% Nb; (ii) that during cold reductions of >99.9%, tensile-strength increases of up to 80% can be expected. This last result is seen more clearly in Figure

Table 3-8. Strengths of

Nb Content at.%	References	Specimen Details	Condition	Test Temperature K[a]	Yield Strength 10^8 N m^{-2}
25	[BYC70]	0.25 mm$^{\varnothing}$ wire	99.86% cw	300	
25	[BAR70]	0.25 mm$^{\varnothing}$ wire	(i) 99.95% cw	300	
		0.25 mm$^{\varnothing}$ wire	(ii) Above + recrystallized 800°C	300	
30	[BYC70]	0.25 mm$^{\varnothing}$ wire	99.86% cw	300	
30	[REA78]	Tensile sample	800°C/WQ	300	4.5
				76	4.1
				4	8.3
30	[REA78]	Tensile sample	800°C/WQ + extruded 86% + 48 h/375°C + drawn 56%	300	10.3
				76	15.8
				4	20.3
34	[HIL76]	Tensile sample	Recrystallized 1 h at 1100 and 1200°C	298	5.2
				77	7.2 ± 0.2
				20	9.2 ± 1.2
				4.2	11.6 ± 0.5
			51% cw	298	5.9
				77	8.3 ± 0.2
				20	10.1 ± 0.7
				4.2	11.9 ± 0.4
			89.6% cw	298	5.7
				77	8.8
				20	14.5 ± 0.5
				4.2	12.8 ± 0.1
			cw + two heat treatments at 380°C	298	6.1
				77	11.5
				20	10.7 ± 1.9
				4.2	10.9 ± 0.2
			cw + two heat treatments at 380°C/final cold deformation	298	6.0
				77	11.7 ± 0.6
				20	11.3 ± 1.7
				4.2	12.4 ± 0.5

Some Ti–Nb Alloys

Tensile Strength 10^8 N m^{-2}	Reduction in Area %	Elongation %	Data Code for Figures 3-22 and 3-23	Comments
6.1			1	UTS also given as function of 1-h aging temperature from 300–900°C.
6.2			2A	Starting with conditions (i) and (ii)—UTS given as function of 1-h aging temperature from 350–500°C and also as function of aging time, up to 100 h at 450°C.
4.9			2B	
6.7			3	UTS given as function of 1-h aging temperature from 300–900°C.
4.9	87	22	4	
10.3	56	8		
11.6	38	9		
10.0	48	3		
16.5	30	3		
20.3	13	3		
5.2	61	3.0	5A	Data extracted from plotted temperature dependence curves. Plotted data available in all cases for the temperatures 4.2, ~26, 77, and 300 K.
11.5	33	2.1		
15.2 ± 0.5	8	0.3		
14.3 ± 1.1	5	0.5		
6.1	80	2.6	5B	
11.5 ± 0.1	47	2.6		
16.0 ± 0.4	16	0.8		
15.3 ± 0.2	15	0.5		
5.9	80	4.9	5C	
11.7	41	3.1		
16.1 ± 0.2	22	0.5		
14.4 ± 1.2	12	0.4		
6.2	78	5.5		
11.7	43	2.5		
15.8 ± 0.3	10	0.5		
14.0 ± 1.7	2	0.6		
6.4	76	10		
13.4 ± 0.7	43	2.1		
15.5 ± 0.6	14	0.6		
16.2 ± 0.3	5	0.5		

Continued

Table 3-8

Nb Content at.%	References	Specimen Details	Condition	Test Temperature K[a]	Yield Strength 10^8 N m^{-2}
34	[HIL73[a]]	0.4–0.6 mm∅	cw	300	8.7
		Wire	30 min/600°C	300	
		Wire	Above + 50% cold reduction	300	
		Wire	30 min/600°C + cold deformed ~45% +(2–4 h)/400°C +cold deformed up to 45%.	300	
		Tensile sample	50% cw	300	4.0 (*l*)
				4.2	12.3 (*l*)
			Above + 20 stress cycles between 2.1 and 12.6 × 10^8 N m^{-2}.	4.2	13.2 (*l*)
			50% cw + 1 h/600°C (argon) +20 stress cycles between 2.1 and 12.5 × 10^8 N m^{-2}.	4.2	13.4 (*l*)
35	[BYC70]	0.25 mm∅ wire	99.86% cw	300	
36	[EAS75]	0.213 mm∅ wire	cw	300	
				4.2	
36	[EAS75]	0.267 mm∅ wire	cw	300	
				77	
				4.2	
36	[EAS75]	0.318 mm∅ wire	cw	300	
				4.2	
36	[EAS75]	1.98 mm∅ tensile sample	800°C anneal	300	
				4.2	
37	[CUR79]	14.22 mm∅	vacuum anneal 1.5 h/800°C +furnace cool	300	4.5
		8.99 mm∅	anneal + 60% cw	300	5.1
		5.59 mm∅	anneal + 85% cw	300	5.1
		2.97 mm∅	anneal + 96% cw	300	5.7
		0.76 mm∅	anneal + 99.7% cw	300	7.8
		0.76 mm∅	anneal + 99.7% cw +24 h/380°C	300	5.3

(cont.)

Tensile Strength 10^8 N m^{-2}	Reduction in Area %	Elongation %	Data Code for Figures 3-22 and 3-23	Comments
8.8		3.0–3.5		
5.7_5				
6.5	—	—	—	UTS measured as function of cold reduction 0–50%.
6.5_5	—	—	—	UTS independent of final cw in this case.
7.2		9		In this series of studies (*l*) indicates "elastic limit." At 300 K this is simply the stress at 0.01% plastic strain; at low temperatures it is taken as the stress at the first (serrated yield) load drop.
14.5		0.5		
14.5				
15.3				
6.4	—	—	6	UTS given as function of 1-h aging temperature from 300–900°C.
11.1	22	2.3		
21.6	7	4.5		
10.3	20	1.2		Referred to in Easton [EAS77].
17.3	17	2.4		
22.1	8	4.8		
13.8	15	4.1		
21.6	12	4.8		
5.0	71	8.8	7	
15.1	11	<0.2		
4.6	70	25	8A	
5.3	74	12	8B	
5.5	74	10.5	8C	
6.1	73	6	8D	
8.3		1	8E	
5.8		10.5		

Continued

Table 3-8

Nb Content at.%	References	Specimen Details	Condition	Test Temperature K[a]	Yield Strength 10^8 N m^{-2}
38	[REE77]	180 twisted 32 μm$^{\varnothing}$ strands coated in formvar	Commercial optimization	298 76 4	
39	[EAS75]	1.98 mm$^{\varnothing}$ tensile sample	800°C anneal	300 77 4.2	
39	[REA78]	Tensile sample	800°C/WQ + extruded 86% + 48 h/375°C +drawn 56%	300 76 4	7.0 11.0 17.5
39	[REA78]	Tensile sample	800°C/WQ	300 76 4	4.6 11.0 15.4
39	[VER76]	0.14–0.20 mm$^{\varnothing}$		300 77.3 21–26 4.2	8.8 15.2
40	[BYC70]	0.25 mm$^{\varnothing}$ wire	99.86% cw	300	
44	[ELB67, ELB68]	0.28 mm$^{\varnothing}$ wire	cw and heat treated 1½h/450°C	300	
50	[BYC70]	0.25 mm$^{\varnothing}$ wire	99.86% cw	300	
50	[REA78]	Tensile sample	800°C/WQ	300 76 4	4.1 9.3 11.1
50	[REA78]	Tensile sample	cw	300 76 4	5.2 9.9 14.3
50	[VER76]	0.12–0.18 mm$^{\varnothing}$ wire		300 77 4.2	16.2 18.6 17.6

[a] Room temperature, if not specified, is quoted here as 300 K.

(*cont.*)

Tensile Strength 10^8 N m^{-2}	Reduction in Area %	Elongation %	Data Code for Figures 3-22 and 3-23	Comments
7.8		2.2		Commercial monolith with the Cu chemically removed—all strands remained twisted 1.3 cm pitch. Sample fractured prior to yield.
10.2		2.2		
12.8		2.2		
5.0	77	9.6	9	
11.0	47	3.3		
15.1	7	<0.2		
7.2	62	8		
12.5	53	8		
17.5	7	5		
4.6	93	23	10	
11.5	48	9		
15.4	0	3		
9.8		0.3		Soviet wire type T 60.
15.7		0.55		
21.9		0.3		
16.7		0.1		
6.2			11	UTS given as function of 1-h aging temperature from 300-900°C.
10.3				
7.1			12	UTS given as function of 1-h aging temperature from 800-900°C.
4.3	94	24	13	
9.5	71	14		
11.6	65	6		
5.4	86	13		
10.0	64	8		
14.3	0	0		
16.2		0.2		Soviet wire type NT 50.
19.8		0.45		
19.1		0.3		

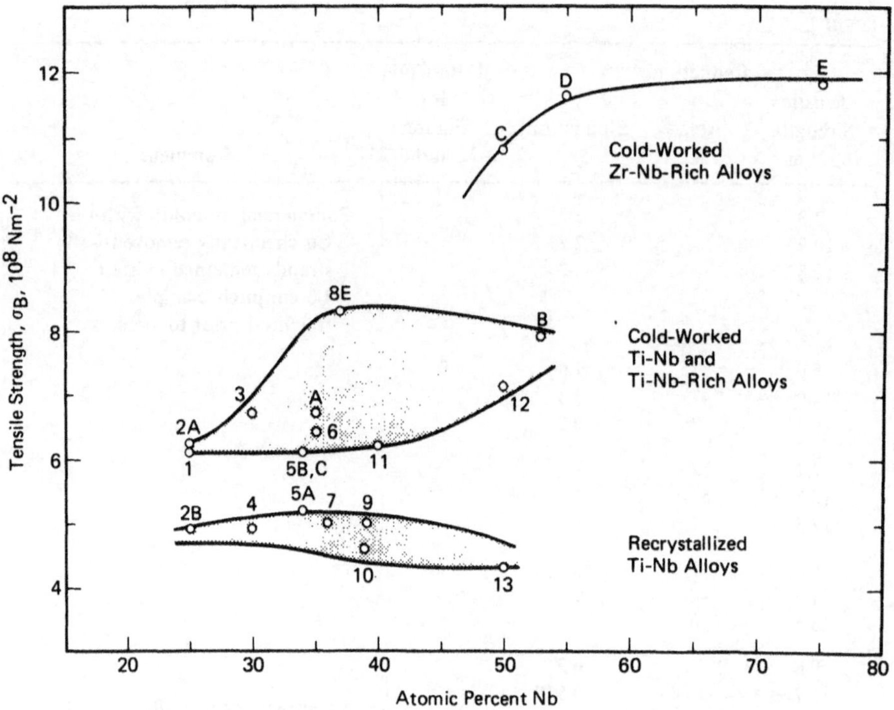

Figure 3-22. Room-temperature tensile strengths of Ti-Nb and Ti-Zr-Nb alloys in various conditions as function of Nb content. The numbers refer to data from Table 3-8, the letters to data from Table 3-11.

3-23, in which tensile strength is plotted *versus* percent area reduction by cold work. Two important sets of data used in the construction of this diagram are those of HILLMANN *et al.* [HIL76], [ALB76] for Ti-Nb(34 at.%) (recrystallized plus 0, 51, and 89.5% cold work) and CURTIS and McDONALD [CUR79] for Ti-Nb(37 at.%) (recrystallized plus 0, 60, 85, 96, and 99.7% cold work). Taken together, as in Figure 3-23, they indicate that although tensile strength is a weak function of cold work in the range 50-90%, a very rapid increase takes place as soon as the cold deformation exceeds about 99%.

In order to obtain estimates for the tensile strengths of optimized Ti-Nb wire, Table 3-9 has been constructed from data selected from Table 3-8. After rejecting "unacceptable" values, it is estimated that the strengths of technically interesting Ti-Nb alloys are independent of concentration and that $\sigma_{B,300\,K} \cong 10 \pm 1$ and $\sigma_{B,4\,K} \cong 21 \pm 2$ in the units 10^8 N m^{-2}.

The important study by HILLMANN *et al.* [HIL76], [ALB76] of the temperature dependence of the strengths of Ti-Nb(34 at.%) is notable not only for the number of measuring temperatures investigated but also for

Table 3-9. Tensile Strengths of Ti–Nb Wires

Nb Content at.%	Tensile Strength σ_B, 10^8 N m^{-2} 4.2 K	300 K	References
36	21.6	11.1	[EAS75]
36	22.1	10.3	[EAS75]
39	16.7[a]	9.8	[VER76]
50	19.1	16.2[a]	[VER76]
Means	21 ± 2	10 ± 1	

[a] Rejected from averaging.

the number of data points acquired at each temperature. Detailed numerical results for five conditions, extracted from the plotted curves of HILLMANN *et al.* [HIL76], were given in Table 3-8 and the complete set of results is summarized for discussion purposes in Figure 3-24. By taking data within the temperature range *between* 4.2 and 77 K, HILLMANN *et al.* discovered some significant features of the tensile-property temperature dependence of the superconducting Ti–Nb alloys, which had been overlooked by the other investigators who had confined their attention to the usual pair of cryogenic temperatures—4.2 and 77 K. With regard to room-temperature properties, scrutiny of Figures 3-24(a) through (e) shows that very little change of tensile strength accompanies cold work of up to 90% and the

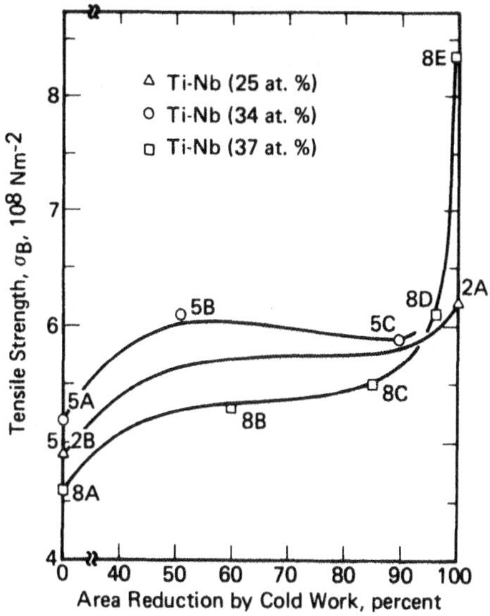

Figure 3-23. Influence of cold work on the room temperature tensile strengths of Ti–Nb alloys. Cold reductions considerably greater than 96% are needed to produce significant increases in tensile strength. The numbers refer to data from Table 3-8.

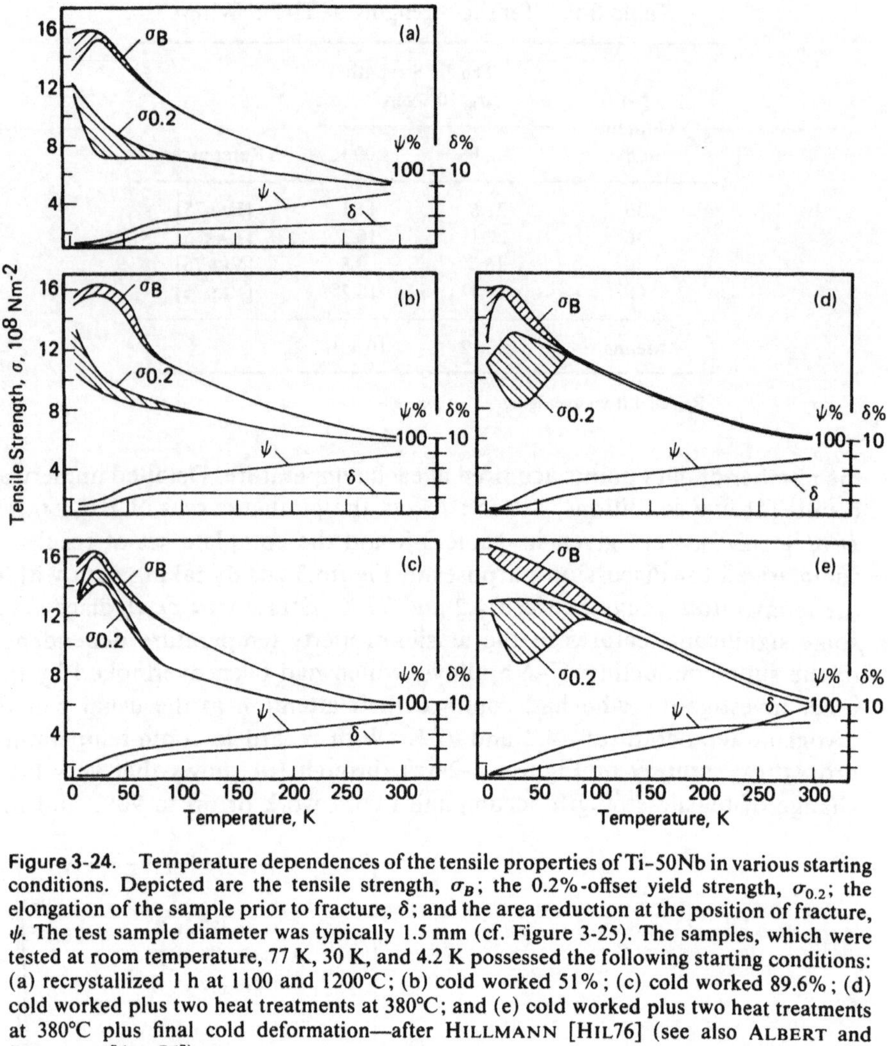

Figure 3-24. Temperature dependences of the tensile properties of Ti–50Nb in various starting conditions. Depicted are the tensile strength, σ_B; the 0.2%-offset yield strength, $\sigma_{0.2}$; the elongation of the sample prior to fracture, δ; and the area reduction at the position of fracture, ψ. The test sample diameter was typically 1.5 mm (cf. Figure 3-25). The samples, which were tested at room temperature, 77 K, 30 K, and 4.2 K possessed the following starting conditions: (a) recrystallized 1 h at 1100 and 1200°C; (b) cold worked 51%; (c) cold worked 89.6%; (d) cold worked plus two heat treatments at 380°C; and (e) cold worked plus two heat treatments at 380°C plus final cold deformation—after HILLMANN [HIL76] (see also ALBERT and PFEIFFER [ALB76]).

application of the two thermomechanical processing schedules mentioned. On the other hand, work hardening raises both $\sigma_{0.2}$ and $\sigma_{0.01}$, and for the deformations in excess of about 90%, maxima appear in these quantities for the first time. Elongations at fracture vary from about 3% to 6%, except for the heat-treated-and-final-cold-worked specimen which elongates about 10% prior to fracture. Clearly the final cold work, administered for electrical reasons, is also beneficial to room-temperature ductility. It is at lower temperatures that spectacular changes in properties take place both as functions of temperature and from sample to sample. All workers agree that the strength of Ti–Nb increases rapidly with decreasing temperature. Tensile specimens tested in the liquid-He regime exhibit strengths that are typically three times the room temperature values. Not previously noted, however, due to lack of data were (i) significant increases in the scatter of all the strength data as the temperature is lowered below about 77 K, and (ii) maxima appearing in one or all of the quantities σ_B, $\sigma_{0.2}$, and $\sigma_{0.01}$, depending on condition, as the temperature is lowered through about 20 K. The *scatter* in the data is related to the occurrence of serrated-yield instability, a phenomenon which eventually leads to slip-like ductile failure; the *maxima* are induced by competition between this failure mechanism and precipitation strengthening mechanisms to be discussed below.

As the temperature of the tensile test is lowered, serrated yielding starts to take place (Figure 3-5); there does not, however, appear to be a characteristic onset temperature for this disturbance. With further decrease in temperature the strength–temperature curves become steeper, and the

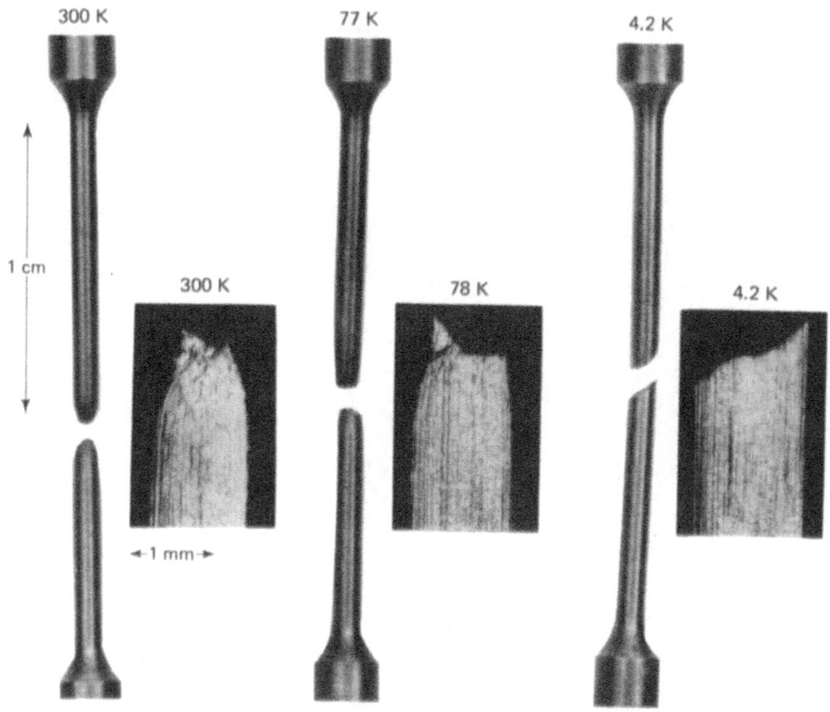

Figure 3-25. Tensile specimens and fracture-ends of samples after testing to fracture at the temperatures indicated. An intercomparison of the results of KOCH and EASTON [Koc77] on Ti–Nb(38.5 at.%) (tensile bars depicted) and ALBERT and PFEIFFER [ALB76] on Ti–Nb(34 at.%) (fracture-ends depicted) demonstrates the reproducibility of the fracture behavior at each of the test temperatures—photographs courtesy of D. S. Easton and H. Hillmann, respectively.

instabilities increase accompanied by an increase in the scatter of the tensile data. In the instability region the elongation, previously several percent, drops to about 0.5%. In Ti–Nb(34 at.%) the instabilities themselves, generally recognized as shearing processes, are thought to be solely responsible for plastic deformation at temperatures below about 20 K [ALB76]. They are, presumably, the principal cause of the fracture that ultimately takes place, with less than 10% area reduction, at an angle of 45° to the wire axis (in the plane of maximum resolved shear stress) as shown in Figure 3-25.

3.17.1 The Fracture of Ti–50Nb and Ti–55Nb

At 300 K, necking due to plastic flow precedes fracture, which then takes place at some 80–90% reduction in cross-sectional area [HIL76] leaving the familiar "cup-and-cone" broken-end pairs. Then at 4.2 K, as mentioned above, a macroscopically smooth 45° fracture takes place with almost no area reduction [ALB76], [HIL76], [KOC77]. In the absence of evidence to the contrary, a fracture surface such as that depicted in Figure 3-26 would be diagnosed as brittle. But the results of intermediate-magnification SEM (Figure 3-27—[ALB76], [KOC77], [SUM80]) all agree that, for the most part,

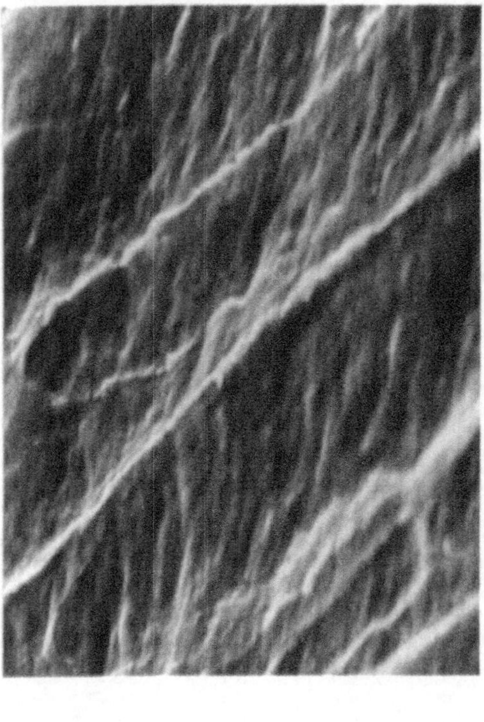

Figure 3-26. Scanning electron micrograph (original magnification, 10,000×) of a small portion of the 4.2 K fracture surface of Ti–Nb(38.5 at.%) showing some evidence of brittle fracture—after KOCH and EASTON [KOC77]. Photograph courtesy of D. S. Easton. Copyright © 1977, IPC Science and Technology Press, reproduced with permission.

the 4.2 K fracture surfaces have the finely dimpled (or "honeycombed" [ALB76]) appearance generally recognized as signifying microscopic *ductile* failure.

The microstructures of the fracture surfaces of the recrystallized and cold-worked samples differ only in the diameters of the dimples, which can be as large as 38 μm. The macroscopic and microscopic surface-structure evidence point to a "ductile fracture of the shear-rupture type" [KOC77]. Information is lacking on the substructure of the fracture ends, but work by READ [REA78] on recrystallized Ti–Nb(30 at.%), strained to fracture at 4.2 K, showed that some twinning had taken place—Figure 3-28—which is not unusual for Ti alloys at low temperatures [MOS70] (see Figure 3-21).

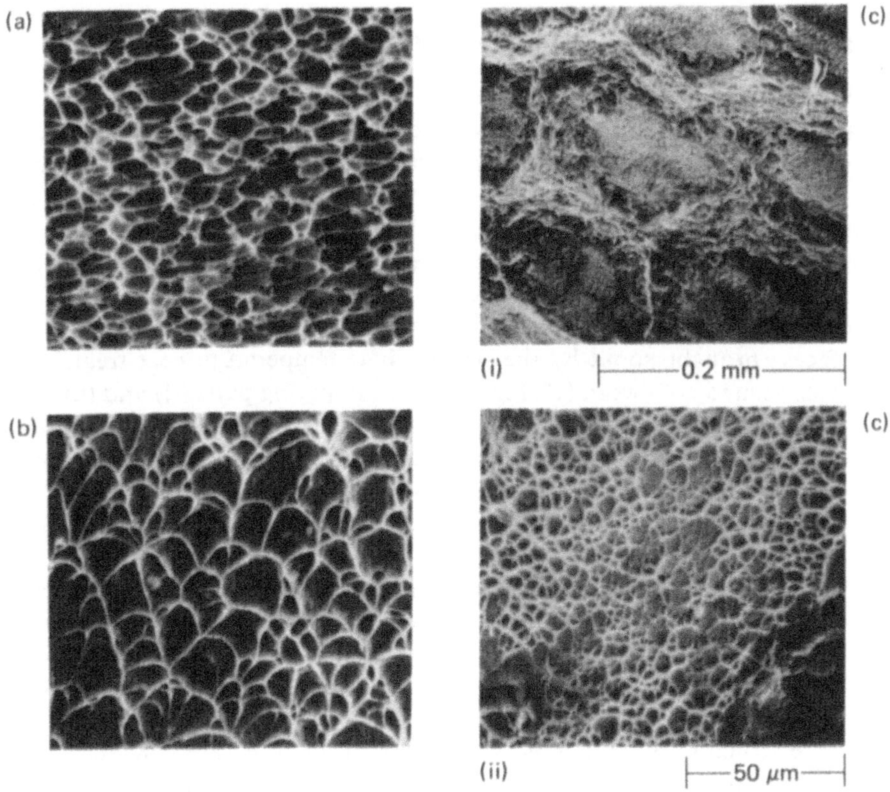

Figure 3-27. Dimpled or "honeycombed" appearance of the 4.2 K shear-fracture surfaces of Ti–Nb samples at intermediate magnifications. SEM results of: (a) KOCH and EASTON [KOC77] (original magnification, 3000×); (b) ALBERT and PFEIFFER [ALB76] (original magnification, 1200×): (c) NAGATA *et al.* [SUM80] (original magnifications (i) 200×, (ii) original magnification, 500×). Photographs courtesy of D. S. Easton, H. Hillmann, and M. Nagata, respectively.

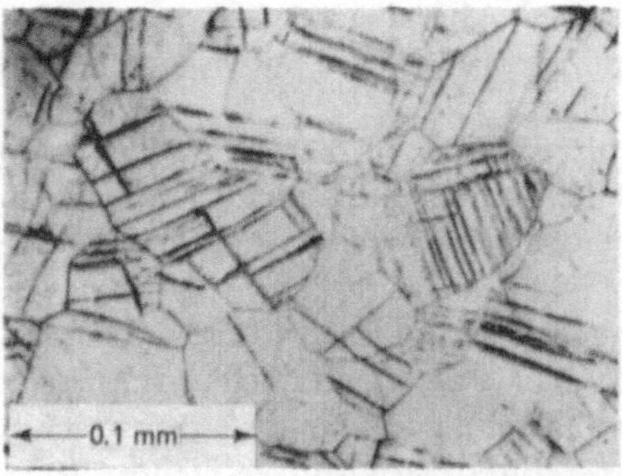

Figure 3-28. Photomicrograph of the subsurface microstructure (original magnification, 600×) of Ti-45Nb (recrystallized 800°C/water quenched) strained to fracture at 4 K showing parallel bands, interpreted as indicating "systematic lattice transformation"—after READ [REA78]. Photograph courtesy of D. T. Read. Copyright © 1978, IPC Science and Technology Press, reproduced with permission.

3.17.2 Influence of the Thermomechanical Processing on the Fracture of Ti-50Nb

Aging at 380°C, a temperature favored by many groups for electrical-property optimization, results in α-phase precipitation. In cold-deformed wire the precipitates are about 0.1-0.2 μm in diameter and are inhomogeneously distributed [ALB76]. Final cold deformation of the heat-treated structure has been claimed to result in a more even distribution of precipitates accompanied by a reduction in particle diameter by about a factor of ten. Changes in the 4.2 K tensile properties (and the fracture surfaces) brought about by the intermediate-temperature heat treatments and associated cold work (cf. Figure 3-24, comparing parts (d) and (e) with (a), (b), and (c)) can be traced to the presence of either coarse or fine precipitation. The influences that coarse and fine, respectively, α-phase precipitate particles exert on the tensile properties of Ti-50Nb can be systematized in the following way.

(a) Large Inhomogeneously Distributed α-Phase Particles. These serve as stress-intensifying nuclei for local brittle fracture during low-temperature tensile testing. In their presence the characteristic drop in the low-temperature tensile strength still takes place. Again, localized crystalline brittle-fracture or cleavage facets have been observed within the dimpled surface structure of Ti-Nb(34 at.%) [ALB76] fractured at 4.2 K and 28 K. Similar evidence of local brittle fracture was noted in Ti-Nb(38.5 at.%) (annealed 800°C) by KOCH and EASTON [KOC77]. There are, of course, room-temperature examples of dimpled "brittle" fracture. For example, Ti-Cr(14 at.%), aged $2\frac{1}{2}$ h/350°C to produce the cubic aged-ω precipitates, failed in a brittle manner with only 2% tensile elongation, but nevertheless showed a finely dimpled fracture surface indicative of limited plastic flow prior to fracture; also Ti-Cr(9 at.%), quenched from 950°C and showing

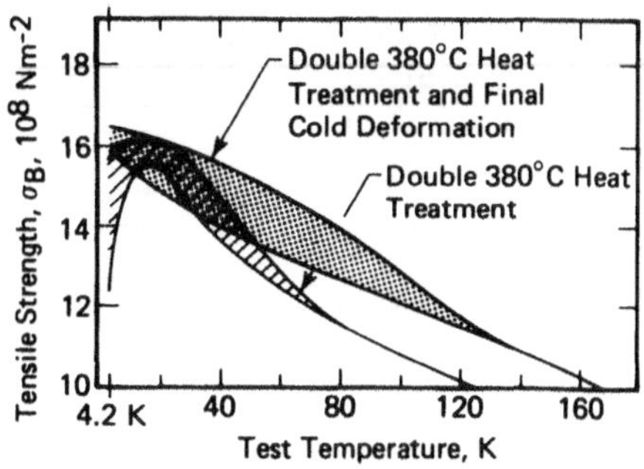

Figure 3-29. Influence of final cold deformation on the low-temperature tensile strength of a 380°C heat-treated Ti-50Nb alloy—after HILLMANN [HIL76].

the β-plus-diffuse-ω microstructure [POL71], revealed a dimpled surface with cleavage facets after fracture.

A full interpretation of fracture modes, of course, falls within the purview of "fracture mechanics," a discipline in its own right and outside the scope of the present work.

(b) Finely Dispersed α-Phase Precipitates. The fine dispersion of α-precipitates that results from the application of final cold deformation to the 380°C-aged alloy brings about a reduction of the brittle component of fracture, a less sudden drop in low-temperature ductility, an improvement in room-temperature ductility, and a maintenance of low-temperature strength by the elimination of the tensile-stress maximum.

The influences of the above-specified two classes of thermomechanical process on the tensile properties of Ti-50Nb are compared in Figure 3-29. It is interesting to note that the process that terminates in the cold-deformation step, and which thereby yields a finely divided homogeneous α-phase precipitate in association with a fine fibrous subband structure, hence an optimal critical current density, also deactivates the slip mechanisms thereby increasing the low-temperature tensile and yield strengths. Thus, as pointed out by ALBERT and PFEIFFER [ALB76], electrical and mechanical optimization can be simultaneously satisfied.

3.18 Strengths of Titanium–Niobium-Base Multicomponent Alloys

3.18.1 Ternary Alloys

The effects on the tensile strength of Ti–Nb of additions of C, Al, Zr, Mo, La, Hf, Ta, W, and Re have been investigated. Table 3-10 is an index to data sources. Solid-solution strengthening in response to very small additions of C, Al, Zr, La, Hf, and W to Ti–Nb(25 and 90 at.%) bases have been studied by BARON *et al.* [BAR65a]. The plotted data indicate clearly

Table 3-10. Strength Studies of Ti–Nb-Base Ternary Alloys

Alloy System						
Alloy Base (at.%)		Added Element				
Ti	Nb	Concentration (at.%, except first entry)	Identity	Data Presentation	Remarks	Literature
75	25	<1 wt% (C, Al, Zr, La, Hf, W)		Plotted	Room temperature dilute-solution-strengthening study.	Baron *et al.* [BAR65]
10	90					
75.1	24.9	—		Tabulated	Room temperature tensile strengths mostly after 99.9% cold deformation.	Bychkova *et al.* [BYC70[a]]
65	25	10	Hf			
50	25	25				
74.2	25.4	0.4	Mo			
70	25	5				
63	28	9				
46	28	26				
74	25	1	Re			
70	25	5				
65	25	10				
50	25	25				

63.1	36.9	—	Ta	Tabulated	Room temperature tensile strengths as function of small additions of Ta to Ti-Nb.	Curtis and McDonald [CUR79]
62.5	37.3	0.18				
62.9	36.8	0.35				
62.5	36.7	0.72				
60	35	5	Zr	Plotted	Tensile strength of sheet aged (0–200 h)/(350–500°C).	Kitada [KIT72]
40	53	7	Zr	Tabulated	Strength of 65 BT at 293, 77, and 20 K as 0.25 mm$^{\varnothing}$ wire (typical composition listed).	Alekseevskii *et al.* [ALE67[a]]
10	50	40	Zr	Plotted	Strength *versus* annealing at 1 h/(700, 800, 1100°C)	Doi *et al.* [DOI66[b]]
				Plotted	Strength *versus* aging (0–500 h)/700°C	Doi *et al.* [DOI68[a]]
4/7	55/50	41/43	Zr	Plotted	σ_B and δ *versus* aging (0–83 h)/700°C for two starting conditions: (i) 5 h/1200°C/WQ + 75% cold draw (ii) 5 h/1200°C/WQ	Soeno and Kuroda [SOE69]

the very rapid strengthening influence of C, the rapid strengthening influence of Al, and the gradual strengthening influences of the other elements Zr, La, Hf, and W. These results are in accord with the hardness data reported in Table 3-2 (see also Figure 3-2), which showed hardening rate to increase in the sequence TM < Al < (C < O < B < N). In fact, if qualitative estimates of relative strengthening effects are desired, instead of direct tensile testing use can be made of the results of hardness measurements (which can also be interpreted in terms of $\sigma_{0.2}$—Sections 3.6 and 3.7) and also of a principle that equates the strengthening rate in β-Ti alloys (i.e., Ti–TM alloys) by a given solute species to that conferred by the same solute on α-Ti itself [CoL73]. As a semiquantitative example we first note that, according to BARON *et al.* [BAR65[a]], C, Al, and Zr strengthen Ti–Nb(25 at.%) (which has an initial room-temperature σ_B of 6.2×10^8 N m^{-2}) at approximate average linear rates of 9.6, 0.5, and 0.14×10^8 N m^{-2} per 0.1 wt.% solute addition, respectively. Secondly, with reference to Table 3-2, in which the hardening rates of α-Ti by C and Al and that of Ti–Nb(33 at.%) by Zr are listed, and bearing in mind the relationship $\sigma_{0.2} \cong H_V/3$, we note that the corresponding *hardness-derived* strengthening rates for C, Al, and Zr, are, respectively, 8.8, 0.5, and 0.1×10^8 N m^{-2} per 0.1 at.% solute.

The purpose of a study conducted by CURTIS and MCDONALD [CUR79] was to investigate the influence of small "controlled" additions of Ta on the superconductive and mechanical properties of Ti–Nb(37 at.%) in a simulated test of a possible deterioration of those properties in the presence of Ta as a tramp element. The full significance of the ability to include a few percent Ta in Ti–Nb alloys is considered in Section 28.1. The results of the tests pointed to a slight overall increase in the tensile strengths of Ti–Nb(37 at.%) acquired in response to additions of Ta to levels of up to 2 wt.%. Under close inspection, however, the data reveal local maxima in both σ_B and $\sigma_{0.2}$ at all degrees of cold deformation of up to 99.7% upon the addition of the first 0.5 wt.% Ta (0.2 at.%). In general, and for the 99.7% deformed series in particular, Ta-solute strengthening took place at the rate of 0.28×10^8 N m^{-2} per wt.% Ta (omitting the 0.5 at.% data). The rate of strengthening of a Ti–Nb base by Ta thus seems to be more gradual, by a factor of five, than that due to Zr. A small decrease in the ductility was accompanied by no noticeable change in the workability.

The Ti–Zr–Nb alloys referred to in Table 3-10 have in the past been regarded as hard superconductors of technical significance. Some of their tensile properties are listed in Table 3-11, which includes for comparison a result for the tensile strength of Zr–Nb(75 at.%). Figure 3-22 compares the strengths of the alloys containing large amounts of Zr—the Zr–Nb-rich or so-called "X-type" Ti–Zr–Nb alloys—with those of the Ti–Nb binaries and the Ti–Nb-rich, or "Z-type," ternary superconductors. We note first that the substitution of a few percent Zr for Ti in alloys of compositions

Table 3-11. Tensile Strengths of Some Common Ti–Zr–Nb Alloys

Composition at.% Ti	Zr	Nb	Condition	Temperature of Test K (if below room temperature)	Tensile Strength σ_B kg mm^{-2}	10^8 N m^{-2}	Sample Code[a]	Elongation δ, %	Literature
60	5	35	Cold-rolled strip		67.9	6.7	A		Kitada [KIT72]
			Rolled + 50 h/350°C		95.9	9.4			
			Rolled + 50 h/400°C		90.8	8.9			
			Rolled + 50 h/450°C		81.2	8.0			
			Rolled + 50 h/500°C		78.7	7.7			
40	~7	~53	0.25 mm$^{\varnothing}$ wire from 65 BT		81	7.9	B		Alekseevskii *et al.* [ALE67]
				77	140	13.7			
				20	192	18.8			
10	40	50	5 h/1100°C $\xrightarrow[\text{cw}]{98.4\%}$ 0.25 mm$^{\varnothing}$		110.5	10.8	C	3.2	Doi *et al.* [DOI66[b]]
			Above +1 h/700°C		99.5	9.8		2.8	
			5 h/1100°C/WQ		96.0	9.4			Doi *et al.* [DOI68[a]]
			Above +10 h/700°C		91.4	9.0			
4	41	55	5 h/1200°C/WQ		87.0	8.5		8.5	Soeno and Kuroda [SOE69]
			Above +1 h/700°C		88.9	8.7		10.5	
			5 h/1200°C/WQ $\xrightarrow[\text{cw}]{75\%}$		118.0	11.6	D	6	
			Above +1 h/700°C		105.4	10.3		8.5	
0	25	75	5 h/1100°C $\xrightarrow[\text{cw}]{98.4\%}$ 0.25 mm$^{\varnothing}$		120.0	11.8	E		Doi *et al.* [DOI66[c]]

[a]Refers to data plotted in Figure 3-22.

near 35 and 50 at.% Nb makes little difference to the strength, in agreement with earlier results on the low-level doping of Ti-Nb alloys by other transition elements. But when substantial Zr levels are introduced to form alloys such as Ti_{10}-Zr_{40}-Nb_{50}, the tensile strengths become commensurate with that of Zr-Nb(75 at.%) itself, and lie on an upper branch of the strength *versus* Nb content diagram, Figure 3-22.

3.18.2 Quaternary Alloys

The tensile strengths of multicomponent Cu-clad composite superconductors based on the quaternary alloy Ti_{61}-Zr_{6}-Nb_{27}-Ta_{6} (a close relative of the binary Ti-Nb(33 at.%)) have been studied by HORIUCHI *et al.* and the results reported in the literature [MON78] and in the technical data sheets of Kobe Steel Ltd. [KOB78, KOB80]. As part of a test of the influence of Cu/SC ratio on the tensile strength, bundles of bare filaments prepared from a multifilamentary Cu/Ti-Nb composite by the etching away of the Cu matrix have been measured by MONJU and TATARA [MON78]. The results of this work are summarized in Table 3-12.

3.19 Modulus and Strength of Composite Superconductors

3.19.1 Relationships Between the Mechanical Properties of the Components

In the selection of materials for fiber-reinforced composites, the intention is generally to stiffen (i.e., increase the Young's modulus) and strengthen the matrix. These goals are not achieved in simple superconductive composites, the components of which are selected for their electrical rather than

Table 3-12. Tensile Strength of a Ti-Zr-Nb-Ta Alloy[a]—After Monju and Tatara [MON78]

Number of Filaments	Filament Diameter μm	Temperature of Test K	Tensile Strength	
			kg mm^{-2}	10^8 N m^{-2}
61	37	RT	150	14.7
		77	207	20.3
		4.2	230	22.6
361	64	RT	115	11.3
		77	151	14.8
		4.2	180	17.6

[a]Test specimens of the alloy Ti_{61}-Zr_{6}-Nb_{27}-Ta_{6} were obtained by removing Cu from two samples of Cryozitt commercial multifilamentary superconductor.

mechanical performances. The moduli and strengths of the several possible components of a superconducting composite are listed separately in Tables 3-4, 3-8, and 3-13. Figure 3-30, constructed from Tables 3-4 and 3-13, shows that the Young's moduli of all Ti–Nb(20–50 at.%) alloys are very much less than that of pure Cu, and in particular that at 4 K the modulus of Cu is almost double those of the practically important Ti–Nb alloys. On the other hand, Ti–Nb can be said to *strengthen* the Cu; stress-strain curves for a typical composite (18 filaments of Ti–Nb(36 at.%); Cu/SC-ratio, 2.8 : 1) at 4.2 K and 300 K, respectively, are given in Figures 3-11 and 3-31. Figure 3-32 shows how the stress-strain curves of a composite and its individual constituents interrelate at low stress levels; it indicates that if the strain can be kept below 0.15% the composite may behave as an elastic unit, with advantage actually being taken of the higher modulus of Cu to stiffen the superconductor.

The plastic properties of Ti–Nb are influenced by the presence of the Cu matrix and cladding in different ways depending on the temperature: (i) *At room temperature* the presence of the Cu improves the strength and ductility of the Ti–Nb. According to PIEHLER [PIE65], the markedly improved mechanical properties of Cu-clad Ti–Nb filaments, as compared to those of the bare alloys, were attributable to the inhibition of localized necking by the presence of the matrix. (ii) *At 4.2 K*, on the other hand, the composite is apparently much less ductile than the bare superconductor. As pointed out in Section 3.9.1(b), interpretation of this effect is again in

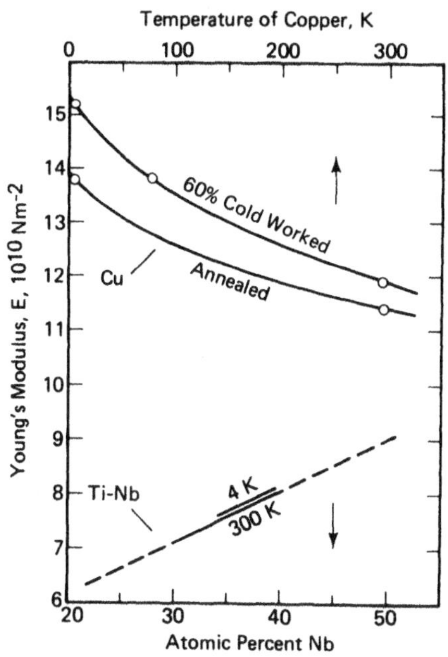

Figure 3-30. Composition dependence of the Young's moduli of Ti–Nb(20–40 at.%) alloys at 4 K and 300 K (data from Table 3-4); also Young's modulus temperature dependences of annealed and 60%-cold-drawn OFHC Cu (data from Table 3-13).

Table 3-13. Tensile Properties of Some Composite-Conductor Components

Material	Condition	Test Temperature[a] K	Young's Modulus 10^{10} N m^{-2} E	Strengths, 10^8 N m^{-2} Tensile σ_B	Yield $\sigma_{0.2}$	Area Reduction ψ, %	Elongation δ, %	References
OFHC Cu	Vacuum annealed 1 h/650°C	298	11.4	2.1	0.30			[MET74][c]
		77	11.6	3.3	0.44			
		4	13.8	4.2	0.48			
	1 h/400°C	300			0.5_1[b]			[TAY68]
	Above + cold drawn 10% (AR)	300			2.2_0[b]			
	Cold drawn 60% (AR)	295	11.9	3.3	3.2	77	17	[REE67]
		76	13.8	4.6	3.8	78	29	
		20		5.1	4.0	76	42	
		4	15.2	5.1	4.0	75	41	
Cu–10Ni	Annealed 40 min/677°C	295	12.2	3.4	1.5	79	37	[REE67]
		76	13.4	5.0	1.7	77	50	
		20		5.7	2.1	73	50	
		4	14.1	5.6	1.7	73	53	
Cu–30Ni	Commercial anneal	295	15.2	4.0	1.3	68	47	[REE67]
		76	15.9	6.2	2.2	70	52	
		20		7.1	2.6	66	51	
		4	16.0	7.2	2.8	65	48	

Cu-0.15Zr	Solution treated +50% cw +aged 1 h/600°C (dispersion hardened)	300			2.7			[TAY68]
Cu-0.18Zr ("Am zirc")	Annealed 950°C/85-90% cold drawn/aged 1 h/450°C (dispersion hardened)	295	10.9	4.4	4.1	62	16	[REE67]
		76	11.9	5.3	4.5	71	26	
		20		5.9	4.6	72	37	
		4	11.9	5.9	4.5	69	36	
304 L Stainless Steel	Annealed	297	19.0	6.6	4.1	81	78	[MET74][d]
		77	20.5	15.1	4.6	66	43	
		20		18.8	5.2	41	42	
		4	20.1	16.6	5.5	56	34	
310 Stainless Steel	Annealed	297		5.9	2.2	71	59	[MET74][e]
		77	20.6	10.9	5.2	50	68	
		20		12.7	6.9	36	46	
		4	20.6	13.0	7.2	41	50	

[a] Room temperature is given as 300 K, unless otherwise stated.
[b] 0.1% yield strength.
[c] Table 5.1.3—ME9.
[d] Table 8.1.3—ME2.
[e] Table 8.1.4—ME4.

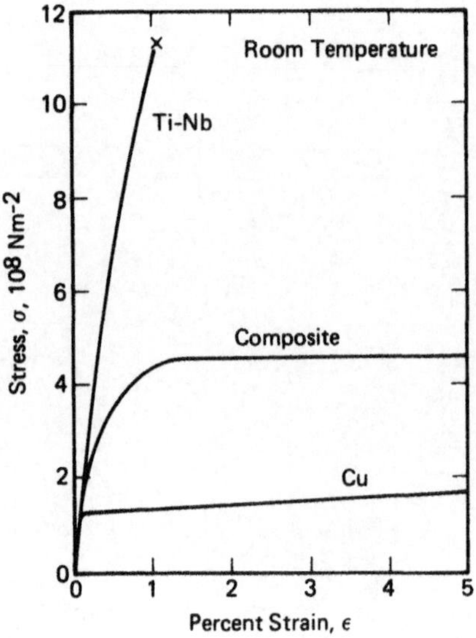

Figure 3-31. Room-temperature stress-strain characteristics of a Cu/Ti-Nb(36 at.%) (Cu/SC, 2.8:1; 18 filaments) superconductive composite compared with those of its individual components. The increase in elongation of the Ti-Nb after cladding with Cu is attributed to the restraint that the matrix exerts on the tendency of the filaments towards localized necking (cf. the 4.2 K behavior, Figure 3-8)—after EASTON and KOCH [EAS77], [KOC77].

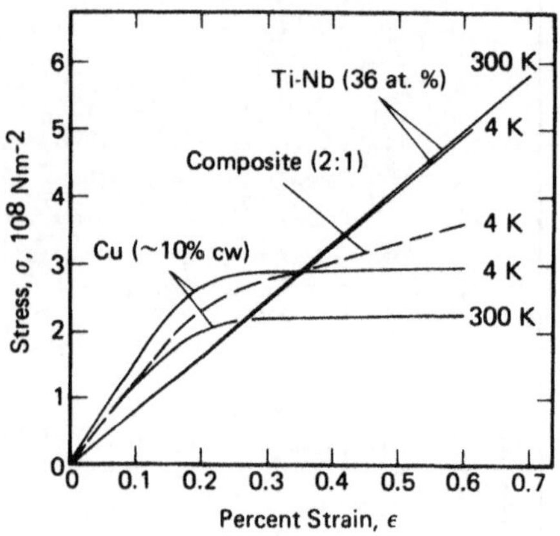

Figure 3-32. Stress-strain characteristics of a Ti-Nb(35 at.%) alloy, unalloyed Cu (~10% cold worked), and a Cu/SC composite of volume ratio, 2:1 (whose σ's were computed using the mixture rule). The diagram is based on the following data:

Material	Temperature K	E 10^{10} N m^{-2}	σ, 10^8 N m^{-2} $\sigma_{0.1}$	σ_B	Sources
Ti-Nb(36 at.%)	300	8.2		5.0	Tables 3-4 and 3-8
	4	8.3		(15.1)	
Cu(60% cw)	300	11.9			Table 3-13
	4	15.2			
Cu(10% cw)	300	11.9			Assumed as for Cu(60% cw)
	4	15.2			Assumed as for Cu(60% cw)
	300		2.2		Table 3-13
	4		2.9		Assume 300 K value + 30%

terms of the constraining influence of the matrix. The elongation noted in bare Ti–Nb is a result of reversible pseudoelastic transformation, an effect which is accommodated by surface rumpling. This is inhibited by the cladding, whose presence restricts the motion of atomic planes at the surface of the superconductor which then has no recourse but to rupture.

3.19.2 Tensile Data

Table 3-14, which summarizes some modulus and strength data for several composite superconductors and braids or cables fabricated from them, includes a large body of useful information on the tensile properties of some U.S. technical superconductors acquired by D. L. EASTON of the Oak Ridge National Laboratory [EAS76, EAS77].

3.19.3 Influence of Heat Treatment, Cold Work, and Cu/SC Ratio on the Tensile Strength

HILLMANN [HIL73[a]] has performed some interesting experiments to determine the factors which influence the strength of composite superconductors. The results of three representative studies were reported, in which:

(1) A Cu/Ti–Nb monofilament was subjected to a recrystallization anneal (for both the Cu and the Ti–Nb) of 30 min/600°C followed by (i) zero to 50% deformation, (ii) 45% deformation, heat treatment for (2–4 h)/400°C, followed by zero and 45% deformation

(2) The Cu/Ti–Nb ratio in a 120-filament conductor was varied from 50 to 90%

(3) A 61-filament Cu/Ti–Nb conductor was heat treated 1 h/400°C and deformed 5 to 45%

In (1), although the 400°C heat treatment increased the strength and reduced the elongation of the unclad Ti–Nb as a result of α-phase precipitation, in the presence of Cu the effect on the tensile strength was naturally less pronounced. In (2) the strength of the composite increased with the increasing volume fraction of the Ti–Nb, and in (3) its strength increased with the degree of final cold work.

These results may be partially understood with the aid of the so-called "rule of mixtures," to be discussed in the following section. The general subject of precipitation strengthening in Ti–Nb alloys and the strengthening of composite superconductors is considered below in Sections 3.21 and 3.22.

3.20 Determination of the Tensile Properties of Composites

Four theoretical methods of dealing with the elastic and thermal properties of fiber-reinforced composites have been introduced, with adequate documentation, by SUN and GRAY [SUN75]. Presumably some or all of the approaches they referred to could also be adapted to the problem of normal-state electrical transport. The thermal expansions of particulate and fibrous composites are considered in Section 9.8.

Exact theoretical solutions to problems in elasticity can be obtained with the aid of the finite-element or finite-difference methods familiar to mechanical metallurgists and engineers [SUN76].

Experimentally, the longitudinal, L, and transverse, T, (referring to the *fiber* direction) elastic properties can be obtained using suitably instrumented tensile specimens. If z represents the *tensile* direction and x, y the plane transverse to it, E_z, of course, is σ_z/ε_z and the Poisson's ratio is

Table 3-14. Tensile Properties

Conductor Type	References	Volume Ratio	Dimensions or Diameter mm	Number of Ti-Nb Filaments	Ti-Nb Filament Diameter μm	Twist Pitch mm	Superconductive Alloy wt.% (at.%) Nb
Cu/Ti-Nb	[REE77]	1.8/1	0.70 × 0.55	180	32	12.5	55 (39)
Cu/Ti-Nb	[SEI79]	2/1	1.45	1710	24	24	50 (34)
Flat cable from above wire, steel reinforced	[SEI79]		17.6 × 3.9	8 of the above wires		200 (transposition length)	
Cu-clad cable from wires of Soviet type T 60 alloy	[VER76]		0.82	6 twisted wires	bare wires ~170 μm		55 (39)
Cu-clad cable from wires of Soviet type NT 50 alloy	[VER76]		1.5	24 wires	bare wires ~150 μm		67 (50)
Cu/Ti-Nb	[HIL76]	2/1	0.4	60		20	50 (34)
Cu/Ti-Nb	[HIL76]	2/1	1.0 × 0.6	125		50	50 (34)
Cu/Ti-Nb	[EAS76]		4.32 × 1.52	3036		25	54 (38)
5056Al/Ti-Nb	[EAS76]		1.6	54 tubes	160 μm O.D. 80 μm I.D.	25	45 (30)
Cu/Cu-Ni/Ti-Nb	[EAS76]	0.5/1/1	2.54 × 1.27	1591			

$\nu_z \equiv -\varepsilon_{x,y}/\varepsilon_z$. Thus with the fibers parallel to the z axis, $E_L = \sigma_L/\varepsilon_L$ and $\nu_L = -\varepsilon_T/\varepsilon_L$. Then if a suitable tensile specimen, with fibers oriented normal to the axis of tension, can be prepared, $E_T = \sigma_T/\varepsilon_T$; while $\nu_T = \varepsilon_{x,\,y}/\varepsilon_z$ in general, and $-\varepsilon_L/\varepsilon_T$ in particular. As an internal check, ν_T should be the reciprocal of ν_L.

If a sample of the composite material can be prepared with the fibers oriented at 45° to the z direction, the longitudinal shear modulus, G_L, can be obtained. This is scarcely possible with a superconducting wire unless a sample can be removed from an unusually large superconductor. As an alternative it is possible to obtain G_L from rosette-strain-gauge readings [SUN75] with the sample mounted in the more convenient longitudinal orientation.

As a rapid semiquantitative approach to the elastic or plastic tensile properties, or transport properties, of composites fabricated from known materials, the "rule of mixtures" is recommended.

of Some Composite Conductors

Temperature K	Young's Modulus, E 10^{10} N m^{-2}	Strengths, σ 10^8 N m^{-2}: Yield $\sigma_{0.2}$	Tensile σ_B	Area Reduction ψ, %	Elongation δ, %	Manufacturer[a]	Remarks
300 4	5.8 to 13.0	 3.6	 8.9		 1.9		Wide spread in modulus data. Same as conductor # 1 in Table 29-2.
4.2	12.0 ± 0.6	5.4 ± 0.2	7.4				
4.2	13.6 ± 0.9	5.8 ± 0.2	6.6				Reinforced with Cu-plated stainless steel strip 15.0 × 1.5 mm^2. Assume wire and cable tested to failure.
300 77.3 21-26 4.2		3.3 5.3 6.0 6.8	4.0 5.9 7.9 7.8		14 25 28 1.7	R	Wire is brittle at 4.2 K.
300 77.3 4.2		3.9 4.5 5.0	5.7 8.1 8.7		6 11 3.5	R	This cable has a larger Cu content than the above and is more plastic at 4.2 K.
300		3.1 ± 0.1	6.18 ± 0.03		7.8 ± 0.3	V	
300		2.7 ± 0.04	5.28 ± 0.03		6.6 ± 0.3	V	
4.2	9.4	5.5	13.1			I	Cable.
4.2	6.3	6.7	11.1			L	Ti-Nb tubes with Al cores imbedded in 5056 Al alloy.
4.2	7.4	6.6	12.3			S	Includes the high-resistivity Cu-Ni-alloy barriers.

Continued

Table 3-1

Conductor Type	References	Volume Ratio	Dimensions or Diameter mm	Number of Ti-Nb Filaments	Ti-Nb Filament Diameter μm	Twist Pitch mm	Superconducti[ng] Alloy wt.% (at.%) M
Cu/Ti-Nb	[EAs76]	1.6/1	2.54 × 1.27	1530		25	54 (38)
Cu/Ti-Nb	[EAs76]	1.6/1	2.54 × 1.27	1530	31	12.5	54 (38)
Cu/Ti-Nb	[EAs76]	1.6/1	3.05 × 1.02	1530	28	25	54 (38)
Cu/Ti-Nb	[EAs76]	1.6/1	2.54 × 1.27	1530	31	25	54 (38)
Cu/Ti-Nb	[EAs76]	2.6/1	2.54 × 1.4	132		25	55 (39)
Cu/Ti-Nb	[EAs76]	2.6/1	2.54 × 1.4	132		12.5	55 (39)
Cu/Ti-Nb	[EAs76]	2.6/1	2.54 × 1.4	234		25	55 (39)
Cu/Ti-Nb	[EAs76]	2.6/1	2.54 × 1.4	234		12.5	55 (39)
Cu/Ti-Nb	[EAs76]	2.6/1	3.15 × 1.04	132		25	55 (39)
Cu/Ti-Nb	[EAs76]	2.6/1	3.15 × 1.04	234		25	55 (39)
Cu/Ti-Nb	[EAs76)	2/1	2.9 × 1.37	160	160	25	55 (39)
Cu/Ti-Nb	[EAs77]	3.5/1	3.12 × 1.30	15	279	∞	(36)
Cu/Ti-Nb	[EAs77]	2.8/1	2.90 × 1.44	18	279	∞	(36)
Cu/Ti-Nb	[EAs77]	2.3/1	3.12 × 1.30	24	203	77	(36)
Cu/Ti-Nb	[EAs77]	3.4/1	6.68 × 1.32	84	188	256	(36)
Cu/Ti-Nb	[EAs77]	3.4/1	10.06 × 0.99	84	173	323	(36)
Cu/Ti-Nb	[EAs77]	3.4/1	12.09 × 0.81	84	183	500	(36)
Cu/Ti-Nb	[EAs76]	2.7/1	2.54 × 1.27	18	220	25	(36)
Cu/Ti-Nb	[EAs76]	2.7/1	2.54 × 1.27	18	220	12.5	(36)
Cu/Ti-Nb	[EAs76]	2.7/1	2.54 × 1.27	18	220	25	(36)
Cu/Ti-Nb	[EAs76]	2.7/1	2.67 × 0.89	18	200	12.5	(36)
Cu-Ni/Cu-Ti-Nb	[EAs77]	3/3/1	0.76	397	14	8.5	(36)
Cu-Ni/Cu/Ti-Nb	[EAs77]	3/3/1	0.51	397	9	6.4	(36)
Cu-Ni/Cu/Ti-Nb	[EAs77]	3/3/1	0.51	397	9	4.2	(36)
Cu-Ni/Cu/Ti-Nb	[EAs77]	3/3/1	0.51	397	9	3.2	(36)
Cu-Ni/Cu/Ti-Nb	[EAs77]	1.5/2/1	0.38	156	14.5	3.6	(36)
Cu-Ni/Cu/Ti-Nb	[EAs77]	1.5/2/1	0.38	156	14.5	3.2	(36)
Cu-Ni/Cu/Ti-Nb	[EAs77]	1.5/2/1	0.38	156	14.5	2.3	(36)
Cu/Ni/Cu/Ti-Nb	[EAs77]	1.5/2/1	0.51	156	19	5.1	(36)
Cu-Ni/Cu/Ti-Nb	[EAs77]	1.5/2/1	0.51	156	19	4.2	(36)
Cu-Ni/Cu/Ti-Nb	[EAs77]	1.5/2/1	0.51	156	19	3.2	(36)

[a] Manufacturers

A Airco
I IGC (Intermagnetics General)
L Alcoa
M MCA (Magnetic Corporation of America)
R Soviet technical alloy
S Supercon
V Vacuumschmelze

3.20.1 The Mixture Rule for Modulus and Strength Estimation

The mixture-rule method is nothing more than the weighted averaging approach adapted to the calculation of the modulus or strength of a composite from those of its components. Implicit in its use are the validities of several assumptions relating to the details of the internal distributions *nt.)*

Temperature K	Young's Modulus, E 10^{10} N m^{-2}	Strengths, σ 10^8 N m^{-2} Yield $\sigma_{0.2}$	Tensile σ_B	Area Reduction ψ, %	Elongation δ, %	Manufacturer[a]	Remarks
4.2	11.4	1.9_3	11.0			I	Group A
4.2	13.5	2.5_7	9.8			I	
4.2	9.2	2.4_4	10.3			I	
4.2	10.9	2.7_4	11.9			I	
4.2	9.7	1.9_4	7.7			M	Group B
4.2	12.8	1.7_7	8.0			M	
4.2	11.1	1.5_9	7.8			M	
4.2	9.7	2.0_9	8.4			M	
4.2	9.9	1.6_6	7.9			M	
4.2	11.5	1.5_7	8.0			M	
4.2	10.8	1.8_4	8.9			A	
4.2	10.8	1.8_3	7.9				
300	9.3		3.7		15	S	
300	10.3		4.3		17	S	
77			6.9		26		
4.2	11.7		7.9		2		
300	11.0		4.0		17	S	
300	9.7		5.1		1.8	S	Group C
4.2	11.0		7.7		2.7		
300	10.3		5.4		9.0	S	
4.2	11.0		7.6		2.2		
300	9.7		5.4		2.0	S	
4.2	11.0		8.3		2.5		
4.2	9.8	1.6_8	7.6			S	Group D
4.2	12.0	2.3_8	8.0			S	
4.2	11.5	2.1_6	6.9			S	
4.2	10.9	1.5_7	6.7			S	
4.2			10.6	11.2	7.7	S	Group E. The properties of Cu-Ni alloys are given in Table 3-13. The alloy referred to here is Cu-30Ni.
4.2	11.7		11.2	21.3	5.7	S	
4.2	13.8		10.6	17.8	5.0	S	
4.2			10.6		8.8	S	
4.2			12.1	18.0	4.6	S	Group F
4.2			12.3	15.8	5.1	S	
4.2	11.7		11.7	20.4	5.9	S	
4.2			11.8	11.0	5.3	S	
4.2			11.7	14.7	5.8	S	
4.2			11.5	17.7	6.8	S	

of stress and strain. Departures from the rule of mixtures are attributable not only to the possible existences of unknown internal stress states, but also to other variables such as preferred orientations in either the fibers or the matrix and the degree of coherency of the fiber-matrix interfaces [LED77].

(a) The Mixture Rule for Modulus. An instantaneous Young's modulus, according to the mixture rule, could be expressed in the form

$$E_{inst} = \frac{\partial \sigma}{\partial \varepsilon} = \left.\frac{\partial \sigma}{\partial \varepsilon}\right|_f + \left.\frac{\partial \sigma}{\partial \varepsilon}\right|_m (1-f) \tag{3-12}$$

where f is the fiber volume fraction, and subscripts f and m refer to "fiber" and "matrix," respectively. Equation (3-12) is defined in two regimes:

(1) The fiber and matrix both deform elastically, in which case $\partial\sigma/\partial\varepsilon|_{f,m} \equiv E_{f,m}$ throughout.

(2) The fiber continues to deform elastically while the matrix deforms plastically, in which case $\partial\sigma/\partial\varepsilon|_f \equiv E_f$ as before, but $\partial\sigma/\partial\varepsilon|_m$ is irreducible.

(b) The Mixture Rule for Strength. By setting $\Delta\varepsilon$ = constant, (3-12) implies a mixture rule for stress. In other words, it implies that tensile strength also obeys the mixture rule provided that the stress-at-fracture of one component (i.e., its actual tensile strength) is combined with the stress in the other at a common level of strain.

In order to reduce electrical coupling between the current-carrying elements of a composite superconductor, a twist is introduced after the final wire-drawing process (Section 25.11). The resulting pitch-dependent reduction of the effective tensile strength and modulus has been studied experimentally and theoretically by several authors [Koc77]. Twist-related corrections have not, however, been introduced into the following examination of the mixture rule.

The degree of validity of the mixture rule for modulus and strength is examined with the aid of data, selected from Table 3-14, grouped according to the matrix/SC volume ratio. The means and standard deviations of modulus and strength within each group are presented in Table 3-15, which also lists, for the various components of the composite, tensile data brought forward from other tables in this chapter.

3.20.2 Young's Modulus

In spite of some evidence to the contrary [Eas77], reliable sources indicate that the Young's modulus of Cu is temperature dependent in the manner shown in Figure 3-30 [Ree67], [Met74]. That of Ti–Nb is composition dependent and only weakly temperature dependent within the entire 4–300 K range (Figure 3-30). The Cu–30Ni data are due to Reed *et al.* [Ree67]. Table 3-15 shows that for Cu/Ti–Nb superconductors the calculated moduli agree rather well with the experimental values—in some cases to within the experimental uncertainty, defined in this case as the modulus

Table 3-15. Test of the Mixture Rule for Modulus and Strength

Conductor Group from Table 3-14	Conductor Type	Nb Content at.%	Volume Ratio	Test Temperature K	Modulus, E 10^{10} N m^{-2} Experiment	Modulus, E Theory	Tensile Strength, σ_B 10^8 N m^{-2} Experiment	Tensile Strength, σ_B Theory
A	Cu/Ti-Nb	38	1.6/1	4.2	11.3 ± 1.8	12.0	10.8 ± 0.9	10.0
B	Cu/Ti-Nb	39	2.6/1	4.2	10.8 ± 1.2	12.7	8.0 ± 0.2	8.3
C	Cu/Ti-Nb	36	3.4/1	300	9.9 ± 0.3	10.8	5.3 ± 0.2	4.2
				4.2	11.0 ± 0.0	13.0	7.9 ± 0.4	7.4
D	Cu/Ti-Nb	36	2.7/1	4.2	11.1 ± 0.9	12.7	7.3 ± 0.6	8.1
E	Cu-Ni/Cu/Ti-Nb	36	3/3/1	4.2	{11.7, 13.8}	14.2	10.8 ± 0.3	7.5
F	Cu-Ni/Cu/Ti-Nb	36	1.5/2/1	4.2	11.7	13.4	11.9 ± 0.3	8.4

Component Data

Material	Data Sources		E	σ_B
OFHC Cu	Estimated values of E for moderate cold work based on Table 3-13, see Figure 3-30. Value of σ_B (4.2 K) selected to fit data (Figure 3-33); σ_B(300 K) chosen to be 23% lower—see Table 3-13	300	11.7	—
		4.2	14.5	3.3
		300	—	2.5
Cu-30Ni	Table 3-13	4.0	16.0	7.2
Ti-Nb	Values of E as function of temperature and composition taken from Figure 3-30; values of σ_B—assumed composition independent in this range—taken from Table 3-9	300	7.9–8.3	10
		4.2	8.0–8.4	21

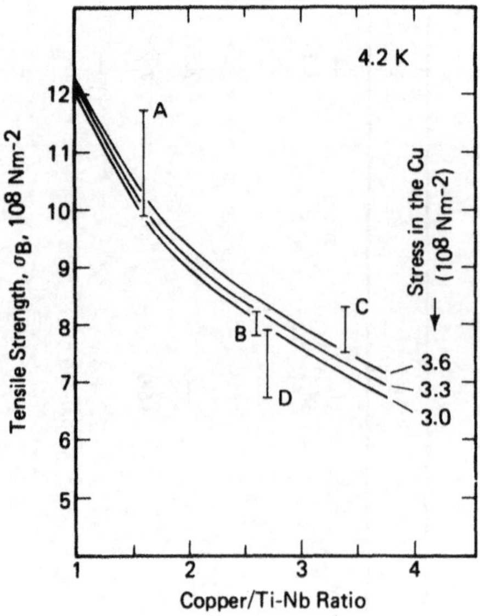

Figure 3-33. Experimental tensile strengths at 4.2 K of Cu/Ti–Nb composites (A, B, C and D from Table 3-15) compared with three curves of σ_B *versus* Cu/SC ratio calculated using the mixture rule and the data σ_B(Ti–Nb) = 21 × 10^8 N m^{-2} (see Table 3-9), the assumption of a common strain-at-fracture, and three values of σ_B(Cu) centered on 3.3 × 10^8 N m^{-2} (see Table 3-15).

scatter for different conductors with the same Cu/SC volume ratio. On the other hand, for reasons that are not well understood, the three-component conductors yield rather poor agreement between the results of mixture-theory and experiment.

3.20.3 Tensile Strength

It is not possible, using the mixture rule, to predict the tensile strength of a Cu-matrix composite without having prior knowledge of the elongation at fracture, since, as explained above, for the Cu component it is an "effective strength" (i.e., the stress in the Cu at the common level of strain) that is needed. This is a number lying somewhere between $\sigma_{0.2}$ (hence 0.5–5 × 10^8 N m^{-2}, depending on the starting degree of cold work) and σ_B (hence 4–5 × 10^8 N m^{-2}). In this analysis we have worked backwards, extracting $\sigma_{B(\text{eff,Cu})}$ from the experimental data, and looked for consistency. The results of so doing are summarized in Figure 3-33, which shows that to obtain reasonable consistency with the experimental results a $\sigma_{B(\text{eff,Cu})} \cong 3.3 \times 10^8$ N m^{-2} is called for; a value which, moreover, is comfortably lower than the tensile strength of Cu alone.

But if this value for $\sigma_{B(\text{eff,Cu})}$ is taken and substituted into a mixture expression for the strength of the Cu–Ni/Cu/Ti–Nb composite, a result is obtained that is much lower than the experimental value (see Table 3-15). Admittedly, available data indicate that the elongations in the ternary

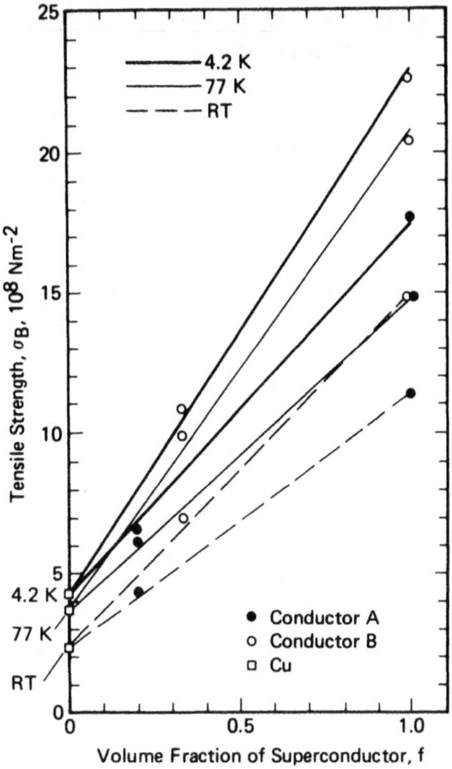

Figure 3-34. Tensile strengths of Cu/Ti-Zr-Nb-Ta composite superconductors in comparison with those of the individual components (i.e., the end points; see also Table 3-12). The conductor specifications are:

Code	Conductor Dimensions or Diameter mm	Cu/SC Ratio	Filament Number	Filament Diameter μm	Twist Pitch mm
A	3.8 × 1.6	4:1	361	64	100
B	0.5	2:1	61	37	10

The bare-superconductor data were obtained on the filament bundle after etching away the Cu from all except the grip ends—after MONJU and TATARA [MON78] and HORIUCHI [HOR80[a]].

composite (7 ± 2%, group E; 6 ± 1%, group F) are three times greater than those in the binary (2.5 ± 0.3%, group C), but a corresponding adjustment of the stress level in the Cu to 4.5×10^8 N m^{-2} in an attempt to take this into account raises the calculated composite tensile strength by only about 0.5×10^8 N m^{-2} and still leaves it short of the experimental value.

An example of the applicability of the mixture rule to the tensile strengths of two Cu/Ti-Zr-Nb-Ta multifilamentary composites at room temperature, 77 K, and 4.2 K, is given in Figure 3-34.

3.21 Strengthening Principles in Alloys and Composite Conductors

It has been stated that typical values of maximum stress likely to be encountered in a large superconducting magnet are in the range 7×10^7 to 2×10^8 N m^{-2}, with strain levels in the range 0.1–0.2% [Koc77]. By way of example: the maximum hoop stress in the European TESPE coils is 2×10^8 N m^{-2} [Sei79] and the MFTF coil design anticipates a maximal tensile strain, upon initial charging, of 0.25% followed by 0.15% residual and 0.10% cyclic strains [Sca80]. Since, according to this chapter, the 4.2 K tensile strengths of Ti–Nb and related alloys are an order of magnitude higher than the values quoted above, it is clear that superconductor strength poses no problem to the magnet designer. Increasing the strength of the superconductor itself is worth considering only if, by so doing, microyield instabilities could be reduced. The improvements in the low-temperature mechanical properties of Ti–Nb(34 at.%) obtained in response to the presence of fine homogeneous α-phase precipitation [Alb76] suggest that this might well be the case. On the other hand, of paramount importance is the task of increasing the effective modulus of the magnet windings (i.e., that of the conductor plus any substructure to which it is attached). This is especially true if the magnet is to be pulsed or cycled, since—as has been discussed in length in Section 3.12.2—serious heat losses are associated with the cyclic yielding of the Cu matrix, whose elastic strain limit at 4 K is typically 0.035% (referring to annealed OFHC Cu). Various approaches to the strengthening of Ti alloys and to the stiffening of composite superconductors are outlined below.

3.22 Strengthening of Alloys

Solution strengthening is considered in sections dealing with alloying effects. In superconducting β-Ti alloys, *precipitation* strengthening is also a possibility. In that the same fine dispersion of precipitate particles that pin dislocations may also pin the Abrikosov flux lattice, heat-treatment-induced-precipitation strengthening may also be compatible with an optimization of the critical current density [Alb76].

3.22.1 Duplex Aging

If an alloy isothermally aged to $\omega+\beta$ is upquenched to a temperature that places it just above the $\omega+\beta$-transus, the ω-phase reverts to a solute-lean β-phase similar to, but more finely distributed than, the β'-phase formed by the phase-separation reaction (Section 2.3). The reversion reaction is not reversible; i.e., although formed from isothermal ω-phase, the new β'-phase

Table 3-16. Strengths of Heat-Treated Ti-15Cr Alloys—After Polonis *et al.* [LUH70], [POL71], [CHA72]

Heat Treatment	Microstructure	Yield Strength $\sigma_{0.2}$, 10^8 N m^{-2}	Elongation δ, %
45 min/950°C/WQ (i.e., HT/Q)	β	8.9	17
HT/Q + 2.5 h/350°C	$\omega+\beta$	11.0	2
above + 5 min/450°C/quench to RT	$\beta'+\beta$	11.7	6

remains in place as the alloy is returned to lower temperatures. The β'-phase, whether produced by the down-quenching of β followed by aging or the upquenching of aged $\omega+\beta$, is a site for further aging-induced α-phase precipitation. The β-reversion route results in finer, more evenly distributed precipitation. All three precipitates: ω, β', and α, are strengtheners. Strengthening by ω-phase is accompanied by embrittlement, a disadvantage that is avoided when it is converted to β', or when β' is subsequently converted to α [LUH70].

An example of strengthening by the upquenching process—a so-called "duplex aging" process—has been provided by POLONIS, LUHMAN, and CHANDRASEKARAN [LUH70], [POL71], [CHA72] in studies of Ti-10Cr and Ti-15Cr. Some of the results are summarized in Table 3-16. In evidence are the equivalent strength and superior ductility of the $\beta'+\beta$-alloy as compared to those of the $\omega+\beta$-alloy.

With regard to flux pinning, although the effectiveness of the $\beta'+\beta''$-modulated structure, as developed in Zr-Nb and Zr-Nb-rich Ti-Zr-Nb alloys by appropriate heat treatment, has been studied and discussed (*structure* in Section 2.7 and *critical current* in Section 21.13.4), an equivalent investigation of flux pinning by β' has not been specifically undertaken.

Using an alloy known as "β-III" (Ti-11.5Mo-6Zr-4.5Sn) as model system, BOYER, TAGGART, and POLONIS [BOY74] discovered that the β-reversion reaction does not always proceed to completion. In addition they were able to show that moderate deformations (of up to 20%) promoted α-phase precipitation along with ω-phase precipitation during aging. They noted, furthermore, that the α-phase remained stable as the ω-phase reverted upon upquenching, and by so doing contributed to a high level of room-temperature strength (typically 14×10^8 N m^{-2}) accompanied by reasonable ductility (typically 6–8%).

3.22.2 Cold Working and α-Phase Precipitation

As implied above and discussed in Sections 2.10 and 2.11, the $\beta \rightarrow \alpha+\beta$ reaction is accelerated by the presence of lattice strain. Specifically, dislocations and deformation-cell walls provide nucleation sites for α-phase precipitation. As formed in heavily cold-worked Ti-Nb(34 at.%) during moderate-temperature aging, the unevenly distributed 0.1-0.2 μm$^{\varnothing}$ α-phase particles actually promote brittle failure. However, a final stage of cold deformation corrects this situation. The resulting finer (~200 Å), more evenly distributed precipitates increase the 4 K tensile and yield strengths and enhance the room-temperature ductility, which rises from $\delta = 6\%$ (as-heat-treated) to $\delta = 10\%$ (following final cold deformation).

As mentioned at the outset, the primary goal of superconductor mechanical-property enhancement should be to increase the Young's modulus. Improvements to the yield strength are useful only in so far as they result in concomitant improvements in resistance to microyield. Having optimized the mechanical and electrical properties of the superconducting alloy itself, the next step is to attend to the mechanical properties of the *composite* conductor and any reinforcing elements with which it is associated.

3.23 Strengthening of Composite Conductors

Three possibilities for the strengthening of composite superconductors are offered for consideration [TAY68]. They consist of strengthening the Cu matrix by (i) work hardening, (ii) dispersion hardening, or (iii) the addition of stainless steel reinforcement. Some representative mechanical properties of these three classes of material are listed in Table 3-13.

3.23.1 Mechanical Properties of the Cu Matrix

With regard to the data for OFHC Cu several points are noteworthy: (i) Young's modulus, which is independent of whether the Cu is in the annealed or 50%-cold-worked condition, increases 25 ± 4% between 300 and 4 K; (ii) the yield strength of OFHC Cu increases by a factor of 10 in response to 60% cold deformation by drawing; (iii) 60% cold work has only a small effect on the temperature dependence of the yield strength, which, between 300 K and 4 K, increases by 60% in the case of annealed Cu and by 25% in the case of cold-worked OFHC Cu.

Texturization has an important influence on the Young's modulus of Cu. During wire drawing, the ⟨100⟩ texture in the direction of the wire axis that first develops at low strains [REE77] is replaced at larger strains by a mixture of ⟨100⟩ and ⟨111⟩ textures. Although other crystal orientations have been shown to have Young's moduli ranging from 15.8×10^{10} to 20×10^{10} N m^{-2}, the Young's moduli of annealed Cu crystals pulled along ⟨100⟩ are typically about 7.6×10^{10} N m^{-2} at room temperature [REE77]. The

Young's moduli presented in Table 3-13 are clearly representative of mixed-texture conditions. The observation that annealing in the temperature range 130–530°C does not significantly change the wire microstructure [REE77] is confirmed by the agreement, to be seen in Table 3-13, between the "vacuum-annealed" and "60%-cold-drawn" room-temperature Young's modulus values.

3.23.2 Electrical Resistivity of the Cu Matrix

In designing a strengthened superconductor, electrical resistivity is another important consideration. Accordingly (and in anticipation of some subsequent discussions), electrical resistivity data for some typical composite-superconductor components are provided in Table 3-17.[6] Assuming the existence at the windings of stresses between 0.7×10^8 and 2×10^8 N m^{-2} (10–30 ksi), vacuum-annealed Cu, with a yield strength at 4 K of 0.5×10^8 N m^{-2}, is in need of reinforcement. Cold drawing to a 10% reduction in cross-sectional area work hardens the OFHC Cu sufficiently to raise its ambient-temperature yield strength to 2.2×10^8 N m^{-2} (enough to cope with the stresses developed under static conditions of magnet excitation) under penalty of some reduction in electrical conductivity. Alternatively, the dispersion-hardened alloy, "Am-zirc" (Cu–0.18Zr), may be considered as a replacement, again at the expense of reduced electrical conductivity. Table 3-17 shows that the resistance ratio of a good OFHC Cu may be reduced from 240 to 125 with 10% of cold work, and that that of aged (dispersion-hardened) Am-zirc is about 30. The implications of this in terms of loss of low-temperature electrical conductivity (and consequently stability) might be regarded as serious until it is realized that the resistance ratio, even of annealed OFHC Cu, is already reduced under magnet operating conditions by the effect of magnetoresistance [TAY68]. The situation is fully dealt with in Section 28.2, whose Figure 28-4 shows that in fields greater than about 40 kOe the resistivities of all Cu's are about the same provided that their starting resistance ratios are greater than about 150. Thus, whereas the benefits that might arise from the use of Am-zirc with its higher yield strength is open to question, it seems that the advantage in mechanical terms that accompanies the use of work-hardened copper is not affected to any appreciable extent by a deterioration in the at-field electrical conductivity. The mechanical properties of the Japanese LCT coil conductor, with particular reference to its strengthening through the use of a 21-% work-hardened Cu stabilizer, has been discussed by YOSHIDA *et al.* [YOS82].

[6]The alloy Cu–Ni, included in these tables for completeness, is, of course, the high-resistivity Cu-base solid solution used as an eddy-current barrier rather than a strengthener (see Section 27.13).

Table 3-17. Electrical Resistivities of Some Composite Conductor Components

Material	Condition	Test Temperature K	Resistivity 10^{-6} Ω cm	Residual Resistance Ratio	References
OFHC Cu	Annealed	273	1.55	97	[MET74] Table 5.1.3-TR1
		50	0.051		
		20	0.0172		
		4	0.0160		
	1 h/400°C	293	1.68	240 ± 10	[TAY68]
	Above + cold drawn 10%	—	—	125 ± 10	
Cu–10Ni	Annealed	273	16.4	1.2	[MET74] Table 5.1.3-TR1
		50	14.6		
		20	14.1		
		4	14.0		
Cu–0.15Zr	Solution treated + 50% cw + 1 h/450°C	—	—	30	[TAY68]
	Unknown (interpolated from value for Cu–0.5Zr)	~270	1.69	—	[TOU67, p. 186]
304 Stainless Steel	Annealed	273	70	1.4	[MET74] Table 8.1.2-TR1
		50	50.5		
		20	49		
		4	49		
310 Stainless Steel	Annealed	273	88	1.3	[MET74] Table 8.1.4-TR1
		50	70		
		20	68		
		4	67		

3.23.3 Properties of Stainless Steel Reinforcement

Two representative 300-series stainless-steel alloys have been selected for inclusion in Table 3-17. AISI Type 310 is a highly alloyed (24–26 wt.% Cr, 19–22 wt.% Ni) austenitic alloy of high structural stability (i.e., low M_s and M_d temperatures), while AISI Type 304L has been listed to represent the class of 304-series alloys (characterized by 18–20 wt.% Cr, 8–12 wt.% Ni and various levels of C and N), whose members are finding numerous cryogenic structural applications. Obviously, stainless steels with their high-field residual resistivities (~zero-field resistivities, ~50 $\mu\Omega$ cm) of more than 1000 times that of Cu cannot be regarded as electrical components of the composite superconductor.

3.23.4 Concluding Note on Composite Strengthening

Ideally, the volume fraction of reinforcement required would be calculated by designing to the impressed-stress level for the structure, bearing in mind the elastic strain limit of the Cu, and using the mixture-rule for Young's modulus. Unfortunately, as a simple calculation will show, no amount of metallic reinforcement will prevent plastic deformation of *annealed* Cu within the stress range referred to above. According to Table 3-13, the elastic strain limit for annealed Cu at 4 K is $(0.5 \times 10^8)/(14 \times 10^{10}) = 0.035\%$, and at the limit of 100% steel reinforcement stresses of 0.7×10^8 to 2×10^8 N m^{-2} will produce strains in the structure of 0.035 to 0.1%. With the work-hardened Cu, the strain limit is a factor of four higher.

3.24 Workability of Titanium-Alloy Superconductors

3.24.1 Principles of Workability

Workability is the ability of a material to withstand deformation during a metalworking process without fracture or void formation. Of the four metalworking processes commonly employed in the production of a superconductor for either experimental investigation or technical application—*viz.* swaging, rolling, extrusion, and wire drawing—the latter two are the most important. Workability is more than just a materials property. Thus, for example, the workability of a material under hydrostatic extrusion, in which compressive forces tend to inhibit cracking, is greater than that of the same material during wire drawing, in which the deformation forces are tensile [KUH78]. The combined influence of materials and process variables is conveniently expressed in the form of a relationship such as

$$\text{workability} \equiv f[f_1(\text{material}), f_2(\text{process})] \qquad (3\text{-}13)$$

in which f_1 is a function of the intrinsic (strain-rate-dependent) ductility of the material and f_2 is a function of the spatial and temporal dependences of the stress system development during working [GIT75], [KUH78]. The process function, f_2, includes the type of process (forging, swaging, rolling, extrusion, drawing) subdivided into parameters such as: (i) the spacial distribution, and rate of change of stress and temperature; (ii) adiabatic heating; (iii) heat transfer; (iv) the properties of the lubricant. The materials-function, f_1, includes: (i) strength; (ii) strain-rate-dependent yield strength (ductility and work hardenability); (iii) the stress- and temperature-dependence of microstructural changes such as phase transformations and precipitation. In terms of theoretical metallurgy, which is outside the scope of this book, (3-13) expresses the existence of a "forming-limit curve" determined by applying the techniques of finite-element or finite-difference mathematics to the physical process, using the "constitutive function" f_1 as materials-data input. This technique, well-known in the metalworking of homogeneous isotropic materials, has yet to be applied to anisotropic materials and composites.

In the present context, interest centers around the workability of Ti-Nb superconductors. The currently known quantitative materials aspects of metalworking have already been presented in terms of room-temperature hardness, modulus, yield strength, percent elongation, and area-reduction-ratio data. As for the full "workability function" as defined above, for the most part only descriptive subjective information is available. Out of necessity, therefore, the following discussion of the manner in which the workability of Ti-Nb is influenced by interstitial and substitutional third-element additions is limited to qualitative comment.

3.24.2 Influence of Temperature on the Workability of Ti-Nb Alloys

The tensile strengths of Ti-Nb(0-50 at.%) alloys tend to decrease monotonically as the temperature rises between 0 and 600°C as shown in Figure 3-35. The strength of Ti-Nb(50 at.%) is least affected by temperature; its ductility also remains practically independent of temperature in the range considered, whereas those of the alloys Ti-Nb(0-30 at.%) increase extremely rapidly as the temperature approaches 600°C. These materials properties are reflected in the temperature-dependence of area reduction ratio achievable by a composite Cu/Ti-Nb billet under hydrostatic extrusion at a given extrusion pressure. Figure 3-36 groups the extrusion data into three workability bands, characterized by temperatures of 20, 350-500, and 625-675°C [MEY76]. Based on the upper limits of these bands, estimates of the maximum-achievable extrusion ratios, at arbitrarily selected extrusion pressures, can be made; some are listed in Table 3-18. The constancy of the extrusion pressure in the temperature range 350-500°C, wherein a rapid decrease with increasing temperature would have been anticipated, is

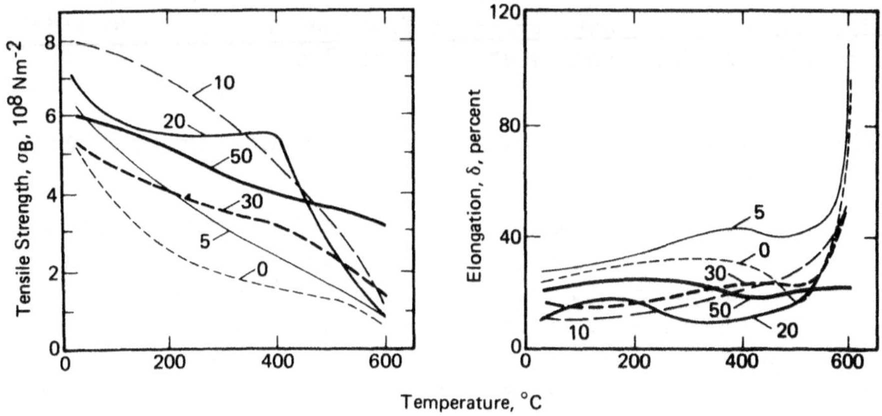

Figure 3-35. Elevated-temperature tensile properties of Ti–Nb alloys. The numbers refer to Nb content in at.%. Of the alloys represented, all of whose tensile strengths decrease with increasing temperature, Ti–Nb(50 at.%) possess the greatest "heat resistance." The elongation of this alloy also remains practically independent of temperature within the range portrayed—after MOLCHANOVA [MOL65, p. 23].

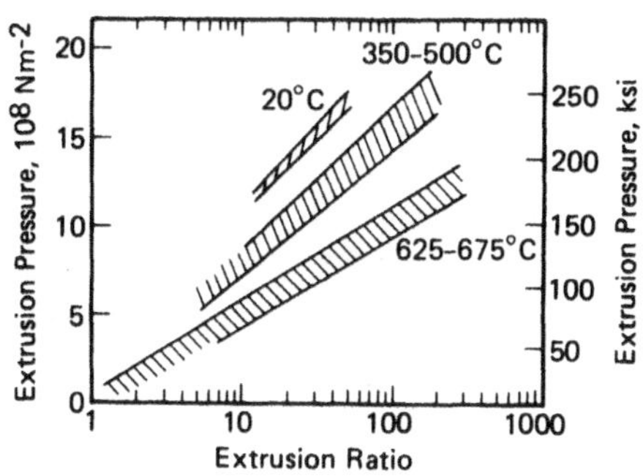

Figure 3-36. Hydrostatic extrusion pressure for Cu/Ti–Nb composite billets as function of extrusion ratio and billet insertion temperature. The hydraulic fluid/lubricant was castor oil. Data acquired at Battelle-Columbus Laboratories—after MEYER *et al.* [MEY76].

Table 3-18. Conservatively Estimated Maximum Achievable Hydrostatic-Extrusion[a] Ratios of Experimental Composite Billets[b] at 17.3×10^8 N m^{-2} as a Function of Temperature—After Meyer *et al.* [MEY76]

Temperature °C	Area-Reduction Ratio
20	48 : 1
350–500	170 : 1
625–675	>1000 : 1

[a] Liner diam.: 6.0 cm; Die angle: 45°; Stem speed: 34 mm s^{-1}; Fluid: AA Castor oil.

[b] Billet size: $15.2\ cm^{l} \times 5.7\ cm^{\varnothing}$; Cu/SC ratio: 1.43 : 1; No. of "filaments": 19.

attributable to a competition between the material variable *ductility* and the processing variable *lubrication efficiency* [MEY76].

3.24.3 Influence of Solute Additions on the Workability of Ti–Nb Alloys

Contemporary Ti–Nb-base superconducting alloys may be regarded as the culmination of a sequence of efforts to improve the electrical and mechanical properties of Zr–Nb, the original technical alloy superconductor. Figure 3-37 traces the development of Ti-alloy superconductors in these terms and also provides a guide to the associated literature. The several stages of development indicated therein are amplified in the following discussion:

1. The alloy Zr–Nb(50 at.%), for example, undergoes a ductile-brittle transition at a temperature of about 100°C [DOI68]. By substituting Ti for some of the Zr, this transition temperature may be significantly reduced: to about room temperature with 1 at.% Ti, and about −50°C with 10 at.%.

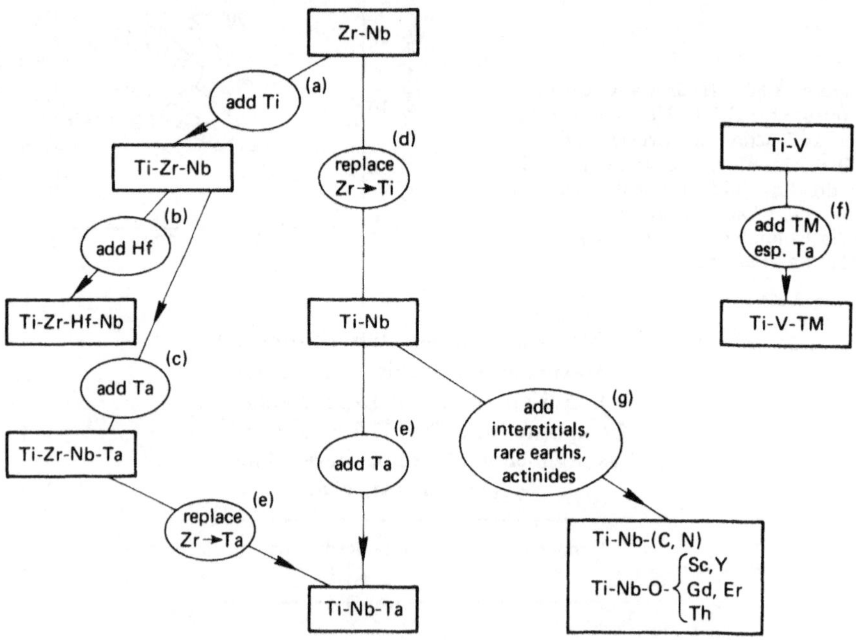

Figure 3-37. Development of the original Zr–Nb-alloy superconductor into contemporary Ti–Nb-base alloy superconductors in terms of improvements to workability. The several branches (a) through (f) are discussed in the text, and the relevant literature is as follows: (a) [IMP65], [WES67], [DOI68], [ZWI70]; (b) [RAY69]; (c) [HOR74]; (d) [COL66], [WES67]; (e) [REI67], [MIT68], [CUR79]; (f) [EFI70], [SAV71a]; and (g) [REU66], [WES67], [COU67, COU69], [KUN70], [RIC70], [RAS72a].

This search for improved workability led to the well-known alloy Ti_{10}-Zr_{40}-Nb_{50} [Doi68]. In other studies it was noted that the addition of a minimum of 10 at.% Ti greatly improved the forgability of Zr-Nb(75 at.%) [Imp65], and that the addition of Ti also improved the workability of Zr-Nb alloys containing either O [Ale67[a]], [Zwi70] or C [Wes67], with regard to which Ti additions of up to about 45 at.% were recommended.

2. Hf, which is isoelectronic with Ti and Zr, has been added to Ti-Zr-Nb in an attempt to further improve its superconductive and mechanical properties. But although it was determined that alloys of the systems Zr-Hf-Nb(70 at.%) and Ti-Zr-Hf-Nb(70 at.%) all forged better than Zr-Nb(70 at.%) itself [Ray69], the use of Hf is not recommended on account of its generally deleterious effect on the superconductive properties.

3. Ta, which is isoelectronic with Nb, has been added with considerable success to Ti-Zr-Nb. Horiuchi *et al.* [Hor74] (Kobe Steel Ltd., Japan) noted that the addition of 1 to 20 at.% Ta improves both the superconductive properties and the workability of Ti-Zr-Nb. The preferred alloy, finally selected for its high upper critical field, was Ti_{61}-Zr_6-Nb_{27}-Ta_6. In summarizing the foregoing results we note that improvements to the workability of Zr-Nb have resulted from the replacement of some of the Zr by Ti, and some of the Nb by Ta, and that workability increases in the sequence Zr-Nb < Ti-Zr-Nb < Ti-Zr-Nb-Ta.

4. The complete replacement of the Zr in Zr-Nb by Ti resulted in alloys that were significantly easier to work [Wes67] and possessed superior superconducting properties. Ti-Nb possesses an additional advantage over Zr-Nb in that its work-hardenability is relatively insensitive to composition between 30 and 80 at.% Nb [Col66].

5. Following the observation that an improvement in workability resulted from the substitution of Ta or Ti into Zr-Nb, Reinbach [Rei67] discovered that both good superconductive properties and workability accompanied the complete replacement of the Zr in Ti-Zr-Nb by Ta. Good cold deformabilities were noted for members of the group Ti-(15-70)Nb-(10-65)Ta and in particular for: Ti-45Nb-30Ta and Ti-25Nb-55Ta, which were able to be cold reduced 99.8% and 99.94%, respectively, without intermediate heat treatment. Similar conclusions were drawn by members of a research group at Mitsubishi Ltd. (Japan) [Mit68], who noted that Ti-Nb-Ta alloys could possess workabilities superior not only to Zr-Nb but also to Ti-Nb. For reasons to be discussed elsewhere (see Section 28.1) Curtis and McDonald became interested in the effects of small ($\lesssim$2 wt.%) additions of Ta to Ti-Nb(38 at.%) [Cur79]. It was found that small increases in strength, accompanied by small decreases in ductility, took place. Furthermore, and not unexpectedly in view of the above results for high-Ta-content alloys, no fabrication difficulties were experienced during heavy cold reductions of up to 99.7%.

6. In a series of studies, which were only marginally related to the properties of technical superconductors, researchers at the Baikov Institute of Metallurgy, Moscow, investigated the effects of third-element (mostly TM) additions to Ti-V alloys. In general it was noted that the potency of the transition-element strengtheners decreased in the sequence Zr, Nb, Re, Mo, Hf, Ta, and W and that alloys containing 10 at.% of W, Hf, Re, Ta, or Sn ruptured when cold deformed greater than 10-30% [EFI70]. On the other hand, by restricting the Ta concentrations in Ti-V(20-80 at.%) alloys to less than 5 at.%, the resulting considerable increases in tensile strength (of from 8-10 $\times 10^8$ to 13-17 $\times 10^8$ N m^{-2}) were accompanied by adequate ductility.

7. A great deal has been written about the effects of additions of the interstitial elements B, C, N, and O, separately and in combination, on the properties of superconductive alloys. It is well known (see Table 3-2) that they are rapid hardeners of Ti and Ti-Nb. Following observations by BETTERTON *et al.* (see [KUN70]) that improvements to the superconductive properties of Zr-Nb(75 at.%) accompanied additions of O (up to 0.025%) and C (up to 0.02%), various workers sought for comparable improvements to the properties of Ti-Nb, especially since it is known that Ti-Nb and Ti-Ta alloys can tolerate appreciably higher percentages of O [REU66]. The electrical results are discussed elsewhere (Sections 21.13.4 and 21.13.5). It was noted that C strengthened Ti-Nb alloys and increased, for example, the rolling force necessary to attain a given reduction [WES67]. The Ti-rich alloy Ti-32Nb, with 0.11 wt.% C, could not be drawn down to wire [RAS72[a]], and even in Ti-66Nb, C levels in excess of 0.1-0.15 wt.% had an embrittling effect on wire [KUN70], presumably in both cases as a result of the effect of carbide precipitation. Oxygen had comparable effects on the workability. COURTNEY and WULFF [COU67] found that O levels above about 0.3 wt.% in Ti-40Nb, and above about 0.35 wt.% in Ti-60Nb caused cold embrittlement. In order to overcome this difficulty, in case high O levels are unavoidably present rather than deliberately introduced, COURTNEY and WULFF studied the effect of additions of oxygen-scavengers (elements that have a greater affinity for O than does Ti) on the electrical and mechanical properties. It was shown that O could be partially removed from solution, accompanied by a useful increase in critical current density and some dispersion strengthening due to oxide particles, by the internal oxidation of added small amounts of the Group-IIIB elements Y [COU67] and Sc [RIC70], the rare-earth elements Gd [COU67] and Er [RIC70], and the actinide element Th [COU67]. In particular, it was noted that an ingot containing 0.45 wt.% O plus 1.0 wt.% Y was readily workable from about 0.45 in.$^{\varnothing}$ to 0.010 in.$^{\varnothing}$ (99.95% reduction) without intermediate annealing [COU67].

II

Physics

Chapter 4

Dynamic Elastic Modulus

4.1 Determination of Dynamic Moduli

4.1.1 Definitions and Interrelationships

In that the elastic constants, C_{ij}, which may be used to calculate the bulk, shear, and Young's moduli, as well as the Debye temperatures of polycrystalline materials, are related to the electronic properties and in some cases have first been calculated from principles, they form a bridge between the mechanical properties of materials and their physical properties. These considerations have directed the placement of this chapter between Chapter 3 (Mechanical Properties) and the normal-state physical-property group—Chapters 5 through 9. The accompanying diagram, Figure 4-1, serves as a reminder of the definitions of the engineering quantities: bulk modulus, K; shear modulus, G; Young's modulus, E; and Poisson's ratio, ν. These, in isotropically elastic solids (fine randomly oriented grains), are simply related to each other according to

$$K = \frac{E}{3(1 - 2\nu)} \qquad (4\text{-}1)$$

$$G = \frac{E}{2(1 + \nu)} \qquad (4\text{-}2)$$

The static Young's moduli of Ti-base alloys, as determined from the results of conventional tensile testing, have been discussed in Section 3.13. To be considered in this chapter are the dynamically determined moduli of both polycrystalline samples and monocrystals—quantities that are obtained from the velocities of longitudinal or transverse sound waves, either audible or ultrasonic (whichever is appropriate), with the aid of the

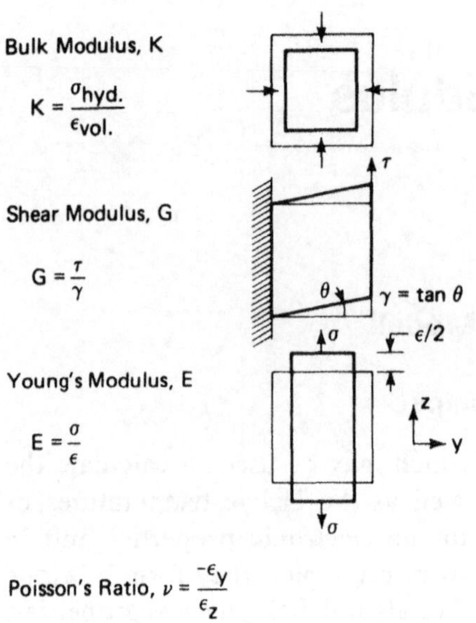

Figure 4-1. Standard definitions of the elastic moduli of solids—see for example HAYDEN *et al.* [HAY65, p. 23].

general relationship: velocity = $\sqrt{\text{modulus/density}}$. Elastic modulus has the dimensions of pressure, which is expressed in the literature in one or other of the units: N m^{-2}, dyn cm^{-2}, Pa, bar, atm, ksi, and kg mm^{-2}. The interrelationships between them are:

$$\begin{aligned} 1\ \text{N m}^{-2} &= 10\ \text{dyn cm}^{-2} \\ &= 1\ \text{Pa} \\ &= 10^{-5}\ \text{bar} \\ &= 0.9869 \times 10^{-5}\ \text{atm} \\ &= 1.4505 \times 10^{-7}\ \text{ksi} \\ &= 1.0197 \times 10^{-7}\ \text{kg mm}^{-2} \end{aligned}$$

The many techniques that are available for the measurement of elastic modulus employ frequencies ranging from zero (as in the static tensile test) to the tens of MHz (often referred to in this context as the "ultrasonic" frequency range). Measurements of the monocrystalline elastic constants usually take place in the ultrasonic frequency range. Examples of the various techniques that have been employed in studying the dynamic elastic properties of Ti–TM alloys and superconductors are given in Table 4-1.

4.1.2 Terminology

Although some authorities (e.g., FISHER [FIS75]) unequivocally prefer the term "modulus" for the parameter coupling linear elastic stress and strain while others (e.g., LEDBETTER [LED82]) prefer to associate the word "constant" with all the elastic parameters except the traditional static engineering quantities such as Young's modulus, all authors tend to agree that no terminological distinction should be drawn between the linear elastic parameters of monocrystalline and polycrystalline materials. This is a pity, since some economy in expression could derive from so doing. According to HEARMON [HEA46]:[1] "The terminology of the quantities specifying the elastic properties of the anisotropic materials is very confused and ill-defined in spite of various attempts that have been made to rationalize it." HEARMON recommended the term "coefficient" for the compliance parameters S_{ij}, the term "parameter" for the C_{ij}, "modulus" for quantities such as the Young's and rigidity moduli, and "constant" to generally cover all of the above.

In this chapter, following FISHER (e.g., [FIS75]), the term "modulus," with appropriate qualifiers, will be used throughout, except that, again following FISHER [STE81] and most other authorities and texts, the monocrystalline elastic stiffness moduli, C_{ij}, will also be referred to as the "elastic constants."

4.1.3. Long-Wavelength Measurement Techniques

When a bar is executing kilocycle-frequency vibrations, the wavelength is commensurate with the length of the sample under investigation. In either its longitudinal (L) or transverse (T) vibrational modes, since unconstrained stretching and relaxation is taking place in either case, the wave velocity is controlled simply by the "static" Young's modulus, E, according to $v_{L,T}$ (long-wavelength) $= \sqrt{E/\rho_d}$, where ρ_d is the density. Torsional vibrations are required for the determination of the shear modulus, G, according to $v_{\text{Torsion}} = \sqrt{G/\rho_d}$. Commercially available equipment often used for this type of measurement is the "Elastomat" originally designed by F. Förster (see [FED63]).

In recent modulus studies of Ti–Nb samples by LEDBETTER *et al.* [LED81] use was made of the "Marx oscillator" technique. In it, a cylindrical-rod specimen, about 5 mm in diameter, was cemented to a matched pair of Y-cut, X-plated, rectangular-rod 50-kHz quartz crystals. Using suitable electronics, this three-component composite Marx oscillator was swept in frequency in a search for the half-wave resonance, which was to be detected by an oscilloscope and measured by a frequency meter. The Young's

[1] I am indebted to Dr. E. S. Fisher for drawing my attention to this article.

Table 4-1. Index of Dynamic Modulus and Elastic Constant Investigations in Ti–TM Alloys and Superconductors

Measurement Technique	Material	Properties Measured	Literature
Use of the "Elastomat," in which the frequencies of transverse, longitudinal, and torsional forced or self-excited oscillations of a bar within the range 0.6–25 kHz yield E (transverse and longitudinal oscillations) and G (torsional oscillations)	Ti–V, -Nb, -Mo	E and G	Fedotov and Belousov [FED63, FED66]
	Ti–Nb	E and G	Fedotov and Belousov [FED64]
	Ti–Cr, -Mn, -Fe, -Co, -Ni Ti–V, -Nb, -Mo	E and G	Fedotov [FED73]
"Dynamic-impact" method	Ti–Nb	E, G, K and ν	Nedyukha and Chernyy [NED66]
Resonant longitudinal oscillation of a rod excited at about 20 kHz using an electrostatic transducer	Ti–Nb	E	Albert and Pfeiffer [ALB72]
Resonant piezoelectrically excited longitudinal or torsional oscillations at 60 and 100 kHz	Small samples from fiberglass insulated coil fabricated from Cu/Ti–Nb composite wire	Elastic compliance moduli S_{ij}, with $ij = 11, 13, 33, 44, 66$	Weston [WES75]

Resonant piezoelectrically excited longitudinal or torsional oscillations at 60 and 100 kHz	Ti–Nb(4.7, 8.9, 14.3, 19.0, 19.4, 24.7 at.%)	E	Ledbetter [LED78, LED82] see also Ledbetter *et al.* [LED81]
Ultrasonic (5 and/or 10 MHz) pulse-echo technique	Rectangular bar of Ti–Nb/Cu superconductive composite	Elastic compliance moduli S_{ij}, with $ij \equiv 11, 22, 33; 44, 55, 66; 12, 13, 23$	Ledbetter and Read [LED77]
Ultrasonic (5 and/or 10 MHz) pulse-echo technique	Ti–Nb(35.7, 38.7 at.%)	E	Ledbetter [LED77, LED82] see also Ledbetter *et al.* [LED81]
Ultrasonic (20–100 MHz) velocities of: (i) ⟨110⟩ longitudinal mode (ii) {110}⟨001⟩ shear mode (iii) {110}⟨110⟩ shear mode	Monocrystals of Ti–Cr	Analyzed to yield:	Fisher and Dever [FIS70, FIS70[a]]
	Monocrystals of Ti–V		Fisher [FIS75]; Katahara *et al.* [KAT79, KAT79[a]]
	Monocrystals of Ti–Nb		Reid *et al* [REI73]

modulus could then be determined from the relationship

$$E = 4\rho_d f^2 l^2 \tag{4-3}$$

in which f is the resonant frequency of the rod and l is its length.

4.2 Ultrasonic (MHz) Methods in Elastic Modulus Measurement

4.2.1 Cubic Monocrystals

(a) Basic Theory. Using ultrasonic techniques in the 20–100 MHz frequency range it is possible to generate sound waves of wavelength $\sim 10^{-4}$ m, very much smaller than the physical dimensions of the specimen. By applying the high-frequency (MHz) ultrasonic method to monocrystalline samples, it is possible to isolate the individual elements of the stiffness matrix $[C_{ij}]$ defined by

$$[\sigma] = [C_{ij}][\varepsilon] \tag{4-4a}$$

or the compliance matrix $[S_{ij}]$, obtained by matrix inversion and defined by

$$[\varepsilon] = [S_{ij}][\sigma] \tag{4-4b}$$

where σ and ε have their usual meanings of stress and strain, respectively. The compliance moduli may also be obtained directly using kHz resonance methods [LED82]. Although, as will be shown later, an ultrasonic pulse-echo or similar method may substitute for low-frequency techniques in the measurement of the elastic moduli of polycrystalline materials, its use is almost mandatory if those of small single crystals are to be obtained conveniently and accurately. In general linear elastic theory, the $[C_{ij}]$ or $[S_{ij}]$ may be represented, in VOIGT's contracted notation, by a 6×6 matrix:

$$\left[\begin{array}{ccc|ccc} 11 & 12 & 13 & 14 & 15 & 16 \\ 21 & 22 & 23 & 24 & 25 & 26 \\ 31 & 32 & 33 & 34 & 35 & 36 \\ \hline 41 & 42 & 43 & 44 & 45 & 46 \\ 51 & 52 & 53 & 54 & 55 & 56 \\ 61 & 62 & 63 & 64 & 65 & 66 \end{array}\right]$$

which for crystals of cubic symmetry reduces to

$$\left[\begin{array}{ccc|ccc} 11 & 12 & 12 & & & \\ 12 & 11 & 12 & & 0 & \\ 12 & 12 & 11 & & & \\ \hline & & & 44 & & \\ & 0 & & & 44 & \\ & & & & & 44 \end{array}\right]$$

leaving only three independent elastic moduli. In the inverse pair of symmetrical matrices $[C]$ and $[S]$, the vanishing of any C_{ij} implies the vanishing of the corresponding S_{ij}; secondly, if the off-diagonal elements of the kth row and the kth column are zero, then $C_{kk} = S_{kk}^{-1}$. An important derived stiffness modulus is $C' = (C_{11} - C_{12})/2$. This, together with C_{11} and C_{44}, are the elastic constants associated with the following modes of ultrasonic wave propagation:[2]

(1) C_{11}: a longitudinal wave in a $\langle 100 \rangle$ direction
(2) C_{44}: a transverse wave in a $\langle 100 \rangle$ direction, polarized along $\langle 010 \rangle$; a transverse wave in a $\langle 110 \rangle$ direction, polarized along $\langle 001 \rangle$
(3) C': the other transverse wave in a $\langle 110 \rangle$ direction, polarized along $\langle \bar{1}10 \rangle$

Since two of these involve propagation in a $\langle 110 \rangle$ direction, it is most convenient to obtain them all by grinding and polishing a pair of $\{110\}$ surfaces on the crystal and after exciting transverse waves, one of polarization $\{110\}\langle 001 \rangle$ and the other of polarization $\{110\}\langle \bar{1}10 \rangle$, to measure $v_{T1} = \sqrt{C'/\rho_d}$ and $v_{T2} = \sqrt{C_{44}/\rho_d}$, respectively. Next, by exciting longitudinal vibrations (parallel to $\langle 110 \rangle$, of course), the velocity $v_L = \sqrt{C_L/\rho_d}$ (where $C_L \equiv (C_{11} + C_{12} + 2C_{44})/2$) is measured to complete the set of three independent elastic constants for cubic symmetry (less symmetric crystals have up to 21 independent elastic constants).

Transverse (shear) and longitudinal waves can be excited by "Y-cut" and "X-cut" piezoelectric transducers, respectively. The transducer, a thin quartz wafer typically 3–5 mm in diameter, is attached to one of an opposite pair of flat ground surfaces either with a grease (for longitudinal waves) or resins or low-melting-point organic salts such as phenyl salicylate [FUL74] (for shear waves).

(b) The Significance of C' for Transition-Metal Alloys. Since C_{44} governs the transverse velocity in a $\langle 100 \rangle$ direction and C' the transverse

[2]Further details dealing with wave and particle vibration directions in crystallographic terms are provided by FISHER *et al.* [FIS75], [KAT79], [STE81].

velocity in a ⟨110⟩ direction, in an isotropic solid $C_{44} = C'$. Departure from isotropy is gauged by the value of an anisotropy parameter such as the Zener "anisotropy ratio," $A \equiv C_{44}/C'$. In bcc simple metals, A tends to be of the order 10, (for example $A_{Na} = 7.5$) whereas for the bcc transition elements of the sixth group—Cr, Mo, and W—$A = 0.713, 0.721$, and 1.008, respectively [FIS75], indicating the influence of the d-band electrons in stabilizing the bcc lattice. An alternative approach to elastic anisotropy, particularly useful when $A < 1$, has been offered by CHUNG and BUESSEM [CHU67].

FISHER discussed in detail the physical significances of A and C' in relationship to electronic factors which govern bcc stability in transition-metal alloys [FIS70, FIS75], [KAT79]. An examination of the bcc structure reveals that, whereas the C_{44} shears are resisted primarily by nearest-neighbor repulsion, the C' shear resistance depends primarily on the next-nearest-neighbor forces, which explains its low values in simple metals (e.g., $0.0065 \times 10^{11}\ \mathrm{N\,m^{-2}}$ for Na) [FIS75]. Additional cohesive contributions from d electrons are thought to be responsible in transition elements for the observed large values of C', which, it appears, may be interpreted as a bcc stability parameter. Thus for the highly stable bcc transition elements Cr, Mo, and W, $C' \cong 1.5 \times 10^{11}\ \mathrm{N\,m^{-2}}$, but with decreasing e/a ratio C' decreases rapidly, tending to zero at room temperature for alloys that possess ω-phase instabilities or that transform martensitically to hcp at ordinary temperatures [COL73]. This continuous decrease of bcc lattice stability with decreasing e/a ratio is illustrated in Figure 1-1 with data for the group-VI elements (Cr, Mo, and W), the group-V elements (V, Nb, and Ta), and members of the alloy systems Zr–Nb, Ti–Nb, Ti–Cr, Ti–V, and Mo–Re. The remarkable feature displayed by this figure is that C' seems to be a universal function of e/a ratio [FIS70].

4.2.2 The Isotropic Solid

Since all directions are equivalent in an isotropic solid, the only variables remaining are the vibrational modes of which there are two—transverse and longitudinal. Accordingly, only two elastic moduli are needed; for example the Lamé parameters, designated λ and μ [JAE62]. Referring to the 6×6 elastic stiffness constant matrix, these are

$$\begin{bmatrix} \lambda + 2\mu & \lambda & \lambda & & & \\ \lambda & \lambda + 2\mu & \lambda & & 0 & \\ \lambda & \lambda & \lambda + 2\mu & & & \\ & & & \mu & & \\ & 0 & & & \mu & \\ & & & & & \mu \end{bmatrix}$$

which implies that sound propagation in an isotropic medium is controlled by the following moduli:

(1) The modulus for *longitudinal-wave* propagation (cf. C_{11}), *viz.* $M \equiv \lambda + 2\mu$;
(2) the modulus for *transverse-wave* propagation (cf. C_{44}), *viz.* μ.

Finally we note that the shear modulus, $C' = (C_{11} - C_{12})/2$, is replaced by μ. But recognizing that C_{44} has already been replaced by μ it follows that the transverse-wave velocity is isotropic, as required.

The Lamé constants must be relatable to the familiar macroscopic elastic moduli—K, E, and G—and they are in the following way [JAE62]:

$$K = \lambda + \tfrac{2}{3}\mu \tag{4-5}$$

$$G = \mu \tag{4-6}$$

$$E = \frac{9KG}{3K + G} = \frac{\mu(3\lambda + 2\mu)}{\lambda + \mu} \tag{4-7}$$

$$\nu = \frac{E}{2G} - 1 = \frac{\lambda}{2(\lambda + \mu)} \tag{4-8}$$

In terms of the macroscopic elastic moduli, the velocities of *ultrasonic* longitudinal and transverse waves in isotropic solids are, therefore,

$$v_L = \sqrt{(\lambda + 2\mu)/\rho_d} \tag{4-9a}$$

$$= \sqrt{(K + \tfrac{4}{3}G)/\rho_d} \tag{4-9b}$$

and

$$v_T = \sqrt{G/\rho_d} \tag{4-10}$$

Note that: (i) these formuli are distinctly different from those for the long-wavelength velocities of the longitudinal and transverse vibrations of rods, *both* of which are governed by E; (ii) long-wavelength measurements of G required a torsional, rather than transverse, vibration.

4.2.3 The Anisotropic Solid

With polycrystalline materials, the particular class of macroscopic anisotropy of special interest in the present context is the orthorhombic symmetry exhibited by some multifilamentary superconductive composites [LED77], for a complete description of whose elastic properties the following nine independent macroscopic stiffness moduli are required:

$$\begin{bmatrix} C_{11} & C_{12} & C_{13} & & & \\ C_{12} & C_{22} & C_{23} & & 0 & \\ C_{13} & C_{23} & C_{33} & & & \\ & & & C_{44} & & \\ & 0 & & & C_{55} & \\ & & & & & C_{66} \end{bmatrix}$$

It has turned out that in such materials deviations from isotropy about the conductor-direction axis may be quite small, sufficiently so that the assignment to the filamentary composites of other classes of axial symmetry (e.g., hexagonal) can be made without introducing serious error [LED77].

4.3 Calculation of Polycrystalline Elastic Moduli from the Monocrystalline Compliance Moduli and Stiffness Moduli (i.e., the Elastic Constants)

4.3.1 The VRH Approximation

If the elastic compliance and stiffness moduli are available from experiments on single crystals, it is possible to compute a set of macroscopic elastic moduli for the corresponding isotropic polycrystalline solid. A procedure for so doing—the VRH approximation, named for the three contributors VOIGT, REUSS, and HILL—has been fully described and discussed by ANDERSON [AND63]. For a polycrystal with no preferred orientation (i.e., "untextured"), it is, of course, sufficient to express two of the moduli (say K and G) independently in terms of the monocrystalline moduli. Expressions for K and G in terms of the C_{ij} were obtained by VOIGT, and in terms of the S_{ij} by REUSS. HILL's contribution was to show thermodynamically that the Voigt and Reuss expressions represented upper and lower bounds, whose means (arithmetic, according to ANDERSON [AND63]) yielded even better estimates of the polycrystalline moduli. For all crystal classes, the aggregate isotropic shear modulus, G_V, according to VOIGT (upper bound) is

$$G_V = \tfrac{1}{15}(C_{11} + C_{22} + C_{33}) - \tfrac{1}{15}(C_{12} + C_{23} + C_{13}) + \tfrac{1}{5}(C_{44} + C_{55} + C_{66}) \tag{4-11a}$$

while the Reuss (lower bound) modulus G_R, is given by

$$\frac{15}{G_R} = 4(S_{11} + S_{22} + S_{33}) - 4(S_{12} + S_{23} + S_{13}) + 3(S_{44} + S_{55} + S_{66}) \tag{4-11b}$$

Similarly, for the bulk modulus, K, according to VOIGT,

$$\left.\begin{aligned} K_V &= \tfrac{1}{9}(C_{11} + C_{22} + C_{33}) + \tfrac{2}{9}(C_{12} + C_{23} + C_{13}) \\ &\text{or more usually} \\ K_V &= \tfrac{1}{9}(C_{11} + C_{22} + C_{33} + 2C_{12} + 2C_{23} + 2C_{13}) \end{aligned}\right\} \quad (4\text{-}12a)$$

According to REUSS

$$\frac{1}{K_R} = (S_{11} + S_{22} + S_{33}) + 2(S_{12} + S_{23} + S_{13}) \quad (4\text{-}12b)$$

Of course for a quasi-isotropic solid composed of crystallites of cubic symmetry the Voigt and Reuss equations reduce to

$$5G_V = C_{11} - C_{12} + 3C_{44} \quad (4\text{-}13a)$$

$$\frac{5}{G_R} = 4S_{11} - 4S_{12} + 3S_{44} \quad (4\text{-}13b)$$

and

$$3K_V = C_{11} + 2C_{12} \quad (4\text{-}14a)$$

$$\frac{1}{3K_R} = S_{11} + 2S_{12} \quad (4\text{-}14b)$$

Thus, following HILL's general conclusions, the G_V and G_R bounds[3] can be averaged either arithmetically or geometrically for insertion into expressions for the Young's modulus, E, and the Poisson ratio, ν. Taking the arithmetic route,

$$G = (G_V + G_R)/2 \quad (4\text{-}15)$$

and

$$K = (K_V + K_R)/2 \quad (4\text{-}16)$$

After this it is possible to derive

$$E = 9KG/(3K + G) \quad (4\text{-}7)$$

$$\nu = (E/2G) - 1 \quad (4\text{-}8)$$

[3]References to bounds more sophisticated than VRH are given by LEDBETTER [LED80].

and a MHz-frequency longitudinal modulus, M, which differs from Young's modulus in that no Poisson dilatation is permitted, given by

$$M = K + \tfrac{4}{3}G$$

$$= \lambda + 2\mu \qquad \text{(see Section 4.2.2)} \qquad (4\text{-}17)$$

In conclusion it is noted that when performing the VRH calculations for cubic crystals, use may be made of the identity set

$$S_{11} \equiv (C_{11} + C_{12})/C \qquad (4\text{-}18a)$$

$$S_{12} \equiv -C_{12}/C \qquad (4\text{-}18b)$$

$$S_{44} \equiv 1/C_{44} \qquad (4\text{-}18c)$$

where

$$C \equiv (C_{11} - C_{12})(C_{11} + 2C_{12}) \qquad (4\text{-}18d)$$

(and a similar set with the S_{ij} and C_{ij} interchanged). For the bulk modulus this leads to

$$K_V = K_R = K = (C_{11} + 2C_{12})/3 = C_{11} - 4C'/3 \qquad (4\text{-}19)$$

and for the shear modulus it leads to

$$5G_V = 2C' + 3C_{44} \qquad (4\text{-}20a)$$

and

$$\frac{5}{G_R} = \frac{2}{C'} + \frac{3}{C_{44}} \qquad (4\text{-}20b)$$

4.3.2 The VRHG Approximation—The Debye Temperature

The Debye temperature, θ_D, obtainable experimentally from the lattice contribution to the low-temperature specific heat, may be regarded as the ultimate embodiment of all the monocrystalline elastic moduli. ANDERSON [AND63] has shown how a reliable value of θ_D for an elastically isotropic solid can be calculated from the monocrystalline elastic moduli using the computed values of the macroscopic moduli as intermediate steps. Proceeding, by way of explanation, in the reverse direction:

$$\theta_D = \frac{h}{k_B}\left(\frac{3\mathcal{N}}{4\pi}\right)^{1/3}\left(\frac{\rho_d}{M}\right)^{1/3} v_m \qquad \text{(K)} \qquad (4\text{-}21)$$

where the product of the first two factors, consisting of atomic constants, is 2.514×10^{-3} in cgs units, ρ_d is the density (g cm^{-3}), M is the mean atomic weight of the alloy, and v_m is an average sound velocity (cm s^{-1}). This polycrystalline sound velocity may be calculated using

$$v_m = \left(\frac{1}{3}\left[\frac{2}{v_T^3} + \frac{1}{v_L^3}\right]\right)^{-1/3} \tag{4-22}$$

after the v_L and v_T, functions of K and G according to (4-9) and (4-10), have been calculated. If so, ANDERSON [AND63] suggested that the appellation VRHG be assigned to the resulting approximation for θ_D in recognition of the contributions that GILVARRY, in addition to the other authors mentioned, was stated to have made to the subject.

Equation (4-22) is, of course, not the only averaging procedure for v_m. According to recent studies by LEDBETTER, who discussed the problem of sound-velocity averaging in detail [LED80], it may in fact not be the best procedure. For Cu, for example, it gives a Debye temperature about 3% too high as compared with the calorimetric value. The discrepancy is expected to increase with increasing elastic anisotropy, A [LED82].

4.4 The Elastic Moduli of Titanium–Transition-Metal Alloys

Using techniques such as the pulse-superposition method and the pulse-echo-overlap method [KAT79], FISHER and colleagues measured the ultrasonic elastic constants of Ti–V [KAT79, KAT79a], Ti–Cr [FIS70, FIS70a], and Ti–Nb [REI73]. The results are summarized in Table 4-2. In some cases [KAT79], as a self-check on precision, the use of the five propagation modes listed in Table 4-3 enabled the three basic constants to be overdetermined.

The 25°C elastic constants of Ti–Cr alloys, whose quenched microstructures vary from $\omega+\beta$ to β with increasing e/a ratio, are plotted in Figure 4-2. The heavy lines connect data points resulting from tests performed on as-quenched alloys, whose structures are $\omega+\beta$ in the case of Ti–Cr (6.98 at.%) and β for all of the others. The finely delineated branches, and their dashed extrapolations, indicate the elastic constants of the single-phase-β structure. After extrapolation, the curves for C_{11} and C_{12} intersect at $e/a \cong 4.1$, which corresponds approximately to the composition below which the β-phase is absolutely unstable at room temperature [COL72] and transforms spontaneously to hcp—the martensitic transformation (see Section 1.7). The elastic-constant composition dependence of Ti–V, mentioned above, is qualitatively similar to that of Ti–Cr, but unfortunately the data do not extend into the $\omega+\beta$-regime. Figure 1-1 has shown that for numerous transition elements and alloys, C' appears to be a universal function of the

Table 4-2. Elastic Constants of Ti-V, Ti-Cr, and Ti-Nb Alloys[a]

Crystal	e/a Ratio	Condition	Elastic Constants 10^{10} N m^{-2} (10^{11} dyn cm^{-2}) C_{11}	C_{44}	C'	Anisotropy Ratio $A \equiv C_{44}/C'$	Molar Weight	Density ρ_d, g cm^{-3}	Derived VRH and VRHG Quantities: Young's Modulus, E 10^{10} N m^{-2}	Debye Temperature[b] θ_D, K
Ti-V(29.4 at.%)	4.29		14.002	3.966	2.026	1.96	48.79	4.929	8.34	328
Ti-V(38.5 at.%)	4.39		14.896	4.095	2.421	1.69	49.07	5.044	9.09	340
Ti-V(53 at.%)	4.53		16.760	4.129	3.125	1.32	49.51	5.328	10.09	355
Ti-V(73 at.%)	4.73		19.227	4.148	4.060	1.02	50.12	5.640	11.22	369
Ti-Cr(6.98 at.%)	4.14	Extrapolated from >900°C to 20°C	12.50	4.10	1.24	3.30	48.19	4.677	7.08	305
Ti-Cr(6.98 at.%)	4.14	Brine quench	15.59	5.54	3.67	1.51	48.19	4.677	12.29	409
Ti-Cr(9.36 at.%)	4.19	Brine quench	13.31	4.27	1.90	2.25	48.28	4.725	8.45	334
Ti-Cr(13.81 at.%)	4.28	Brine quench	13.99	4.42	2.18	2.02	48.47	4.834	9.08	345
Ti-Cr(28.37 at.%)	4.57	Brine quench	15.91	4.77	3.25	1.47	49.06	5.027	10.98	377
Ti-Nb(40.4 at.%)	4.40	Annealed 65 h/1700°C	15.65	3.963	2.247	1.76	66.08	6.189	8.74	291
Nb	5.00		24.74	2.80	5.69	0.49	92.91	8.578	10.45	268[c]

[a]Ti-V data — Katahara *et al.* [KAT79]
Ti-Cr data — Fisher and Dever [FIS70, FIS70[a]]
Ti-Nb data — Reid *et al.* [REI73]
Nb data — Fisher *et al.* [FIS75[a]]

[b]Calculated from Eq. (4-21) with the assistance, in turn, of Eqs. (4-22), (4-9b), (4-10), (4-15) and (4-16) and with $K_V = K_R$ given by Eq. (4-19) and G_V and G_R given by Eqs. (4-20a) and (4-20b). In this c.g.s. representation, the C_{ij} are in dyn cm^{-2}.

[c]cf. Table 8-3.

Table 4-3. Elastic Wave Propagation Modes Available for Use in Measuring the Elastic Constants of Cubic Crystals—After Katahara *et al.* [KAT79]

Mode Number	Direction of Wave Propagation	Direction of Particle Motion	Elastic Constant Derived
1	[001]	[001]	C_{11}
2	[001]	⊥[001]	C_{44}
3	[110]	[001]	C_{44}
4	[110]	[1$\bar{1}$0]	$C' \equiv (C_{11} - C_{12})/2$
5	[110]	[110]	$C_L \equiv (C_{11} + C_{12} + 2C_{44})/2$

e/a ratio. The importance of C' as an index of the bcc-lattice stability of transition metals has been discussed in detail by FISHER and DEVER [FIS70], according to whom the vanishing of C' seems to be responsible for the $\beta \rightarrow \alpha^m$ martensitic transformation in low-concentration Ti–TM alloys. It has also been claimed to be responsible for the fcc→fct and bcc(A15)→bct transformations exhibited by several A_3B-type intermetallic compounds [DEW79, p. 140].

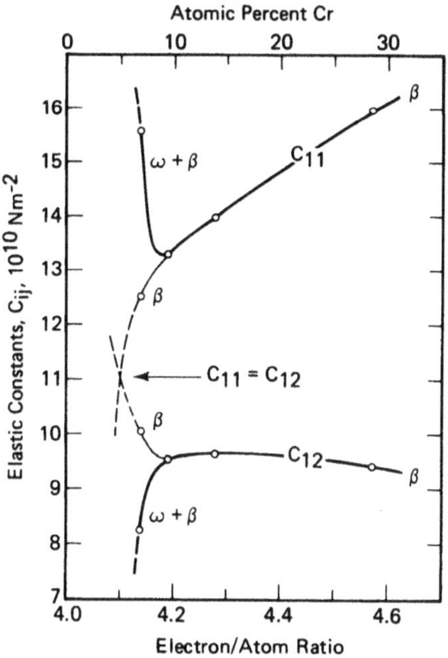

Figure 4-2. Elastic constants C_{11} and C_{12} of Ti-Cr. The heavy lines represent data for the quenched $\omega+\beta$- and β-phase alloys; the fine full lines represent extrapolations to room temperature of high-temperature β-phase data—after FISHER and DEVER [FIS70[a]]; see also COLLINGS *et al.* [COL72].

4.5 Systematic Variation of Elastic Moduli with Composition and Microstructure in Titanium–Transition-Metal Alloys

4.5.1 The β-Isomorphous Alloys Ti–V, Ti–Nb, and Ti–Mo

Using the long-wavelength transverse, longitudinal, and torsional resonances of alloy rods, FEDOTOV and colleagues [FED63, FED64, FED66, FED73] have measured, as functions of composition, the Young's moduli, E, and shear moduli, G, of quenched (from 24 h/900°C) and quenched-plus-aged (200 h/700°C + 500 h/600°C) Ti-V, Ti-Nb, and Ti-Mo alloys. Their results are summarized in Figure 4-3. The remarkable parallelism exhibited by the three pairs of curves is a result of the common responses of the microstructures of the three alloy systems to variation of the e/a ratio. The similarity between Figures 4-3(a) and 4-3(b) is a consequence of the proportionality of E to G as expressed by (4-2), for example. In discussing these figures reference is made to the known composition- or e/a-ratio-dependences of Ti-TM-alloy microstructure as described in Section 1.4.

Near the Ti-rich margin, the quenched alloys are martensitic—first α' and then α'' at higher solute levels (Table 1-6). The rapid softening noted in each case could be thought of as a consequence of the approach to the

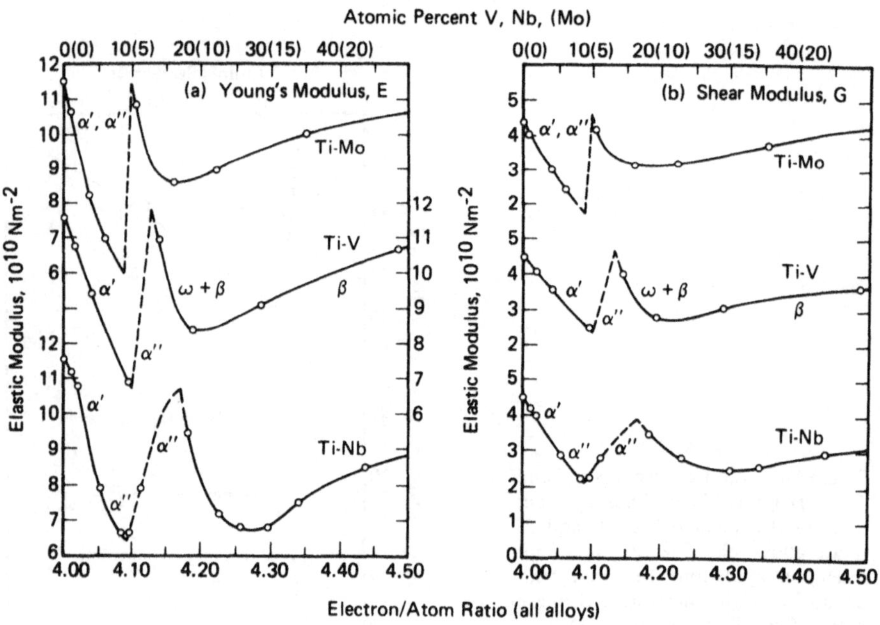

Figure 4-3. Elastic moduli *versus* e/a ratio for three β-quenched, β-isomorphous Ti–TM alloys; (a) Young's modulus, and (b) shear modulus—after FEDOTOV [FED73].

martensitic phase boundary.[4] As pointed out in Section 1.7, the boundaries between the hcp-α' and orthorhombic-α'' martensitic phases are 2 at.% Mo, 8.9 at.% V, and 5.7 at.% Nb. Thus, according to Figure 4-3(a), the α'/α'' transformation appears not to interrupt the martensitic-E *versus* e/a ratio curves for Ti–Mo and Ti–Nb, whereas the converse is true for Ti–V.

Transferring attention to the β-phase alloys on the right-hand side of the diagram, as the solute content is *decreased* the β-phase becomes continuously softer until the produce of that instability, ω-phase precipitation, eventually makes its presence felt by stiffening the lattice. According to BAGARIATSKII *et al.* [BAG59] (Table 1-4) the solute concentrations for which (athermal) ω-phase is formed on quenching are: 4.5 at.% Mo, 13 at.% V, and 18 at.% Nb, values which agree quite well with the positions of the E- and G-modulus peaks in Figures 4-3(a) and (b). The dashed lines represent the transitions from α^m to $\omega+\beta$.

Numerous other physical properties respond to the composition-controlled quenched microstructure in a manner similar to that depicted in Figure 4-3. Since, as pointed out above, the Debye temperature can be synthesized from the macroscopic elastic moduli, a parallelism between θ_D and E or G is expected and, according to Figure 4-4, is indeed observed. The behavior of the Vickers hardness curve also exemplifies the connection (albeit somewhat more tenuous) between hardness and strength or modulus. As solute content decreases in the bcc field, the composition dependence of the four parameters plotted in Figure 4-4 respond in comparable ways to the stiffening and hardening influences of ω-phase precipitation. On the α^m side only the hardness data, particularly at low solute concentrations, exhibit departures from parallelism in a manifestation of some kind of competition between solution strengthening and intrinsic lattice softening.

The influence of heat treatment on modulus is considered in Figure 4-5 with reference to the equilibrium and metastable-equilibrium phase diagrams of Ti–Nb, constructed from data sources referred to in the caption. The form of the E *versus* composition curve for the 24 h/900°C/WQ alloys is easily understood in the light of the occurrence of α'-martensite and ω-phase. After the water-quenched (WQ) alloys have been aged according to 200 h/700°C + 500 h/600°C, the new curve A–B–C–D develops. The segment A–B, which represents the equilibrium α-phase, naturally follows the "as-quenched" data; the segment C–D, which is in the 600°C equilibrium-β field also follows the old data; finally, B–C, for the equilibrium $\alpha+\beta$-field, is simply a "tie-line."

The temperature dependence data depicted in Figure 4-5(c) are particularly interesting. With a metastable alloy, and depending on the

[4] An isothermal approach to a phase boundary may be regarded as being equivalent to increasing the "homologous temperature" of the material.

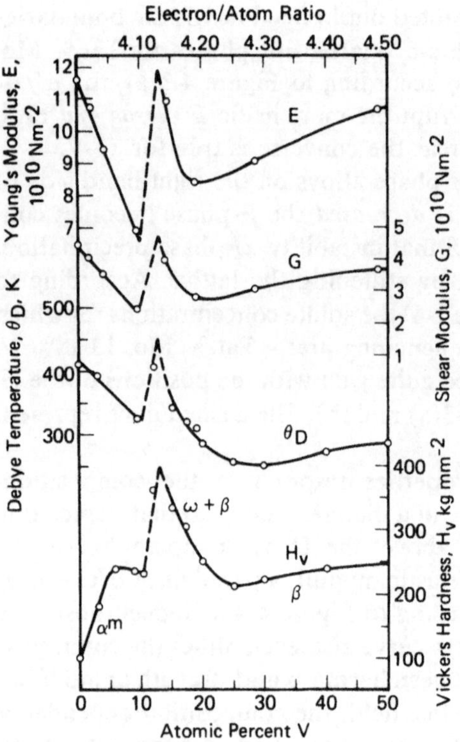

Figure 4-4. Comparison of the composition dependences of Young's modulus, E, shear modulus, G, Debye temperature, θ_D, and Vickers hardness, H_V, in quenched Ti-V alloys—after FEDOTOV (E and G) [FED73] (cf. Figure 4-3); COLLINGS *et al.* (θ_D) [COL75]; and COLLINGS (5-kg Vickers diamond pyramid hardness, H_V) [COL84] (cf. Figure 3-1).

decomposition kinetics, a temperature dependence experiment may also be a short-time aging experiment. For example, quenched Ti-Nb(18 at.%) possesses a high volume fraction of ω-phase; upon heating through 300°C solute diffusion becomes active and additional isothermal precipitation takes place resulting in a modulus peak centered about 400°C, an optimal temperature for isothermal ω-phase precipitation. At higher temperatures some α-precipitation commences, thereby enriching the matrix with Nb and lowering E; although within the 1 h that elapses as the temperature is raised from 400 to 800°C thermodynamic equilibrium is not achieved. The experiment illustrates the relatively rapid kinetics of the ω-phase reaction.

The second alloy, Ti-Nb(34 at.%), is in the $\alpha+\beta$-field at room temperature. But since the reaction kinetics at ordinary temperatures are quite sluggish (hence the need for the 500-h equilibration time at 600°C), the experiment actually measured the E-modulus temperature dependence of the β-phase. This turned out to be very small, in accordance with: (i) the results of ALBERT and PFEIFFER [ALB72] for both recrystallized (1 h/1000°C) and 95%-cold-worked Ti-Nb(34 and 44 at.%); (ii) the very slight (~1%) change in E modulus noted between 300 and 4.2 K for several technical alloys (Table 3-4).

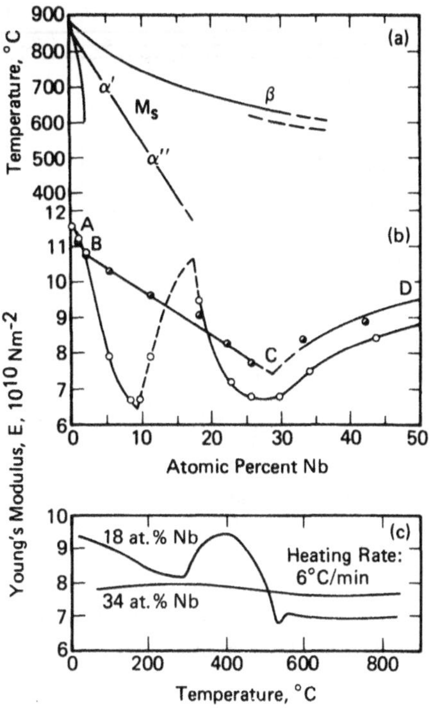

Figure 4-5. Influence of heat treatment on the Young's modulus of Ti–Nb: (a) equilibrium transi are from conventional sources (cf. Figure 1-7) and this study (*viz.*, the $\alpha+\beta/\beta$-transus near 600°C), the M_s line is due to JEPSON *et al.* [JEP70]; (b) Young's moduli of β-quenched (from 24 h/900°C), and quenched-plus-aged (100 h/800°C + 200 h/700°C + 500 h/600°C) alloys—after FEDOTOV [FED73] (see also FEDOTOV and BELOUSOV [FED64]); and (c) change of Young's modulus during heating at the rate of about 6°C/min—after FEDOTOV and BELOUSOV [FED64].

4.5.2 The β-Eutectoid Alloys: Ti–Cr, Ti–Mn, Ti–Fe, Ti–Co, and Ti–Ni

The elastic properties of the β-eutectoid alloys of Ti with first-row transition elements (see Section 1.2) have been studied by FEDOTOV and colleagues [FED73]. The modulus-composition dependences of alloys quenched from 1000°C are compared in Figure 4-6. As before, with decreasing solute content, they all exhibit a rapid increase in stiffness in the composition range where athermal ω-phase is expected.

With regard to the quenched-α^m regime, two effects are noteworthy: first, the Young's modulus composition dependences of the eutectoid alloys are generally much smaller than those noted above for the β-isomorphous alloys and second, an interesting trend is noted within the eutectoid series itself—particularly for the Ti–Fe, Ti–Co, and Ti–Ni trio. For these, a diminished composition dependence is believed to be due to departures from the ideal martensitic structure as a consequence of partial decomposition of either (i) the β-phase above M_s, or (ii) the martensitic phase below M_s during the quenching process, as a result of the very high diffusivities in Ti of Fe, Co, and Ni. The special positions occupied by these three elements with regard to diffusivity are shown in Figure 1-12. For what is presumed to be the same reason, dilute "martensitic" Ti–Fe, Ti–Co, and

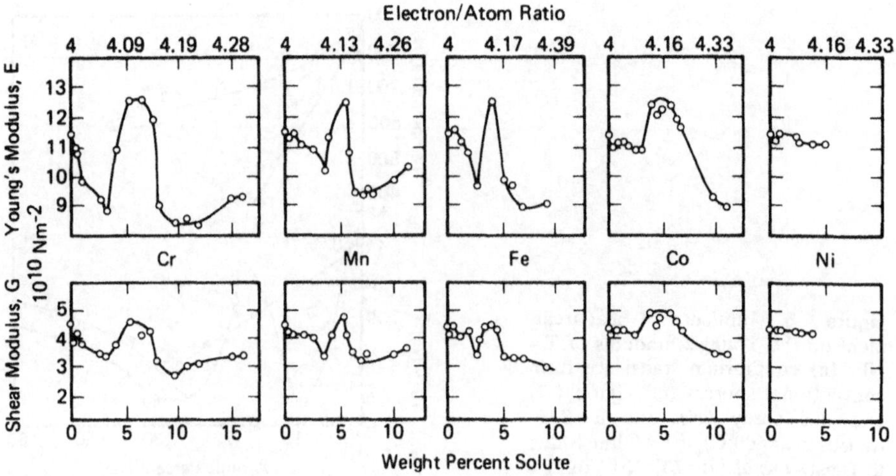

Figure 4-6. Composition dependences of the Young's and shear moduli of β-quenched, β-eutectoid alloys of Ti with the first-period transition elements Cr, Mn, Fe, Co, and Ni—after FEDOTOV [FED73].

Ti-Ni alloys exhibit anomalous calorimetrically detected superconducting transitions (see Section 10.3).

4.6 The Dynamic Modulus of Titanium–Niobium

Table 4-4 lists the results of known dynamic Young's modulus investigations of Ti-Nb alloys. The work of FEDOTOV and colleagues [FED63, FED64, FED66, FED73] focussed attention on the interesting $\alpha' \rightarrow \alpha'' \rightarrow \omega+\beta \rightarrow \beta$ microstructural range, while NEDYUKHA and CHERNYY [NED66] completed the data set across to unalloyed Nb. LEDBETTER [LED78], using a piezoelectrically excited rod-resonance technique in the 60–100 kHz frequency range, determined E and G for a series of six Ti-Nb(4.7–24.7 at.%) alloys and, with the aid of an ultrasonic (5–10 MHz) pulse-echo method, investigated the two technically interesting alloys Ti-Nb(35.7 and 38.7 at.%). Also of particular interest from a technical standpoint is the work of ALBERT and PFEIFFER [ALB72], who measured, using a rod-resonance technique (longitudinal oscillations at ~20 kHz), the Young's moduli of Ti-Nb(22, 34 and 44 at.%) in conditions relevant to those encountered in superconductive wires (*viz.*, 95% cold work (cw) and cw plus ($\lesssim$100 h)/($\lesssim$450°C)) and after recrystallization for 1 h/1000°C.

As shown in Table 4-4, *recrystallization* generally results in an increase in Young's modulus. That this is traceable to a sharpening of the ⟨110⟩ deformation texture can be demonstrated with the aid of the VOIGT approximation (Section 4.3.1), by means of which it can be proved that the

Table 4-4. Dynamic Young's Modulus Data for Ti–Nb Alloys

Nb Concentration at.%	Condition	Young's Modulus, E 10^{10} N m^{-2}	Literature	Symbol in Figure 4-7
0	24 h/900°C/WQ	11.7	Fedotov *et al.* [FED63, FED64, FED66]	○
1.1		11.2		
2.2		10.8		
5.4		7.9		
8.3		6.7		
9.5		6.7		
11.4		7.9		
18.1		9.4		
22.7		7.2		
25.6		6.8		
29.7		6.8		
34.0		7.5		
43.6		8.4		
42.5		9.1	Nedyukha and Chernyy [NED66]	□
49		9.4		
61		10.1		
75		10.4		
82		10.6		
91		10.5		
100		10.4		
22	95% cold work	6.3	Albert and Pfeiffer [ALB72]	▽
34		7.1		
44		8.1		
22	Recrystallized 1 h/1000°C	7.9	Albert and Pfeiffer [ALB72]	△
34		7.9		
44		8.7		
4.7		8.1	Ledbetter [LED78]	×
8.9		6.7		
14.3		10.0		
19.0		6.5		
19.4		6.4		
24.7		7.2		
35.7		7.4		
38.7		7.7		
40.4		8.74	Reid *et al.* [REI73]	◪
100		10.45	Fisher *et al.* [FIS75[a]]	◩

Table 4-5. Temperature Dependence of the Young's Modulus, *E*, of Ti–Nb—After Albert and Pfeiffer [ALB72]

Nb Concentration at.%	Condition	Change in *E* Between 25 and 200°C
22	95% cw	+3%
	Recrystallized	+2%
34	95% cw	+0.4%
	Recrystallized	+2%
44	95% cw	−0.2%
	Recrystallized	−0.1%

longitudinal ⟨110⟩ elastic constant, *viz.*, $C_L = (C_{11} + C_{12} + 2C_{44})/2$, is greater than $9KG/(3K + G)$, the random-polycrystalline Young's modulus (provided that $A > 1$).[5] ALBERT and PFEIFFER paid particular attention to the temperature dependences of the *E*-modulus within the temperature range 25–300°C. As indicated in Table 4-5 (see also Table 3-4), these were generally small. Any anomalies observed in the temperature dependences were attributed to lattice defects and/or precipitates, present as a result of the thermomechanical processing to which the specimens had been subjected.

The only single-crystal result to date is that of REID *et al.* [REI73] for Ti–Nb(40.4 at.%).

The results of all reported dynamic *E*-modulus experiments on Ti–Nb are assembled in Figure 4-7. In the "technically interesting" composition range of 30–40 at.% Nb, the results of FEDOTOV *et al.*, for alloys heat treated 24 h/900°C/WQ, lie midway between the two sets of data acquired by ALBERT and PFEIFFER on 1 h/1000°C heat treated and 95% cold worked

[5]As pointed out by LEDBETTER [LED82], for crystals of cubic symmetry, Young's modulus varies with direction according to

$$\frac{1}{E} = S_{11} - 2[(S_{11} - S_{12}) - \tfrac{1}{2}S_{44}](l_1^2 l_2^2 + l_2^2 l_3^2 + l_3^2 l_1^2)$$

where the l_i are direction cosines. Thus,

$$\frac{1}{E_{110}} = S_{11} - \tfrac{1}{2}(S_{11} - S_{12} - \tfrac{1}{2}S_{44})$$

so that E_{110} exceeds the VOIGT Young's modulus as calculated from (4-7) by way of (4-19) and (4-20a).

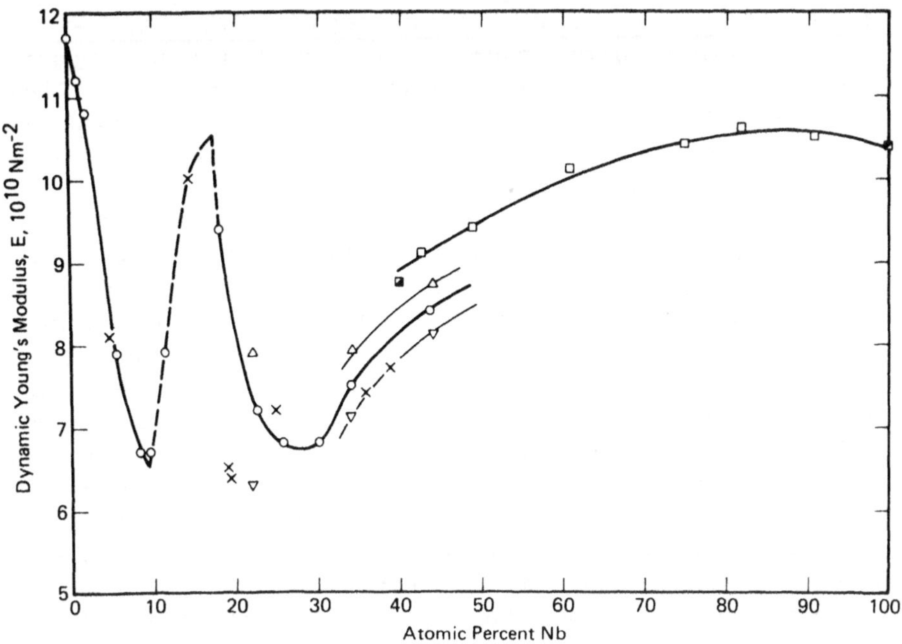

Figure 4-7. Composite plot of collected room-temperature dynamic-modulus data for Ti-Nb. Contributing to the figure are the elastomat data of FEDOTOV *et al.* [FED63, FED64, FED66, FED73] (○), "dynamic-impact" data of NEDYUKHA and CHERNYY [NED66] (□), the 20-kHz rod-resonance results of ALBERT and PFEIFFER [ALB72] (▽, △), the 60–100-kHz rod-resonance results of LEDBETTER [LED82] (×), and the ultrasonic monocrystalline data of REID *et al.* [REI73] (◪) and FISHER *et al.* [FIS75^a] (◩). Within the composition range 30–50 at.% the composite curve may be approximated by a straight line connecting the points (30, 7.1) and (50, 9.0); such lines have been plotted in Figures 3-16 and 3-30.

material, respectively. Although they lead to E-modulus values higher than all the other data, in the region where overlap occurs the results of NEDYUKHA and CHERNYY [NED66] are in excellent agreement with the moduli of Ti-Nb(40 at.%) and pure Nb as computed, using the VRH approximation, from the single-crystal elastic constant results of REID *et al.* [REI73] and FISHER *et al.* [FIS75^a], respectively. For use in the engineering-type calculations performed elsewhere (Section 3.19), an "average" E-modulus composition dependence based on the data of Figure 4-7 has been assigned. This is the line segment connecting the data-pairs (30 at.% Nb, 7.1×10^{10} N m^{-2}) and (50 at.% Nb, 9.0×10^{10} N m^{-2}).

4.7 The Dynamic Moduli of Composite Superconductors

Two interesting investigations of the elastic moduli of composite materials have been conducted at the National Bureau of Standards, in

Boulder, Colorado. One of these, herein designated "study A," was performed on small cylindrical samples ($\sim$0.4 cm$^{\varnothing}$ by $\sim$1.5 cm^{l}) removed from a fiberglass-insulated epoxy-impregnated Cu/Ti–Nb composite coil (24 cm O.D. $\times$ 19 cm I.D. $\times$ 10 cm^{l}); while the other, "study B," was conducted upon a rectangular bar (1 $\times$ 0.5 $\times$ 20 cm^3); i.e., a 20 cm length of Cu/Ti–Nb composite superconductor.

In study A [WES75], samples with the cylinder axis machined parallel to the filament direction were regarded as isotropic in the transverse direction—i.e., "orthotropically" symmetrical. As such their elastic properties could be described in terms of five independent moduli, indexable according to ij = 11, 33, 44, 12, and 13. The moduli were measured by exciting resonant 50 and 100 kHz longitudinal and torsional oscillations in samples cut with their cylinder-axes parallel, and at angles of 50° and 90°, to the fiber direction. The results of the measurements are listed, for comparison with those of study B [LED77], in Table 4-6. Containing as it does 25% volume fraction of epoxy, the coil composite possesses relatively low elastic stiffnesses.

Study B [LED77] describes an ultrasonic (5 and 10 MHz) investigation of a Cu/Ti–Nb monolithic composite. Acting on the prior assumption that the material would exhibit orthorhombic symmetry about each of three mutually perpendicular twofold axes, the superconductive strand direction being one of them, LEDBETTER and READ [LED77] set out to cut samples that would be suitable for the measurement of nine independent stiffness moduli. Their experiment, which of course involved the propagation of acoustic waves parallel, normal, and at 45° to the fiber direction, is discussed in detail in the original article. The results, summarized in Table 4-6 for

Table 4-6. Elastic Stiffness Moduli, C_{ij}, 10^{10} N m^{-2}, for a Representative Coil Composite (Weston [WES75]) and a Representative Conductor Composite (Ledbetter [LED77]) at Room Temperature

Index ij	Coil Composite (Orthotropic)	Conductor Composite (Orthorhombic)
11	3.82	18.34
22		18.78
33	9.35	18.80
12		10.56
13	1.77	10.39
23		10.41
44	0.74	4.16
55		4.07
66	1.22	4.14

comparison with data on the coil composite referred to above, turned out to agree reasonably well with the static data of SUN and GRAY [SUN75], and also indicated that the monolithic composite could be reasonably well described in terms of cubic symmetry for which only the three moduli defined by $ij = 11$, 12, and 44 would be needed.

Chapter 5

Electrical Resistivity

5.1 Electrical Resistivity of Titanium-Alloy Superconductors

In studies of the physics of metals and alloys a wealth of useful information has been obtained from properly planned and interpreted electrical resistivity measurements. One of the most easily measurable of the physical properties, electrical resistivity can be determined to a precision of 10^{-6} [Riz74] and in carefully prepared specimens to an accuracy of typically ±0.2%, most of the error being associated with cross-sectional area determination [Col74]. Low-temperature resistivity, or relative-resistivity, measurements have occasionally been used to determine the superconducting transition temperature, T_c, and, of course, to study flux-flow resistivity (e.g., [Kim65]); and critical current density, J_c, measurements have generally made use of the four-probe experimental arrangement.

In association with J_c investigations, the upper critical field has frequently been determined resistometrically. For example, the resistive critical fields, H_r, of the alloy systems Ti–V, Ti–Nb, Ti–Ta, Ti–Mo, and several others have been measured at a current density of 10 A cm^{-2} in a classical series of experiments by Berlincourt and Hake [Ber63]. J_c is, of course, a function of the applied magnetic field, H_a; conversely, H_r is sensitive to the choice of measuring current density [Col66], [Haw80], which should, therefore, always be specified as a condition of the measurement. Not only is resistivity a valuable aid to critical field determination, normal-state resistivity is also an ingredient of a *calculated* H_{c2}. Following the development of Ginzburg–Landau theory, as discussed in Chapter 14, the nonparamagnetically limited zero-K upper critical field, H^*_{c20}, was able to be expressed in the Gor'kov–Goodman–Maki–Eilenberger form

$$H^*_{c20} = 3.06 \times 10^4 \rho_n \gamma_v T_c \qquad \text{(Oe)} \qquad (5\text{-}1)$$

where ρ_n is the residual resistivity of the alloy (in Ω cm), as measured either

Table 5-1. Electrical Resistivities of

Alloy or Alloy System	Variable: Composition	Variable: Temperature	Variable: Aging Time and Temperature	Data Presentation: Numerical	Data Presentation: Graphical
Comparison of Simple Metal and Transition Metal Solutes					
Ti-Al(0-8 at.%)	×				×
Ti-Zr, -Nb					
Ti-Al, -Ga, -Si, -Ge, -Sn, or -Bi	×				×
Ti-V					
Ti-Al, -Ga, -Ge, or -Sn	×			×	×
$Ti-Mo_{25}$-Al, -Ga, -Ge, or -Sn					
Binary Transition Metal Alloys					
Ti with Group III					
Ti-Sc					
Ti-La					
Ti with Group V (excluding Nb)					
Ti-V					
Ti-V(60-100 at.%)	×			×	
Ti-V(15-96 at.%)	×			×	×
Ti-V(0-100 at.%)	×	×		×	×
Ti-V(20-100 at.%)	×	×			×
Ti-V(0-100 at.%)	×	×			×
Ti-Ta					
Ti-Ta(10-90 at.%)	×			×	×
Ti-Ta(30-90 at.%)	×				×
Ti-Ta(~20-100 at.%)	×				
Ti with Group VI					
Ti-Cr					
Ti-Cr(13 at.%)			×		×
Ti-Cr(10, 13, 15, 20 at.%)		×	×	×	×
Ti-Mo					
Ti-Mo(12-30 wt.%)		×			×

Titanium Alloys—Relevant Data Sources

Comments (Results, Discussion)	Literature
Scattering strengths of simple-metal and transition-metal additions to α-Ti compared	Collings *et al.* [COL71[a]]
Same as above	Collings and Gegel [COL75[d]]
Scattering strengths of simple-metal additions to α-Ti and β-Ti-Mo compared	Collings and Gegel [COL75[d]] Gegel *et al.* [GEG73]
See Alloy Series C below	
See Alloy Series B below	
See Alloy Series C below	
ρ_n measured just above T_c	El Bindari and Litvak [ELB63[a]]
ρ_n at 1.2 K in as-cast alloys studied as ingredient of H_{c2}^*	Berlincourt and Hake [BER63]
Study of anomalous ρ *versus* c and anomalous (negative) $d\rho/dT$	Collings [COL74]
Study of anomalous ρ *versus* c and anomalous (negative) $d\rho/dT$	Prekul *et al.* [PRE74]
Reversible anomalous $d\rho/dT$ compared with reversible anomalous magnetic susceptibility (χ) temperature dependence	Collings [COL78]
See Alloy Series B below	
ρ_n at 1.2 K in as-cast alloys studied as ingredient of H_{c2}^*	Berlincourt and Hake [BER63]
ρ_n at 4.2 K for cold-worked alloys	Colling *et al.* [COL66]
Anomalous concentration dependence of residual resistivity	Prekul *et al.* [PRE74]
See Alloy Series A, B and C below	
$\rho_{77\,K}$ *versus* aging time (up to 1 h) at 300°C of previously heat-treated (1 h/1000°C/WQ) alloy	Luhman [LUH70]
A study of negative $d\rho/dT$ discussed with reference to a review of the literature of negative $d\rho/dT$ in Ti-TM and Zr-Nb alloys	Chandrasekaran *et al.* [CHA74] see also [CHA73]
See Alloy Series B and D below	
A study of negative $d\rho/dT$	Yoshida and Tsuya [YOS56]

Continued

Table 5-1.

Alloy or Alloy System	Variable: Composition	Variable: Temperature	Variable: Aging Time and Temperature	Data Presentation: Numerical	Data Presentation: Graphical
Ti–Mo(6–50 at.%)	×			×	×
Ti–Mo(5–20 at.%)	×	×	×	×	×
Ti with Group VII					
Ti–Mn					
Ti–Re					
Ti with Group VIII					
Ti–Fe					
Ti–Fe(4, 9.5 at.%)			×	×	×
Ti–Fe(1.5–20 at.%)	×	×		×	×
Ti–Ru					
Ti–Os					
Ti with Group IX					
Ti–Co					
Ti with Group X					
Ti–Ni					
Ti–Pd					
Alloy Series A					
Ti–{Cr, Mn, Fe, Co, Ni}	×			×	×
Alloy Series B					
Ti–{La, Ta, Cr, Mo, Re, Pd}		×			×

(cont.)

Comments (Results, Discussion)	Literature
ρ_n at 1.2 K in as-cast alloys studied as ingredient of H^*_{c2}	Berlincourt and Hake [BER63]
Study of anomalous ρ *versus* c and anomalous (negative) $d\rho/dT$	Ho and Collings [HO72]
See Alloy Series A, C and D below	
See Alloy Series B below	
See Alloy Series A, C and D below	
$R(T)/R(77\text{ K})$ *versus* aging time at 400 and 500°C (T = 193, 273, 373 K)	Nishimura and Zwicker [NIS68]
A study of resistivity minima in low-concentration alloys, the negative $d\rho/dT$ effect in Ti-Fe(6.5-12 at.%), and the transition to the superconductive state	Prekul *et al.* [PRE76]
See Alloy Series D below	
See Alloy Series A and C below	
See Alloy Series A and C below	
See Alloy Series B below	
ρ(4.2 K, 273 K) and ρ(2-24 K) for Ti-TM alloys containing about 1 at.% of TM additions in a search for low-temperature resistance anomalies	Hake *et al.* [HAK62]
Temperature dependence of resistivity (−200 to $\gtrsim$1000°C) for alloys with solute concentrations of generally only a few percent	Ul'yanov and Kovtun [ULY64]

Continued

Table 5-1.

Alloy or Alloy System	Variable			Data Presentation	
	Composition	Temperature	Aging Time and Temperature	Numerical	Graphical
Alloy Series C					
Sc		×		×	
V					
Cr					
Ti– Mn					
Fe					
Co					
Ni					
Alloy Series D					
(V + Cr)					
Mo					×
Ti– Mn					
Fe					
Ru					
Os					
Ti–Nb Alloys					
Ti–Nb(20–100 at.%)	×	×			×
Ti–Nb(10–95 at.%)	×			×	×
Ti–Nb(22 at.%)			×	×	
Ti–Nb(22 at.%)			×		×
Ti–Nb(25 at.%)			×		×
Ti–Nb(20–50 at.%)			×		×
Ti–Nb(4, 9, 20, 25 at.%)			×	×	×
Ti–Nb(50, 60, 80 wt.%)			×		×
Ti–Nb(0–100 at.%)	×				×
Ti–Nb(0–50 at.%)	×				×

(*cont.*)

Comments (Results, Discussion)	Literature
$\rho_{300\,K}$, $\rho_{300\,K}/\rho_{77\,K}$ and $\rho_{300\,K}/\rho_{4.2\,K}$ for dilute alloys (solute concentrations ~1 at.%)	Agarwal [AGA74]
Discussion of anomalous resistivity temperature dependence $\alpha^T = (\partial\rho/\partial T)_H/\rho$ (one intermediate-concentration alloy each system)	Hake *et al.* [HAK75]
Emphasis on negative $d\rho/dT$	Ames and McQuillan [AME54]
ρ_n at 1.2 K in as-cast alloys studied as ingredient of H^*_{c2}	Berlincourt and Hake [BER63]
ρ_{RT} as-drawn, and ρ_{RT} after heat treatment at 200–650°C	Salter [SAL64]
Relative resistivity (post-age/pre-age) plotted isochronally *versus* aging temperature	Salter [SAL66]
Specific resistivity as function of annealing temperature; various starting conditions	Baron *et al.* [BAR70]
Specific resistivity as function of annealing temperature and aging temperature	Bychkova *et al.* [BYC70]
$R(T)/R(77\text{ K})$ *versus* aging time at 400 and 500°C (T = 193, 273, and 373 K)	Nishimura and Zwicker [NIS68]
Relative resistivity *versus* 1-h aging at temperatures up to 750°C	Rauch [RAU68]
Specific resistivity at 77 and 300 K	Ricketts [RIC69]
Specific resistivity of alloy series in the conditions: as-cast cold-deformed 6 h/900°C/WQ	Zwicker *et al.* [ZWI70]

Continued

Table 5-1.

Alloy or Alloy System	Variable: Composition	Variable: Temperature	Variable: Aging Time and Temperature	Data Presentation: Numerical	Data Presentation: Graphical
Ti–Nb(33–60 wt.%)	×				×
Ti–Nb(30, 55 at.%)		×	×	×	
Ti–Nb(22 at.%)			×	×	
Ti–Nb(35 at.%)			×		×
Ti-Nb(25–100 at.%)	×	×			×
Ti–Nb(33 at.%)		×			×
Ti–Nb(45 at.%)				×	
Ti–Nb(50, 56 wt.%)		×			×
Ti–Nb(22, 32, 35, 38, 45 at.%)	×	×			×
Ti–Nb(22 at.%)		×			×
Doped Ti–Nb Alloys					
Ti–40Nb–{$Sc_{0.65,\ 0.82}$ / $Er_{0.66,\ 0.72,\ 1.33}$}			×	×	×
Ti–Nb_{33}–$Cu_{3,\ 5}$			×		×
Ti–(44–49)Nb–(0.6–4.7)Ge			×		×
Ternary Transition Metal Alloys					
Ti–V–Ta (wide range)	×				×

(*cont.*)

Comments (Results, Discussion)	Literature
Specific resistivity of alloy series in the conditions: 97% cold worked; 97% cold worked plus 6 h/900°C/WQ	Heller [HEL71]
$\rho_{77,300\,K}$ as function of aging temperature (200–500°C). Prior condition: generally 1 h/250°C; small measured amounts of Fe and O	Easton *et al.* [EAS71]
$\rho_{300\,K}$ in the as-drawn condition; $\rho_{11,300\,K}$ after precipitation-heat-treatment	McInturff and Chase [MCI73]
Resistivity as-received and after 1 h/400°C, also as function of heat-treatment time at 350 and 750°C in vacuum (10^{-4} torr) and argon	Witcomb and Dew-Hughes [WIT73]
Anomalous composition-dependence; anomalous $d\rho/dT$	Prekul *et al.* [PRE74]
Study of low-temperature $\rho(T)$ as a companion to thermal conductivity investigation	Morton *et al.* [MOR75]
$\rho_{11,RT}$ in as-swaged and heat-treated condition as companion to thermal conductivity investigation	Ikebe *et al.* [IKE77]
Resistivity temperature dependence in range 77–300 K	Feher *et al.* [FEH77]
Electrical resistivities studied as functions of temperature and concentration	Liedtke [LIE81]
$\rho(T)$ measured with the alloy as-quenched (negative $d\rho/dT$) and after tensile deformation ($\beta\rightarrow\alpha''$; $d\rho/dT$ positive). Resistivity measured during the execution of a stress-strain cycle	Hochstuhl and Obst [HOC82] see also Hochstuhl [HOC81]
Tabulated values of $\rho_{77,300\,K}$; relative resistivity *versus* 1-h aging at temperatures up to 600°C. O-levels of doped samples 2900 to 4500 ppm	Ricketts [RIC69]
$\rho_{77\,K}$ and $\rho_{20\,K}/\rho_{77\,K}$ *versus* aging time to 50 h/400°C of previously deformed wire	Löhberg [LOH71]
Specific electrical resistivity of quenched alloys as function of aging time at 400°C	Heller [HEL71]
Resistivity at 12 and 77 K plotted on composition triangle	Savitskii *et al.* [SAV71[a]]

Continued

Table 5-1.

Alloy or Alloy System	Variable: Composition	Variable: Temperature	Variable: Aging Time and Temperature	Data Presentation: Numerical	Data Presentation: Graphical
Ti-Nb-{Hf, Mo, Re}	×			×	
Ti-{$Nb_{4.9}$-$Fe_{8.2}$, Nb_{10}-Fe_4, Nb_{25}-Fe_3}			×	×	×
Ti-Zr-Nb and Ti-Zr-Nb-Base Alloys					
Ti_{0-77}-Zr_{0-46}-Nb_{18-99}	×				×
Ti_{18-74}-Zr_{7-31}-Nb_{19-64}			×		×
$Ti_{4,7}$-$Zr_{41,43}$-$Nb_{55,\,50}$			×		×
Ti-7.5Zr-(30, 48)Nb			×		×
Ti-10Zr-65Nb		×			×
Soviet Alloy 65 BT				×	
Soviet Alloy 65 BT				×	
Ti_{24-65}-Zr_{4-10}-Nb_{5-35}-Ta_{5-10}	×				×
Ti_{33}-Zr_{30}-Nb_{30}-Ta_7			×		×

at $T > T_c$, or at $T < T_c$ but with the superconductivity suppressed magnetically, and γ_v is the electronic specific heat coefficient (erg cm^{-3} K^{-2}). Frequent use has been made of ρ_n, in association with (5-1), for estimating the upper critical field from "normal-state" parameters [Ber63], [Bra72], [Hor73].

Table 5-1 is a listing of Ti-alloy-resistivity data sources, systematized, for the most part, according to the position in the periodic table occupied by the solute elements.

(*cont.*)

Comments (Results, Discussion)	Literature
Specific resistivity data for a few compositions in each system	Bychkova *et al.* [BYC70[a]]
$R(T)/R(77\ \text{K})$ *versus* aging time at 400 and 500°C ($T = 193, 273, 373$ K)	Nishimura and Zwicker [NIS68]
$\rho_{77,\,293\ \text{K}}$ for 0.25 mm$^{\varnothing}$ wire plotted on composition triangle	Doi *et al.* [DOI66]
$\rho(T)/\rho(77\ \text{K})$ for various metallurgical conditions such as: annealed/quenched/deformed as-cast/deformed, etc. temperature = −200 to +100°C	Zwicker *et al.* [ZWI68]
Relative resistivity used to monitor structural decomposition in T-T-T-diagram construction	Soeno and Kuroda [SOE69]
Specific resistivity as function of annealing temperature after quenching or cold deformation	Kadykova [KAD71]
Resistivity temperature dependence (77-300 K)	Feher *et al.* [FEH77]
Specific electrical resistivity at 20, 77 and 293 K	Alekseevskii *et al.* [ALE67]
Specific electrical resistivity at 20 and 300 K	Gorina *et al.* [GOR69]
Residual resistivity determined as an ingredient of the calculated H_{c2}	Horiuchi *et al.* [HOR73]
Resistivity used as an indicator of phase transformation in T-T-T-diagram construction	Horiuchi *et al.* [HOR73[a]]

5.2 Resistometrically Monitored Transformation and Aging

In metallurgical-property diagnosis, measurements of room-temperature resistivity, in association with appropriate microstructural studies, have frequently been used to monitor the progress of solid-state reactions during the aging of metastable structures. Some Ti-TM systems for which data are available are: Ti-Cr [LUH70], [CHA74], Ti-Fe [NIS68], Ti-Nb [WIT73], Ti-Nb-Cu [LOH71], and Ti-Nb-Ge [HEL71]. The resistivity changes that occur during aging, in response to the generally complicated metallurgical processes that take place, must, of course, be "calibrated" from time to time by means of direct microstructural evaluation. Typical effects expected in a Ti-Nb alloy may include: (i) precipitation of α-phase with an accompanying Nb enrichment of the β-matrix [HEL71, p. 62], which, according to the $\rho(c)$ curves to be considered below (Figure 5-1), would result in a decrease in matrix resistivity; (ii) the diffusion of O and rare earths, if they are present, to the newly-forming Ti-rich α-precipitates, again contributing to a reduction in the resistivity of the matrix [RIC69].

In a related class of investigation, the effects of temperature (usually up to about 600°C) on the results of isochronal aging (frequently for 1 h) have been studied by several workers. Numerous such tests have been carried out on Ti-Nb alloys [SAL66], [RAU68], [BAR70], [BYC70], [EAS71], some of which were doped with O or rare earth elements [RIC69]. The results for both doped [RIC69] and undoped [SAL66], [RIC69] Ti-Nb alloys agree in revealing a rather sharp minimum, formed by the dropping of the relative resistivity to about 0.6 within the temperature range of 400-500°C [COL83, p. 224]. This drop in resistivity in the temperature range of the minimum is attributable to the precipitation effects described above. Its subsequent increase, to form the high-temperature boundary of the minimum, is attributable to the dissolving of the α-phase and a redistribution of the interstitials upon the alloy's entry into the single-phase-β field.

Finally, an important and useful semiquantitative application of relative-resistivity monitoring has to do with the construction of time-temperature-transformation (T-T-T) diagrams. Examples are those developed by SOENO and KURODA [SOE69] for $Ti_{4.4}$-$Zr_{40.7}$-$Nb_{54.9}$ (cold drawn 0, 25, 50, 75, and 99%—see Figure 2-14), and HORIUCHI *et al.* [HOR73[a]] for Ti_{33}-Zr_{30}-Nb_{30}-Ta_7 (cold-worked 90 and 99%—see Figure 2-15). In the former investigation, for example, the relative resistivity was measured as a function of time ($\lesssim$100 h) during isothermal annealing at

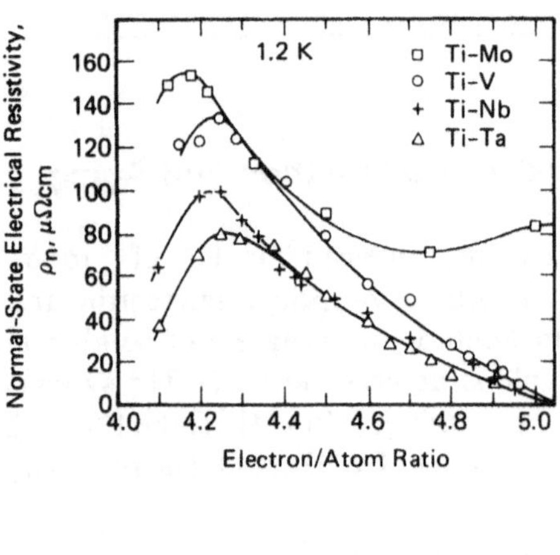

Figure 5-1. Resistivity composition dependences of Ti-TM alloys. Normal-state electrical resistivities at 1.2 K of representatives of the alloy systems Ti-V, -Nb, -Ta and -Mo (wires or arc-melted and cold-rolled strips) plotted *versus* e/a ratio (see Figure 4-3 for alloy phase information)—after BERLINCOURT and HAKE [BER63] (see [PRE74] for further data on Ti-V, -Nb and -Ta).

temperatures between 650 and 800°C. For each sample (i.e., each starting condition), T-T-T diagrams were constructed representative of consecutive stages (designated $f_R = 0.25$, 0.5, and 0.75 where f_R is the fractional change in resistivity) in the β-phase-decomposition process.

5.3 The Resistivity of Alloys—Composition Dependence

5.3.1 Simple Models of Alloy Resistivity

The electrical resistivity of an alloy (especially at low solute concentrations) is frequently expressed in the form

$$\rho_{\text{total}}(c, T) = \rho_i(T) + \rho_s(c) \tag{5-2a}$$

where ρ_i is the "ideal" resistivity of the host (the temperature-dependent component of ρ_{total}) and ρ_s is the resistive contribution of "impurities," such as precipitates or solute atoms, etc. Under Matthiessen's rule, ρ_i is solute independent, the only result of changing solute species or concentration being a vertical displacement of the $\rho_{\text{total}}(T)$ curve. In the more likely event that the presence of solute perturbs both the phonon spectrum *and* the electronic band structure of the host, ρ_i itself becomes solute dependent and the equation must be written as

$$\rho_{\text{total}}(c, T) = \rho_i(c, T) + \rho_s(c), \tag{5-2b}$$

where c represents the solute concentration in, say, at.%. In some alloy systems, especially Ti–TM systems in certain e/a ratio ranges, the modified phonon spectrum is related to lattice stability. Then whether or not impurity scattering, ρ_s, dominates the total resistivity, will depend on the relative strength of the electron-impurity scattering (which is then expressible in terms of a "specific solute resistivity" or specific resistivity per at.% solute) as compared to the influence of solute on the intrinsic resistivity of the host.

Mott and Jones [Mot58, p. 300], in discussing the resistivities of alloys containing transition elements, pointed out that the scattering of the conduction s electrons into d states, the cross section for which is proportional to the Fermi density of d states, $n^d(E_F)$, could make an important contribution to the resistivity. In accordance with this, Chandrasekhar and Hulm [Cha58] suggested that $\rho_s(c)$, for an alloy in which this so-called s-d scattering is the dominant mechanism, should be expressible in the form

$$\rho_s(c) \propto n^d(E_F)c(100 - c) \tag{5-3a}$$

Then if $n^d(E_F)$ is able to be approximately equated to the total $(s + d)$

density of states, $n(E_F)$, which in turn is proportional to the electronic specific heat coefficient, γ (see Section 8.1), the impurity resistivity may be written as

$$\rho_s(c) \underset{\sim}{\propto} n(E_F)c(100 - c) \tag{5-3b}$$

$$\propto \gamma c(100 - c) \tag{5-3c}$$

Thus, in a binary transition-metal alloy in which γ *versus* c is roughly parabolic in shape, it would be expected that ρ_s *versus* c would also tend to maximize near the middle of the concentration range. The term "anomalous composition dependence" is used herein to signify departures from this simple picture. As we shall see, the resistivity composition dependences of binary transition-metal alloys are generally anomalous [Ho72].

5.3.2 Residual Resistivities of Binary Transition-Metal Alloys

Figure 5-1, from the work of BERLINCOURT and HAKE [BER63], intercompares the normal-state resistivities, $\rho_{n,1.2\,K}$, of the alloy systems Ti–V, Ti–Nb, Ti–Ta, and Ti–Mo. Other systems studied at the time, but not represented in the figure, were Nb–Zr, Nb–Hf, Nb–U, Hf–Ta, and U–Mo. The curves, which are in general agreement with the earlier results of

Table 5-2. Residual Resistivities of Ti–Nb Alloys[a] at 1.2 K—After Berlincourt and Hake [BER63]

Nb Concentration at.%	Thickness Reduction Ratio[b]	Residual Resistivity ρ_n, $\mu\Omega$ cm
10.0[c]	9.7	63.8
20.0[c]	14.3	97.2
25.0[c]	13.6	98.5
30.0	215.0	79.4
30.0	wire	85.7
34.1	wire	78.3
38.7	wire	63.0
42.5	193.0	58.7
43.7	wire	55.8
52.0	wire	49.3
60.0	269.0	42.2
70.0	15.6	30.6
85.0	17.3	19.0
90.0	45.3	12.0
95.0	56.4	6.8

[a] Condition: arc melted and cold worked.
[b] Reduction by cold rolling; thickness reduction ratio ≅ARR.
[c] Mainly bcc but contained some second phase.

ADENSTEDT *et al.* [ADE52] for Ti-V, AMES and MCQUILLAN [AME54] for Ti-Nb, and the subsequently acquired results of PREKUL *et al.* [PRE74] for Ti-V, Ti-Nb, and Ti-Ta, indicate resistivities that rise monotonically as the solute concentrations *decrease* on their way to low values of e/a. In so doing, they are in sharp contrast to the predictions of (5-3). The maxima, which eventually develop [PRE74] when data are taken to sufficiently small solute concentrations, occur at e/a ratios of roughly 4.2–4.3 in alloys of Ti with the group-V transition elements. In arguments to be presented below, with particular reference to Ti-V [COL74] and Ti-Mo [HO72] alloys, the anomalously high residual resistivities and the displaced resistivity maxima will be identified with electron scattering associated with ω-phase precipitation [COL78].

The 1.2 K residual resistivities of Ti-Nb alloys are listed in Table 5-2.

5.3.3 Relative Scattering Strengths of Simple-Metal and Transition-Metal Solutes in Ti

It has been emphasized repeatedly by COLLINGS and coworkers [COL73, COL73[b]], [GEG73], that Ti-SM and Ti-TM represent two distinct classes of Ti-base binary alloys, the former being distinguished by relatively high solution-strengthening rates, large *negative* values of the regular-solution interaction coefficient, and finally, as indicated in Figure 5-2, by relatively large values of specific solute resistivity. For several reasons, not the least among which has to do with the relationship between specific resistivity and rates of solution strengthening [COL73, COL75[d]], [GEG73],

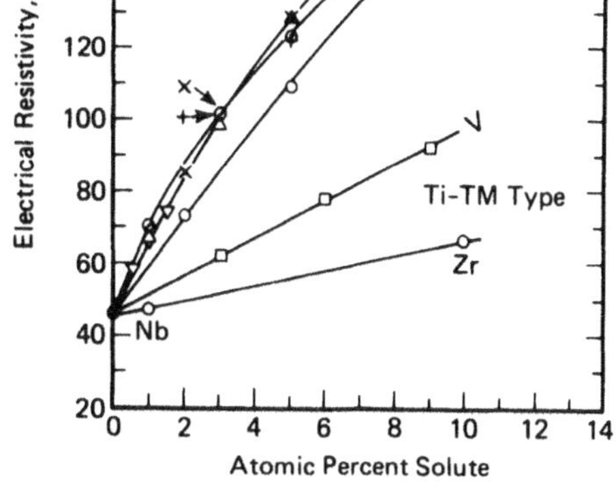

Figure 5-2. Resistivity composition dependences of Ti-SM alloys. A comparison is also made between various low-concentration Ti-SM and Ti-TM alloys. Alloys with Sn (△), Ga (∅), and Al (○) were measured as-cast, those with Ge (×), Bi (+), Si (▽), and V (□) had been annealed 1 h/1000°C and quenched—after COLLINGS and GEGEL [COL73, COL75[d]].

Table 5-3. Specific Solute Resistivities of Simple-Metal Additions to α-Ti and β-Ti-Mo(25 at.%)—After Collings and Gegel [COL75[d]], [GEG73]

Alloy System	Specific Solute Resistivity $\mu\Omega$ cm per at.%
Ti-Al	1220
Ti-Ga	1918
Ti-Ge	2254
Ti-Sn	2093
$Ti-Mo_{25}-Al$	330
$Ti-Mo_{25}-Ga$	400
$Ti-Mo_{25}-Ge$	529
$Ti-Mo_{25}-Sn$	770

it is interesting to inquire into the change that takes place in the specific resistivity of a simple-metal solute during its transfer from α-Ti to a β-stabilized Ti-TM alloy. The selected solute is Ti-Mo(25 at.%), a bcc alloy whose composition falls outside of the range where ω-phase forms or where the lattice is unstable. The results of the resistivity comparison [GEG73], [COL75[d]] are summarized in Table 5-3, which reveals that simple metals when dissolved in β-Ti-Mo(25 at.%) are on the average four-times weaker in electron scattering strength than they are when α-Ti is the host lattice. The reason for this will become clearer during a discussion, to be presented below, of anomalous resistivity temperature dependences in strong-scattering concentrated binary alloys. For the time being it is sufficient to note that since the mean-free-path in the comparable alloy Ti-Mo(16 at.%) is already only 5 Å [HAK67[b]], any further lowering of the mean-free-path by simple-metal additions will tend to be much less severe than would be the case if the same solutes were added to unalloyed α-Ti. The specific resistivities at room temperature of Ti-Mo(15, 16, and 20 at.%), according to the measurements by HO and COLLINGS [HO72], are 120, 117, and 118 $\mu\Omega$ cm, respectively, and that of Ti-Mo(25 at.%) is estimated to be even higher: $\sim$140 $\mu\Omega$ cm [HO72].

5.4 The Resistivity of Alloys—Temperature Dependence

5.4.1 Dilute Alloys at Low Temperatures

HAKE *et al.* [HAK62], in studying the low-temperature transport properties of dilute ($\sim$1 at.%) alloys of Ti with Cr, Mn, Fe, Co, and Ni (Figure

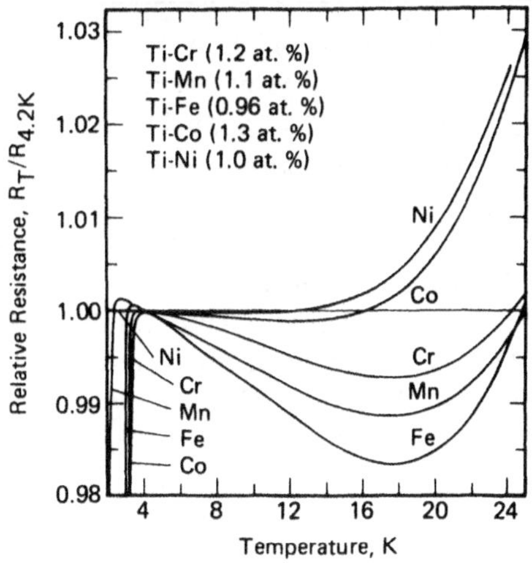

Figure 5-3. Relative resistivity temperature dependences, $\rho(T)/\rho(4.2\text{ K})$, of dilute (i.e., ~1 at.%) Ti-TM alloys. Resistance minima are exhibited by only Ti-Cr, Ti-Mn and Ti-Fe—after HAKE *et al.* [HAK62].

5-3) noted the occurrence of pronounced low-temperature minima in the electrical resistivities of Ti-Cr, Ti-Mn, and Ti-Fe, but not in those of any of the others. Of these three alloys, only Ti-Mn exhibited negative magnetoresistivity.

Magnetic susceptibility experiments by CAPE [CAP63] showed that whereas dilute Ti-Mn possessed almost ideal Curie paramagnetism ($\chi = C_m/T$), Ti-Fe(0.96 at.%) and Ti-Co(1.30 at.%) exhibited practically temperature-independent magnetic susceptibilities, χ. These results are in agreement with some related observations made by COLLINGS *et al.*, namely that: (i) while α-Ti-Mn alloys (quenched Ti-Mn (0.5, 1, 2, 3, and 4 at.%)) were Curie-Weiss paramagnetic, the ω- and β-phase higher-Mn-concentration members of the same alloy series possessed magnetic susceptibility temperature dependences comparable to those of Ti-TM alloys in general; (ii) the room-temperature susceptibilities of Ti-Fe, even when hcp (e.g., quenched-α^m Ti-Fe(2.5 at.%)), were comparable to those exhibited by the unequivocally "Pauli paramagnetic" (see Section 7.3) members of the entire Ti-Mo series [HO74].

By way of summary, and as indicated originally by HAKE [HAK62], it seems that all properties point to a localized magnetic moment on Mn in α-Ti-Mn alloys, and that while Ti-Cr and Ti-Fe exhibit resistivity minima, other results suggest that their solute moments are not sufficiently "localized" (see spin fluctuation discussions by COLES [COL72[d]] and others) to enable them to yield such characteristically "magnetic" effects as negative magnetoresistance, Curie-Weiss paramagnetism, suppression of the superconducting transition temperature [RIZ74], and so on.

5.4.2 Ti-Alloy Resistivity at Moderate-to-High Temperatures

UL'YANOV and KOVTUN [ULY64] measured the specific resistivities of the iodide and commercial grades of unalloyed Ti, as well as those of α-phase Ti-SM and Ti-TM alloys, in the temperature range 77–1500 K. The solute concentrations in the alloys studied were Al: 2, 4, 6 wt.%; La; 0.4, 1, 1.8, 2.2 wt.%; Cr: 2, 4 wt.%; Mo: 2, 4 wt.%; Pd: 0.2, 0.8, 2 wt.%; Ta: 1.7, 6.5, 13 wt.%; and Re: 0.4, 0.7, 1, 3 wt.%. Also measured were Ti-2Al-2Cr-2Mo and Ti-5Al-2Cr-2Mo.

The resistivity temperature dependences of Ti-Al(0, 3, 6, 11, and 33 at.%) have been measured over the temperature range 90–1075 K by MOOIJ [MOO73]. The phases assumed by these alloys at ordinary temperatures are: α-phase—Ti-Al(0, 3, and 6 at.%); α_2-phase—Ti-Al(33 at.%) [COL70[a]]; mixed $\alpha + \alpha_2$—Ti-Al(11 at.%) [COL70[a]]. The interesting feature common to all the temperature-dependence curves, and exemplified by Figure 5-4, is the tendency for the resistivity to "saturate" once it

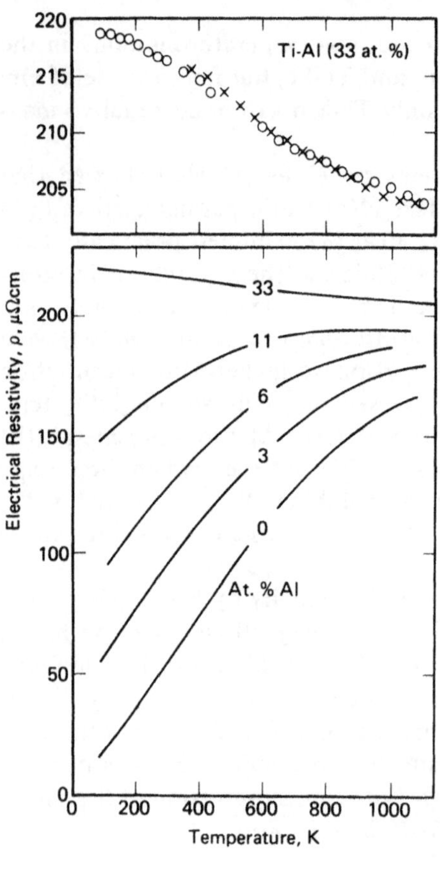

Figure 5-4. Resistivity temperature dependences of a series of typical Ti-SM alloys. Electrical resistivities of Ti-Al alloys containing 0, 3, 6, 11, and 33 at.% Al showing the gradual tendency from positive to negative $d\rho/dT$ with increasing Al content—after MOOIJ [MOO73].

has risen to values within the range 180–200 $\mu\Omega$ cm. Applying an argument similar to that used above in connection with the effect of the large background *impurity* resistivity on the incremental resistivity (i.e., the "specific solute resistivity") associated with simple-metal additions to Ti-Mo(25 at.%), it is suggested, with reference to the lower-concentration Ti-SM and Ti-TM alloys, that once the *ideal* resistivity itself becomes sufficiently large further increases in temperature become less and less effective in raising it further. The alloy could be said to be moving into a regime of anomalous temperature dependence. Similarly, again with reference to Figure 5-4, it is noted that as the *solute concentration* increases, the resistivity temperature dependence decreases, until finally with Ti-Al(33 at.%) the temperature dependence is actually negative. The parameter $\alpha^T \equiv (1/\rho)(d\rho/dT)$ has been used by MOOIJ [MOO73] and HAKE *et al.* [HAK75] as a measure of normal-state resistivity temperature dependence and in particular of the anomalous behavior frequently exhibited by concentrated alloys.

5.4.3 Ti-Alloy Resistivity at Moderate to Low Temperatures—Gross Features

Specific electrical resistivities below room temperature for the representative α-stabilized systems Ti-Al and Ti-Sn [COL84] and the representative β-stabilized systems Ti-V [COL74] and Ti-Mo [HO72] are listed in Table 5-4. A typical set of α-alloy resistivity temperature dependence curves was depicted in Figure 5-4; that for a typical β-alloy series, Ti-Mo, is given in Figure 5-5. The alloy resistivities exhibit several noteworthy features, not all apparent in the figures, only some of which are able to be

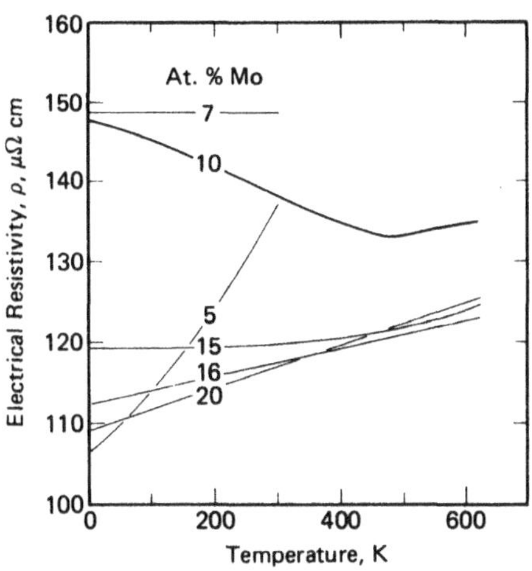

Figure 5-5. Resistivity temperature dependences of a series of typical Ti-TM alloys. Electrical resistivities of Ti-Mo alloys containing 5, 7, 10, 15, 16, and 20 at.% Mo showing the region of zero, and even negative, $d\rho/dT$ for alloys of compositions near 7–10 at.% Mo—after HO and COLLINGS [HO72].

Table 5-4. Specific Electrical Resistivities of Representative α-Stabilized (Ti–SM) and β-Stabilized (Ti–TM) Alloys

A. α-Stabilized Alloy, Ti–Al [CoL84]

Al Concentration at.%			Specific Resistivity, ρ 10^{-6} Ω cm			Specific Resistivity Temperature Dependence 10^{-8} Ω cm K^{-1}	
Nominal	Analyzed	Condition	4.2 K	77.3 K	RT^a	$d\rho/dT\|_{4.2}^{77}$	$d\rho/dT\|_{77}^{RT}$
0	—	As-cast	1.68	5.22	47.8	4.8	19.3
2	2.2	As-cast	25.7	32.8	72.7	9.7	18.1
5	5.4	As-cast	64.5	70	109	7.5	17.7
10	10.6	As-cast	129	132	161	4.1	13.1
15	14.7	As-cast	170	174	186	5.5	5.4
20	19.8	As-cast	169	171	195	2.7	10.9
23	22.4	As-cast	177	182	209	6.8	12.2
23	22.4	330 h/900°C/furnace cool	127	131	165	5.5	15.4
25	25.6	As-cast	78.4	84.7	141	8.6	25.5
25	25.6	330 h/900°C/furnace cool	43.9	51.0	110	9.7	26.7
28	27.2	As-cast	199	203	220	5.5	7.7
28	27.2	330 h/900°C/furnace cool	145	150	176	6.8	11.8
30	30.9	As-cast	203	203	209	0.0	2.7

B. α-Stabilized Alloy, Ti–Sn [Col84]

Sn Concentration at.%			Specific Resistivity, ρ $10^{-6}\ \Omega$ cm			Specific Resistivity Temperature Dependence $10^{-8}\ \Omega$ cm K^{-1}	
Nominal	Analyzed	Condition	4.2 K	77.3 K	RT^a	$d\rho/dT\|_{4.2}^{77}$	$d\rho/dT\|_{77}^{RT}$
3	2.90	8 h/1300°C/IBQ	49.9	56.7	97.4	9.3	18.4
5	5.20	8 h/1300°C/IBQ	84.7	91.3	129	9.0	17.1
7	6.78	8 h/1300°C/IBQ	111	117	146	8.2	13.1
10	10.2	8 h/1300°C/IBQ	146	150	169	5.5	8.6
13	12.8	8 h/1300°C/IBQ	143	147	163	5.5	7.2
17	17.3	8 h/1300°C/IBQ	117	124	148	9.6	10.9
23	22.6	8 h/1300°C/IBQ	92.4	101	138	11.8	16.7
25	24.3	As-cast	17.8	31.1	100	18.2	31.2
25	24.3	As-cast + 330 h/900°C/furnace cool	16.2	29.4	99.7	18.1	31.8
27	27.1	8 h/1300°C/IBQ	19.1	36.4	122	23.7	38.7
30	28.7	As-cast	43.3	61.9	136	25.4	33.5

Table 5-4. (*cont.*)

C. β-Stabilized Alloy, Ti–V [CoL74]

Nominal V Concentration, at.%	Condition	Specific Resistivity, ρ $10^{-6}\ \Omega$ cm				Specific Resistivity Temperature Dependence $10^{-8}\ \Omega$ cm K^{-1}
		77.3 K	200.0 K	273.2 K	300.0 K	$d\rho/dT\|_{77}^{300}$
0	—	5.9_9	26.7_5	40.9_2	46.5_3	18.2
3	1 h/1050°C/IBQ	18.5_3	42.1_9	56.5_0	62.4_0	19.7
6	1 h/1050°C/IBQ	33.3_2	58.1_8	72.5_2	78.1_5	20.1
9	1 h/1050°C/IBQ	48.7_0	73.1_2	87.1_1	92.1_2	19.5
10.5	1 h/1050°C/IBQ	55.5_0	79.5_0	92.9_0	97.9_9	19.1
12	1 h/1050°C/IBQ	89.6_7	109.8_2	121.6_0	125.4_5	16.1
15	1 h/1050°C/IBQ	120.0_7	132.0_1	137.5_2	139.6_2	8.8
17.5	1 h/1050°C/IBQ	125.7_8	134.2_2	137.7_7	139.0_4	6.0
20	1 h/1050°C/IBQ	136.8_4	136.3_0	136.9_6	136.7_8	0.0
25	1 h/1050°C/IBQ	127.6_4	123.2_8	212.4_7	121.1_3	−2.9
30	1 h/1050°C/IBQ	114.1_1	112.0_0	111.7_2	112.0_8	−0.9
40	1 h/1050°C/IBQ	92.3_7	95.5_1	97.4_6	98.3_7	2.7
50	1 h/1050°C/IBQ	—	—	84.9_5	86.3_0	—
50	Above +8h/1350°C/IBQ	76.7_1	82.2_3	86.0_5	86.8_3	4.5
50	New sample 8h/1350°C/IBQ	82.8_1	87.6_1	91.0_4	92.0_3	4.1
70	1 h/1050°C/IBQ	—	—	63.0_0	65.0_3	—
70	Above +8h/1350°C/IBQ	46.7_6	56.2_4	61.2_8	62.8_7	7.2
70	New sample 8h/1350°C/IBQ	47.6_4	56.3_9	60.9_8	62.9_0	6.9
80	1 h/1050°C/IBQ	—	—	49.0_8	51.1_3	—

80	Above +8h/1350°C/IBQ	34.9	45.1_0	50.5_1	52.3_2	7.8
100	—	2.4_3	12.9	18.3	20.3	8.0

D. β-Stabilized Alloy, Ti-Mo [Ho72]

Mo Concentration, at.%			Specific Resistivity, ρ $10^{-6}\ \Omega$ cm			Specific Resistivity Temperature Dependence $10^{-8}\ \Omega$ cm K^{-1}	
Nominal	Analyzed	Condition	4.2 K	77.3 K	RT^a	$d\rho/dT\vert_{4.2}^{77}$	$d\rho/dT\vert_{77}^{RT}$
5	5.09	8 h/1000°C/IBQ	107	112	137	6.8	11.3
7	6.86	8 h/1000°C/IBQ	149	149	149	0.0	0.0
10	10.30	2 h/1000°C/IBQ	148	146	138	−2.7	−3.6
10	10.30	2 h/1000°C/IBQ + 160 h/350°C	120	123	130	4.1	3.2
15	14.92	2 h/1000°C/IBQ	120	120	120	0.0	0.0
15	14.92	2 h/1000°C/IBQ + 160 h/350°C	119	120	120	1.4	0.0
16	—	2 h/1000°C/IBQ	112	115	117	4.1	0.9
16	—	2 h/1000°C/IBQ + 160 h/350°C	112	114	117	2.7	1.4
20	19.38	2 h/1000°C/IBQ	110	111	118	1.4	3.2
20	19.38	2 h/1000°C/IBQ + 160 h/350°C	109	111	118	2.7	3.2

[a] Assume 298 K.

discussed in the present context:

(1) The concentrated disordered α-phase and β-phase alloys are characterized by high specific resistivities (generally of the order 100 $\mu\Omega$ cm) and low $d\rho/dT$s.
(2) In the α-phase alloys, as a function of composition, the resistivity passes through a local minimum in the vicinity of the ordered α_2 phase (i.e., in the Ti_3SM-based regime [Col82[a]]) and the $d\rho/dT$, in the same composition range, rises to values comparable to those expected in low-concentration alloys.
(3) The β-phase alloys possess what appear to be anomalous resistivity maxima near $e/a \cong 4.2$.
(4) $d\rho/dT$ is negative: (i) in Ti-Al alloys having Al concentrations greater than 30 at.% (see Figure 5-4) over the entire 90–1075 K temperature range; (ii) for a narrow range of Ti–Mo concentrations centered about Ti–Mo(10 at.%) over the temperature range 0–470 K, within which the quenched $\omega+\beta$-phase is moderately stable (Section 1.9). The occurrence of negative $d\rho/dT$ within the observed concentration range associates it with the resistivity maximum.

Negative resistivity temperature dependence is a branch of alloy physics that has attracted considerable attention over a prolonged period of time. Depending on temperature range and alloy classification, the phenomenon has been attributed to: (i) the Kondo effect (dilute alloys at low temperatures [Riz74]); (ii) the increase with decreasing temperature of spin-disorder scattering from local-moment clusters (e.g., concentrated Cu–Ni alloys [Hou70]); (iii) an increase with decreasing temperature of the density of ω-phase-related soft phonons [Col74] or the thermally reversible ω-phase precipitation itself [Ho72] (as in Ti–V and Ti–Mo alloys and related alloy systems—see also references in [Cha74]); (iv) a smearing-out with increasing temperature of the density-of-states structure near E_F in certain classes of strong-scattering concentrated binary alloys [Che72] (such as Ti-Al [Moo73]).

The relevance of the last two mechanisms to anomalous resistivity temperature dependence in Ti–TM and Ti–SM alloys is considered below.

5.5 Anomalous Resistivity Concentration Dependence and Temperature Dependence in Titanium-Base Alloys

5.5.1 Anomalous Concentration Dependence

According to (5-2), the total resistivity may be regarded as the sum of an intrinsic or ideal component, $\rho_i(c, T)$ and an impurity- or scattering-

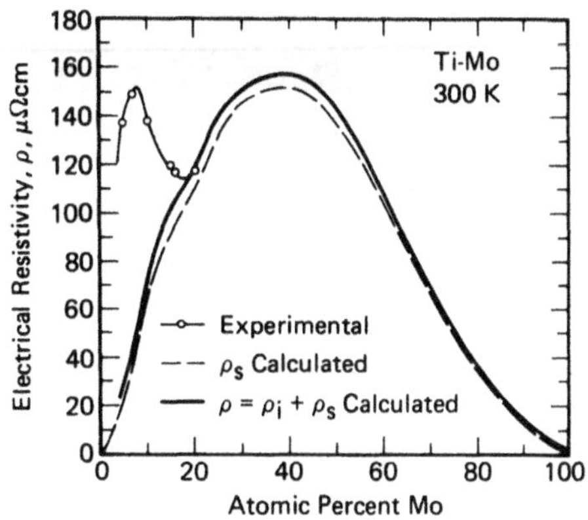

Figure 5-6. Comparison of experimental (300 K) and calculated resistivity composition dependences of Ti–Mo alloys. The ideal resistivity, ρ_i, given by $\rho_i \propto n(E_F)/\theta_D^2$ (in turn based on calorimetric data, see COLLINGS *et al.* [COL72]) was normalized to ρ_i(Ti-Mo(20 at.%), 300 K) = 8 $\mu\Omega$ cm. The solute-scattering (impurity) resistivity, ρ_s, given by $\rho_s \propto n(E_F)c(100 - c)$ was normalized to ρ_s(Ti-Mo(20 at.%)) = 110 $\mu\Omega$ cm—after HO and COLLINGS [HO72].

dependent component, $\rho_s(c)$. In discussing anomalous electrical resistivity in Ti–Mo alloys, both these components have been taken into consideration [HO72]. After arguing that at a fixed temperature

$$\rho_i(c) \propto n(E_F)/n_e\theta_D^2 \tag{5-4}$$

where n_e is the number-density of conduction electrons, and recalling from (5-3c) that $\rho_s(c) \,\tilde{\propto}\, n(E_F)c(100 - c)$, a $\rho(c)$ curve for the Ti–Mo system has been calculated semiempirically using as input data:

(1) ρ_s(Ti-Mo(20 at.%)) $\cong \rho_{4\,\mathrm{K}} = 110\ \mu\Omega$ cm,
ρ_i(Ti-Mo(20 at.%)) $\cong \rho_{300\,\mathrm{K}} - \rho_{4\,\mathrm{K}} = 8\ \mu\Omega$ cm

(2) low-temperature-specific-heat-derived values of $n(E_F)$ and θ_D [COL71, COL72]

Comparing the results of this calculation with those of direct measurement, as in Figure 5-6, it is seen that the resistivity (either at 4.2 K or 300 K) is anomalously large for alloys in a concentration range centered near 7 to 10 at.% Mo (i.e., near an e/a ratio of between 4.14 and 4.20).

The resistivity composition dependences of Ti–V alloys at 77, 200, and 300 K [COL74] are plotted in Figure 5-7. Again the resistivity appears to be anomalously large in the vicinity of 15-20 at.% V (e/a = 4.15-4.20). Indeed using a fully empirical approach [COL74], it was determined that a resistivity increment $\Delta\rho_{300\,\mathrm{K}}$, in excess of what would normally be expected from the sum of ρ_i and ρ_s, rose to values larger than 20 $\mu\Omega$ cm within the concentration range 12–17 at.% V. After the same approach was applied to the 77 K resistivity isothermal, the values of $\Delta\rho_{77\,\mathrm{K}}$ within the composition interval 13-29 at.% V were found to exceed 40 $\mu\Omega$ cm. In fact, since within the concentration range 20–32 at.% V the increase in $\Delta\rho$, which accrued as the temperature decreased from 300 to 77 K (Figure 5-8), was larger than the decrease in ρ_i within the same temperature interval, the temperature

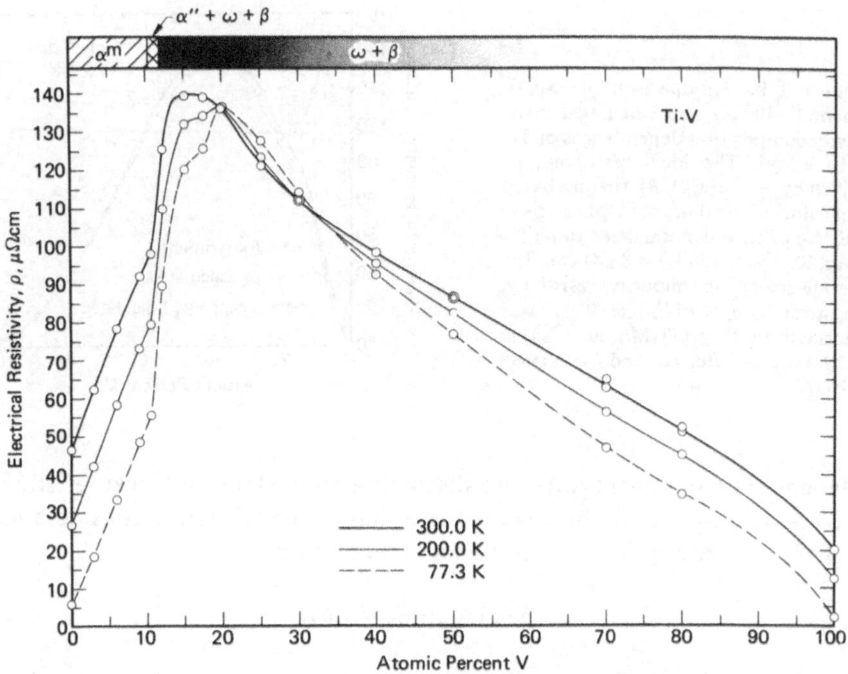

Figure 5-7. Resistivity composition isothermals for Ti–V alloys together with a representation of the corresponding quenched microstructural states. The actual liquid-N data are plotted; the 200.0 K data were obtained from the ~200 ± 1 K results after correction using measured $[d\rho/dT]_{77\,K}^{\sim200}$; the 200.0 K data were obtained from the ~298 ± 1 K results after correction using measured $[d\rho/dT]_{\sim273\,K}^{\sim300\,K}$—after COLLINGS [COL74].

coefficient of resistivity within the composition range was obliged to be negative.

The occurrence of ω-phase in quenched Ti–TM alloys has been considered in Section 1.8, and its occurrence in Ti–V alloys in particular has been discussed in detail by MCCABE and SASS [MCC71]. According to those authors, ω-phase is seen as a submicroscopic precipitate in the concentration range 13–25 at.%; then as the V concentration increases beyond 13 at.% the precipitate changes character, as evidenced by the electron-diffraction pattern, which changes from sharp to diffuse spots and finally to halos that persist in gradually decreasing intensity all the way across to pure V. This situation, in which there is no sharp line of demarcation between the $\omega+\beta$- and β-fields, is represented by the graded shading in Figure 5-7. The Ti–Mo system has comparable microstructural properties, although the range of occurrence of the ω-phase is more restricted. In the diffraction studies referred to in Section 1.8, DE FONTAINE *et al.* [DEF70, DEF71] have noted the presence of athermal ω-phase in Ti–V(19 at.%), an alloy in the center

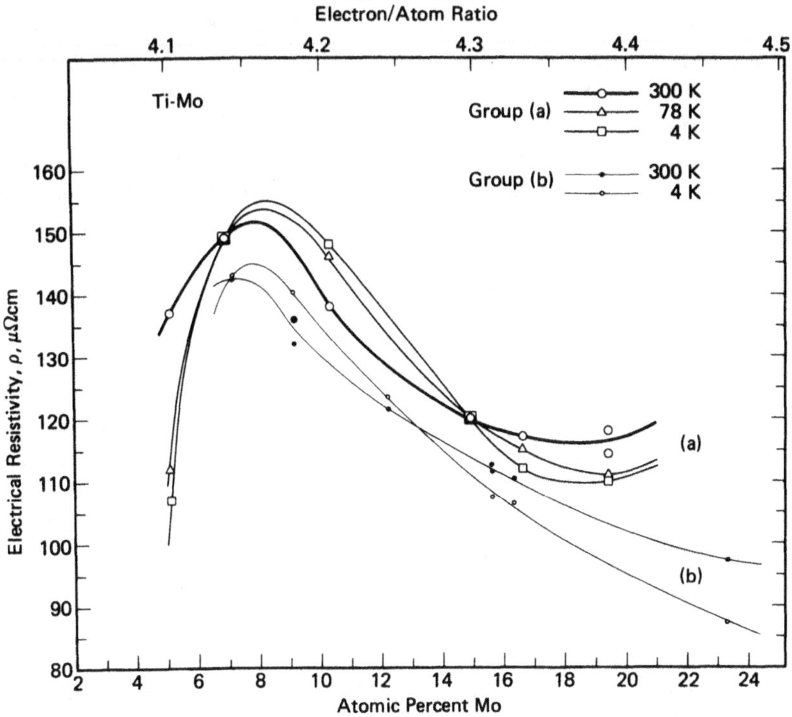

Figure 5-8. Resistivity composition isothermals for Ti–Mo alloys. Group (a) was ground from annealed-and-quenched material. Group (b) was ground from cast ingots (see Figure 5-1 for cast-and-cold-rolled 1.2 K data). The occurrence and extents of the regimes of negative $d\rho/dT$ are in good agreement—after Ho and Collings [Ho72] (Group (a)) and Hake *et al.* [Hak61] (Group (b)).

of the anomalous range under discussion. It is likely, therefore, that both crystalline and phonon-induced structural anomalies exist out to 25 at.% V. The obvious conclusion is that the anomalous excess isothermal resistivity is coupled to the occurrence, within the same concentration range, of both the isothermal (crystalline) and the athermal (thermally reversible, soft-phonon generated) ω-phases.

If athermal ω-phase is partially responsible for the excess resistivity component $\Delta\rho$, it is obviously a close contender in the choice of a mechanism for negative $d\rho/dT$. But the lack of a *direct* experimental link between ω-phase and negative $d\rho/dT$ has been responsible for much controversy over the years [Cha74].

5.5.2 Negative Temperature Dependence

Negative $d\rho/dT$ in Ti–TM alloys is a frequently encountered phenomenon usually occurring over a wide temperature range but over a narrow composition range. In this review, which is intended to be representative rather than exhaustive, it is noted that negative $d\rho/dT$ has been detected in Ti-V and Ti-Nb but not in Ti-Ta [PRE74]; it has also been observed in Ti-Cr, Ti-Mo, and Ti-Fe (see below), as well as in Zr-Nb [CHA74], Zr-Mo, Hf-Mo, U-Mo, and U-Nb [HAK61]. The temperature range of measurement has usually been 4.2-300 K or 77-300 K, although if present below room temperature the negative $d\rho/dT$ can generally be expected to persist to higher temperatures (e.g., ~200°C) depending on the stability of the quenched structure. The following systems have been subject to extensive investigation.

(a) Ti–V. Although undetected in the initial investigation of ADENSTEDT *et al.* [ADE52], a negative $d\rho/dT$ has been noted by BROTZEN *et al.* in quenched alloys of compositions of 15-20 wt.% (14-19 at.%) over the temperature range 77-300 K. Negative $d\rho/dT$ was reported by HAKE *et al.* [HAK61] in Ti-V(29 at.%), by COLLINGS [COL74] in quenched Ti-V(20, 25, and 30 at.%) within the temperature range 77-300 K, and by PREKUL *et al.* [PRE74], also in quenched Ti-V(20, 25, and 30 at.%), at temperatures between T_c and 273 K.

(b) Ti–Nb. AMES and MCQUILLAN [AME54] first noted that in Ti-Nb alloys $d\rho/dT$ decreased with increasing Ti content, becoming negative for Ti-Nb(25 and 20 at.%) over the temperature range 73-473 K, within which the quenched alloys were stable. PREKUL *et al.* [PRE74], after studying a wide range of Ti-Nb alloys, noted that only in Ti-Nb(25 at.%) quenched from 900°C was it possible to obtain a negative $d\rho/dT$, and furthermore that the tendency to achieve negative $d\rho/dT$ decreased in the order Ti-V→Ti-Nb→Ti-Ta, the effect being absent from the resistivity temperature dependence of the latter alloy. The most recent investigations of the concentration and temperature dependences of the electrical resistivity of Ti-Nb alloys are those of OBST and LIEDTKE [LIE81], some of whose results are depicted in Figure 5-9. It can be seen there that, in agreement with the observation of AMES and MCQUILLAN, the concentration threshold for negative resistivity temperature dependence lies somewhere between 22 and 32 at.% Nb. In a subsequent study, HOCHSTUHL and OBST [HOC82] were able to associate negative $d\rho/dT$ with bcc-lattice instability. Focusing attention on Ti-Nb(22 at.%), they recognized that: (i) $d\rho/dT$ in the as-quenched alloy was negative throughout the temperature range from 300 K down to T_c; (ii) although Ti-Nb(22 at.%) does not spontaneously undergo $\beta\rightarrow\alpha''$ transformation during cooling through M_s (see Figure 3-7 and associated discussion) such a transformation can be induced by deformation; (iii) partial transformations, of the kind $\beta\rightarrow\omega$ or $\beta\rightarrow\alpha''$ (which preempts the former), serve to stabilize the crystal. By performing resistivity measurements

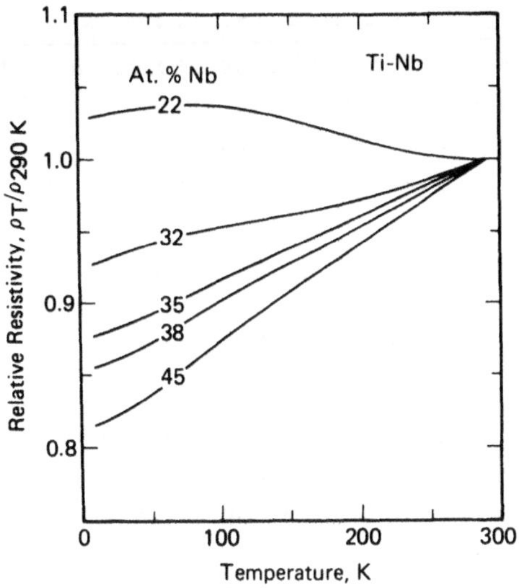

Figure 5-9. Relative resistivity temperature dependences of a series of Ti-Nb(22, 32, 35, 38, and 45 at.%) alloys. Condition: annealed 2 h/900°C and water quenched—after LIEDKE [LIE81].

in a tensile-test machine, HOCHSTUHL and OBST were able to show that, although the *as-quenched* Ti-Nb(22 at.%) alloy exhibited the expected negative $d\rho/dT$, after it was subjected to 2.4% tensile deformation at 180 K, in response to which partial $\beta \rightarrow \alpha''$ transformation took place, during subsequent cycling down to 10 K and back up to 300 K its resistivity temperature dependence was positive.

(c) Ti–Cr. LUHMAN *et al.* [LUH68] noted, as a result of measurements taken in the temperature range 4.2-473 K, that $d\rho/dT$ was negative for Ti-Cr alloys within the composition range of $\lesssim$10 at.% to 20 at.% Cr, and CHANDRASEKARAN *et al.* [CHA73, CHA74] have shown in another detailed study that $d\rho/dT$ is negative in Ti-Cr(10, 13, 15, and 20 at.%).

(d) Ti–Mo. YOSHIDA *et al.* [YOS56], the first to study resistivity temperature dependence (in the temperature range −190° to 900°C) in Ti-Mo alloys, reported the existence of negative $d\rho/dT$, at temperatures below about 200°C, in Ti-Mo(12, 14, and 17 wt.%) (i.e., $6._4$, $7._5$, and $9._3$ at.% Mo). Subsequently, negative $d\rho/dT$ was noted in: (i) as-cast Ti-Mo(9 and 12 at.%) by HAKE *et al.* [HAK61]; (ii) a series of quenched Ti-Mo alloys bounded by Ti-Mo(7 at.%) and Ti-Mo(15 at.%) (whose temperature coefficients within the 4.2-300 K range were themselves zero) by HO and COLLINGS [HO72]; (iii) Ti-20Mo (i.e., 11 at.% Mo), within the temperature range −196 to 0°C, by CHANDRASEKARAN *et al.* [CHA73, CHA74].

(e) Ti–Fe. HAKE *et al.* observed negative $d\rho/dT$ within the temperature range 4.2-300 K in Ti-Fe(9.1 and 11.9 at.%) [HAK61]. This result was confirmed by PREKUL *et al.* [PRE76], who, in measurements of the resistivity temperature dependences of ten quenched Ti-Fe(1.46-20 at.%) alloys within the temperature range 4.2-300 K, noted that those within the composition range 4 to 12 at.% Fe all possessed negative $d\rho/dT$s, which were claimed to persist to fairly high temperatures.

5.6 Three Case Studies of Negative $d\rho/dT$

5.6.1 Negative $d\rho/dT$ in Ti–V and Ti–Mo

The electrical resistivities of a series of Ti-V alloys were measured and corrected to 77.3, 200.0, and 300.0 K. The corresponding resistivity-composition isothermals are seen in Figure 5-7 to intersect in such a way as to reverse their order and establish a negative $d\rho/dT$ within the composition interval 20-32 at.% V. Similar behavior is to be seen in Figure 5-8 for the Ti-Mo alloy series. Presented in this way, it appears that negative $d\rho/dT$ occurs as a minor temperature-dependent perturbation to a much larger effect—the anomalous composition dependence. In discussing the Ti-Mo [HO72] and Ti-V [COL74] systems, it was concluded that the negative $d\rho/dT$ was a manifestation of some reversible effect associated with ω-phase precipitation, either the presence of crystalline ω-phase itself [HO72] or the soft phonons associated with its formation [COL74]. As pointed out in Section 1.8, DE FONTAINE *et al.* [DEF71] have demonstrated that ω-phase will precipitate athermally and reversibly, between room temperature and 77 K, from very rapidly quenched Ti-Mo(8 at.%). This clearly favorable composition for ω-phase formation coincides, according to Figure 5-8, with the peak of the anomalous-resistivity function. Maximal negative $d\rho/dT$ occurs at somewhat higher Mo concentrations, suggesting that it may be associated with an ω-phase precursor-effect (such as the presence of soft phonons, see [DEF70, DEF71]) rather than with the crystalline precipitate itself. This argument was further developed in a discussion of the resistivity composition and temperature dependences of Ti-V [COL74], in which it was pointed out that since athermal ω-phase had already been noted in Ti-V(19 at.%), it was likely that both phonon-induced anomalies *and* crystalline precipitates coexisted out to 25 at.% V, the upper limit, according to MCCABE and SASS [MCC71], of particulate precipitation.

In studies to be considered below, the magnetic susceptibility temperature dependences of Ti-V alloys have been measured in the temperature range 77 to about 400 K. With the aid of an analysis based on the deduced susceptibility temperature dependences of ω-phase and β-phase Ti-V, it was concluded that part of the observed total *susceptibility* temperature dependence in low-concentration alloys was a result of the reversible precipitation of ω-phase, obviously in the form of particles sufficiently large to

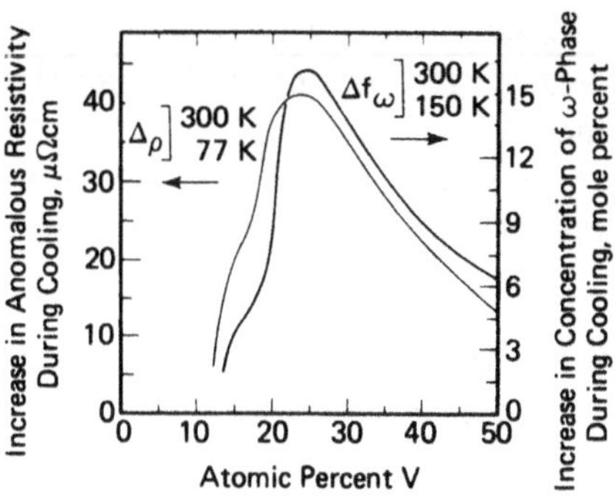

Figure 5-10. Increase in anomalous resistivity, $\Delta\rho$, incurred on lowering the temperature of quenched Ti-V alloys from 300 to 77 K, compared with a magnetically-derived estimate (see Chapter 7) of the increase in ω-phase abundance, which takes place as the temperature is lowered from 300 to 150 K—after COLLINGS [COL74, COL78].

have definite density-of-states properties [COL78]. The magnetically deduced volume fraction of reversible ω-phase precipitation is plotted in Figure 5-10 alongside the anomalous reversible resistivity component, $\Delta\rho|_{77}^{300}$. Represented according to the scales indicated in the figure, the $\Delta\rho$ and Δf_ω curves agree remarkably well, suggesting rather strongly that reversible crystalline precipitation is accompanied by the anomalous reversible electrical resistivity, which, if sufficiently large over a limited concentration range, yields a negative $d\rho/dT$. Whether the $\Delta\rho$ (or its extreme manifestation, negative $d\rho/dT$) is the direct result of reversible particulate ω-phase precipitation, or of the soft-phonons (i.e., the fluctuations) that accompany it, was not determinable by these experiments alone.

5.6.2 Negative $d\rho/dT$ in Ti–Cr

In studies of Ti-Cr alloys by LUHMAN *et al.* [LUH68] and CHANDRASEKARAN *et al.* [CHA73, CHA74], reversible $d\rho/dT$ has been disassociated from athermal ω-phase precipitation. Nevertheless the results do support an interpretation in terms of the properties of the metastable β-phase, or, in other words, in terms of the soft-phonon modes of DE FONTAINE *et al.* [DEF70, DEF71]. The role of athermal ω-phase itself as a source of negative $d\rho/dT$ was denied by the following observations: (i) the alloys Ti-Cr(15 and 20 at.%), both of which yielded negative $d\rho/dT$, failed to exhibit reversible electron-diffraction intensity maxima during cold-stage transmission electron microscopy down to −180°C [CHA74]; (ii) according to LUHMAN *et al.* [LUH68], Ti-Cr(12 at.%) exhibited negative $d\rho/dT$ during heating to 473 K in spite of the fact that ω_s[1] for this alloy was supposed to be below room temperature; (iii) according to [CHA74], the resistivity temperature coefficient of quenched Ti-Cr(20 at.%) was $d\rho/dT|_{77}^{273} = -2.76\times10^{-2}$ $\mu\Omega$ cm K^{-1}, then after aging 1140 min/300°C to develop $\omega+\beta$, whose $d\rho/dT$ is positive, and reverting 4 min/435°C, the $d\rho/dT|_{77}^{273}$ of the

[1] ω_s represents the "ω-start" transus; cf. M_s for martensitic transformation (Figure 1-11).

resulting phase-separated structure, $\beta'+\beta$, (see Section 2.3) was again -2.67×10^{-2} $\mu\Omega$ cm K^{-1}. Since during β-reversion the ω-phase is transformed to, and retained as, a solute-lean β [CHA78], this final observation alone was regarded as convincing evidence for the association of negative $d\rho/dT$ with the metastable β-phase rather than the reversible ω-phase itself.

5.7 Mechanisms of Anomalous Resistivity Temperature Dependence

Both the ideal, ρ_i, and impurity, ρ_s, components of total resistivity are candidates for anomalous resistivity temperature dependence.

5.7.1 Impurity-Scattering Mechanisms

As mentioned in Section 5.4.3, spin-disorder scattering in magnetic alloys, under suitable conditions, is purported to give rise to a negative $d\rho/dT$ [HOU70]. In Ti-TM alloys at sufficiently high temperatures $d\rho/dT$ is positive, even in alloys for which it is negative in the lower-temperature regime. The resistivity of Ti-Mo(10 at.%) (Figure 5-5) is an example of this. The existence of a $\rho(T)$ minimum separating regions of low-temperature negative $d\rho/dT$ and high-temperature positive $d\rho/dT$, led PREKUL *et al.* [PRE74, PRE76] to advocate the use of a localized-spin-fluctuation model to explain the occurrence of anomalous resistivity temperature and/or composition dependences in alloys such as Ti-V, Ti-Nb, Ti-Ta, and Ti-Fe. The validity of such an approach can be judged from the foregoing discussions. Clearly, the preferred scattering model for anomalous resistivity in such alloys has to do with phonon scattering associated with the β-phase instability. As such it should be categorized (with reference to the θ_D of (5-4)) as a phonon contribution to ideal resistivity.

5.7.2 Ideal- (i.e., Phonon-) Scattering Mechanisms

The other ideal resistivity term to be taken into account (referring again to (5-4)) is $n(E_F)$. In one of the earliest attempts to identify a mechanism for the anomalous resistivity temperature dependence in Ti-TM alloys, AMES and MCQUILLAN [AME54] compared the negative $d\rho/dT$ of Ti-Nb to that of a semiconductor. The deficiencies in the comparison are obvious, but nevertheless their proposed model, which attributed the effect to structure in the $n(E)$ function rather than electron-scattering, anticipated, in a way, the results of CHEN *et al.* [CHE72], who showed with the aid of the coherent-potential approximation (CPA) [FAU82] that negative $d\rho/dT$

could be associated with suitably located minima in the density-of-states curves of strong-scattering disordered binary alloys. The model is clearly not applicable to the Ti–TM alloys discussed above. On the other hand it could well be applicable to concentrated Ti–SM; these are strong-scattering alloys (e.g., [GEG73], [COL75d]), which, as Figure 5-4 suggests, may have positive, zero, or negative $d\rho/dT$s.

5.7.3 Anomalous $d\rho/dT$ in Strong-Scattering Disordered Binary Alloys

CHEN *et al.* [CHE72], using the CPA method, performed a model calculation for a concentrated binary alloy system, and have watched the changes in density-of-states that occur in response to: (i) change of concentration from f (mole fraction) = 0.1 to $f = 0.5$; (ii) change of atomic-scattering strength, represented in Figures 5-11 and 5-12 by δ; (iii) change of temperature, represented in Figure 5-11 by t. They also calculated the relative electrical conductivity as a function of band filling and have been able to predict, in a semiquantitative way, the manner in which resistivity may change with temperature in two classes of concentrated binary alloy: (i) virtual-crystal or weak-scattering alloys; (ii) strong-scattering alloys. Ti–Al seems to be an example of the latter.

The total density-of-states of a binary alloy consisting of components A and B may be represented by

$$n(E) = fn_A(E) + (1 - f)n_B(E) \qquad (5\text{-}5)$$

leading, in the simple model situation defined by CHEN *et al.* [CHE72], to

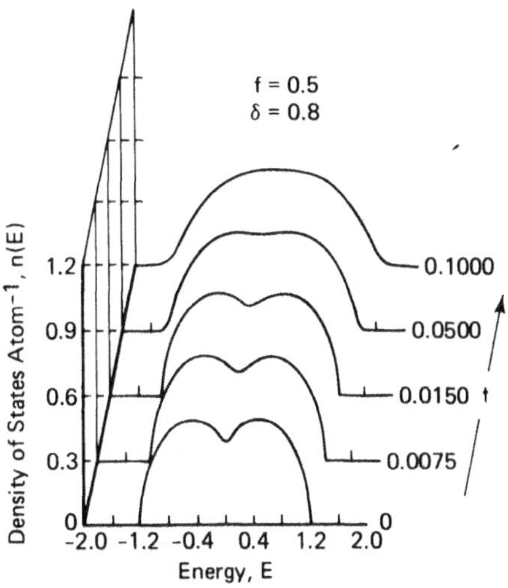

Figure 5-11. Coherent-potential-approximation calculated density-of-states, $n(E)$. Curves for five reduced temperatures (t) of a model strong-scattering ($\delta = 0.8$ on a scale of 0 to 1) equiatomic ($f = 0.5$) binary alloy—after CHEN *et al.* [CHE72].

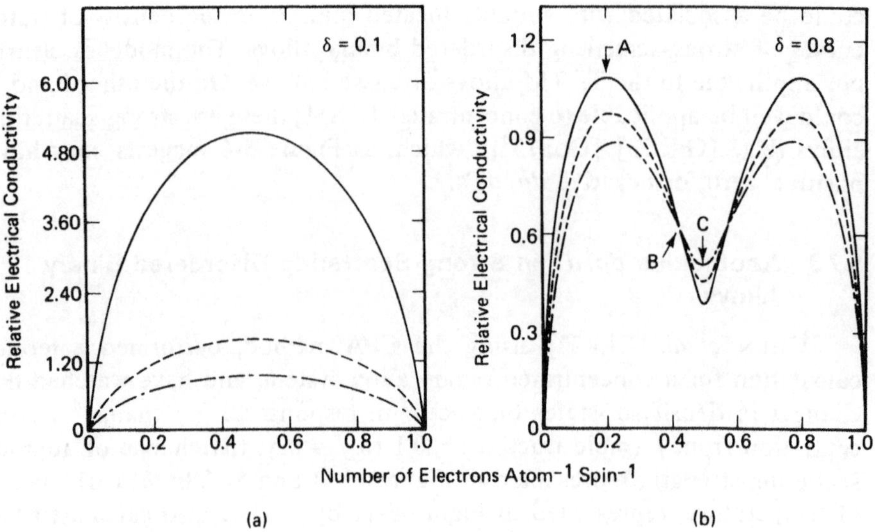

Figure 5-12. Relative electrical conductivity as function of band filling for three reduced temperatures (t) of a model equi-atomic ($f = 0.5$) binary alloy: (a) weak-scattering case ($\delta = 0.1$), $t = 0.000$ (—), 0.006 (---), 0.012 (—-—), (b) strong-scattering case ($\delta = 0.8$), $t = 0.000$ (—), 0.0075 (---), 0.015 (—-—). In the strong-scattering case three $d\rho/dT$ signatures are possible depending on the level of band filling: $d\rho/dT$ is positive at A, zero at B, and negative at C—after CHEN *et al.* [CHE72].

the double-humped density-of-states function represented in Figure 5-11, in which one maximum derives from component A, and the other from B.

Electrical conductivity is plotted as a function of band filling in Figures 5-12(a) and (b) for alloys that have equal electron–phonon interaction parameters, λ_A and λ_B, but which differ in electron-scattering strengths, δ_A and δ_B (δ for each alloy representing the difference between the atomic potentials of the constituents [COL73[b]], [GEG73]). Figure 5-12(a), for weak electron-scattering, represents an alloy in the "virtual-crystal limit," for which perturbation theory is valid. For such an alloy, thermal disorder increases the scattering of Bloch-like conduction electron states; its resistivity increases rapidly with temperature according to the usual formulae, which reflect that, under the conditions stated, the change in density-of-states with temperature is inconsequential and the increase in resistivity is simply the result of electron–phonon scattering.

In strong-scattering alloys on the other hand, the dominant mechanism controlling resistivity temperature dependence has to do with the effect of temperature on the structure of the total density-of-states function. Thermal disorder broadens and smears the density-of-states. While this has no noticeable effect in the virtual-crystal limit, in strong-scattering alloys, as depicted in Figure 5-11, the minimum (or "pseudogap") begins to close and will eventually fill up completely at high temperatures. In Figure 5-12(b), which is a direct descendant of the density-of-states data displayed in Figure 5-11, the influence of temperature on the electrical conductivity of strong-scattering alloys, as a function of position of the Fermi level, is clearly visible. If for example the Fermi level is at A, $d\rho/dT$ is positive, if at B it is zero, while if at C it is negative.

Ti–Al, a typical TM–SM alloy, like numerous other combinations of very dissimilar elements, is a strong-scattering alloy. The large specific resistivity of Al in Ti, and the relatively rapid strengthening coefficients are confirmations of this [GEG73], [COL75[d]]. It is, therefore, suggested that the concentrated-Ti–Al resistivity temperature dependences portrayed in Figure 5-4 are explicable in terms of a suitable specialization of the model of CHEN *et al.* [CHE72].

As pointed out above, thermal disorder can lead to one of two possible effects: (i) in virtual-crystal alloys it can scatter electrons between Bloch-like states and in so doing contribute resistivity, or (ii) in strong-scattering alloys it can assist in the hopping of highly damped quasilocalized states thus aiding conductivity. It is interesting to conclude this review by recalling that the latter effect is precisely the metallic equivalent of the mechanism initially invoked by AMES and MCQUILLAN to explain negative $d\rho/dT$ in Ti–TM alloys [AME54].

5.8 Magnetoresistivity in Normal Metals

The electrical resistivities of metals usually increase in the presence of a magnetic field, H_a, applied either transversely or longitudinally to the direction of current flow. The change, $\Delta\rho(H_a)$, is referred to as the magnetoresistivity. In free-electron theory both the longitudinal and transverse magnetoresistivities are zero: (i) in the former case this is clearly so, since the longitudinal field influences only the transverse motion of the conduction electrons; (ii) in the latter case the Hall e.m.f. (see Section 23.11) exactly compensates for the effect on the conduction electrons of a transverse magnetic field, and again restores their longitudinal motion, and hence scattering, to its field-free state. Thus, Na, with its practically spherical Fermi surface, has the smallest magnetoresistance of all the measured metallic elements [ROS63, p. 107]. The nonzero magnetoresistivity that emerges with departures from the free-electron model for metals is explicable in terms of the usual circular motion of the electron between scattering points on the Fermi surface combined with scattering effects associated with the nonsphericity of that surface. A simple scaling-law for magnetoresistivity is Kohler's rule, which states:

$$\Delta\rho/\rho(0, T) = \text{function}[H_a/\rho(0, T)] \qquad (5\text{-}6)$$

where $\rho(0, T)$, the zero-field resistivity at temperature T, is a normalizing factor that may be changed either by changing the temperature or the purity of the sample. Under variation of both of these parameters it is usually found that all the experimental points, when plotted in the format $\Delta\rho/\rho(0, T)$ *versus* $H_a/\rho(0, T)$, but more usually as $\Delta R(H_a, T)/R(0, T)$ *versus* $H_a/R(0, T)$, fall on a universal curve.[2] An example of such a Kohler plot, as it is called, is presented later in Figure 28-3. The transverse magnetoresistivity is usually greater than the longitudinal, is usually positive (except when scattering is by magnetic ions), and is usually proportional to H_a^2.

Some special effects that occur in galvanomagnetic studies of *monocrystalline* samples are explicable in terms of Fermi-surface topologies. In very strong applied fields the electron orbit may be sufficiently small as to permit its completion without eliciting a scattering event; any further increase in H_a is accompanied by no further increase in the then-saturated magnetoresistivity. But in certain crystals, such as Cu, whose Fermi surface reveals an extended path through **k**-space when sectioned parallel to one or other of the (111) planes, the electron may execute open orbits thereby permitting the magnetoresistivity to continue to increase indefinitely with applied field strength.

These effects are reflected in the *polycrystalline* results: thus, whereas the transverse magnetoresistivity of Cu increases monotonically with H_a to fields of at least 90 kOe [FIC72], that of Al tends to saturate in fields of more than about 10 kOe [FIC71]. In the present context, Cu and Al are important stabilizers of multifilamentary superconducting composites. As such their magnetoresistivities are discussed in Sections 28.2.1 and 28.2.2, respectively.

[2]Alternatives to the quotient $H_a/R(0, T)$ are $H_a \times \dfrac{R(0, 273)}{R(0, T)} \equiv H_a \times RRR$.

Chapter 6

Thermal Conductivity

6.1 Thermal Conductivity in Insulators and Normal Metals

As illustrated by two classic results from the literature (Figure 6-1), the thermal conductivities of both insulators and pure metals increase rapidly as their temperatures are decreased (within the cryogenic range) and pass through maxima before dropping to zero at zero K. They do so for entirely different sets of reasons, to be outlined below with reference to Figure 6-2. A useful basic equation, and one which makes an excellent

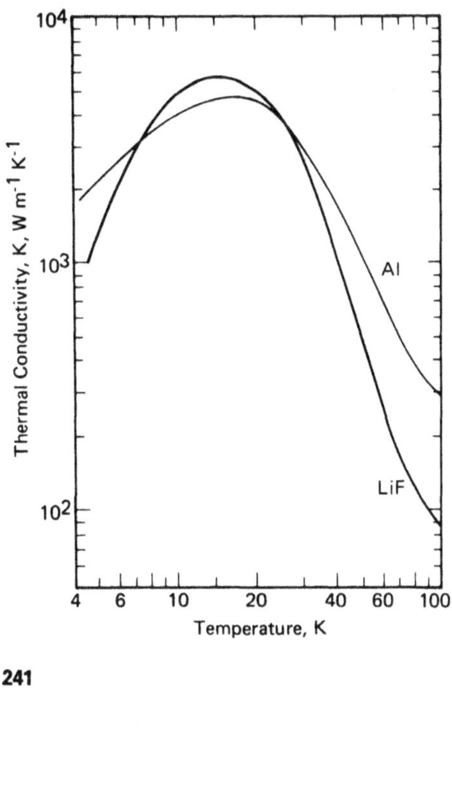

Figure 6-1. Intercomparison of the low temperature thermal conductivities of a typical metal (Al) and a typical insulator (the ionic solid LiF). The particular Al referred to is Johnson Matthey >99.995%, RRR = 467—data from CHILDS *et al.* [CHI73, p. 123]. The LiF is specified as ORNL 97.2% ^{7}Li in LiF—data from [CHI73, p. 439].

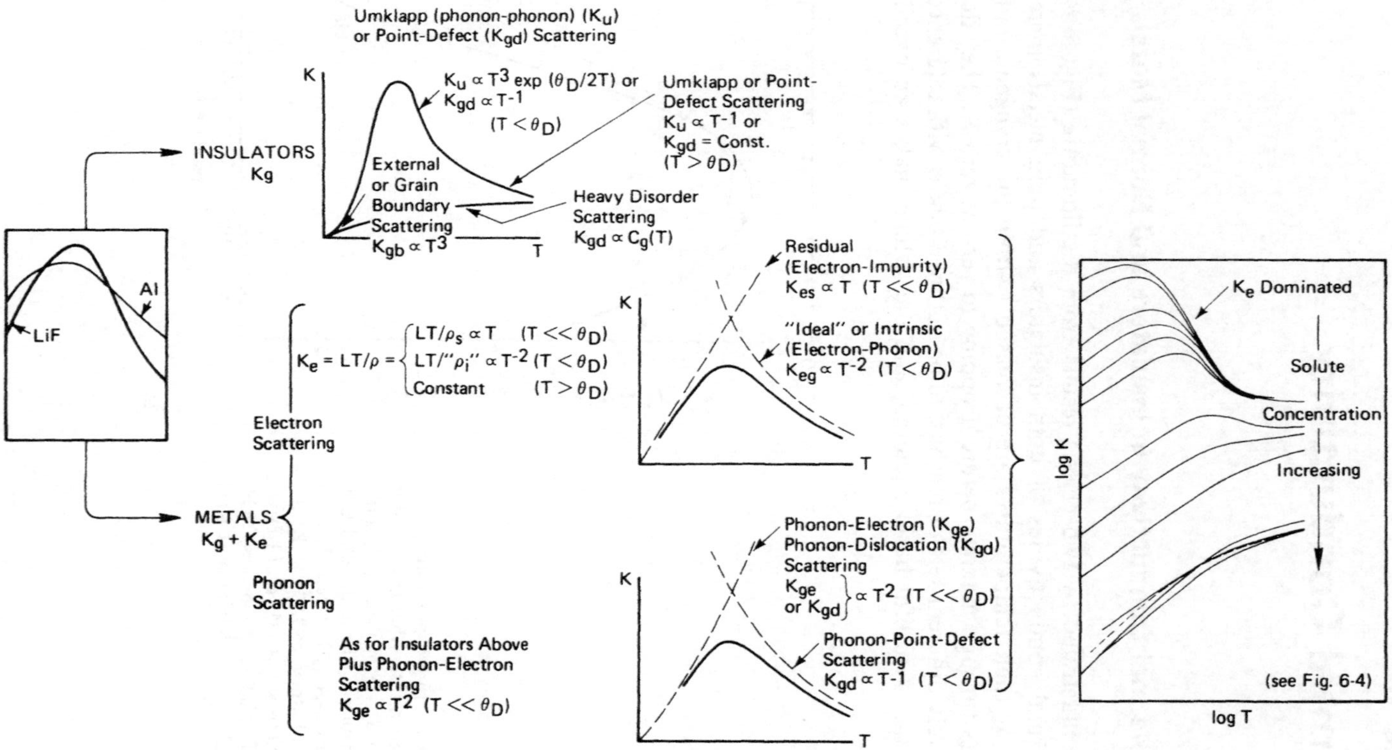

Figure 6-2. Components of the thermal conductivities of insulators and metals.

starting point for discussion of thermal conductivity, is

$$K(T) = \tfrac{1}{3}C^v(T)vl(T) \tag{6-1}$$

In this equation, originally due to Debye, $C^v(T)$ is the specific heat per unit volume, and v and l are the velocity and mean-free-path length, respectively, of the "carriers" of heat. All four symbols may be subscripted with either a "g" or an "e" indicating the applicability of (6-1) to either the phonon (lattice) or electronic heat conductivities, respectively. The known temperature dependence of $C^v(T)$ being taken for granted, that of $K(T)$ is discussed primarily in terms of the influence of temperature on the numerous possible phonon- and electron-scattering processes.

In the following discussions thermal conductivity is denoted by K and thermal resistivity (the reciprocal of conductivity) by W, to each of which subscripts are appended to denote the following types of scattering process:

eg electron-lattice (i.e., intrinsic or "ideal" scattering)
es electron-impurity
ge phonon-electron
gd phonon-defect
gi phonon-phonon (intrinsic or "ideal" scattering again)

6.2 Insulators

6.2.1 Intrinsic Lattice Conductivity

In large, perfect, insulating crystals the only impediment to the propagation of phonons is their mutual scattering. But even within the context of phonon-phonon scattering, momentum can be extracted from an otherwise conservative three-phonon interaction only if the mechanism referred to by Peierls as the "umklapp" process is operative.[1] Umklapp scattering is weak at very low temperatures—K_g is large. As the temperature increases, K_g decreases at a rate described by $K_u \propto (T^3/\theta_D)\, e^{\theta_D/2T}$ [KLE58]. At high temperatures ($T > \theta_D$), the operation of umklapp phonon-phonon scattering confers a T^{-1} dependence upon K_g. Returning to the low-temperature regime, we note that a maximum develops as K_g tends to zero at absolute zero. On the low-temperature side of the maximum, the phonon mean-free-path length, l_g, being limited by crystal size, may be regarded as constant. In this so-called "boundary-scattering regime," $K_g(T)$ is proportional to $C_g(T)$, which in turn is proportional to T^3 in the liquid He temperature

[1] A conservative collision process in which the total wavevector of the phonons is conserved is represented by $\mathbf{K}_1 + \mathbf{K}_2 = \mathbf{K}_3$. Thermal resistance in a perfect crystal requires the intervention of the lattice, as represented by $\mathbf{G}$ (a reciprocal lattice vector), such that $\mathbf{K}_1 + \mathbf{K}_2 = \mathbf{K}_3 + \mathbf{G}$ [KIT68, p. 190]. This is the "umklapp" (literally "doubling back") process.

range. Thus, for the intrinsic or "ideal" lattice thermal resistivity, W_{gi}, we have, according to KLEMENS [KLE56],

$$W_{gi} = A_{gi}T^{-3} + B_{gi}T^{-3}\, e^{-\theta_D/2T} \text{ (low to intermediate temperatures, } T < \theta_D) \tag{6-2a}$$

$$W_{gi} = C_{gi}T \text{ (high temperatures, } T > \theta_D) \tag{6-2b}$$

where the A, B, and C are constants of proportionality.

6.2.2 Influence of Impurities

The scattering of phonons by defects depends on their size and nature (i.e., whether they are point defects, dislocations, etc.) and on the phonon wavelength in relationship to these factors. KLEMENS [KLE56] has discussed in detail phonon scattering from point defects, single dislocations, grain boundaries, and external boundaries. In specifically dealing with the temperature dependence of such scattering KLEMENS [KLE58] pointed out that at low temperatures ($T < \theta_D$) the individual contributions to phonon scattering from crystalline imperfections of various kinds can be expressed in the form:

$$W_{gd} \propto T^m \tag{6-3a}$$

$$\text{where } m = \begin{Bmatrix} 1 \\ 0 \\ -1 \\ -2 \\ -3 \end{Bmatrix} \text{ for } \begin{Bmatrix} \text{point defects} \\ \text{dislocation cores} \\ \text{stacking faults} \\ \text{dislocation strains} \\ \text{sample or grain boundaries} \end{Bmatrix} T < \theta_D \tag{6-3b}$$

In particular, for an annealed impure crystal $W_{gd} \propto T$; but at high temperatures ($T > \theta_D$) all imperfections yield W_{gd} = constant. In imperfect crystals these terms replace the previously mentioned umklapp resistance, which is only resolvable in (i) sufficiently near-perfect crystals, and (ii) large crystals such that boundary scattering is deferred to sufficiently low temperatures (long phonon wavelengths). Thus, for crystals with structural defects in addition to sample and grain boundaries the thermal resistance is given, also according to KLEMENS [KLE58], by

$$W_{gd} = A_{gd}T^{-3} + \sum_n A_{gn}T^m \tag{6-4a}$$

and in general

$$W_{gd} = \sum_n B_{gn}T^m \tag{6-4b}$$

(low to intermediate temperatures, $T < \theta_D$)

where the As and Bs are constants of proportionality, while

$$W_{gd} = \text{constant at high temperatures } (T > \theta_D) \tag{6-4c}$$

6.2.3 Influence of Grain Boundaries and Lattice Disorder

KITTEL [KIT49] has discussed the thermal conductivity of oxide glasses, whose structures might, in a sense, be regarded as limiting examples of polycrystalline disorder. The theory focuses attention on the limitation of phonon mean-free-path length, $l(T)$, that takes place when the dominant phonon wavelength, λ_d, which is a decreasing function of temperature according to

$$\lambda_d = hv_g/k_BT \tag{6-5}$$

(where v_g is the acoustic velocity), drops below some value d which, for glass, was a dimension of the unit cell. At temperatures below some T_0 (defined such that $\lambda_d = d$) the longer wavelength phonons do not "see" the lattice distortion, and $l_g(T)$ responds to temperature in the normal way by increasing with decreasing temperature at ordinary temperatures. At higher temperatures, such that λ_d is now less than d, $l_g(T)$, being constrained by d, is constant; under this condition the thermal-conductivity temperature dependence, which according to (6-1) is now directly proportional to $C_g(T)$, follows a monotonically increasing Debye-like curve. Similar behavior is expected of fine-polycrystalline, densely precipitated, or heavily deformed alloys, the temperature range of the effect being determined by the relationship cited above in the form

$$d = \lambda_d = hv_g/k_BT = (h/k_B)(E/\rho_d)^{1/2}/T_0 \tag{6-6}$$

where E is the Young's modulus and ρ_d is the density.

If mean-free-path limitation occurs at temperatures below the peak in the conductivity curve (see Figure 6-1), since $K_g(T)$ is proportional only to $C_g(T)$, the peak will be suppressed. Evidence in support of this argument is seen in the results of the measurements by BERMAN [BER51] of the influence of successively heavier neutron dosages on the low-temperature thermal conductivity of crystalline quartz.

KITTEL [KIT49] cited as further evidence in support of the path-length-limitation theory some results from acoustic attenuation studies, according to which, if in a polycrystalline solid the wavelength is much smaller than the grain size, the free path is constant and independent of frequency. This, of course, imposes only an *upper limit* on free-path length—if the grains are sufficiently large as to contain another level of lattice defects, the free

Table 6-1. Solutions of the Equation $k_BT = hv_g/\lambda_d$ for Ti-Nb(~30 at.%)

T, K	λ_d, Å
100	15
50	30
20	75
10	150
5	300

$E \equiv 7 \times 10^{11}$ dyn cm^{-2} (Table 4-4).
$\rho_d \equiv 6.5$ g cm^{-3}, and hence
$v_g \equiv 3.3 \times 10^5$ cm s^{-1}.

path will be appropriately reduced by them. In order to investigate whether an effect comparable to this might be seen in alloys of interest in the present context, values of E (from Table 4-4) and ρ_d (estimated) for Ti-Nb(30 at.%) were inserted into (6-6). The results, listed in Table 6-1, suggest that, in the presence of ω-phase precipitation (see Section 1.8), critical temperatures for free-path limitation may occur in the 5–20 K temperature range.

6.3 Conductors

The foregoing discussion applies to all solids—insulators or conductors. But in the latter, the presence of conduction electrons results in: (i) an additional scattering opportunity for the phonons, and hence an increase in the phonon resistivity; (ii) a competitive mechanism for heat flow in the form of electronic thermal conductivity which "short-circuits" the phonon component. Thus, to the expressions for *phonon resistivity*, (6-2) and (6-4), must be added fourth terms representing scattering via the electron–phonon interaction. The *electronic component* of thermal conductivity, K_e, is responsive to the same electron-scattering mechanisms that control electrical conductivity—the electron–phonon interaction again, plus impurity scattering.

Before proceeding it is useful to note that for a given conduction type, phonon or electronic, the respective aggregates of all scattering processes are the *sums* of the individual *thermal-resistivity* components: thus, $W_g = \sum_m W_{g,m}$ for the phonons, as in (6-2) and (6-4); likewise $W_e = \sum_n W_{e,n}$ for the electrons. On the other hand, for a metal, in which both processes take place concurrently and hence in "parallel," the combined effect of phonon and electron scattering is best described in terms of *conductivities*, thus,

$$K = K_e + K_g \tag{6-7a}$$

$$\equiv W_e^{-1} + W_g^{-1} \tag{6-7b}$$

$$\cong K_{es} + K_{ge} \tag{6-7c}$$

6.3.1 Relative Magnitudes of Insulator and Conductor Conductivity

The thermal conductivity temperature dependence of a typical pure metal (e.g., Cu) and a typical ionic insulator (e.g., LiF) are juxtaposed for comparison in Figures 6-3(a) and (b). Clearly it is fruitless to make any evaluation on the quotient $K_{\text{metal}}/K_{\text{insulator}}$ at temperatures much below about 40 K since in that regime the low-temperature conductivities of different purities of a given material differ by factors as large as 30. But at temperatures above about 100 K, wherein the conductivities of each class of material coalesce into a single curve, a valid comparison of K_{metal} and $K_{\text{insulator}}$ is possible. Accordingly, in the construction of Table 6-2, in which the conductivities of various metals, alloys, and insulators are compared, 273 K was selected as the temperature of the comparison. In general it

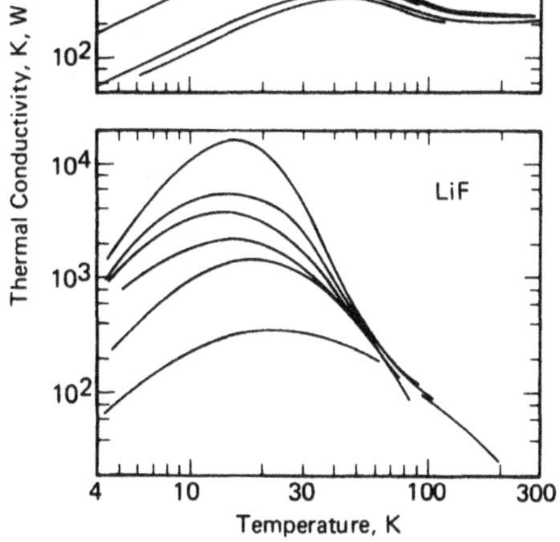

Figure 6-3. Temperature dependences of thermal conductivity of various samples of a typical unalloyed metal (Al) and a typical insulator (LiF). One curve from each group had been selected for use in Figure 6-1. The curves show that differentiation between the thermal conductivities of unspecified "Al" and "LiF" is only meaningful at temperatures above about 100 K—after CHILDS *et al.* [CHI73, pp. 123, 439].

appears that the room temperature thermal conductivities of pure metals and low-concentration alloys are some 30 times greater than those of some common insulators. Deliberately omitted from the table are the insulators Al_2O_3 (sapphire) and C (diamond), which possess exceptionally high room-temperature thermal conductivities, and unalloyed Ti and Nb, whose conductivities are rather low for supposedly pure metals.

If conduction electrons could be somehow added to an insulator, its lattice thermal conductivity would be lowered on account of the additional phonon–electron scattering. This possibility, together with the data of Table 6-2, suggests that in a metal the electronic component of thermal conductivity will, at room temperature, be at least 30 times greater than the lattice component. Indeed a comparison based on (6-1) combined with a knowledge of the relative specific heats, velocities, and relaxation times, leads to a similar conclusion.

Table 6-2. Thermal Conductivities at 273 K of Selected Pure Metals, Alloys and Insulators[a]

Material		Thermal Conductivity $W\,m^{-1}\,K^{-1}$	
Pure Metals	Related Alloys		
Al		235	
	Al 110		220
	Al 2024		130
	Al 3003		160
Cu		401	
	Free cutting leaded		380
	Phosphorus deoxidized		220
	Leaded brass		120
Fe		84	
	SAE 1020		65
	SAE 1059		45
Insulators			
AgCl		1.2	
$CuCo_3$	∥ to c-axis	5.5	
	⊥ to c-axis	4.6	
KCl		7.0	
NaCl		6.4	
SiO_2 (quartz)	∥ to c-axis	12	
	⊥ to c-axis	6.8	
TiO_2 (rutile)	∥ to optic axis	13	
	⊥ to optic axis	9	

[a]American Institute of Physics Handbook, Third Edition, 1972, McGraw-Hill, Inc. Used with permission of McGraw-Hill Book Company.

By way of illustration, in a moderately concentrated swaged-and-annealed Al–Mg alloy, *viz.*, Al–Mg(7 at.%), theory and experiment applied to thermal conductivity data for the temperature interval 4–60 K [KLA75] have shown that at 10 K, $K_e/K_g \cong 9.51/0.52$ (in W m^{-1} K^{-1}) $\cong 18$. But the situation is quite different for concentrated disordered alloys, such as stainless steels, in which the phonon and electronic contributions may be similar in magnitude.

6.3.2 The Electronic Component

The title of this paragraph implies the existence of a relationship of some sort between the electronic and thermal conductivities of metals. It can be shown by quite simple free-electron arguments, as in introductory texts or more generally as in a work by KLEMENS [KLE56], that if σ is the electrical conductivity and e the electronic charge:

$$\frac{K_e}{\sigma} = \frac{\pi^2}{3}\left(\frac{k_B}{e}\right)^2 T \tag{6-8}$$

This is known as the Wiedemann–Franz–(Lorenz) law. The temperature-premultiplier in the above expression is the Lorenz constant, L, which when evaluated in c.g.s.-practical units is 2.447×10^{-8} W Ω K^{-2}. The breakdown of the validity of the Wiedemann–Franz law, when it occurs, may be expressed in terms of the departure of L from this value. Thus, whereas for eleven pure metals at 0°C $L = 2.48 \pm 0.22$, in good accord with expectation, pronounced departures from the theoretically predicted and experimentally verified (at room temperature) values are noted as the temperatures of *pure* metals are lowered into the liquid-He range [BER52]. This deviation has to do with the development of different scattering lengths for electrical transport and heat transport, respectively, [KLE56]. On the other hand for *alloys* at low temperatures, in which both K_e and σ (then designated σ_s) are controlled by impurity scattering, (6-8) is again valid. The Wiedemann–Franz law always breaks down at intermediate temperatures.

Equation (6-8) is $W_e = \rho/LT$; thus, we would expect the thermal resistivity to be the sum of "impurity" and "ideal" components proportional, respectively, to the electrical-resistivity components ρ_s and ρ_i of (5-2), according to

$$W_e = (\rho_s + \rho_i)/(LT) \begin{cases} T \ll \theta_D \\ T > \theta_D \end{cases} \tag{6-9}$$

We will consider in turn each of the two components of this equation.

(a) Thermal Resistivity due to Electron–Impurity Scattering, W_{es}. Since ρ_s is temperature independent, from (6-9)

$$\left.\begin{aligned} W_{es}(T) &= (\rho_s/L)T^{-1} \propto T^{-1} \\ K_{es}(T) &\propto T \end{aligned}\right\} \tag{6-10a}$$

hence

Alternatively, in dealing with the same problem, one could invoke an electronic form of (6-1), put $C_e(T) = \gamma T$, $v \equiv v_F$ (the Fermi velocity), and $l \equiv l_i$ (the distance between fixed impurity centers), to find that

$$K_{es}(T) = \tfrac{1}{3}\gamma T v_F l_i \tag{6-10b}$$

which, since v_F is a constant and l_i, the distance between impurity centers, is fixed, is again proportional simply to T.

(b) Thermal Resistivity due to Electron–Phonon Scattering, W_{eg}. At intermediate temperatures, defined by $T < \theta_D$, yet above the "residual" regime, electronic thermal resistivity is controlled by scattering against phonons. In keeping with (6-10), but with ρ_s replaced by ρ_i, and after recognizing that $\rho_i \propto T^5/\theta_D^6$ (the well-known "T^5 law" for ideal electrical resistivity), it would appear that $W_{eg}(T) \propto T^4$. It turns out, however, that because the phonon processes that control electrical transport in this temperature regime are *different* from those responsible for the scattering of electrons in heat transport, the above proportionality and consequently the Wiedemann-Franz law from which it stemmed are invalid for $T < \theta_D$. Formal arguments have demonstrated that

$$W_{eg} = A_{eg}T^2 \qquad (T < \theta_D) \tag{6-11}$$

where A_{eg} is a constant of proportionality, and furthermore that since the same phonon processes control both thermal *and* electron transport at *high temperatures,* in which regime $\rho_i \propto T$,

$$W_{eg}(T) = (\rho_i/L)T^{-1} = \text{constant} \qquad (T > \theta_D) \tag{6-12}$$

(c) Summary. The temperature dependence of *electronic* thermal resistivity over the entire temperature range is describable in the form:

$$W_e(T) = W_{es}(T) + W_{eg}(T) \tag{6-13a}$$

$$= (\rho_s/L)T^{-1} + A_{eg}T^2 \qquad (T < \theta_D) \tag{6-13b}$$

$$= (\rho_s/L)T^{-1} + \text{constant} \qquad (T > \theta_D) \qquad \text{(6-13c)}$$

$$\cong \text{constant} \qquad (T \gg \theta_D) \qquad \text{(6-13d)}$$

all terms except $W_{eg}(T < \theta_D)$ being in accord with the Wiedemann-Franz law.

6.3.3 The Lattice Component—Thermal Conductivity under Phonon-Electron and Phonon-Impurity Scattering

All the phonon-scattering mechanisms operative in insulators also apply to conductors, in which, however, the presence of conduction electrons offers an additional process—phonon-electron scattering. The subject has been dealt with in detail by KLEMENS [KLE56, KLE58]. The form of the temperature dependence, a signature of great importance to the experimentalist, can be predicted by the following simple argument: the mean-free-path for phonon-electron scattering will be inversely proportional to the number density of electrons available for scattering, i.e., those within $\sim k_B T$ of the Fermi surface—thus, $l_{ge} \propto T^{-1}$. Next, with reference again to (6-1), and making the substitutions $C_g(T) \propto T^3$ and v_g (the phonon velocity) = constant,

$$K_{ge} = a_{ge}T^2 \qquad \text{(6-14a)}$$

or

$$W_{ge} = A_{ge}T^{-2} \qquad \text{(6-14b)}$$

where a_{ge} and A_{ge} are constants of proportionality. Other opportunities for phonon scattering in metals are similar to those that exist in insulators and have already been dealt with.

The full expression for lattice resistivity, the phonon counterpart of (6-13a), is

$$W_g(T) = W_{gd}(T) + W_{ge}(T) \qquad \text{(6-15)}$$

Whether or not the full expression needs to be retained in a given situation depends, of course, on relative magnitudes. Thus, while point defects are extremely effective in scattering electrons at low temperatures and contribute significantly to the residual resistivity, ρ_s, they are not "seen" by the long-wavelength phonons. Dislocations, which scatter both phonons and electrons, contribute according to (6-3), a T^2 temperature dependence to the phonon conductivity. Thus, the scattering of phonons by dislocations on one hand or by electrons on the other are indistinguishable, as gauged by their temperature dependences.

6.4 Thermal Conductivity of Alloys

6.4.1 Influence of Solute Concentration

The thermal conductivity processes outlined above are summarized and illustrated in Figure 6-2. As a result of the foregoing discussion, it is possible to appreciate the changes in thermal scattering processes that underlie the trends observed in Figure 6-4, in which the thermal conductivities of various samples of Cu, some Al alloys, and some stainless steels are intercompared. With reference to Figure 6-4, the addition of solute to the initially pure metal results in the steady washing out of the low-temperature peak in $K(T)$, which eventually becomes reduced to a monotonic increase with respect to temperature. For alloys at low temperatures ($T \ll \theta_D$), K_e is dominated by impurity scattering and K_g by phonon–electron and phonon–dislocation scattering; the operative form of (6-7) is, therefore,

$$K = K_e + K_g \tag{6-7a}$$

$$\cong K_{es} + K_{ge} \tag{6-7c}$$

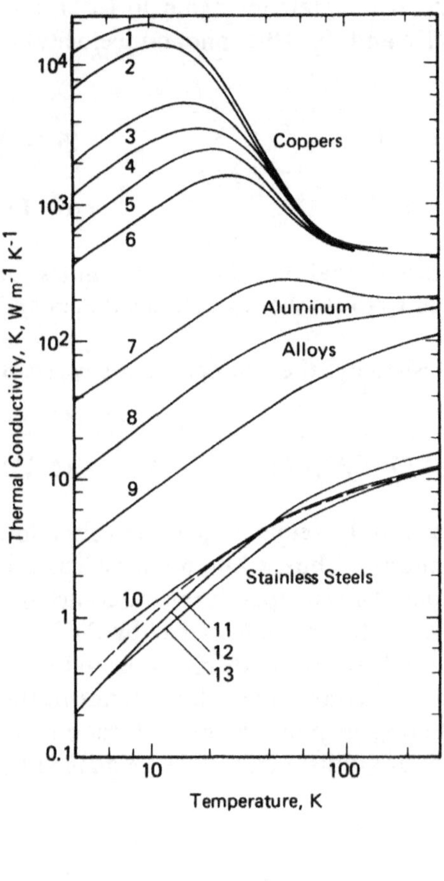

Figure 6-4. Thermal conductivities of various pure metals and alloys: *Copper*: (1) RRR = 2000, (2) RRR = 1530, (3) Johnson Matthey 5N-Cu, vacuum annealed, (4) another batch of the same, (5) RRR = 110, (6) specimen no. 3 plus cold work. *Aluminum Alloys*: ALCOA Al with: (7) 0.65% Mg, 0.38% Si, etc., (8) 1.23% Mn, 0.43% Fe, 0.15% Si, etc., (9) 4.58% Cu, 1.70% Mg, etc. *Stainless Steels*: (10) AISI 310, (11) A-286, (12) AISI 304, 316, (13) Kromarc 58. Data sources: Cu and Al alloys, CHILDS *et al.* [CHI73, pp. 97, 347]; stainless steels, [MET74, Section 8.0].

which, with the aid of (6-10a) and (6-14a), becomes

$$K = (L/\rho_s)T + a_{ge}T^2 \qquad (T \ll \theta_D) \qquad (6\text{-}16)$$

whose constants are determinable from the intercept and slope of a plot of K/T *versus* T. At higher temperatures, but still such that $T < \theta_D$, K_e is replaced by $K_{eg} \propto T^{-2}$ (see (6-11)), and K_g becomes dominated by point-defect scattering such that, according to (6-3), $K_{gd} \propto T^{-1}$. These effects serve to counteract the initially quadratic increase in K with temperature. A maximum may be induced by this succession of scattering processes (or the pronounced peak exhibited by a pure solvent, more or less-retained). At high temperatures ($T > \theta_D$) in the disordered alloy, both K_{eg} ((6-13c)) and K_{gd} ((6-4c)) tend to become independent of temperature.

6.4.2 Separation of the Electronic and Lattice Components

Useful information regarding the nature of lattice defects can be obtained from the temperature dependence of lattice thermal conductivity, $K_g(T)$, a quantity that can be separated away from the measured total thermal conductivity, K, with the aid of (6-7), (6-13b), and (6-16) in the following way: The range of temperatures within which the electrical resistivity has dropped to its residual value, ρ_s, is first determined. Thermal conductivity data taken within this same range, hence describable by (6-16), and plotted in the format K/T *versus* T, yields values of (L/ρ_s) and a_{ge}. The separation of $K(T)$ into its two components, for $T \ll \theta_D$, has now been achieved. To extend the range of this separation, the full $T < \theta_D$ expression for electronic thermal resistivity ((6-13b)) is next constructed with the aid of an A_{eg} value, which has to be taken from measurements on the pure solvent or calculated from first principles [Mor73]. Finally, a reapplication of (6-7) in the form

$$K_g(T) = K(T) - [(\rho_s/L)T^{-1} + A_{eg}T^2]^{-1} \qquad (T < \theta_D) \qquad (6\text{-}17)$$

leads to the desired result.

A recent example of the application of this procedure to Al alloys is given in a paper by Klaffky *et al.* [Kla75]. The results showed that for numerous alloys, such as Al-Mg(6 and 7 at.%), K_g closely followed a T^n-law, with $n = 2$, as prescribed by phonon-electron and phonon-dislocation scattering, for $T \lesssim 10$–15 K. They also showed that above these temperatures, n began to drop below 2 as the effects of phonon-point-defect scattering began to make themselves felt.

6.5 Thermal Conductivity Data

Cryogenic thermal conductivity data for materials such as Cu-base and Al-base alloys are readily available. Not so for the stainless steel alloys of use in superconducting machinery, whose thermal conductivities are, therefore, listed in Table 6-3. These data for stainless steels have also been included in Figure 6-4 as examples of concentrated alloys with very low thermal conductivities. The temperature dependences of AISI types 304, 316, and 310, based on the plotted data of [MET74], are proportional to $T^{1.4}$(4–8 K), $T^{1.4}$(4–8 K), and $T^{1.0}$(6–30 K), respectively, indicating the early influence of general phonon–defect scattering mechanisms that tend to reduce $K_g(T)$ below its expected T^2 low-temperature temperature dependence. Also included in the table are some data for unalloyed Ti and Nb in the normal state.

Table 6-3. Thermal Conductivities of Some Representative Stainless Steels; also Unalloyed Titanium and Niobium

Material	Cr/Ni Weight-% Ratio[a]	Thermal Conductivity, K (W m^{-1} K^{-1}), Temperature, K: 10	20	50	100	273	Reference
AISI Type Stainless Steels							
304	(18–20)/(8–10.5)	0.78	1.95	5.7	9.5	14.7	[MET74]
316	(16–18)/(10–14)	0.78	1.90	5.6	9.4	14.7	[MET74]
310	(24–26)/(19–22)	1.15	2.30	5.4	8.1	12.4	[MET74]
Proprietary Type Stainless Steels							
Kromarc 58	15/22	0.72	1.55	4.2	6.8	11.5	[MET74]
A-286	15/26	1.01	2.20	5.0	7.6	11.2	
Unalloyed Ti	99.99%	{11.5	21	30.5	30	—}	[CHI73][b]
		14	28	38	35	—	
Unalloyed Nb	RRR = 10	16.5	31	46	(50)[e]	—	[CHI73][c]
	RRR = 33	55	85	66	(54)	—	[CHI73][d]

[a] Stainless steels also include stabilizing and strengthening additions of C, N, Mo, Mn, etc.
[b] Data of (i) White Woods (RRR = 16–22) and (ii) Rosenberg, together make up this spread.
[c] Data of (i) Mendelssohn and Rosenberg and (ii) Rosenberg } "low-" and "high-purity" Nb
[d] Data of White and Woods } respectively—see [CHI73].
[e] ( ) indicates extrapolated from ~90 K.

6.6 Thermal Conductivity in a Magnetic Field

As pointed out by ROSENBERG [ROS63, p. 145], several workers, using monocrystalline simple metals, observed large variations in thermal resistance, and concurrent variations in electrical and thermal resistances, in response to changes of applied magnetic field strength. A magnetic field, when applied to metals with the usual nonspherical Fermi surfaces, influences the electron scattering processes, and is thereby responsible for the magnetoresistive effect discussed in Section 5.8. By the same token it should also influence the electronic component of thermal conductivity, and indeed an effect known as "magnetothermal conductivity" has been observed in the past and has also been the subject of several investigations by SPARKS [SPA77, SPA78] and FEVRIER and MORIZE [FEV73]. Since the phonons are unresponsive to the applied field, its influence on thermal conductivity should diminish as the dominance of the electronic contribution to the conduction process decreases. Thus, whereas a field of 80 kOe was able to reduce the 5 K thermal conductivity of unalloyed Cu by 340% (for RRR = 1520) and 138% (for RRR = 107), the alloys 1100-Al (annealed, RRR = 33), Ti-6Al-4V, and Fe-47.5Ni experienced reductions of only 41%, 16%, and 4.1%, respectively [SPA78].

Those experiments were performed in a longitudinal applied field. It is, of course, important to enquire whether the influence of a transverse field would be smaller or larger. Magnetoresistivity results suggest that the latter would tend to be the case. Indeed, experiments by FEVRIER and MORIZE [FEV73] on two grades of unalloyed Cu (RRR = 62 and RRR = 163) demonstrated that the reduction in thermal conductivity due to an applied magnetic field was significantly larger when it was applied in the transverse direction; at 50 kOe, for example, the drop in magnetothermal conductivity accompanying a 90° rotation of a previously longitudinal field was 17% at RRR = 62 and 32% in the RRR = 163 sample of Cu.

Pure grades of Cu and Al are needed for the stabilization of multifilamentary superconducting composites. In these materials K_e is the dominant thermal-conductivity component and the magnetothermal conductivity effect is correspondingly pronounced. It would be useful to be able to predict its magnitude from the results of magnetoresistivity measurements alone. This could certainly be accomplished if the Wiedemann-Franz law, which usually governs the relationship between ρ and K_e in the absence of a magnetic field ((6-8) and (6-9)), was equally valid in its presence. FEVRIER and MORIZE [FEV73], using the two grades of Cu referred to above, examined this problem and, as indicated in Table 6-4, came to the conclusion that the Wiedemann-Franz law remained applicable in fields of up to at least 50 kOe, thus, enabling the thermal conductivity in applied transverse or longitudinal fields to be deduced directly from measured values of the

Table 6-4. Influence of Transverse and Longitudinal Magnetic Fields of up to 50 kOe on the Lorenz Numbers[a] of Two Samples of Unalloyed Copper—After Fevrier and Morize [FEV73]

Sample	RRR	Thermal Conductivity, K at 4.2 K $\mathrm{W\,m^{-1}\,K^{-1}}$	$L(H_a)/L(H_a = 0)$	
			Transverse	Longitudinal
Cu-I	62	330	1.13 ± 0.10	1.06 ± 0.03
Cu-II	163	950	1.03 ± 0.03	1.06 ± 0.04

[a] The isothermal Lorenz-number ratio, according to Eq. (6-8) is $K_e(H)\rho(H)/K_e(0)\rho(0)$, and this was the ratio actually computed.

resistivity in those fields coupled with the zero-field Lorenz number. A similar conclusion was reached by SPARKS [SPA78] as a result of studies on unalloyed Al (RRR = 100).

Magnetothermal conductivity and its companion parameter, magnetoresistivity, are both important ingredients of composite conductor *design*, which is the subject of Chapter 27. In that they influence the choice of stabilizer material, the magnetoresistivities and magnetothermal conductivities of Cu and Al are intercompared in Chapter 28, which discusses the *fabrication* of composite superconductors.

6.7 Superconductors

Early treatments of heat capacity and heat transport in superconductors were based on the assumption that the conduction electrons could be regarded as constituting a pair of immiscible "fluids" whose volume ratio depended on the reduced temperature, $t \equiv T/T_c$ [MEN50]. One of the fluids was supposed to exhibit normal electronic properties, while the other consisted of the superelectrons. While it is recognized by now that the two-fluid idea is microscopically incorrect it can with some justification, according to BARDEEN, RICKAYZEN, and TEWORDT [BAR69], be employed to describe the phenomenological calorimetric and thermal transport properties of the superconducting state.

In the following discussion superscripts n and s are used to distinguish between normal-state and superconductive-state parameters.

6.7.1 The Electronic Thermal Conductivity of Superconductors, K_e^s

According to the two-fluid model, as the temperature decreases so does the fraction of normal electrons. The calorimetric consequences of this are

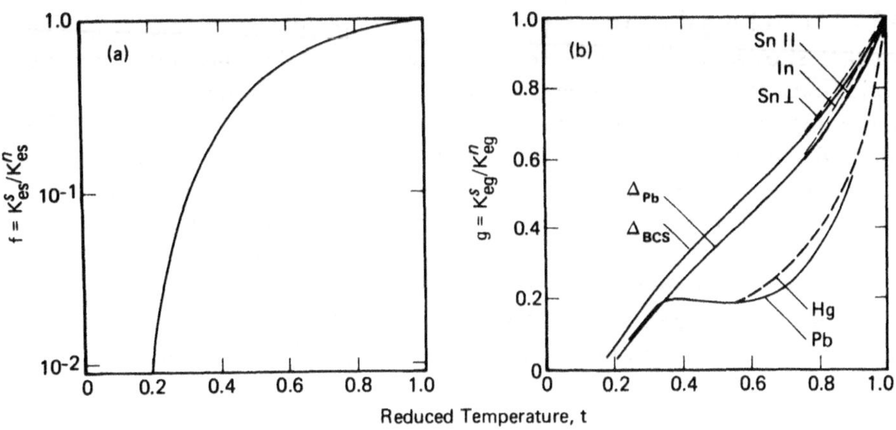

Figure 6-5. Three components, in reduced form, of the thermal conductivity of superconductors plotted *versus* reduced temperature $t \equiv T/T_c$: (a) the component $f = K^s_{es}/K^n_{es}$ representing impurity-scattered electronic thermal conductivity—after BARDEEN *et al.* [BAR59]; (b) the component $g = K^s_{eg}/K^n_{eg}$ representing phonon-scattered electronic (i.e., "intrinsic" electronic) thermal conductivity. The theoretical curves (full lines) are based on the BCS gap energy and that for Pb, respectively; the broken lines are experimental results—after TEWORDT [TEW63]; (c) the component $h = K^s_{ge}/K^n_{ge}$ representing the phonon (-electron) thermal conductivity—after BARDEEN *et al.* [BAR59].

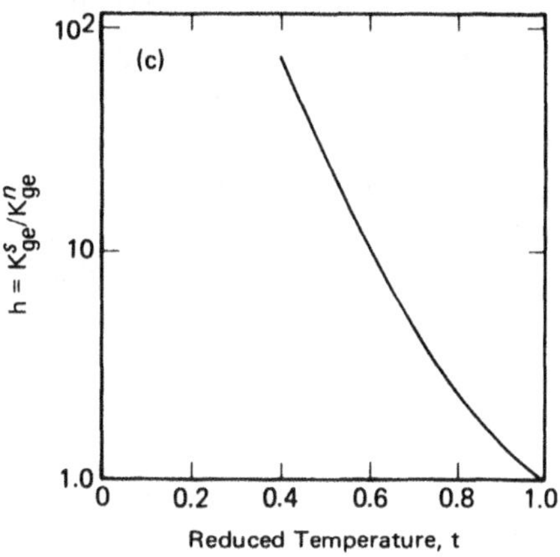

dealt with in Chapter 8. If the heat transport in the normal state is dominated by K_e (see Section 6.3.2), it can be anticipated that K^s will begin to drop markedly below K^n-projected as T decreases below T_c. These predictions were quantitatively confirmed by BARDEEN *et al.* [BAR59], who treated the electronic thermal conductivity of superconductors within the framework of BCS theory [BAR57][2] assuming, in addition to weak electron–phonon coupling, that the dominant thermal scattering mechanism was electron-impurity in nature (see Section 6.3.2(a)). The results of the calculation, expressed in the form of a plot of K^s_{es}/K^n_{es} *versus* t, are given in Figure 6-5(a), a particular feature of which is the manner in which the ratio approaches unity—namely, with a continuously decreasing slope that vanishes at $t = 1$. In a subsequent calculation, TEWORDT [TEW63] considered the intrinsic electronic thermal conductivity of superconductors (see Section

[2]The well-known theory of BARDEEN, COOPER, and SCHRIEFFER [BAR57].

6.3.2(b)) and inquired into the temperature dependence of K^s_{eg}/K^n_{eg}. The results, shown in Figure 6-5(b), indicate that under phonon-scattering conditions, the electronic thermal conductivity ratio approaches $t = 1$ with a limiting slope of about 1.6. As indicated by the authors [TEW63], the thermal conductivities of In and Sn were in excellent accord with the theory in the important temperature range above $t = 0.8$. But with Hg and Pb, both of which are strong-coupled superconductors (see Sections 8.5 and 8.6), the ratio K^s_{eg}/K^n_{eg} quickly dropped below the predicted curve with decreasing t. Nevertheless theory and experiment, as represented by Figures 6-5(a) and (b), agree that under conditions of either electron-impurity [BAR69] or electron-phonon [TEW63] scattering K^s_e/K^n_e decreases to zero with decreasing t.

6.7.2 The Phonon Thermal Conductivity of Superconductors, K^s_g

Again invoking the two-fluid idea for the purpose of conceptualization, we note with brief reference to a subsequent discussion (Section 8.4) that the fraction of normal electrons decreases to zero as t^4. Since these are the only electrons that can exchange net momentum with the lattice (a BCS concept) electron-phonon scattering (and, by reciprocity, phonon-electron scattering) drops to zero with decreasing t, thus depriving the phonons of

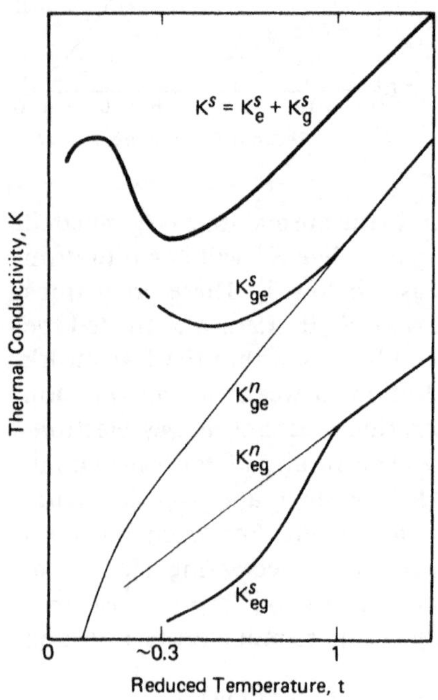

Figure 6-6. Development of a minimum followed, with decreasing t (<1), by a maximum in the superconductive thermal conductivity as a consequence of the competition between the phonon (K^s_{ge}) and electronic (K^s_{eg}) thermal conductivities.

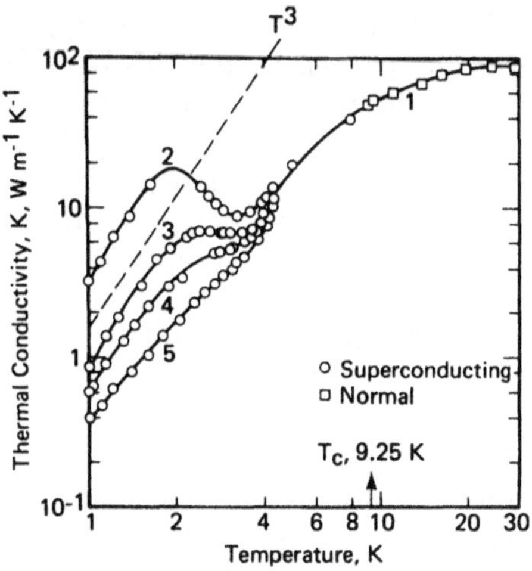

Figure 6-7. Thermal conductivity temperature dependence of pure Nb: (1) Fansteel 99.9%, RRR (22 K) = 33; (2) Johnson-Matthey zone refined, $H_V = 66$; (3) the same; (4) the same, 51% strain; (5) the same, 195% strain—after CHILDS *et al.* [CHI73, p. 53].

their principal low-temperature scattering mechanism, thereby allowing the phonon conductivity to increase. This prediction was also confirmed by BARDEEN *et al.* [BAR59], who calculated the lattice thermal conductivity by solving a Boltzmann equation within the framework of BCS theory. The results of this work are illustrated in Figure 6-5(c), in which we see K^s_{ge}/K^n_{ge} *increasing* rapidly (at a rate $\propto t^{-5}$) with decreasing t. This phonon-electron enhancement factor is usually represented by h (e.g., [KLE64], [MOR77]); while MORTON [MOR73] adopted the symbols f and g for the ratios K^s_{es}/K^n_{es} and K^s_{eg}/K^n_{eg}, respectively. The increasing K^s_{ge} and decreasing K^s_{es} or K^s_{eg} are in competition; in the event that the ratios K_e/K_{eg} are suitable, it is to be expected that the total thermal conductivity $K^s = K^s_{es,g} + K^s_{ge}$ (e.g., the additivity relationship, (6-7)) should develop a minimum at some intermediate value of t ($\cong 0.3$ according to experiment) as illustrated in Figure 6-6. At very low temperatures in even the purest metal, with the freezing out of the conduction electrons, heat conductivity is transferred completely to the lattice; the metal then assumes the character of an insulator, whose conductivity drops to zero according to T^3 (see (6-2a) and (6-4a)). These statements are amply substantiated by the results of numerous studies of the thermal conductivity of high-purity Nb, Figure 6-7. Particularly spectacular examples of the "superconductive phonon peak,"[3] generated in the manner just described, are to be seen in the low-temperature thermal conductivity temperature dependences of monocrystalline Nb [MEN61], [UML67] and Ta [MEN61]. In commenting on their observations, MENDELSSOHN and ROSENBERG [MEN61] noted that such enhanced phonon conduction was generally to be expected in transition metals with their already low values

[3] "Normal phonon peaks," when they occur, usually do so at temperatures near 20 K—Figures 6-1, 6-3, 6-4, and to a lesser extent Figure 6-7 (normal data), are typical in this regard.

of intrinsic thermal conductivity, K^s_{es}. However, with metals such as Nb, with its propensity towards contamination by O (a strong electron scatterer) and Ta (a weak scatterer), both of which are extremely difficult to remove (see, for example, [CoL72^b]), the influences of point-defect scattering on K^s_{es} and K^s_{ge} come into play. The effect is to reduce K^n_{es}, and consequently K^s_{es}, for a given K^s_{es}/K^n_{es} ratio. But possibly more important is the influence of point-defect scattering on K^s_{ge}. As predicted by KLEMENS and TEWORDT [KLE64], and amply confirmed by MORTON *et al.* in studies of Nb-Zr(25%) [MOR73, MOR74], point-defect scattering reduces the rate of increase of K^s_g/K^n_g with decreasing t to values much smaller than those predicted on the basis of the initially offered [BAR59] phonon–electron scattering theory. Thus, the presence of point defects in the form of interstitial or substitutional solute atoms, radiation damage, or even the Abrikosov flux lattice, reduces the phonon peak first to a shoulder and finally at sufficient density to a monotonic continuation of the higher temperature curve. This peak suppression is exemplified by the results of the early investigations of Nb by MENDELSSOHN and OLSEN [MEN50] and MENDELSSOHN and ROSENBERG (Nb, $RRR_{20\,K} = 10.4$) [MEN52] (Figure 6-8(a)), and by the more recent results of SCHMIDT [SCH75] on a Nb sample of somewhat higher purity ($RRR_{10\,K} = 44$) (Figure 6-8(b)).

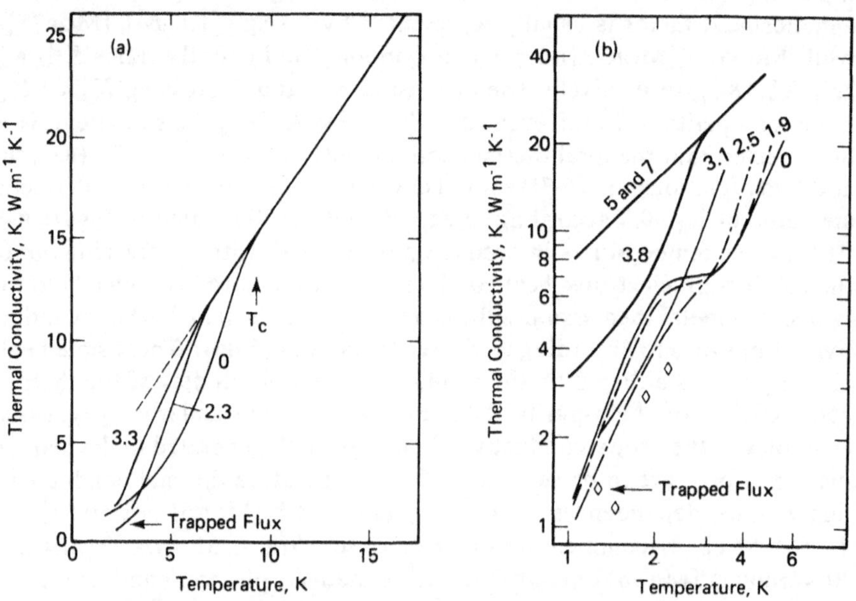

Figure 6-8. Influence of applied magnetic fields on the thermal conductivity of two specimens of pure Nb: (a) $RRR_{20\,K} = 10.4$—after MENDELSSOHN and ROSENBERG [MEN52]; (b) $RRR_{10\,K} = 44$—after SCHMIDT [SCH75]. The numbers against the curves are the applied field strength, H_a, in kOe.

Reduction of the thermal conductivity anomaly is also expected in strained high-purity Nb when, although the decrease in K_{eg}^{s} is substantial, the already point-defect-diminished increase in K_{ge}^{s} is further reduced by the $K_{gd}^{n} \propto t^2$ decrease due to phonon-dislocation scattering followed again by the $K_{gb} \propto t^3$ boundary-scattering effect ((6-3)). Examples of such lattice-defect suppression mechanisms operating in sequence are to be found in the results of the low-temperature thermal conductivity studies (0.5 to 1.5 K) by IKEBE, KOBAYASHI, and MUTO [IKE74] of unalloyed Ta, both as-received and cold rolled, and the alloys Ti–Nb(20, 40, and 80 at.%). The influences of point defects and strain on the low temperature thermal conductivity of superconductors has been discussed by MENDELSSOHN and ROSENBERG [MEN61] whose early review of the low-temperature thermal conductivities of metals serves as a useful background to the recent literature.

6.8 The Mixed State

Thermal conductivity in the superconducting state, although independent of the strengths of small applied magnetic fields, H_a, eventually tends to pass through a minimum before rising to its normal value above the (upper) critical field. This behavior, although first observed by MENDELSSOHN and OLSEN [MEN50] in Nb and dilute Pb–Bi alloys, was later shown to be exhibited by the pure elemental type-I superconductors In, Sn, and Pb. Thus, although the following qualitative discussion will be couched within the terminology of mixed-state type-II superconductivity, it would seem to apply equally well to the type-I intermediate state, which is otherwise of no concern to us in this treatise. Magnetothermal conductivity in superconductors has been briefly reviewed by UMLAUF [UML67], with reference to the pre-1966 literature, as part of a detailed study of monocrystalline ($RRR_{10\,K} = 22$) and wire ($RRR_{10\,K} = 250$) samples of Nb. The results of a study of Nb ($RRR_{20\,K} = 10.4$) by MENDELSSOHN and ROSENBERG [MEN52] are summarized in Figure 6-8(a) and those of the most recent investigation of Nb, that undertaken by SCHMIDT [SCH75] during a study of the Kapitza resistances against Cu of Nb and Ti–Nb, are to be seen in Figure 6-8(b). Data taken from this latter group of curves, and plotted as *K versus* H_a isothermals, as in Figure 6-9, clearly exhibit the characteristic minimum within certain ranges of field and temperature. The existence of the mixed state is a necessary condition for the occurrence of the conductivity minimum. As the field increases, thermal conductivity first decreases due to the appearance of additional scattering sites, presumably in the form of Abrikosov fluxoids. This decrease eventually comes into competition with an increase in K_e^n due to electronic transport through the increasing volume fraction of the essentially normal cores of those fluxoids, following the

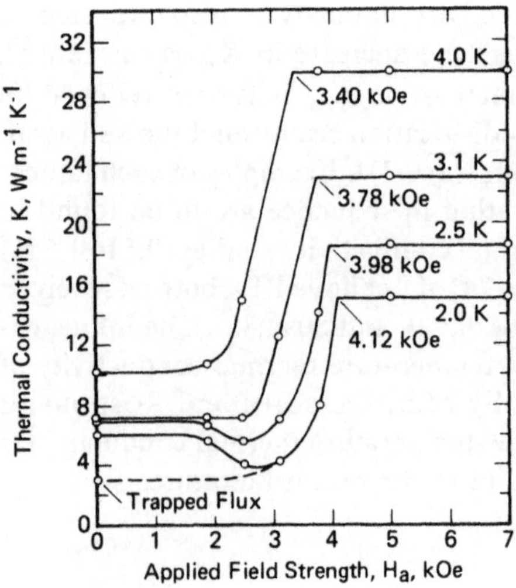

Figure 6-9. *K versus* H_a isothermals plotted from the data of Figure 6-8. The minimum in the magnetothermal conductivity is a consequence of phonon-scattering from fluxoids.

ideas of BARDEEN and STEPHEN [BAR65] on that subject (see Section 23.8.1). The increase stops abruptly at H_{c2},[4] at which point normal thermal conductivity resumes.

With regard to the initial conductivity decrease, the "additional scattering" referred to above could be a result of either phonon or electron interaction with the matrix/fluxoid boundaries. UMLAUF [UML67] proposed an electron–fluxoid scattering mechanism but a comparison of Figures 6-8(a) and 6-9 suggests that phonon–fluxoid scattering is the more likely cause for the following reasons: (i) the minimum in $K(H_a)$ occurs only if $K(H_a = 0)$ corresponds to a position on the "phonon peak;" (ii) application of field (i.e., the insertion of mixed-state fluxoids) washes out this peak in much the same manner as does the addition of impurities or solute atoms. This argument is strengthened by the behavior of the magnetothermal conductivity at 2 K, with regard to which we note that the presence of trapped flux which removes the phonon peak, as in Figure 6-8(b), also eliminates the minimum, as in Figure 6-9.

6.8.1 General Conclusion

The increase of magnetic field in the mixed state can be expected to result in a quadratic-like increase of K from $K^s(H_a)$ to $K^n(H_a = H_{c2})$, excepting that in very pure metals, if the zero-field starting point is located

[4]The values of H_{c2} shown in Figure 6-9 were computed from the relationship $H_{c2} = \sqrt{2}\,\kappa_{GL} H_{c0}(1-1.18\,t^2)$, where κ_{GL} is the Ginzburg–Landau parameter and H_{c0} is the zero-K thermodynamic critical field. With the literature value of $H_{c0,\,Nb} = 1.95$ kOe and the derived value of $\kappa_{GL} = 1.58$, this equation represents a least-squares fit of $H_{c2} = A + Bt^2$ (correlation coefficient 0.999) through three datum points provided by [SCH75].

on the phonon maximum, the ultimate increase will be preceded by a small dip.

6.9 Transition-Metal-Alloy Superconductors

As indicated in Table 6-5 studies have been made of the thermal conductivities of a representative sampling of transition-metal-alloy superconductors, including Ti–Nb for which an abundance of recent information is available. In analyzing the results, attempts are usually made to interpret the data in terms of: (i) the relative importance of the roles played by electron and phonon conductivities in various temperature regimes; and (ii) the most probable electron-scattering and phonon-scattering processes and the influences of metallurgical microstructures on them.

The analysis by MORTON *et al.* [MOR73] of the low-temperature conductivity of Zr–Nb(75%), which commences with a full expression for thermal conductivity above the boundary-scattering limit, makes an interesting case study: The general expression for conductivity is, of course

$$K = K_e + K_g \tag{6-7a}$$

$$\equiv W_e^{-1} + W_g^{-1} \tag{6-7b}$$

regarding which it is understood that both normal and superconducting states are to be included, and in which

$$W_e = W_{es} + W_{eg} \tag{6-13a}$$

and

$$W_g = W_{ge} + W_{gd} \tag{6-15}$$

with

$$W_{es} = f^{-1}(\rho_s/L)T^{-1} \qquad \text{(cf. (6-10a))} \tag{6-18}$$

$$W_{eg} = g^{-1}A_{eg}T^2 \qquad \text{(cf. (6-11))} \tag{6-19}$$

$$W_{ge} = h^{-1}A_{ge}T^{-2} \qquad \text{(cf. (6-14b))} \tag{6-20}$$

$$f = g = h = 1 \quad \text{for } t > 1 \qquad \text{(cf. Figure 6-5)}$$

and

$$W_{gd} = \sum_n B_{gn}T^m \tag{6-4b}$$

where m responds to lattice defects according to (6-3b).

Table 6-5. List of Representative Transition-Metal-Alloy Superconductors for Which Thermal Conductivity Data Are Available

Alloy	Temperature Range of Measurement K	Condition	Literature
Nb–25Zr	3–18	Cold rolled; Above plus 1 h/1500°C/oil quench	Morton *et al.* [MOR74]
		Above plus 10 min/800°C/quench or 1 h/800°C/quench or 5 h/800°C/quench	Morton *et al.* [MOR73]
Ta–Nb(20, 40, 80 at.%)	0.5–1.5	Annealed	Ikebe *et al.* [IKE74]
Ta–Nb(80 at.%); Nb–Mo(20, 25 at.%)	0.5–1.2	Annealed (several large grains)	Ikebe *et al.* [IKE74[a]]
Nb–V(20 at.%)	0.7–1.5	Swaged; Annealed	Ikebe *et al.* [IKE77]
Ti–V(39, 59, 79, 89 at.%)	2.5–14	Annealed 1 h/950°C	Morton *et al.* [MOR77]
Zr–Nb(20 at.%)	5–27		Bischof *et al.* [BIS78]
Ti–Nb Alloys, at.% Nb			
33	2.2–4.2	Forged and cold rolled	Dubeck *et al.* [DUB68]
35	4.2–20	Cold rolled; Annealed $\frac{1}{2}$ h/550°C/cooled	Morton *et al.* [MOR75]
34	7.4–220	Unknown	Bischof *et al.* [BIS78]
22	5–95	Cold worked; Annealed at 850°C	Moore [MOO80]
34; 28, 34; 45; 34; 36	Refer to Table 6-6 for details		Schmidt [SCH75]; Schmidt [SCH79]; Ikebe *et al.* [IKE77]; Flachbart *et al.* [FLA78]; Frederking *et al.* [FRE79]

In (6-18) and (6-19), f and g represent the electronic-conductivity suppression factors of Figures 6-5(a) and (b) ($f, g \lesssim 1$, $t \lesssim 1$). In (6-20), h is the phonon–electron enhancement factor of Figure 6-5(c) ($h \gtrsim 1$, $t \lesssim 1$). The analysis of the experimental data, according to MORTON *et al.* [MOR73], then proceeds in the manner outlined in the following subsections.

6.9.1 Normal-State Electronic Resistivity

W^n_{es} is obtained from low-temperature resistometric results, A_{eg} hence W^n_{eg} from high-temperature resistometric and other data. According to MORTON [MOR73], $A_{eg} \cong (70\, N_{\text{eff}}^{2/3})/(K^{\infty}_{eg}\theta^2_D)$, where N_{eff} is the effective e/a ratio and $K^{\infty}_{eg} \equiv L\sigma_{\infty}T$, since the Wiedemann–Franz law is again valid at high temperatures where the electrical conductivity is σ_{∞}.

6.9.2 Normal-State Lattice Resistivity

Commencing with (6-7a) and making the assumption that $W_{es} \gg W_{eg}$, it follows that

$$(K^n - K^n_{es})^{-1} \cong W^n_g \tag{6-21}$$

the right-hand side of which, according to (6-14b) and (6-4b) may then be written as

$$W^n_g = W^n_{ge} + W^n_{gd} \tag{6-15}$$

$$= A_{ge}T^{-2} + \sum_n B_{gn}T^m \tag{6-22}$$

It follows that under the assumption of point-defect scattering ($m = 1$, from (6-3b)) a plot of $T^2/(K^n - K^n_{es})$ *versus* T^3 will be approximately linear with slope B_{g1}, say, and intercept A_{ge}. After some iterations of this procedure, as outlined in [MOR73], W^n_{ge} and W^n_{gd} were able to be separately evaluated.

6.9.3 Superconducting State

W^n_{es} and W^n_{eg} having been determined in the manner indicated in Section 6.9.1, W^s_e is able to be calculated directly with the aid of f and g values from Figures 6-5(a) and 6-5(b). Subtraction of the resulting $K^s_e = (W^s_{es} + W^s_{eg})^{-1}$ from the measured K^s yields a K^s_g for comparison with $(h^{-1}W^n_{ge} + W^s_{gd})^{-1}$. Finally, having already obtained W^n_{ge}, and after recognizing that W_{gd} is the same in both normal and superconducting states, a "satisfactory" determination of h as a function of t concludes the analysis.

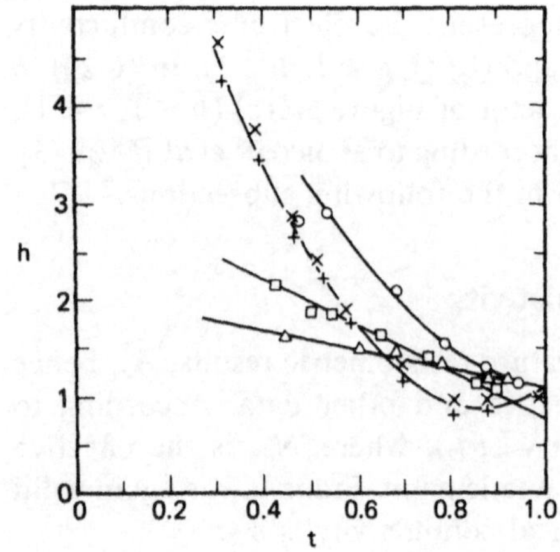

Figure 6-10. Variation of $h = K^s_{ge}/K^n_{ge}$ (cf. Figure 6-5) with reduced temperature for Zr-75Nb in the following conditions: cold rolled (□); annealed 1 h/1500°C/oil quenched (△); annealed 10 min/800°C (×); 1 h/800°C (+); 5 h/800°C (○)—after MORTON *et al.* [MOR73, MOR74].

Plots of the phonon-electron-scattering enhancement factor for Zr-Nb(75%) in four heat treatment conditions are given in Figure 6-10. The values of *h* obtained were orders of magnitude lower than the theoretically predicted ratios (Figure 6-5(c)), in keeping with the absence of phonon peaks from all but the highest purity metals.

In treating the low-temperature thermal conductivity data for (i) cold-rolled and (ii) heat-treated 1 h/1500°C/oil-quenched Nb-Zr(25%), MORTON [MOR74] noted that since the resistometically derived K_{es} was no more than 3% of the total *K*, the thermal conductivity was primarily due to phonons and could, therefore, be dealt with in terms of the following simplified relationship, obtained with the aid of (6-20) and (6-4b):

$$K \cong K_g = \left(h^{-1}A_{ge}T^{-2} + \sum_n B_{gn}T^m\right)^{-1} \tag{6-23}$$

A similar approach was adopted with (i) cold-rolled and (ii) annealed ~0.5 h/550°C/cooled Ti-Nb(35 at.%) [MOR75], for which $K_{es}/K \cong 1\%$. But in both cases the decision to regard phonon conduction (with phonon-electron plus phonon-defect scattering) as the dominant mechanism of thermal transport stemmed from what appeared to be, in the light not only of other results for transition-metal alloys but also for pure Nb itself, inordinately large measured values of the total conductivity.

6.10 Thermal Transport in Titanium–Niobium Alloys

6.10.1 Thermal Conductivity of Ti–Nb

The results of all known studies of the thermal conductivities of Ti-Nb(28-45 at.%) alloys, excepting those of MORTON *et al.* [MOR75] referred to above, are summarized in Figure 6-11. The results of FREDERKING and CHANG [FRE79], depicted there, are also presented in numerical form in Table 6-6.

FLACHBART *et al.* [FLA78] measured the thermal conductivity of Ti-50Nb over the temperature range 1.3–300 K using a technique that had reproduced the thermal conductivity of stainless steel (NBS standard reference material No. 735) to better than ±3.5%. In analyzing the data, it was assumed that in the normal state both electrons (scattered by static defects) and phonons (scattered by electrons) contributed to the thermal conductivity which could then be represented simply by

$$K^{n,s} = K^{n,s}_{es} + K^{n,s}_{ge} \qquad \text{(cf. (6-7c))}$$

$$= fK^{n}_{es} + hK^{n}_{ge} \qquad (\text{with } f = h = 1,\ T \geqslant T_c)$$

$$= f(L/\rho_s)T + ha_{ge}T^2 \qquad \text{(cf. (6-16))}$$

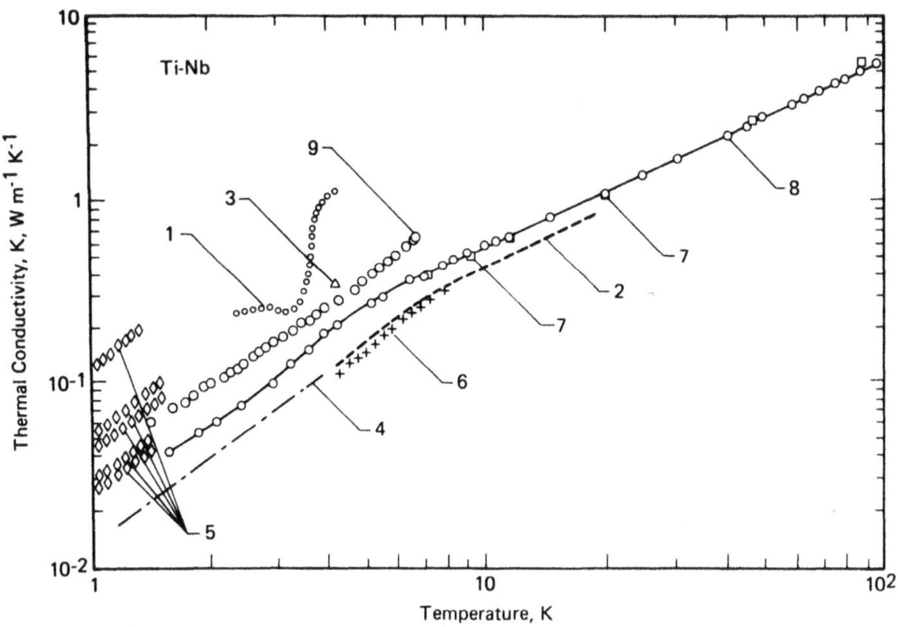

Figure 6-11. Thermal conductivity of Ti–Nb alloys. Samples represented are:

1. Ti–49Nb [DUB68]
2. — [MON69]
3. Ti–58Nb [SCH70]
4. Ti–50Nb [SCH75]
5. Ti–61Nb, various conditions [IKE77]
6. Ti–43.5Nb, Ti–50Nb [SCH79]
7. Ti–50Nb [BIS78]
8. Ti–50Nb [FLA78]
9. Ti–52Nb [FRE79]

Table 6-6. Thermal Conductivity of Ti–Nb(36 at.%) (i.e., the Commercial Alloy Nb–48Ti) in the Temperature Range 1.4 to 6.7 K—After Frederking and Chang [FRE79]

Temperature K	Thermal Conductivity 10^{-1} W m^{-1} K^{-1}	Temperature K	Thermal Conductivity 10^{-1} W m^{-1} K^{-1}
1.4	0.61	3.80	2.33
1.6	0.71	3.99	2.5
1.7	0.77	4.05	2.6
1.82	0.84	4.1	2.7
1.92	0.93_7	4.2	2.75
2.18	1.05	4.25	2.78
2.20	1.16	4.5	3.08
2.35	1.22	4.55	3.15
2.40	1.32	4.75	3.4
2.54	1.38	4.75	3.45
2.6	1.45	4.9	3.48
2.7	1.52	4.93	3.6
2.75	1.58	5.1	3.78
2.85	1.69	5.1	3.9
2.9	1.71	5.2	3.95
3.05	1.80	5.38	4.25
3.15	1.89	5.45	4.3
3.24	1.92	5.6	4.68
3.3	2.0	5.9	4.9
3.4	2.09	6.05	5.2
3.5	2.15	6.3	5.6
3.55	2.19	6.5	6.0
3.6	2.20	6.7	6.2
3.65	2.25		
3.70	2.30		

In the opinion of FLACHBART *et al.* [FLA78], *electronic* conduction was the major contributor above 0.3 T_c. With regard to the temperature range below that value: phonon conduction is always dominant as f drops below 0.1 (Figure 6-5(a)), and h increases by one order of magnitude (e.g., Figures 6-5(c) and 6-10). Thus, in Ti–Nb for $t \lesssim 0.3$, $K^s = K^s_{ge} = hK^n_{ge} = ha_{ge}T^2$. But, of course, in the phonon-scattering regime a quadratic temperature dependence (see (6-3b) or Figure 6-2) could equally well signify phonon-dislocation-strain scattering—a rather likely event in cold-worked material.

The problem of low-temperature phonon scattering was discussed by IKEBE *et al.* [IKE77] in connection with the observation of anomalous phonon resistivity in Ti–Nb(45 at.%), for which TEM investigations

Table 6-7. Thermal Conductivity Temperature Dependences at Low Temperatures, $K^{s'} = K_0 T^m$, of Ti–Nb Superconductors

Nb Concentration at.%	Temperature Range of Measurement[a] K	Condition	Prefactor K_0 10^{-3} W m^{-1} K^{-1}	Exponent m	Thermal Conductivity at 4.2 K W m^{-1} K^{-1}	Literature
34	1-4	Heat treated 850°C	13	1.54	0.12	Schmidt [SCH75]
45	0.7–1.5	Swaged 99%	50	1.70	0.57	Ikebe *et al.* [IKE77]
		Swagged 99% plus 1 h/500°C/water quench	41	1.65	0.44	
		Swaged 99% plus 5 h/500°C/water quench	28	1.65	0.30	
		Swaged 99% plus 25 h/500°C/water quench	30	1.65	0.32	
		Swaged 99% plus 25 h/1600°C/vacuum cool	125	1.80	1.65	
34	1.6–6[a]	Forged 90% at 900°C plus cold forged 90%	20	1.58	0.18	Flachbart *et al.* [FLA78]
28	4.4–9	Commercial	7.5	1.85	0.11	Schmidt [SCH79]
34			7.7	1.82	0.10	
36	1.4–6.7	Incomplete commercial cold work and heat treatment.[b]	35	1.47	0.27_5	Frederking *et al.* [FRE79]

[a] Range of this analysis; the full experiment covered the temperature range 1.6–300 K.
[b] Commercial (Supercon) Cu/Ti–Nb monofilament at the 0.71-cm$^{\varnothing}$ stage of manufacture.

revealed evidence of diffuse-ω fluctuations. It was postulated that the anomalous phonon scattering could be associated with either (i) the dynamical $\omega \rightleftharpoons \beta$ fluctuations referred to in Section 1.8 (see also Table 1-7), or (ii) phonon-electron scattering associated with the normal electrons of the inherently low-T_c ω-phase. The latter is indeed a possibility unless the ω-regions are sufficiently small, compared to a coherence length, to guarantee the operation of a complete proximity effect (see Section 11.9). The argument offered by DUBECK and SETTY [DUB68], on the other hand, in support of a nonelectronic phonon-scattering mechanism in cold-rolled Ti-Nb(33 at.%), an alloy for which they claimed that $K_e^s \ll K_g^s$ in the temperature range below about 4 K, was based on the observed insensitivity of the thermal conductivity to the introduction of further "structural" scattering sites in the form of mixed-state fluxoids.

In conclusion, it is notable that in no case has a full quadratic temperature dependence been found in the superconducting temperature range; values of the temperature exponent obtained by or derivable from the results of the various research groups are listed in Table 6-7.

The importance of metallurgical condition in controlling the low-temperature thermal conductivities of Ti-Nb alloys was emphasized by SCHMIDT [SCH79], who pointed out, in comparing his own results with those of IKEBE *et al.* [IKE77], that although technical alloys of differing Nb contents (28 and 34 at.%, respectively) when processed under similar conditions possessed similar thermal conductivities [SCH79], a single alloy, Ti-Nb(45 at.%), yielded a range of low temperature conductivities in response to various post-cold-work (99%) heat treatments [IKE77].

6.10.2 Thermal Diffusivity of Ti-Nb

When heat flow must be treated as a nonsteady state diffusional process (e.g., Table 18-2), as in flux jumping (Section 18.14 *et seq.*) and normal-zone propagation (Section 25.37 *et seq.*), solution of the diffusion equations

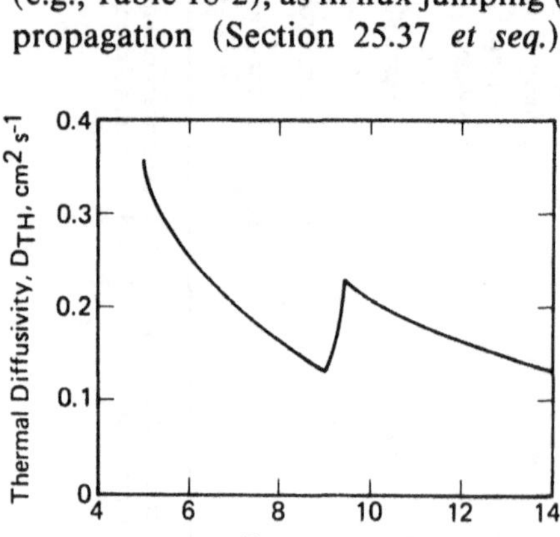

Figure 6-12. Thermal diffusivity as function of temperature $5 \lesssim T \lesssim 14$ K for Ti-Nb(34 at.%) (condition: hot forged 90% AR at 900°C plus cold forged 90% AR)—after FLACHBART *et al.* [FLA78].

requires a knowledge of the thermal diffusivity, $D_{TH} = K/\rho_d C$, where K is the thermal conductivity, ρ_d the density, and C the specific heat. Using thermal-conductivity data for Ti-50Nb from Figure 6-11 and a set of unpublished specific heat data for the same sample, FLACHBART *et al.* [FLA78] calculated thermal-diffusivity temperature dependence within the interval 5–14 K. The result of the investigation, as summarized in Figure 6-12, shows that although the superconducting thermal conductivity, K^s, drops rapidly with temperature (see Figure 6-11 and Table 6-6) the accompanying decrease in specific heat (cf. Figure 8-5) is even faster such that the thermal diffusivity actually increases as the temperature is lowered.

6.11 Thermal Resistance of Superconductor/Normal Interfaces

6.11.1 Occurrence of the Thermal-Boundary Effect—Kapitza Resistance

When heat is released in a superconducting filament imbedded in a composite conductor, it must be transported to the coolant not only through the other filaments of the composite and several layers of stabilizer, the eddy-current barriers, and perhaps a strengthening sheath, but also through a corresponding succession of metal–metal interfaces, and, most importantly, the final interface between the external surface of the conductor and liquid-He bath.

The properties of intermetallic thermal interfaces are familiar to mechanical and heat-transfer engineers. In this regard, the study by VAN SCIVER and HARTWIG [VAN80] of the thermal-boundary resistance between a rod of high-purity ("5N") Al and a cold-mechanically-bonded outer cylinder of lower purity ("3N6") Al is of some interest in the present context. But of principal concern in this section is the special, but nonetheless practically important, class of heat transfer that takes place between the *superconducting* and *normal* components of a composite.

KAPITZA was the first to notice that a temperature discontinuity occurred at the boundary between a normal metal and liquid He (see for example CHALLIS *et al.* [CHA74[a]]). In commemoration of this discovery, the heat transfer coefficient of the interface is referred as the Kapitza conductance, h_K. Its reciprocal is the Kapitza resistance R_K and the defining equation is

$$R_K^{-1} = h_K = \dot{Q}/\Delta T_K \tag{6-24}$$

where ΔT_K is the temperature drop across the interface produced by $\dot{Q}$,

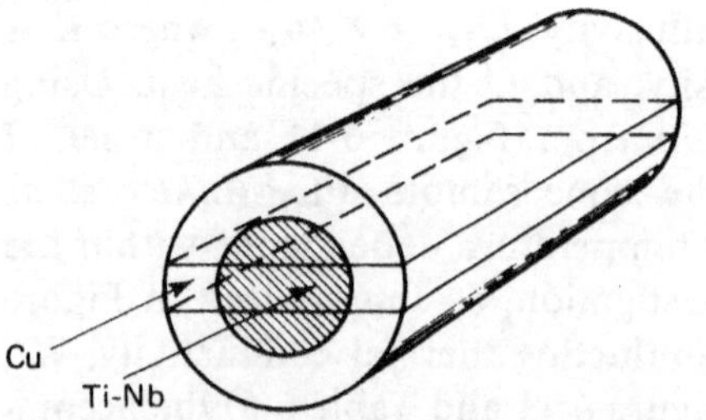

Figure 6-13. Preparation of a pair of Cu/Ti–Nb(35 at.%) junctions from a partially drawn single-core composite for use in thermal-boundary-resistance studies—after FREDERKING and CHANG [FRE79].

the unit-area heat flux. The effect is recognized as an acoustic mismatch phenomenon and as such is a perfectly general property of the passage of phonons across interfaces between appropriate pairs of substances. It is, therefore, encountered at the interface between a normal metal and a superconductor operating in the phonon-conduction mode, but not between a pair of *normal* metals for which thermal transport is predominantly electronic in nature.

The importance of R_K to the release of heat from a Ti–Nb filament imbedded in a Cu matrix is obvious. As an acoustic-mismatch phenomenon, it can, of course, be studied ultrasonically, but it is also amenable to direct measurement using low-temperature thermal conductivity test equipment. SCHMIDT [SCH75] performed just such a measurement using a thin superconducting disc of Ti–50Nb sandwiched between a pair of instrumented Cu bars ("meter bars" in the terminology of thermal conductivity measurement). The test sample was fabricated in the laboratory using diffusion bonding at 850°C to achieve intimate metal–metal contact. The unfortunate result of the use of this technique is that the metallurgical state of the alloy no longer duplicates that of a technical superconductor; the sensitivity of phonon conductivity to heat treatment has already been demonstrated. This objection does not apply to the sample used by FREDERKING and CHANG [FRE79], which was in the form of a slab (Figure 6-13) cut from a monofilamentary composite rod which had been taken out of an early phase of the wire-drawing process. The authors claimed, during a detailed discussion of the interface heat transfer problem, that interface conductance measurements make a useful, but unfortunately not nondestructive, test of the bond quality in a composite conductor.

6.11.2 Kapitza Resistance of the Cu/Ti–Nb Interface

SCHMIDT [SCH75] measured the Kapitza resistance of his Cu/Ti–Nb couple as a function of both temperature (1–4 K) and field (0–10 kOe). In the zero-field tests, it was found that R_K decreased by over an order of magnitude as the temperature was raised from 1 to 2 K, while as the applied field was increased to 10 kOe at 1.01 K it was noted that $R_K(H_a)/R_K(0)$

decreased by a factor of 2. The temperature dependence of R_K, as determined by SCHMIDT [SCH75], was represented closely by

$$R_K = (6 \pm 0.5) \times 10^{-4} / T^{2.75\pm0.2} \qquad (\mathrm{m^2\,K\,W^{-1}}) \tag{6-25}$$

The result is, of course, specific to the particular pair of metals referred to and undoubtedly includes contributions from non-Kapitza heat-transfer mechanisms [FRE79].

6.11.3 Temperature Drop at the Cu/Ti–Nb Interface in a Composite Conductor—A Simple Model Calculation

Suppose heat to be dissipated at a constant rate uniformly throughout the volume of a superconducting cylinder of radius r clad in a Cu jacket. If ΔT is the temperature difference between the axis and surface of the superconductor, then according to CARSLAW and JAEGER [CAR47, p. 168] the radial heat flux per unit area at the surface is

$$\dot{Q} = (2K^s/r)\Delta T \tag{6-26}$$

where K^s is the thermal conductivity of the superconductor. Since this is also the heat flux across the interface, which is controlled by (6-24), the ratio of the temperature discontinuity at the boundary to the radial temperature drop in the superconductor is given by

$$\Delta T_K/\Delta T = 2K^s R_K/r \tag{6-27}$$

Thus, the relative importance of Kapitza resistance is proportional to K^s and inversely proportional to r. It is, of course, also temperature dependent. After insertion of the appropriate expression for R_K from (6-25), and the $K^s = 20 \times 10^{-3}\ T^{1.58}$ from Table 6-7 (FLACHBART [FLA78] data), (6-27), with r expressed in μm, becomes

$$\Delta T_K/\Delta T|_{\mathrm{Cu/Ti-Nb}} = 24/rT^{1.17} \tag{6-28}$$

Equation (6-28) shows that the Kapitza effect decreases in relative importance with increasing temperature and increasing filament radius.

As a demonstration of the significance of Kapitza resistance in impeding heat flow from a superconductor to the stabilizing matrix, we put $\Delta T_K = \Delta T$ in (6-28), after which the filament diameter, w_{eq}, for which the internal and surface temperature drops are equal, can be seen to be

$$w_{eq} = 48/T^{1.17} \qquad (\mu\mathrm{m}) \tag{6-29}$$

For example, at temperatures of 1.0 and 4.2 K, $w_{eq} = 48$ and 9 μm, respectively—values which are encountered in commercial superconducting composites. The presence of a magnetic field must, of course, also be taken into consideration: $R_K(H)$ decreases with field and reduces the effect, while according to DUBECK and SETTY [DUB68] (see Section 6.10.1) the (phonon) conductivity of Ti–Nb(33 at.%) is insensitive to the establishment of a mixed state.

Chapter 7

Magnetic Susceptibility

7.1 Introduction

7.1.1 Magnetic Susceptibilities of Solids

(a) Quantities and Units Employed. The susceptibility of a body to the acquisition of a magnetic moment when exposed to a magnetic field, H_a, is its magnetic susceptibility, $\chi \equiv M/H_a$, where M is its magnetization. In the nonrationalized c.g.s. system of units usually employed by physicists and chemists when studying this subject, the induction, B, within the body is given by $B = H_a + 4\pi M$, such that its permeability, $\mu \equiv B/H_a$, is $1 + 4\pi\chi^v$. The superscript, v, is here used to denote a "per unit volume (cm^3) quantity." Other commonly employed bases for calculation and measurement are the gram and the mole, the susceptibilities corresponding to which are here denoted by χ^g and χ, respectively. The latter, unsuperscripted, symbol will also be used in the *general* case (just as *heat capacity* is the general counterpart of *specific heat*, to be discussed in the following chapter). In the c.g.s.–e.m.u. system, in which magnetic susceptibility is expressed in e.m.u. per unit mass, volume, etc., the unit of e.m.u. is cm^3; thus, χ^v is dimensionless.

Some symbols denoting the subdivision of χ into its various components and other quantities relevant to the following discussion, expressed in non-c.g.s. units convenient for the evaluation of experimental results, are presented in Table 7-1(a). A chart enabling conversions to be made between the c.g.s.–e.m.u. (i.e., Gaussian) and the SI systems of units is provided in Table 7-1(b).

(b) Classes of Magnetic Behavior. Useful descriptions of the ways in which magnetic susceptibility manifests itself in various classes of solids are to be found in several standard texts (e.g., [Cus58, Chapters 12 to 15])

Table 7-1(a). Definitions of Frequently-Used Magnetic and Calorimetric Quantities

Symbol	Quantity	Units
χ	Total magnetic susceptibility	$cm^3\ mol^{-1}$[a]
$\chi_p^{s,d}$	Pauli spin paramagnetism of *s* and *d* electrons	$cm^3\ mol^{-1}$[a]
χ_p^0	Pauli spin paramagnetism of free noninteracting electrons	$cm^3\ mol^{-1}$[a]
$\chi_L^{s,d}$	Landau diamagnetism of *s* and *d* electrons	$cm^3\ mol^{-1}$[a]
χ_i	Ion-core diamagnetism	$cm^3\ mol^{-1}$[a]
χ_{orb}	Orbital paramagnetism	$cm^3\ mol^{-1}$[a]
χ_{so}	Spin-orbit paramagnetism	$cm^3\ mol^{-1}$[a]
ϕ	Electron-electron interaction potential	eV
γ	Measured electronic specific heat coefficient	$J\ mol^{-1}\ K^{-2}$
γ^0	Above after correction for electron-phonon interaction (i.e., "bare")	$J\ mol^{-1}\ K^{-2}$
V	Electron-(phonon)-electron pairing potential	eV
λ	Electron-phonon coupling constant	Number
$n(E_F)$	Fermi density-of-states (states per atom considering one direction of spin)	$eV^{-1}\ atom^{-1}$

[a] Also expressed in the units $cm^3\ g^{-1}$ and $cm^3\ cm^{-3}$, the latter being dimensionless.

and in the books by MULAY and BOUDEREAUX [MUL76, MUL76[a]], upon the second of which the following synopsis is based. (i) *Diamagnetism in General*: The magnetic induction, *B*, tends to avoid the interiors of samples prepared from diamagnetic substances. The permeability, μ, is less than unity, and the magnetization, M or $4\pi M$ (hence χ), is negative. In an inhomogeneous magnetic field, a diamagnetic material experiences a force

Table 7-1(b). Conversion from Gaussian to SI units[a]

Multiply the number for (below)		by (below)	to obtain the number for (below)	
Gaussian Quantity	Unit		SI Quantity	Unit
Flux density, B	G	10^{-4}	Flux density, B	$T \equiv Wb\ m^{-2}$
Magnetic field strength, H	Oe	$10^3/4\pi$	Magnetic field strength, H	$A\ m^{-1}$
Volume susceptibility,[b] χ^v	emu[c] cm^{-3}	4π	Rationalized volume susceptibility, κ	dimensionless
Mass susceptibility, χ^g	$emu\ g^{-1}$ ($\equiv cm^3\ g^{-1}$)	$4\pi \times 10^{-3}$	Rationalized mass susceptibility, κ_ρ	$m^3\ kg^{-1}$
Molar susceptibility, χ	$emu\ mol^{-1}$ ($\equiv cm^3\ mol^{-1}$)	$4\pi \times 10^{-6}$	Rationalized molar susceptibility, κ_{mol}	$m^3\ mol^{-1}$
Magnetization, M or I (per cm^3)	G or Oe	10^3	Magnetization, M	$A\ m^{-1}$

[a] Abstracted from [MUL82[a]] with the permission of L. N. Mulay.
[b] Generally represented by κ.
[c] The emu is cm^3.

in the direction of (or down) the field gradient. Diamagnetism can be an atomic (e.g., metallic ion cores, χ_i), molecular, or energy-band property. Most organic and inorganic compounds, with the exception of those involving some of the transition- or rare-earth elements, are diamagnetic. As pointed out below, metallic conduction electrons also possess a diamagnetic component of the total susceptibility. (ii) *Paramagnetism in General*: Paramagnetism, the converse of diamagnetism, is associated with $\mu > 1$ and a positive value of M, hence χ. The magnetic induction tends to become concentrated as it approaches and enters a paramagnetic substance, which is always attracted to the stronger part of a graded magnetic field. (iii) *Temperature-Dependent Paramagnetism*: Quantum mechanically, paramagnetism can be described in terms of a field-induced splitting of energy states coupled with a related imbalance of the spin populations. Suppose Δ to be the induced energy gap. Then if $\Delta \ll k_BT$, a thermal redistribution of spins among the states is possible, and the paramagnetism is temperature dependent [Kit68, p. 628]. Commonly the Curie–Weiss law, $\chi = C_m/(T - \theta_W)$, where C_m is the Curie constant and θ_W is the Weiss temperature, describes the paramagnetic susceptibility temperature dependence. (iv) *Van Vleck Paramagnetism*: Not unusual in molecular systems is the condition $\Delta \gg k_BT$. In this case, named for J. H. Van Vleck who analyzed the situation in detail, the field-induced moment (the paramagnetism) is unable to respond to change of temperature. (v) *Orbital Paramagnetism*: If the field-induced polarization is being experienced by the electronic energy bands of a metal, and if as before $\Delta \gg k_BT$, the Van Vleck type temperature-independent paramagnetism that results is referred to as "orbital paramagnetism," χ_{orb}. (vi) *Other Components of Metallic Magnetization*: The spins of conduction electrons at the Fermi surface give rise to another temperature-independent paramagnetism, which is proportional to $n(E_F)$, the Fermi density-of-states; this is known as the Pauli paramagnetism, χ_p. Associated with the quantization of angular momentum of the conduction electrons throughout the entire band (not just at E_F) is the Landau diamagnetism, χ_L. Cooperative interactions between conduction-electron spins, either with or without the assistance of those of the ion cores, lead to spin-density waves, spin waves, antiferromagnetism, and ferromagnetism. Spin interactions, if only partially cooperative, can lead to the superparamagnetism encountered among systems of independent giant moments and to spin-glass effects [Col79[a], Col79[c], Col80].

7.1.2 The Role of Magnetic Susceptibility in Ti-Alloy Physics

Considered in this chapter are the magnetic susceptibilities of Ti as well as those of its binary alloys with V, Nb, Mo, and Fe; passing reference is also made to the magnetic susceptibility of Ti–Al, as an example of the

Ti–SM class of alloys. The room-temperature composition dependences of susceptibility of quenched Ti–V [CoL75[b]], Ti–Nb [CoL76], Ti–Mo [CoL71, CoL72], and Ti–Fe [Ho74] are presented, with reference to their quenched microstructures as determined either by electron or optical microscopy or as described in the literature. The susceptibility temperature dependence of β-Ti–Mo [CoL71] and those of both the α- and β-phase modifications of Ti–V [CoL75[b], CoL84] are discussed here, together with those of members of the Ti–Al system [CoL71[e]], whose susceptibility temperature dependences are associated (via orbital paramagnetism) with lattice thermal expansion. Also to be considered is Ti–Mn, a most unusual system in that of all the microstructural variants (α^m, ω, and β) only the hcp structure (α^m in this case) is associated with a Curie–Weiss susceptibility temperature dependence [CoL84]. As a result of magnetic susceptibility temperature dependence studies in Ti–V, it has been possible to quantify the thermally reversible occurrence of athermal ω-phase [CoL75[b]] and to correlate the results with the anomalous temperature-dependent resistivity component, $\Delta\rho|_{77}^{300}$ [CoL78], considered above in Section 5.6.1. Magnetic susceptibility has also been found to be a useful tool in the investigation of the equilibrium phase diagram—recent studies of the Ti–Al system [CoL79], [SwA81] are examples of this—and in following the precipitations of ω-phase in alloys such as Ti–V [CoL75[b]] and Ti–Fe [Ho74] during prolonged moderate-temperature (175 and 300°C) aging.

Magnetic susceptibility results have also been found to be useful and in some cases essential adjuncts to low-temperature specific heat results when values of the Fermi density-of-states are required for alloys which are stable only at elevated temperatures. Several published studies of the magnetic and calorimetric properties of the Ti–Mo system [CoL71, CoL72, CoL72[a]] testify to the utility of the joint approach in this regard. Furthermore, since the electronic specific heat coefficient is directly related to the Pauli spin susceptibility (see below), provided suitable values can be assumed for the other small residual components of total susceptibility, it is possible, again from the conjoint results of magnetic and calorimetric studies, to extract a value for the orbital paramagnetism. In this way, χ_{orb} has been deduced for the pure elements Ti, Zr, and Hf [CoL70, CoL71[d]], V and Nb [CoL76], and several Ti–Mo alloys [CoL70[d], CoL71, CoL72]. Conversely, armed with values of χ_{orb} and χ_i (the ion-core diamagnetism) for the pure-element endpoints of the systems Ti–V, Ti–Nb, and Ti–Mo, and assuming a linear interpolation between these endpoints, it has been possible to estimate the composition dependences of the Pauli spin susceptibility in those alloys [CoL76]. In superconductivity, the normal-state Pauli paramagnetism, χ_p, is a particularly important quantity in that, as discussed in Chapter 15, the value of the upper critical field is strongly influenced by paramagnetic lowering of the normal-state free energy. For example, in an alloy such as Ti–Nb(30 at.%), the paramagnetic energy, $\frac{1}{2}\chi_p^v H^2$, in a field, H, of say 90 kOe is already equal to about half of the zero-K condensation energy.[1]

The manner in which the total transition metal susceptibility can be resolved, with the aid of calorimetric and other data, into its individual components is also treated in this chapter.

7.2. Components of the Total Magnetic Susceptibility of Transition Metals and Their Alloys

It will be necessary here to anticipate some of the low-temperature specific heat results to be treated in Chapter 8, particularly those dealing with the semiempirical determination of the electron–phonon coupling constant, a knowledge of which is essential for extracting a value for the Fermi density-of-states from the measured electronic specific heat coefficient. It is an artificial, although conceptionally useful, convention to subdivide the measured total susceptibility of a material into the sum of individually identifiable components. Thus,

$$\chi = \chi_p + \chi_L + \chi_i + \chi_{so} + \chi_{\text{orb}} \qquad (7\text{-}1)$$

in which, when dealing with transition metals, the first two terms may be subdivided even further into contributions (indicated by appropriate superscripts) from s band and d band electrons. It is one of the goals of joint magnetic susceptibility and low-temperature specific heat investigations to accomplish this subdivision experimentally. Relevant metals and alloys for which approximations to this form of separation have been attempted are: Ti [CoL70, CoL71[d]], Zr and Hf [CoL71[d]], V, Nb, and Mo [CoL76], Ti–V and Ti–Nb [CoL76], and Ti–Mo [CoL70[d], CoL72, CoL76].

[1]The difference between the zero-K n and s free energies, according to (8-6) is: $g_n - g_s = H_{c0}/8\pi$; in which, according to (13-48), $H_{c0,\text{BCS}} = 2.43\gamma_v^{1/2}T_c$. Taking for γ_v a published value attributed to Ti–Nb(25 at.%) [SUK68], *viz.*, $\gamma_v(=5.45\times10^{-3}\,\text{J mol}^{-1}\,\text{K}^{-2}) = 5.82\times10^3\,\text{erg cm}^{-3}\,\text{K}^{-2}$, $T_c = 8$ K, and for χ_p^v a value appropriate to Ti–Nb(30 at.%) [COL76], *viz.*, $\chi_p^v(=110\times10^{-6}\,\text{cm}^3\,\text{mol}^{-1}$, see below$) = 11.75\times10^{-6}$, and comparing the magnetic free energy $g_m = \frac{1}{2}\chi_p^v H^2$ at say 90 kOe with the zero-K condensation energy, $\Delta g_{n/s} = H_{c0}^2/8\pi$, we find that $g_m = 0.53\Delta g_{n/s}$.

This calculation serves to demonstrate that the field-dependent normal-state Pauli paramagnetic energy can significantly influence the point at which the applied field brings about a transition to the normal state (i.e., the upper critical field H_{c2}). This effect, which is generally referred to as "Pauli paramagnetic limitation" [CHA62], [CLO62[a]], is dealt with further in Chapter 15.

7.3 Pauli Paramagnetic Susceptibility

Pauli paramagnetism has been discussed in detail by authors such as MORRISH [MOR65] and PEIERLS [PEI55]. For free noninteracting electrons, the Pauli spin paramagnetic susceptibilities, $\chi_p^{v,0}$ and χ_p^0, are related to the Fermi density-of-states according to

$$\chi_p^{v,0} = 2\mu_B^2 n_v(E_F) \qquad \text{(dimensionless)} \tag{7-2a}$$

and

$$\chi_p^0 = 64.7 \times 10^{-6} n(E_F) \qquad \text{in the units of Table 7-1} \tag{7-2b}$$

where $n_v(E_F)$ is the "bare" density-of-states per cm^3 for *one* direction of spin (implicitly allowing two electrons, ↑↓, per state) and $n(E_F)$ is its per-atom value.

7.3.1 Many-Body Effects in Pauli Paramagnetism

Strictly speaking, through the influence of exchange and correlation the "observed" spin susceptibility, χ_p, is larger than that represented in (7-2) by an amount that is given by the well-known expression due to SEITZ (see also GARDNER *et al.* [GAR65, GAR65[a]])

$$\chi_p = \chi_p^0/(1 - \alpha^e) \tag{7-3}$$

where $\alpha^e \equiv 2n(E_F)\phi$, in which ϕ is an electron–electron interaction potential. As an example of the expected magnitude of this correction we cite the work of GARDNER *et al.* [GAR65[a]], who concluded that for Sc and Y ϕ was equal to 0.091, and 0.029 eV, respectively, values which were consistent with the observation [GAR65] that $0.03 < \phi < 0.30$ eV for all the paramagnetic transition elements. Next, using the simply calculated $n(E_F)$ values, it may be shown that $\alpha_{\text{Sc}}^e = 0.4$ and $\alpha_{\text{Y}}^e = 0.1$. In order to obtain crude estimates of α^e for Ti, Zr, and Hf we take into consideration: (i) the trend $\alpha_{\text{Sc}}^e \rightarrow \alpha_{\text{Y}}^e$; (ii) the position of these elements with respect to those of Sc and Y in the periodic table, bearing in mind that ϕ must also increase somehow in going towards the Pd/Pt corner of the transition-metal block; and (iii) that the $n(E_F)$s of Ti, Zr, and Hf are much smaller than those of Sc and Y. Accordingly it is suggested that α^e for Ti, Zr, and Hf might lie in the vicinity of 0.05 and as such would result in a negligible correction to the magnetic density-of-states expression, (7-2), enabling the superscript "0" to be dropped. Furthermore, in all subsequent treatments of the magnetic susceptibilities of Ti-base transition-metal alloys, the possibility of many-body corrections to χ_p is ignored.

When properly interpreted, the results of low-temperature specific heat experiments enable χ_p to be computed independently of all the other components of magnetic susceptibility. Thus, again in the absence of many-body effects, the bare electronic specific heat coefficients, γ^0 and γ_v^0, may be written as[2]

$$\gamma_v^0 = \tfrac{2}{3}\pi^2 k_B^2 n_v(E_F) \qquad (\text{erg cm}^{-3}\text{K}^{-2}) \tag{7-4a}$$

and

$$\gamma^0 = 4.72 \times 10^{-3} n(E_F), \text{ in the units of Table 7-1.} \tag{7-4b}$$

7.3.2 Many-Body Effects in Electronic Specific Heat

Although corrections for electron–electron exchange and correlation interactions may be regarded as negligible [PIN55], the calorimetrically measured γ is enhanced with respect to γ^0 according to

$$\gamma = \gamma^0(1 + \lambda) \tag{7-5}$$

due to the effect of electron–phonon interactions parameterized by λ, the electron–phonon coupling constant. Naturally, this is particularly important for a superconductor. According to MCMILLAN [MCM68], the expression

$$\lambda = \frac{0.10 \ln + 1.04}{0.94 \ln - 1.04}$$

with

$$\ln \equiv \ln(\theta_D/1.45\, T_c) \tag{7-6}$$

enables an electron–phonon correction to be applied to the measured electronic specific heat coefficient using the parameters θ_D and T_c generally obtainable (provided T_c is not too low) from the same specific heat measurement. For Ti, Zr, and Hf, MCMILLAN found λ to be 0.38, 0.41, and 0.34, respectively [MCM68].

If calorimetric data are available for an *alloy series* it may also be possible, using an alternative approach to be dealt with in Section 8.7, to make such data self-correcting for the effects of electron–phonon interactions. In a case study of Ti–Mo [COL72, COL72[a]], it has been determined that $V = 0.26$ eV for the entire series; then recognizing that

$$\lambda = n(E_F)V \tag{7-7}$$

[2]Provided, of course, that the energies are all in ergs.

Table 7-2. Magnetic Susceptibility Components of Ti, Zr, Hf, Nb,

Element	Electronic Specific Heat Coefficient, γ mJ mole^{-1} K^{-2}	McMillan[a] Electron-Phonon Coupling Constant λ	Pauli Spin[b] Paramagnetism χ_P	Landau[c] Diamagnetism χ_L
Ti	3.36	0.38	33.4	−11.1
Zr	2.80	0.41	27.2	−9.1
Hf	2.15	0.34	22.0	−7.3
V	9.82	0.60	84.2	−28.1
Nb	7.80	0.82	58.8	−19.6
Mo	1.85	0.41	18.0	−6.0

[a] Reference [McM68].
[b] Eq (7-8)—all magnetic susceptibilities in the units 10^{-6} cm^3 mol^{-1}.
[c] Eq. (7-13).

values of λ have been computed for the Ti endpoint ($\lambda_{\alpha\text{-Ti}} = 0.18_5$) and the individual members of the Ti-Mo alloy series. It should be noted that this is an oversimplification of a fully empirical method to be presented in greater detail below.

The McMillan-calculated pure-element λ-values are used in the construction of Table 7-2, while for supplying λ-parameter data for inclusion in Table 7-3, which deals with Ti-Mo alloys, use has been made of the "alloy-series method" referred to in the previous paragraph.

Table 7-3. Determination Orbital

Nominal Mo Concentration at.%	Actual Mo Concentration at.%	Electronic Specific Heat Coefficient, γ mJ mol^{-1} K^{-2}	Electron-Phonon Coupling Constant "Alloy Series Method"[a] λ
100	100.0	1.8_5	0.102
70	71.0	2.6_5	0.146
40	39.8	6.1	0.336
25	25.4	7.0	0.386
20	19.4	7.1	0.391
0 (hcp)	0.0	3.3_6	0.185
0 (bcc)	0.0	—	—

[a] References [Col71, Col71[b], Col72]—see also below.
[b] Eq. (7-8)—all magnetic susceptibilities in the units 10^{-6} cm^3 mol^{-1}.
[c] $\chi_L = 1/3\chi_p$.
[d] Calculated by linearly interpolating between the end-points.

and Mo and Determination of the Orbital Paramagnetism

Mean[d] Ion-Core Diamagnetism χ_i	Magnetic Susceptibility[e] (Measured) and Corrected to 70 K χ		Orbital[f] Paramagnetism χ_{orb}	Reference
−6	(151.8)	140.1	123.8	[CoL70, CoL71[d], CoL76]
−12	(116.8)	108.5	102.4	[CoL71[d]]
−19	(73.7)	68.4	72.7	[CoL71[d]]
−4	(285.7)	291.0	238.9	[CoL76]
−9	(214.1)	220.3	190.1	[CoL76]
−7	(84.4)	83.9	78.9	[CoL76]

[d] Data from [ANG32] and [SEL56].
[e] (Room temperature values) corrected to 70 K using the $d\chi/dT$ listed in Table 7-4.
[f] $\chi_{orb} = \chi - (\chi_p + \chi_L + \chi_i)$; see Eq. (7-14).

7.3.3 Calorimetrically Determined χ_p

A value for λ having first been determined or estimated, $n(E_F)$ may be eliminated between (7-2b), (7-4b), and (7-5) yielding

$$\chi_p = 13.71 \times 10^{-3}\gamma/(1+\lambda) \tag{7-8}$$

with χ_p in the units $cm^3\,mol^{-1}$, and γ in the units $J\,mol^{-1}\,K^{-2}$, thereby

Paramagnetism in Ti–Mo Alloys

Pauli Spin[b] Paramagnetism χ_p	Landau[c] Diamagnetism χ_L	Ion-Core[d] Diamagnetism χ_i	Measured[e] Magnetic Susceptibility χ	Orbital[g] Paramagnetism χ_{orb}
23.0	−7.7	−7	84.4	76.1
31.7	−10.6	−6.7	132.2	117.8
62.6	−20.9	−6.4	190.4	155.1
69.2	−23.1	−6.3	200.3	160.5
70.0	−23.3	−6.2	202.8	162.3
38.9	−13.0	−6	151.8	131.9
108.6	−36.2	−6	231.4[f]	165[h]

[e] Room temperature values are adequate since $d\chi/dT$ is small.
[f] Obtained by extrapolating elevated-temperature bcc data to room temperature.
[g] $\chi_{orb} = \chi - (\chi_p + \chi_L + \chi_i)$; see Eq. (7-14).
[h] The boxes indicate derived values, see p. 288.

illustrating the use of low-temperature specific heat as an adjunct to magnetic susceptibility measurement in accomplishing the first step in separating the total χ into its components.

7.4 Landau Diamagnetism

The Landau term, χ_L, represents the diamagnetic component of conduction-electron susceptibility, which in a $(d+s)$-band metal, may be written:

$$\chi_{\text{conduction electron}} = \chi_P^s + \chi_P^d + \chi_L^s + \chi_L^d \tag{7-9}$$

The Pauli and Landau terms are generally related according to

$$\chi_L^{s,d} = -\tfrac{1}{3}(m/m^*)\chi_P^{s,d} \tag{7-10}$$

where m^* is the effective mass of the conduction electrons in the crystal [MOR65]. Equation (7-10) is then specialized according to the requirements of the particular model under investigation. With regard to Pt for example, CLOGSTON *et al.* [CLO64] assumed a high density-of-states for d holes near the top of the band, in which case $n^d(E_F) \gg n^s(E_F)$ and $(m^*/m)^d \gg 1$. In addition, by assuming that for the s electrons $(m^*/m)^s = 1$, (7-9) reduces to

$$\chi_{\text{conduction electron}} \cong \chi_P^d + \tfrac{2}{3}\chi_P^s \tag{7-11}$$

For the group-IV transition metals Ti, Zr, and Hf, although we may start out with a two-band model, since the d bands are rather wide (calculated values for Ti and Zr being 6.5 and 3.9 eV, respectively [ALT67]), it is justifiable to take $(m^*/m)^d \cong (m^*/m)^s = 1$ (a result which is in accord with the de Haas–van Alphen data of THORSEN and JOSEPH [THO63]). This leads to the conclusion that a single-band model in which

$$\chi_{\text{conduction electron}} \cong \chi_P + \chi_L \tag{7-12}$$

and

$$\chi_L = -\tfrac{1}{3}\chi_P = -13.71 \times 10^{-3}\gamma/3(1+\lambda) \tag{7-13}$$

in the units of (7-8), would adequately describe the magnetic properties. In the following treatment of groups IV to VI transition metals and alloys we continue to rely on (7-13) for calorimetrically estimated values of χ_L.

7.5 Ion-Core Diamagnetism

The diamagnetic susceptibilities of most of the elements have been tabulated by ANGUS [ANG32], SELWOOD [SEL56, p. 78], and DEHN and MULAY [DEH68]. For use in the calculation of metallic susceptibility, it is obviously best to select a value corresponding to the highest chemically excited states (e.g., "4 + " for the group-IV transition elements).

7.6 Orbital Paramagnetism

KUBO and OBATA [KUB56] seem to have been the first to point out, in a general discussion of paramagnetism in metals, that for partly filled non-s bands the Pauli spin susceptibility is augmented by a term comparable to it in magnitude resulting from a type of Van Vleck magnetic polarization. In general, the band *paramagnetism* is $\chi_p + \chi_{so} + \chi_{orb}$, in which, according to KUBO and OBATA, χ_{so} (the so-called "spin-orbit" component) is usually negligible compared to χ_p, and $\chi_{orb} \cong \mu_B^2/\Delta$ (where Δ refers to the mean separation of occupied and unoccupied $3d$ subbands). That χ_{orb} should be an important contributor to the paramagnetism of transition metals was later independently suggested by ORGEL,[3] who estimated that its magnitude would be typically about $125 \times 10^{-6}\ \text{cm}^3\ \text{mol}^{-1}$. Numerical estimates of χ_{orb} for specific metals were subsequently made by CLOGSTON *et al.*[3] ($\chi_{orb,V} \cong 210 \times 10^{-6}\ \text{cm}^3\ \text{mol}^{-1}$), DENBIGH and LOMER[3] ($\chi_{orb,Cr} \cong 100 \times 10^{-6}\ \text{cm}^3\ \text{mol}^{-1}$), BUTTERWORTH[3] (V–Cr), CLOGSTON *et al.*[4] (Pt), MORI[3] (several transition metals), GARDNER and PENFOLD[3] (Sc), and others.

Following KUBO and OBATA [KUB56], we will assume χ_{so} to be negligible thereby enabling the total susceptibility to be expressed in the form

$$\chi = \chi_p + \chi_L + \chi_i + \chi_{orb} \tag{7-14}$$

and χ_{orb} to be taken as the difference between the measured susceptibility and the other "known" components χ_p, χ_L, and χ_i.

The experimentally derived orbital susceptibilities of Ti, Zr, Hf, V, Nb, and Mo are listed in Table 7-2 and discussed below. DENBIGH and LOMER [DEN63] derived a theoretical expression for χ_{orb}, the exact evaluation of which requires a detailed knowledge of the band structure; but in view of the other approximations involved, it is satisfactory to use the frequently quoted simplified expression

$$\chi_{orb} \cong 4\mathcal{N}\mu_B^2 n(10 - n)/10\Delta \tag{7-15}$$

[3]The references are given in COLLINGS and HO [COL71[d]].
[4]The papers cited are listed in COLLINGS and GEHLEN [COL71[e]].

where $\mathcal{N}$ is Avogadro's number, n is the number of occupied states per atom in the band, and Δ has the same meaning as before. Assuming a spectroscopic state for Ti of [Ar]$3d^2 4s^2$, we find $\Delta \cong 1.7$ eV, while for the other group-IV transition elements Δ has the values 2.1 eV(Zr) and 2.8 eV(Hf). With $\Delta \gg k_B T$, χ_{orb} is temperature independent; so also is χ_p. It follows that the *intrinsic* temperature dependences of transition elements must be practically zero. A clue to a possible mechanism for the *observed* temperature dependences is provided by the observation that the susceptibilities of Ti, Zr, and Hf are anisotropic. Of the various susceptibility components considered above only χ_{orb}, which depends on the band structure and properties of the crystal lattice, can be anisotropic. Accordingly, it has been suggested [CoL70, CoL71[d], CoL71[e]] that it is this lattice-coupled susceptibility component that is responsible, via lattice expansion, for the observed susceptibility temperature dependences of some transition metals. In conclusion, it should be mentioned that spin–orbit coupling may also provide a lattice-property related susceptibility component, particularly in simple metals for which χ_{orb}, as described above, is not operative [MuL82].

7.7 Magnetic Susceptibilities of Some Pure Transition Elements

The magnetic susceptibility components of the pure elements Ti, Zr, Hf, V, Nb, and Mo, deduced in the manner outlined above, are listed in Table 7-2.

7.8 Susceptibility Composition Dependences in Binary Transition-Metal Alloys

7.8.1 Total Magnetic Susceptibility

The composition dependences of total magnetic susceptibility of the alloys Ti–V [CoL75[b]], Ti–Nb [CoL76], Ti–Mo [CoL71, CoL72], and Ti–Fe [Ho74] are depicted in Figure 7-1.

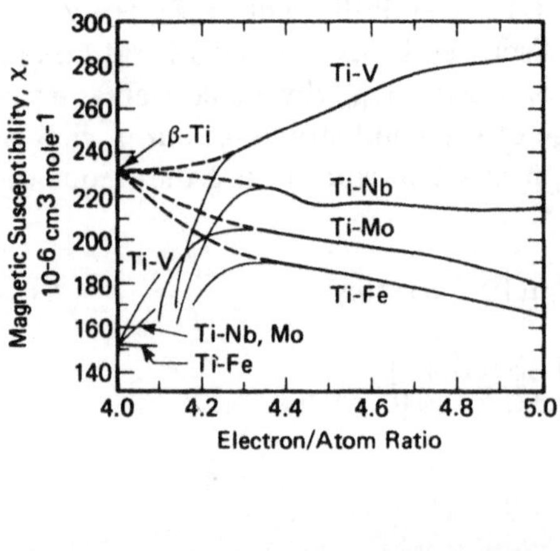

Figure 7-1. Experimentally determined e/a ratio dependences of the room temperature magnetic susceptibilities of quenched Ti–V, Ti–Nb, Ti–Mo, and Ti–Fe alloys. Extrapolations of the principally-bcc data lead to an estimate of the hypothetical room-temperature susceptibility of β-Ti—after Collings *et al*: Ti–V [CoL75[b]], Ti–Nb [CoL76], Ti–Mo [CoL71, CoL72], Ti–Fe [Ho74].

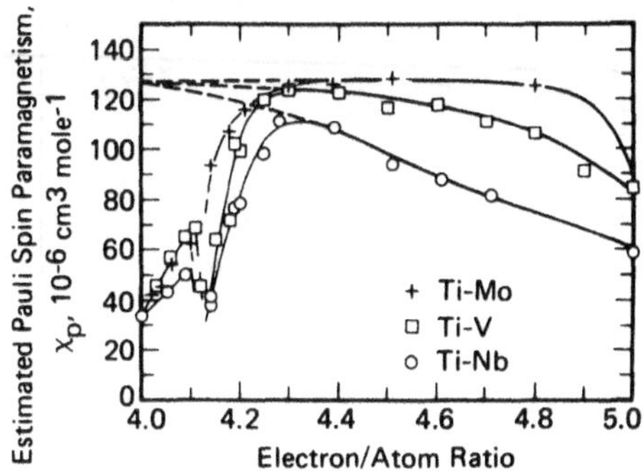

Figure 7-2. Estimated e/a ratio dependences of the Pauli spin susceptibilities of Ti–V, Ti–Nb, and Ti–Mo. The data points correspond to values of χ_p calculated according to (7-16) and are based on experimental *room temperature* χ and interpolated values of $(\chi_i + \chi_{orb})$—after COLLINGS [COL76].

7.8.2 Pauli Paramagnetism

Based on the alloy data represented in Figure 7-1 and those for the pure elements Ti, V, Nb, and Mo listed in Table 7-2, the e/a ratio dependence of χ_p has been computed using (7-14) in the form

$$\chi_p = \tfrac{3}{2}[\chi - (\chi_i + \chi_{orb})_{alloy}] \tag{7-16}$$

in which $(\chi_i + \chi_{orb})_{alloy}$ has been obtained by linear interpolation between the pure-element endpoints. The results are plotted in Figure 7-2. All three data sets extrapolate with little distortion towards a common point suggesting that the value $\chi_p^\beta \cong 127 \times 10^{-6}\,\text{cm}^3\,\text{mol}^{-1}$ is acceptable as the Pauli paramagnetic susceptibility of β-Ti were it stable at room temperature.

7.8.3 Orbital Paramagnetism

As pointed out above, provided that calorimetric data are available, it is possible to extract χ_p directly from the electronic specific heat coefficient γ, and go on to determine χ_{orb} as a function of alloy composition. This has been done for the Ti–Mo system [COL70[d], COL71, COL72], whose data is collected in Table 7-3, and the results are plotted in Figure 7-3. The values of χ_{orb} presented here represent a correction to those given in the articles cited above, from which χ_L and χ_i had, for several reasons valid at the time, been deliberately neglected. The most important quantities to be derived from the magnetic aspects of the joint magnetic/calorimetric studies are χ_p and an associated Fermi density-of-states. Consequently, in most of the alloy work, χ_{orb} has been relegated to the role of residual to be separated from the total susceptibility and, in a sense, discarded. Under these circumstances we are with some degree of justification, able to write

$$\chi = \chi_p + (\chi_L + \chi_i) + \chi_{orb} \tag{7-14}$$

$$\equiv \chi_p + \chi_R \tag{7-17}$$

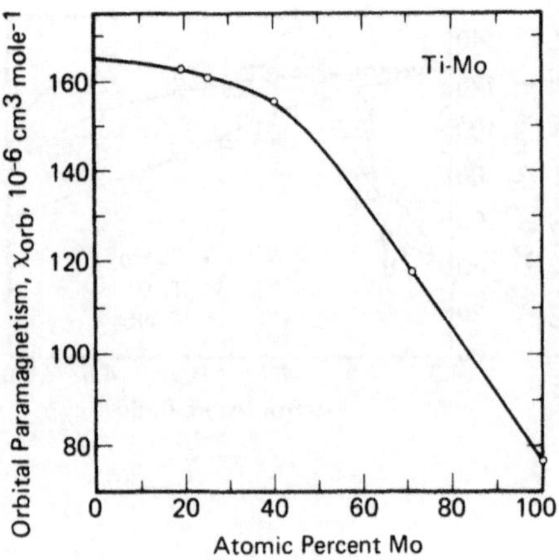

Figure 7-3. Composition dependence of χ_{orb} for Ti–Mo alloys from the data of Table 7-3; an extrapolation of the semiempirical curve yields $\chi_{orb}^{\beta\text{-Ti}} = 165 \times 10^{-6}\ cm^3\ mol^{-1}$.

where, provided no physical significance is to be attached to χ_{orb}, χ_L and χ_i may be operationally absorbed into it. Thus, in articles such as [CoL71] and [CoL72], the values of "χ_{orb}" listed are some $30 \times 10^{-6}\ cm^3\ mol^{-1}$ lower than the "exact values" reported in Figure 7-3.

The curve in Figure 7-3 extrapolates to $\chi_{orb}^{\beta\text{-Ti}} \cong 165 \times 10^{-6}\ cm^3\ mol^{-1}$, which, when combined with the "measured" value of $\chi_{total}^{\beta\text{-Ti}}$ (i.e., the elevated-temperature value extrapolated to room temperature [CoL71, CoL72]), and the value of χ_i listed above, yields a Pauli spin paramagnetism for β-Ti of about $109 \times 10^{-6}\ cm^3\ mol^{-1}$. This differs by only 13% from that indicated in Figure 7-2, a plot of χ_p *versus* composition derived using (7-16) without the aid of any calorimetric data for the alloys themselves.

The derived values of $\chi_{orb}^{\beta\text{-Ti}}$ and $\chi_{p,L}^{\beta\text{-Ti}}$ are specially indicated (rectangular boxes) in Table 7-3.

7.9 Susceptibility Temperature Dependences of Pure Transition Elements

The temperature dependence of susceptibility of metals has been considered by many workers since Stoner first introduced the subject in 1935. The subject has been developed in each of two important directions, one of which might be termed the "two-band approach" and the other the "density-of-states approach." In the latter, which has been applied mainly to the *d* bands of transition metals, the susceptibility temperature dependence is ascribed to the Pauli paramagnetism expressed as a function of the magnitude, slope, and curvature of the density-of-states function in the

Table 7-4. Magnetic Susceptibility Temperature Dependences of Some Pure Transition Elements

Group Number	Element	Temperature Range K	$\partial\chi/\partial T$ 10^{-8} cm^3 mol^{-1} K^{-1}	Reference
IV	α-Ti[a]	150–400	+5.1	[CoL70, CoL71[d]]
	β-Ti	1213–1384	−0.18	[CoL84]
	Zr	300–400	+3.6	[CoL71[d]]
	Hf	300–400	+2.3	[CoL71[d]]
V	V	77–297	−2.3[b]	[CoL84]
	Nb	77–160	−1.9	[CoL72[b]]
		160–300	−3.4	
VI	Mo	77–300	+0.22	[CoL84]

[a] At temperatures below about 70 K the susceptibility of pure Ti tends to be temperature independent.
[b] Data for Ti–V(90 at.%).

vicinity of E_F. This model, first introduced in 1936 by STONER in the form

$$\chi_p^v = 2\mu_B^2 n_v(E_F)\left\{1 + \tfrac{1}{6}(\pi k_B T)^2\left[\frac{1}{n}\frac{\partial^2 n}{\partial E^2} - \left(\frac{1}{n}\frac{\partial n}{\partial E}\right)^2\right]_{E_F}\right\} \quad (7\text{-}18)$$

with

$$n \equiv n_v(E)$$

(cf. (7-2a)) was subsequently developed by KRIESSMAN and CALLEN (1954),[4] KOJIMA *et al.* (1971),[4] TANIGUCHI *et al.* (1962),[4] and by SHIMIZU and coworkers (1963–1966)[4] in a series of nine papers. According to (7-18), the temperature dependence of χ_p is positive or negative depending on the sign of the term in square brackets, which in turn depends on the sign and relative magnitude of $(\partial^2 n/\partial E^2)_{E_F}$. Thus, on proceeding from group IV along a given row of the transition-metal block of the periodic table, the remarkable correlation that is encountered between the sign of $d\chi/dT$ and the relative magnitude of the electronic specific heat coefficient (an indication of whether E_F for an individual transition element is situated at a maximum or minimum in its $n(E)$ curve) has frequently been cited as evidence in support of the density-of-states model for susceptibility temperature dependence as embodied in (7-18).

Recent values of the susceptibility temperature dependences of samples of very high purity Ti, Zr, Hf, and Nb, and commercially pure Mo, which turn out to be linear over quite wide temperature ranges, are listed in Table 7-4. Noticeable is the alternation of sign, alluded to above, which appears

[4] The papers cited are listed in COLLINGS and GEHLEN [CoL71[e]].

upon proceeding from one group to the next, and which would seem to confirm the validity of the density-of-states model.

However, for Ti and presumably for other transition elements, a close examination of the curvature needed to achieve the measured temperature dependence reveals that it is actually several orders of magnitude greater than that provided by the rigid-band $n(E)$ function upon which the initial argument was based. There is little doubt that the density-of-states curve of a *pure element* could provide the necessary conditions for Pauli paramagnetic susceptibility temperature dependence, *viz.*, a turning point in $n(E)$, of width about 5×10^{-3} Ryd, with E_F situated within it. But such fine structure could scarcely be maintained at elevated temperatures or survive alloying with the associated smearing of the electron states (see Figure 5-11), situations with which any theory for transition-metal susceptibility must cope. For example, the susceptibility temperature dependences of hcp α-Ti-alloys (for which $d\chi/dT$ is very much greater than that in the β-phase) form sets of almost parallel curves out to relatively high solute concentrations. This has been graphically demonstrated with Ti–Al alloys by COLLINGS and GEHLEN [COL71[c]], and again in the $\chi(T)$ curves for α'-Ti–V referred to below (Figure 7-8). The numerical comparison between these

Table 7-5. Magnetic Susceptibility Temperature Dependences of Representative α-Phase Ti-SM and Ti-TM Alloys [COL84]

Solute	Nominal[a] Solute Concentration at.%	Condition time(h)/ Temperature(°C)	Temperature Range K	$\partial\chi/\partial T$ 10^{-8} cm^3 mol^{-1} K^{-1}
Al	0	As-cast	150–300	+5.1
	2	As-cast	124–454	+4.9
	3	As-cast	105–380	+4.5
	5	As-cast	112–412	+4.5
	10	48/900/IBQ	108–382	+3.5
	15	24/1000/IBQ	123–396	+3.7
V	0	—	150–300	+5.1
	3	1/1000/IBQ	150–300	+4.9
	6	1/1000/IBQ	150–300	+4.0
	9	1/1000/IBQ	150–300	+3.2
	10.5	1/1000/IBQ	150–300	+3.0
	12	1/1000/IBQ	150–300	+2.7

[a]The analyzed Al concentrations are: 9, 2.3, 3.2, 5.2, 10.7, and 14.8 at.%, respectively; the estimated (from weight-loss during melting) V concentrations are: 0, 3.08, 6.04, 9.09, unknown, and 12.10 at.%, respectively.

systems is made in Table 7-5. Faced with these results, it must be concluded that the susceptibility temperature dependences of transition elements or their alloys are not governed by the density-of-states model or in fact by any simple electronic thermal excitation mechanism.

Since χ_{orb} is the principal contributor to the total susceptibility of Ti and many of its alloys, it cannot be ignored when considering the mechanisms of both magnetocrystalline anisotropy and susceptibility temperature dependence. A clue to the possible coupling between χ_{orb}, anisotropy, and $\chi(T)$ has been provided by the observation that anomalies in the susceptibility temperature dependences are often accompanied by corresponding anomalies in the temperature dependences of the lattice parameters [CoL71[e]] or in the thermal expansion coefficients [CoL72[b]].

Noting that $\chi_{\mathrm{orb}} \propto \mu_B^2/\Delta$, as in (7-15), where the energy gap Δ is much greater than k_BT, we see that χ_{orb} fulfills two essential experimental requirements: (i) an insensitivity to changes of temperature range or solute concentration; and (ii) a responsiveness to changes in certain of the crystal-lattice properties—in particular, lattice symmetry and lattice parameter. Accordingly, we offer χ_{orb} as the seat of susceptibility temperature dependence of transition elements and alloys. After further verification, this approach may turn out to be applicable to nontransition metals, with χ_{so} (the spin-orbit component, see Section 7.6) replacing χ_{orb} as the lattice-sensitive susceptibility component.

7.10 Curie–Weiss Paramagnetism in Titanium–Transition-Metal Alloys

7.10.1 Dilute Alloys

In response to a study of superconductivity in alloys of Ti with Cr, Mn, and the Fe-group transition elements by MATTHIAS *et al.* [MAT59], in which some questions (now refuted) were raised about the possibility of a "magnetic interaction" for superconductivity CAPE [CAP63] instituted a study of the magnetic susceptibility temperature dependences of some dilute as-cast Ti-Cr, -Mn, -Fe, and -Co alloys with a view to examining the existences of localized magnetic moments. The results, which are summarized in Figure 7-4, show that the susceptibilities of the alloys Ti-Cr, -Fe, and -Co are just as temperature independent as that of the host element itself and that only Ti-Mn supports a localized magnetic moment. From an analysis of the Curie–Weiss plot for Ti-Mn (Figure 7-4(iii)) based on the equation

$$(\chi - \chi_0)_{\mathrm{alloy}} = C_m/(T - \theta_W) \tag{7-19}$$

where χ_0 is some constant to be determined, C_m is the Curie constant, and

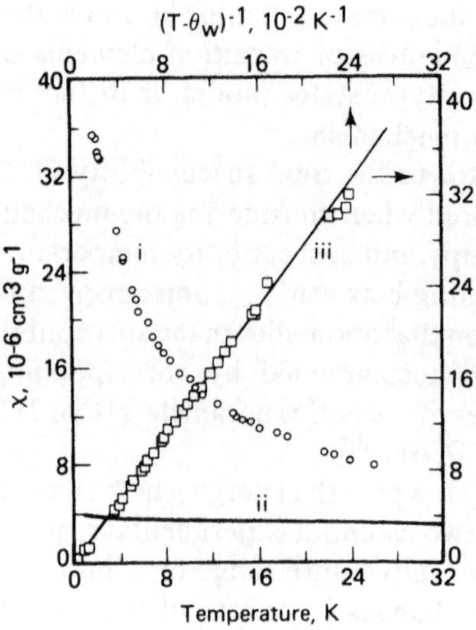

Figure 7-4. Magnetic susceptibility temperature dependences of some dilute Ti–TM alloys plotted in the format χ *versus* T in the case of (i) Ti–Mn(1.0 at.%) and (ii) unalloyed Ti, Ti–Cr(1.15 at.%), Ti–Fe(0.96 at.%), Ti–Co(1.30 at.%), and in the format $\chi - \chi_0$ *versus* $(T - \theta_W)^{-1}$ in the case of (iii) Ti–Mn(1.0 at.%) with $\chi_0 = 3.4 \times 10^{-6}$ cm^3 g^{-1} and $\theta_W = -2.2$ K. Evidently, of the solutes studied, only Mn carries a local moment—after CAPE [CAP63].

θ_W is the Weiss temperature [CUS58, p. 298], it was found that $\chi_0^g = 3.4 \times 10^{-6}$ cm^3 g^{-1}—a value adequately close to that of the solvent, Ti—and that $C_m = np_{\text{eff}}^2 \mu_B^2/3k_B = 1.39 \times 10^{-4}$ cm^3 K g^{-1} (where n is the number of Mn atoms per gram, and p_{eff} is the "effective magneton number" of the Mn ions) leading to the quite substantial localized moment per Mn ion of about 1.5 μ_B.

Associated investigations led to the interesting conclusion that: (i) the localized moment was confined to Ti–Mn alloys of compositions less than 4 at.% Mn whose quenched structures were α^m (and probably hcp); (ii) although superconductivity was present in as-cast Ti–Mn at 1 K, presumably due to the presence of a continuous filamentary path of retained β-phase, the same was not true of dilute quenched alloys, which remained normal and revealed no evidence of any incipient superconducting transition down to ~1 K, the lowest available temperature; (iii) in the nonmagnetic, apparently bcc, Ti–Mn alloys the transition temperature responded in the usual way to an increasing Fermi density of states, while in the hcp alloys, T_c actually seemed to be depressed by the presence of the localized moments.

7.10.2 Concentrated Ti–Mn Alloys

The magnetic susceptibilities of a series of quenched Ti–Mn alloys of nominal compositions 0.5 through 25 at.% have been measured in the temperature range 77–400 K. The composition dependence of room-temperature (297 K) susceptibility of alloys quenched from $\frac{1}{2}$ h and 5 h heat treatments at 1000°C are given in Figure 7-5. Also shown in that figure are the effects of aging, for up to 1000 h at 300°C, on Ti–Mn(5 and 7 at.%,

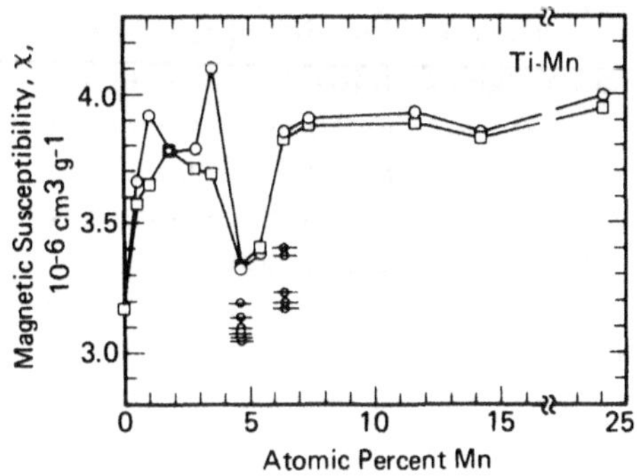

Figure 7-5. Composition dependence of room-temperature magnetic susceptibility of Ti–Mn alloys quenched from heat treatments of $\frac{1}{2}$ h/1000°C (○) and 5 h/1000°C (□). Also shown (⊖) are the decreases in susceptibility which accompany the aging at 300°C of Ti–Mn(4.6 at.%) for $\frac{1}{2}$, 1, 10, 100, 300, and 1000 h and Ti–Mn(6.4 at.%) for $\frac{1}{2}$, 1, 10, 100 and 800 h—after COLLINGS [COL84].

nominal), the results of which were to develop large volume fractions of ω-phase precipitates [WIL69] in both systems. The ω-phase generally has a lower susceptibility than the β-phase, consequently, as will be considered in greater detail below, the effect of aging in the $\omega+\beta$-field is to continuously lower the average susceptibility, which in turn makes an excellent indicator of the progress of the aging reaction. Evidently ω-phase precipitation was already well advanced in the as-quenched Ti–Mn(5 at.%) sample, while heat treatment was necessary for its development in the higher concentration alloy.

The results of the magnetic susceptibility temperature dependence measurements of the first six alloys of the series are shown in Figure 7-6. All the quenched alloys with Mn concentrations less than about 4 at.% (nominal) exhibit Curie–Weiss susceptibilities, in agreement with the earlier conclusions of CAPE [CAP63]. But the new result emerging from this study is the observation that the localized magnetic moment is absent not only from the β-phase alloy but also from ω-phase Ti–Mn, in spite of its being hexagonal.

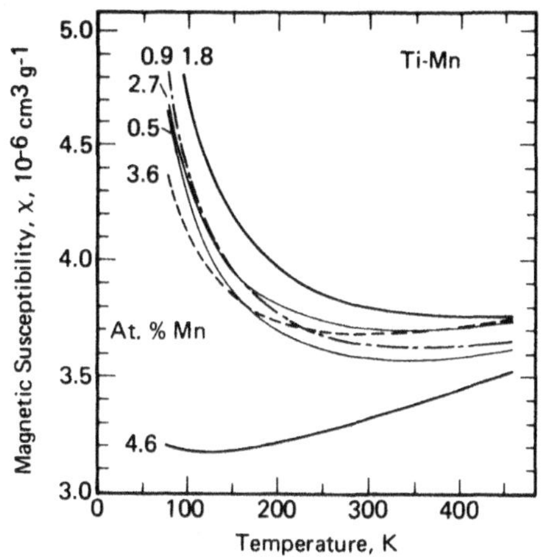

Figure 7-6. Magnetic susceptibilities of Ti–Mn(1–5 at.%) alloys over the temperature range 77–450 K. Only the hcp alloys show Curie–Weiss temperature dependences. The numbers refer to the analyzed compositions in at.%—after COLLINGS [COL84].

The complete disappearance of localized magnetic moment in response to change of matrix crystal structure in the manner demonstrated in Figure 7-6 is quite remarkable, especially in view of the fact that the classic models of localized-moment formation, such as those of FRIEDEL [FRI56] and ANDERSON [AND61[a]], are structure independent, and also in the light of the results of various other magnetic susceptibility studies that have demonstrated that the localized moments in Cu [NAK59] and certain simple metals [TAM67], [COL70[c], COL71[c]] persist through the melting point.

7.11 Susceptibility Temperature Dependence in Concentrated Titanium-Base Alloys—Case Studies of Ti–Al, Ti–V, and Ti–Mo

7.11.1 β-Phase Alloys

The β-Ti–Mn alloys in all concentrations have very slight susceptibility temperature dependences. That of unalloyed β-Ti itself has been given in Table 7-4, those of Ti–Mo(5 to 25 at.%) are listed in Table 7-6, while the susceptibility temperature dependences of a series of Ti–V alloys are depicted in Figure 7-7, in which the effect of the $(\alpha+\beta)\rightarrow\beta$ transformation is to be seen and the β-phase susceptibility is graphically extrapolated back to 150 K for subsequent use.

7.11.2 α-Phase Alloys

In contrast to β-phase Ti–TM alloys, the susceptibility temperature dependences of α-phase Ti-base alloys are relatively strong, at least above

Table 7-6. Magnetic Susceptibility Temperature Dependences of β-Ti–Mo Alloys [COL84]

Nominal[a] Mo Concentration at.%	Temperature Range K	$\partial\chi/\partial T$ 10^{-8} cm^3 mol^{-1} K^{-1}
0	1213–1384	−0.18
5	1030–1417	−0.19
9	827–1413	−0.19
10	907–1305	−0.19
15	879–1298	−0.19
20	576–1358	+0.09
25	298–1414	−0.01

[a] The analyzed compositions are: 0, 5.2, 8.9, 10.3, 14.9, 19.4, and 25.4 at.% Mo, respectively.

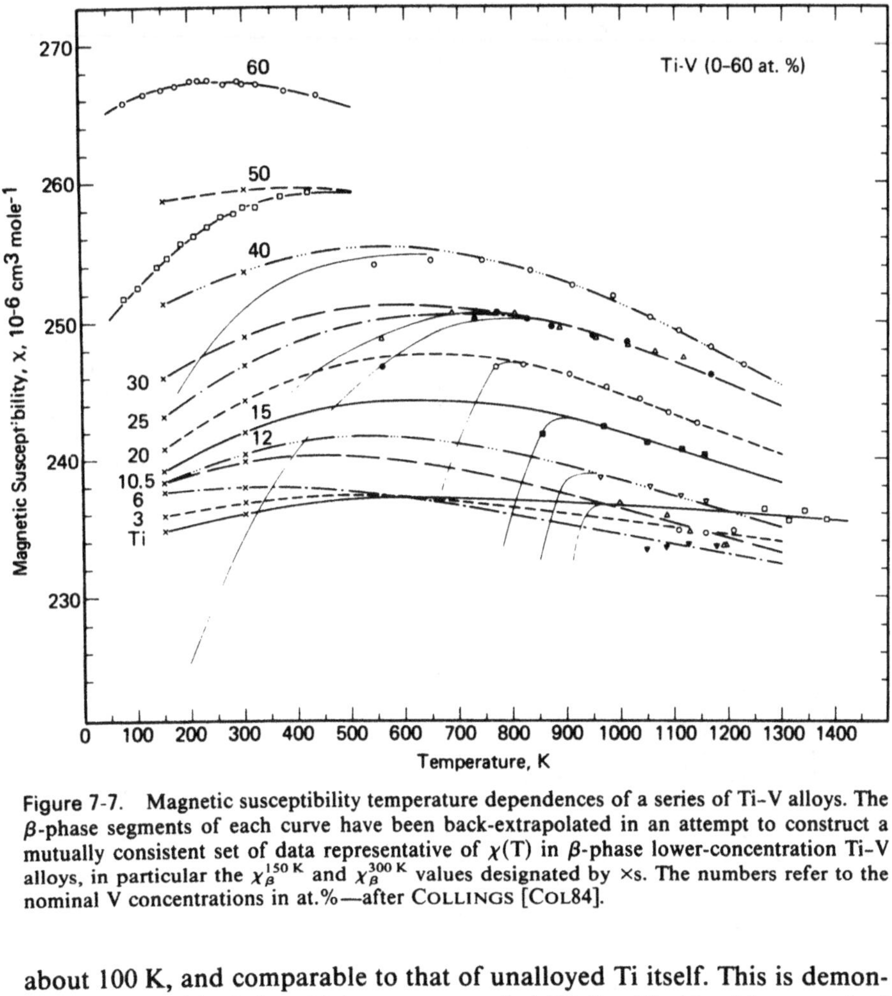

Figure 7-7. Magnetic susceptibility temperature dependences of a series of Ti–V alloys. The β-phase segments of each curve have been back-extrapolated in an attempt to construct a mutually consistent set of data representative of $\chi(T)$ in β-phase lower-concentration Ti–V alloys, in particular the $\chi_\beta^{150\,K}$ and $\chi_\beta^{300\,K}$ values designated by ×s. The numbers refer to the nominal V concentrations in at.%—after COLLINGS [COL84].

about 100 K, and comparable to that of unalloyed Ti itself. This is demonstrated in Table 7-5, which compares $d\chi/dT$ for Ti–Al with quenched α^m-Ti–V(0–10.5 at.%) alloys, and in Figure 7-8, which compares the temperature dependence of α^m- and β-Ti–V(0–10.5 at.%) alloys within their appropriate temperature ranges of stability.

At concentrations greater than 12 at.% V, the upper limit in Table 7-5, $d\chi/dT$ begins to increase as the alloy, initially in the quenched $\omega+\beta$ condition, experiences a change in the $\omega{:}\beta$ volume ratio followed by a transformation to $\alpha+\beta$ and finally to β-phase, with continuously increasing temperature. This behavior, which is illustrated in Figure 7-9, introduces a new feature into the susceptibility temperature dependence—one involving the change of susceptibility that results from temperature-dependent phase transformation. The effect is particularly interesting and useful when reversible ω-phase precipitation is involved. To ensure reversibility in this case

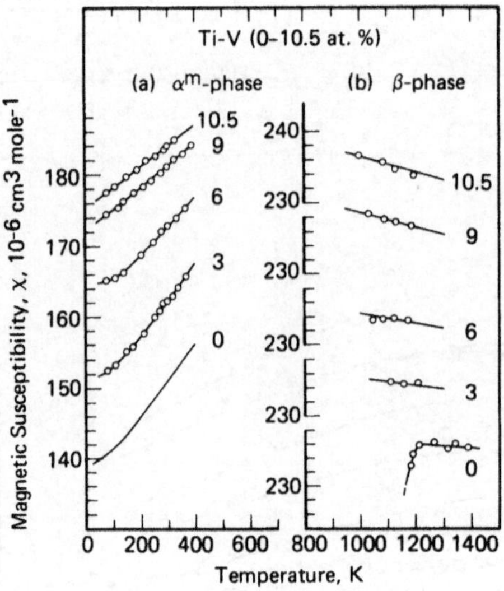

Figure 7-8. Magnetic susceptibility temperature dependences of Ti–V(0–10.5 at.%) alloys in the (a) as-quenched α^m-phase state, and (b) elevated-temperature β-phase. The numbers refer to the nominal V concentration in at.%—after COLLINGS [COL84].

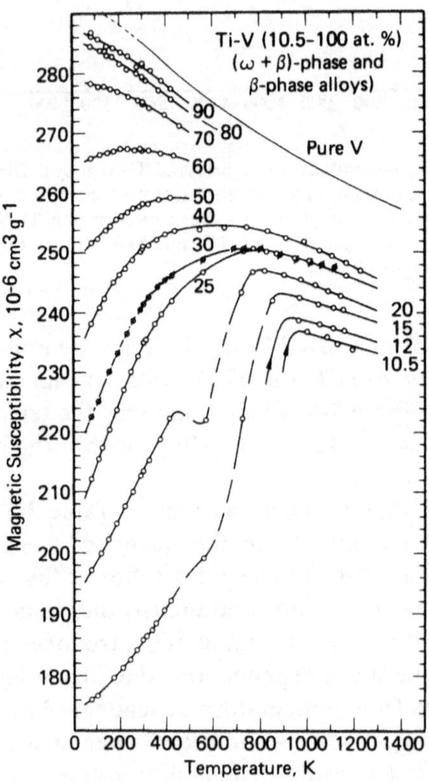

Figure 7-9. Magnetic susceptibility temperature dependences of Ti–V(10.5–100 at.%) alloys. The curves display the combined effects of intrinsic temperature dependence and phase transformation. The numbers refer to the nominal V concentration in at.%—after COLLINGS [COL84].

we confine ourselves to the temperature range below room temperature. By considering temperatures between 150 and 300 K and the composition range of 15 to 50 at.% V, within which a reversible ω-phase effect of one kind or another is expected (see Section 1.8), an *anomalous susceptibility temperature dependence* is revealed, analogous to the anomalous $d\rho/dT$ of Section 5.6.1.

7.12 Concentration, Microstructure, and Temperature Dependences of Magnetic Susceptibility—A Case Study of Titanium–Vanadium

7.12.1 Concentration and Microstructure Dependence

The magnetic susceptibilities of Ti–V(0–100 at.%) alloys at 150 K and 300 K, in all their microstructural variants, have been plotted in Figure 7-10 using data already contained, for the most part, in the previous three figures: (i) Figure 7-7—β-phase data (indicated by the crosses) obtained by back-extrapolation from the β-stable elevated-temperature field; (ii) Figure 7-8(a)—data from the quenched α^m-phase; (iii) Figure 7-9—data for the $\omega+\beta$- and equilibrium-β-phase (high concentration range) susceptibilities. Additional data not included in the earlier figures have to do with the estimated ω-phase isothermals. These have been developed using an argument that commences with the identification of a special point, $\chi_\omega^{300\,K}$ (14 at.% V) $= 168.8 \times 10^{-6}\,cm^3\,mol^{-1}$ acquired from the results of the experiment on the aging of Ti–V(19 at.%) referred to below. Next, according to COLLINGS [COL75[b]], in which it was concluded that near 14 at.% V $(d\chi/dT)_\omega^{Ti\text{-}V} \cong 0.9 \times 10^{-6}\,cm^3\,mol^{-1}\,K^{-1}$, the composition dependence of $\chi_\omega^{Ti\text{-}V}$ could be reasonably well represented by a straight line connecting the special point to $\chi_{Ti}^{300\,K} = 151.8 \times 10^{-6}\,cm^3\,mol^{-1}$. If $\chi_\omega^{300\,K}(c)$ is to lie on this line, then it is reasonable to assume that $\chi_\omega^{150\,K}(c)$ will lie on a lower, parallel line commencing at $\chi_{Ti}^{150\,K} = 143.4 \times 10^{-6}\,cm^3\,mol^{-1}$.

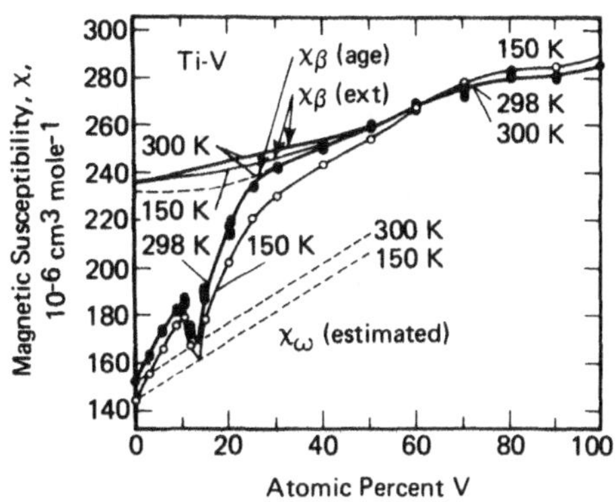

Figure 7-10. Experimental (● and ○) and estimated composition-dependence isothermals (150 K and ~300 K) for Ti-V alloys in all their microstructural variants. The curves χ_β(ext) are the loci of the points designated by ×s in Figure 7-7; the dashed curve designated χ_β (age) relies on an estimated χ_β(300 K) of $232 \times 10^{-6}\,cm^3\,mol^{-1}$ for Ti-V(15 at.%) based on an aging study—after COLLINGS [COL75[b], COL84].

This method, whose results appear to be in conflict with those of another approach to the susceptibility composition dependence of low-solute-concentration ω-Ti–TM alloys, to be considered below, is offered as a simple model construction intended for use in analyzing the susceptibility components of $\omega+\beta$-phase alloys over the composition range 15–50 at.% V, within which no other relevant data exist. With their relatively small slopes, the curves suggested in Figure 7-10 lead to very conservative estimates of the fraction of ω-phase required to produce given depressions of average susceptibility in the high concentration Ti–V alloys.

7.12.2 Anomalous Temperature Dependence

The experimentally determined susceptibility temperature dependences of Ti–V alloys evaluated *at* 150 K and *at* 300 K, respectively, are plotted in Figure 7-11(a), where they may be compared with values of $d\rho/dT$ averaged over the interval 77–200 K (i.e., "at" ~150 K) and again at 300 K, Figure 7-11(b). Clearly $d\chi/dT$ is anomalously large within the same composition range that $d\rho/dT$ is anomalously small or negative.

Susceptibility temperature dependence, including the effect of reversible $\omega \rightleftharpoons \beta$ conversion, is dealt with by first differentiating the continuity equation

$$\chi = f_\omega \chi_\omega + f_\beta \chi_\beta \tag{7-20}$$

and writing

$$f_\omega = 1 - f_\beta$$

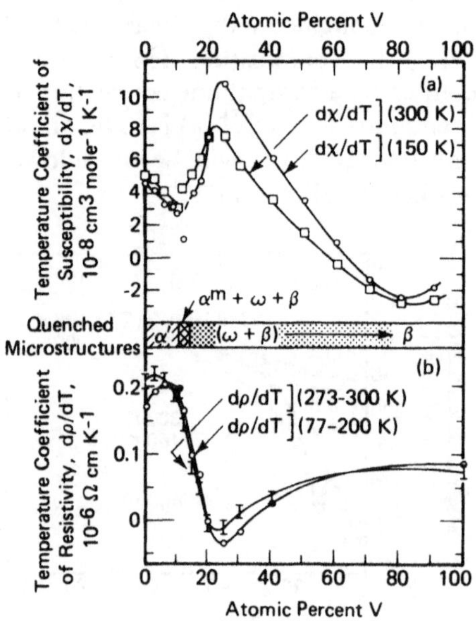

Figure 7-11. (a) Magnetic susceptibility temperature dependences at 150 K and 300 K of Ti–V alloys showing anomalous behavior at compositions near 25 at.% V, a range within which, according to (b), the resistivity temperature dependence is negative (see Figures 5-7 and 5-10)—after COLLINGS [COL78, COL84].

after which

$$\frac{d\chi}{dT} = \underbrace{f_\omega \frac{\partial \chi_\omega}{\partial T} + f_\beta \frac{\partial \chi_\beta}{\partial T}}_{\substack{\text{Intrinsic temperature} \\ \text{dependence of} \\ \text{average } \chi}} + \underbrace{(\chi_\omega - \chi_\beta)\frac{\partial f_\omega}{\partial T}}_{\substack{\text{Reversible } \beta \rightleftharpoons \omega \\ \text{transformation}}} \tag{7-21}$$

In (7-21), the first group of terms represents the normal susceptibility temperature dependence of a stable two-phase $\omega+\beta$-mixture, while the second group takes into account the effect of $\omega \rightleftharpoons \beta$ conversion. If the temperature increases, $\partial f_\omega/\partial T$ decreases as the ω-phase dissolves. But $\chi_\omega < \chi_\beta$, consequently the second term represents a positive contribution to $d\chi/dT$, as required by the isothermals of Figure 7-11(a).

A calculation of the mole fraction ($\cong$ volume fraction) of ω-phase, which exists, at any given temperature T within the selected interval 150–300 K, again relies upon continuity equation (7-20) this time in the form

$$f_{\omega,T} = \left.\frac{\chi_\beta - \chi}{\chi_\beta - \chi_\omega}\right|_T \tag{7-22}$$

Applications of (7-22) at each of two temperatures T_1 and T_2 enable the volume fraction of ω-phase,

$$\Delta f_\omega\Big|_{T_2}^{T_1} = f_{\omega, T_2} - f_{\omega, T_1} \tag{7-23}$$

which exchanges reversibly with β-phase within the temperature interval $T_2 - T_1$, to be easily calculated from the results of a susceptibility temperature dependence experiment provided that $\chi_\beta(T)$ and $\chi_\omega(T)$ have already been estimated in the manner outlined above in connection with the construction of Figure 7-10.

The results of applying this procedure were plotted in Figure 5-9 for comparison with the anomalous resistivity that also accrues reversibly as the temperature of Ti–V is lowered from 300 to 77 K. As pointed out in connection with that comparison (Section 5.6.1), the fact that the reversible $\Delta\rho$ is accompanied by a reversible susceptibility response attributable to crystalline ω-phase suggests that the anomalous electrical scattering, even if not a direct consequence of particulate precipitation (as claimed by POLONIS and colleagues [LUH68], [CHA73, CHA74], see Section 5.6.1), is at least very closely associated with it.

7.13 Magnetic Susceptibility as a Function of Microstructure in Titanium-Base Alloys

7.13.1 Quenched Ti–TM Alloys

The results of numerous susceptibility composition dependence investigations of quenched Ti–TM alloys are available in the literature. Systems which have been studied over wide composition (or rather, e/a ratio) ranges include: Ti-V [CoL75[b]], Ti-Nb [CoL76], Ti-Mo [Ho71], [CoL76[a]], Ti-Mn [CoL84], and Ti-Fe [Ho74]. The susceptibilities of the quenched alloys respond to composition- and heat-treatment-related microstructure in three major regimes: (i) low-concentration—the martensitic α' and α'' structures; (ii) intermediate-concentration—$\omega+\beta$; (iii) high concentration—single-phase-β. Several curves have already been presented above, both individually (as in Figures 7-5 and 7-10) or as a group, such as in Figure 7-1 where the susceptibility composition dependences of Ti-V, -Nb, -Mo, and -Fe are compared. The occurrence and structural properties of the nonequilibrium quenched phases α', α'' and ω have been considered in Section 1.6 *et seq.* With regard to the martensitic structures, we note that the compositions at which the hexagonal α' gives way to the orthorhombic α'' are listed in Table 1-6.

As a typical example of susceptibility composition/microstructure dependence in an important alloy system some results for Ti-Nb [CoL76] are presented in Figure 7-12. The conditions for the occurrence of the α', α'', ω, and β phases in quenched Ti-Nb alloys have already been discussed in Chapter 1, providing the microstructural information necessary for the interpretation of the figure in metallurgical terms. The susceptibility of the homogeneous, solution-treated-and-quenched (8 h/1350°C/IBQ, i.e., STQ), material is almost independent of composition above 50 at.% Nb. This fact is alone sufficient to explain the insensitivity of the results in that composition range to the existence of coring, as demonstrated by the excellent agreement between the data for as-cast β-quenched (1 h/1000°C/IBQ) and STQ materials (which for Ti-Nb(50 at%), for example, are 215.8, 215.8, and 215.0×10^{-6} cm^3 mol^{-1}, respectively). Turning to the low-concentration alloys, the two distinct branches in the susceptibility data for the martensitic (α', α'') regimes indicate (i) that quenching yields α' for Nb concentrations

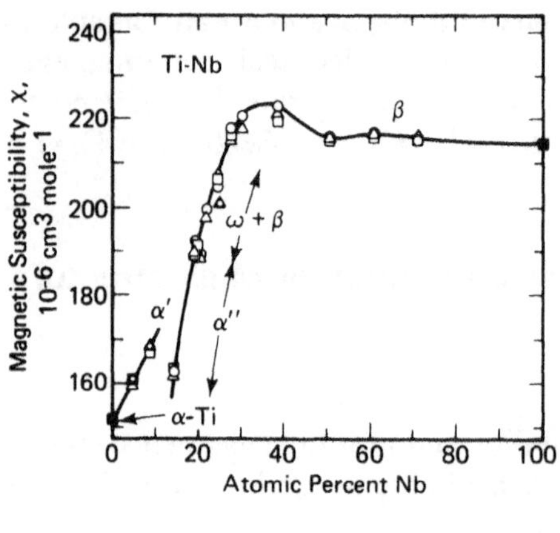

Figure 7-12. Composition dependence of room-temperature magnetic susceptibility of Ti-Nb alloys measured in the as-cast condition (△) or after quenching from heat treatments at 1000°C (○) or at 1350°C (□)—after COLLINGS and SMITH [CoL76].

up to 8.9 at.% and (ii) that the hexagonal (α') and orthorhombic (α'') structures have different electronic densities-of-states (at equivalent Nb concentrations), that for α'' being the lower. Based on the microstructural data presented in Section 1.9, it may be concluded that the structures beyond 20 at.% Nb are $\omega+\beta$ followed in turn by "diffuse-ω"+β and essentially single-phase-β, as indicated in Figure 7-12.

Summarizing these observations, it may be concluded that as Ti is added to Nb ($\chi = 214.1 \times 10^{-6}\ \mathrm{cm^3\ mol^{-1}}$) the susceptibility increases very slowly with composition and would head towards 230–240 $\times 10^{-6}\ \mathrm{cm^3\ mol^{-1}}$ for $\chi_{\beta\text{-Ti}}^{300\,K}$ (see Table 7-3) excepting that increasing amounts of ω-phase, and finally α''-phase, intervene to severely depress the average susceptibility. The presence of ω-phase is always such as to depress the values of measured density-of-states-related properties such as χ and γ, the latter to be considered in the following chapter. The results of both direct and indirect measurements of the susceptibilities of ω-phase Ti and some of its alloys are treated below.

7.13.2 Magnetic Susceptibility of ω-Phase

As mentioned above, the magnetic susceptibility of ω-Ti–V(14 at.%) at 300 K has been deduced from the results of magnetic studies of aging [COL75[b]]. Likewise, a value for ω-Ti–Mo(4.3 at.%) has been extracted from the results of calorimetric studies of aging in the Ti–Mo system [COL72[a]]. The results of these rather indirect measurements on boundary-concentration ω-phase alloys are summarized in Table 7-7.

In the pure group-IV elements (and, presumably, their dilute alloys), ω-phase, which is the equilibrium phase at high pressure [KOU70[b]], is also retainable in the metastable state after the pressure has been removed. DEGTYAREVA *et al.* [DEG74] investigated the superconducting transition temperatures and the susceptibility temperature dependences (4.2–300 K)

Table 7-7. Magnetic Susceptibilities of Some ω-Phase Metals and Alloys

Sample	Magnetic Susceptibility at 300 K χ $10^{-6}\ \mathrm{cm^3\ mol^{-1}}$	Magnetic Susceptibility Temperature Dependence $d\chi/dT$ $10^{-8}\ \mathrm{cm^3\ mol^{-1}\ K^{-1}}$	Reference
Ti	109.2	+5.8	[DEG74]
Zr	107.6	+1.6	[DEG74]
Ti–V(14 at.%)	169	—	[COL75[b]]
Ti–Mo(4.3 at.%)	149	—	[COL72[a]]

of ω-Ti and ω-Zr obtained by subjecting samples of the high-purity metals to pressures of up to 120 kbar for 30 min at 300 K. $\chi_\omega(T)$ essentially parallelled that of pure α-Ti over the temperature range of measurement, but at $\chi_\omega^{300\,K} = 109.2 \times 10^{-6}$ cm^3 mol^{-1}, lay some 42.6×10^{-6} cm^3 mol^{-1} below it, (Figure 7-13). Since graphite was used as the pressurizing medium the sample conditions were not exactly hydrostatic. The magnetocrystalline anisotropy of pure α-Ti is 25.0×10^{-6} cm^3 mol^{-1} [CoL70], but the anisotropy of ω-Ti is unknown, as was the influence of the pressurization on the sample texture. Consequently, it is expected that some uncertainty may be associated with the value of 109.2×10^{-6} cm^3 mol^{-1} quoted above, a reservation that should be borne in mind when considering Figure 7-13, in which the room-temperature susceptibilities of the three ω-phase samples referred to here are plotted. Nevertheless, it seems abundantly clear that the susceptibility of ω-Ti is significantly lower than that of its α-phase allotrope. On the other hand, the band-structure calculations of VOHRA *et al.* [VOH79] showed that the Fermi density-of-states in ω-Ti is a factor of two higher than that of the α-phase, a result which leads to a theoretical (bare) electronic specific heat coefficient of $\gamma_\omega^0 = 5.20$ mJ mol^{-1} K^{-2}, to be compared with a theoretical (bare) electronic specific heat coefficient for α-Ti of 2.56 mJ mol^{-1} K^{-2} ($\gamma^0 = \gamma/(1+\lambda)$, (7-5)) derivable from the data of Table 7-2. The corresponding Pauli paramagnetic susceptibility is $\chi_p^\omega = 71.3 \times 10^{-6}$ cm^3 mol^{-1};

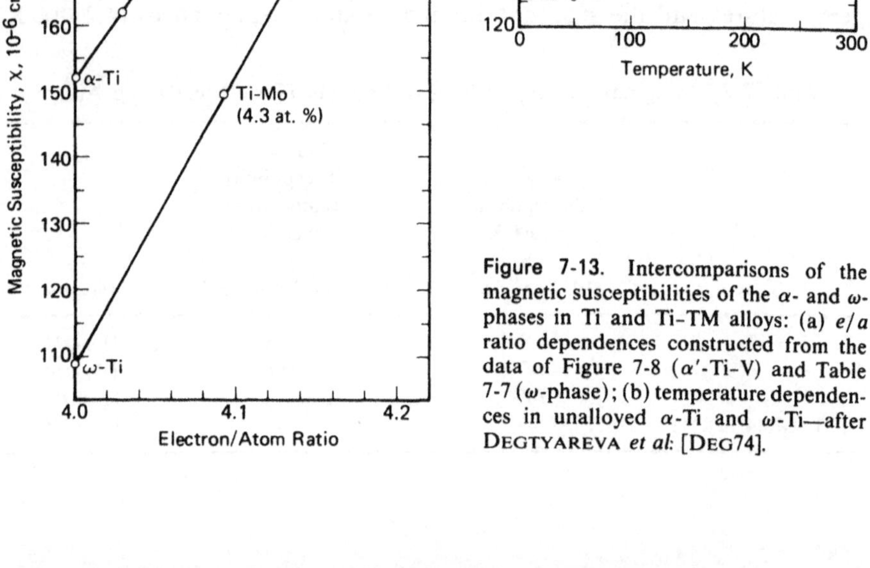

Figure 7-13. Intercomparisons of the magnetic susceptibilities of the α- and ω-phases in Ti and Ti–TM alloys: (a) e/a ratio dependences constructed from the data of Figure 7-8 (α'-Ti–V) and Table 7-7 (ω-phase); (b) temperature dependences in unalloyed α-Ti and ω-Ti—after DEGTYAREVA *et al*: [DEG74].

then taking $\chi^{\omega}_{70\,K} = 95.9 \times 10^{-6}\,cm^3\,mol^{-1}$ (Table 7-7) for the low-temperature total susceptibility of ω-Ti, it follows that $\chi^{\omega}_{orb} = 54.4 \times 10^{-6}\,cm^3\,mol^{-1}$, less than half that of α-Ti (*viz.*, $123.8 \times 10^{-6}\,cm^3\,mol^{-1}$). It would be advisable to confirm the susceptibility data, but if the results of experiment and calculation presented above are correct, they would suggest that a severely reduced χ_{orb} (enhanced Δ, see (7-15)) is responsible for the relatively low total susceptibility of ω-Ti. Discussions of the electronic and related properties of ω-phase are continued in Section 11.9.

7.14 Magnetic Studies of Precipitation and Aging in Titanium–Transition-Metal Alloys

In addition to equilibrium-phase-diagram analysis (such as has been applied elsewhere to studies of the Ti–Al system [CoL79]), "metaequilibrium" phase relationships,[5] as well as the kinetics of moderate-temperature aging leading to the metaequilibrium condition, are amenable to study using magnetic susceptibility and low-temperature specific heat as diagnostic properties. The agings of Ti–V, Ti–Fe (Section 10.5), and Ti–Mo (Sections 10.5 and 11.9), have been monitored using both of these techniques. In studies of Ti–V(15 and 19 at.%) using magnetic susceptibility alone it has been possible to examine the time dependence of isothermal ω-phase precipitation from the original quenched $\omega+\beta$-phase alloys and, with the aid of some results from the x-ray investigations by HICKMAN [HIC68, HIC69], to determine the magnetic susceptibility of ω-Ti–V at the phase boundary where $[V] \cong 14$ at.%.

7.14.1 The Aging Process in the $\omega+\beta$-Field

As the following analysis will demonstrate, the effect of aging can be regarded as equivalent to an allotropic $\beta \rightarrow \omega$ transformation in association with a compositional change (i.e., a transfer of solute from ω to β).

In N moles of alloy, if ΔN moles of solute are transferred from ω to β during aging, the compositional changes would be $\Delta c_\omega = -\Delta N/f_\omega N$ and $\Delta c_\beta = +\Delta N/f_\beta N$, respectively. The susceptibility change is, of course, $\Delta\chi = (\partial\chi/\partial c)\Delta c$ with appropriate subscripts, thus

$$f_i \Delta\chi_i = (\partial\chi/\partial c)_i(\Delta N/N) \qquad i = \omega \text{ or } \beta \tag{7-24}$$

[5]The prolonged aging of Ti–TM alloys within a suitable concentration range at a temperature below some 400°C leads to the equilibration of $\omega+\beta$ within the lower-energy $\alpha+\beta$-field (see, for example, Figure 2-4).

This is next substituted in the differential form of (7-20), *viz.*,

$$\Delta\chi = \Delta(f_\omega \chi_\omega) + \Delta(f_\beta \chi_\beta) \tag{7-25}$$

which, after taking into account that $\Delta f_\beta = -\Delta f_\omega$ then becomes (cf. (7-21))

$$\Delta\chi = \underbrace{-\frac{\Delta N}{N}\left[\left(\frac{\partial \chi}{\partial c}\right)_\omega - \left(\frac{\partial \chi}{\partial c}\right)_\beta\right]}_{\substack{\text{Composition dependences} \\ \text{of } \chi_\omega \text{ and } \chi_\beta}} + \underbrace{(\chi_\omega - \chi_\beta)\Delta f_\omega}_{\substack{\text{Allotropic} \\ \beta \to \omega \\ \text{transformation}}} \qquad (\Delta N > 1) \tag{7-26}$$

As indicated at the outset, one of the components of $\Delta\chi$ (*viz.*, the second term in (7-26)) refers to the allotropic transformation, while the other term contains the composition dependences of χ_ω and χ_β. Recognizing that $\chi_\omega < \chi_\beta$, the equation shows that if $\chi_\omega(c)$ and $\chi_\beta(c)$ are exactly parallel, the susceptibility change with aging (invariably a decrease) will be a direct result of the allotropic $\beta \to \omega$ transformation component. Otherwise this decrease will be augmented if $(\partial\chi/\partial c)_\omega$ is more positive than $(\partial\chi/\partial c)_\beta$, as it is in Ti–V (see Figure 7-10), or will be partially offset if the converse is true.

The effect of moderate-temperature aging on the magnetic susceptibility of Ti–V(19 at.%) has been investigated [COL75[b]]. After being quenched from a heat treatment of 1 h/300°C, the susceptibility specimen was aged at 300°C for various times of from 0.5 to 2500 h as indicated in the semi-logarithmic plot of Figure 7-14. Data for the as-quenched alloys, after being positioned in such a way as to be consistent with the aging results, indicate that the ice-brine quenching of a vial of small encapsulated samples is equivalent to an ideal quench followed by a 2 min anneal at 300°C. Finally,

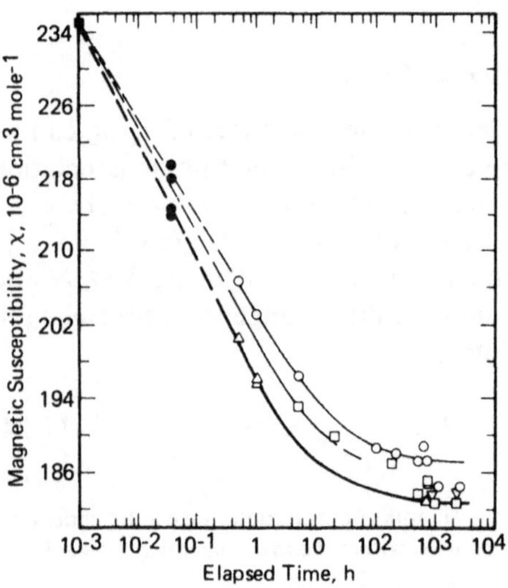

Figure 7-14. Response of the room-temperature susceptibility of Ti–V(19 at.%) to aging at 300°C; the four samples measured are indicated by ○, □, △, and ▽. The as-quenched data (●), which have been inserted so as to fit the aging curve, suggest that the quenching of these encapsulated samples is equivalent to aging at 300°C for about 2 min. The datum point for β-Ti–V(19 at.%) (■, see the χ(age) curve in Figure 7-10), has been arbitrarily placed at an elapsed time of 0.001 h—after COLLINGS [COL75[b]].

we note that the 300 K susceptibility of β-Ti–V(19 at.%) (taken, for example, from Figure 7-10), if positioned at $t = 0.001$ h, also falls naturally on a back-extrapolation of the aging results.

7.14.2 Properties of a "Saturation-Aged" $\omega+\beta$-Phase Ti–TM Alloy

After the $\omega+\beta$-phase alloy has been aged for a sufficiently long time it eventually reaches a state of metastable equilibrium in which the compositions of the ω- and β-phases are those of the phase boundaries at the aging temperature. According to HICKMAN [HIC68, HIC69], if Ti–V(15–25 at.%) alloys are aged at temperatures below 400°C (i) the V content of the ω-phase approaches a final value of 13.5–14.5 at.%, (ii) for Ti–V(19 at.%), as the aging time approaches 1000 h, the volume fraction of ω-phase tends towards 0.84. Inserting these values into a continuity equation for composition of the form

$$[V]_\beta = \frac{[V] - f_\omega [V]_\omega}{1 - f_\omega} \tag{7-27}$$

yields $[V]_\beta^{sat} = 45.3 \pm 2.6$ at.%, associated with which, according to Figure 7-10, is a room-temperature susceptibility of $\chi_\beta^{sat} = (255.3 \pm 2.3) \times 10^{-6}\,\text{mol}^{-1}$, the "error bars" resulting from the ±0.5% uncertainty in $[V]_\omega$.

As shown in Figure 7-14, during aging at 300°C the magnetic susceptibility of a typical $\omega+\beta$-phase alloy also approaches saturation at sufficiently long aging times. In this case the smoothed results suggest that $\chi^{1000\,\text{h}} = 182.6 \times 10^{-6}\,\text{cm}^3\,\text{mol}^{-1}$. This, and the above data, when inserted into (7-20) yield $\chi_\omega = (168.8 \pm 0.45) \times 10^{-6}\,\text{cm}^3\,\text{mol}^{-1}$ for the susceptibility of ω-Ti–V(14 at.%) at about 300 K, the value already listed in Table 7-7 and one that has played an essential role (see the discussion associated with Figure 8-10) in determining the susceptibility-composition dependence of ω-Ti–V.

Chapter 8

Low-Temperature Specific Heat

8.1 Low-Temperature Specific Heat of Solids

8.1.1 Specific Heat of Insulators

(a) Quantities and Units Employed. Numerous discussions of lattice specific heat are available in the literature; for descriptive introductory summaries, the reader is directed to standard texts, such as that of KITTEL [KIT68], and a cryogenic monograph in this series by GOPAL [GOP66]. In developing the Debye equations, it is usual to begin by calculating the thermal energy associated with the vibrations of N atoms of a solid. The resulting quantity is conveniently referred to as the "heat capacity" of the assembly. The Debye calculation takes place implicitly at constant volume, and leads accordingly to the constant-volume heat capacity, C_v, which is, therefore, associated with some fixed volume, V, of material. If the solid is heated at constant pressure it generally expands, and in so doing absorbs some extra energy. The difference between the associated constant-pressure heat capacity, C_p, and C_v, at some temperature T, is $C_p - C_v = TV\beta_v^2 K$, where β_v is the volume expansion coefficient and K is the bulk modulus. Unless lattice expansion and its consequences are of special interest, as in Chapter 9, the difference $C_p - C_v$ may be neglected, particularly below room temperature [KIT68, p. 165], and the bare symbol, C, may be assigned to the general heat capacity. In what follows that symbol will also be assigned to the molar heat capacity or the molar specific heat, while the superscripts v and g will designate specific heats per unit volume (cm^3) and per gram, respectively. A subscript g may be employed to indicate lattice[1] specific heat if a distinction between it and the electronic specific heat component of metals (see below) is to be made.

[1]"Lattice," in German, is Gitter, hence the subscript g.

(b) Temperature Dependence of the Lattice Specific Heat. At very low temperatures, the lattice heat capacity is simply βT^3, where the lattice specific heat coefficient, β, is a function of the Debye temperature, θ_D; the addition to this term of a series of higher odd powers of T usually describes $C_g(T)$ accurately throughout the entire He-temperature range.

8.1.2 Low-Temperature Specific Heat of Metals

To take into account the thermal excitation of those conduction electrons that lie within an energy range of width about $k_B T$ at the Fermi surface, simple free-electron theory assigns to metals an electronic specific heat contribution, proportional to T and with a coefficient, γ, which is proportional to the Fermi density-of-states ((7-4)). In what follows, this bare symbol will imply the molar quantity, while γ_v will denote a per unit volume (cm^3) electronic specific heat coefficient. The total molar low-temperature specific heat of a metal may thus be written [BAR57[b]]

$$C = \gamma T + \sum_{i=1}^{n} \beta_i T^{2i+1} \tag{8-1}$$

In a study of Cu, for example, MARTIN [MAR73] has shown (8-1), with $n = 3$, to be valid over the temperature range of 0.4 to 30 K. The form more usually employed, and one which turns out to be very satisfactory at temperatures below about $\theta_D/12$ (i.e., $\lesssim$4 K for many metals), is simply

$$C = \gamma T + \beta T^3 \tag{8-2}$$

Equation (8-2) indicates that γ and β can be conveniently obtained as the intercept and slope, respectively, of a plot of C/T *versus* T^2, a format which is still used to display the experimental results even when the data are analyzed by computer.

The low temperature specific heat measurement is a useful technique in metal physics, providing as it does (i) *electronic-property* data which, *via* $n(E_F)$ and the BCS theory, is connected to the superconducting transition temperature, T_c (see below), and (ii) *metallurgical information,* in that θ_D is a kind of bulk stiffness constant or, by virtue of its relationship to the C_{ij}s and in particular to C' (Sections 4.3 and 4.4), a lattice-stability parameter. The Fermi density-of-states, $n(E_F)$, has been shown to be related to the free, noninteracting (i.e., "bare") electronic specific heat coefficient, γ^0, according to (7-4). But as mentioned above, the measured γ has been enhanced by electron–phonon interactions parameterized by λ, the electron–phonon coupling constant, such that γ and γ^0 are connected by (7-5). An experimentally convenient relationship between $n(E_F)$ and the measured electronic specific heat coefficient is, therefore,

$$n(E_F) = 212\gamma/(1+\lambda) \qquad (\text{states eV}^{-1}\,\text{atom}^{-1}) \tag{8-3}$$

with γ expressed in the units J mol^{-1} K^{-2}.

The Debye temperature is derivable from the low-temperature specific heat data with the aid of the formula

$$\theta_D = \left(\frac{1.944 \times 10^3}{\beta}\right)^{1/3} \qquad (\text{K}) \tag{8-4}$$

provided β is expressed in J mol^{-1} K^{-4}.

The normal-state low-temperature specific heats of the unalloyed transition elements have been comprehensively reviewed by HEINIGER *et al.* [HEI66], who also tabulated a set of critically selected values of γ and θ_D. The low-temperature specific heats of high-purity samples of Ti, Zr, and Hf have been measured in this laboratory [COL70, COL71^d]; the results of this work, together with the critical data of HEINIGER *et al.* for the transition elements of groups IV and V, are presented for reference in Table 8-1.

Table 8-1. Normal State Low-Temperature Specific Heat Parameters for the Pure Transition Elements of Groups IV and V

Element	Electronic Specific Heat Coefficient γ, mJ mol^{-1} K^{-2}	Lattice Specific Heat Coefficient β, mJ mol^{-1} K^{-4}	Debye Temperature θ_D, K	Literature
Ti	3.32		425	Heiniger *et al.* [HEI66]
	3.36	0.026	420	Collings and Ho [COL70]
Zr	2.78		290	Heiniger *et al.* [HEI66]
	2.80	0.079	291	Collings and Ho [COL71^d]
Hf	2.16		252	Heiniger *et al.* [HEI66]
	2.15	0.127	248	Collings and Ho [COL71^d]
V	9.9		399	Heiniger *et al.* [HEI66]
Nb	7.8		277	Heiniger *et al.* [HEI66]
Ta	6.0		258	Heiniger *et al.* [HEI66]

8.1.3 Interrelationships Between γ and θ_D

Already present in Table 8-1 (in particular, the data from [HEI66]) is the suggestion of an inverse relationship between γ and θ_D. This effect becomes obvious when data for the entire transition-metal block of the periodic table are scanned, and even more so when γ and θ_D for the members of a closely spaced alloy series are intercompared using, say, the e/a ratio as a mapping parameter [COL72c]. Similar inverse trends have also been noted in the dual plots of calorimetrically derived values of $n(E_F)$ (using (8-3)) and θ_D. Figure 8-1, an excellent example of this, represents the alloy systems Ti–Al and Ti–Mo [COL73a]. The remarkable inverse scaling indicated in the figure is claimed to be an expression of the factors that influence the variation of bcc stability with band filling [COL72c, COL73a], in regard to which it is interesting to note the parallelism that exists between the variation of θ_D with e/a referred to here and that of C' *versus* e/a as depicted in Figure 1-1. An important observation that emerges from these comparisons is that the bcc lattice is softest when $n(E_F)$ is a maximum (i.e., when a high Fermi density of conduction-electron states is available for screening the ions).

As implied by a combination of (7-2) and (8-3) in the form

$$10^6\chi_p/64.7 = n(E_F) = 212\gamma/(1+\lambda) \tag{8-5}$$

with χ_p in $cm^3\ mol^{-1}$, $n(E_F)$ in $eV^{-1}\ atom^{-1}$, and γ in $J\ mol^{-1}\ K^{-2}$, $n(E_F)$ couples the Pauli spin paramagnetic susceptibility to the electronic specific heat coefficient. It is to be expected, therefore, that some degree of parallelism might also exist between the total magnetic susceptibility and γ. Figure

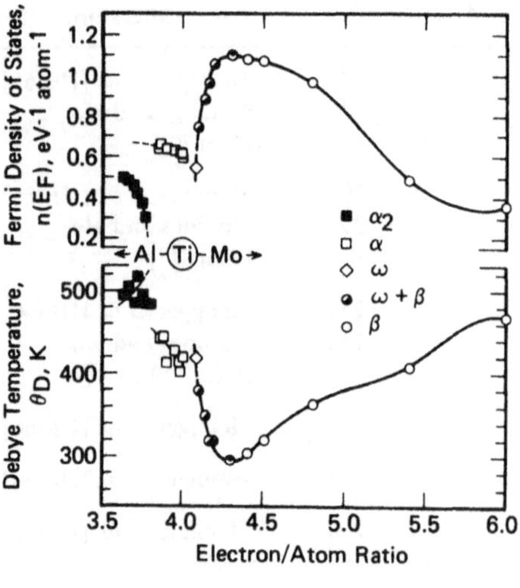

Figure 8-1. Calorimetrically derived Fermi density-of-states, $n(E_F)$ and the directly measured Debye temperature, θ_D, as functions of e/a ratio (and microstructure) in Ti-Al ($e/a < 4$) and Ti-Mo ($e/a > 4$) alloys. Data for the α^m-martensitic Ti-Mo alloys are omitted. The properties of the α-, ω-, and β-structures are considered throughout this book; the α_2-structure has been discussed elsewhere [COL82a]—after COLLINGS *et al.* [COL73a].

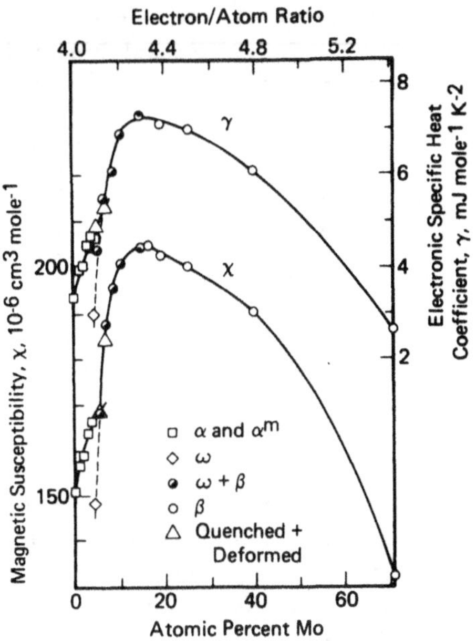

Figure 8-2. Intercomparison of the composition dependences of room-temperature magnetic susceptibility, χ, and electronic specific heat coefficient, γ, for quenched Ti–Mo alloys—after Ho and COLLINGS [HO71, HO73], [COL72[a]].

8-2 for Ti–Mo provides a good example of this prediction, which, as we shall see, turns out to be a fairly general property of Ti–TM alloys.

8.2 Composition and Microstructure Dependence of Low-Temperature Specific Heat in Titanium–Transition-Metal Alloys

8.2.1 General Description

The normal-state low-temperature specific heats of transition-metal binary alloys in general have been comprehensively reviewed by HEINIGER, BUCHER, and MÜLLER [HEI66]. Confining our attention now to Ti-TM systems, low-temperature specific heat data are available for alloys of Ti with both 3*d* and 4*d* transition elements such as Ti–V(0–80 at.%) [UPT72], [COL75], Ti–Fe(0–20 at.%) [HO74], Ti–Zr(0–90 at.%) [BUC64], [BET68], Ti–Nb(0–90 at.%) [SUK71], [SAV73, p. 230], Ti–Mo(0–70 at.%) [HAK61[a]], [SIN68], [COL71, COL72, COL73[a]], and Ti–Rh(0–10 at.%) [DUM68], [DAN69].

The essential points which need to be made concerning the interrelationships between γ and composition-controlled microstructure are already implied in Figures 8-1 and 8-2 for Ti–Mo, which, as a prototype of Ti-TM alloys, has been studied in considerable detail (e.g., [COL72,

Table 8-2. Low-Temperature Specific Heat Parameters for Several Structural States of Ti-Mo[a]

Alloy	Structure	γ mJ mol^{-1} K^{-2}	θ_D K
Ti	α	3.36	420
Ti	β	(10.1	<200)
Ti-Mo(4.3 at.%)	ω	(3.0	420)
Ti-Mo(25 at.%)	β	7.0	320
Mo[b]	β	1.83	460

[a]The quantities in parentheses were estimated using the techniques described in Sections 10.4 and 11.9.
[b]Reference [HEI66].

COL73[a]]). As Ti is added to Mo (a group VI element), γ increases smoothly towards a maximum between about 15 and 20 at.% Mo ($e/a = 4.3 \sim 4.4$); with further increases in Ti level, the onset of ω-phase precipitation rapidly reduces the average γ. In the quenched martensitic ($\alpha^m \equiv \alpha'$ or α''—Section 1.6.1) regime, γ increases steadily with Mo concentration within the range concerned (0–4½ at.% Mo). The γ-values for α^m are generally greater than those for ω-phase but less than those for β-Ti-Mo if projected into the low-concentration regime. A set of measured and estimated low-temperature specific heat data for several microstructural forms of the Ti-Mo prototype alloy is given in Table 8-2. Of particular interest is the extremely low value predicted for γ_ω; one which is in accord, however, with the relatively low value of total magnetic susceptibility, χ_ω, obtained by DEGTYAREVA *et al.* [DEG74] from a study of pressure-transformed unalloyed Ti (Section 7.13.2, Figure 7-13).

8.2.2 Low-Temperature Specific Heats of Ti–V, Ti–Mo, and Ti–Fe

The electronic specific heat coefficients and Debye temperatures of quenched alloys selected from the systems Ti-V [UPT72], [COL75], Ti-Mo [COL72], and Ti-Fe [HO74] are intercompared as functions of e/a ratio in Figure 8-3, which serves to summarize several of the important points mentioned above: (i) The curves of γ *versus* e/a bear a resemblance to those of χ_p *versus* e/a given in Figure 7-2, as required by (8-5). (ii) The γ and θ_D curves exhibit inverse scaling, a drop in γ, and a stiffening of θ_D occurring within the ω-phase precipitation regime. (iii) The solutes V, Mo, and Fe are from groups V, VI, and VIII, respectively, of the periodic table. Nevertheless their electronic and microstructural properties vary in unison when displayed as functions of e/a; in particular the discontinuities in their γ and θ_D curves occur at $e/a \cong 4.15$, in excellent agreement with the

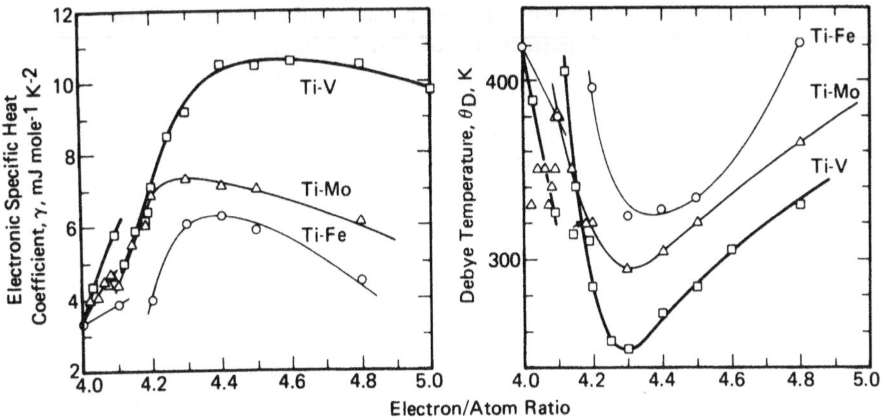

Figure 8-3. Intercomparison of the e/a-ratio dependences of electronic specific heat coefficient, γ, and Debye temperature, θ_D, for three series of β-quenched Ti-TM alloys. Data sources: Ti-V—COLLINGS *et al.* [UPT72], [COL75]; Ti-Mo—COLLINGS *et al.* [COL72, COL73[a]]; Ti-Fe—HO and COLLINGS [HO74].

$e/a = 4.15 \pm 0.03$ given in Table 1-3 as the average value of the $M_s(200°C)$ position for ten Ti-TM($3d$, $4d$, or $5d$) alloys.

8.2.3 Normal-State Low-Temperature Specific Heat of Ti–Nb

Because of its practical importance, the specific heat of the Ti-Nb system is singled out for individual attention. Low-temperature specific heat

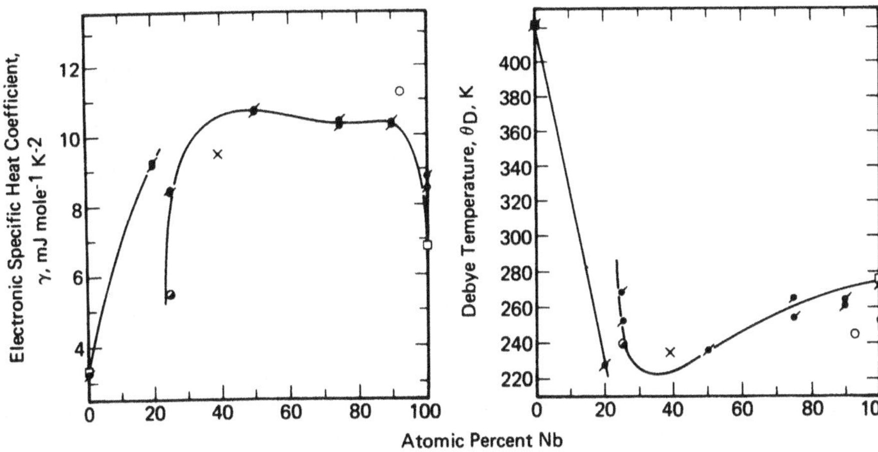

Figure 8-4. Electronic specific heat coefficient, γ, and Debye temperature, θ_D, as function of composition in Ti-Nb alloys. Pure endpoint data (□) from standard sources (cf. Table 8-3). Alloy data sources: ●, SAVITSKII *et al.* [SAV73, p. 320]; ●, SUKHAREVSKII *et al.* [SUK71] and VERKIN [VER80]; ◑, SUKHAREVSKII *et al.* [SUK68]; ×, ELROD *et al.* [ELR82]; ○, SHCHETKIN and KHARCHENKO [SHC73].

Table 8-3. Electronic Specific Heat Coefficients and Debye Temperatures of Ti–Nb Alloys

Nominal Nb Concentration[a] at.%	Electronic Specific Heat Coefficient γ, mJ mol^{-1} K^{-2}	Debye Temperature θ_D, K	Literature[b]
0	3.32	425	[HEI66]
	3.36	420	[COL70]
	3.24	422	(a)
	3.25	422	(b)
20	9.21	228	(a)
	9.13	228	(b)
25	5.45	239	[SUK68]
	8.40	268	(a)
	8.45	252	(b)
39.0	9.49	235	[ELR82]
50	10.76	236	(a)
	10.66	236	(b)
75	10.40	254	(a)
	10.22	265	(b)
90	10.33	264	(a)
	10.25	261	(b)
92.8	11.2	245	[SHC73]
100	8.80	252	(a)
	8.46	273	(b)
	7.80	275	[LEU64]
	7.80	277	[HEI66]

[a] As plotted, the compositions of Reference (a) appear to be 20, 25.5, 50, 75, and 90 at.% Nb; those of Reference (b) appear to be 20, 24.5, 49.5, 74, and 89.5 at.% Nb.
[b] Reference (a)—[SAV73, p. 230]—these data have the symbol ● in Figure 8-4. Reference (b)—[SUK71], [VER80]—these data have the symbol ● in Figure 8-4.

data for the entire alloy series obtained by workers at the Baikov Institute, Moscow, have been presented in graphical form by SAVITSKII *et al.* [SAV73, p. 230], and data for the individual alloys Ti–Nb(97.77 at.%) and Ti–Nb(25 at.%) have been reported by SHCHETKIN and KHARCHENKO [SCH73] and SUKHAREVSKII *et al.* [SUK68], respectively, of the Physico-Technical Institute, Kharkov. SUKHAREVSKII, SHCHETKIN, and FAL'KO [SUK71] reported having measured a series of Ti–Nb(10, 25, 50, 75, and 80 at.%) alloys. Their data, originally published in a work by SUKHAREVSKII *et al.* [SUK69], have been extracted from a set of graphs supplied by VERKIN [VER80] for presentation in Table 8-3 and Figure 8-4 along with the other data referred

Table 8-4. Low-Temperature Specific Heat Coefficients γ and β for Ti–Nb Extracted from the Smooth Curves of Figure 8-4. The Total Specific Heat, C, at Temperature, T, Is Given by (8-2)

Nb Concentration	Molar Weight	Density[a]	Electronic Specific Heat Coefficient, γ		Temperature	Lattice Specific Heat Coefficient, β	
at.%	g	g cm^{-3}	10^{-1} J kg^{-1} K^{-2}	10^{-3} J cm^{-3} K^{-2}	θ_D, K	10^{-3} J kg^{-1} K^{-4}	10^{-5} J cm^{-3} K^{-4}
25	59.2	6.08	1.27	0.77	252	2.05	1.25
30	61.4	6.31	1.56	0.98	226	2.74	1.73
35	63.7	6.53	1.61	1.05	222	2.79	1.82
40	65.9	6.73	1.61	1.08	224	2.62	1.76
45	68.2	6.92	1.57	1.09	229	2.38	1.65
50	70.4	7.10	1.52	1.08	236	2.10	1.49
55	72.7	7.26	1.47	1.08	240	1.94	1.41

[a]Obtained by linear extrapolation (wt.% basis) between 4.58(Ti) and 8.4(Nb) from standard sources.

to above. From the smoothed curves of Figure 8-4, "average" values of γ, θ_D, and β for Ti–Nb alloys within the composition range of technical interest (25–55 at.% Nb) have been obtained; they are presented in Table 8-4 in both mass and volume units enabling values of the total low-temperature specific heats at any temperature within the cryogenic range to be conveniently calculated with the aid of (8-2). A further set of data, that due to ELROD *et al.* [ELR82] for a sample of annealed commercial Nb–44.6 Ti (i.e., Ti–Nb(39 at.%)), agrees reasonably well with the earlier results.

In Figure 8-4, both γ and θ_D for alloys within the composition range 0–20 at.% Nb, whose quenched microstructures are α', followed by α'' (Section 1.7.2), are both confined to single highly-composition-dependent branches. Since this is in contrast to the susceptibility results of Figure 7-12, which shows the α' and α'' data on separate branches, the latter being an extension of the $\omega+\beta$-phase curve, some confirmation would be desirable. Since both γ and θ_D vary rapidly with composition in the ω-rich end of the $\omega+\beta$-field (~25 at.% Nb), precise composition control in this region is essential. The results of the Baikov Institute and Physico-Technical Institute groups are in good general agreement in the mid-to-high composition range except for the isolated data for Ti–Nb(92.8 at.%). The collected electronic specific heat data for Ti–Nb, taken together with a crucial data point for pure Nb [HEI66], indicate that an unusually rapid increase in γ accompanies the additions of small amounts of Ti to Nb. No such anomalous change is reflected in the Debye temperature.

8.3 Low-Temperature Specific Heats of Superconductors

8.3.1 Experimental Observations

The temperature dependence of normal-state low-temperature specific heat is adequately described by (8-2). If the metal is a superconductor, its second-order transition into the superconductive state at T_c is accompanied by a jump, ΔC, in the electronic specific heat. The height of this jump is about 50% of the normal-state electronic specific heat γT_c at the critical temperature. There are many examples of this in the literature. Numerous detailed calorimetric comparisons of the normal and superconductive states of pure transition elements have been undertaken and the results compared with current theory. For example, the specific heats of V have been measured by CORAK *et al.* (pre-BCS[2]) [COR56] and RADEBAUGH and KEESOM [RAD66], those of Nb by BROWN *et al.* (pre-BCS) [BRO53], HIRSHFELD *et al.* [HIR62], LEUPOLD and BOORSE (monocrystalline Nb) [LEU64], and VAN

[2]BARDEEN, COOPER, and SCHRIEFFER [BAR57].

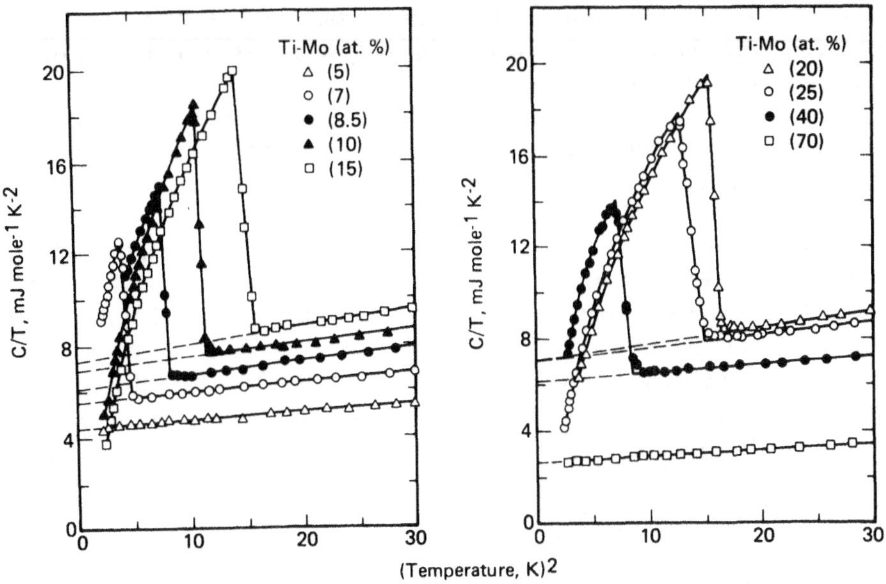

Figure 8-5. Low-temperature specific heat results for quenched Ti–Mo(5–70%) alloys plotted in the usual format C/T *versus* T^2. A sharp superconducting specific heat "jump" is observed for most specimens—after COLLINGS and HO [COL70[a], COL71, COL72[a]], [HO73].

DER HOEVEN and KEESOM [VAN64], and those of Mo by RORER *et al.* [ROR65].

Transition metal alloy superconductivity is typified by the behavior of the Ti–Mo system, which has been subjected to intense study particularly by HAKE and colleagues. A useful calorimetric investigation of Ti–Mo(16 at.%) in applied magnetic fields of up to 29 kOe, during which the transition temperature decreases from 4.25 to 3.22 K, was undertaken by BARNES and HAKE [BAR66]. The results of a set of *zero-field* measurements of a series of quenched Ti–Mo(7–40 at.%) alloys, taken through the superconducting transition, are presented in Figure 8-5. Some important features of the calorimetric superconductive transition to be considered below are: (i) the degree of sharpness (or rounding) of the transition; (ii) the relative height of the unrounded (or deconvoluted) specific heat jump at T_c; (iii) the magnitude and functional form of the lattice specific heat below T_c; (iv) the functional form of the electronic specific heat below T_c.

8.3.2 Lattice Specific Heat in the Normal and Superconducting States

A topic of considerable interest has been the superconductive-state lattice specific heat and its relationship to that in the normal state. In order to achieve this comparison it is necessary to take one of the specific heat data sets with the sample's superconductivity suppressed by means of an applied magnetic field.

In studies of Nb, BROWN *et al.* [BRO53] used a transverse magnetic field of 5 kOe and noted that in a plot of C/T *versus* T^2 the field-on and field-off data above T_c, and the field-on data below it, were colinear with slope equivalent to 252 K. The subsequent results of LEUPOLD and BOORSE [LEU64], in the vicinity of T_c, were essentially in agreement with this (albeit with θ_D (3 K to $\sim T_c$) = 241 K); but when the field-on (10 and 17 kOe) normal-state measurements were extended to temperatures below about 3 K, the data fell on a line corresponding to $\theta_D = 275$ K, a value which agreed with the results of ultrasonic measurements. This, and the equality of the slope of this very low temperature field-on line to that of the tangent to the superconductive C/T *versus* T^2 curve below $T^2 = 1\ \text{K}^2$ [LEU64], were taken as further evidence for the equality of the normal-state and superconductive-state lattice specific heats.

The shift in θ_D from 241 K to 275 K at temperatures below about 3 K was thought not to be artifactual but rather an inherent property of the Nb crystal. In Ti–Mo, which appears to be devoid of anomalous lattice specific heat temperature dependence, the interpretation of field-on/field-off results is less complicated. As mentioned above, BARNES and HAKE [BAR66] have reduced the calorimetrically measured T_c of Ti–Mo(16 at.%) from 4.25 K to 3.22 K through the application of magnetic fields of strengths of up to 29 kOe, and in so doing have detected no discontinuities or changes in slope of the lattice specific heat base line.

8.4 The Superconductive Electronic Specific Heat

8.4.1 The Gorter–Casimir Two-Fluid Relationships

With the aid of a simple thermodynamic argument based on the following equation for the difference in unit-volume zero-field free energies between the superconductive and normal states at some reduced temperature $t = T/T_c$

$$g_n(0) - g_s(0) = H_c^2/8\pi \qquad (\text{erg cm}^{-3}) \tag{8-6}$$

(with H in Oe), and the empirical relationship

$$H_c/H_{c0} = 1 - t^2 \tag{8-7}$$

for the thermodynamic critical field temperature dependence, it can easily be shown [RIC65, p. 32] that the *superconductive* unit-volume electronic specific heat, C_{es}^{v}, is cubic in temperature according to

$$C_{es}^{v} = \frac{3H_{c0}^{2}}{2\pi T_c} t^3 \qquad (\text{erg cm}^{-3}\,\text{K}^{-1}) \tag{8-8}$$

i.e.,

$$C_{es}^{v}/\gamma_v T_c = 3t^3 \tag{8-9a}$$

where $\gamma_v (= H_{c0}^2/2\pi T_c^2)$ is the electronic specific heat coefficient per unit volume. More generally, of course,

$$C_{es}/\gamma T_c = 3t^3 \tag{8-9b}$$

As pointed out by RICKAYZEN [RIC65, p. 38], departure from the t^3 law is observed and indeed expected at low temperatures as the underlying critical-field relationship, (8-7), breaks down. Nevertheless, the simplicity of the analytical form of (8-8) has led to its continued use as a descriptor of superconductive electronic specific heat (e.g., [BUC65]), at least as a first approximation.

8.4.2 The Exponential Form

According to BARDEEN, COOPER and SCHREIFFER (BCS) [BAR57], an analytical expression for C_{es} that agrees well with the experimental data over limited temperature ranges is the exponential form

$$C_{es}/\gamma T_c = a \exp(-b/T) \tag{8-10}$$

where a and b are constants; in particular, for two representative temperature ranges well removed from T_c, RICKAYZEN [RIC65, p. 205] recommended the following relationships:

$$0.4 > t > 0.17; \qquad a = 8.5, \qquad b = 0.82\Delta_{00}/k_B$$

$$0.14 > t > 0.09; \qquad a = 26, \qquad b = 0.92\Delta_{00}/k_B$$

where Δ_{00} is the BCS half-energy-gap at zero K. For V and Nb, for example, these equations in the two temperature ranges listed above reduce to

$$\text{V:} \qquad C_{es}/\gamma T_c = \begin{cases} 8.5 \exp(-1.39/t) \\ 26 \exp(-1.56/t) \end{cases} \tag{8-11a}$$

$$\text{Nb:} \qquad C_{es}/\gamma T_c = \begin{cases} 8.5 \exp(-1.15/t) \\ 26 \exp(-1.29/t) \end{cases} \tag{8-11b}$$

8.4.3 The Full BCS Electronic Specific Heat

The BSC electronic specific heat as a function of temperature must be handled numerically, for which purpose MÜHLSCHLEGEL [MUH59] has listed values of the quotient $C_{es}/\gamma T$ (as well as those of the other normalized BCS quantities: energy gap, entropy, free energy, critical field, and penetration depth) for t between 0.14 and 1.00 at intervals of 0.02.

8.4.4 The Electronic Specific Heat at T_c—Height of the Specific Heat Jump, ΔC

According to BARDEEN, COOPER, and SCHREIFFER [BAR57] the relationship between the superconductive and normal state electronic specific heats at T_c itself is given by

$$C_{es}/\gamma T_c = 2.43 \tag{8-12}$$

Then presupposing that the lattice specific heat is the same in both states (Section 8.3.2), the height of the jump at T_c is given by

$$\Delta C \equiv (C_s - C_n)|_{T_c} = (C_{es} - C_{en})|_{T_c} \tag{8-13}$$

It follows that the relative jump height at T_c is

$$\frac{\Delta C}{\gamma T_c} = \left.\frac{C_{es} - \gamma T}{\gamma T}\right|_{T_c} = 1.43 \tag{8-14}$$

Provided weak-coupling BCS-type superconductivity can be assumed, a comparison of the calorimetrically measured relative jump height with the value 1.43 can be used to gauge the fraction of the material that becomes superconducting at T_c. For example, in exploring the operation of a proximity effect in alloys containing fine precipitates, such as ω-phase, the jump-height makes a useful indicator of the completeness of the transition. Of course, most materials depart more or less from the ideal BCS model; and as we shall see, an increase in jump height accompanies any increase in the coupling strengths.

Figure 8-6(a) shows the variation of $\Delta C/\gamma T_c$ with e/a ratio for the alloy systems Ti–V and Ti–Mo. The curves, whose shapes tend to mimic those representing the electronic specific heats of the same alloy systems (Figure 8-3), indicate that maximum $\Delta C/\gamma T_c$ is achieved for the most Ti-rich all-bcc alloys—i.e., those for which the Fermi density-of-states is highest and the bcc lattice most unstable (low C', etc.) but yet for which ω-phase has not yet appeared in sufficient abundance to stabilize the lattice, reduce the average $n(E_F)$, and suppress the superconductivity.

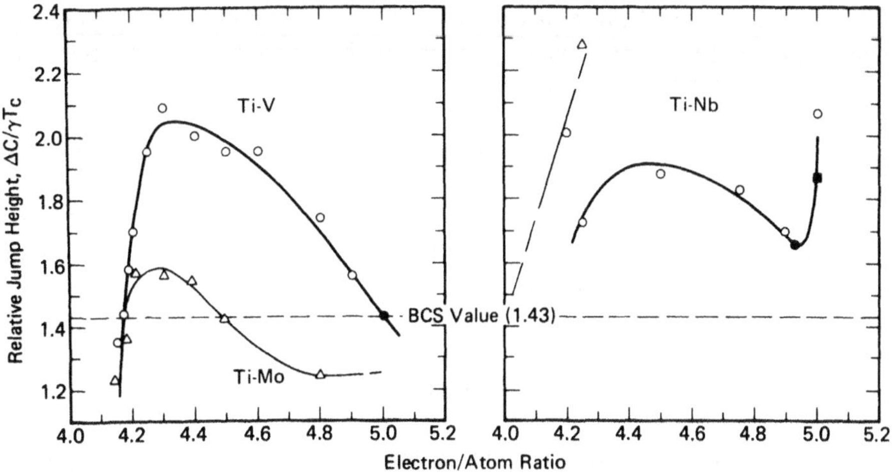

Figure 8-6. Relative height of the specific heat jump at T_c, $\Delta C/\gamma T_c$, plotted *versus* e/a ratio for (a) Ti-V, Ti-Mo, and (b) Ti-Nb alloys. Data sources: Ti-V alloys—○, COLLINGS *et al.* [COL75]; V—●, RADEBAUGH and KEESOM [RAD66]; Ti-Mo alloys—△, COLLINGS *et al.* [COL84] (see also Figure 8-5); pure Mo ($e/a = 6.0$, not shown, $\Delta C/\gamma T_c = 1.30$)—RORER *et al.* [ROR65]; Ti-Nb alloys—○, VERKIN [VER80]; ●, SHCHETKIN and KHARCHENKO [SHC73]; △, SUKHAREVSKII *et al.* [SUK68]; Nb—■, LEUPOLD and BOORSE [LEU64].

Figure 8-6(b) depicts the composition dependence of $\Delta C/\gamma T_c$ for Ti-Nb, plotted using data from four sources. The characteristic maximum near $e/a = 4.4$ again appears; but particularly interesting is what seems to be an anomalously high value for the relative jump height of unalloyed Nb. This remains an unexplained phenomenon; although it has been tentatively suggested [CAP82] that it may be traceable to uncertainties associated with the extraction of γ and θ_D values from Nb's anomalous low-temperature C/T *versus* T^2 curve (Section 8.3.2).

8.5 The Electron–Phonon Coupling Strength

8.5.1 Coupling Strength and the Temperature-Ratio T_c/θ_D

This review of low-temperature specific heat concludes with a discussion of some empirical indicators of the strength of the superconductive electron–phonon coupling and its relationship to the relative height of the specific heat jump at T_c. According to the weak-coupling BCS theory [BAR57, p. 1186], the superconducting transition temperature is expressible in the well-known form

$$k_B T_c = 1.14 h\nu \exp[-1/n(E_F)V] \qquad (8\text{-}15a)^3$$

[3]Although the prefactor 1.14 appears in BCS [BAR57, p. 1186] and is commonly quoted, integration of BCS equation (3-28) actually gives 1.134 (V. E. Wood, personal communication).

where ν is a typical phonon frequency and V is the interaction potential between Cooper pairs (see (7-7)); or, in the format preferred by some subsequent authors (e.g., [McM68]),

$$T_c = 1.14\langle\omega\rangle \exp[-1/n(E_F)V] \tag{8-15b}$$

where the average phonon "energy" is now $\langle\omega\rangle$, a temperature. Next, starting the Eliashberg theory, McMillan [McM68] derived a strong-coupling expression for T_c in the form

$$T_c = \omega_0 \exp\left[\frac{-(1+\lambda)}{\lambda - \mu^* - (\langle\omega\rangle/\omega_0)\lambda\mu^*}\right] \tag{8-16}$$

where ω_0 is the maximum phonon "energy," μ^* is the Coulomb coupling constant or pseudopotential, and λ is the usual electron–phonon coupling constant—a direct measure of the "coupling strength." Intermediate coupling strengths are characterized by $\lambda \cong 1$; such values of λ are an order of magnitude greater than μ^*, which has often been set equal to 0.1. As pointed out by McMillan, in the weak-coupling limit ($\lambda \ll 1$) (8-16) reduces to the BCS form with $\lambda - \mu^*$ substituting for the $n(E_F)V$ of (8-15). Subsequently, Allen and Dynes [Dyn76] after reanalyzing Eliashberg theory in the strong-coupling limit, pointed out that (8-16) was valid for $\lambda \lesssim 1.5$—i.e., for weak to intermediate coupling strengths—and recommended, for strong coupling, the form:

$$T_c = f_1 f_2 \frac{\omega_{\log}}{1.2} \exp\left[\frac{-1.04(1+\lambda)}{\lambda - \mu^* - 0.62\lambda\mu^*}\right] \tag{8-17}$$

where f_1 is a "strong-coupling correction" and f_2 is a "shape correction" [Dyn76], and in which $\omega_{\log} \cong 2\langle\omega\rangle - \langle\omega^2\rangle^{1/2}$. For small λ, in which case $f_1 \cong f_2 \cong 1$, this reduces to (8-16); for large λ, according to Allen and Dynes [Dyn76],

$$T_c \to (\lambda\langle\omega^2\rangle)^{1/2} \tag{8-18}$$

indicating that T_c can be expected to increase continuously as λ increases, in contrast to McMillan's prediction of a maximum in T_c at $\lambda = 2.6$.

For transition metals and their alloys, McMillan theory itself is adequate, following which, and assuming $\mu^* = 0.1$, λ is given by (7-6). McMillan has plotted λ as a function of e/a ratio for $3d$, $4d$, and $5d$ transition elements and binary alloys [McM68, Figure 8] and obtained a set of three curves whose forms are qualitatively similar to those of $n(E_F)$ or γ *versus* e/a (cf. [Hei66, Figure 4] for $4 \lesssim e/a \lesssim 6.5$). He was also able

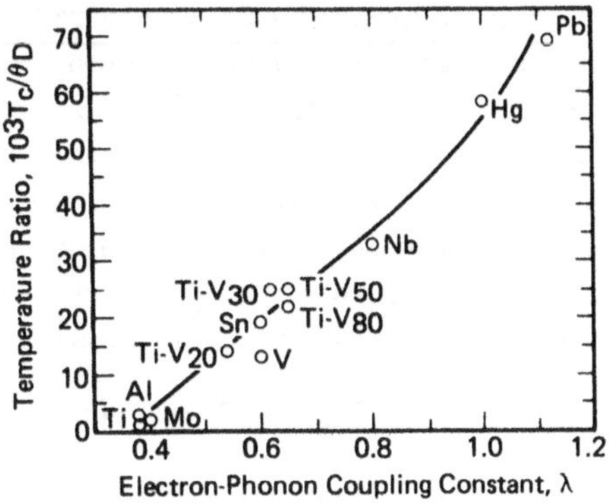

Figure 8-7. Transition-temperature/Debye-temperature ratio, $10^3 T_c/\theta_D$, plotted *versus* the electron–phonon coupling constant λ in a demonstration of the validity of the statement that T_c/θ_D is a measure of coupling strength. The subscripts to Ti-V indicate compositions in at.%. Data source—McMillan [McM68].

to show rigorously that

$$\lambda = \frac{n(E_F)\langle I^2\rangle}{M\langle\omega^2\rangle} \equiv \frac{C_M}{M\langle\omega^2\rangle} \qquad (8\text{-}19)^4$$

where $\langle I^2\rangle$ was referred to as an "average squared electronic matrix element" and M was the atomic mass, and empirically that for the transition metals and alloys (i.e., with the aid of experimentally derived values for λ and $\langle\omega^2\rangle$) C_M was roughly constant. As a consequence it was possible to make the statement that λ, which under BCS was governed by $n(E_F)$,[5] was in the intermediate-coupling approximation primarily controlled by $M\langle\omega^2\rangle$ (i.e., by the lattice stiffness). It was, however, pointed out that $n(E_F)$ still came into the picture, in the sense that density-of-states and lattice softness were positively correlated (see Figure 8-1). It follows that T_c/θ_D should be a measure of coupling strength, a conclusion that is verified in Figure 8-7 (constructed using some tabulated data from McMillan [McM68, Table III] and other sources), in which T_c/θ_D is plotted *versus* λ for selected pure metals and alloys.

The temperature quotient has the value 2.7×10^{-3} for Al, a weak-coupling superconductor to which BCS theory is applicable; at the other extreme, $T_c/\theta_D = 68 \times 10^{-3}$ for Pb to which strong-coupling extensions of BCS must be applied. Among the pure transition elements, examples of weak-to-intermediate strength superconductors (including values of their T_c/θ_D) are Mo (2.0×10^{-3}), V (12.8×10^{-3}), and Nb (33.3×10^{-3}). Various Ti–TM alloys achieve maximal values of coupling strength at solute compositions near (i) 15 to 20 at.% Mo ($T_c/\theta_D = 13.1 \times 10^{-3}$), (ii) 40 at.% V ($25.7 \times 10^{-3}$), and (iii) 50 at.% Nb (40.3×10^{-3}).

[4]Here, C_M is the same as the η of Hopfield theory—see Chapter 12.
[5]$\lambda \cong n(E_F)V$, with $V \cong$ constant—see below.

As pointed out by LAIBOWITZ *et al.* [LAI70], the ratio[6] $2\Delta'_{00}/k_BT_c$, which is equal to 4.2_5 (from tunnelling experiments) for Pb and 4.6 (infrared absorption) for Hg [LYN64, p. 99] but which assumes the BCS value 3.52 in the weak-coupling limit, can also be used as an indicator of coupling strength. It is moreover, according to GOODMAN [LYN64, p. 134], positively correlated with T_c/θ_D, again drawing attention to the temperature-quotient as a coupling-strength parameter. As the result of a theoretical study, GEILIKMAN and KRESIN [GEI66] found

$$2\Delta'_{00}/k_BT_c = 3.5[1 + 3.5(T_c/a\theta_D)^2 \ln(a\theta_D/T_c)] \tag{8-20}$$

in which a is a constant relating the "characteristic phonon frequency" to the Debye temperature. Subsequently, as the result of an experimental study of superconductivity in Nb_xN_{1-x}, whose T_c could be varied between about 4 and 16 K, LAIBOWITZ *et al.* [LAI70] offered the empirical formula

$$2\Delta'_{00}/k_BT_c = 3.5[1 + b\exp(cT_c/\theta_D)] \tag{8-21}$$

with $b = 0.78 \times 10^{-3}$ and $c = 114$, as an almost universal relationship between $2\Delta'_{00}/k_BT_c$ and T_c/θ_D for numerous superconductors over a wide range of values of the temperature quotient. The influence of strong coupling on Pauli paramagnetic limitation, and hence the upper critical field, is discussed in Section 15.8.1.

8.5.2 Coupling Strength and the Deviation Function, $D(t)$

Coupling strength may be reliably expressed in terms of the reduced-temperature dependence of the so-called "deviation function:"

$$D(t) = h_c(t) - (1 - t^2) \tag{8-22}$$

where $h_c(t) \equiv H_c/H_{c0}$ is the reduced thermodynamic critical field and, as usual, $t \equiv T/T_c$. According to standard texts (e.g., [LYN64]), the reduced critical field can be very closely approximated by the first term in the expansion

$$\left.\begin{aligned} h_c(t) &= 1 - \sum_{i=2}^{n} a_i t^i \\ \sum_n a_n &= 1 \end{aligned}\right\} \tag{8-23}$$

i.e., by the parabolic relationship of (8-7), frequently regarded as a feature of the Gorter–Casimir two-fluid model (e.g., [BAR57, p. 1187]).

[6] Δ'_{00} is the half gap energy for arbitrary coupling strength.

Implicit in the extension of the above series to $i > 2$ is the expectation that the critical field temperature dependence of most superconductors, when closely examined, will deviate slightly from the parabolic law. It turns out that the extent to which they do deviate, as gauged by the so-called "deviation function" defined by (8-22), is a useful measure of the coupling strength. In strong-coupled superconductors such as Pb, $D(t)$ is positive, while in the weak-coupling limit it is negative. According to BARDEEN, COOPER, and SCHREIFFER [BAR57], for $t \ll 1$,

$$H_c^2 = H_{c0}^2[1 - \tfrac{2}{3}\pi^2(k_B T/\Delta_0)^2] \tag{8-24}$$

then substituting $2\Delta_0(t \ll 1) \cong 2\Delta_{00} = 3.53\ k_B T_c$,

$$H_c \cong H_{c0}(1 - 1.06t^2) \qquad (t \ll 1) \tag{8-25}$$

Extending this approximation to larger values of t, for the BCS superconductor at, say, $t^2 = 0.5$ the deviation would be expected to be of the order

$$\left.\begin{aligned} D(t) &\cong (1 - 1.06t^2) - (1 - t^2)\big|_{t^2=0.5} \\ &= -0.03 \end{aligned}\right\} \tag{8-26}$$

Finally, by combining (8-22) and (8-24) it is possible to write

$$D(t) \cong \left[1 - \tfrac{4}{3}\pi^2\left(\frac{1}{\delta_0}\right)^2\right]t^2 \tag{8-27}$$

which serves to emphasize the relationship, in terms of coupling strength, between the deviation function and the zero-K gap width, $2\Delta'_{00}$, whose relative value is herein defined as $\delta_0 \equiv 2\Delta'_{00}/k_B T_c$. With $\delta_0 = 3.52$, as for a BCS superconductor, the second term in the brackets is 1.06 as before and $D(t)$ is negative. As δ_0 increases, the maximum deviation, $D(t^2 = 0.5)$, tends to become positive; according to (8-27), it reaches its greatest value of +0.09 in Pb for which δ_0 is about 4 [LYN64, p. 99].

In Figure 8-8, the BCS deviation function is compared to those of the simple metals Pb, Hg, and In from [LYN64, Figure 23] and those of the transition elements V, Nb, and Mo from the work of CORAK *et al.* [COR56], LEUPOLD and BOORSE [LEU64], and RORER *et al.* [ROR65], respectively. With Pb, the discrepancy between the theoretically predicted $D(t^2 = 0.5) = +0.09$ and the experimentally obtained 0.025 implies a breakdown of simple theory for very strongly coupled superconductors. For further discussions of the deviation function [RIC65, p. 201] and [SCA69, p. 539] are recommended.

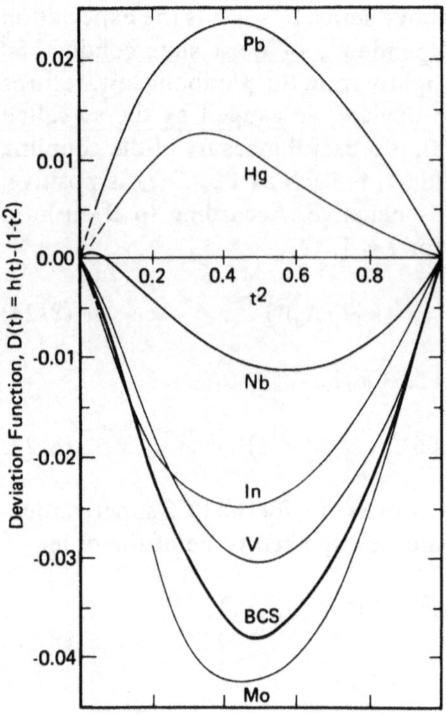

Figure 8-8. Deviation functions, $D(t) \equiv h_c(t) - (1 - t^2)$, plotted *versus* t^2. The BCS deviation function is compared with those of the simple metals Pb, Hg, and In and those of the transition elements V, Nb, and Mo. Data sources—simple metals, LYNTON [LYN64, p. 84]; V, CORACK *et al.* [COR56]; Mo, RORER *et al.* [ROR65]; Nb and BCS, LEUPOLD and BOORSE [LEU64].

8.6 Relative Height of the Specific Heat Jump at T_c as a Function of Coupling Strength

8.6.1 Jump Height in Terms of the Deviation Function

For the difference between the superconductive- and normal-state molar specific heats, a standard thermodynamic argument (e.g., [LYN64, p. 15]) yields

$$C_s - C_n = \frac{VT}{4\pi}\left[H_c\left(\frac{d^2H_c}{dT^2}\right) + \left(\frac{dH_c}{dT}\right)^2\right] \qquad (8\text{-}28)$$

where V is the molar volume. This transforms to

$$\Delta C \equiv (C_s - C_n)|_{T_c} = \frac{V}{\pi}\frac{H_{c0}^2}{T_c}(1+\beta)^2 \qquad \text{as } t\to 1 \qquad (8\text{-}29)$$

and reduces to

$$C_n = \gamma T = \frac{V}{2\pi}\left(\frac{H_{c0}^2}{T_c}\right)(1-\alpha)T \qquad \text{as } t\to 0 \qquad (8\text{-}30)$$

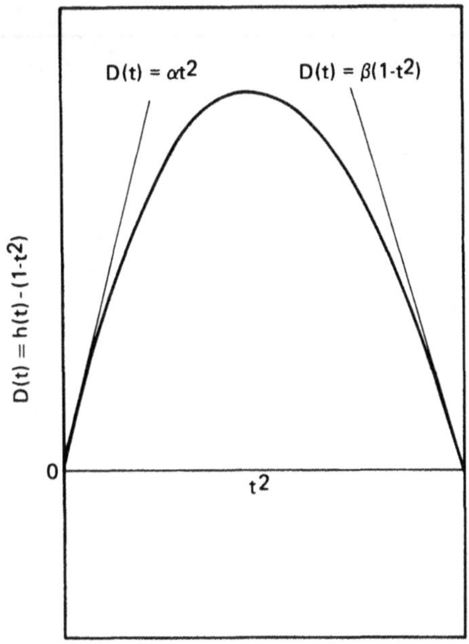

Figure 8-9. Schematic of $D(t)$ *versus* t^2 illustrating the definitions of the "relative jump height parameters" α and β in terms of the limiting slopes of $D(t)$ at $t = 0$ and $t = 1.0$—after SCALAPINO [PAR69, p. 449].

where α and $-\beta$ are the limiting slopes of $D(t)$ at $t = 0$ and 1, respectively, as in Figure 8-9. It follows from (8-29) and (8-30) that in general

$$\Delta C/\gamma T_c = 2(1 + \beta^2)/(1 - \alpha) \tag{8-31}$$

Clearly in the Gorter-Casimir model, with $\alpha = \beta = 0$, $\Delta C/\gamma T_c = 2.0$; in the weaker coupling situation, with $\alpha, \beta < 0$, $\Delta C/\gamma T_c < 2.0$ (in particular 1.43 for BCS); and in strong-coupling metals such as Pb and Hg, with $\alpha, \beta > 0$, $\Delta C/\gamma T_c > 2.0$ (in particular 2.6 for Pb). The relative jump heights for several pure elements are listed in the standard texts, for example in LYNTON [LYN64, p. 135] and in RICKAYZEN [RIC65, p. 204]. Those for the alloy series Ti-V, Ti-Nb, and Ti-Mo, were compared with the BCS prediction and presented as functions of e/a in Figure 8-6, which indicated that the coupling strengths became maximum in such systems for $e/a = 4.2$–4.3.

8.6.2 Jump Height in Terms of T_c/θ_D

As indicated above (see Figure 8-7), the quotient T_c/θ_D may be used as a qualitative indicator of coupling strength. Since the relative jump height is also a measure of coupling strength, a positive correlation is expected between $\Delta C/\gamma T_c$ and T_c/θ_D. Indeed such a correlation has been noted in the series Ti-Mo(7-40 at.%), which included not only single-phase β-quenched alloys but also a two-phase Ti-Mo(10.3 at.%) sample that had been subjected to aging at 350°C (Figure 8-10).

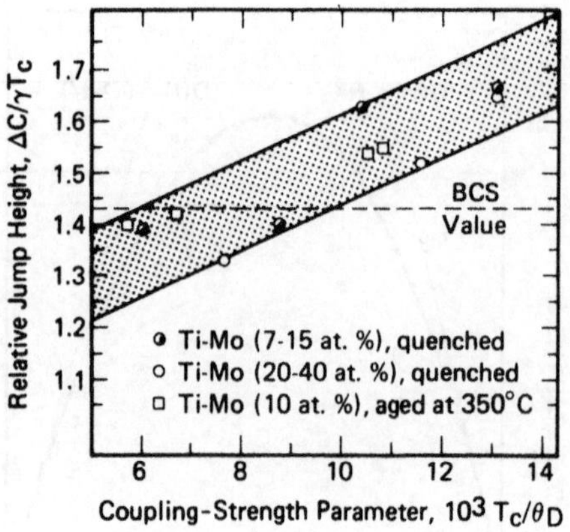

Figure 8-10. Relative height of the specific heat jump at T_c, $\Delta C/\gamma T_c$, as a function of a coupling-strength parameter which, following Figure 8-7, is taken as $10^3 T_c/\theta_D$. Alloys depicted are quenched Ti–Mo(7–40 at.%) and Ti–Mo(10 at.%) aged at 350°C for up to about 1000 h [CoL84].

In an attempt to place the subject on a quantitative basis, KRESIN and PARKHOMENKO [KRE75], as the result of an analysis based on Eliashberg theory, showed that

$$\Delta C/\gamma T_c = 1.4\left[1 + 1.8\left(\ln\frac{\langle\omega\rangle}{T_c} + 0.5\right)\left(\frac{\pi T_c}{\langle\omega\rangle}\right)^2\right] \tag{8-32}$$

in which $\langle\omega\rangle$, referred to in [KRE75] as the "characteristic phonon-frequency" has, for Pb, been taken as one half of the Debye temperature. The validity of (8-32), with $\langle\omega\rangle = \theta_D/2$, for various pure simple metals and transition metals is demonstrated in Table 8-5. However, in comparing theory with experiment for two sequences of Ti–TM alloys, as in Figure 8-11, it is noticeable that (8-32) as it stands, although capable of yielding values of $\Delta C/\gamma T_c$ of about the correct magnitude (i.e., somewhat larger than 1.4), is not sufficiently responsive to variation of T_c/θ_D.

Table 8-5. Relative Height of the Specific-Heat Jump At T_c, $\Delta C/\gamma T_c$, Computed Using (8-32) from the Theory of Kresin and Parkhomenko [KRE75]

Element	T_c K	θ_D K	$\Delta C/\gamma T_c$: (8-32) with $\langle\omega\rangle = \theta_D/2$	$\Delta C/\gamma T_c$: Published Values	Literature
Pb	7.19	105	2.58	2.65	[RIC65, p. 204]
Hg	4.16	72	2.28	2.18	[RIC65, p. 204]
Sn	3.72	200	1.53	1.60	[RIC65, p. 204]
Al	1.16	428	1.40	1.60	[RIC65, p. 204]
V	5.30	399	1.47	1.43	[RAD66]
Nb	9.22	277	1.75	2.07	[BRO53]
Mo	0.92	460	1.40	1.30	[ROR65]

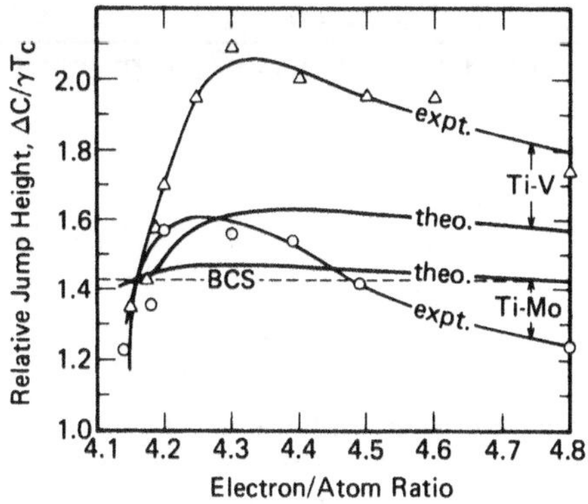

Figure 8-11. Relative height of the specific heat jump at T_c *versus* e/a ratio for Ti–V and Ti–Mo alloys; a comparison of experimental values with those calculated from (8-32) [CoL84].

8.7 Empirical Determination of the Electron–Phonon Coupling Constant—A Case Study of Ti–Mo Alloys

8.7.1 Electron–Phonon Enhancement of the Density-of-States—Theoretical

It was pointed out above, in a discussion of (7-4), (7-5), and (8-3) that an average Fermi density-of-states is not directly obtainable from the experimentally measured electronic specific heat coefficient, γ, since the latter is considerably larger than the band-structure or "bare" value, γ^0, as a consequence of electron–phonon (as well as electron–electron) interaction. The enhancement factor (see (7-5)) is $(1 + \lambda_t)$, where λ_t, the total interaction constant, contains both electron–electron and electron–phonon components. Ashcroft and Wilkins [Ash65] have succeeded in calculating these contributions for Na, Al, and Pb, have shown the electron–phonon component to be dominant, and have demonstrated that the enhancement factors lay between 1.2 and 2.1. This author is not aware of any similar *ab initio* calculations of λ_t having been carried out for transition metals—certainly not for their alloys. In view of the dominance of electron–phonon interaction, it is assumed here that $\lambda_t \cong \lambda_{\text{electron–phonon}} \equiv \lambda$.

8.7.2 Electron–Phonon Effects—Semiempirical

(a) BCS Theory. A semiempirical value of the enhancement factor can be derived if the metal is a superconductor. As pointed out by Clogston *et al.* [Clo62], the λ of the above discussion is equivalent to the $n(E_F)V$ of the BCS expression for T_c (from (8-15) above) where V is the net pairing potential that takes into account the effects of both electron–phonon and screened-Coulomb interactions. For the $\langle\omega\rangle$ of (8-15b), Morel [Mor59] has recommended the approximation $C_0\theta_D$, with C_0 being a universal numerical constant equal to 3/4.[7] After making this substitution into (8-15b) CLOGSTON *et al.* [CLO62] were able to express a BCS version of λ in the form

$$\lambda^{-1} \equiv [n(E_F)V]^{-1} = \ln(1.14C_0\theta_D/T_c) \qquad C_0 = 3/4 \qquad (8\text{-}33)$$

But MORIN and MAITA [MOR63] subsequently showed that 3/4 was not a good choice for C_0. As the result of a series of measurements they demonstrated experimentally that, for various alloy systems, C_0 fell within the range of 0.08 to 0.44. Based on this (and recognizing that λ was logarithmically related to C_0) it appeared that $C_0 = 1/4$ was a much better assumption than that proposed earlier, and that it could be expected to lead to values of λ correct to within about 25%. Thus, in the absence of some other more reliable value of C_0 for some particular metal or alloy, an application of this version of (8-33) to the results of a single specific heat measurement had been shown to yield a satisfactory correction factor.

(b) McMillan Theory. Another approach available to superconductors is by way of the McMillan formula [MCM68], (7-6), which according to ALLEN and DYNES [DYN76] is valid for $\lambda \lessapprox 1.5$. By substituting values of T_c and θ_D from the results of low-temperature specific heat measurements into that expression, λ is directly obtainable provided μ^*, the Coulomb pseudopotential, is known. With $\mu^* \cong 0.1$, which was taken to be appropriate for transition metals, λs for the transition elements of groups IV and V, for example, were found to lie between 0.25 (Hf) and 0.72 (Nb).

8.7.3 Electron–Phonon Effects—Empirical Method for an Alloy Series

Both of the above methods are useful for estimating the enhancement factor $(1+\lambda)$ as a step towards computing $n(E_F)$ from the results of low-temperature specific heat measurements on single specimens. But when a *family* of alloys is under study, a third possible approach enables the specific heat data to be *self-corrected* for the many-body effects. The technique to be described is based on an approach first introduced by MORIN and MAITA [MOR63], who pointed out that since the substituted form of (8-15b) (cf. (8-33)) is equivalent to

$$T_c/\theta_D = 1.14C_0 \exp[-1/n(E_F)V] \qquad (8\text{-}34)$$

a semilog plot of T_c/θ_D *versus* $1/n(E_F)$ would be linear with a slope of $-1/V$ if the latter were constant for the alloy sequence considered. If

[7] Actually, at the time, the substitution was couched within the framework of (8-15a).

successful the method would have certain advantages over the two described in the previous section: (i) the requirement $\langle\omega\rangle = C_0\theta_D$, upon which (8-33) depends, is reduced to a proportionality, i.e., a prior knowledge of C_0 is no longer required; (ii) the necessity of obtaining or assuming a value for μ^* for use in the McMillan equation (7-6) (see also Section 8.5.1) is avoided.

But the calorimetrically derived T_c, θ_D, and γ values, although allowing semilog plots of T_c/θ_D *versus* $1/\gamma$ to be constructed, do not themselves provide values of V since a calculation of the accompanying $n(E_F)$ ((8-34)) requires a prior knowledge of the enhancement factor, the quantity to be determined. In spite of this, it turns out, however, that the slope of the experimental semilog plot is just what is needed to extract a value of λ from the measured electronic specific heat coefficient.

Let us first define an "apparent pairing potential," V_{app}, as

$$V_{app} \equiv V/(1+\lambda) \qquad \text{(eV)} \tag{8-35}$$

whose $(1+\lambda)$-factor also appears in

$$\gamma = \gamma^0(1+\lambda) \tag{7-5}$$

Then with V in eV, $n(E_F)$ in $\text{eV}^{-1}\,\text{atom}^{-1}$, and γ in $\text{J mol}^{-1}\,\text{K}^{-2}$, since

$$n(E_F) = 212\gamma/(1+\lambda) \tag{8-3}$$

Table 8-6. Empirically Determined Electron–Phonon Coupling Constants in a Series of Quenched Ti–Mo Alloys—Based on $V_{app} = 0.26$ eV

Nominal Mo Concentration at.%	Electronic Specific Heat Coefficient[a] γ, $\text{mJ mol}^{-1}\,\text{K}^{-2}$	Electron–Phonon Coupling Constant λ
5	4.4	0.24
7	5.5	0.30
8.5	6.1	0.34
10	6.9	0.38
15	7.3	0.40
20	7.1	0.39
25	7.0	0.39
40	6.1	0.34
70	2.6_5	0.15
100	1.8_5	0.10

[a] Reference—[Col72].

it follows that

$$\lambda \equiv n(E_F)V = 212\gamma V_{app} \tag{8-36}$$

This, when substituted into (8-34), yields

$$\left.\begin{aligned} \ln(T_c/\theta_D) &= \ln(1.14C_0) - 1/212\gamma V_{app} \\ \log_{10}(T_c/\theta_D) &= 0.434\ln(1.14C_0) - (0.434/V_{app})(1/212\gamma) \end{aligned}\right\} \tag{8-37}$$

Thus, if V_{app} is constant for a family of alloys, a plot of $\log_{10}(T_c/\theta_D)$ *versus* $(212\gamma)^{-1}$ will be linear with slope $-0.434/V_{app}$; and the V_{app} so derived is just what is needed to extract λ from the measured γ (using (8-36)) and hence to convert γ to $n(E_F)$ using (8-3). Figure 8-12 illustrates the applicability of this technique to a series of quenched Ti–Mo alloys. The semilog plot is indeed linear. Its slope yields $V_{app} = 0.26$ eV, with the aid of which values of λ may be calculated, as in Table 8-6, for the individual members of the Ti–Mo alloy series.

The intercept corresponds to $(T_c/\theta_D) = 0.164$, from which, going back to (8-34), it may be concluded that for the system under discussion $C_0 = 0.164/1.14 = 0.14$. This is much smaller than the MOREL [MOR59] suggestion of $\frac{3}{4}$, but falls well within the MORIN–MAITA [MOR63] experimentally derived range of 0.08 to 0.44 (Section 8.7.2).

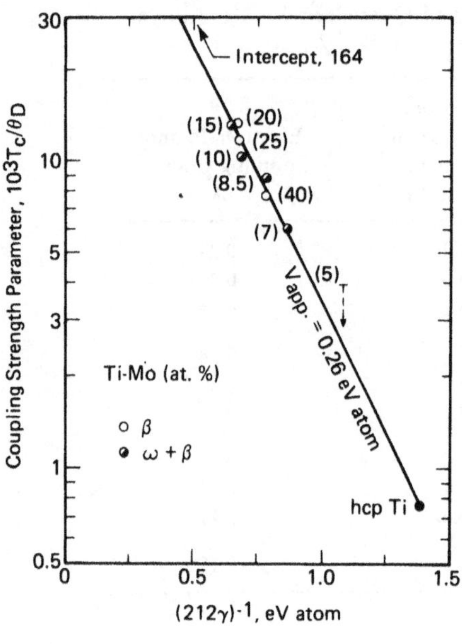

Figure 8-12. Semilog plot based on (8-37). The slope of the line, $-0.434/V_{app}$, yields a practical value for the apparent pairing potential of $V_{app} = 0.26$ eV atom. The experimental line, extrapolated to the axis, yields a value for C_0. As discussed elsewhere, points on the extrapolated line itself aid in developing data for low-concentration β-Ti–Mo alloys and even for β-Ti itself [COL72].

In conclusion it is useful to note that (7-5) and (8-36) may be synthesized into the reciprocal relationships

$$\gamma = \gamma^0(1 + 212\gamma V_{\text{app}}) \tag{8-38}$$

and

$$\gamma^0 = \gamma(1 - 212\gamma^0 V_{\text{app}}) \tag{8-39}$$

which, together with (8-5), are useful in the analysis of joint magnetic and calorimetric data.

Chapter 9

Low-Temperature Thermal Expansion

Both this and the preceding chapter form a transition between the normal-state mechanical and physical properties of Ti alloys and their superconductive ones, detailed discussion of which begins in Chapter 10. If treated phenomenologically as simply a "temperature-dependent property," thermal expansion could be considered almost anywhere among the physical-property chapters of this book; if treated as a "thermal property," it could logically be associated with the "thermal conductivity"-"specific heat" group. Mechanistically, however, the thermal expansion of a crystal is the result of an interplay between the thermal excitation of lattice vibrations (responsible for its heat capacity), the anharmonicity of those vibrations (which is also responsible for thermal resistivity), and the elastic properties of the lattice. "Thermal expansion," a tensor, is considerably more complicated than "heat capacity," the subject of the preceding chapter; thermodynamically, both can be regarded as deriving from the entropy (or free energy)—thermal expansion as its volume derivative and heat capacity as its temperature derivative. From the elastic-property standpoint, the following discussion of low-temperature thermal expansion can also be regarded as an extension of Chapter 4.

9.1 Thermal Expansion of Insulators and Metals

9.1.1 Harmonicity and Anharmonicity in Thermal Expansion

Discussions of thermal expansion generally begin by quoting the work of GRÜNEISEN, according to whom

$$\frac{\beta_v V K_T}{C_v} = \frac{\beta_v V K_S}{C_p} = \text{constant}^1 \tag{9-1}$$

[1] GRÜNEISEN, in a detailed discussion of the components of γ in various crystalline directions and its average value among crystals of various kinds, recognized that although γ was not strictly constant, its variability was surprisingly small from crystal to crystal as compared with, say, the compressibility which can vary over several orders of magnitude [GRU26, p. 29].

where V is the molar volume, K_T and K_S are the isothermal and adiabatic bulk moduli, C_v and C_p are the constant-volume and constant-pressure molar heat capacities,[2] and β_v is the coefficient of the volumetric expansion, defined by

$$\beta_v = (1/V)(\partial V/\partial T)_P \equiv (\partial \ln V/\partial T)_P \tag{9-2}$$

where P is pressure [GRU26].

In his analysis, GRÜNEISEN adopted the Debye isotropic continuum model and showed that the above constant was γ_G, given by

$$\gamma_G \equiv -(\partial \ln \theta_D/\partial \ln V)_T \tag{9-3}$$

The thermodynamic "Grüneisen gamma" is frequently unsubscripted. A subscript *is* used here in order to avert confusion with the electronic heat capacity coefficient that frequently occurs in association with it when metals are under discussion. The numerous specializations of the Grüneisen gamma are distinguished by other subscripts and superscripts. Equation (9-3) shows that because γ_G, and hence β_v, is nonzero for real materials, θ_D must be volume dependent.

In the classical Einstein solid, an increase in temperature translates into an increase in oscillation amplitude. If the oscillation is harmonic (i.e., if the corresponding potential is quadratic in the displacement, r) the mean position of each oscillating atom is independent of its amplitude of oscillation and expansion does not take place. Only if the potential is asymmetrical (i.e., if it is perturbed by a small term in r^3, in the simplest case) can expansion occur. It follows, therefore, that (i) vibrational anharmonicity is an essential ingredient of thermal expansion, and (ii) the anharmonicity is embodied in the parameter γ_G. In all crystals each atom oscillates in an asymmetric potential well created by its interaction with the surrounding atoms. An amplitude-dependent expansion, based on an interatomic potential function of the form

$$V(r) = \sum_{i=2}^{\infty} a_i r^i \tag{9-4}$$

and the expansion associated with it, is illustrated in Figure 9-1. Although an $i = 4$ term would represent a softening at large amplitudes, only the odd-i terms contribute asymmetry to $V(r)$, hence expansion. Of course, in

[2]It is not necessary to specify "molar" quantities, but in an experimental discussion it is convenient to do so since the results of specific heat measurements are frequently expressed on a per-mole basis.

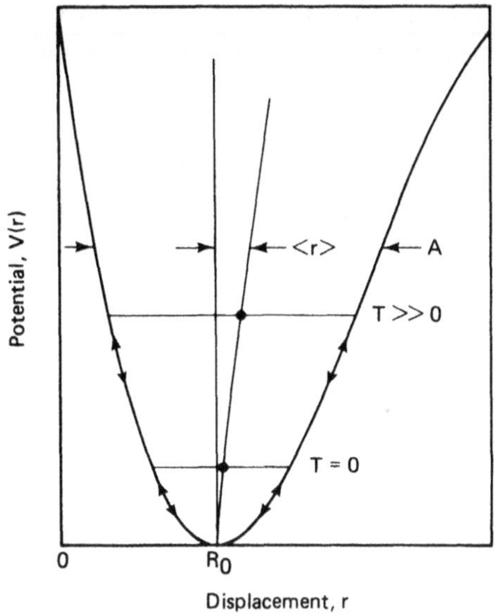

Figure 9-1. Diagram indicating that the time-averaged displacement from rest, $\langle r \rangle$, of a particle oscillating in a potential well increases with vibration amplitude, A, provided first of all that the well is asymmetric in the manner indicated.

the presence of the odd-i terms, the even-i terms control *how much* the solid expands [Kit68, p. 184]. This is also illustrated in Figure 9-1. As a simple example [Tou77, p. 5a], consider a particle of mass m oscillating in an anharmonic potential well of the form

$$V(r) = \tfrac{1}{2}Sr^2 - \tfrac{1}{3}Ssr^3 \tag{9-5}$$

Its equation of motion, $m(d^2r/dt^2) = -dV(r)/dr$, would then be

$$m\frac{d^2r}{dt^2} + S_{eq}r = 0 \tag{9-6a}$$

with

$$S_{eq} = S(1 - sr) \tag{9-6b}$$

It follows [Tou77, p. 5a] that the time-averaged position of a particle oscillating with amplitude A is

$$\langle r \rangle = \frac{sA^2}{2} \tag{9-7}$$

Thus, for a classical Einstein solid, in which A^2 is proportional to T ($A^2 = aT$, say), the expansion is proportional to T and to s, an anharmonicity parameter. Finally, also within this one-dimensional-oscillator model, it is easy to see how anharmonicity is equivalent to a volume-dependent frequency. The equivalent harmonic-oscillator stiffness constant, according to (9-6b), is $S_{eq} = S(1 - sr)$, the mean value of which is

$$\langle S_{eq} \rangle = S(1 - s\langle r \rangle) \tag{9-8}$$

It follows that, as the volume ($R_0 + \langle r \rangle$ in this case, where R_0 is the rest position of the particle) increases, the restoring force softens and the linear-oscillator frequency, ω_l, decreases. Writing $\omega_l = \sqrt{\langle S_{eq} \rangle / m}$, we deduce from (9-7) and (9-8) that

$$-\frac{d \ln \omega_l}{d \ln \langle r \rangle} = \frac{s^2}{4} aT$$

or

$$-\left(\frac{\partial \ln \omega_l}{\partial \ln \langle r \rangle}\right)_T = bs^2 = \text{constant} \tag{9-9}$$

where, it will be recalled, s is the anharmonicity constant. The analogy between (9-3) and (9-9) is clear, and serves to illustrate the role played by γ_G as a descriptor of lattice-vibrational anharmonicity. The simple model just described also illustrates another concept central to most treatments of thermal expansion, *viz.*, the so-called "quasiharmonic" approximation [CoL64, p. 452], in which, although the atomic oscillators vibrate harmonically (9-6), they do so with a volume-dependent frequency.

9.1.2 Development of the Subject

The thermodynamic approach coupled with expressions for the average or isotropic elastic properties is adequate to describe the volume expansions of isotropic materials and cubic crystals. For other crystals (hexagonal, tetragonal, and trigonal being the simplest departures from cubic), thermal expansion is more complicated [WHI74[a]], [BAR80] and must be treated as the trace of an expansion tensor formed from the specific heat, molar volume, and the components of the Grüneisen tensor. The latter are calculable using the monocrystalline elastic constants (C_{ij}) and their strain derivatives [BAR67[a]], whereas for cubic solids only the (isotropic) pressure derivatives of the C_{ij} are needed [SHE58], [COL63].

Theoretical advances have, of course, been made over the years, but particularly rapid progress was made during the late 1950s and early 1960s when BARRON (see [BAR57[a], BAR61, BAR64]) and BLACKMAN authored an important series of papers dealing with the thermodynamics and lattice dynamics of thermal expansion and departures from the simple Grüneisen model (see [COL64] and [BAR80] for listings). At the same time, innumerable measurements of thermal expansion at temperatures down to about $\theta_D/20$ were being carried out. Adequate testing of the 1950s calculations required experimental data for the liquid-He temperature range where the expansion of solids is extremely small and detection sensitivities of the order 1 Å are needed. The development of high-sensitivity, low-temperature dilatometry during that period, as reported for the first time in the VIIth International Low Temperature Physics Conference (Toronto, 1960), provided the needed experimental impetus [WHI61], [ABB61], [FRA61]. In particular, the invention by WHITE of the three-terminal capacitance dilatometer was a major experimental breakthrough, which, by enabling accurate data to be reliably taken in the temperature range of liquid He and its vapor, revitalized the experimental side of the subject. Low-temperature thermal expansion measurement of insulators, metals, magnetic materials, and superconductors commenced immediately and has continued up to the present day. Some idea of the extent to which the advent of three-terminal capacitance dilatometry stimulated new work in the field of low-temperature thermal expansion can be obtained by recognizing that in 1959 the discipline, experimentally at least, was at about the same stage of development as low-temperature specific heat had been some thirty years previously. Due at least in part to WHITE's work, other dilatometers, some with greater sensitivity, were quickly devised and a wide range of both insulating and metallic materials began to be measured at low temperatures. As a result, and considering only metals, already by 1963 some five rare-earth elements, seven simple metals, and thirteen transition metals had been measured at liquid-He temperatures [COL64], [AND64[a]], [BAR80].

9.1.3 The Electronic, Magnetic, and Other Contributions to Low-Temperature Thermal Expansion

According to the Mie–Grüneisen theory for an isotropic crystal, the volume expansion coefficient, $\beta_v(T)$, is given by (9-1), (9-2), and (9-3) combined in the form

$$\beta_v(T) = \gamma_G(T)\frac{C_v(T)}{K_T(T)V(T)} = \gamma_G(T)\frac{C_p(T)}{K_S(T)V(T)} \tag{9-10}$$

Since the temperature dependences of $K(T)$ and $V(T)$ are both weak compared to those of C_v or C_p,[3] and γ_G for most materials usually varies

[3]For NaCl, for example, between 280 K and 4 K, $V(T)$ decreases by 1.8% and $K(T)$ increases by 5.3% [WHI73]. On the other hand, over a comparable temperature range, C_p decreases from 49.6 J mol^{-1} K^{-1} (260 K) to 14.9 mJ mol^{-1} K^{-1} (5 K) [BAR64], a ratio of more than 3000:1.

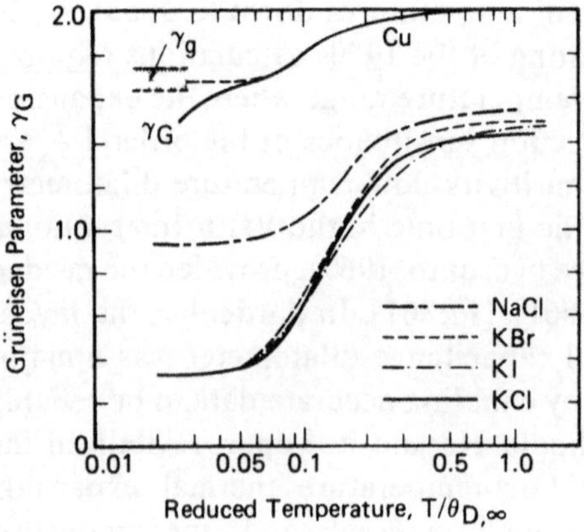

Figure 9-2. Measured variation of the Grüneisen parameter, γ_G, with reduced temperature for some alkali halides and a metal (Cu). In the latter case, the dashed extrapolation indicates the lattice component, in this case γ_g—after WHITE [WHI63].

by no more than a factor of 1.5–3 between very high and very low temperatures [WHI63] (Figure 9-2), $\beta_v(T)$ is dominated by the temperature dependence of the heat capacity (see previous footnote). One could, therefore, write $\beta_v(T) \mathrel{\widetilde{\propto}} C(T)$, or preferably

$$\beta_v(T) \mathrel{\widetilde{\propto}} \gamma_G(T)C(T) \tag{9-11}$$

in which the Grüneisen parameter is retained (i) to indicate that its presence as a finite quantity is an essential prerequisite for the *existence* of thermal expansion, and (ii) to enable its slight variation with temperature to be taken into consideration. Equation (9-11) indicates that the thermal expansion and the specific heat are formed from corresponding physical components, and that any property or effect that influences the one will also be reflected in the other. Thus, contributions to the thermal expansion of metals, particularly at low temperatures, are to be expected from the conduction electrons—both normal and superconducting—and anomalies in the linear thermal expansion ($\Delta l/l_0$) and linear expansion coefficient ($d \ln l/dT$) are expected as a superconductor is cooled through T_c. Likewise, expansion anomalies are expected to be associated with magnetic transitions in both insulators and metals.

9.1.4 Literature Sources—Plan of the Chapter

Thermal expansion has been the subject of eight symposia, the proceedings of many of which have been published (see for example [PEG78], the sixth, and [LAR82], the seventh). It has also been the subject of several important reviews since its renaissance as a topic for scientific study in the late 1950s. Two reviews, authored by COLLINS and WHITE [COL64] and ANDRES [AND64[a]], quickly followed the development and use of high-sensitivity low-temperature dilatometry. Recent books on the subject have been authored by YATES [YAT72] and KRISHNAN, SRINIVASAN, and DEVANARAYANAN [KRI79], and a wealth of data and a useful summary of the theory are to be found in Volumes 12 and 13 of the *Thermophysical Properties of Matter* series edited by TOULOUKIAN *et al.* [TOU75, TOU77]. Finally, attention is drawn to a recent comprehensive and authoritative review of low-temperature thermal expansion appropriately authored by BARRON, COLLINS, and WHITE [BAR80] which, it is fairly safe to predict, is likely to remain the last word on the subject for many years to come.

This chapter is based on material drawn from the above-mentioned reviews and other sources, as cited. It commences with a brief summary of the usual thermodynamic and lattice-dynamical aspects of thermal expansion, first of insulators, and then of metals in which the electrons play an important role at low temperatures. Experimental measuring techniques are considered next, especially the high-sensitivity methods required for expansion measurement in the liquid-He temperature range. After presenting some thermal expansion data for selected Ti-base and other structural alloys, the chapter goes on to consider the thermal expansion of superconductors both as a function of temperature below T_c as well as through the n/s transition. The chapter concludes with a discussion of the thermal expansion of composite materials with reference both to nonmetallic composites of interest in superconducting machinery construction and to the superconductive metal-matrix composite Cu/Ti–Nb.

9.2 Thermal Expansion of Insulators

9.2.1 Thermodynamics of the Debye Isotropic Continuum

(a) Internal Energy Formulation of the Expansion Equation. According to the Debye theory, the heat capacities of all solids can be described by a universal function provided that the temperature is scaled by θ_D, a characteristic temperature for each solid. The atomic oscillations are assumed to be harmonic. In the quasiharmonic Debye model for thermal expansion this is still assumed to be true, except that θ_D is permitted to be volume dependent. It is the assignment of a non-zero, but negative, value to $\partial \ln \theta_D / \partial \ln V$ that enables the Debye solid to expand.

The standard thermodynamic treatment [KIT68], [YAT72], [TOU77] usually begins with the statement[4]

$$F = U - TS \tag{9-12}$$

[4]As before this is a general statement, but in this context molar quantities are assumed.

for the Helmholtz free energy, F, expressed in terms of the internal energy, U, and the entropy, S, and with the following differential form of the first law of thermodynamics:

$$TdS = dU + PdV \tag{9-13}$$

From these it follows that

$$P = -\left(\frac{\partial F}{\partial V}\right)_T \tag{9-14}$$

At this point, following standard approaches (e.g., [KIT68, p. 183], [YAT72, p. 33]), the Debye model is introduced by requiring that F be the sum of the zero-K internal energy, U_0, and the Debye vibrational free energy, F_D, from which it follows that

$$P = -\left(\frac{\partial U_0}{\partial V}\right)_T - \left(\frac{\partial F_D}{\partial \theta_D}\right)_T \left(\frac{\partial \theta_D}{\partial V}\right)_T \tag{9-15}$$

Next, recognizing that the Debye internal energy component, U_D, may be expressed as the product of T and a function of θ_D/T, consequently that F_D must also be of that form, it can be shown [YAT72, p. 34] that $\partial F_D/\partial \theta_D = U_D/\theta_D$. With these provisions (9-14) yields

$$P = -\left(\frac{\partial U_0}{\partial V}\right)_T - \frac{U_D}{\theta_D}\left(\frac{\partial \theta_D}{\partial V}\right)_T \tag{9-16a}$$

$$= -\left(\frac{\partial U_0}{\partial V}\right)_T + \gamma_G \frac{U_D}{V} \tag{9-16b}$$

with

$$\gamma_G \equiv -(\partial \ln \theta_D/\partial \ln V)_T = -(V/\theta_D)(\partial \theta_D/\partial V)_T \tag{9-17}$$

Equation (9-16a) or (9-16b) is the Debye equation of state. Since, in the Debye model, θ_D is independent of temperature, so also is γ_G. Practical departures from the model generally force some temperature dependence on those characteristic quantities. In each case they tend to have constant "high-temperature" and "low-temperature" values, with transitions over a range of intermediate temperatures that differ from one material to another.

That the nonvanishing of γ_G leads to thermal expansion follows by associating

$$(\partial P/\partial T)_V = \gamma_G C_v/V \tag{9-18}$$

from (9-16b) with the identity $(\partial P/\partial T)_V \equiv -(\partial V/\partial T)_P(\partial P/\partial V)_T$, the definition $\beta_v \equiv (1/V)(\partial V/\partial T)_P$, and the bulk modulus $K_T \equiv -V(\partial P/\partial V)_T$, the result of which is

$$\beta_v = \gamma_G \frac{C_v}{VK_T} = \gamma_G \frac{C_p}{VK_S} \tag{9-19}$$

as in (9-1) and (9-3). As mentioned above, under the Debye approximation γ_G is temperature independent. Since K_T is temperature independent, this leaves $\beta_v \propto C_v$ (as mentioned in the introduction), which is Grüneisen's rule [YAT72].

(b) Entropy Formulation of the Grüneisen Relation. Following VARLEY [VAR56] and BARRON *et al.* [BAR80], it is useful to recall the thermodynamic identity relating entropy, *S*, to pressure and volume,

$$\left(\frac{\partial S}{\partial V}\right)_T = \left(\frac{\partial P}{\partial T}\right)_V \tag{9-20}$$

which with the aid of (9-18) leads to

$$\left(\frac{\partial S}{\partial V}\right)_T = \gamma_G \frac{C_v}{V} \tag{9-21a}$$

$$= \beta_v K_T \tag{9-21b}$$

It follows that, for a given solid, any contribution to the entropy also contributes to the volume expansion. As BARRON *et al.* [BAR80] pointed out, since the bulk modulus is usually dominated by the static lattice energy there is no need to consider the possibility of *its* being separable into components. Thus, it is satisfactory to write

$$\beta_v = \sum_k \beta_{v,k} \tag{9-22a}$$

$$= \left(\frac{1}{K_T}\right) \sum_k \left(\frac{\partial S_k}{\partial V}\right)_T \tag{9-22b}$$

where k is nothing more than a running index. Components to the total entropy arise from numerous well-known physical sources, such as the presence of conduction electrons and the existence of magnetic and superconductive interactions, to be considered below.

9.2.2 Lattice-Dynamical Approach

(a) Cubic Crystals. From (9-21), (9-20), and (9-14) it follows that

$$\beta_v K_T = -\frac{\partial^2 F}{\partial V\,\partial T} \tag{9-23}$$

a relationship that is frequently used as the starting point for the development of lattice-dynamical theories of thermal expansion. Following BLACKMAN [BLA57] and COLLINS and WHITE [COL64], if the temperature dependent part of the crystal's free energy is entirely vibrational

$$F = U_0 + k_B T \sum_{k=1}^{3N} \ln\{1 - \exp(-\hbar\omega_k/k_B T)\} \tag{9-24}$$

where U_0 is the zero-K internal energy, as before, and the sum is taken over all of the $3N$ normal modes of the crystal. After performing the substitutions, (9-23) becomes

$$\beta_v K_T = \frac{1}{V}\sum_k \gamma_k C_k(\hbar\omega_k/k_B T) = \frac{1}{V}\sum_k \gamma_k C_k(\theta_k/T) \tag{9-25}$$

where

$$\theta_k \equiv \hbar\omega_k/k_B$$

$$\gamma_k = -\frac{V}{\omega_k}\left(\frac{\partial\omega_k}{\partial V}\right)_T = -(\partial \ln \omega_k/\partial \ln V)_T \tag{9-26}$$

and

$$C_k(\theta_k/T) \equiv k_B\left(\frac{\theta_k}{T}\right)^2 \frac{\exp(\theta_k/T)}{\{\exp(\theta_k/T) - 1\}^2}$$

which is the Einstein heat capacity of the kth vibrational mode. The identification of (9-25) with the Grüneisen relation (9-19), after recognizing that $C_v \equiv \sum_k C_k$, implies for the aggregate γ_G the definition

$$\gamma_G = \sum_{k=1}^{3N} \gamma_k C_k \Big/ \sum_{k=1}^{3N} C_k \tag{9-27}$$

Alternatively, (9-25) can now be simply rewritten in the form

$$\beta_v = \sum_k \gamma_k C_k / K_T V \tag{9-28}$$

Because of its central importance in thermal expansion theory, calculations usually focus upon the γ_G-parameter. It is useful to first consider its high- and low-temperature limits.

(i) $\gamma_\infty \equiv \gamma_G$ *at High Temperatures* ($T \gg \theta_D$). At high temperatures, when all modes contribute equally to the heat capacity, the aggregate γ_G is simply an average given by

$$\gamma_\infty = \langle \gamma_k \rangle = \frac{1}{3N} \sum_{k=1}^{3N} \gamma_k = \text{constant} \tag{9-29}$$

(ii) $\gamma_0 \equiv \gamma_G$ *at Low Temperatures* ($T \ll \theta_D$). At low temperatures the heat capacity obeys a Debye law and (9-27) reverts to the original Grüneisen definition (9-3). Calculations by BARRON [BAR55] for a model fcc lattice with central forces between nearest neighbors yielded $\gamma_\infty - \gamma_0 \equiv 0.3$ (see Figure 9-2).

The above equations, and others related to them, play important roles in the study of crystal lattice dynamics. The purpose of calculating γ_G, and hence β_v, is not to derive the expansion coefficient per se, since it is much easier to measure it directly if only data are needed, but to use the comparison between the calculated and measured β_vs as a test of lattice-dynamical model validity.

Early calculations of γ_G considered a cubic solid as an anisotropic continuum [SHE58], [COL63] supporting normal modes of frequency, $\omega_k(\mathbf{k})$, where $\mathbf{k}$ is the wave vector of the kth lattice mode which is associated with a stiffness modulus, say m_k. According to COLLINS [COL63], the γ_k for use in (9-27) was given by

$$\gamma_k = -\tfrac{1}{6} + \tfrac{1}{2} K_T (\partial \ln m_k / \partial P)_T \tag{9-30}$$

in which the derivative is measured directly or may be calculated from the measured third-order elastic moduli (the $\partial C_{ij}/\partial P$).

Equation (9-19) is frequently inverted in order to express γ_G explicitly in terms of the other quantities, all of which are amenable to direct measurement. The thermal expansion then becomes part of the experimental input to the calculation of $\gamma_G(T)$, the acquisition of which has been the goal of a large number of studies, such as those by BARRON *et al.* (e.g., [BAR64]) and subsequently WHITE and COLLINS [WHI73], of the alkali halides. The function $\gamma(n)$, defined as the dimensionless volume dependence of the nth moment of the normal-mode frequency distribution, can be used to derive the volume dependence of a number of crystal properties [YAT72]. The effects of high frequency modes were not represented in the early calculations. This deficiency was corrected by subsequent workers, for example

VETELINO *et al.* [VET70], who calculated the thermal expansions of some alkali halide and Zn-blend-type crystals, and NAMJOSHI *et al.* [NAM72], who investigated the applicabilities of the "modified rigid ion," the "simple shell," and the "breathing shell" models to the thermal expansions of crystals with the NaCl structure.

The starting point for the calculation of γ_G is (9-26) rewritten in the form

$$\gamma_k(\mathbf{k}) = -d[\ln \omega_k(\mathbf{k})]/d[\ln V] \tag{9-31}$$

hence

$$\gamma_k(\mathbf{k}) = [K_T/\omega_k(\mathbf{k})][\Delta\omega_k(\mathbf{k})/\Delta P] \tag{9-32}$$

where in this case $\omega_k(\mathbf{k})$ denotes the frequency of the normal mode belonging to the kth branch at wave vector $\mathbf{k}$ in the first Brillouin zone [VET70], and $\Delta\omega_k(\mathbf{k})/\Delta P$ is its pressure dependence. Normal mode frequencies and their pressure derivatives can be model-calculated using the values and pressure-derivatives of the elastic constants, the long-wavelength transverse and longitudinal optic phonon frequencies, and the high- and low-frequency dielectric constants as experimental inputs. The thermal expansion can then be calculated via (9-28), using no additional data except the crystal volume, for comparison with experiment as a test of the model.

(b) Noncubic Crystals. In cubic crystals or isotropic solids the linear thermal expansion, α, is isotropic, thus $\beta_v = 3\alpha$. Otherwise, α is generally different along each of any three mutually perpendicular directions and β_v is equal to their sum (i.e., the trace of the expansion tensor).[5] The anisotropic case has been treated by BARRON and MUNN [BAR67] and by COLLINS and WHITE [COL64] (see also [TOU77], [DEL80]). Referring to the principal crystallographic directions, the three principal expansion coefficients are given by

$$\alpha_i = \frac{C_v}{V}\sum_{j=1}^{6} S_{ij}\gamma_j \qquad i = 1, 2, 3, \tag{9-33}$$

where S_{ij} are the usual elastic compliances in Voigt notation (see Section 4.2) and γ_j is now the derivative of the lattice entropy with respect to uniaxial strain rather than to volume as in the isotropic case (see Section 9.2.1(b)). β_v is of course $\sum_{i=1}^{3} \alpha_i$. For axial crystals, j does not exceed 3; furthermore, for such crystals (hexagonal, tetragonal, and trigonal) only two linear coefficients are needed: $\alpha_\perp$ in the basal plane, and $\alpha_\parallel$ in the

[5]Not to be confused with the thermal expansion of an anisotropic polycrystal (see Section 9.8).

c-axis direction. These, according to BARRON and MUNN [BAR67[a]], are given by

$$\alpha_\perp = \frac{C_v}{V} \frac{C_{33}\gamma_\perp - C_{13}\gamma_\parallel}{C_{33}(C_{11} + C_{12}) - 2C_{13}^2} \tag{9-34a}$$

$$\alpha_\parallel = \frac{C_v}{V} \frac{(C_{11} + C_{12})\gamma_\parallel - 2C_{13}\gamma_\perp}{C_{33}(C_{11} + C_{12}) - 2C_{13}^2} \tag{9-34b}$$

A reciprocal relationship for $\gamma_\parallel$ and $\gamma_\perp$ formally closes the equation set. To break the loop and make further progress it would be necessary to perform independent theoretical-model calculations of the γ_js themselves by calculating the uniaxial strain variation of lattice entropy. It appears that this has not yet been done [BAR80, p. 619].

9.3 Thermal Expansion of Metals

9.3.1 The Electronic Expansion Coefficient

(a) The Heat-Capacity Analogy. In general, the coefficient of thermal expansion is proportional to the heat capacity. This is illustrated in Figure 9-3 using data for pure Al as an example. With regard to insulators, only the lattice heat capacity need be considered, but for a metal, whose heat capacity contains both lattice and electronic components (particularly noticeable at low temperatures, Section 8.1), an electronic component of thermal expansion, β_e, defined by

$$\beta_e \propto \gamma_e C_e = \gamma_e \gamma T \tag{9-35}$$

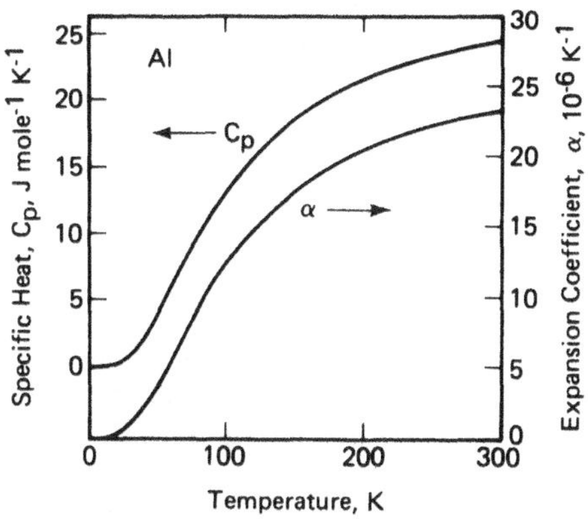

Figure 9-3. Coefficient of linear thermal expansion, α, for pure Al juxtaposed against its molar constant-pressure heat capacity, C_p,—after standard sources.

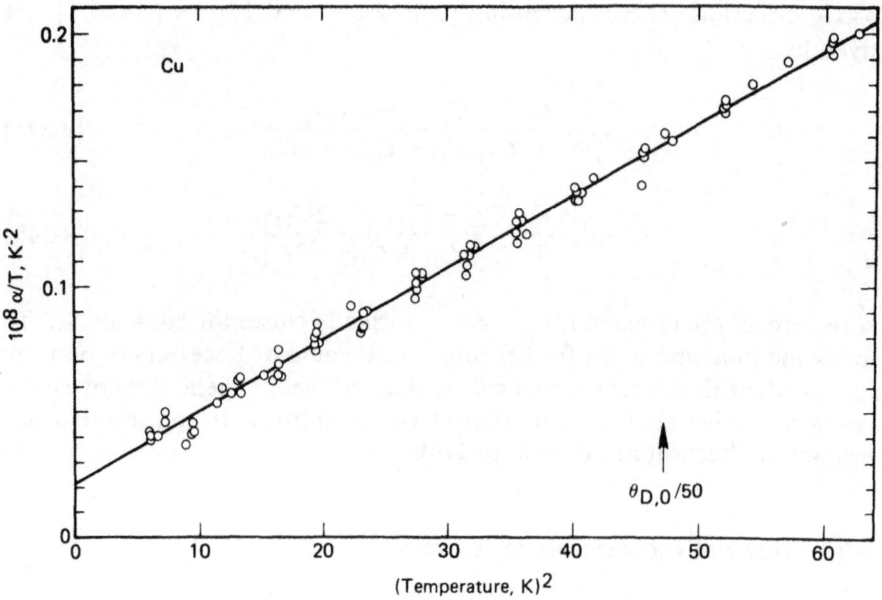

Figure 9-4. α/T *versus* T^2 for Cu ($\theta_{D,0} = 344.5$ K) and the fitted polynomial $10^{10}\alpha = 2.1_5 T + 0.284\ T^3 + 0.5 \times 10^{-4}\ T^5$, associated with which are $\gamma_e = 0.93$ and $\gamma_g = 1.78$—after WHITE and COLLINS [WHI72ᵃ], see also [PER72].

is to be expected. Again by analogy with the low-temperature heat capacity, this should become equal to, and eventually dominate, the lattice thermal-expansion component as the temperature is reduced in the liquid-He regime [WHI61ᵇ]. Indeed, for most metals, β_e and β_v start to become comparable in magnitude, and therefore separable experimentally, at temperatures between about $\theta_D/100$ and $\theta_D/50$ (i.e., between about 1 and 10 K).

The expansion coefficient of a metal is the sum of the electronic and lattice terms. According to (9-35), β_e is proportional to $C_e = \gamma T$, while β_v, being proportional to C_v (9-24b), varies as T^3 at low temperatures. For cubic and isotropic systems it is, therefore, possible to write (in terms of linear thermal expansion coefficients) [WHI61ᵇ]:

$$\beta_v/3 = \alpha = \alpha_e + \alpha_g \qquad (9\text{-}36a)^6$$

$$= AT + BT^3 \qquad (9\text{-}36b)$$

where the A and B are unnamed constants. Low-temperature metallic expansion data are often tabulated in terms of $A\ (=\alpha_e/T)$ and $B\ (=\alpha_g/T^3)$, thereby enabling α to be calculated for any selected temperature below

[6]The subscript g (for "Gitter") is traditionally used to denote lattice properties.

about $\theta_D/30$ (see Section 9.6). A plot of α/T *versus* T^2 for pure polycrystalline Cu illustrating the validity of (9-36) at low temperatures is given in Figure 9-4. Obviously γ_G is the weighted mean of γ_e and γ_g according to

$$\gamma_G = (\gamma_e C_e + \gamma_g C_g)/C \tag{9-37}$$

In plots of γ_G *versus* T, γ_e will begin to dominate at very low temperatures. In Cu, $\gamma_{g,\,293\,\mathrm{K}} = 1.98$ [BAR80, p. 682] and $\gamma_{g,\infty} - \gamma_{g,0} \cong 0.35$,[7] while γ_e (free electron) $= \frac{2}{3}$ and $\gamma_{e,\,\mathrm{Cu}} = 0.9$ [BAR80, p. 682]. Thus, the electronic component will cause the aggregate γ_G to decrease steeply as the temperature falls below about 17 K ($\sim\theta_D/20$). This effect is seen in Figure 9-2.

(b) Entropy Formulation of the Electronic Expansion Coefficient. According to (9-21), any contribution to the entropy of the crystal supplies an additional term to the thermal expansion coefficient. The first to discuss in detail the electronic thermal expansion of non-free-electron metals was VARLEY [VAR56], who commenced his analysis with an expression for the electronic entropy of the form

$$S_e = \tfrac{2}{3}\pi^2 k_B^2 T V n_v(E_F) = \gamma_v V T \tag{9-38}$$

where $n_v(E_F)$ is the usual Fermi density-of-states per unit energy per unit volume *for one direction of spin*,[8] and $\gamma_v \equiv \frac{2}{3}\pi^2 k_B^2 n_v(E_F)$ is the unit-volume electronic specific heat coefficient. Following BARRON *et al.* [BAR80], it is, however, better to think in terms of the total (in this case molar) Fermi density-of-states $N(E_F) = Vn(E_F)$ and the electronic heat capacity coefficient, $\gamma_v V \equiv \gamma$, such that the electronic heat capacity is $C_e = \gamma T$ $(=S_e)$, in which case the electronic form of (9-21) yields

$$\beta_e K_T = \left(\frac{\partial S}{\partial V}\right)_T = \left(\frac{\partial \ln \gamma}{\partial \ln V}\right)_T \left(\frac{C_e}{V}\right) \tag{9-39}$$

Identification of this with the expected $\gamma_e(C_e/V)$ reveals that

$$\gamma_e = \left(\frac{\partial \ln \gamma}{\partial \ln V}\right)_T \tag{9-40a}$$

that is,

$$\gamma_e = \left(\frac{\partial \ln N(E_F)}{\partial \ln V}\right)_T \tag{9-40b}$$

[7] A model fcc lattice yields $\gamma_{g,\infty} - \gamma_{g,0} \cong 0.3$ [BAR55].

[8] The *total* density-of-states is then $2n_v(E_F)$.

Equation (9-40b) is of course equivalent to the Varley expression

$$\gamma_e = \left[1 + \frac{\partial \ln n_v(E_F)}{\partial \ln V}\right] \tag{9-40c}$$

9.3.2 The Free-Electron Expansion Coefficient

According to the standard texts (e.g., [KIT68, p. 208]) for a free electron gas, $N(E) \propto E^{1/2}$, where $N(E)$ is the (molar) state density at energy E, and the Fermi energy, E_F, is proportional to $(N/V)^{2/3}$. It follows that $\partial \ln N(E_F)/\partial \ln V = \frac{2}{3}$, or $\partial \ln n_v(E_F)/\partial \ln V = -\frac{1}{3}$.

In either case

$$\gamma_e = \tfrac{2}{3} \tag{9-41}$$

It is useful to compare the electronic thermal expansion coefficient to that of the lattice, β_v, whose value is closer to common experience. The ratio is obviously given by $\beta_e/\beta_v = (\gamma_e/\gamma_G)(C_e/C_g)$. By comparing Figure 9-2 and (9-41) it is apparent that the γ_es and the low-temperature γ_Gs are of the same order of magnitude—and this seems to be generally true (e.g., [BAR80, p. 682]). It follows that $\beta_e \cong \beta_v$ within the same temperature range that $C_e \cong C_g$ (i.e., within the liquid-He range). According to the Debye theory of specific heats, C_g at low temperatures is given by $\beta T^3 = 1.944 \times 10^3 (T/\theta_D)^3 \text{ mol}^{-1} \text{ K}^{-1}$ (from (8-4)). For the nearly free electron metal Al, for example, for which $\theta_D = 426$ K (low-temperature calorimetric) and $\gamma = 1.36 \text{ mJ mol}^{-1} \text{ K}^{-2}$, $C_e = \gamma T$ begins to exceed C_g when T drops below about $\theta_D/58$ (i.e., below about 7.3 K).

9.3.3 Relative Linear Expansion at Low Temperatures

The quantity that is usually measured experimentally is the incremental expansion, dl, associated with a small temperature interval, dT, which as a general rule is preferred to be less then about $T/20$ [BAR80, p. 646]. The expansion coefficient, $\alpha = (1/l_0)(dl/dt)$, where l_0 is the sample length at some reference temperature, can be acquired numerically or otherwise from the slope of the Δl $(=\int dl)$ *versus* T curve.

The equipment sensitivity needed to acquire low-temperature thermal expansion data can be judged from a consideration of the expansion coefficient of, for example, Al. At 10 K, $\alpha_{Al} = 3.6 \times 10^{-8} \text{ K}^{-1}$. In order to measure the expansion of a sample of length $l_0 = 1$ cm in response to a temperature increase of 1 K to a precision of 1%, a length resolution of 0.03 Å would be needed. Such a remarkable degree of sensitivity is achievable using three-terminal capacitance dilatometry and other techniques to be described in Section 9.5.

9.3.4 Further Calculations of the Electronic Thermal Expansion Coefficient—Departures from the Free Electron Model

During the expansion of a free electron gas, $N(E_F)$ increases as $V^{2/3}$. But in a real metal whose conduction electron energies are confined to bands, which in turn depend on the lattice parameter and in which $N(E)$ may be a complicated function of E, additional components must be included in the expression for γ_e. Take, for instance, the case of a single energy band. As V increases, the bandwidth will narrow, causing the density-of-states to increase more than the simple volume dependence would predict. γ_e would then be expected to be larger than the $\frac{2}{3}$ of (9-31), and this has certainly been found to be the case for a large number of metals. VARLEY [VAR56] showed that for band electrons

$$\gamma_e = \left[1 + \left(\frac{\partial \ln n}{\partial \ln V}\right)_T - \frac{n_e}{n^2}\left(\frac{\partial n}{\partial E}\right)_V \left(1 + \frac{\partial \ln n_e}{\partial \ln V}\right)_T + \frac{C_e}{\alpha V n^3}\left(\frac{\partial n}{\partial E}\right)_V^2\right]_{E_F} \tag{9-42}$$

with $n \equiv n_v(E)$ and n_e = number density of the conduction electrons. This relationship, which is similar in spirit to that previously introduced by STONER to account for the magnetic susceptibility temperature dependences of transition metals, (7-18), suggests that important contributions to γ_e could arise out of the structure of the $n_v(E)$ *versus* E curve in the vicinity of E_F. In particular, large positive or negative values of $(\partial n_v(E)/\partial E)_{E_F}$ in the third term of (9-42) could significantly decrease or increase, respectively, the $\frac{2}{3}$ arising from the previous two. However, the same objections that were levelled against the magnetic susceptibility function in Section 7.9 also apply here. Of course, as a function of $n_v(E_F)$, the electronic Grüneisen parameter should also respond to the same many-body effects that influence the electronic specific heat coefficient, Section 8.1.

As pointed out by BARRON *et al.* [BAR80, p. 642], standard band-structure calculations have been applied to the calculation of γ_e values for particular metals. With Cu, for which the KKR method was appropriate, a poor value, $\gamma_e \cong 0.4$, was obtained. The pseudopotential method, when applied to the almost-free-electron metal Al, yielded $\gamma_e = 1.63$ in excellent agreement with experiment.

9.4 Thermal Expansion of Magnetic Solids

When a solid undergoes a temperature dependent magnetic anomaly or transformation, its specific heat *versus* temperature profile is often duplicated by that of the thermal expansion coefficient, appropriately scaled. If

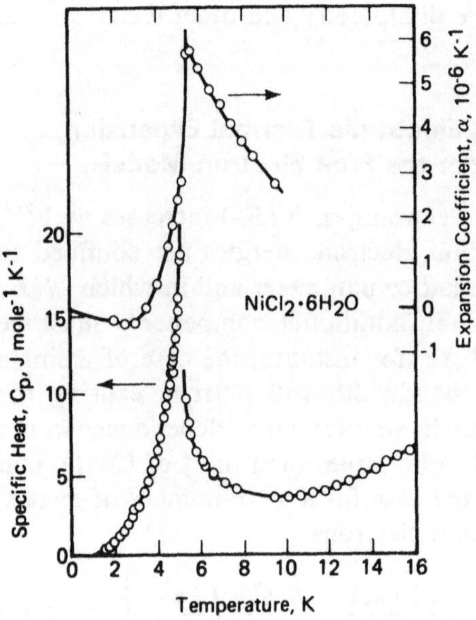

Figure 9-5. The *b*-axis thermal expansion coefficient of $NiCl_2 \cdot 6H_2O$ (after LANCHESTER [LAN70]) juxtaposed against its specific heat (after GOPAL [GOP66, p. 100]).

C_m is the magnetic contribution to the total specific heat, then a magnetic component of thermal expansion is coupled to it via a magnetic Grüneisen parameter, γ_m. Thus, in antiferromagnetic crystals such as the hydrated halides of Mn, Co, and Ni characteristic λ-anomalies are observed in both the specific heat [GOP66, p. 100] and the various monocrystalline linear thermal expansion coefficients [LAN70] as the temperature is varied through the magnetic order-disorder transformation point (Figure 9-5). Similarly in dilute "local-moment" alloys, such as Cu with a few percent Mn, the "excess" linear thermal expansion coefficient almost exactly parallels the "excess" specific heat when both are plotted *versus* temperature [WHI62[a]], Figure 9-6.

Magnetic contributions to specific heat have been discussed by GOPAL [GOP66, Chapter 4]. The effects of ferromagnetism, spin waves, antiferromagnetism, and paramagnetism (in particular, the Schottky effect) were considered. According to (9-11) these should all contribute to β_v, and indeed

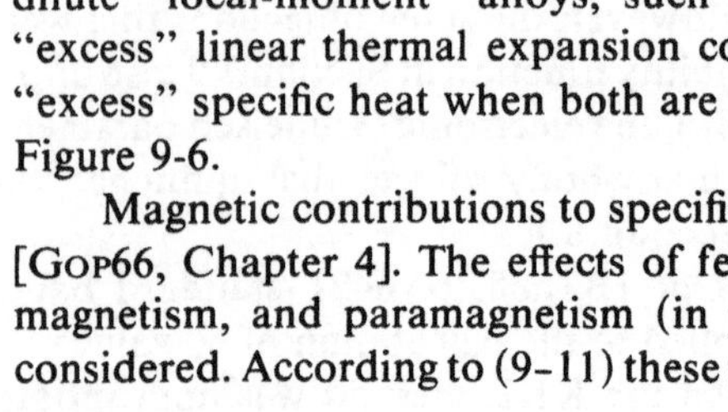

Figure 9-6. Excess linear thermal expansion coefficient, $TE \equiv \Delta\alpha/T$, compared to the excess specific heat, $SH \equiv \Delta C/T$ (both relative to pure Cu), for dilute Cu-Mn alloys of concentrations (i) 0.56, (ii) 0.23, and (iii) 0.15 at.% Mn—after WHITE [WHI62[a]].

they do according to BARRON *et al.* [BAR80, pp. 642–646], who discussed, with the aid of numerous examples, the influences on thermal expansion of magnetic transformations and the thermal excitation of magnetic energy states.

It is clear that several classes of magnetic solid remain unmeasured and that much work remains to be done. Of particular interest in this context is the thermal expansion of austenitic stainless steels. It is known as a result of magnetic susceptibility [COL79[a], COL80] and sound velocity [COL79[b]], [LED79] measurements that austenitic stainless steels containing appreciable amounts of Mn exhibit magnetic and associated acoustic anomalies in the intermediate-temperature range. Magnetic susceptibility, magnetization, and low-temperature specific heat measurements [COL77, COL78[b], COL79[c]] performed on austenitic Fe-Cr-Ni stainless steels have also revealed the existence of low-temperature cluster superparamagnetism and possible spin-glass ordering. All of these magnetic effects should have subtle influences, as yet unmeasured, on the thermal expansion coefficients.

9.5 Thermal Dilatometry

9.5.1 Introduction

Experimental techniques for the measurement of thermal expansion—either the relative expansion (with respect to some reference temperature, T_0, and length, l_0), $\Delta l/l_0|_{T_0}^{T}$, or the thermal expansion coefficient, $(1/l_0) \times (dl/dT)$—have been compared and discussed in several review articles and books [COL64], [YAT72], [TOU75, TOU77], [BAR80]. The methods used may be loosely subdivided into two categories—"low sensitivity," in which $\delta l/l = 10^{-5}$–10^{-6} where δl is the smallest measurable length change, and "high sensitivity," in which $\delta l/l = 10^{-8}$–10^{-9} or less [YAT72]. Since various levels of sensitivity can be attained within the framework of a given measuring principle, the terms "low" and "high" have as much to do with experimental goals as with technical limitations [YAT72]. Thus, for example, in measuring relative expansion, $\Delta l/l_0$, over the temperature range 4–300 K for the purpose of acquiring engineering design data, a low-sensitivity technique or the low-sensitivity range of a general-purpose technique will be adequate. On the other hand, if accurate expansion coefficient data are required over a narrow temperature range, particularly at liquid-He temperatures, very high sensitivities would be needed. Thus, as pointed out in Section 9.3.3, in order to measure the $\Delta T = 1$ K length change of an $l_0 = 1$ cm bar of Al at 10 K to an accuracy of 1%, a sensitivity of $\delta l/l_0 = 3.6 \times 10^{-10}$ would be needed. Some of the ways in which this can be achieved, and even exceeded, are mentioned below.

It would be inappropriate, if not impossible in this space, to attempt to compete with the reviews already in existence; thus, only a few representative techniques are to be mentioned. In view of the overlap between the categories "particular measuring technique," "measuring principle," and "detection sensitivity" it is also impossible to describe the subject properly in terms of a single linear categorization scheme. A systems approach, with appropriate diagrams, would be more appropriate. In its absence it is hoped that the framework for description adopted here will at least be instructive.

9.5.2 Interference Methods

(a) Light Waves. The Fizeau interferometer relies on the self-interference between incident and reflected beams of collimated light in their two-way passage across a wedge-shaped gap [LEV68, p. 107]. A set of straight, parallel fringes is produced, whose movement is a magnified result of the change of gap thickness. Since for normal incidence on a narrow gap [TOU77]

$$\delta l = (\lambda/2)\delta N \tag{9-43}$$

where λ is the wavelength of the light and δN is the number of fringes that pass a given point when the gap thickness changes by an amount δl, one's ability to detect movement of 10^{-2} to 10^{-3} of a fringe corresponds, in an $l_0 = 1$ cm sample, to $\delta l/l_0 \cong 10^{-7}$ to 10^{-8}.

If the gap is made parallel, and the light beam convergent or divergent, the result is the Fabry–Perot interferometer in its simplest form [TOU77]. The diffraction pattern is now a set of concentric rings whose diameters change in response to changes in the gap thickness—and hence in response to the expansion of an attached specimen. Using such a device the expansion of several metals down to 10–20 K have been measured [FRA61].

If the reflectivity of the parallel-sided optical cell is increased, the light becomes partially trapped inside what has now become a "multiple-beam Fabry–Perot etalon" [LEV68, p. 115]. As an optical resonator, this device will be considered in the next section.

(b) X-Rays. Bragg diffraction is another example of interference between electromagnetic waves. Equation (9-43) as it stands leads to the expectation that the use of x-rays, with wavelengths three orders of magnitude less than those of light, would result in very high sensitivities. But in x-ray diffraction the basis of measurement is one lattice parameter ($\cong \lambda$) rather than a film of thickness $\sim 10^3\lambda$, and the equivalent δN is not measured

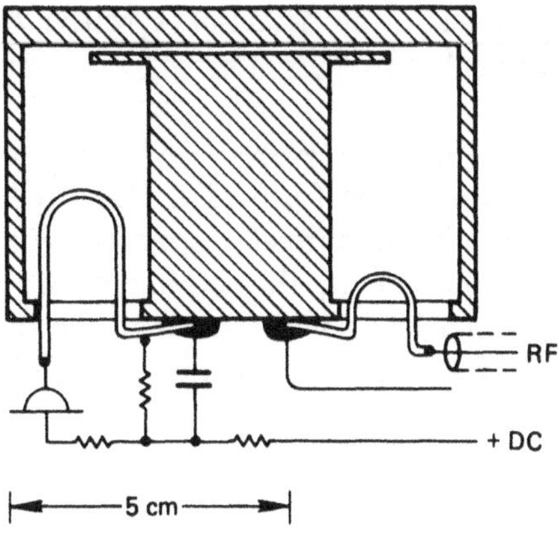

Figure 9-7. Toroidal L-C resonator for 84 MHz and associated tunnel diode circuitry. The 0.0037-cm gap at the top of the center-post has a capacitance of 470 pF; the inductance of the toroid is 7.7 nHy. The diode-coupling and output-coupling loops project through 0.6-cm wide radial slots—after VAN DEGRIFT [VAN74].

to the same degree of precision.[9] As a result, the sensitivity is correspondingly reduced to $\delta l/l_0 \cong 10^{-5}$, rising to 10^{-7}–10^{-8} only in special situations [TOU77], [BAR80, p. 647].

9.5.3 Resonance Methods

(a) Radio Waves. Resonant circuits, either self-oscillating or driven, have been used to detect small changes in the gap thicknesses of their capacitive elements. A series L-C circuit mounted within a cryostat, and resonating at 1 MHz, has been used to measure the expansions of the noble metals in the temperature range 4.2 to 15 K [KOS69]. Assisted by frequency multiplication and counting circuitry the device had a sensitivity of $\delta l/l_0 \cong 10^{-11}$. The electronics in this case were complicated by the need to couple the L-C circuit through coaxial leads to the rest of the oscillator circuit outside the cryostat. This disadvantage was avoided by VAN DEGRIFT [VAN74] who mounted the entire oscillator in the He bath. In this case a toroidal resonator was coupled to the negative resistance of a tunnel diode (Figure 9-7). The arrangement was compact and capable of operating at high frequency (e.g., 84 MHz). It was estimated that, if suitably transmitted to the capacitance gap, a length change of 7×10^{-12} cm could be detected. The oscillating circuit is self-shielded and, in the unpolished and nonsuperconducting condition, possessed a Q of 16,000.

Higher frequencies still were employed by PUDALOV and KHAIKIN [PUD69], who used a coaxial microwave resonator (9.5 GHz) as a dilatometric transducer. Employing an x-cut quartz crystal as a calibrating

[9]Actually for a cubic crystal of lattice parameter, a, using the Debye-Scherrer x-ray method, the expansion is given by $\Delta a/a = -\cot\theta\Delta\theta$, independent of λ, where θ is the angle between the x-ray beam and the crystal planes.

element,[10] the detection sensitivity for unity signal/noise ratio was found to be 1×10^{-12}.

(b) Light Waves. As mentioned above, a pair of parallel reflective plates, in the form of a Fabry–Perot multiple-beam "etalon" [LEV68, p. 115], is an optical resonater. When collimated monochromatic light is incident on it, the exiting beam intensity peaks sharply each time the geometrical condition for phase reinforcement is met. Comparable to the Q of an RF circuit is the "finesse," F, of an optical resonator given by

$$F = \frac{\pi\sqrt{\rho}}{1-\rho} \tag{9-44}$$

(where ρ is the reflectance of the surfaces) and defined as the *ratio of the phase shift between two maxima to the half-intensity width phase shift* [LEV68, p. 114].

After tuning a stable HeNe red laser to such an etalon,[11] JACOBS *et al.* [JAC72] were able to achieve a length change sensitivity, $\delta l/l_0 = \delta\nu/\nu_0$, of 10^{-9}, limited only by laser stability. It was estimated that the application of improved stabilization techniques could increase the detection sensitivity to 10^{-10} or even 10^{-13}.

9.5.4 Push-Rod, Optical-Lever, SQUID, and Capacitive Techniques

(a) Push-Rod Dilatometry. This method is suitable for expansion measurement over a wide temperature range. In it, the sample resides in a furnace or cryostat at the bottom of a long tube of some low-expansivity and/or low-thermal-conductivity material. A long push-rod of the same material rests on the sample. Differential movement between the top of the tube and the push-rod is measured at room temperature using, typically, a dial gauge [FOW73], [CLA68, CLA81] or linear differential transformer (LVDT) [EVA66]. The typical length-change sensitivities (of 5×10^{-5} cm and 3×10^{-6} cm, respectively) characteristic of these techniques are usually adequate for most wide-temperature-range expansion measurements. If improvements to the usual LVDT are made, however, length-change sensitivity, $\delta l/l_0$, as high as 10^{-10} is achievable [TOU77]. Thus, the method itself can be included in either the "low-sensitivity" or the "high-sensitivity" category depending on the purpose for which its particular variant is intended.

[10] Normal to the plane of a 2-mm′ disc, the piezoelectric constant was 2×10^{-2} Å V^{-1} at room temperature and 2.5×10^{-2} Å V^{-1} in He.

[11] In the absence of a stable tunable laser, tuning may be achieved by varying the frequency of an amplitude-modulation-produced side-band.

As might be expected, the greatest problems with push-rod dilatometry enter through what might be termed the "displacement-transfer" mechanism. In the usual low-sensitivity applications, the temperature gradient in the push-rod gives rise to systematic error which can be overcome only by calibration or through the use of a reference sample. In high-sensitivity measurements, mechanical couplings limit the reproducibility whatever the choice of displacement sensor (optical lever, capacitance, inductance, etc.) [WHI83].

(b) Optical Lever Methods. With the upsurge of interest in high-sensitivity low-temperature dilatometry in the late 1950s, several optical lever techniques were developed, the later versions of which were intended to improve upon the sensitivity of WHITE's three-terminal capacitance method (considered below). In the method used by ABBISS *et al.* [ABB61], which seems to be an adaptation of the push-rod approach, the relative motion of the sample and a reference material was first amplified by a twin-strip flexure mechanism, or "reed comparator" (Figure 9-8), before being passed on, in the form of a moving light beam, to an optical level. The sensitivity was about 2×10^{-7}. Subsequently, GRAHAM *et al.* [SHA64] used an "Ayrton strip" as the movement transducer. In this, the slight unravelling of a doubly twisted strip bearing a mirror is converted into the motion of a light beam that is further amplified by an optical lever. The sensitivity of the first device used was 3×10^{-10} cm; applied to an 8-cm sample this yielded a $\delta l / l_0$ of about 2×10^{-9}. A second version, improved by the placing of the Ayrton strip in line with the specimen rather than parallel to it thereby enabling it to be immersed in the liquid He, and by associating it with a better optical lever, had a sensitivity, $\delta l / l_0$, of 10^{-11}. An interesting feature of a further modification of this device [PER72] was the inclusion of a quartz crystal in series with the specimen; its piezoelectric response of 2.32×10^{-10} cm V^{-1} was used as a displacement reference.

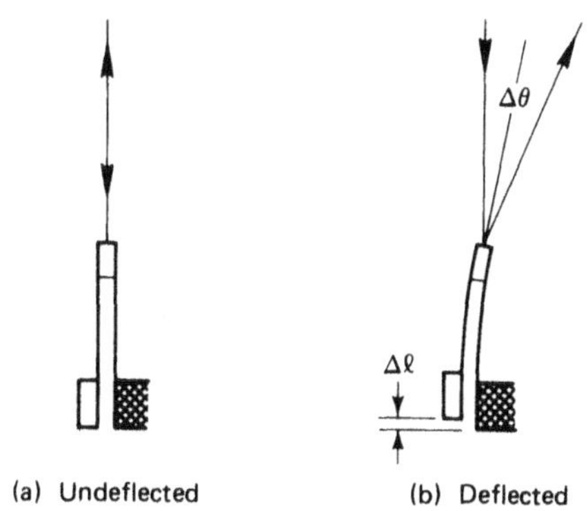

Figure 9-8. Reed-comparator-type flexure mechanism for converting differential linear motion into angular motion (of an end-mounted mirror)—after GEARY [GEA54] (see also [ABB61]).

In the above-mentioned optical lever, whose sensitivity was $\sim 10^{-8}$ rad (signal/noise = 1:1), the moving image of one grid, cast onto a second, caused the output light intensity to vary strongly with the angle of the beam. The dilatometer devised by ANDRES [AND61, AND64a] also gauged the motion of an expanding sample in terms of the variation of the intensity of a light beam as it passed through a pair of grids in relative motion on its way to a photocell.[12] In this case the grids were mounted physically side by side and close to the specimen in the tail of a cryostat. The sensitivity of this apparatus, which was used extensively in a series of low-temperature dilatometric studies [AND63a, AND64a], was such that an expansion $\delta l = 10^{-8}$ cm would produce a signal/noise ratio of 5:1 in the electrical output of the photocell amplifier.

(c) SQUID Technology. It was inevitable that the SQUID technique would eventually be applied to thermal dilatometry. ACKERMAN and ANDERSON reported on the construction of such a dilatometer for the measurement of thermal expansion coefficients in the temperature range 0.1–10 K [ACK82]. The principle of the method is straightforward: the expansion of a sample causes a pick-up loop to advance into a magnetic field, the SQUID-measured change of enclosed flux being interpreted in terms of linear displacement. In practice the method employs temperature-modulation (hence displacement-modulation), enabling signal averaging to improve the signal/noise ratio. The resolution of the dilatometer was claimed to be 2×10^{-12} cm.

(d) The Three-Terminal Capacitance Method. As pointed out by WHITE [WHI61a], the three-terminal capacitance method of measuring small displacements stemmed from the development at CSIRO, Sydney, Australia, of a precision small-capacitance measurement technique using three-terminal capacitance standards in association with transformer-bridge circuitry. Any "two-terminal" capacitance is a three-terminal network when the unavoidable stray capacitances to ground from each of the terminals are taken into consideration (Figure 9-9). However, with a transformer bridge as an essential component of the measuring circuit, the effect of stray capacitance can be eliminated and the "direct" capacitance, C_{12}, measured. Based on this principle, a capacitance cell such as that depicted schematically in Figure 9-10 was constructed. In association with the proper detection circuitry [WHI61a, WHI72a], [BAR80] the 0.1-mm gap capacitance of 10 pF can be measured to 10^{-6} pF (hence $\delta l = 10^{-9}$ cm) with the usual commercial bridge, or 10^{-7} pF ($\delta l = 10^{-10}$ cm) with a high-sensitivity one.

[12] Actually a differential arrangement was used.

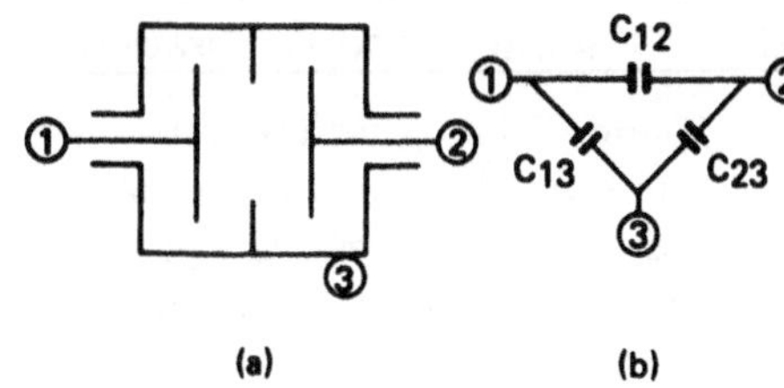

Figure 9-9. (a) Schematic representation of a shielded capacitor; and (b) its equivalent circuit showing the direct capacitance, C_{12}, and the two stray capacitances to ground, C_{13} and C_{23}, forming a three-terminal network. The transformer bridge measures only C_{12}—after WHITE [WHI61[a]].

Although some exotic techniques possess higher nominal sensitivities, because its precision is combined with robustness, lack of vibration sensitivity, versatility, and other advantages, the three-terminal capacitance technique has, in effect, become the standard method of measuring the low-temperature thermal expansion coefficient.

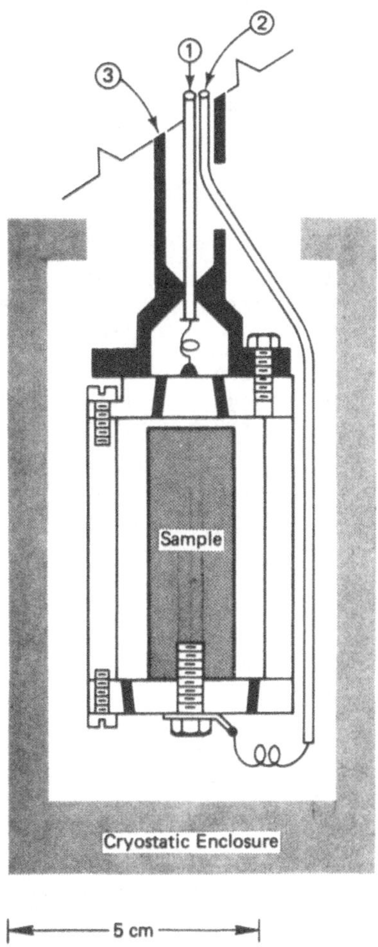

Figure 9-10. Schematic representation of the three-terminal capacitance differential-expansion cell. The numbers correspond to the terminals represented in Figure 9-9—after WHITE [WHI72[a]], [BAR80].

9.5.5 Standards and Calibration Techniques

Using Fizeau interferometry over the temperature range 20–800 K, the U.S. National Bureau of Standards has measured samples taken from a single drawing of OFHC Cu in order to establish it as Standard Reference Material 736 [HAH70]. Subsequently, a set of reference data enabling the

Table 9-1. Linear Thermal Expansion of Coppers above 20 K

	Linear Expansion Coefficient $(1/l_{293})(dl/dT)_T$, 10^{-6} K^{-1}			Relative Thermal Expansion[c]	Normalized Contraction Factor, $N(T)$[d]
T, K	a	b	c	$\Delta l/l_{293}$, %	$\frac{l_{293} - l_T}{(dl/dT)_{293}}$, K
20	0.23		0.27	−0.3250	195.3
25	0.64				
30	1.04				
40	2.28	2.31	2.29	−0.3229	194.1
50		3.87	3.87	−0.3198	
60	5.40	5.50	5.48	−0.3151	189.4
70		7.00	6.98	−0.3089	
80	8.21	8.33	8.30	−0.3012	181.0
90		9.50	9.46	−0.2923	
100	10.33	10.50	10.46	−0.2823	169.7
120		12.04	12.05	−0.2597	
130	12.63		12.67	−0.2474	148.7
150		13.66	13.64	−0.2210	
160	14.09		14.01	−0.2072	124.5
200	15.18	15.18	15.14	−0.1487	89.4
250	15.98	16.05			
280		16.39	16.50	−0.0215	12.9
283		16.43			
293	16.54		16.64	0	0.0
300	16.61	16.61	16.71	0.0117	
400	17.58		17.53	0.1831	
500	18.32		18.25	0.3620	
600	18.92		18.95	0.5480	
700	19.48		19.69	0.7411	
800	20.09		20.51	0.9420	

[a] Data of HAHN [HAH70] for NBS Standard Reference Material (SRM) 736 (OFHC Cu)—Method: Fizeau interferometry.
[b] Data of WHITE [WHI74], see also [WHI72]—Method: capacitance cell.
[c] Based on SRM 736, revised certificate, by R. K. Kirby and T. A. Hahn. This SRM is available as a rod 6.4 mm ($\frac{1}{4}$ in.) in diameter. SRM 736 L1 is 51 mm (2 in.) long; SRM 736 L2 is 102 mm (4 in.) long; SRM 736 L3 is 152 mm (6 in.) long. These data are in close agreement with [KRO77[a]].
[d] These data can provide the basis for a universal contraction curve, valid for all metals to better than ~8% such that: $(l_{293} - l_T)/l_{293}(\%) = [10^{-4}N(T)]\cdot(10^6\alpha_{293})$.

elements Si, Cu, Ag, Al, and Ge to be used as standard reference materials has been supplied and discussed by WHITE [WHI74]. Data for vitreous silica were also presented. Some results for Cu are given in Table 9-1.

As indicated in Sections 9.5.3(a) and 9.5.4(b) the piezoelectric effect can be used to generate a standard small displacement. With piezoelectric constants of typically 2×10^{-2} Å V^{-1}, quartz crystals are eminently suitable for the calibration of high-sensitivity dilatometers.

9.6 Thermal Expansions of Selected Metals and Alloys

Presented below are the thermal expansions of Cu, Al, Ti, and some technical alloys.

9.6.1 Thermal Expansion of Cu

The low-temperature thermal expansion coefficient of Cu at temperatures below about $\theta_D/30$ (where $\theta_D = 345$ K) is given by: $\alpha = \alpha_e + \alpha_l$ such that $10^{10}\alpha = (2.15 \pm 0.1)T + (0.284 \pm 0.005)T^3$ (K^{-1}) [BAR80, p. 683]. The thermal expansion of Cu at higher temperatures is given in Table 9-1. References to further literature and data are provided in the works of WHITE [WHI74] and BARRON *et al* [BAR80].

9.6.2 Thermal Expansion of Al

The low-temperature thermal expansion coefficient of Al at temperatures below about $\theta_D/30$ (where $\theta_D = 426$ K (low-temperature calorimetric)) is given by $10^{10}\alpha = (9.3 \pm 0.1)T + (0.271 \pm 0.003)T^3$ (K^{-1}) [BAR80, p. 688]. The thermal expansion of Al within the temperature interval 12–320 K, from the work of KROEGER and SWENSON [KRO77[a]] whose data were very reproducible over that range, is given in Table 9-2. References to numerous other data sets are to be found in the works of TOULOUKIAN *et al.* [TOU75] and BARRON *et al.* [BAR80, p. 688].

9.6.3 Thermal Expansion of Ti

The low-temperature thermal expansion of Ti has been measured by ANDRES [AND63[a], AND64[a]], COWAN *et al.* [COW68], ILIYEV *et al.* [ILI80] and others, [BAR80]. The results are rather scattered, presumably on account of the difficulty that is usually encountered in obtaining truly random polycrystalline samples of hcp metals. The results of x-ray lattice parameter measurements over the range 4.2–300 K [COL71[e]] enable the two principal thermal expansion components to be calculated.

Table 9-2. Linear Thermal Expansion of Aluminum[a]

T, K	Linear Expansion Coefficient $(d \ln l/dT)$, 10^{-6} K^{-1}	Relative Thermal Expansion $\Delta l/l_{293.15}$, %
12	0.0575	−0.4140
16	0.1214	−0.4140
18	0.1690	−0.4139
20	0.2303	−0.4138
25	0.4626	−0.4137
30	0.8365	−0.4134
35	1.360	−0.4129
40	2.027	−0.4120
45	2.805	−0.4109
50	3.662	−0.4092
60	5.502	−0.4047
70	7.354	−0.3983
80	9.105	−0.3901
90	10.714	−0.3802
100	12.157	−0.3688
120	14.553	−0.3421
140	16.415	−0.3111
160	17.882	−0.2769
180	19.054	−0.2400
200	20.009	−0.2010
220	20.803	−0.1602
240	21.475	−0.1180
273.15	22.395	−0.0453
280	22.563	−0.0299
293.15	22.870	0
300	23.024	+0.0157
320	23.466	+0.0622

[a] From data of KROEGER and SWENSON [KRO77a].

There is considerable uncertainty in the polycrystalline expansion coefficient below $\theta_D/30$ (where $\theta_D = 427$ K) presumably on account of the texturization just referred to. According to BARRON *et al.* [BAR80, p. 682], $10^{10}\alpha = (15 \text{ to } 23)T + (0.03 \text{ to } 0.12)T^3$ (K^{-1}), while according to ILIYEV *et al.* [ILI80], for temperatures below 15 K, $10^{10}\alpha = (21 \pm 3)T + (0.15 \pm 0.03)T^3$ (K^{-1}). The thermal expansion coefficient at higher temperatures is given in Table 9-3.

Table 9-3. Linear Thermal Expansion Coefficient of Titanium above 20 K[a]

T, K	Linear Thermal Expansion Coefficient, α $(d \ln l / dT)$, $10^{-6}\,K^{-1}$
20	0.167
22	0.220
24	0.283
26	0.363
28	0.465
30	0.595
35	0.963
40	1.33
45	1.70
50	2.07
60	2.83
70	3.54
80	4.15
90	4.72
100	5.20
150	6.92
200	7.75
250	8.29
300	8.61

[a] Data of ILIYEV *et al.* [ILI80]—optical lever methods.

9.6.4 Thermal Expansions of Some Selected Technical Alloys

The thermal expansions of two stainless steel alloys, AISI 304L and AISI 310, and several α-phase and $\alpha+\beta$-phase Ti-base alloys, useful in cryogenic applications, are given in Table 9-4. A typical expansion curve, that for Ti–6Al–4V, is given in Figure 9-11.

9.6.5 Estimation of Thermal Expansion (Contraction) Curves

(a) Thermal Contractions of Metals and Alloys Based on Their Room-Temperature Thermal Expansion Coefficients. Assume that the length, l, of a sample varies with temperature, T, according to

$$l = a_0 + a_1 T + a_2 T^2 + \cdots \tag{9-45}$$

Table 9-4. Thermal Expansion of Some Stainless Steels and Technical Titanium-Base Cryogenic Alloys

	Relative Thermal Expansion, $\Delta l/l_0$, %				
	Austenitic Stainless Steels		Technical Ti-Base Alloys		
T, K	AISI 304L[a]	AISI 310[b]	Ti-5Al-2.5Sn[c]	Ti-6Al-4V[d]	Ti-8Al-1Mo-1V[e]
20	−0.306	−0.288	−0.175	−0.173	−0.162
30			−0.174	−0.172	−0.161
40	−0.303	−0.286	−0.172	−0.171	−0.160
50			−0.170	−0.169	−0.158
60	−0.294	−0.281	−0.168	−0.167	−0.156
70			−0.165	−0.165	−0.153
80	−0.281	−0.271	−0.162	−0.162	−0.150
90			−0.158	−0.158	−0.147
100	−0.265	−0.256	−0.153	−0.154	−0.143
120	−0.245	−0.235	−0.143	−0.141	−0.133
140	−0.222	−0.212	−0.130	−0.126	−0.122
160	−0.195	−0.188	−0.116	−0.110	−0.110
180	−0.168	−0.163	−0.100	−0.094	−0.096
200	−0.140	−0.135	−0.084	−0.078	−0.081
220	−0.113	−0.106	−0.067	−0.062	−0.065
140	−0.083	−0.077	−0.050	−0.045	−0.049
260	−0.052	−0.048	−0.031	−0.027	−0.031
280	−0.021	−0.019	−0.012	−0.009	−0.012
293	0.000	0.000	0.000	0.000	0.000

[a] Data of Arp *et al.* (1962) [Tou75, p. 1150]—annealed sample.
[b] Data of Arp *et al.* (1962) [Tou75, p. 1151]—annealed sample.
[c] [Sal79, p. 166]—annealed sample, α-phase alloy, normal interstitial content.
[d] [Sal79, p. 112]—annealed sample, $\alpha+\beta$-phase alloy, extra low interstitial content.
[e] [Sal79, p. 203]—annealed sample, $\alpha+\beta$-phase alloy.

where a_0 is the length at zero K. Then the contraction from length, l_0, on cooling to temperatures, T, below some reference temperature, T_0, (usually 293 K) is

$$\Delta l \equiv l_0 - l = (T_0 - T)[a_1 + a_2(T_0 + T) + \cdots] \tag{9-46}$$

It follows that

$$\frac{\Delta l}{(dl/dT)_{T_0}} = (T_0 - T)\left[\frac{a_1 + a_2(T_0 + T) + \cdots}{a_1 + a_2(2T_0) + \cdots}\right] \tag{9-47a}$$

$$= \text{say, } N(T) \tag{9-47b}$$

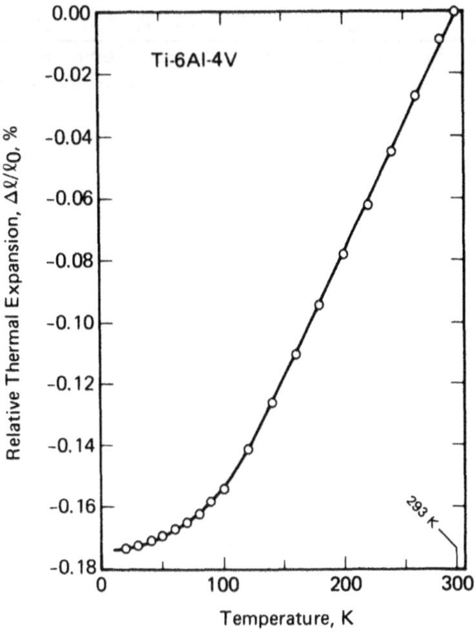

Figure 9-11. Thermal expansion, $\Delta l/l_0$, *versus* temperature for Ti-6Al-4V (annealed; extra-low interstitial content)—after SALMON [SAL79, p. 112].

Two conclusions can be drawn from this:

(1) The ratio, $N(T)$, of the relative contraction, $\Delta l/l_0$, to the expansion coefficient at the reference temperature, $(1/l_0) \times (dl/dT)_{T_0}$, is approximately $(T_0 - T)$, especially at intermediate temperatures, although departures from this value occur when $T \ll T_0$.

(2) If the expansion curves for all metals are the same "shape" (hence all similar to Figure 9-11, for example, as indeed they are), such that they can be generated from (9-45) by the application of a metal-dependent but otherwise constant scale factor to each of the a_i coefficients, the ratio $N(T)$, although temperature dependent, is *metal-independent.*

The extent to which these statements are true can be judged from an inspection of Table 9-5. According to this table, $N(200\text{ K})$ and $N(20\text{ K})$ for some 15 pure metals are $89 \pm 3\%$ and $193 \pm 8\%$, respectively. According to CLARK, who was the first to draw attention to (9-47) as an empirical rule [CLA70], $N(4\text{ K}) = 185.5 \pm 8\%$ for 85 different metals and alloys [CLA81]. A tabulation of $N(T =$ "near absolute zero") for 26 Al-base, Ni-base, Cu-base, and Fe-base technical alloys, also by CLARK [CLA68], yielded a value of $181 \pm 6(3\%)$. Table 9-1, for NBS standard OFHC Cu (SRM 736) [HAH70] enables the rule to be applied at suitable intervals of temperature over the range 20–293 K in order to predict the low-temperature thermal

Table 9-5. $N(T)$, the Ratio of Thermal Expansions $(\Delta l/l_{293})|_{20}^{293}$ and $(\Delta l/l_{293})|_{300}^{293}$ to the Expansion Coefficient, α_{293}, for Several Metallic Elements—Data from [TOU75]

Element	Linear Expansion Coefficient, $10^6\alpha_{293}$, K^{-1}	Linear Expansion Ratio, $N(T)$: $N(200) \equiv \frac{(l_{293} - l_{200})}{(dl/dT)_{293}}$ K	$N(20) \equiv \frac{(l_{293} - l_{20})}{(dl/dT)_{293}}$ K
Al	23.1	87.8	181
Cu	16.5	89.7	196
Ag	18.9	90.5	216
Au	14.2	91.6	229
Cd	30.8	91.9	239
Zn	30.2	90.4	225
Pb	28.9	90.3	242
Ti	8.6	86.0	178
V	8.4	85.7	185
Cr	4.9	104.1	200
Fe	11.8	86.4	172
Co	13.0	86.2	182
Ni	13.4	85.1	172
Zr	5.7	89.5	202
Nb	7.3	87.7	201
Mo	4.8	93.8	192
W	4.5	88.9	191
Mean and standard deviation		89 ± 2.5(3%)[a]	193 ± 16(8%)[b]

[a] Excluding Cr, which has a Néel point at about 310 K.
[b] Excluding Cd, Zn, and Pb.

contractions of metals and alloys on the basis solely of their room-temperature thermal expansion coefficients.

(b) Empirical Relationship Between the Melting Point and the Room-Temperature Thermal Expansion Coefficient. According to (9-1), the expansion coefficient is approximately proportional to the molar heat capacity. According to the Dulong–Petit law, the room-temperature molar heat capacities of all the elements are about the same and equal to 26–27 J per gram atom (mole). Room-temperature departures from the Dulong–Petit law are gauged by the relative magnitude of θ_D. If the Debye temperature is low (e.g., 158 K, as in Na), room temperature is "high" and the heat capacity should conform to the Dulong–Petit prescription. On the other hand, should θ_D be high (e.g., ~450 K, as in the transition elements), room temperature is considerably lower than θ_D and the heat capacity would fall well below its "high-temperature-limiting" value. There should therefore be a correlation between θ_D and the room-temperature molar heat capacity, hence the room-temperature thermal expansion coefficient, α_{RT}. Furthermore, the existence of the Lindemann melting formula, $\theta_D = \text{const}\cdot(T_M/AV^{2/3})^{1/2}$, where A and V are the atomic weight and volume, respectively, suggests that there should also be a correlation of sorts between the expansion coefficient, α_{RT}, and the absolute melting point, T_M. Indeed, VAN UITERT *et al.* have noted the existences of correlations between the α_{RT}s and the centigrade melting points of a wide range of crystalline substances [UIT77]. Based on data listed in their metallic tabulation, it appears that for twenty-five metals (mostly close-packed, fcc and hcp, and with $10^6\alpha_{RT}$s ranging from 6 to 26) $10^3\alpha_{RT}T_M = 19.6 \pm 14\%$.

It follows from this and the preceding subsection that knowledge of a metal's melting point together with the $N(T)$ function for a reference material, say Cu, is sufficient to provide a rough estimate of α_{RT}, and hence of the entire contraction curve below room temperature.

9.7 Thermal Expansion of Superconductors

9.7.1 Thermal Expansion Through the Superconducting Transition

As shown in Figure 8-5 and discussed in Section 8.4, as the temperature of a superconductor is decreased through T_c, the electronic specific heat undergoes a jump of height, $\Delta C = 1.43\ T_c$ (in BCS theory, (8-14)), as a result of the increase in electronic entropy. It follows from an uncritical acceptance of equations such as (9-11) that the thermal expansion coefficient should mimic this behavior, and in some cases it does [SMI76[a]], [SIM78] [POT81]. The conjoint plots of C/T *versus* T^2 and α/T *versus* T^2 for La_3S_4 provided in Figure 9-12 are an excellent example of this. It will be shown below, however, that although the jump in the specific heat is always "positive-going," that in β_v or α may be either positive or negative depending on the sign of dT_c/dP.

Experimental evidence suggests that the change in the expansion coefficient is entirely "electronic" in origin: (i) Elastic moduli, when measured, change by only a few parts per million during passage through T_c [BAR80, p. 712]; (ii) as discussed in Section 8.3.2, the lattice specific heat of Nb seems to be the same in both the normal and superconducting states.

Since the linear expansion coefficient is the derivative of the relative expansion, $\Delta l/l_0$, it responds much more strongly than the latter to the change of temperature through T_c (cf. Figures 9-12 and 9-13). Nevertheless, results are frequently presented in the latter format which provides, in a more direct way, information concerning the change in volume which accompanies the s/n transition at constant temperature, $T < T_c$.

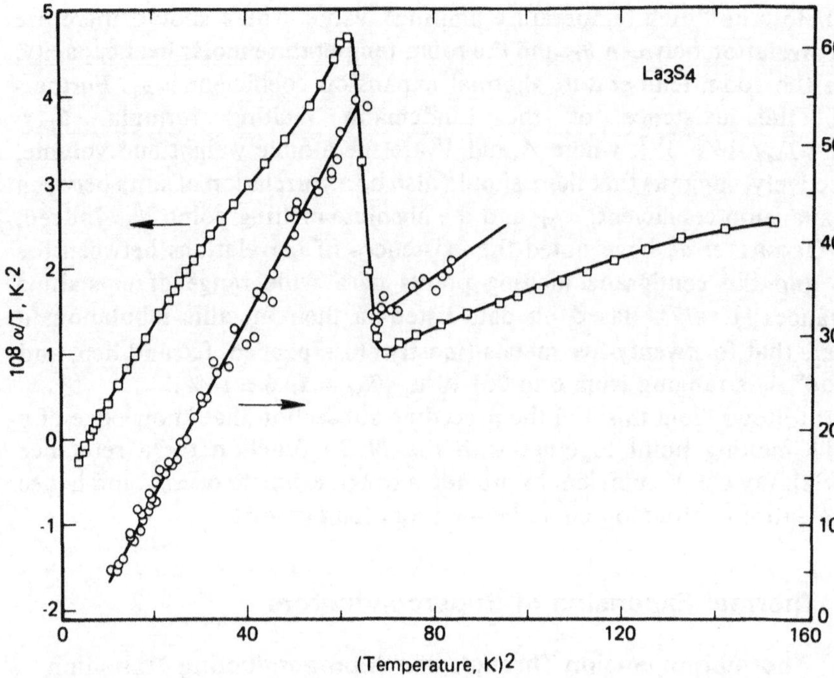

Figure 9-12. α/T *versus* T^2 for La_3S_4 (□, after POTT *et al.* [POT81]) juxtaposed against its C/T *versus* T^2 curve (○, after WESTERHOLT *et al.* [WES79]).

It is shown below that during the normal to superconducting transition the changes in volume at constant arbitrary temperature, and in the expansion coefficients at $T = T_c$, are thermodynamically related to the pressure dependences of the superconducting transition temperature and the thermodynamic critical field, H_c (see [AND62[a]], [SWE62], [WHI62[b]]). Thus, just as in normal solids, for which the expansion coefficient is used to calculate the fundamental thermodynamic quantity γ_G, the results of thermal expansion measurements when applied to superconductors are used

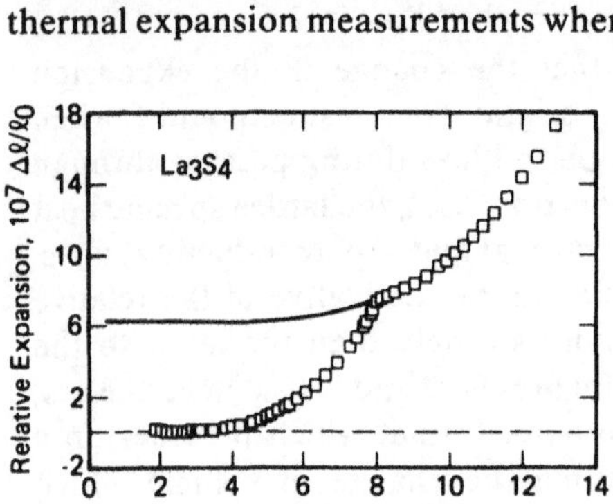

Figure 9-13. Relative length change $\Delta l/l_0$ of La_3S_4 as a function of temperature—after POTT *et al.* [POT81].

to provide information concerning the thermodynamic quantities $\partial T_c/\partial P$ and $\partial H_c/\partial P$ mentioned above [BAR80, p. 710] as well as the electron coupling mechanisms, especially in the case of strong-coupled superconductors [POT81].

9.7.2 Phenomenological Thermodynamic Relationships

The superconducting transition is usually dealt with in terms of the Gibbs free energy, G. Ignoring changes of volume at the transition, and following standard treatments of the subject (e.g., [LYN64, pp. 14ff]), (8-6) in the form

$$G_n(0) - G_s(0) = VH_c^2/8\pi \tag{8-6}$$

after two differentiations with respect to T leads to

$$C_s - C_n = (VT/4\pi)(dH_c/dT)^2|_{T=T_c, H_c=0} \tag{9-48}$$

Introduction of the effect of volume change begins with

$$G_s(H_a) - G_s(0) = (V_s/8\pi)H_a^2 \tag{9-49}$$

Combining the P derivative of this with the P derivative of (8-6) leads, via the relationship $V = (\partial G/\partial P)_{T,H}$, to

$$V_n(H_c) - V_s(H_c) = (V_sH_c/4\pi)(\partial H_c/\partial P)_T \tag{9-50}$$

for the isothermal volume change at the transition caused by raising the applied field to H_c, the thermodynamic critical field. Obviously this only applies to type-I superconductors (which undergo s/n transitions at $H_a = H_c$) such as, for example, the pure simple metals In, Pb, and Hg, and the pure transition elements La, V, and Ta. Nb and Tc are elemental type-II superconductors. Equation (9-50), or its equivalent in terms of length change $\Delta l = l_n - l_s$ [BAR80, p. 708], shows that as the temperature is lowered below T_c the expansion curve should split into a pair of branches, that for the normal state (either numerically extrapolated or measured at $H_a > H_c$) being uppermost if $(\partial H_c/\partial P)_T$ is positive. Examples of such $\Delta l/l$ splitting are to be found in the work of ANDRES on La [AND64] and Ta [AND63[a], AND64[a]] and WHITE [WHI62] on V, Nb, and Ta (Figure 9-14). Analysis of the results of such experiments tends to lead to a discussion of the pressure dependence of H_c.

As a next step in the thermodynamic analysis, (9-50) is differentiated by T and P, yielding for the difference in the volume expansion coefficients

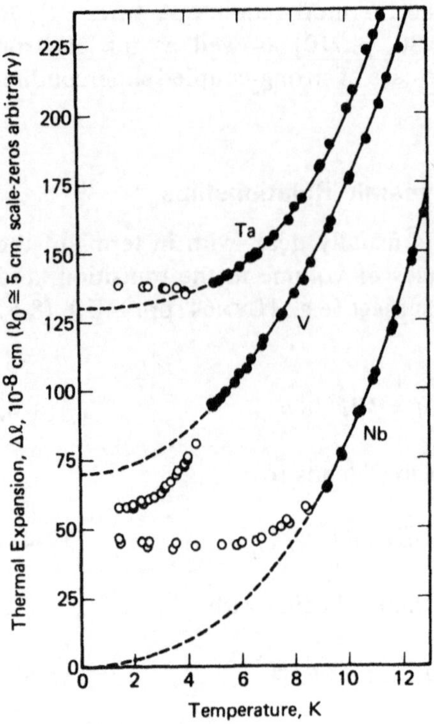

Figure 9-14. Linear thermal expansions of V, Nb, and Ta in the normal (●) and zero-field superconducting (○) states. The dashed lines have been fitted to the normal-state $T_c < T < \theta_D/20$ data—after WHITE [WHI62].

[LYN64, p. 16]

$$\beta_{v,n} - \beta_{v,s} = \frac{1}{4\pi}\left(\frac{\partial H_c}{\partial T}\right)_P \left(\frac{\partial H_c}{\partial P}\right)_T \Bigg|_{H_c=0} \tag{9-51}$$

Replacing $(\partial H/\partial P)_T$ by $-(\partial H/\partial T)_P(\partial T/\partial P)_H$, and introducing (9-48) yields (now for $T = T_c$, $H_c = 0$)

$$\beta_{v,n} - \beta_{v,s} = \frac{C_n - C_s}{V} \cdot \frac{\partial \ln T}{\partial P}\Bigg|_{T=T_c,\ H_c=0} \tag{9-52}$$

which is an Ehrenfest equation for the superconducting phase transition. HAKE [HAK69] has derived a similar equation for type-II superconductors, valid for the transition surface where $T = T_c$, and $H_a = H_{c2}$. For further discussion of the thermodynamics of type-I and type-II superconductors HAKE's work [HAK69] should be consulted.

9.7.3 Thermal Expansion At and Below T_c

(a) Modified Weak-Coupling Approach. SIMPSON and SMITH [SIM78] have calculated the temperature dependence of the electronic thermal expansion coefficient using BCS theory, and fitted it to the experimental results for Nb (a type-II superconductor of intermediate strength, see Figure 8-8), Nb–Zr, and Mo–Re with the assistance of a variably magnified BCS deviation function, σD_{BCS}. In this way an excellent fit to the α_e data could be obtained over a wide temperature range below T_c (i.e., down to about $t = 0.5$). Using standard BCS theory, the temperature dependence of the electronic volume expansion coefficient in the superconducting state ($\beta_{es} \equiv$ the β_s of the thermodynamic treatment) was found to be

$$\beta_{es} = \beta_{en}\frac{S_{es}}{\gamma T} + \frac{\gamma T}{V}\cdot\frac{1}{T_c}\cdot\frac{dT_c}{dP}\cdot\frac{C_{es} - S_{es}}{\gamma T} \tag{9-53}$$

In (9-53) γT is the electronic molar heat capacity ($=C_{en}$); and S_{es}, the electronic molar entropy, is a function of $N(E_F)$, T, and the BCS deviation function. The results of the analysis, as applied to Zr–Nb(71.3 at.%) and pure Nb, are shown and contrasted in Figures 9-15(a) and (b). As implied in the figures, good fits to the temperature dependences of β_s can be obtained using as experimental input data the normal-state electronic specific heat and the expansion coefficients, dT_c/dP, and a value of σ deduced from the height of the specific heat discontinuity at T_c.

At $T = T_c$, (9-53) reduces to

$$\Delta\beta_e|_{T_c} = \frac{\gamma}{V}\left(\frac{\Delta C}{\gamma T_c}\right)\left(\frac{dT_c}{dP}\right) \tag{9-54}$$

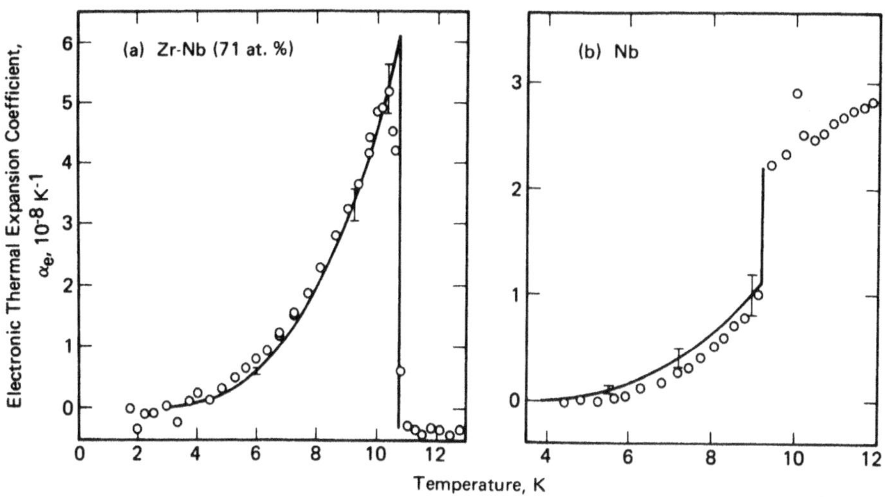

Figure 9-15. Electronic thermal expansion coefficients for (a) Zr–Nb(71.3 at.%) and (b) unalloyed Nb, demonstrating the possibility of both "positive" and "negative" discontinuities at T_c. The solid curves (with uncertainty bars) are the results of calculation (see Section 9.7.3(a)); in (a) $\sigma = 1.48$ and in (b), $\sigma = 1.27$—after SIMPSON and SMITH [SIM78].

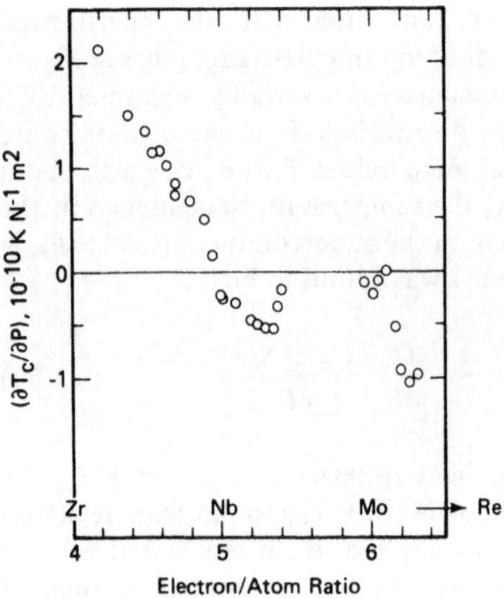

Figure 9-16. Pressure dependence of T_c for twenty-nine pure metals and binary alloys forming the sequence Zr, Zr-Nb, Nb, Nb-Mo, Mo, Mo-Re—after SMITH *et al.* [SMI75, SMI76[a]].

which is again the Ehrenfest relation, (9-52). Expressed in this way, the expansion-coefficient discontinuity is seen to be proportional to the product of the relative specific heat jump (see Section 8.6) *and* dT_c/dP. A plot of directly measured dT_c/dP for the alloy series Zr-Nb-Mo-Re [SMI76[a]] is given in Figure 9-16. The results presented correlate with the observations that $\Delta\beta_e|_{T_c}$ is "positive-going" (i.e., in the same direction as ΔC) for Zr-Nb and "negative-going" for Nb (Figure 9-15), while for Nb-Mo(50 at.%) α was found to be a smooth function of T with *no discontinuity at all* at T_c [SMI76[a]].

(b) Strong-Coupling Approach. According to SMITH *et al.* [SMI76[a], SMI77] and the references therein, numerous studies have been undertaken of the pressure dependence of T_c. In recent times, using McMillan theory as a starting point, several attempts have been made to calculate dT_c/dP, and to understand the physics that controls its magnitude and sign [GAR72], [POT81], [SMI72, SMI76[a], SMI77]. GARLAND and BENNEMANN [GAR72] and subsequently SMITH *et al.* (e.g., [SMI76]) started with a slightly modified McMillan equation for the superconducting transition temperature [MCM68] (see (8-16) and (8-17)), *viz.*,

$$\left.\begin{aligned} T_c &= \frac{\langle\omega^2\rangle^{1/2}}{1.2}\exp(-1/g) \\ \text{where}\qquad g &= \frac{0.96\lambda - \mu^*(1+0.6\lambda)}{1+\lambda} \end{aligned}\right\} \qquad (9\text{-}55\text{a})$$

with

$$\lambda = \frac{C_M}{M\langle\omega^2\rangle} \equiv \frac{n(E_F)\langle I^2\rangle}{M\langle\omega^2\rangle} \qquad [8\text{-}19]\,(9\text{-}55b)$$

the quantities there having been defined in Section 8.5.1.

The calculation of $\partial T_c/\partial P$ then proceeded by way of several intermediate variables, the first of which was $(\partial \ln T_c/\partial \ln V)_P \equiv (K_T/T_c)(\partial T_c/\partial P)_V$. Next from (9-55a), $\partial \ln T_c/\partial \ln V$ varies as $\partial \ln g/\partial \ln V$ which in turn varies as $\partial \ln \lambda/\partial \ln V$. This, according to (9-55b), is given by

$$\left.\begin{aligned} \frac{\partial \ln \lambda}{\partial \ln V} &= \frac{\partial \ln C_M}{\partial \ln V} - \frac{\partial \ln\langle\omega^2\rangle}{\partial \ln V} \\ \text{with}\quad \frac{\partial \ln C_M}{\partial \ln V} &= \left[\frac{\partial \ln n(E_F)}{\partial \ln V} + \frac{\partial \ln\langle I^2\rangle}{\partial \ln V}\right] \end{aligned}\right\} \qquad (9\text{-}56)$$

Thus, as pointed out by SMITH and SHELTON [SMI75], the sign of $\partial T_c/\partial P$ depends upon the balance between $\partial \ln C_M/\partial \ln V$ and $\partial \ln\langle\omega^2\rangle/\partial \ln V \cong 2\gamma_G^{300\text{ K}}$.

Further development of the analysis along these lines then devolved to a discussion of the volume dependences of $n(E_F)$ and $\langle I^2\rangle$ and, at the hands of SMITH *et al.* [SMI72[a], SMI75, SMI76[a]] and others [POT81], became a test of the strong-coupling theory of superconductivity and the degree to which its components could be quantified using experimental data from various sources.

(c) Prognosis for Ti–Nb. Although an abundance of data is available for the Zr–Nb system, no thermal expansion or dT_c/dP measurements seem to have been made on superconducting Ti–Nb. However, assuming that its behavior parallels that of the former system, it is predicted that the dT_c/dP for the technically important alloy Ti–Nb(34 at.%) would be about +15 mK per 10^8 N m^{-2}. This value, within the context of Figure 9-16, being a relatively large positive quantity, determines that the α-discontinuity at T_c should be similar to that of Zr–Nb as depicted in Figure 9-15(a) and mimic that of the low-temperature specific heat. From a practical standpoint, recognizing that the tensile strength of Ti–50Nb at liquid-He temperatures is about 15×10^8 N m^{-2} (Figure 3-24), the pressure dependence of T_c is insignificant.

9.7.4 Normal-State Thermal Expansions of Ti–Nb and Ti–Zr–Nb

Although no high-sensitivity low-temperature measurements have been made of the thermal expansion (or its coefficient) of technically important Ti–Nb-base superconductors, some low-sensitivity expansion data for the

Table 9-6. Linear Thermal Expansions of Some Research and Technical Ti–Nb Alloys

	Relative Thermal Expansion, $\Delta l/l_0$, %					
	Atomic Percent Nb[a]				Weight Percent, Nb[b]	
T, K	5	10	15	20	52	55
300	0.019	0.018	0.017	0.014	0.025	0.026
280	0.005	0.005	0.005	0.004	0.006	0.007
273	0.000	0.000	0.000	0.000	0.000	0.000
260	−0.009	−0.009	−0.008	−0.007	−0.012	−0.011
240	−0.023	−0.021	−0.020	−0.018	−0.029	−0.026
220	−0.036	−0.034	−0.032	−0.028	−0.046	−0.044
200	−0.050	−0.047	−0.044	−0.039	−0.063	−0.059
180	−0.066	−0.062	−0.057	−0.050	−0.080	−0.074
160	−0.080	−0.076	−0.068	−0.062	−0.095	−0.090
140	−0.095	−0.090	−0.080	−0.073	−0.110	−0.105
120	−0.109	−0.101	−0.093	−0.084	−0.125	−0.120
100	−0.122	−0.113	−0.105	−0.093	−0.140	−0.137
90	−0.129	−0.120	−0.113	−0.098	−0.146	−0.143
80	−0.133	−0.121	−0.116	−0.102	−0.151	−0.148
78	−0.133	—	−0.116	—	—	—
77	—	−0.123	—	−0.103	−0.153	−0.151
60	−0.142	−0.131	−0.124	−0.110	−0.162	−0.159
50	−0.146	−0.134	−0.128	−0.113	−0.163	−0.163
40	−0.149	−0.137	−0.131	−0.115	−0.168	−0.165
30	−0.152	−0.140	−0.133	−0.117	−0.170	−0.167
20	−0.154	−0.141	−0.134	−0.118	−0.171	−0.168
10	−0.155	−0.142	−0.134	−0.119	−0.172	−0.169
5	−0.155	−0.142	−0.135	−0.120	—	—
4	—	—	—	—	−0.174	−0.169

[a] After JELINEK [JEL74].
[b] After JELINEK [JEL75].

entire liquid-He to room-temperature range are available thanks to the work of JELINEK [JEL74, JEL75] and GOLIK *et al.* [GOL69]. Presented in Table 9-6 are data for the technical alloys Ti-52Nb and Ti-55Nb and (for comparison only) for research alloys with 5, 10, 15, and 20 at.% Nb. In Table 9-7 are given the results of GOLIK *et al.* for Ti–25% Zr–50% Nb (probably atomic percent, in which case the composition corresponds to that of Soviet alloy type SS-2 [LAZ72]).

Table 9-7. Linear Thermal Expansion of Ti-Zr-Nb[a]—After Golik *et al.* [GOL69]

T, K	Relative Thermal Expansion, $\Delta l/l_0$, %
300	0.0189
275	0.0000
250	−0.0185
200	−0.0538
150	−0.0867
100	−0.1160
50	−0.1366
25	−0.1402
0	−0.1405

[a] Specified Ti-25%Zr-50%Nb (probably atomic percent, hence corresponding to Soviet Type SS-2).

9.8 Thermal Expansion of Metallic and Nonmetallic Composites

9.8.1 Parallel Strips

The simplest example of a metallic composite is the bimetallic strip. Consisting of a pair of strips differing in thermal expansion coefficient (α_i), Young's modulus (E_i), and thickness (t_i), and bonded side by side, when heated through a temperature interval ΔT it bends into the arc of a circle of radius R given by [TOU77, p. 15a]

$$|R| = \frac{3(t_1 + t_2)^2 t_1 t_2 E_1 E_2 + (t_1 E_1 + t_2 E_2)(t_1^3 E_1 + t_2^3 E_2)}{6(\alpha_1 - \alpha_2)\Delta T(t_1 + t_2) t_1 t_2 E_1 E_2} \tag{9-57}$$

A symmetrical laminate (or sandwich) does not of course bend; instead its length changes to an extent prescribed by an average expansion coefficient, α_c, given by [TOU77, p. 15a]

$$\alpha_c = \sum_i \frac{\alpha_i E_i t_i}{1 - \nu_i} \Big/ \sum_i \frac{E_i t_i}{1 - \nu_i} \tag{9-58}$$

where ν_i, the Poisson's ratio, is given by $K_i = E_i/3(1 - 2\nu_i)$, (4-1). If the expansion over a given temperature interval is required, it can, according to Section 9.6.5, be obtained simply by substituting $\Delta l_i/l_0$ for α_i in (9-58); and of course the same rule will apply to all the expressions in α presented below.

9.8.2 Isotropic Solid-State Dispersion

The thermal expansion of a system of noninteracting components is expected to obey the mixture rule

$$\alpha_c = \sum_i \alpha_i f_i \tag{9-59}$$

where f_i is the volume fraction of the ith component. However, when the components are mutually bonded, since those with the higher expansivities will tend to stretch those with the lower to which they are mechanically coupled, and conversely, the net dilatation is controlled by a combination of α_i, f_i, and the elastic properties. In reviewing the thermal expansions of binary composites, TUMMALA and FRIEDBERG [TUM70] referred to the results of four previous theoretical analyses (those due to TURNER, KERNER, BLACKBURN, and THOMAS (see [TUM70]). They also derived a formula, based on a model system consisting of elastic spheres dispersed in an infinite elastic continuum, and considered its applications to the systems: ceramic/glass, metal/metal, and metal/organic. At about the same time, FAHMY and RAGAI [FAH70] produced a somewhat similar formula to describe the expansion of dispersions such as fused-silica particles in a Pb matrix and a Si-precipitated Al alloy. The formulae that issued from these analyses are as follows. The subscripts d, m, and c within the formulae signify "disperse phase," "matrix," and "composite," respectively:

$$\alpha_c = \frac{\alpha_d f_d K_d + \alpha_m f_m K_m}{f_d K_d + f_m K_m} \qquad \text{(Turner)}^{13} \tag{9-60}$$

$$\alpha_c = \alpha_m f_m + \alpha_d f_d + 4\frac{G_m}{K_c}\left[\frac{K_c - K_d}{3K_d + 4G_m}(\alpha_m - \alpha_d) f_d\right]$$

with

$$K_c = \frac{K_m f_m/(3K_m + 4G_m) + K_d f_d/(3K_d + 4G_m)}{f_m/(3K_m + 4G_m) + f_d/(3K_d + 4G_m)} \qquad \text{(Kerner)}^{14} \tag{9-61}$$

[13] Turner's original analysis was quite broad in its applications; see P. S. Turner, *J. Res. NBS*, **37**, 239–50 (1946).

[14] Note that, according to (4-17), the quantity $K + (4/3)G$ is a longitudinal modulus similar to E but without the Poisson dilatation, and the third term acts as a correction to the rule-of-mixtures formulation, i.e., to (9-59). Equation (9-61) is identical to the following expression from TOULOUKIAN *et al.* [TOU75, p. 15a]:

$$\alpha_c = \alpha_m + \frac{(\alpha_d - \alpha_m) f_d}{\dfrac{12K_m G_m}{3K_m + 4G_m}\left[\dfrac{f_d}{3K_m} + \dfrac{1}{4G_m} + \dfrac{f_m}{3K_d}\right]}, \tag{9-61a}$$

but not to that listed by TUMMALA and FRIEDBERG [TUM70] and ascribed to KERNER [KER69].

$$\alpha_c = \frac{[\alpha_d + f_m(\alpha_m - \alpha_d)]\frac{3}{2}(1-\nu_c)}{\frac{1}{2}(1-\nu_d) + f_m(1-2\nu_d) + f_d(1-2\nu_m)} \frac{E_d}{E_m} \quad \text{(Blackburn)} \tag{9-62}$$

$$\alpha_c = \alpha_d^{f_d} \alpha_m^{f_m} \quad \text{(Thomas)} \tag{9-63}$$

$$\alpha_c = \alpha_m - \frac{f_d(\alpha_m - \alpha_d)(1+\nu_m)}{(1-2\nu_d)(2E_m/E_d) + (1+\nu_m)} \quad \text{(Tummala)} \tag{9-64}$$

$$\alpha_c = \alpha_m - \frac{3f_d(\alpha_m - \alpha_d)(1-\nu_m)}{(1-2\nu_d)(2E_m/E_d)f_m + (1-2\nu_m)2f_d + (1+\nu_m)} \quad \text{(Fahmy)} \tag{9-65}$$

9.8.3 Granular Compacts

KERNER's result for a densely packed aggregate of i types of granular particles is [KER69]

$$\alpha_c = \left(\frac{4G_c}{K_c} + 3\right) \sum_i \frac{\alpha_i f_i}{(4G_c/K_i) + 3} \tag{9-66}$$

It follows that an untextured polycrystal of anisotropic crystallites, each with principal expansion coefficients α_1, α_2, and α_3, regarded as a three-component compaction, has a coefficient that differs from the simple arithmetic mean of the α_i.

9.8.4 Fiber-Reinforced Composites

Fiber-reinforced composites generally exhibit orthorhombic symmetry (e.g., a fabric-reinforced laminate, see also Section 4.7) but are frequently uniaxial (e.g., a length of Cu/Ti–Nb multifilamentary superconductor). In the latter case, the expansion can be described in terms of only two coefficients—the longitudinal and the transverse, say α_{cl} and α_{ct}, respectively. These components of thermal expansion can of course be exactly calculated by "brute force" using the techniques of finite-element computation [FAH74]; otherwise, useful approximate values can be obtained with the aid of simple algebraic formulae.

(a) The Longitudinal Thermal Expansion. If the Poisson's ratio of the components are identical, which is certainly very close to the truth in a metal–metal composite for which, in general, $\nu_m \cong \nu_f \cong 0.3$ (where sub-f

refers to the fiber), TURNER's formula, (9-60), reduces to

$$\alpha_{cl} = \frac{\alpha_f f_f E_f + \alpha_m f_m E_m}{f_f E_f + f_m E_m} \tag{9-67}$$

As pointed out by FAHMY and RAGAI [FAH70], the expansion coefficients given by (9-67) may fall either above or below the mixture-rule line. However, in practice, they are generally found below the line. The reason for this is embodied in the fundamental expansion equation, (9-10), which shows that the component with the lower expansion coefficient will have

Table 9-8. Longitudinal Thermal Expansions of Some Superconducting Coil Composites[a]

	Longitudinal Relative Thermal Expansion, $\Delta l/l_0$, %			
T, K	(a) Cu/Ti-Nb	(b) Cu/Ti-Nb	(c) Cu/Ti-Nb/Fiberglass	(d) Fiberglass/Epoxy
300	+0.040	—	$+0.0009_9$	$+0.0008_2$
293	—	0.000	0.0000	0.0000
280	+0.010	−0.016	-0.0017_0	-0.0015_2
273	0.000	−0.025	-0.0026_4	-0.0024_4
260	−0.018	−0.041	-0.0043_9	-0.0038_5
240	−0.043	−0.066	-0.0070_8	-0.0061_9
220	−0.067	−0.091	-0.0097_7	-0.0085_3
200	−0.093	−0.117	−0.0125	−0.0109
180	−0.103	−0.142	−0.0152	−0.0132
160	−0.138	−0.167	−0.0176	−0.0154
140	−0.157	−0.190	−0.0198	−0.0175
120	−0.176	−0.212	−0.0220	−0.0195
100	−0.192	−0.231	−0.0240	−0.0213
90	−0.199	—	−0.0248	−0.0221
80	−0.206	−0.245	−0.0256	−0.0229
70	−0.212	—	−0.0262	−0.0235
60	−0.218	−0.256	−0.0268	−0.0242
50	−0.220	—	−0.0272	−0.0248
40	−0.223	−0.262	−0.0275	−0.0252
30	−0.225	—	−0.0277	−0.0256
20	−0.226	−0.265	−0.0278	−0.0260
10	−0.227	−0.265	−0.0278	−0.0262
4	−0.231	−0.265	−0.0278	−0.0262

[a]Refer to text for materials details.

the higher elastic modulus and therefore dominate the expansion of the composite. It was this observation that prompted FAHMY and RAGAI [FAH70] to develop an improved version of the Turner equation. Other authors found the original equation to be satisfactory for their purposes [NAK74], [CLA81] (see also references included in the latter); CLARK *et al.*, for example, used it to predict the longitudinal low-temperature thermal expansions of fiberglass/epoxy and Cu/Ti-Nb composites [CLA81].

(b) Transverse Thermal Expansion. CLARK *et al.* [CLA81] found the mixture rule to be satisfactory for calculating the transverse expansion (or its coefficient) when the thermomechanical stress is acting perpendicular to the bond between the fibers and the matrix. In studies of boron/epoxy and graphite/epoxy composites, NAKAMURA and LARSEN [NAK74[a]] preferred to use the following modification of the mixture rule, which they attributed to KREIDER and PATARINI:

$$\alpha_{ct} = (1 + \nu_f)\alpha_f f_f + (1 + \nu_m)\alpha_m f_m - \alpha_{cl}\nu_c \qquad (9\text{-}68)$$

(c) Thermal Expansion Data for Some Superconducting Magnet Composites. Expansion data for a representative selection of metal-matrix and organic-matrix composites are presented in Table 9-8.

Reported there are the thermal expansions of:

(a) a multifilamentary Cu/Ti-Nb superconducter (MCA 233-1) with Cu/SC ratio = 4:1 [JEL75]
(b) a multifilamentary Cu/Ti-Nb superconducter with Cu/SC ratio = 1.8:1 containing 180 Ti-Nb filaments twisted to a pitch of 2 per inch [CLA81] (for mechanical properties, see Table 3-14)
(c) a "coil composite" consisting of layers of the Cu/Ti-Nb (1.8:1) wires just referred to imbedded in epoxy and reinforced with fiberglass cloth between the layers [FOW73] (the results of the measurements were then analyzed within the framework of a binary composite although each component—Cu/Ti-Nb and FG/epoxy—was itself a composite [CLA81])
(d) a fiberglass/epoxy composite intended to model the "matrix" of the above [FOW73], [CLA81]

III
The Superconducting Transition

Chapter 10

Calorimetric Studies of the Superconducting Transition and the Mixed State

10.1 The Calorimetrically Determined Transition Temperature

The preceding general discussion of low-temperature specific heat (Chapter 8) naturally leads into a consideration of its use in experimental investigations of the superconducting transition. Calorimetry, a particularly powerful tool in this regard, possesses numerous advantages: (i) as a bulk-property measurement technique it is superior to electrical resistivity for the study of multiphase or otherwise inhomogeneous samples; (ii) the specific heat result yields not only T_c but also two other important solid-state parameters, γ and θ_D; (iii) if the transition for one reason or another is sharp but incomplete, the relative magnitude of the specific heat jump at T_c, *viz.*, $\Delta C/\gamma T_c$ (see Sections 8.4 and 8.6), in comparison to some expected value (such as the BCS value, 1.43, or perhaps some more appropriate experimentally calibrated ratio) yields a measure of the fraction of the specimen that participates in the observed transition; (iv) if on the other hand the transition is broad, a knowledge of the form of the unrounded specific heat temperature dependence below T_c enables one to derive a transition temperature distribution function related to the metallurgical condition of the sample.

To be considered in this chapter are the responses of the calorimetrically determined superconducting transition to variations of composition, composition-related microstructure, aging, and deformation. In so doing this chapter sets the stage for the consideration of other experimental approaches to the problem and for further discussions of T_c from theoretical, phenomenological, experimental, and materials standpoints.

Most of the equations used in this chapter are dimensionless or independent of units; but in equations such as (10-7) and (10-10), where a "per unit volume" basis is employed, the symbol v is used as a superscript for C and a subscript for γ, as explained in Section 8.1.1.

10.2 Calorimetric Studies of T_c as a Function of Composition-Related Microstructure—General Descriptions

The classic noncalorimetric work relating T_c to composition and structure in transition metal binary alloys was undertaken by HULM and BLAUGHER [HUL61] in the early 1960s. Their work was the forerunner of numerous investigations to be made into the superconducting transition temperatures of Ti–TM alloys within both low and high concentration ranges. Calorimetric studies of superconductivity have, of course, considerable scientific value, but disregarding any microscopic interpretation (i.e., if used diagnostically) the calorimetrically observed superconducting transition can provide a wealth of information relating to the microstructural states of quenched alloys, solute diffusion rates, and the effects of aging and deformation.

The calorimetrically measured T_cs of the three series of quenched β-isomorphous alloys (see Figure 1-2), *viz.*, Ti–V [COL75], Ti–Nb [SUK68],

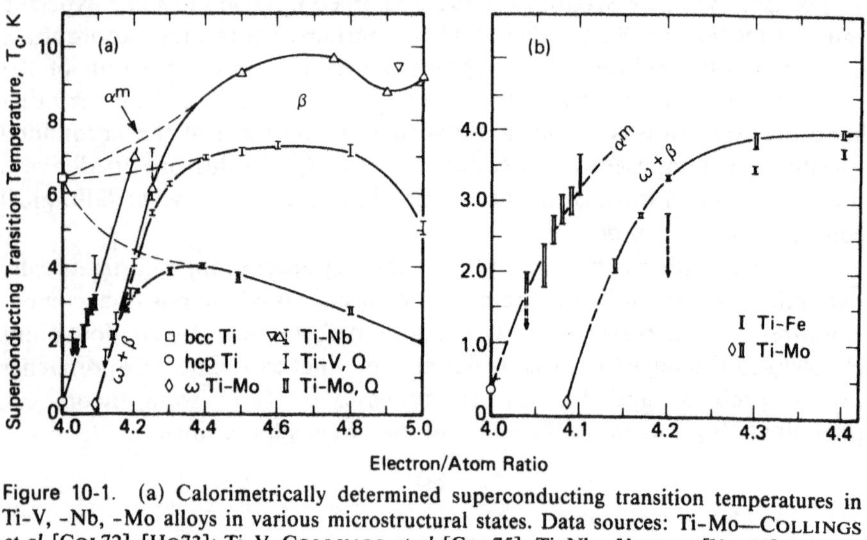

Figure 10-1. (a) Calorimetrically determined superconducting transition temperatures in Ti–V, -Nb, -Mo alloys in various microstructural states. Data sources: Ti–Mo—COLLINGS *et al.* [COL72], [HO73]; Ti–V, COLLINGS *et al.* [COL75]; Ti–Nb—VERKIN [VER80] (△) (see also SUKHAREVSKII *et al.* [SUK71]), SHCHETKIN and KHARCHENKO [SCH75] (▽), and SUKHAREVSKII *et al.* [SUK68] (I). (b) Detail of Ti–Mo data from above figure together with some calorimetric results for Ti–Fe—after HO and COLLINGS [HO74].

[SHC75], [VER80], and Ti–Mo [COL72], [HO73], are displayed as functions of e/a ratio in Figure 10-1. Interpreting Figure 10-1(a) from right to left, as Ti is added to V, Mo, or Nb, the transition temperature rises to a broad plateau or maximum before dropping suddenly to very low values for $e/a \lesssim 4.3$. This rapid final decrease in T_c is due to the influences of the diffuse-, and finally the athermal ω-phases (Sections 1.6 and 1.8) on the bulk or average T_c. It is important to note that, in spite of the presence of ω-phase, the transitions remain sharp and bulk ($\Delta C/\gamma T_c \not< 1.4$) over the composition range within which athermal ω-phase is expected (e.g., Figure 8-5) indicating that both the precipitate and its matrix undergo the superconducting transition simultaneously in a manifestation of a complete "proximity effect." At lower values of e/a, the transition temperatures of the quenched alloys, now α^m-martensitic in structure (i.e., α' or α'', see Section 1.7.2), reside on separate strongly composition-dependent branches lying more or less parallel to their respective $\omega+\beta$-phase segments.

10.2.1 T_c in the Martensitic Alloys

The rapid increase of T_c with composition exhibited by the heavily dislocated martensitic structures has been previously interpreted as a manifestation of a disorder-induced enhancement of T_c attributable to a localized-soft-phonon mechanism of the kind proposed earlier by GARLAND *et al.* [GAR68] and STRONGIN *et al.* [STR68] to explain T_c enhancement in structures such as amorphous and microcrystalline films or in granules [COL69, COL70[b], COL71[f], COL75[a]], [HO71]. However, the absence of any noticeable T_c enhancement in arc-melted (hence martensitically transformed during hearth cooling) as compared to annealed (hence polycrystalline hcp) samples of unalloyed Ti does militate against such an interpretation. Superconductivity in both quench- and deformation-induced martensitic structures is considered in more detail below.

10.2.2 T_c in the $\omega + \beta$-Phase Alloys

The rapid decrease of T_c with increasing Ti content in the $e/a \lesssim 4.3$ regime has been interpreted in terms of a low intrinsic T_c for the ω-phase, in association with the operation of a proximity effect between the ω-precipitate and the β-matrix [HO69[a], HO71[a]], [COL76[b]].

Superconductive proximity effects, the subject of Chapter 11, have generally been studied using thin-film sandwiches prepared by vacuum evaporation or sputtering, in which case the transition temperature must be measured resistively. But with certain alloy systems it is possible to obtain lamellar or globular precipitates, sufficiently fine compared to a coherence length, so that it is meaningful to consider the operation of a proximity effect between adjacent platelets or between a precipitate and its matrix. As will be demonstrated in the next chapter, the as-quenched and quenched-plus-aged ω-phases provide examples of such *in situ* proximity-effect arrangements.

10.2.3 T_c and Other Properties of ω-Ti–Mo

A calorimetric and magnetic investigation of aging in a Ti-Mo(10.3 at.%) alloy is to be discussed in Section 11.9. The results of the aging experiments, analyzed using a conventional continuity-equation approach, yielded values for the room-temperature magnetic susceptibility, χ_ω (see Section 7.14), and the calorimetrically determined properties γ_ω, $\theta_{D,\omega}$, and $T_{c,\omega}$. These are listed in Table 10-1 for comparison with corresponding data for pure α-Ti, Ti–Mo(19.4 at.%) (a representative β-Ti–Mo alloy), and unalloyed β-Ti, whose physical properties are analyzed below (Section 10.4.1). It is immediately seen that the properties of α-Ti and ω-Ti-Mo(4.3 at.%), both hexagonal crystals, are fairly similar; their density-of-states properties are very much lower than those of the β-phase alloy and, as gauged by θ_D, their crystal lattices are considerably stiffer. The data show how the presence of ω-phase in an $\omega+\beta$-phase alloy can reduce χ and γ by simple averaging, and severely suppress T_c if the precipitate size is favorable for the operation of a complete proximity effect.

It is instructive to compare the presently estimated values of T_c for ω-Ti–Mo(4.3 at.%) with those of the ω-phase allotropes of Zr and Ti. The superconducting transition temperature of ω-Zr has been measured directly

Table 10-1. Properties of ω-Ti–Mo(4.3 at.%)[a] Compared with Those of α-Ti, a Typical β-Phase Ti–Mo Alloy, and β-Ti

Property			Structure			
			α-Ti	ω	β	β-Ti
Mo Concentration	[Mo]	(at.%)	0.00	4.3	19.4	0.00
Magnetic Susceptibility At Room Temperature	χ	(10^{-6} cm^3 mol^{-1})	151.8	148.5	202.8	231.4
Electronic Specific Heat Coefficient	γ	(10^{-3} J mol^{-1} K^{-2})	3.36	3.00	7.1	11.6
Debye Temperature	θ_D	(K)	420	420	305	200
Superconducting Transition	T_c	(K)	0.38	0.15	3.9_8	6.4

[a]Together with Ti–Mo(21 at.%) this is the "metaequilibrium" product of aging Ti–Mo(10.3 at.%) for 880 h/350°C (Section 11.9.3).

on pressure-transformed material by TITTMANN *et al.* [TIT64]. Those of ω-Ti and ω-Zr have been measured by DEGTYAREVA [DEG74] in a study of both T_c and the magnetic susceptibility temperature dependences (see Section 7.13.2). After subjecting samples of high-purity Zr and Ti to quasihydrostatic pressures (in graphite) of 120 kbar for 30 min, it was noted that the transition temperatures of the resulting ω-phases were lower than those of the α-phase starting materials. It may be concluded that the effect of the $\alpha \rightarrow \omega$ transformation is to lower the T_c of Zr by about 0.06 K (either about 0.72 to 0.65 K [TIT64] or 1.1 to 0.95 K [DEG74]); and that of Ti from 0.45 to less than 0.06 K [DEG74].

10.3 Calorimetrically Determined Superconducting Transitions in Quenched Low-Concentration α^m-Phase Ti–TM Alloys

As discussed in Section 1.6.3, quenched low-concentration Ti–TM alloys characterized by $e/a \lesssim 4.15$ (see Table 1-3) assume a martensitic structure (α^m in general), designated either α' or α'' depending on e/a within this range (see Table 1-6). Interest in the superconductivity of such alloys stemmed from the experimental observations of MATTHIAS *et al.* [MAT59] who noted, with particular reference to comparisons between the T_c composition dependences of (i) Ti–Fe and Ti–Ru, and (ii) Ti–Co and Ti–Rh, that small amounts of the "magnetic" (i.e., late 3*d*) transition elements Fe and Co when added to Ti appeared to raise T_c with exceptional rapidity as compared with the nonmagnetic 4*d* transition elements. Subsequent experiments by HEINIGER, MÜLLER, and others (e.g., [HEI64]), who used more closely spaced compositional sequences, failed to confirm MATTHIAS' conclusion, which, however, had sparked considerable interest in the superconductive behavior of dilute Ti–TM alloys in general.

Second-phase precipitation, if it takes place in dilute quenched equilibrium-two-phase systems, usually does so at grain boundaries, thereby establishing an extensive multiply connected net of solute-rich and potentially higher T_c material. Thus, in measuring the superconducting transition temperature of such an alloy by the usual resistometric or magnetization techniques, that of the major constituent may be masked by the effects of short-circuit or shielding currents flowing along the precipitation paths. In order to distinguish between homogeneous and nonhomogeneous transitions, and to make quantitative evaluations of the latter, it is necessary to employ low-temperature specific heat. Numerous such studies have been made on quenched martensitic Ti–TM alloys, notably Ti–V and Ti–Nb [HEI64], Ti–Mn and Ti–Co [HAK64], Ti–Fe [HEI64], [BAT64], and Ti–Mo [COL69, COL70[b]]. BUCHER *et al.* [BUC65], using Ar-gas-quenched samples,

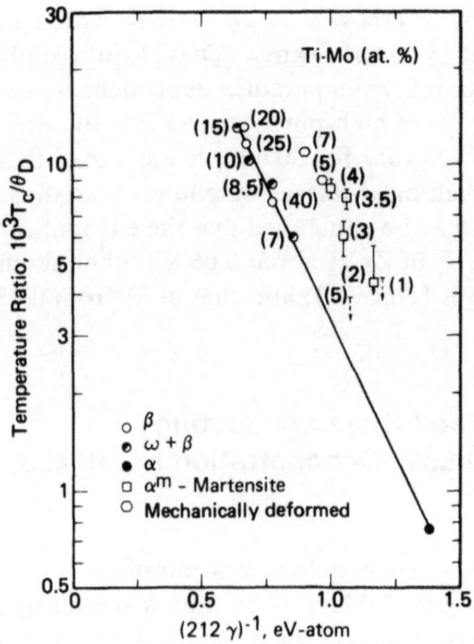

Figure 10-2. Semilog plot based on the expression $\log_{10}(T_c/\theta_D) \propto (212\gamma V_{app})^{-1}$. The alloys with the martensitic structures (either as-quenched or quenched-plus-deformed) have enhanced values of T_c/θ_D, hence T_c, as compared to the $\omega+\beta$-phase or β-phase alloys. See Table 10-2 for units, etc.—after Ho and Collings [Ho74].

have undertaken a comparative study of the superconducting transition in representative members of the Ti-(3d)TM, Ti-(4d)TM, and Ti-(5d)TM sequences. Alloy systems measured were (i) Ti–V, Ti–Cr, Ti–Mn, and Ti–Fe, (ii) Ti–Zr, Ti–Nb, Ti–Mo, and Ti–Ru, and (iii) Ti–W and Ti–Re. In order to further examine, confirm, and clarify some of the results obtained, Agarwal [Aga74] conducted a calorimetric investigation of ice-brine-quenched representatives of the systems: Ti–V, Ti–Cr, Ti–Mn, Ti–Fe, Ti–Co, Ti–Ni, and Ti–Hf.

10.3.1 Typical Results

Figures 10-1(a) and (b), from the results of low-temperature specific heat experiments on the Ti–V, Ti–Nb, and Ti–Mo systems, show the α^m-alloy transition temperatures increasing at the rate of about 30 K per unit e/a ratio. A similar rate of increase is evidenced by the $\alpha+\beta$-phase alloys. But a comparison of Figure 10-1 with Figure 8-2 shows the relative rate of increase of T_c with α^m-composition to be significantly greater than that of the electronic specific heat coefficient, γ, which in turn is a measure of $n(E_F)$ according to (8-3). Anomalies in the expected relationship between T_c and γ can best be expressed with the aid of a BCS-Morel-Morin-Maita[1] plot of $\log_{10}(T_c/\theta_D)$ *versus* $(212\gamma)^{-1}$ such as Figure 10-2 (see also Figures

[1]See Morel [Mor59], and Morin and Maita [Mor63]; also see (8-37).

8-12 and 11-7). From the heights of the α^m points above the full line in Figure 10-2 it appears that T_c in a martensitic alloy is some $2\frac{1}{2}$ times higher than that of the "corresponding" (i.e., same γ) $\omega+\beta$-structured phase. Although other explanations (see below) are possible, this could be interpreted as the result of an atomic-disorder-induced enhancement mechanism [CoL69, CoL75[a]], [Ho71].

The latter approach demands an explanation consistent with that offered to justify T_c augmentation in other structurally disordered systems, such as: (i) homogeneous amorphous films, (ii) amorphous metallic granules, (iii) microcrystalline films [GAR68], and (iv) layered metallic films [STR68]. In such systems the observed enhancements of T_c were interpreted within the framework of McMillan theory (see Section 8.5 and in particular (8-19)) as being due to increases in electron–phonon coupling, λ, through reductions in mean-square phonon frequency, $\langle\omega^2\rangle$, as a consequence of the coupling of lattice modes to low-frequency localized vibrational modes associated with loosely bound lattice defects. In such a context the martensitic structure, densely faulted on an atomic level, is a possible fifth example of the operation of the localized-soft-phonon-mode T_c enhancement mechanism.

10.3.2 Atypical Results—Ti–Mn

As pointed out in Section 7.10.2, in which the magnetic susceptibility of Ti–Mn is considered as a function of both composition (Figure 7-5) and temperature (Figure 7-6), only the hcp alloys of Ti with Mn, i.e., the quenched martensitic alloys (and presumably the dilute equilibrium α-phase alloys—although these have not been studied), are Curie paramagnets and, as such, tend to have suppressed superconducting transition temperatures. The low-temperature specific heats of Ti–Mn(0.17, 0.36, 0.85, 1.7, and 14 at.%) and of a Ti–Zr–Mn alloy have been considered in detail by HAKE and CAPE [HAK64] and those of a pair of dilute quenched α^m alloys, Ti–Mn(0.39 and 1.71 at.%), have been reinvestigated by AGARWAL [AGA74]. Both groups agree that provided the structure is hcp and, therefore, according to Figure 7-6, amenable to the support of a localized Mn magnetic moment, no superconductivity is observed down to the lowest temperature obtainable in the usual calorimetry experiment (1.2–1.3 K); in fact when plotted in the usual format, C/T *versus* T^2, a curve is obtained that rises with accelerating steepness as the temperature decreases. In examining the mechanism underlying this specific heat anomaly, defined as $\Delta C \equiv C_{\text{alloy}} - C_{\text{host}}$, plots of $\log_{10}(\Delta C/T)$ *versus* $\log_{10}T$ and $\Delta C/T$ *versus* T have been constructed. From the former it has been ascertained that $\Delta C/T \propto T^{-1.46\pm0.1}$ for Ti–Mn(0.17, 0.36 at.%) [HAK64] and $\Delta C/T \propto T^{-1.6}$ for Ti–Mn(0.39, 1.71 at.%) [AGA74]. These temperature dependences are characteristic of neither nuclear hyperfine interaction, for which $\Delta C/T \propto T^{-3}$, nor cluster superparamagnetism, a more attractive candidate (see [COL77, COL78[b], COL79[c], COL80]) for which $\Delta C/T \propto T^{-1}$.

It is, of course, always possible to adopt a thermodynamic approach and simply compare the area under the $\Delta C/T$ *versus* T curve with

$$S = fR \ln(2s + 1) \tag{10-1}$$

the decrease in total spin entropy, S, being associated with some local-moment ordering postulated to take place during cooling through the temperature range of the experiment, and where f is the mole fraction of magnetic solute of spin quantum number, s, and R is the gas constant. Taking $s = \frac{3}{2}$ for the Mn solute and introducing a relative total spin entropy, S_R, defined as $100 \int (\Delta C/T)\, dT/S$, it was estimated that the observed specific heat anomalies could be associated with S_R values of typically 26%; in particular, 27% for Ti-Mn(0.17, 0.36 at.%) [HAK64] and 25% for Ti-Mn(0.39 at.%) [AGA74].

As additional evidence for the depressive influence of Mn, as a localized magnetic moment, on the calorimetrically determined superconducting transition temperature, we cite a further experiment due to HAKE and CAPE [HAK64] whose results showed that the addition of 0.2 at.% Mn to Ti-Zr(50 at.%) reduced its T_c from 1.6 K to less than 1.24 K, the lower temperature limit of the experiment.

10.3.3 Distributed Calorimetric Transitions

(a) An Early Analysis of Superconducting Transition Rounding. As a result of the work of BUCHER *et al.* [BUC65] and the subsequent investigation of AGARWAL [AGA74], information is available on the low-temperature specific heat temperature dependences of low-concentration alloys of Ti with: (i) all the other 3*d* transition elements from Sc to Ni, (ii) the 4*d* elements from Zr to Ru, and (iii) the 5*d* elements Hf, W, and Re. Of the alloys for which specific heat transitions were noted it is interesting to compare the properties of dilute members of the "β-isomorphous" class of alloy such as Ti-V, Ti-Nb, and Ti-Mo with those of α^m-Ti-Cr, Ti-Fe, and Ti-Co, which are members of the "β-eutectoid" class (Figure 1-2). The specific heat jumps of the former are reasonably sharp, while those of the latter are exceptionally broad. A subdivision of quenched α^m-Ti-TM alloys into two groups consisting of Ti-V, Ti-Nb, and Ti-Mo on one hand and Ti-Fe on the other, with Ti-Cr occupying an intermediate position, is apparent in Figure 10-3.

For a detailed comparison of the properties of Ti-V and Ti-Nb with Ti-Fe, we turn to the work of HEINIGER and MÜLLER [HEI64], who discussed the transition smearing in terms of a distribution of transition

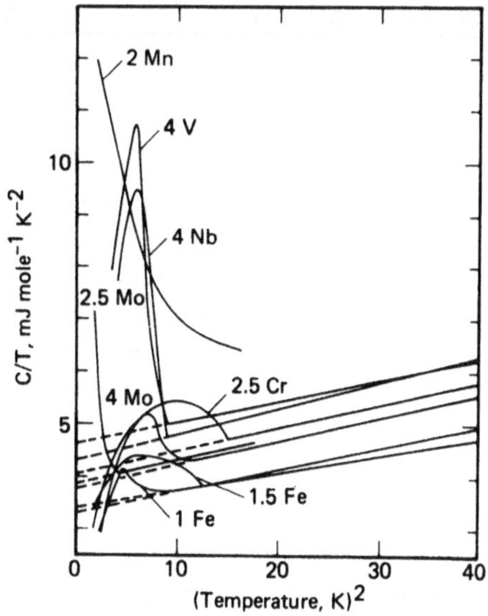

Figure 10-3. Low-temperature specific heats in the vicinity of their superconducting transitions for low-concentration martensitic Ti–V, Ti–Cr, Ti–Mn, Ti–Fe, and Ti–Mo alloys indicating a decrease in the abruptness of the specific heat jump on proceeding from Ti–V, Ti–Nb, Ti–Mo, towards Ti–Fe—after BUCHER *et al.* [BUC65].

temperatures, $F(T_c)$, normalized according to the somewhat truncated, although experimentally realistic, procedure described by

$$\int_0^{T_m} F(T_c)\,dT_c \cong \int_{-\infty}^{+\infty} F(T_c)\,dT_c = 1 \tag{10-2}$$

where T_m is the estimated maximal observed transition temperature. The analytical procedure adopted was to subdivide the rounded electronic specific heat contribution, $C_{er}(T)$, into two temperature regimes: (i) $T < T_c$ and (ii) $T > T_c$, where T_c is treated as a variable distributed according to (10-2). Thus,

$$C_{er}(T) = \underbrace{\int_0^T \gamma T F(T_c)\,dT_c}_{\substack{\text{Normal Region}\\ 0<T_c<T}} + \underbrace{\int_T^{T_m} C_{es}(T, T_c)F(T_c)\,dT_c}_{\substack{\text{Superconducting Region}\\ T<T_c<T_m}} \tag{10-3}$$

This equation can be rearranged into the following form with the aid of (10-2):

$$\left[\frac{C_{er}(T)}{\gamma T} - 1\right] = \int_T^{T_m} F(T_c)\left[\frac{C_{es}(T, T_c)}{\gamma T} - 1\right] dT_c \tag{10-4}$$

and with the unrounded electronic specific heat function, C_{es}, replaced by the following version of the Gorter–Casimir two-fluid result (see (8-9)):

$$\frac{C_{es}(T, T_c)}{\gamma T} = 3t^2 \tag{10-5}$$

yields an equation from which the transition temperature distribution function $F(T_c)$ may be estimated by means of a numerical trial and error process. Applying this method of analysis to the alloys mentioned it was then possible to compute the fraction, f, of material that underwent transition within a certain selected temperature interval, say $T_m/2 < T < T_m$. Thus, after defining

$$f = \int_{T_m/2}^{T_m} F(T_c)\, dT_c \tag{10-6}$$

HEINIGER and MÜLLER [HEI64] were able to show that Ti–V and Ti–Nb exhibited sharp transitions for which $0.95 < f < 1$. With Ti–Fe(1.5 at.%), on the other hand, they found that $0.3 < f < 0.35$ indicating that only about 30% of the sample was able to participate in the superconducting transition within the temperature range of the analysis.

A rationalization of the behavior illustrated in Figure 10-3, and specified above, is possible in metallurgical terms. The α^m-alloys under discussion are all prepared by rapid quenching, from an elevated-temperature anneal, into the equilibrium $\alpha+\beta$-field. If solute diffusion is sufficiently rapid some decomposition of the original β-alloy will take place during quenching. Figure 1-12, which intercompares the tracer diffusivities of several transition elements in β-Ti [ZWI74, p. 108], indicates that of the alloys for which data are available only Ti–V, Ti–Nb, and Ti–Mo have the opportunity to transform athermally to α^m during quenching from the β-phase; while in alloys such as Ti–Fe, Ti–Co, and Ti–Ni, whose diffusivities are almost two orders of magnitude higher, significant levels of solute redistribution and Widmanstätten growth can be expected during the quenching of the moderately massive samples needed for the usual kind of low-temperature specific heat measurement.

(b) Later Examples of Calorimetric Transition Rounding in General. Impurities, dislocations, and their strain fields may influence the properties of any long-range ordered material, be it superconducting, ferromagnetic, antiferromagnetic, and so on. If so, their effects are most strongly felt in the vicinity of the calorimetrically detected order-disorder transition where a cusp in the heat capacity is coupled with small strain- or impurity-induced changes in the critical temperature. The result is a measurable rounding of an otherwise sharp calorimetric anomaly. With either theoretical or experimental results for an instrained parent sample in hand, any rounded transition, detected in the same material after deformation, can be interpreted in terms of a strain-induced distribution of transition temperatures.

This technique has been applied by WILLIAMS, GOPAL, and STREET in interpreting the strain-induced Néel-temperature distribution in deformed Cr [WIL79] and the strained-induced Curie-temperature distribution in deformed Gd [WIL81]. Further examples of its applicability to the superconducting transition temperatures of strained Ti–TM alloys are to be presented in Sections 10.7 *et seq.*

10.4 Transition Temperatures of Unstable bcc Alloys—A Case Study of Titanium–Molybdenum

10.4.1 The T_c of bcc–Ti

As indicated in the foregoing discussion of magnetic susceptibility and low-temperature specific heat (see, in particular, (8-5)), the Pauli paramagnetic susceptibility, χ_p, and the measured electronic specific heat coefficient, γ, are coupled through the Fermi density-of-states, $n(E_F)$. Next with the aid of the BCS–Morel–Morin–Maita relationship between T_c/θ_D and γ (see Section 10.3.1 and footnote) it is possible to extract values for T_c provided some additional information relating to θ_D is available. χ_p is, of course, not directly measurable but may be extracted from the experimentally determined total magnetic susceptibility with the aid of extrapolation techniques [COL71, COL72] involving (7-17) and (8-5) and a series of joint magnetic and calorimetric experiments performed on cryogenically stable alloys.

Commencing with a series of magnetic susceptibility measurements, and using such an approach, it has been possible to make some semi-quantitative estimates of the density-of-states properties [COL71], including the superconducting transition temperature [COL72], of single-phase-bcc Ti–Mo alloys in the compositional regime in which they are unstable at ordinary temperatures. Of particular interest is the estimated superconducting transition temperature of unalloyed β-Ti. In attempting to determine the cryogenic properties of dilute Ti–TM alloys, since martensitic transformation and/or the formation of ω-phase precipitates cannot be suppressed by quenching, experiments on low-temperature-unstable single-phase bcc alloys must be performed at elevated temperatures. This automatically excludes low-temperature calorimetry as a technique; however, the elevated-temperature single-phase region can be explored using magnetic susceptibility, which has already been shown to make a useful substitute (after suitable

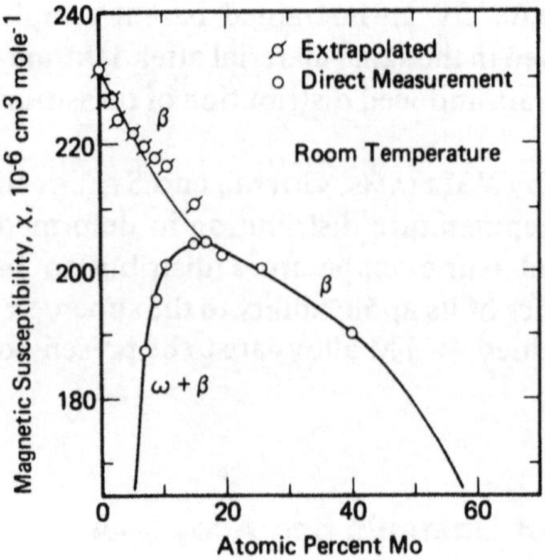

Figure 10-4. Room-temperature magnetic susceptibilities of as-quenched intermediate-to-high concentration $\omega+\beta$-phase and β-phase Ti–Mo alloys and low-concentration β-phase alloys. The latter data were obtained by extrapolating to room temperature the high-temperature results in the manner portrayed in Figure 7-7 for the β-Ti–V data—after HO and COLLINGS [HO69[a]], [COL71].

corrections have been applied) for the electronic specific heat coefficient in experimental density-of-states studies.

10.4.2 The T_c of Dilute bcc Ti–Mo Alloys

The elevated temperature magnetic susceptibilities of β-Ti–Mo alloys are only slightly temperature dependent (Table 7-6) and may be extrapolated back to room temperature with considerable confidence in order to supply a set of $\chi_\beta^{300\,\mathrm{K}}$ data for the entire alloy series. The results of so doing, which are shown in Figure 10-4, are the starting point of the empirical procedure for estimating the electronic and superconductive properties of the unstable[2] β-phase alloys as outlined in Figure 10-5 and summarized below.

A knowledge of $\chi_R = \chi - \chi_p$ (from (7-17)), already obtained by back extrapolation (cf. χ_{orb}, Figure 7-3) from the results of magnetic and calorimetric experiments on cryogenically stable alloys, next enables a set of values of $\chi_p^{300\,\mathrm{K}}$ to be extracted for the dilute β-Ti–Mo members of the series. At this point, the procedure, initially applied to the concentrated (>20 at.%) Ti–Mo alloys, is reversed and a corresponding set of values of γ_{dilute} is synthesized for insertion into the equation for T_c/θ_D referred to in Section 8.7.3, upon the assumption that V_{app} is still equal to 0.26 eV-atom. In other words, the T_c/θ_D are obtained by applying the estimated γs to an

[2]COLLINGS *et al.* [COL72] referred to single-phase-bcc alloys at room temperature in the composition range 0–15 at.% Mo as "unstable" and divided them into two classes: (i) those in the range $5 <$ at.% Mo < 15, which, although normally unstable do not yield unphysical elastic moduli, were designated "virtual" β-Ti–Mo; (ii) those within the martensitic regime ($\lesssim 4\frac{1}{2}$ at.% Mo) were referred to as "absolutely unstable," since for these the extrapolated elastic constants yield $C_{11} < C_{12}$ (see Figure 4-2). Data obtained by extrapolation into this latter region must be interpreted with considerable caution.

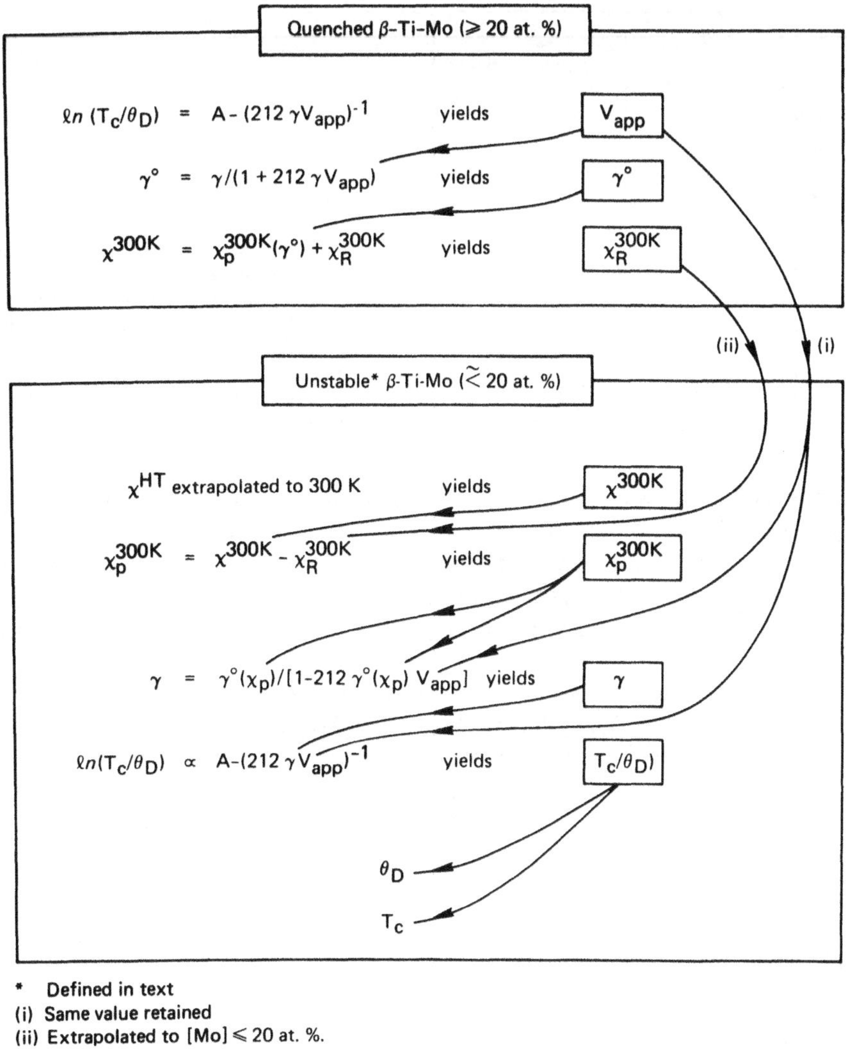

Figure 10-5. Algorithm for the empirical procedure followed in estimating the electronic and superconductive properties of low-concentration β-Ti–Mo alloys. At concentrations below about 15 at.% the β-phase of such alloys is unable to be retained during quenching to room temperature; its properties can be extracted only with the aid of some extrapolation procedure. See also Table 10-2—after COLLINGS *et al.* [COL72].

extension of the previously constructed BCS–Morel–Morin–Maita plot (Figure 8-12) as in Figure 10-6. The equations used in performing the calculations are assembled for convenient reference in Table 10-2.

The goal of the analysis was to derive estimates of the T_cs of unstable β-Ti–Mo alloys and in particular that of β-Ti, which necessitated separation of the temperature-ratio, T_c/θ_D, into its individual components. If, as [Mo]

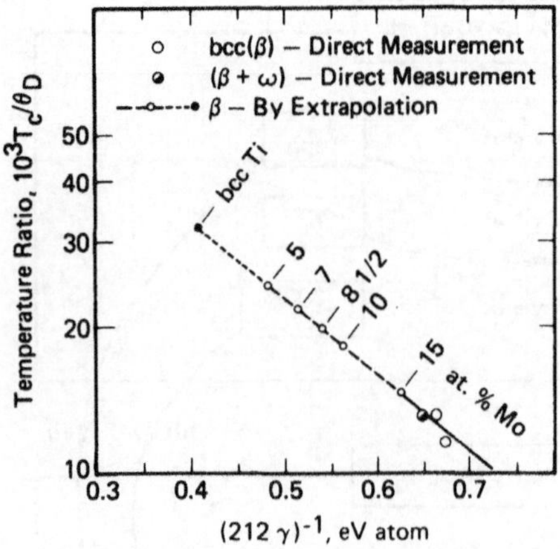

Figure 10-6. Semilog plot based on the expression $\log_{10}(T_c/\theta_D) \propto (212\gamma V_{app})^{-1}$ for use, in conjunction with the procedure outlined in Figure 10-5, in estimating T_c and θ_D for unstable-β low-concentration Ti–Mo alloys. See Table 10-2 for units, etc.—after COLLINGS *et al.* [COL72].

decreases below about 20 at.%, θ_D is permitted to remain constant and equal to the lowest directly measured value (*viz.*, 295 K for Ti–Mo(15 at.%)), T_c rises to what appears to be excessively high values (e.g., 7.2 K for β-Ti–Mo(5 at.%)). To obtain a more conservative estimate for T_c, allowance was made for a possible decrease in θ_D with decreasing solute concentration,

Table 10-2. List of Approximate Relationships[a] Used in the Derivation of T_c/θ_D for "Virtual-β" Ti–Mo Alloys

	T_c and θ_D	Equation Numbers
$\ln(T_c/\theta_D)$	$= \text{const} - (1/V)[n(E_F)]^{-1}$	(8-34)
$n(E_F)$	$= 212\gamma^0$	(7-4b)
γ^0	$\equiv \gamma/(1+\lambda)$	(7-5)
V_{app}	$\equiv V/(1+\lambda)$	(8-35)
$\ln(T_c/\theta_D)$	$= \text{const} - (1/V_{app})(212\gamma)^{-1}$	(8-37)
	Evaluation of γ	
λ	$= n(E_F)V$	
	$= 212\gamma V_{app}$ (using (7-4b), (7-5), and (8-35))	(8-36)
V_{app}	$= 0.26$ eV (experimental value from Figures 8-12 or 10-2)	
λ	$= 55.1\gamma$	
	γ, χ	
γ^0	$= 4.72 \times 10^{-3} n(E_F)$	(7-4b)
χ_P^0	$= 64.7 \times 10^{-6} n(E_F)$	(7-2b)
	$= 1.37 \times 10^{-2} \gamma^0$	
χ	$= \chi_p + \chi_R$	
γ	$= \gamma^0/(1-55.1\gamma^0)$	(8-39)

[a]Units: γ, J mol^{-1} K^{-2}; χ, cm^3 mol^{-1}; $n(E_F)$, eV^{-1} atom^{-1}; V, V_{app}, eV.

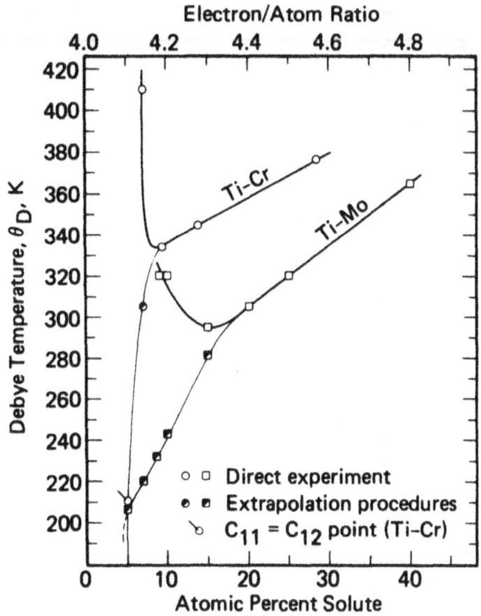

Figure 10-7. Comparison of the temperature dependences of the Debye temperature, θ_D, for Ti-Cr and Ti-Mo alloys. In the case of Ti-Cr the extrapolation was based on extrapolated elastic constant data (cf. Figure 4-2) in association with VRHG calculations of θ_D (Section 4.3.2)—after COLLINGS *et al.* [COL72].

behavior which could itself be interpreted in terms of a softening of the lattice in the unstable regime. The separation was achieved by trial and error in such a way that (i) the extrapolated T_c tended to scale with γ, and (ii) $\theta_{D,\text{Ti-Mo}}$ was consistent with $\theta_{D,\text{Ti-Cr}}$ (Figure 10-7) computed from the extrapolated elastic constant data of FISHER and DEVER [FIS70] (Figure 7-2) with the aid of the VRHG approximation (Section 4.3.2). The results of the extrapolation exercise are displayed in Figure 10-8 together with curves representing the corresponding directly measured properties of the $\omega+\beta$-phase and β-phase alloys. By way of summary the estimated physical properties of β-Ti are listed in Table 10-1.

10.5 Influence of Aging on the Transition Temperatures of Titanium–Transition Metal Alloys

The metallurgical consequence of the aging of quenched Ti–TM alloys have been discussed in Section 2.1. As is well known by now, moderate-temperature aging of the quenched alloys (at temperatures of 150–400°C) results in a continuous increase in the volume fraction of the ω-phase precipitate, accompanied by an increase in the solute content of the β-matrix at the expense of that of the precipitate [HIC68, HIC69[a]]. At the same time, some increase in precipitate particle size may also be anticipated (Table 1 of [BLA68]).

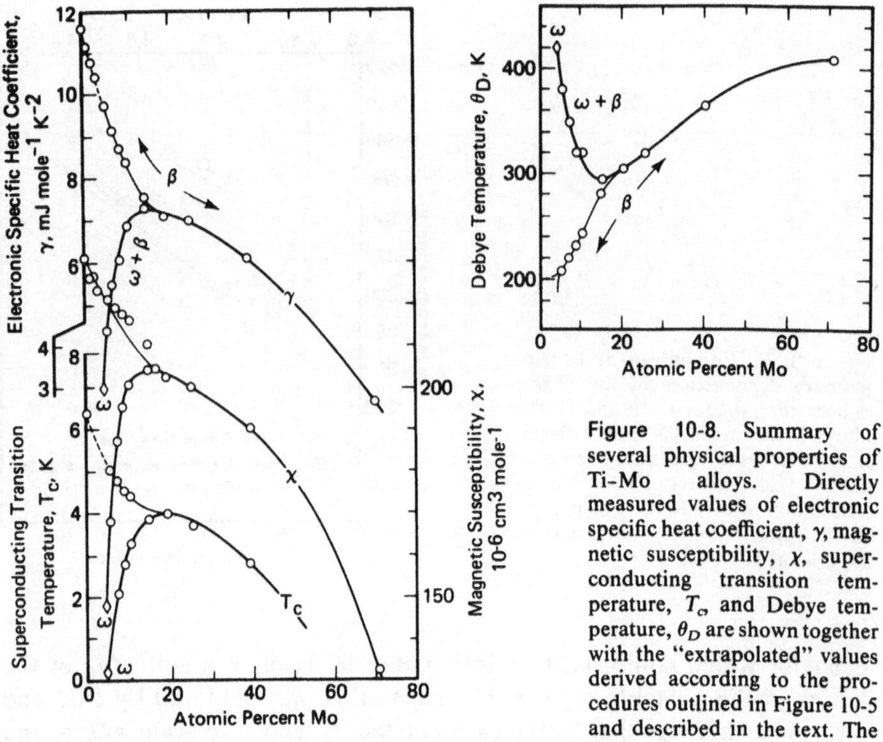

Figure 10-8. Summary of several physical properties of Ti–Mo alloys. Directly measured values of electronic specific heat coefficient, γ, magnetic susceptibility, χ, superconducting transition temperature, T_c, and Debye temperature, θ_D are shown together with the "extrapolated" values derived according to the procedures outlined in Figure 10-5 and described in the text. The value of T_c(β-Ti) so obtained agrees well with the T_c = 6.6 K value recently acquired by Poon [Poo83] from the extrapolated T_c *versus* composition data for rapidly liquid-quenched Ti–Mo and Ti–Pd alloys.

10.5.1 Magnetic Susceptibility and Electronic Specific Heat

The applicability of magnetic susceptibility to the semiquantitative monitoring of ω+β-phase aging of a Ti–V(19 at.%) alloy has been discussed in Section 7.14. As a consequence of its close relationship to magnetic susceptibility, (8-5), low-temperature specific heat could play a similar role. In support of a discussion of the superconductive proximity effect, the influence of aging to metastable ω+β-phase equilibrium on the properties of Ti–Mo(10.3 at.%) is considered in Section 11.9.3. There, with the aid of the relevant continuity equations (e.g., (7-20) *et seq.*), estimates of the magnetic susceptibility and electronic specific heat coefficient of ω-Ti–Mo(4.3 at.%) have been made. Calorimetric studies of aging have been conducted on Ti–V(15 at.%) (up to 1030 h/300°C) and Ti–V(19 at.%) (up to 2200 h/300°C) [Col75, Col76[b], Col82]; Ti–Mo(10 at.%) (up to 880 h/350°C) [Ho73], [Col72[a], Col82]; and Ti–Fe(7.5 at.%) (up to 1170 h/175°C followed by an additional 88 h/300°C) [Ho74]. Magnetic susceptibility measurements were also made on small aged specimens taken

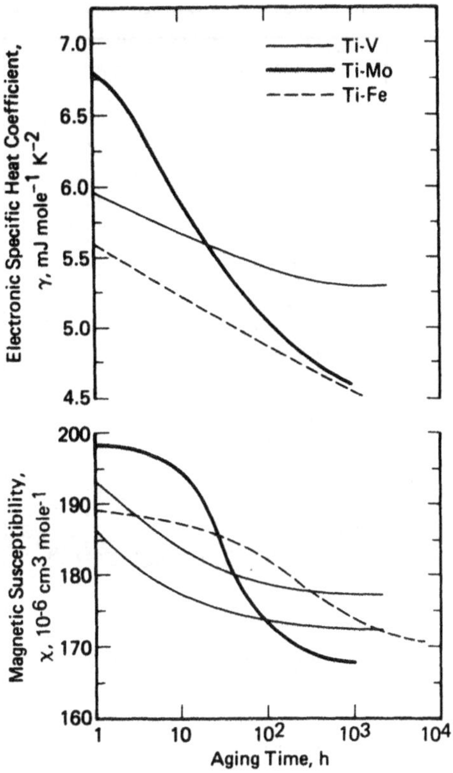

Figure 10-9. Influence of aging on the electronic specific heat coefficient, γ, and magnetic susceptibility, χ, of Ti-V, Ti-Mo, and Ti-Fe alloys. Data sources: Ti-V(19 at.%) (aged at 300°C)—COLLINGS *et al.* [COL75, COL75[b]]; Ti-Mo(10 at.%) (aged at 350°C after 1 h/300°C)—COLLINGS [COL82]; Ti-Fe(7.5 at.%) (aged at 175°C)—HO and COLLINGS [HO74].

from the same starting ingots. On account of the relative simplicity of this measuring technique, data were able to be taken at much closer intervals of aging time than was possible in the calorimetric part of the investigation. It is generally true (see Section 7.14.2, Table 10-1, and Figure 10-8) that χ_ω and γ_ω are substantially less than the χ_β and γ_β with which they are associated and that, as a consequence, both $\chi_{\omega+\beta}$ and $\gamma_{\omega+\beta}$ decrease monotonically with aging time. Examples of this are seen in Figure 10-9 and later on in Figure 10-11.

10.5.2 The Transition Temperature

The influence of aging on the calorimetric superconducting transition of Ti-V(19 at.%) is shown in Figure 10-10(a) and that of Ti-Mo(10.3 at.%) is shown in Figure 10-10(b). A property of Ti-V(19 at.%), also shared by Ti-V(15 at.%) during aging, is a broadening of the transition associated with a continuous *increase* in the threshold T_c [COL75, COL76[b]]. On the other hand with Ti-Mo(10 at.%), as with Ti-Fe(7.5 at.%) [HO74], the transition remains relatively sharp during aging and shifts towards lower temperatures. This behavior is discussed in Section 11.9 in terms of the degree

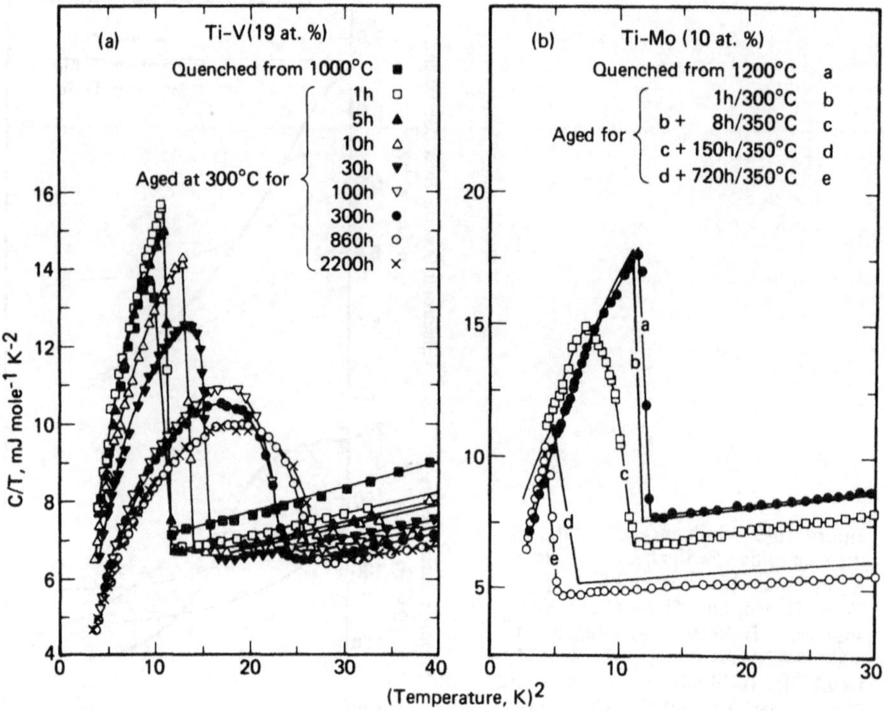

Figure 10-10. Influence of aging on the calorimetric superconducting transitions in (a) Ti-V(19 at.%)—after COLLINGS *et al.* [COL75] and (b) Ti-Mo(10 at.%)—after COLLINGS [COL84].

to which the proximity effect between matrix and ω-phase precipitate is "complete" in each class of alloy.

In aged Ti-V it seems that the proximity effect is "incomplete" (the coherence length, ξ, is presumably less than $\langle r \rangle$, the mean minor radius of the precipitate), as evidenced by: (i) a pronounced increase in the width of the transition, indicative of a distribution of transition temperatures; (ii) a decrease in the relative specific heat jump height, $\Delta C/\gamma T_c$, of from 1.32 to 0.58 (Ti-V(15 at.%)) and from 1.58 to 0.82 (Ti-V(19 at.%)), indicative of decreases with time of the superconductive-component abundances to values as low as 40% and 50%, respectively [COL75]; (iii) *increases* in the threshold T_c, in spite of *decreases* in χ and γ (both bulk average quantities) with ω-phase development, as incomplete superelectron penetration into the precipitate weights the transition temperature towards that of the progressively enriching β-Ti-V phase. On the other hand with Ti-Mo and Ti-Fe the converse seems to be true, in that both jump height and transition sharpness are preserved during aging and T_c decreases with aging on account of the dominance of the low-T_c ω-phase under *complete* proximity-effect conditions. In Ti–Fe(7.5 at.%), for example, $\Delta C/\gamma T_c \cong 1.6$ (compared to the BCS value of 1.43) in both the as-quenched alloy and after aging for 546 h/175°C, during which T_c decreases from 3.5 to 1.9 K [Ho74].

10.5.3 Verification of the Properties of β-Ti–Mo(10.3 at.%)

Based on extrapolations from an elevated-temperature magnetic susceptibility study of low-concentration Ti–Mo alloys, by way of the semiempirical analysis outlined in Figure 10-5, a set of curves claimed to represent the physical properties of bcc low-concentration Ti–Mo alloys (Figure 10-8) were derived. It is interesting to note that these conclusions are substantiated, at least with regard to Ti–Mo(10.3 at.%), by the results of the aging study presently under discussion.

Figure 10-11, based on data from Figures 10-9 and 10-10(b), shows χ, γ, and T_c decreasing monotonically with aging time at 350°C (from an initial aging of 1 h/300°C). These data, extrapolated back three decades in time towards starting points that simulate the results of a hypothetical ultrarapid quench to a fully-retained-β condition, intersect the vertical axis at points which agree with the values of χ_β, γ_β, and $T_{c,\beta}$ found in Figure 10-8 above a solute concentration of 10 at.% Mo.

10.6 Low-Temperature Specific Heat in the Mixed State

10.6.1 The Electronic Specific Heat

The electronic specific heat of the mixed state can be related to that of the normal state in terms of the second derivative with respect to temperature of the molar or unit-volume Gibbs free energy differences between those states. According to MAKI [MAK65], for example, in unit-volume terms,

$$C^{v}_{em} - C^{v}_{en} = -T\frac{\partial^2}{\partial T^2}(g_m - g_n) \tag{10-7}$$

Figure 10-11. Influence of aging at 350°C on T_c (cf. Figure 10-10(b)), γ, and χ in Ti–Mo(10 at.%) (open points). Data back-extrapolated to some very short aging time (half-shaded points representing the nascent β-phase) agree with those deduced using the previously discussed extrapolation procedure, in particular with the values to be found on the extrapolated branches of Figure 10-8 opposite 10 at.% Mo—after COLLINGS [COL84].

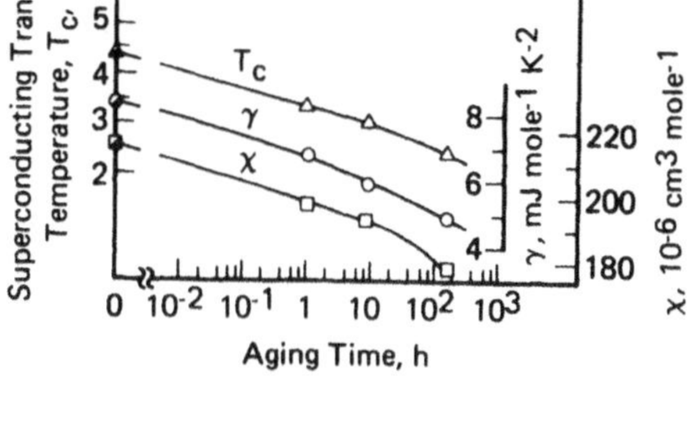

The regimes of validity of the calculation were chosen to be those for which H_a, the applied field, was (i) just below H_{c2}, and (ii) just above H_{c1}. MAKI's work was followed by a detailed study of the mixed state by VAN VIJFEIJKEN [VIJ68], whose calculation of the electronic specific heat, again based on an expression for the Gibbs free energy density, was valid over the entire mixed-state region except near the upper critical field curve, $H_{c2}(T)$.

In developing an expression for the mixed-state magnetization an important simplifying assumption was made. The actual shielding-magnetization $B-H_a$ curve (e.g., Figure 16-7c) was replaced by a straight line connecting $(0, H_{c1})$ on the applied-field axis to the point $(B\ (= H_{c2}), H_{c2})$. The equation of such a line is, of course:

$$\frac{B(H,T)}{H_{c2}(T)} = \frac{H_a - H_{c1}(T)}{H_{c2}(T) - H_{c1}(T)} \tag{10-8}$$

After substituting for $H_{c2}(T)$ and $H_{c1}(T)$ from (14-10) and (14-12), respectively, including the usual parabolic temperature dependence for $H_c(T)$ ((8-7)) and upon the assumption that $\kappa_1(T) \cong \kappa_2(T) \cong \kappa_{GL}$, the common value of those parameters at T_c (Figure 14-5), (10-8) can be rewritten

$$\frac{B(H,T)}{H_{c2}(T)} = \frac{H_a - H_{c0}(1-t^2)\ln\kappa/\sqrt{2}\kappa}{(\sqrt{2}\kappa - \ln\kappa/\sqrt{2}\kappa)H_{c0}(1-t^2)} \tag{10-9}$$

wherein κ implies κ_{GL}. Equation (10-9) is the source of terms in κ_{GL} and $\ln\kappa_{GL}$ to appear in the final expression for the mixed-state specific heat, which is

$$C^v_{em} = \gamma_v T - \frac{H_{c0}T}{2\pi T_c^2}\left[\frac{\sqrt{2}\kappa H_{c0} - H_a}{\sqrt{2}\kappa - \ln\kappa/\sqrt{2}\kappa}\right] + \frac{3H_{c0}^2 T}{2\pi T_c^2}\left[\frac{\sqrt{2}\kappa}{\sqrt{2}\kappa - \ln\kappa/\sqrt{2}\kappa}\right]t^2 \tag{10-10}$$

(cf. [VIJ68, Eq. (3.90)] and [ZUB79, Eq. (4)]).

Implicit in the derivation of this equation is its applicability to only the mixed state. It is, however, natural to inquire whether it reduces to a recognizable form for the electronic specific heat temperature dependence as $H_a \to 0$. Making the usual substitution for H_{c0}/T_c, *viz.*, $\gamma_v = (1/2\pi)(H_{c0}/T_c)^2$, (see [LYN64, p. 21] and [BAR57, Eq. (3.40)]) into (10-10) one obtains, in molar units or quite generally,

$$C_{em} = \gamma T - \gamma T\left[\frac{\sqrt{2}\kappa - H_a/H_{c0}}{\sqrt{2}\kappa - \ln\kappa/\sqrt{2}\kappa}\right] + 3\gamma T\left[\frac{\sqrt{2}\kappa}{\sqrt{2}\kappa - \ln\kappa/\sqrt{2}\kappa}\right]t^2 \tag{10-11}$$

In zero applied field the bracketed quotients are equal and maximize at 1.1 for $\kappa_{GL} = 1.6$. Thus, for all values of $\kappa_{GL} \gtrsim 0.9$, C_{em} is essentially equal to the thermodynamically derived Gorter–Casimir expression for the electronic specific heat temperature dependence, (10-5). Equation (10-10) is of the form

$$C_{em} = AT + BT^3 \tag{10-12}$$

which is comparable to (8-2) for the total *normal-state* low-temperature specific heat. It follows that the *mixed-state electronic* specific heat data, when plotted in the format C_{em}/T *versus* T^2, should also lie on a straight line. As pointed out by VAN VIJFEIJKEN [VIJ68], the data of FERREIRA DA SILVA *et al.* [FER66, FER66[a]] conform rather well to this rule except that the experimental intercept is some 30% higher than that predicted by (10-10) after the insertion of appropriate numerical values for γ, H_{c0}, κ_{GL}, and T_c.

10.6.2 Height of the Specific Heat Jump

The height of the specific heat jump at some temperature, T_s, in a magnetic field, H_a, is

$$\Delta C(T_s) = C_{em}(T_s) - C_n(T_s) = C_{em}(T_s) - \gamma T_s \tag{10-13}$$

After substituting a molar or general form of (10-10), and recognizing[3] that

$$H_a = H_{c2}(T_s) = H_{c20}\left[1 - \left(\frac{T_s}{T_c}\right)^2\right] \cong \sqrt{2}\kappa_{GL} H_c(T_s) \tag{10-14}$$

(10-13) leads to

$$\frac{\Delta C(T_s)}{\gamma T_s} = \frac{2}{\gamma}\frac{H_{c0}^2}{2\pi T_c^2}\left[\frac{\sqrt{2}\kappa}{\sqrt{2}\kappa - \ln \kappa/\sqrt{2}\kappa}\right]\left(\frac{T_s}{T_c}\right)^2 \qquad \kappa \equiv \kappa_{GL} \tag{10-15}$$

again similar, under the conditions on γ and κ_{GL} referred to above, to the Gorter–Casimir relative jump height at T_c, *viz.*, $\Delta C/\gamma T_c = 2$.

It follows from (10-15) that the height of the jump, $\Delta C(T_s)$, varies as T_s^3; also, the normalized relative jump height can be expressed in the form

$$\frac{\Delta C(T_s)/\gamma T_s}{\Delta C(T_c)/\gamma T_c} = \frac{\Delta C(T_s)/T_s}{\Delta C(T_c)/T_c} = \left(\frac{T_s}{T_c}\right)^2 \equiv t_s^2 \tag{10-16}$$

[3]Equations used in this transformation are (8-7) and (14-10); at the same time it is assumed, as before, that $\kappa_1(T) \cong \kappa_{GL}$.

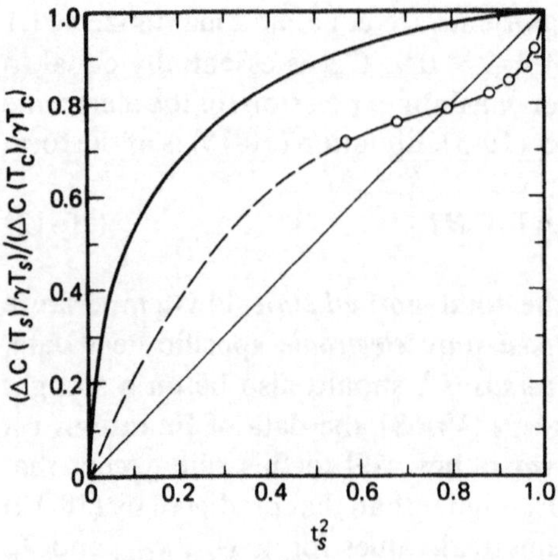

Figure 10-12. Reduced relative height of the specific heat jump at the transition from the mixed to the normal state at temperature T_s plotted *versus* the square of the reduced transition temperature, $t_s \equiv T_s/T_c$. Experimental data, for Ti-Mo(16 at.%), are due to BARNES and HAKE [BAR66]; the fine straight line represents the quadratic relationship; the heavier curve is a result due to MAKI for $\kappa = 2$ (after FERREIRA DA SILVA *et al.* [FER66[a]]).

Experimental investigations of mixed-state specific heat have been undertaken by: (i) FERREIRA DA SILVA *et al.* [FER66, FER66[a]], who measured bundles of unannealed Nb wires in magnetic fields of up to 9.4 kOe; (ii) ZUBECK *et al.* [ZUB79], who also studied Nb but with particular reference to the influence of plastic deformation on the mixed-state transition; (iii) BARNES and HAKE [BAR66], who measured a sample of Ti-Mo(16 at.%) in fields of up to 29 kOe. According to FERREIRA DE SILVA *et al.* [FER66, FER66[a]], the normalized relative jump height in Nb followed the T^2 law prescribed by (10-16), within experimental error over the entire temperature range. On the other hand, as indicated in Figure 10-12, the results for Ti-Mo(16 at.%) departed markedly from the simple quadratic form, partly as a consequence of unexplained excess jump height exhibited by the alloy specific heat in the three strongest fields, see [BAR66, Figure 2] and [BAR67, Figure 1].

10.7 Influence of Deformation Itself, and Deformation- or Solute-Induced Phase Transformations on the Superconducting Transition

10.7.1 Deformation of Pure Elements

Numerous studies have been made of the influence of plastic deformation on the superconducting transition. This and the sections to follow, however, emphasize low-temperature calorimetry, which as pointed out already in Section 10.3.3(b), is a powerful technique for probing the effects of strain fields, not only on the superconducting transition, but on critical transitions in general. In a resistometric investigation, SWANSON and QUENNEVILLE [SWA73] conducted a study of compressionally deformed Sn alloys.

They pointed out, with documentation, that: (i) the introduction of point defects such as impurities and vacancies could bring about *reductions* in T_c; (ii) static dislocations could for similar reasons *reduce* T_c; (iii) increases in the electron–phonon coupling associated with the interaction of electrons with dislocation-vibrations could be responsible for small *increases* in T_c. As pointed out in Section 10.3.1, the latter mechanism had been proposed earlier by GARLAND *et al.* [GAR68] and STRONGIN *et al.* [STR68] to explain some observed T_c enhancements in heavily defected Al films. THOMPSON and FLEWITT [THO75], as the result of a comprehensive magnetic study of the superconducting transition temperatures and deformation microstructures of cold-rolled and cold-drawn Nb, found that for deformations of up to 10% T_c *increased* linearly with the dislocation density, but that at higher levels of deformation during the development and refinement of a cellular structure (Section 2.8) T_c *decreased* linearly with cell size. On the other hand, during the zero-field segment of a study of the influence of deformation on the calorimetrically observed transition of Nb into the superconducting mixed state, ZUBECK *et al.* [ZUB79] found that *compressive* strains of up to 75% (reduction in thickness) resulted in continuous *increases* in both the temperature and width of the superconducting transition. Quite clearly there is no simple prescription linking deformation with either increases or decreases of the superconducting transition temperature; the observed effects depend on the metallurgical details.

Estimates of the transition temperature distributions derived by ZUBECK *et al.* [ZUB79], using an analytical procedure to be outlined below, are given in Table 10-3. As a result of their analysis, those authors noted that, although a symmetrical Gaussian distribution gave an adequate description of the rounding at the 7% and 20% levels of strain, after 75%

Table 10-3. Influence of Plastic Strain on the Calorimetrically Measured Superconducting Transition in Nb—After [ZUB79]

Strain (Reduction in Thickness) ε %	Transition Temperature T_c K	Standard Deviation of Symmetrical Gaussian Distribution σ K
0	9.19	$<0.5 \times 10^{-3}$
7	9.255	4×10^{-3}
20	9.268	11×10^{-3}
75	9.285	Broad and skewed to higher temperatures[a]

[a]The high temperature side appears to be exponential rather than Gaussian.

strain this was not the case, the distribution appearing to be exponential rather than Gaussian and severely skewed towards high temperatures.

10.7.2 Deformation-Induced Transformation in Ti–TM Alloys

(a) Deformation Martensite. It has already been shown that T_c responds rather strongly to changes of phase in Ti–TM alloys. Thus, as indicated in Figure 10-1, the transition temperatures of α^m-Ti–TM alloys are very much higher than those of the corresponding $\omega+\beta$-phases. As a consequence, pronounced changes of T_c can be expected if martensitic transformation can be triggered by the deformation. Figure 10-13 [Ho71], [CoL71[f]] is one example of this. Whereas in the as-quenched state Ti-Mo(5 at.%) is expected to have a transition temperature of about 0.8 K (see Figure 10-1(b)), after some 23% compressive deformation a broad, substantial, superconducting transition distributed between 2.7 and 3.3 K (peak and onset temperatures, respectively) was noted. This spectacular increase in T_c is attributed to a deformation-induced martensitic transformation (Figure 10-14), which, with reference to Figure 10-1(b), transports the T_c-datum point from the $\omega+\beta$- to the α^m-branch of the T_c *versus* composition curve. After transformation to martensite, it is possible that the soft-phonon mechanism, alluded to above in Section 10.3.1, comes into play to enhance the transition temperature.

(b) Deformation Twins. With regard to Ti–Mo(7 at.%) [Ho71], it is believed that microstructural effects associated with a deformation-induced *twinning-transformation* (as distinct from martensitic) are instrumental in raising the midpoint of the transition from 2.11 to 2.59 K, an increase of 23%.

10.7.3 Solute-Induced Transformation in Ti–TM Alloys

Martensitic transformation of a Ti–TM alloy under quenching can also be induced by the addition of small amounts of a strong α-stabilizer. In

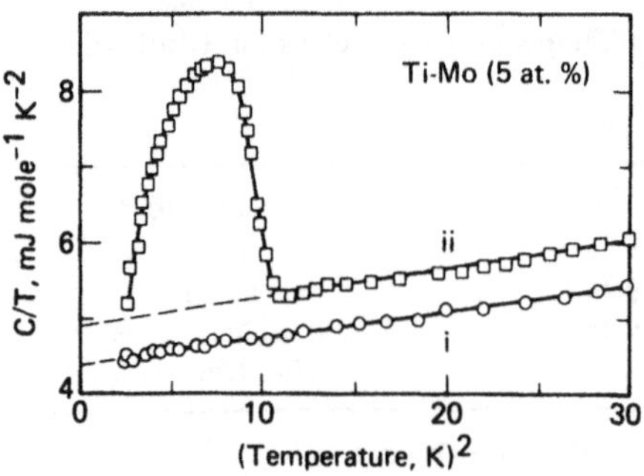

Figure 10-13. Low-temperature specific heats of (i) as-quenched and (ii) quenched and subsequently deformed Ti-Mo(5 at.%). The transformation is responsible for the enhanced T_c—after Ho and Collings [Ho71].

Ti-Mo (5 at. %), Quenched   Quenched + Deformed

Figure 10-14. Photomicrographs of the samples referred to in the previous figure. The as-quenched alloy exhibits a few patches of martensite. During deformation, extensive transformation takes place. Magnification of the original 3.5×3.5 in^2 picture, 50×.

Ti–Mo(5 at.%) this has been achieved by the presence of 1 and 3 at.% Al in solid solution. In calorimetric studies of such alloys [COL76[a]] it has been determined that the "solute induced" martensitic transformation of Ti-Mo(5 at.%) is responsible for elevating the midpoint of the superconducting transition from an estimated 0.8 K to 3.1 K (1 at.% Al) and 2.8 K (3 at.% Al), respectively, Figure 10-15.

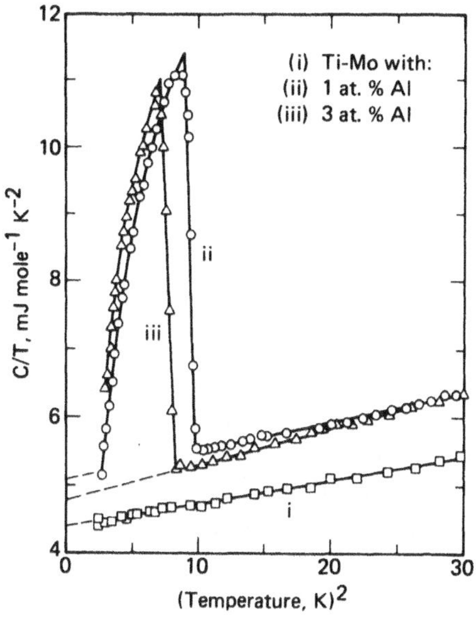

Figure 10-15. Low-temperature specific heats of: (i) quenched Ti–Mo(5 at.%) (as in Figure 10-13, curve (i)); (ii) the same alloy base doped with 1 at.% Al (annealed and quenched); and (iii) the same doped with 3 at.% Al (annealed and quenched). As in the previous figure the effect of martensitic transformation is responsible for the enhanced T_c—after COLLINGS and HO [COL84].

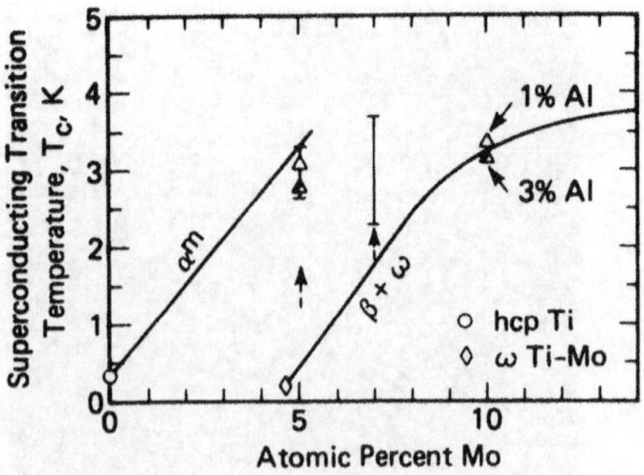

Figure 10-16. Influence on the calorimetrically determined T_cs of Ti-Mo(5, 7, and 10 at.%), which in the as-quenched state lie along the curve labelled $\omega+\beta$, of deformation ([), and alloying with 1 at.% Al (△) and 3 at.% Al (▲)—after COLLINGS and HO [COL76[a], COL84].

These results are summarized in Figure 10-16, which also shows that the addition to Ti-Mo(10 at.%) of 1 at.% and 3 at.% of Al, which does not promote martensitic transformation of the host under quenching, has relatively little influence on the transition temperature.

10.8 Analysis of the Rounded Zero-Field Calorimetric Superconducting Transition

10.8.1 The Transition Temperature Distribution Function

If as a result of plastic deformation or partial structural transformation, or through the presence of precipitated phases, and so on, the superconducting transition is not only incomplete but distributed over a range of temperatures, the specific heat transition will be correspondingly broad and reduced in peak height. In general, the observed rounded total specific heat temperature dependence will be represented by

$$C_r(T) = \int_{-\infty}^{+\infty} C_u(T, T_c)F(T_c)\, dT_c \tag{10-17}$$

with

$$\int_{-\infty}^{+\infty} F(T_c)\, dT_c = 1 \tag{10-18}$$

where $C_u(T, T_c)$ is the unrounded specific heat, and $F(T_c)$ is the normalized transition temperature distribution function introduced in (10-2).

Numerous approaches to the solution of (10-17), with C_u representing either the entire specific heat or only its electronic component, are possible. Variations that are encountered in the literature consist of: (i) taking a theoretical model for $C_u(T, T_c)$ and, with the aid of the experimental electronic specific heat data C_{er}, calculating $F(T_c)$ numerically and from it the fraction f of material that transforms within a selected temperature range [HEI64], [BUC65]; (ii) taking a differential form of (10-17) and, focussing attention on the immediate neighborhood of the transition, deriving a distribution function $F(T_c)$ in terms of the derivative with respect to temperature of the specific heat jump at T_c [ZUB79]; (iii) taking theoretical models for both $C_u(T, T_c)$ and $F(T_c)$ and computing the specific heat parameters together with f, the superconductive mole fraction of a two-component (superconductive/normal) system [WHI76], [COL78[a]].

An analytical procedure was devised by HEINIGER and MÜLLER [HEI64], [BUC65] to estimate transition temperature distributions in quenched-α^m Ti–TM alloys. Their method, which was modelled on Gorter–Casimir theory, has already been described in Section 10.3.3.

In an analysis of specific heat rounding in plastically deformed Nb in the mixed state, with zero-field distributed-T_c rounding as a special case, ZUBECK *et al.* [ZUB79] started out by taking for the $C_u(T, T_c)|_{T<T_c}$ of (10-17) the full van Vijfiejken expression for the mixed-state electronic specific heat, (10-11). Although in zero field, according to foregoing arguments (Section 10.6.1), this reduces to the simple Gorter–Casimir cubic temperature dependence of (8-9b) and (10-5), an even simpler model was in fact selected for the electronic specific heat in the close neighborhood of T_c, *viz.*,

$$\frac{C_{es}}{T} = \begin{cases} \gamma + \Delta C/T & T \leqq T_c \\ \gamma & T > T_c \end{cases} \tag{10-19}$$

where ΔC is the specific heat jump at T_c. On the basis of this model, the function $(C_{es}/T - \gamma)\mathcal{H}(T_c - T)$, where $\mathcal{H}$ is the Heaviside unit function, represents a rectangular step of height $\Delta C/T_c$ (Figure 10-17(a)), while the actual measured electronic specific-heat quotient, C_{er}/T, yields a rounded step, $(C_{er}/T - \gamma)\mathcal{H}(T_c - T)$. According to arguments advanced by ZUBECK *et al.* [ZUB79], a distribution function defined by

$$F_c(T) = \frac{1}{A}\frac{d}{dT}(C_{er}/T - \gamma) \tag{10-20}$$

e.g., Figure 10-17(b), with normalization factor, A, given by

$$A \equiv \int_{T_{\min}}^{T_{\max}} \frac{d}{dT}(C_{er}/T - \gamma)\, dT \qquad T_{\max} > T > T_{\min} \tag{10-21}$$

(both limits of integration being well removed from T_c) is just that required to satisfy (10-17) within the framework of the model (i.e., provided $C_{es}/T(T \neq T_c)$ can be regarded as being independent of temperature over a range of temperatures larger than the anticipated breadth of the rounding). Some results are given in Figures 10-17(c) and (d).

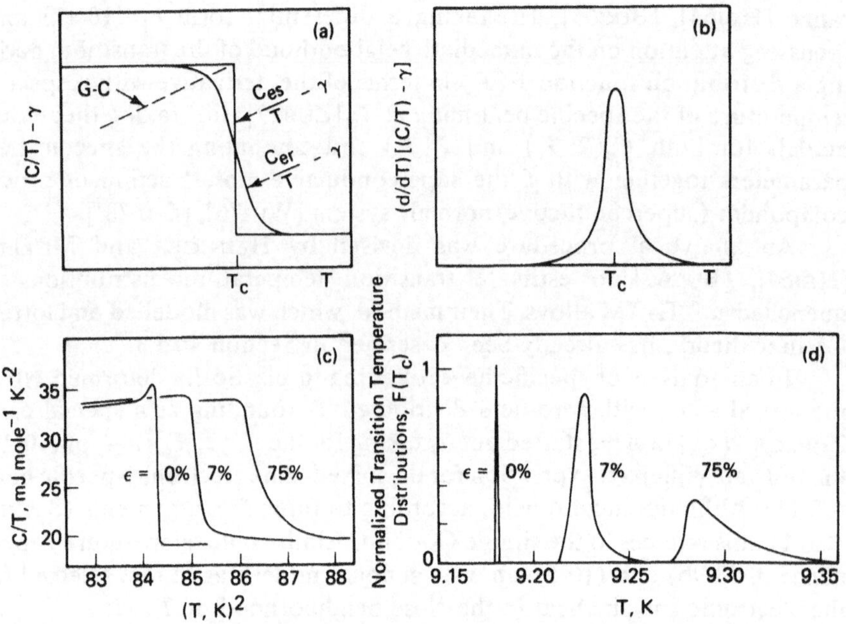

Figure 10-17. (a) Representation of the distributed C/T *versus* T function in terms of the rounding of a rectangular step (as proposed by ZUBECK *et al.* [ZUB79]) valid over a narrow temperature range about T_c; the Gorter-Casimir function (see (10-5)) is also indicated. (b) Derivative of the rounded step function. (c) Zero-field specific heat of Nb at three levels of strain (ε) produced by cold rolling—ZUBECK *et al.* [ZUB79]. (d) Normalized zero-field transition temperature distributions as derived by ZUBECK *et al.* [ZUB79] from the derivatives of the specific-heat functions.

The existence of a temperature-dependent sub-T_c C_{es}/T would have severe consequences for the theory. The difference between the constancy subsumed by the above model and a Gorter–Casimir temperature dependence such as (10-5) varies as $1 - t^2$; if this is to be kept below 0.1, say, then the half-width of the permissible broadening for $T_c = 9$ K must be less than about 0.5 K. Table 10-3 demonstrates that this requirement was adequately met in plastically deformed Nb. Although its inability to deal with broad transitions might under some circumstances be regarded as a disadvantage, the method of ZUBEK *et al.* [ZUB79] enables narrow superconducting transitions, such as those exhibited by single-phase systems, to be elegantly described.

The third approach to the analysis of rounded specific heat transitions *assumes* a transition temperature distribution function such as an asymmetric-exponential, or an "asymmetric-Gaussian," combines it with an exponential or BCS sub-T_c specific heat function, and fits the result to the experimental data using a suitable nonlinear least-squares-fitting code, such as the "NONLIN 4," devised by WHITE [WHI74[b]] for the calorimetric analysis of magnetic transitions. Advantages of this method are: (i) its capacity for dealing with extremely broad transitions in the presence of any nominated unrounded total specific heat function, and transition temperature distributions ranging continuously from symmetric to fully asymmetric (e.g., "half-Gaussian"); (ii) its ability to cope with a two-component, superconductive-plus-normal system and to assign a reliable numerical value to the fraction of superconducting component in such a mixture.

Alloys which have been treated using NONLIN 4 are Ti–Mo(5 at.%) (quenched + deformed—i.e., Q + D) [WHI76], [COL78[a]], Ti–Mo(7 at.%) (Q and Q + D) [COL78[a]], and $Ti\text{-}Mo_5\text{-}Al_{1,3}$. An example of the application of the (asymmetric-Gaussian-rounded)-(exponential-specific-heat) combination to Ti–Mo(5 at.%) (Q + D) is presented in [WHI76], while that of the (asymmetric-Gaussian-rounded)-(BCS-specific-heat) combination is given in [COL78[a]]. Following a brief general description of the procedure, some hitherto unpublished results for Ti–Mo(7 at.%) (Q and Q + D) are presented.

10.8.2 Outline of a Two-Component Model

The specific heat function to be fitted to the data is $C_r(T)$, as given by (10-17) in which (i) $C_u(T, T_c)$, now the unrounded total specific heat of the two-component sample, depends on T as well as the T_cs of the superconductive subcomponents, and (ii) $F(T_c)$ is, as before, a normalized distribution function.

In case a Gaussian distribution is selected

$$F(T_c) = \frac{1}{\sqrt{\pi(\Gamma^2/\ln 2)}} \exp[-(T_c - T_{c0})^2/(\Gamma_s^2/\ln 2)] \qquad (10\text{-}22)$$

where Γ, the half-width of $F(T_c)$ at half maximum, is related to the usual standard deviation σ by $(\Gamma^2/\ln 2) = 2\sigma^2$. The asymmetry of the Gaussian is controlled by the deviation term Γ_s (equal to Γ for a symmetrical distribution characterized by $s = 0$) which assumes different values above and below T_{c0}, the modal temperature, according to

$$\Gamma_s \equiv \begin{Bmatrix} \Gamma(1+s) & T_c > T_{c0} \\ \Gamma(1-s) & T_c < T_{c0} \end{Bmatrix} \quad |s| < 1 \qquad (10\text{-}23)$$

The general form of the unrounded two-component specific heat function is

$$C_u(T, T_c) = \begin{cases} f\gamma_s T_c C_m(T, T_c) + (1-f)\gamma_n T + \{f\beta_s + (1-f)\beta_n\}T^3 & T < T_c \\ \{f\gamma_s + (1-f)\gamma_n\}T + \{f\beta_s + (1-f)\beta_n\}T^3 & T > T_c \end{cases} \qquad (10\text{-}24)$$

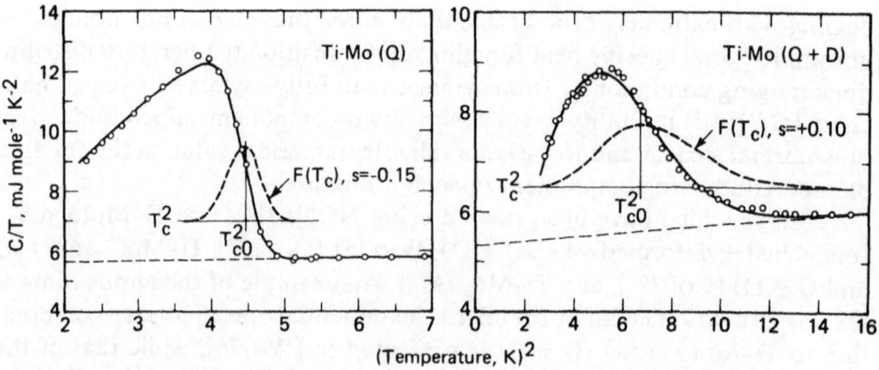

Figure 10-18. Example of the application of the program NONLIN 4 [WHI74b] in the Gaussian–BCS modelling of the calorimetric superconducting transitions in as-quenched and quenched-plus-deformed Ti–Mo(7 at.%). The full line represents the nonlinear least-squares fit of (10-25) to the data points. In order to accomplish the fit, the transition temperatures must be distributed according to the Gaussian, $F(T_c)$, and skewed according to $\Gamma_s = \Gamma(1 \pm s)$ (see (10-23)) where s has been assigned the values indicated. The full sets of fitting parameters are listed in Table 10-4—after WHITE and COLLINGS [WHI76] [COL78a, COL84].

in which the subscripts s and n refer to the superconductive and normal components, respectively, f is the mole-fraction of s, γ and β are the electronic and lattice specific-heat coefficients, respectively, and C_m is a dimensionless model specific-heat function for the superconductive component such as (i) a suitable exponential function [WHI76], (ii) the Gorter–Casimir function as in (10-5), or (iii) the BCS function as tabulated by MÜHLSCHLEGEL [MÜH59].[4] If the latter is selected and if, as is usually the case, it is not possible to separate γ and β into their s and n components, (10-24) simplifies to the operational form

$$C_u(T, T_c) = \begin{cases} f\gamma T_c C_{\mathrm{BCS}}(T/T_c) + (1-f)\gamma T + \beta T^3 & T < T_c \\ \gamma T + \beta T^3 & T > T_c \end{cases} \qquad (10\text{-}25)$$

in which any of the parameters f, γ, β, T_{c0}, Γ, and s may be held fixed or else evaluated by NONLIN 4.

10.8.3 Application of the Two-Component Model

Figure 10-18(a) and (b) and Table 10-4 summarize the application of NONLIN 4 [WHI74b] to the Gaussian–BCS modelling of Ti–Mo(7 at.%) in the as-quenched (Q) and quenched-plus-deformed (Q + D, 47% reduction in thickness) conditions, respectively. In the as-quenched alloy the

[4]This entire analytical approach is restricted by the lack of similarly tabulated numerical results from strong-coupled specific heat theory.

Table 10-4. Two-Part[a] Six-Parameter Gaussian–BCS Fit of the Low-Temperature Specific Heat Data[b] for Ti–Mo and Ti–Mo–Al Alloys

Alloy	Condition[c]	γ, 10^{-3} J mol^{-1} K^{-2}	β, 10^{-5} J mol^{-1} K^{-4}	$100f$ %	T_{c0} K	Γ K	s
Ti–Mo(5 at.%)	Q	4.38 ± 0.01	3.55 ± 0.08		0.8[d]		
Ti–Mo(5 at.%)	Q + D	4.91 ± 0.05	3.73 ± 0.27	68.3 ± 0.8	3.272 ± 0.007	0.233 ± 0.008	−1.0
Ti–Mo(7 at.%)	Q	5.57 ± 0.02	3.68 ± 0.15	92.8 ± 0.7	2.115 ± 0.006	0.632 ± 0.002	−0.15 ± 0.07
Ti–Mo(7 at.%)	Q + D	5.30 ± 0.05	4.54 ± 0.21	73.1 ± 0.7	2.593 ± 0.025	0.444 ± 0.011	−0.10 ± 0.05
Ti_{94}–Mo_5–Al_1	Q	5.07 ± 0.03	4.32 ± 0.12	89.4 ± 0.9	3.103 ± 0.009	0.046 ± 0.002	−0.68 ± 0.16
Ti_{92}–Mo_5–Al_3	Q	4.81 ± 0.03	5.18 ± 0.14	92.1 ± 0.6	2.838 ± 0.002	0.056 ± 0.002	−1.0

[a] A form of the program in which the parameter sets f, γ, β, and T_{c0}, Γ, s are treated separately [WHI76].
[b] Computed and provided by J. J. White III, Battelle-Columbus Laboratories.
[c] Q = quenched.
D = deformed.
[d] Estimated.

transition, centered about $T_{c0} = 2.11$ K, is slightly biased towards lower temperatures ($s = -0.15$); after deformation the transition is considerably broader and biased towards higher temperatures ($s = +0.10$). By way of metallurgical interpretation of the results for the deformed alloy, we first note that a twinned rather than martensitic structure is the more likely product of the deformation. Analysis of the specific heat data yields a 23% T_c enhancement, a 73% complete transition, a small decrease in both γ and θ_D, and a slightly positively skewed distribution of T_c with a large full-width of 0.90 K. Due to the recent finding that the ω-phase in aged samples of this system is depleted near slip planes, a plausible interpretation of the specific heat analysis suggests a structure consisting of up to 10% high-T_c ω-deficient $\omega+\beta$-phase near the twin boundaries, 27% low-T_c twin-boundary and highly defected material, and the remainder being the original $\omega+\beta$-phase. A slightly incomplete proximity effect on the small amount of high-T_c material could explain the high-temperature tail of the T_c distribution, while the presence of appreciable quantities of defect- and proximity-effect-broadened original $\omega+\beta$-phase could account for the large width of the distribution.[5]

Results for other investigated alloys are also summarized in Table 10-4. The types of transition amenable to treatment by the method just described are unrestricted as to width or skewness. For example, the transition depicted in Figure 10-13 turned out to possess a low-temperature ($T_c < T_{c0}$) half-width of $\Gamma_s = \Gamma(1-s) = 0.466$ K yet was unsmeared on the high-temperature side of T_{c0}; the transitions in Ti–Mo_5–$Al_{1,3}$, although skewed, were rather sharp—as such they are possible candidates for treatment by the Zubeck–Barbee technique [ZUB79].

10.9 Rounded Calorimetric Transitions into the Mixed State

10.9.1 General Principles

In a strong applied field, H_a, the superconducting transition occurs at a temperature T_s given by

$$H_a = H_{c2}(T_s) = H_{c20}\left[1 - \left(\frac{T_s}{T_c}\right)^2\right] \cong \sqrt{2}\kappa_{GL} H_c(T_s) \qquad (10\text{-}14)$$

and illustrated in Figure 10-19(a). Thus, T_s depends on H_{c2}, which in turn

[5]The heat capacity of metallic bilayers in superconductive proximity is discussed by ZAITLIN [ZAI82]; proximity effect in general is the subject of Chapter 11.

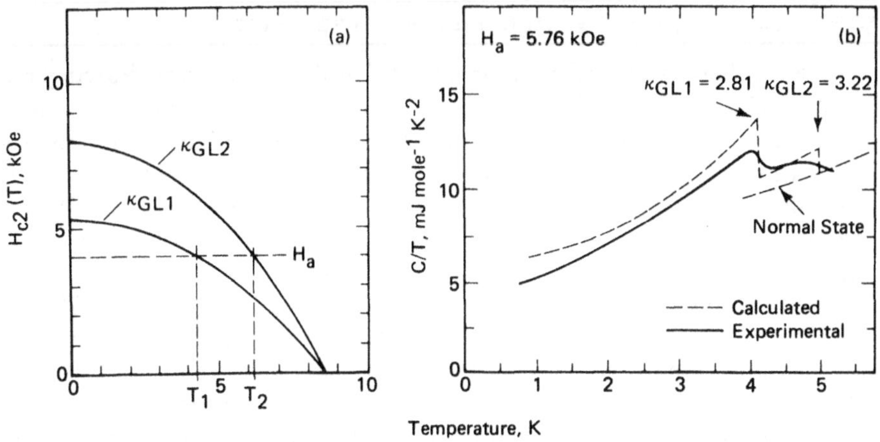

Figure 10-19. (a) Schematic representation of H_{c2} *versus* T for two values of the Ginzburg-Landau parameter, κ_{GL}, indicating separate transitions into the mixed state during cooling in applied field H_a at $T_s = T_2$ and T_1, respectively; (b) mixed state specific heat calculated by BARBEE [BAR69] with the aid of the VAN VIJFEIJKEN [VIJ68] expression ((10-10)) with $H_a = 5.76$ kOe and the two values of κ_{GL} shown, for comparison with an experimental result for cold-drawn Nb wires due to FERREIRA DA SILVA *et al.* [FER66[a]].

depends on the Ginzburg-Landau parameter κ_{GL}. But as considered in detail in Chapter 14, since

$$\kappa_{\text{GL}} = \kappa_{\text{GL}}^{c} + \kappa_{\text{GL}}^{d} \tag{10-26a}$$

$$= \kappa_{\text{GL}}^{c} + 7.49 \times 10^3 \gamma_v^{1/2} \rho_n \tag{10-26b}$$

(the first term referring to the "clean limit" of κ_{GL}), $H_{c2}(T_s)$ is sensitive to metallurgical factors that influence ρ_n, the normal-state residual resistivity. Taken together, (10-14) and (10-26) indicate that T_s responds to variations of T_c at constant H_{c2}, as well as to variations of H_{c2}, κ_{GL}, and ρ_n at fixed T_c. We have already noted that deformation is responsible for small variations in T_c; in the mixed state this will be reflected as a variation in T_s. But this contribution to transition rounding has been found to be dominated by the much more pronounced distribution in T_s that accompanies the modulation of ρ_n (thus κ_{GL} and hence H_{c2}) brought about by the same degree of deformation. Mixed-state transition temperature broadening by the mechanism just discussed is not restricted by size considerations; thus, for example, precipitation of any kind could be expected to have similar consequences.

Figure 10-19(b), from the work of BARBEE [BAR69] and FERREIRA DA SILVA *et al.* [FER66[a]], illustrates the effect of two-κ_{GL} broadening on the specific heat transition in deformed Nb. In practice, of course, a *distribution* of κ_{GL}s is to be expected, a problem that was to be addressed by BARBEE and colleagues in subsequent publications (e.g., [ZUB79]).

10.9.2 Deformation Structure and the Mixed State

The influence of deformation by cold rolling and wire drawing has been considered in Section 2.8. The deformation processes, which lead ultimately to the establishment of the fine fibrous structures favored for high-current alloy superconductors, begin with the assemblage of an initially uniform distribution of dislocations into a cellular network. In Nb, for example, this process takes place as the level of deformation increases within the range of about 3 to 20% (in particular 3–10% [NAR66]; 10–20% [THO75]). Cell diameters are claimed to be typically 0.1–1 μm [ZUB79] (cf. 0.5–1.2 μm, also in Nb [NAR66], [BAR69] and ~0.5 μm, for cold-worked Zr–Nb and Ti–Nb [NAR66]) with cell walls of thickness 0.01–0.2 μm [BAR69], [ZUB79]. Since the cell walls are, in effect, the repositories of the dislocations produced by the deformation processes, it is not surprising that the dislocation density within them, n_w, is relatively high. According to TEM and related studies [BAR69], n_w is in fact twenty times greater than that in the cell interiors. Considering only cell-wall dislocations then, the resulting modulation of H_{c2} would be given by

$$\left.\begin{aligned} \Delta H_{c2} &= \sqrt{2}\Delta\kappa_{\rm GL} H_c(T) \\ \text{with}\quad \Delta\kappa_{\rm GL} &= 7.5\times 10^3 \gamma_v^{1/2}\Delta\rho_n \\ \text{and}\quad \Delta\rho_n &= n_w \rho_D \end{aligned}\right\} \tag{10-27}$$

where ρ_D is the specific dislocation resistivity for the particular material under consideration. A table of data for cold-rolled Nb has been developed by BARBEE [BAR69], from which some typical values are as follows:

$$n_w = 5\times 10^{10}\ \text{cm}^{-2}$$

$$\Delta\rho_n = 3.52\times 10^{-7}\ \Omega\ \text{cm}$$

$$\Delta\kappa_{\rm GL} = 0.225$$

$$\Delta H_{c20} = 628\ \text{Oe}$$

Other estimates of dislocation density in cell walls have been 10^{11} cm^{-2} [BRA72], 10^{12} cm^{-2} [PFE68],[6] and 10^{13} cm^{-2} [FIE69].

[6] In studies of heavily cold worked Ti-50Nb, PFEIFFER and HILLMANN [PFE68] noted subband densities of up to about 5×10^{11} cm^{-2} corresponding to which the dislocation density within the subbands was about 10^{12} cm^{-2}.

The pinning of individual fluxoids (or the entire flux lattice) by dislocation cell walls is regarded as essential for the maintenance of high transport supercurrents in strong transverse magnetic fields. A flux-pinning mechanism in which the κ_{GL}-modulation just introduced translates into energy modulation in the presence of a magnetic field, in the manner initially proposed by NARLIKAR and DEW-HUGHES [NAR66] and developed by subsequent authors (e.g., [HAM72]), is discussed in Chapter 21. As suggested by HAMPSHIRE [HAM74], practically useful flux pinning requires $\Delta\kappa_{GL}/\kappa_{GL}$ to be about 2%. In Ti–Nb(32 at.%) $\kappa_{GL} \cong 78$ [BRA72], which calls for $\Delta\kappa_{GL} \cong 1.6$. This is an order of magnitude greater than that listed above for pure Nb, but so also is the corresponding value of n_w [PFE68].

10.9.3 Calorimetric Studies of κ_{GL} Modulation

In view of the importance of κ_{GL} (cell-wall) modulation in flux pinning an experimental technique was needed, in addition to TEM, whose results could be interpreted in terms of the spacial variation of κ_{GL}. Since, in the presence of a magnetic field, the $\Delta\kappa_{GL}$ is responsible for spread of transition temperatures, the low-temperature specific heat experiment, which as we have seen can be made to yield a transition temperature distribution function, was selected as the experimental probe.

The starting point of the approach used by ZUBECK, BARBEE *et al.* [ZUB79] was an extension of (10-17) in the form

$$C_r(T) = \int\int dT_c \, d\kappa_{GL} \, F(T_c)P(\kappa_{GL})C_u(T, H_a, T_c, \kappa_{GL}) \qquad (10\text{-}28)$$

which takes into account, in addition to the original T_c rounding, a normalized distribution, $P(\kappa_{GL})$, of Ginzburg–Landau parameters. For mathematical convenience C_u may be replaced by $f_1 = (C_{em} - \gamma T)/T$, which vanishes for $T > T_c$ and in which C_{em} is given by (10-10). In developing an expression for $P(\kappa_{GL})$, or equivalently $P_H(H_{c2})$, and during the differentiation of (10-28) it was recognized that the intrinsic T_c distribution, its contribution to the total T_s distribution being less than 5% of that due to the $P(\kappa_{GL})$ or $P_H(H_{c2})$ that was sought, did not contribute significantly to the total rounding and could be treated as a stepfunction. T_c was replaced by some single average value $\langle T_c \rangle$. The final result, in the form

$$\frac{d}{dT}\left[\frac{C_{er} - \gamma T}{T}\right] = \frac{H_a f_1(T, H_a, \langle T_c \rangle, H_{c2})}{\langle T_c \rangle^2[1 - (T/\langle T_c \rangle)^2]} P_H(H_{c2}) \qquad (10\text{-}29)$$

enabled $P_H(H_{c2})$ to be obtained as a function of the derivative with respect to the temperature of the smoothed specific heat data.

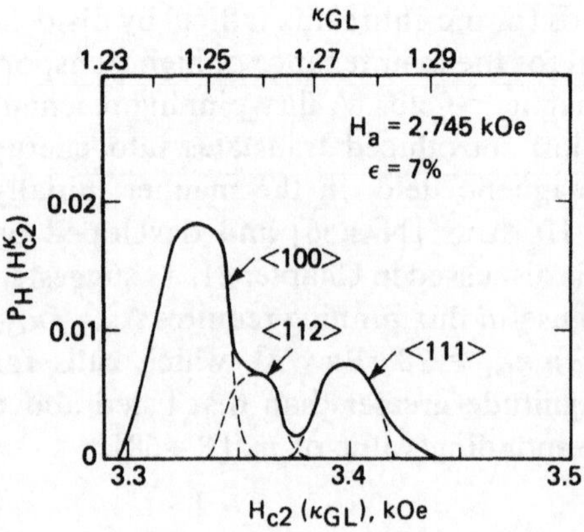

Figure 10-20. Distribution in κ_{GL} (hence $H_{c2}(\kappa_{GL})$) about a mean value of 1.251, as derived from the analysis of mixed-state ($H_a = 2.745$ kOe) specific heat data for Nb, cold deformed 7% by rolling. The prominent features of the distribution were attributed by ZUBECK *et al.* [ZUB79], the authors of this analysis, to grains oriented with respect to H_a (which was parallel to the surface normal) in the manner indicated.

As indicated in Figure 10-20, a typical result of the application of this procedure to deformed Nb, the upper critical field may be the sum of several contributions. Curves such as those depicted were discussed by ZUBECK *et al.* [ZUB79] in terms of distributions of grain orientations, the observed and deduced deformation structures, and the results of electrical resistivity measurements.

Chapter 11

The Superconductive Proximity Effect

11.1 Introduction

11.1.1 Terminology

The superconductive proximity effect is understood to mean the modification of the superconducting properties of one metal when it resides in intimate contact with another. The second metal may itself be superconductive, "normal" (i.e., with a very low or "negative" T_c), or characterized by special magnetic or other properties. The theory of the proximity effect usually presupposes a lamellate sample geometry. The practical bimetallic pair may also conform to this specification, being fabricated in bilayer form using thin-film techniques. Thus, for example, a superconducting film may be deposited by vacuum evaporation or sputtering over a normal-metal *underlayer*, both being supported by an inert *substrate*, whose properties do not enter into the discussion. *In situ* formation of proximity-effect couples is also possible. In such cases, one of the materials may be in the form of either elongated directionally solidified precipitates or of a finely precipitated second-phase dispersion.

Although the deposition of a superconductor *onto* an underlayer or the formation of a precipitate *within* a matrix have been mentioned, it must be recognized that the proximity effect is mutual or reciprocal. Thus, within some characteristic distances from the interface the properties of *both* members of the pair are influenced by proximity. If, for example, a superconductive film (s, of transition transition temperature T_{cs}) is deposited onto a "normal" metal (n, of transition temperature $T_{cn} \gtrless 0$), the transition temperature of the former may be reduced by the presence of the latter; conversely, measurable superconductivity may be induced in the normal metal itself, such that a thin-film bilayer may acquire a new transition temperature, T_{sn}, which is intermediate between T_{cs} and T_{cn}. The characteristic distances referred to above are the coherence lengths ξ_s and ξ_n.

11.1.2 Coherence Length and Literature Survey

The magnitudes of the transition temperature changes depend on the thicknesses D_s and D_n of the participating members of the couple, these thicknesses being gauged in terms of the coherence lengths ξ_s and ξ_n (typically 100–1000 Å) to be defined below. The effect is illustrated by Figure 11-1 from HILSCH's pioneering studies of Pb/Cu double films. As suggested by the figure: with $D_s \gg \xi_s$, the superconducting transition temperature of the bilayer, T_{sn}, decreases as D_n increases, but tends towards a finite limit or saturation value as D_n exceeds ξ_n; on the other hand for $D_s \ll \xi_s$, T_{sn} drops rapidly below the threshold of detection as D_n increases towards ξ_n. As first pointed out by COOPER [Coo61], the proximity effect arises through the ability of the zero-momentum electron pairs (the Cooper pairs) to maintain their coherence over macroscopic distances even after being injected into underlayers of zero or negative T_c (or pairing potential, V). This can be seen in the following expression due to DEUTSCHER and DE GENNES [DEU69] for the probability, P, of finding a Cooper pair a distance x from a s/n interface. According to those authors, if n is "clean," i.e., if the electron mean-free-path, l, is very much greater than ξ_n, to be defined below, then

$$P \propto \exp(-|x|/L) \qquad \text{true for large } x \tag{11-1}$$

with

$$L \equiv \hbar v_F / 2\pi k_B T \tag{11-2}$$

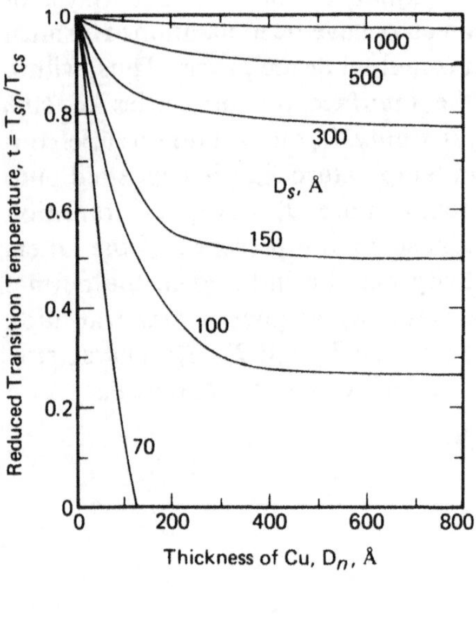

Figure 11-1. Representative results from HILSCH's pioneering study of the superconductive proximity effect in Pb/Cu double films. Depicted are the reduced transition temperatures, T_c/T_{cs}, of Pb films of thickness D_s (=70–1000 Å) in contact with Cu films of thickness D_n—after WERTHAMER [WER63].

where v_F is the Fermi velocity in metal n. It is useful to note that this expression for the characteristic length, L, is similar in form to that of the BCS coherence length, ξ_0, for an isolated superconductor of transition temperature, T_{cs} (see Section 13.4), *viz.*,

$$\xi_0 = \left(\frac{e^{\gamma}}{\pi^2}\right)\frac{\hbar v_F}{k_B T_{cs}} \qquad e^{\gamma} = 1.781 \tag{11-3a}$$

$$\cong \hbar v_F/6k_B T_{cs} \tag{11-3b}$$

If the normal metal is "dirty," that is if the electron mean-free-path, l, is very much less than ξ_n, the leakage of coupled electrons into it takes place by diffusion. Under this condition, the characteristic length for use in (11-1) becomes the normal-metal coherence length, ξ_n. Expressed in c.g.s. units this is

$$\xi_n = (\hbar D_e/2\pi k_B T)^{1/2} \qquad \text{(cm)} \tag{11-4a}$$

where D_e is the single-electron diffusion coefficient, $v_F l/3$. Substituting this value yields[1]

$$\xi_n = (\hbar v_F l/6\pi k_B T)^{1/2} \qquad \text{(cm)} \tag{11-4b}$$

Two other useful expressions for ξ can then be obtained by substituting for either $v_F l$ or $\hbar v_F$ in (11-4b). In the first case, use is made of the relationship [HAK67[b]], [ORL79]

$$v_F l = 9 \times 10^{11}(\pi k_B)^2/e^2 \rho_n \gamma_v \tag{11-5}$$

where γ_v is the electronic volume specific heat coefficient (erg cm^{-3} K^{-2}), ρ_n is the normal-state residual electrical resistivity (Ω cm), e is the electronic charge (e.s.u.), and k_B is Boltzmann's constant (erg K^{-1}). Making the

[1]Occurring throughout the literature are slight variations of this definition depending on the meanings associated with l and T. According to DE GENNES [DEU69], T in (11-2) and (11-4) is an arbitrary temperature but may also represent the transition temperature of the couple [DEG64[a]]. To summarize, the following expressions have appeared in the literature:

$\xi_n = (\hbar v_{Fn} l_n/6\pi k_B T_{sn})^{1/2}$ — HAUSER *et al.* [HAU64]

$\xi = (\hbar v_F l/6\pi k_B T_{sn})^{1/2}$ — WERTHAMER [WER63]

$\left.\begin{array}{l}\xi_s = (\hbar v_{F_s} l_s/6\pi k_B T_{sn})^{1/2} \\ \xi_n = (\hbar v_{Fn} l_n/6\pi k_B T_{sn})^{1/2}\end{array}\right\}$ — JACOBS and GINSBERG [JAC68]

where T_{sn} is the transition temperature of the superposed film pair which, for thick films at least, is not much different from T_{cs} the transition temperature of the isolated superconductor.

substitution yields

$$\xi_n = (9 \times 10^{11})^{1/2}(\hbar\pi k_B/6e^2\rho_n\gamma_v T_{sn})^{1/2} \tag{11-6a}$$

$$= 5.46 \times 10^{-7}(\rho_n\gamma_v T_{sn})^{-1/2} \qquad \text{(cm)} \tag{11-6b}$$

In the second case, $\hbar v_F$ from (11-3a) is substituted into (11-4b) immediately yielding

$$\xi_n = (\xi_0 l/3.4)^{1/2} \cong (\xi_0 l/3)^{1/2} \tag{11-7}$$

Comparing (11-6) and (11-7) with two forms of the dirty-limit zero K Ginzburg–Landau coherence length, ξ^d_{GL0} (Table 14-1), *viz.*,

$$\xi^d_{\text{GL0}} = 8.57 \times 10^{-7}(\rho_n\gamma_v T_c)^{-1/2} \tag{11-8}$$

and

$$\xi^d_{\text{GL0}} = 0.852(\xi_0 l)^{1/2} \tag{11-9}$$

respectively, we note that in either case

$$\xi_n = 0.64\xi^d_{\text{GL0}} \tag{11-10}$$

Some representative values of the (11-6b) form of the coherence parameter (with T_{sn} replaced by the isolated-material T_c) for several representative Ti–TM alloys are given in Table 11-1.[2] Also listed for comparison are corresponding values of the dirty-limit Ginzburg–Landau coherence length appropriate to some intermediate temperature, say $t = 0.5$, computed using

$$\xi^d_{\text{GL}}(t) = \xi^d_{\text{GL0}}(1-t)^{-1/2} \qquad \text{(see Table 14-1)} \tag{11-11a}$$

$$= 1.57\xi_n(1-t)^{-1/2} \tag{11-11b}$$

The early literature of the proximity effect has been reviewed by HILSCH [HIL62]. The development of the currently accepted models of the effect is outlined in Table 11-2, while several important case studies are mentioned in Table 11-3. The lists are, of course, not exhaustive; for further details the reader is referred to sources cited within the papers selected. An early review of the theory of the proximity effect has been provided by DEUTSCHER and DE GENNES [DEU69], who dealt with the classical de Gennes-Werthamer theories referred to in Table 11-2(a), tunnelling implications of

[2] The calculations are confined to the high-concentration bcc alloys; and it may be assumed that $\rho_{\text{bcc}}(c)$ is roughly symmetrical about $c = 0.5$, the asymmetry (i.e., anomalous resistivity) actually observed being a result of the ω-phase effect—see Section 5.5.

Table 11-1. Values of the Coherence Parameters $\xi = 54.6/(\rho_n \gamma T_c)^{1/2}$ Å and $\xi_{GL}^d(0.5)$ for Several Representative Ti–V, Ti–Nb, and Ti–Mo Alloys

Concentration at.%	Molar Weight M g mol^{-1}	Nominal Density[a] ρ_d g cm^{-3}	Residual Resistivity[b] ρ_n $\mu\Omega$ cm	Electronic Specific Heat Coefficient γ mJ mol^{-1} K^{-2}	γ 10^3 erg cm^{-3} K^{-2}	Superconducting Transition Temperature T_c K	$\xi(T_c)$[c] Å	$\xi_{GL}^d(0.5)$[d] Å
Vanadium								
90	50.64	5.9	17.6(a)	10.2(b)	11.9×10^3	5.7_5 (b)	50	111
80	50.33	5.8	28.0	10.5	12.1×10^3	7.2	35	78
70	50.03	5.6	48.8	10.6	11.9×10^3	7.3	26	58
60	49.73	5.5	56.4	10.6	11.7×10^3	7.3	25	56
50	49.42	5.3	78.9	10.5	11.3×10^3	7.1	22	49
Niobium								
90	88.41	8.0	12.0(a)	10.3(c)	9.3×10^3	9.5(d)	53	118
70	79.40	7.2	30.6	10.3_5	9.4×10^3	9.7	33	73
60	74.90	6.8	42.2	10.5_5	9.6×10^3	9.9	27	60
52	71.30	6.5	49.3	10.7	9.8×10^3	9.8	25	56
Molybdenum[b]								
50	71.92	7.4	83.2(a)	5.1_5(e)	5.3×10^3	(1.6)(e)	65	144
37.5	65.92	6.6	70.8	6.2_5	6.3×10^3	2.9	48	107
25.0	59.91	5.9	89.1	7.0	6.9×10^3	3.9	35	78

[a]Calculated by linear interpolation on a mole-fraction basis.
[b]Electrical resistivities of quenched Ti–Mo(5–20 at.%) alloys are listed in Table 5-4.
[c]Eq. (11-6a).
[d]Eqs. (11-8) and (11-11).

Literature
(a) Berlincourt and Hake [BER63]
(b) Collings *et al.* [COL75]
(c) Verkin [VER80]
(d) Collected data
(e) Collings *et al.* [COL72]

Table 11-2. Superconductive Proximity Effect—Sequential Development of Theoretical Models

Analytical Approach	Expressions[a]	Literature
(*a*) *The de Gennes-Guyon-Werthamer Model*		
Modification of the BCS coupling parameter $n(E_F)V$ for a bimetallic sample in the "Cooper limit" of $\xi \gg$ film thickness, $D_{n,s}$ in case $[n(E_F)V]_n = 0$.	$[n(E_F)V]_{\text{eff}} = \frac{D_s}{D_n + D_s}[n(E_F)V]_s$ for use in BCS expressions based on $\Delta_{00} = 2\langle h\nu\rangle \exp[-1/n(E_F)V]$	Cooper [Coo61]
Generalization of Gor'kov's Green's-function treatment, under the assumption $D_{n,s} \gg \xi$, in a discussion of induced superconductivity in "normal metal."	$\exp(-2kD_n) = (T_{ns} - \theta)[-f'_T(\theta, D_s)/2] \quad t_c \ll 1$ enabled k^{-1}, a reciprocal normal-layer coherence length (with limiting value ξ_n (11-4a)), hence $n(E_F)V$ etc., to be estimated. Here, $\theta(D_s)$ is some "limiting temperature" and f is the Fermi function.	de Gennes and Guyon [DEG63]
An extension of the above treatment and the development of expressions for T_{sn} (relative to that of the bulk superconductor, T_{cs}) valid for (a) semiinfinite normal substrate, (b) finite normal layer thickness, (c) the Cooper limit of $D_{n,s} \ll \xi$.	*General Solutions* $\ln(T_{cs}/T_{sn}) = \chi(\xi_s^2 k_s^2)$ $\ln(T_{cn}/T_{sn}) = \chi(-\xi_n^2 k_n^2)$ $[n(E_F)\xi^2 k \tan kD]_s = [n(E_F)\xi^2 k \tanh kD]_n$ *Particular Solutions for Cases* (*a*), (*b*) *and* (*c*) (a) $d_s = (\pi/2)(1 - t_c)^{-1/2} \cot^{-1}[(2/\pi)(t_c^{-1} - 1)^{1/2}]$ (b) $t_c \cong 1 - (t_c/t_\infty)(1 - t_\infty) \tanh D_n/\xi(t_\infty)$ (c) $t_c \cong [1 + (\pi^2/4)(D_n/D_s)]^{-1}$	Werthamer [WER63]

Full development of the Gor'kov Green's-function treatment and the development of explicit expressions for T_{sn} under the conditions (a) the Cooper limit D_n, $D_s \ll \xi_n$, ξ_s, respectively, (b) thick films defined by D_n, $D_s \gg \xi_n$, ξ_s, respectively, (c) thin SC layer ($D_s \ll \xi_s$) on a thick substrate ($D_n \to \infty \gg \xi_n$), (d) SNS junctions.	(a) $[n(E_F)V]_{\text{eff}} = \dfrac{(nV)_n n_n D_n + (nV)_s n_s D_s}{n_n D_n + n_s D_s}$ with $n \equiv n(E_F)$ (c) $D_s = \dfrac{D_t}{1 - \beta(\gamma/2t_c)^{1/2}}$ with $D_t = (n_n/n_s)(2\gamma t_c/\pi)^{1/2}\xi_n$ $\beta = (\sqrt{2} - 2)\zeta(\frac{1}{2}) = 0.86$ γ = Euler's constant = 1.78	de Gennes [DEG64[a]]
Generalized theory of the T_{sn} of superposed films based on the methods of DE GENNES and WERTHAMER. Yields results in good accord with experiment.	The results of WERTHAMER follow analytically in the dirty limit if the T_cs of the films are nearly equal.	Moormann [MOO66]
A refinement of the de Gennes-Guyon-Werthamer calculation of the T_{sn} for superposed metals by the use of coupled integral equations for the order parameters $\Delta_s(x)$ and $\Delta_n(x)$ in the two metals.	The results of MOORMANN as well as those of DE GENNES, GUYON, and WERTHAMER can be derived from the new coupled-integral-equations theory in the dirty limit.	Jacobs [JAC67]
(b) Other Approaches and Properties		
Calculation of the thermodynamic properties of a double layer with the aid of the generalized Ginzburg-Landau equations for such a system in association with standard methods for the determination of the free energy. Results were presented for the height of the specific heat jump and for T_{sn} primarily for the paramagnetic substrate (strong pair-breaking) case.	*Specific Heat Jump, ΔC (Thick Paramagnetic Films)* $\Delta C_{sn}/\Delta C_s = t_c(1 - 4.72\rho_c)/\beta$ with ρ_c, the value at T_{sn} of the pair-breaking parameter $\rho = (\hbar v_F l/12\pi k_B T)(\pi/2D_s)^2$ i.e., $\rho_c = (\pi^2/8)/(\xi/D_s)^2$ and $\beta \cong 3/2$	Fulde and Moormann [FUL67]

Continued

Table 11-2. *(cont.)*

Analytical Approach	Expressions[a]	Literature
	Transition Temperature (Thick Paramagnetic Substrate) (a) Dirty limit ($l \ll \xi$): $\ln t_c^{-1} = (\pi/2)^4/d_s^2$ (b) arbitrary mean-free-path: $\ln t_c^{-1} = (\pi/2)^2[\Lambda(T_{sn})/D_s]^2$ with $\Lambda(T_{sn}) \xrightarrow[l \gg \xi]{} (7/12)\zeta(3)\xi_0^2$	
A BCS approach is extended to a sandwich consisting of normal and superconducting metals and a potential barrier through which tunnelling takes place. The analysis can be made to yield an expression for the T_{sn}.	$\ln t_c^{-1} = (\Gamma_n/\Gamma)[\Psi(-\frac{1}{2} + \Gamma/2\pi T_c) - \Psi(-\frac{1}{2})]$ with $\Gamma = \Gamma_n + \Gamma_s$ $\Gamma_n = \hbar v_{Fn}\sigma/2BD_n$ $\Gamma_s/\Gamma_n = [n(E_F)D]_n/[n(E_F)D]_s$ where σ = tunnel probability B = function of l/D	McMillan [McM68[a]]
The "Bogoliubov–de Gennes" equations for a two-component superconductor were solved analytically and an expression for the transition temperature was developed in case a superconducting layer of thickness D_s is attached to a thick normal layer.	For the system to be superconductive, D_s must be thicker than some threshold value, D_t, given by $D_t = 0.825\hbar v_F/c4\pi k_B T_{cs}$ Near $D_s \gtrsim D_t$, the transition temperature of the bilayer increases according to $t_c^2 \cong (24/5c^2)(1 - D_t/D_s)$ with $c = \exp[-\Psi(\frac{1}{2}) - 1] = 2.62$	Bar-Sagi [Bar78]

[a] $D_{n,s}$ = thickness of superconductive (s) or normal (n) layers

$\xi_{n,s}$ = characteristic coherence distance for electron pairs in either the normal or superconductive components of the couple such that:

$\xi = (\hbar v_F l / 6\pi k_B T_{cn})^{1/2}$

$\cong (\frac{1}{3}\xi_0 l)^{1/2}$, where ξ_0 is the BCS coherence length of the superconductive component, s, of the couple, Eqs (11-3) and (11-7).

$\zeta(3) \equiv 1.202$

$$\chi(z) \equiv \Psi(\tfrac{1}{2} + \tfrac{1}{2}z) - \Psi(\tfrac{1}{2}) \rightarrow \begin{cases} \ln[1 + (\pi^2 z/4)] & (z \gtrsim 0) \\ (\pi^2/4)\ln(1+z) & (z \lesssim 0) \end{cases}$$ Ψ being the digamma function.

$d = D/\xi$

$t_c = T_{sn}/T_{cs}$

$t_\infty = t_c$ in the limit $D_n \rightarrow \infty$, as in [WER63].

The $k_{s,n}$ are defined in Section 11.6.

Table 11-3. Superconductive Proximity Effect—Some Representative Case Studies[a]

Synopsis of Investigation	Literature
Measurement of the T_{sn} of Pb films in contact with the "nonmagnetic" substrates Cu and Pt. The results were analyzed with the aid of an extension of the DE GENNES ($D \gg l$) and WERTHAMER (finite thickness) theories.	Hauser, Theuerer, and Werthamer [HAU64]
Measurements of the T_{sn} of Pb films in contact with "magnetic" substrates: (a) the ferromagnets Fe, Ni and Gd, (b) the antiferromagnet Cr, and (c) the dilute paramagnetic alloys Mo–Fe(1%) and Pb-Gd (2.9%). T_{sn} was measured as a function of film (Pb in most cases) thickness and the results compared with DE GENNES–WERTHAMER proximity effect theory.	Hauser, Theuerer, and Werthamer[b] [HAU66]
Measurements of the T_{sn} of In/Tl and In/Pb superposed films as functions of normal-metal thickness in tests of the validity of the Jacobs-refined DE GENNES-GUYON-WERTHAMER (weak coupling) theory and its applicability to sandwiches in which the strongly-coupled superconductor, Pb, is one of the elements.	Jacobs and Ginsburg [JAC68]
Calorimetric study of the proximity effect in a directionally solidified eutectic Pb–Sn alloy ("*in situ*" two-component composite). The depression of the specific heat jump at T_c gave qualitative support to the general conclusions of the FULDE-MOORMANN theory (Table 11-2(b)).	Lechevet, Neighbor, and Shiffman [LEC72]

Calorimetric study of the proximity effect in a directionally solidified eutectic Pb-Cd alloy. The depression of the specific heat jump at T_{sn} was compared as a function of laminate spacing with predictions of the FULDE-MOORMANN theory (Table 11-2(b)) in its paramagnetic-substrate limit.	Bevolo, Gibson, Verhoeven, and Finnemore [BEV76]
The heat capacity of an s/n bilayer or sandwich was calculated using BOGOLIUBOV theory. It was claimed that this approach, in contrast to the GINZBURG-LANDAU theory approach used by FULDE and MOORMANN [FUL67] and others, was able to take into account the effects of microscopic electronic excitations at the s/n interface. Expressions were also derived for the specific heat jump at T_{ns} as a function of film thickness.	Zaitlin [ZAI82]
Measurements of the T_{sn} of Nb/Cu multilayers (so-called "superlattices") as function of layer thickness using a mutual-inductance bridge technique. Good agreement with DE GENNES-GUYON-WERTHAMER predictions were obtained for $D_n = D_s > 600$ Å. With thinner layers, such that $D_{n,s}/2 \lesssim \xi_s$, T_{sn} was less than the predicted value due, it was presumed, to a lowering of the intrinsic T_{cs} of the Nb as a consequence of a film-thickness (size-effect) restriction of its electronic mean-free-path.	Banerjee, Yang, Falco, and Schuller [BAN82]

[a] For a complete picture the references cited within the works listed should also be consulted.
[b] Submitted in 1964, a few months after the previous paper.

the proximity effect (see Table 11-2(b)), and the proximity effect in strong magnetic fields. At about the same time, CLARKE [CLA68[a]] reviewed the experimental situation, paying particular attention to transition temperatures of *n*/*s* bilayers, critical currents of *s*/*n*/*s* sandwiches, and tunnelling through *s*/oxide/*n* sandwiches. Neither of these reviews included MCMILLAN's contribution to the subject [MCM68[a]]. An excellent up-to-date review of the effect, published in 1977 and which was thereby able to include a comparison of MCMILLAN's tunnelling model with the diffusion model of DE GENNES and WERTHAMER, has been prepared by GILABERT [GIL77]. Also included in that review were discussions on transition temperature, excitation spectra of *n*/*s* bilayers including specific heat and thermal conductivity, depairing processes in the form of applied magnetic fields and magnetic impurities, and supercurrents across what would now be referred to as Josephson junctions. Finally, it should be pointed out, a wealth of useful relevant information is to be found in the proceedings of the conference *Inhomogeneous Superconductors—1979* [GUB80].

11.1.3 Measuring Techniques

For several reasons, electrical transport techniques (either four-terminal resistivity or tunnelling) are far and away the most popular methods with which to study the properties of proximity-effect systems. With regard to the latter, current-voltage isotherms, acquired over a suitable range of temperatures, will readily yield T_{sn} as the temperature at which supercurrent tunnelling characteristics give way to normal resistive linearity. Magnetic techniques are also available; SCHULLER *et al.*, for example, studied multilayers of Nb/Ti [ZHE81] and Nb/Cu [BAN82] using the NBS mutual-inductance-bridge method. As described by SOULEN *et al.* [SOU73], the sample (or several samples with different transition temperatures) is placed within one of two similar coil-pairs connected so as to form an audio-frequency bridge circuit. With the bridge initially balanced, the temperature of the bridge-containing cryostat is swept over a range of several mK so that the sample (or one of them) passes completely through its superconducting transition. Finally, provided that suitable samples and/or measuring equipment are available, low-temperature specific heat is strongly recommended for the study of superconductive proximity effect. As pointed out by FULDE and MOORMANN [FUL67], who analyzed the thermodynamic (particularly calorimetric) properties of superconducting couples, the specific heat jump at T_c (i.e., the ΔC, discussed in Section 8.4.4) is a much more sensitive indicator of proximity effect than is any resistively measured reduction in T_c. SHIFFMAN *et al.* [SHI64], [LEC72] using eutectic Pb-Sn alloys, and subsequently BEVOLO *et al.* [BEV76] with eutectic Pb-Cd alloys, investigated the proximity-effect-induced reduction in the height of ΔC, obtaining results which were then discussed in terms of the Fulde–Moormann theory. For contemporary experimentallists the theoretical results of ZAITLIN [ZAI82] (see Table 11-3(b)) are also available to aid in the analysis of calorimetric proximity-effect measurements.

11.1.4 Influence of the Underlayer on T_{sn}

The influence of the underlayer on the transition temperature of the superimposed superconductor, and hence T_{sn}, the transition temperature of the couple, depends among other things on its pair-breaking ability—i.e., upon whether its pairing potential, V, (hence BCS $n(E_F)V$, see for example (8-15)) is greater than or less than zero. The reduction of T_{cs} by the underlayer thus becomes more pronounced as its paramagnetic susceptibility increases [FUL67], becoming greatest when it is ferromagnetic [HAU66]. The influence of underlayer susceptibility on the transition temperature has been studied experimentally by HAUSER *et al.* Underlayer materials chosen for study were diamagnetic (Cu), Pauli paramagnetic (Pt) [HAU64], dilute magnetic (Mo–Fe, Pb–Gd), antiferromagnetic (Cr) and ferromagnetic (Fe, Ni, and Gd) [HAU66]. In this connection it is interesting to report that proximity-effect T_c-depression was suggested as being a reliable indicator of the existence of a localized magnetic moment in situations where other more conventional techniques might not be applicable [HAU66]. Specific-heat jump-height has been considered theoretically by FULDE and MOORMANN [FUL67], who noted that in case the normal substrate was paramagnetic, jump-height reductions of about 15% were still encountered when the substrate thickness was increased to 100 times the coherence distance; a weaker effect would of course have been expected when the superconductor was in contact with a nonmagnetic normal film.

11.2 Selection of Couples

As indicated above, sample and material variables that have been adjusted during the study of proximity effect have been (i) the relative thickness $d_{n,s} = (D/\xi)_{n,s}$ of the superconductive and normal layers, and (ii) the magnetic properties of the underlayer. Several other materials and geometrical considerations that have been addressed are listed below.

Although the de Gennes–Werthamer model (see Table 11-2(a)) was based on weak-coupling theory, its verification had generally relied on the results of tests in which the strong-coupled superconductor Pb was one of the elements of the junction. Accordingly, JACOBS and GINSBERG [JAC68] designed an experiment in which the properties of In/Tl and In/Pb couples were intercompared and related to de Gennes–Werthamer–Jacobs theory

[JAC67]. The results demonstrated the absence of pronounced strong-coupling effects in the Pb-element couples thus substantiating the conclusions drawn by the earlier investigators.

In the design and preparation of couples, mean-free-path length, l, is also a variable. If a large l is selected, experimental and interpretational difficulties are encountered as unavoidable nonuniformity of the interface changes its reflection and transmission properties in an uncontrolled manner. These difficulties can be eliminated through the use of small-l, or "dirty," materials in which electron diffusion (see (11-4)) is the governing mechanism and interface irregularities are masked by intrinsic bulk scattering. Second, in the preparation of couples it is obviously important, in the interest of good electrical contact between n and s, to avoid the inclusion of foreign atoms, particularly those that would establish a thin insulating layer between the components. Finally, intermetallic diffusion should be avoided; this suggests the use of n and s metals, which, in pairs, form neither solid solutions nor intermetallic compounds [DEU69]—a requirement which, of course, may be relaxed slightly if a small l has been selected [DEU69]. The requirements of a small l, clean interfaces and low interdiffusion are simultaneously satisfied if the techniques of evaporation or sputtering onto a cryogenically cooled substrate are employed. The small l is a result of the quenching-in of atomic defects [DEG64[a]]. But restraint must be exercised in the choice of l—since according to the relationship $\xi \propto l^{1/2}$, (11-4), there are limits below which l should not be reduced if experiments are to be performed in which $D \cong \xi$ and the films are not to be unduly thin.

Both nonequilibrium and equilibrium metallurgical phase transformations may also provide conditions amenable to the operation of superconductive proximity effects between host alloys and precipitated phases. Thus, for example, the presence of athermal ω-phase lowers the bulk T_cs of as-quenched $\omega+\beta$-phase Ti–TM alloys (Section 10.2). Two-phase samples suitable for proximity-effect study may also be prepared by the directional solidification of binary alloys of eutectic compositions. SHIFFMAN *et al.* [SHI64], using a Pb–Sn eutectic (soft solder), were the first to prepare samples in this way. The work was subsequently extended, for comparison with the results of the thermodynamic calculations of FULDE and MOORMANN [FUL67], by LECHEVET *et al.* again with eutectic Pb–Sn [LEC72] and by BEVOLO *et al.* with eutectic Pb–Cd [BEV76]. Since it is possible to prepare rather massive samples of such *in situ* composites, they are able to be studied using conventional low-temperature specific heat techniques. As indicated in the previous chapter, bulk superconducting transition temperature has been employed diagnostically in metallurgical investigations of $\omega+\beta$-phase Ti–TM alloys. On the other hand, the directionally solidified eutectics have been employed as subjects for the study of proximity-effect reductions of the specific-heat jump itself.

11.3 Experimental Materials and Techniques

The experiments listed in Table 11-3 provide several useful examples of sample preparation methods for proximity-effect investigation. Jacobs and Ginsberg [Jac68] prepared samples about 100 Å thick by vacuum evaporation from a pair of Mo boats onto glass slides attached with Apiezon N grease to a He-cooled Cu block. Transition temperatures were measured using AC (97.5 Hz) four-terminal resistometry. Hauser *et al.* prepared samples by getter-sputtering onto a liquid-N-cooled glass substrate. Films of thicknesses ranging from several tens to several thousands of Å were prepared in that way. In the first series of experiments (on Pb/Cu and Pb/Pt films [Hau64]) the sputtering had to be interrupted while the electrodes were changed, exposing the interface to the possibility of contamination. The installation of a twin sputtering chamber prior to the preparation of the second series of double films (Pb deposited on various magnetic substrates [Hau66]) removed this danger. Transition temperatures were measured using DC resistometry. Nowadays, sophisticated ion-beam deposition techniques are available.

Proximity-effect couples may also be prepared using *in situ* techniques which exploit one or another of several possible precipitational processes. Although the danger of interface contamination is averted using such procedures, the possibility of interdiffusion or the formation of interface-phases cannot be ignored. The lamellar product of eutectic solidification (see [Fle74, pp. 93–104] for a full treatment of this topic) is ideally suited for this purpose. Shiffman *et al.* [Shi64], [Lec72], the first to exploit this approach, prepared samples of eutectic Pb–Sn alloy by conventional directional solidification using a travelling-zone furnace, a technique which resulted in the establishment of large colonies of lamellae running parallel to the rod axis. Bevolo *et al.* [Bev76] prepared samples of eutectic Cd–Pb alloy (Pb concentration 28 at.%) for calorimetric investigation using the technique of high-speed (i.e., high temperature-gradient) directional solidification. Two methods of establishing the necessary temperature gradients are in vogue: a furnace, the lower end of which consists of a water-cooled chill block, is raised with respect to a suitably enclosed cylinder of molten alloy [Par78]; alternatively a tube of molten alloy is lowered through the steep temperature gradient established between the furnace and suitably arranged jets of cold water [Bev76]. In the experiments of Bevolo *et al.* [Bev76] the rate of movement, r, of the sample was varied from 3×10^{-1} to 3×10^{-5} cm s^{-1} resulting in laminate repeat distances, $L = 500\, r^{-1/2}$ (where L is in Å and r is in cm s^{-1}), of between 1000 and 85,000 Å. Although the lamellae consisted of Cd-rich (normal) and Pb-rich (superconducting) plates, under the assumption that they were unalloyed, estimates of plate thicknesses were able to be obtained from measurements of the repeat distance, L, aided by density data and a knowledge of the mean alloy composition. For the specimen in which $L = 1000$ Å, for example, $D_n = 190$ Å and $D_s = 810$ Å.

The superconductive properties of the eutectic couples were studied by low-temperature calorimetry, particular attention being focused on the height, ΔC, of the specific heat jump at T_{sn}, which, according to FULDE and MOORMANN [FUL67], is a particularly sensitive probe of the proximity effect. Another, but less controllable, proximity-effect configuration is provided by ω-phase precipitation. In calorimetric studies of quenched $\omega+\beta$-phase Ti-TM alloys to be discussed below, emphasis was placed on the proximity-effect-induced reduction of bulk T_c in the Cooper limit.

11.4 Theoretical Considerations—Cooper's Model

In his pioneering 1961 article, COOPER [COO61] proposed a mechanism for the proximity effect within the framework of BCS theory, placing particular emphasis on the dimensionless BCS interaction parameter $n(E_F)V$ (see, for example, (8-15)). He suggested that when a superconductor (s) is laid over a normal metal (n) the wave function (or order parameter) of the zero-momentum superelectron pairs is able to penetrate a considerable distance into the substrate even when $[n(E_F)V]_n = 0$. Thus, while the electron pairs sample the entire volume of the couple (i.e., $\xi_{n,s} \gg D_n$ and $\xi_s \gg D_n$—subsequently referred to as the "Cooper limit"), the positive electron-electron interaction is confined to n. This renormalization was expressed in terms of an effective pair interaction given by[3]

$$[n(E_F)V]_{\text{eff}} = \frac{D_s}{D_s + D_n}[n(E_F)V]_s \tag{11-12}$$

which, incidentally, may be regarded as a special case of the equation[4]

$$(nV)_{\text{eff}} = \frac{n_s V_s D_s + n_n V_n D_n}{D_s + D_n} \qquad n \equiv n(E_F) \tag{11-13}$$

Finally, as pointed out by WERTHAMER [WER63], (11-12), when inserted

[3]This argument should be compared with that introduced by FULDE and MOORMANN [FUL67] (see Section 11.8), whose simplest model called for a vanishing of the *order parameter* at the interface, $\Delta(x) = 0$, to achieve which a strongly paramagnetic substrate $[n(E_F)V]_n < 0$ was required. Cooper's model, of course, requires only a vanishing of the *coupling parameter*, $[n(E_F)]_n = 0$, at the interface and within the normal layer.

[4]Henceforth, n will be used as an abbreviation for $n(E_F)$, the per-atom Fermi density-of-state in eV.

into the (8-34) form of the BCS equation for the transition temperature, leads to a particularly simple expression for the T_c reduction, *viz.*,

$$t_c \equiv T_{sn}/T_{cs} = (T_{cs}/1.14C_0\theta_D)^{D_n/D_s} \tag{11-14}$$

It could be argued, however, that as a weighting factor for the pair interaction the layer thickness, D, although necessary, is not sufficient, and that, as indicated below with reference to the work of DE GENNES [DEG64[a]], the electronic density-of-states in each medium should also play a role.

The early development of proximity-effect theory in terms of the work of COOPER [COO61], DE GENNES and GUYON [DEG63], WERTHAMER [WER63], DE GENNES [DEG64[a]], and others is outlined in Table 11-2(a). Some of the important contributions made by these authors are considered below.

11.5 The Theory of de Gennes

DE GENNES has presented a full development of Gor'kov's Green's-function treatment and discussed superconductivity in the dirty limit ($l \ll \xi$) of various double-film configurations. Bilayer geometry was expressed in terms of the relationships of film thicknesses $D_{n,s}$ to the respective characteristic (or coherence) distances $\xi_{n,s}$ defined in Section 11.1.2. The various arrangements which were considered are outlined below.

11.5.1 Thick Films in Which $D_{n,s} \gg \xi_{n,s}$

The result of the dirty-limit thick-film calculation was an implicit equation for the transition temperature involving the properties of the superconductive and normal layers. It was recognized, however, that although $n_{n,s}$, $D_{n,s}$, T_{cs} and V_s (or at least $[nV]_s$) were known, and T_{sn} could

Table 11-4. Proximity-Effect Deduced Superconducting Transition Temperature Pure Copper

T_c	Literature	Comments
~1 mK	de Gennes and Guyon [DEG63]	Also deduced critical field (<0.1 Oe)
6 mK	Hauser *et al.* [HAU64]	T_{sn} of n/s
10 mK	Moormann [MOO66]	See Table 11-2(a)
0.02–70 mK	Minnigerode [MIN66]	T_{sn} of n/s
~70 mK	Jacobs [JAC67]	Regarded as a correction to earlier deductions
0.03–400 mK	Clarke [CLA69]	Supercurrents in $s/n/s$

be measured, the normal-layer properties V_n and T_{cn} were generally unknown. Under these conditions, although the theory would not be able to be tested, V_n or T_{cn} could be obtained as results of the experiment itself.

Using such an approach it is possible to produce estimates of the superconducting transition temperatures of metals usually regarded as "non-superconducting," Cu for example. The results of several such measurements are listed in Table 11-4. Evidently the *percentage* errors associated with proximity-effect estimates of T_{cn} (and $[nV_n]$, [CLA69]) are enormous. The actual values obtained are, however, qualitatively correct and heuristically valuable.

11.5.2 Thin Superconductive Film on a Massive Normal Underlayer

Under the conditions $D_s \ll \xi_s$ (i.e., Cooper limit for s) and $D_n > \xi_n$, de Gennes theory has led to the following implicit expression for $t_c \equiv T_{sn}/T_{cs}$:

$$\left.\begin{aligned} &\frac{D_s}{D_t} = \frac{1}{1-\beta(\gamma/2t_c)^{1/2}} \\ &\text{with}\quad \begin{cases} \beta = (\sqrt{2}-2)\zeta(\tfrac{1}{2}) = 0.86 \\ \gamma = \text{Euler's Constant} = 1.78 \\ D_t = (n_n/n_s)(2\gamma t_c/\pi)^{1/2}\xi_n \end{cases} \end{aligned}\right\} \tag{11-15}$$

For values of superconductive layer thickness D_s less than some threshold value D_t the superconductivity of the couple vanishes even in the presence of a nonrepulsive interaction V_n in the normal layer. This comes about, according to DE GENNES [DEG64[a]], as a consequence of the Cooper reduction of V_s by the factor $D_s/(D_s + \xi_n)$ (cf. (11-12)) coupled with the temperature dependence of ξ_n (according to (11-4)) which brings about an accelerating decrease of T_{sn} as D_s decreases.

The recent analysis of BAR-SAGI [BAR78] (see Table 11-2(b)) for a superconductive film in contact with a thick normal underlayer also leads to a film-thickness threshold for superconductivity, but of the form

$$t_c^2 \propto (1 - D_t/D_s) \tag{11-16}$$

11.5.3 Thin Films—The Cooper Limit ($D \ll \xi$)

With the film thicknesses D_s and D_n both much smaller than the respective coherence lengths ξ_s and ξ_n, DE GENNES' analysis [DEG64[a]] leads to an expression for an effective BCS coupling parameter given by

$$(nV)_{\text{eff}} = \frac{n_s^2 V_s D_s + n_n^2 V_n D_n}{n_s D_s + n_n D_n} \tag{11-17}$$

which is the Cooper formula of (11-13) but with the weighting factors D_n and D_s replaced by $n_n D_n$ and $n_s D_s$.

A practical version of this expression is employed in Section 11.9.3 to describe the proximity effect in as-quenched Ti–Mo(10.3 at.%), a β-phase alloy containing a high volume density of a very fine evenly dispersed ω-phase precipitate.

11.6 The Theory of de Gennes and Werthamer

11.6.1 Formulations of the Theory

As indicated in Table 11-2(a), the initial dirty-limit theory of DE GENNES and GUYON [DEG63] was extended first by WERTHAMER [WER63] and then by JACOBS [JAC67], whose more general theory was applicable to all film thicknesses and mean-free-paths. The *Cooper limit* ($D \ll \xi$) of JACOB's theory does not produce the correct result; instead it predicts that $T_{sn} = 0$ when $D_s = 0$ rather than D_t (cf. (11-15)), even for $V_n < 0$ where T_{sn} should clearly drop to zero for small but finite values of D_s. In the *dirty limit* ($l \ll \xi$) the de Gennes–Werthamer equations [WER63], [JAC67] are recovered in the form[5]

$$\ln(T_{cs}/T_{sn}) = \chi(k_s^2 \xi_s^2) \tag{11-18}$$

$$\ln(T_{cn}/T_{sn}) = \chi(-k_n^2 \xi_n^2) \tag{11-19}$$

with

$$[n\xi^2 k \tan kD]_s = [n\xi^2 k \tanh kD]_n \tag{11-20}$$

where $n_{n,s}$ represents, as usual, $n(E_F)|_{n,s}$. The χ function of (11-18) and (11-19) is defined in terms of the digamma function, Ψ, according to

$$\chi(z) = \Psi(\tfrac{1}{2} + \tfrac{1}{2}z) - \Psi(\tfrac{1}{2}) \gtrsim \begin{cases} \ln[1 + (\pi^2 z/4)] & z \geq 0 \\ (\pi^2/4)\ln(1+z) & z \leq 0 \end{cases} \tag{11-21}$$

where, it should be mentioned, the logarithmic approximation is valid only for a small range of Z [BAN82[a]]. The quantity k may be regarded as a decay

[5]In WERTHAMER's version [WER63], (11-18) and (11-19) are formulated using the *common* ξ, defined in the footnote to Section 11.1.2 as $(\hbar v_F l/6\pi k_B T_{sn})^{1/2}$.

constant for superconductive interaction. Thus, k_s^{-1} is essentially the Ginzburg-Landau coherence length and k_n^{-1} represents the depth of penetration of Cooper pairs into the normal component of the bilayer [CLA68[a]]. It follows from (11-18) to (11-21) that

$$k_s = (2/\pi\xi)[(T_{cs}/T_{sn}) - 1]^{1/2} \tag{11-22}$$

$$k_n = \xi^{-1}[1 - (T_{cn}/T_{sn})^{4/\pi^2}]^{1/2} \tag{11-23}$$

Equation (11-23) then enables, at least in principle, a value of T_{cn} to be extracted from experimentally determined values of k_n, ξ_n and T_{sn} [WER63].

WERTHAMER [WER63] specialized these equations to the following situations:

(a) The normal film is very thick, $k_n D_n \gg 1$, in which case

$$d_s = (\pi/2)(1 - t_c)^{-1/2} \cot^{-1}[(2/\pi)(t_c^{-1} - 1)^{1/2}] \tag{11-24}$$

where $d_s \equiv D_s/(\hbar v_{Fs} l_s/6\pi k_B T_{cs})^{1/2}$; $t_c \equiv T_{sn}/T_{cs}$

(b) The film thicknesses are finite, in which case

$$t_c \cong 1 - (t_c/t_\infty)(1 - t_\infty)\tanh D_n/\xi \tag{11-25}$$

where $t_\infty \equiv t_c(D_n \to \infty)$ and ξ (which is the "common" version as defined in footnotes 1 and 5) is evaluated at $t_c = t_\infty$.

(c) The film thicknesses are at the "Cooper limit," $D_{n,s} \ll \xi$, in which case

$$t_c \cong [1 + (\pi^2/4)(D_n/D_s)]^{-1} \tag{11-26}$$

which may be compared with Werthamer's extension of the basic ($n_n V_n = 0$) Cooper formula, *viz.*,

$$t_c = (1.14 C_0 \theta_D/T_{cs})^{-D_n/D_s} \tag{11-14}$$

The application of the de Gennes-Werthamer theory to the proximity effects noted when Pb is laid over various diamagnetic, paramagnetic, and ferromagnetic underlayers [HAU63, HAU64] has yielded some very interesting results.

11.6.2 Proximity Effect Against Normal Metals

HAUSER *et al.* [HAU64] showed how equations (11-18) to (11-21) would lead, under certain simplifying assumptions, to

$$D_n = k_n^{-1}\left[\frac{1}{2t_c^{1/2}}\ln\left(\frac{1+J}{1-J}\right)\right] \tag{11-27}$$

where $J \equiv [(1-t_c)/(1-t_\infty)](t_\infty/t_c)^{1/2}$, and t_c and t_∞ are as defined in parts (a) and (b) of the above display.

In studies of proximity effect in Pb/Cu and Pb/Pt couples, underlayer thickness, D_n, was plotted against the term from (11-27) in square brackets, which involves only the transition temperature. After measuring the slope of this plot, k_n^{-1}, and calculating ξ_n with the aid of (11-4) (equating T to T_{sn} is adequate) sufficient data are available for the calculation of T_{cn} using (11-23). The approach actually taken by HAUSER *et al.* [HAU64], however, made use of the following relationship attributed to DE GENNES [DEG64[a]]:

$$k_n^2 = (1/\xi_n^2)[1 - 2n_nV_n/(1 - cn_nV_n)] \tag{11-28}$$

with

$$c = \ln(1.14C_0\theta_{Dn}/T_{sn}) - 2$$

Having acquired $[nV]_n$ in this way, T_{cn} was obtained from an application of the usual BCS formula for the transition temperature, (8-34). As a result of their experiments with Pb/Sn bilayers HAUSER *et al.* [HAU64] concluded that $T_{c,\mathrm{Cu}} \cong 6 \times 10^{-3}$ K. Other estimates of the transition temperature of Cu are listed in Table 11-4.

11.6.3 Proximity Effect Against Magnetic Metals

HAUSER *et al.* [HAU66] went on to apply de Gennes-Werthamer theory to the results of proximity-effect studies of the Pb/magnetic-metal couples listed in Table 11-3. Recognizing that the transition temperature, T_{ca}, of an alloy formed by the addition of a paramagnetic solute to an otherwise pure metal of transition temperature T_{cp} is given by

$$\ln(T_{cp}/T_{ca}) = \chi(\hbar/2\pi\tau_s k_B/T_{ca}) \tag{11-29}$$

where τ_s is an electron-spin relaxation time in the alloy, and the function χ is as defined by (11-21), HAUSER *et al.* [HAU66] suggested that it might be appropriate to replace (11-19) by

$$\ln(T_{cn}/T_{sn}) = \chi(-k_n^2\xi_n^2 + (\hbar/2\pi\tau_s k_B T_{sn})) \tag{11-30}$$

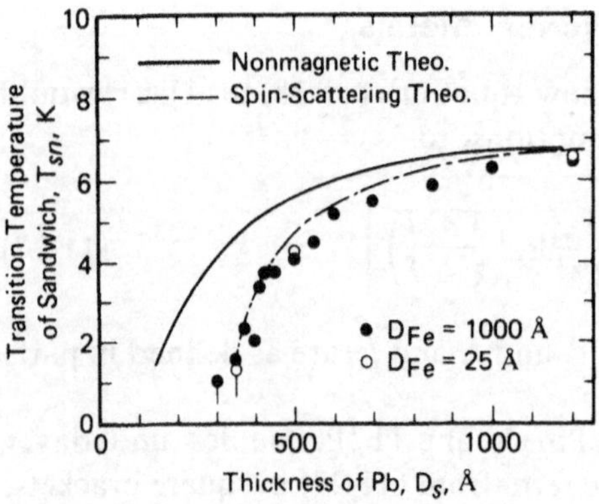

Figure 11-2. Superconducting transition temperature for Pb/Fe sandwiches as function of Pb-film thickness for Fe-film thicknesses of 25 and 1000 Å. Also shown is the theoretical curve based on the de Gennes-Werthamer theory (see equations (11-18), (11-19), and (11-20)) (full line) and that taking into account spin-scattering effects (see (11-30)) (chain line)—after HAUSER *et al.* [HAU66].

After inserting a suitable value for τ_s into this expression it was found that (11-18), (11-20), and (11-30) resulted in plots of T_{sn} *versus* D_s that were in good accord with the experimental data for Pb/Fe sandwiches, Figure 11-2.

Ferromagnetic substrates are exceptionally strong decouplers of the spins of penetrating electron pairs. Thus, it was estimated [HAU66] that whereas k_n^{-1} would have been expected to vary from 23 Å (at 7.2 K) to 50 Å (at 1.5 K) if the layer underlying the Pb were nonmagnetic, in practice the Fe underlayer yielded a k_n^{-1} value of only 6 Å, which is significantly smaller than the nonmagnetic prediction and independent of temperature as predicted by magnetic theory.

11.7 The Evolution of Proximity-Effect Research

The previous section has reviewed the essential results of the de Gennes–Werthamer theory, a dirty-limit approximation developed and originally discussed by DE GENNES and GUYON [DEG63], WERTHAMER [WER63], and DE GENNES [DEG64[a]]. As indicated above (see Table 11-1) this work was subsequently refined by: (i) MOORMANN [MOO66], who produced a generalized theory of the transition temperature of superposed films based on the methods of the above authors whose results could be analytically recovered in the dirty limit; (ii) JACOBS [JAC67], who solved a coupled pair of Gor'kov integral equations for the order parameter in the s and n components of a bilayer. This method, which had general applicability, could also be made to reproduce the de Gennes–Werthamer results in the dirty limit. As indicated briefly in Table 11-1, further contributions to proximity-effect theory have been made by MCMILLAN [MCM68[a]] and by BAR-SAGI [BAR78]. The status of the subject as of 1976 has been reviewed by GILABERT [GIL77].

For the experimentalist, three interesting directions have emerged since the mid-1960s, *viz.*, (i) the study supercurrent tunnelling through $s/n/s$

sandwiches, (ii) the study of proximity effect in samples made up of many alternating thin films of a normal metal and a superconductor (so-called "layered metals" [ZHE81], "modulated structures," or "*s*/*n* superlattices" [BAN82]), and (iii) the application of low-temperature specific heat to the study of proximity-effect bilayers.

(a) Supercurrent Tunnelling. In a pioneering investigation of supercurrent tunnelling, CLARK [CLA69] conducted a series of detailed experiments on thin-film Pb/Cu/Pb sandwiches. The effect of a magnetic field on the tunnelling supercurrent was investigated (see also [HSI80] for some recent studies of tunnelling-current field dependences) whereupon it was discovered that the *s*/*n*/*s* sandwiches behaved phenomenologically as *s*/oxide/*s* Josephson junctions. Today, as "superconducting electronics" the Josephson effect has developed into a branch of applied superconductivity rivalling in magnitude that of "conductor development and large-scale applications." As a consequence of this, the superconductive proximity effect, and especially its supercurrent tunnelling aspects, is acquiring a new importance.

(b) Proximity Effect in Modulated Structures. Superconductivity in layered materials has been the subject of considerable study by SCHULLER and his group at the Argonne National Laboratory (e.g., [ZHE81], [BAN82]). Multiple laminates consisting of equithickness alternate layers of Nb and another metal such as Ti [ZHE81] or Cu [BAN82] were prepared by high-rate sputtering. The individual films generally varied in thickness from less than 10 Å to many tens or even thousands of Å; the total specimen thickness was usually maintained at about 1–2 μm. As a result of resistive and inductive (see [SOU73]) transition temperature measurements, some interesting proximity effects and "anomalous" thin-film-related effects were observed. For example, in the Nb/Cu "superlattice" studies, whereas de Gennes–Werthamer theory was able to describe the transition temperature in samples with layer half-thicknesses, d, greater than ξ_s, deviations from it began to occur as d became comparable to the coherence length [BAN82]. It was pointed out that, since the above-mentioned theory should in fact be able to maintain its validity to $d < \xi_s$, the observed apparent deviations from it were probably a consequence of size-effect-induced changes in the electronic mean-free-path of the Nb.

(c) Low-Temperature Specific Heat in the Study of Proximity Effect. The remainder of this chapter deals with theoretical and experimental explorations of the low-temperature specific heats of proximity-effect-controlled two-phase superconducting systems.

11.8 Low-Temperature Specific Heats of Proximity Effect Couples

11.8.1 Theory of the Specific Heat and Its Discontinuity at T_c

The specific heat jump at T_c, ΔC, and its relationship to coupling strength has been discussed in Section 8.6. If the superconductor is placed in close contact with a metal of lower T_c or with a normal metal, not only is the T_{sn} of the couple depressed but so also is the height of the specific heat jump at that temperature. The magnitude of the suppression depends on the pair breaking ability of the underlayer, the reductions in T_{sn} and ΔC being greatest when it is paramagnetic. As a measure of the proximity effect, calorimetric-jump suppression is more sensitive than transition-temperature shift [FUL67].

In an analysis based on Ginzburg–Landau theory, FULDE and MOORMANN [FUL67] computed the free-energy density difference, Δg, between superconductive and normal states, and from it have calculated $\Delta C_{sn}/\Delta C_s$, the ratio of the specific-heat jump height of the couple, ΔC_{sn}, to that of the isolated superconductor, ΔC_s. In case the underlayer was strongly paramagnetic (i.e., such that the superconductive order parameter, $\Delta(x)$, was zero everywhere within it, and in particular at the interface ($x = 0$) between the layers), the calculations were quite straightforward and led to simple expressions for the jump-height ratio. Under the paramagnetic condition, according to FULDE and MOORMANN,

$$\Delta g = -\left(\frac{mp_F}{2\pi^2}\right)\frac{f_1(\rho)}{(4\pi T)^2}\langle\Delta^4(x)\rangle \tag{11-31}$$

which, since

$$\Delta C^v_{sn} = -T_{sn}\left.\frac{\partial^2 \Delta g}{\partial T^2}\right|_{T=T_{sn}} \qquad [10\text{-}7](11\text{-}32)$$

leads via the generalized Ginzburg–Landau equations to

$$\frac{\Delta C_{sn}}{\Delta C_s} = t_c\frac{[1-\rho_c g_1(\rho_c)]^2}{\beta}\frac{f_1(0)}{f_1(\rho_c)} \tag{11-33}$$

where

$$\rho_c = \rho|_{T=T_{sn}} = (\hbar v_F l/12\pi k_B T)(\pi/2D_s)^2|_{T=T_{sn}}$$

$$= (\pi^2/8)/(\xi/D_s)^2$$

$$\beta = \langle\Delta^4(x)\rangle/\langle\Delta^2(x)\rangle^2 = \tfrac{2}{3}$$

Here f_1 and g_1 are functions of the digamma function Ψ, and the $\langle\,\rangle$ indicates spacial averaging over the sample. For $D_{s,n} \gg \xi_{s,n}$, (11-33) reduces to

$$\Delta C_{sn}/\Delta C_s = t_c(1 - 4.72\rho_c)/\beta \tag{11-34}$$

In case the underlayer is nonparamagnetic or only weakly paramagnetic the expressions for the specific heat jumps are somewhat more complicated. Equation (11-34) is useful because of its simplicity and also as an upper limit to the jump-height suppression.

In calculating the heat capacity and density-of-states of a n/s bilayer, ZAITLIN [ZAI82], instead of adopting the more usual Ginzburg–Landau theory thermodynamic approach, decided to solve the Bogoliubov equation for the bilayer in order to take into account the effects of microscopic quasiparticle scattering near the n/s interface and the resulting establishment of bound states below the energy gap. Expressions for the heat capacities and densities-of-states of very thick (as compared to the coherence length) and very thin bilayers were developed. In the former case the no-proximity-effect rule-of-mixtures result was recovered, while in the latter the density-of-states was BCS-like with an effective pair potential, or energy gap, Δ_{eff}, equal to some volume average over the bilayer. With regard to the specific heat jump, the quantity which is usually of greatest interest to the experimentalist, thick films yielded simply $[\Delta C_{sn}/C_n]_{T=T_{sn}} = 1.43\,f$, the BCS result, with f the superconducting fraction in the bilayer. For thin films, the microscopic analysis also yielded a BCS-like result, but this time as for a homogeneous superconductor with some effective energy gap, Δ_{eff}.

11.8.2 Experimental Studies of Specific Heat in the Proximity Effect Regime

The validity of the Fulde–Moormann theory [FUL67] has been investigated by two research groups [LEC72], [BEV76]. Since massive amounts of well-categorized material are usually required for specific heat determination, samples selected for study have been in the form of directionally-solidified eutectics. Such materials may be regarded as an assemblage of *in situ*-formed proximity effect couples. LECHEVET *et al.* [LEC72] investigated a Pb–Sn eutectic alloy both of whose components (Pb-rich and Sn-rich lamellae) were superconductive in the He-temperature range. Lamellae thicknesses in the sample were characterized by

$$\begin{bmatrix} 4.4 \le D_s/\xi_s \le 44 \\ 2.0 \le D_n/\xi_n \le 18.4 \end{bmatrix}$$

and it is interesting to note that under such circumstances two distinct jumps were recorded, one near 7 K corresponding to the Pb-rich phase and the other in the range 3.8–4.5 K corresponding to the Sn-rich component. Subsequently, BEVOLO *et al.* [BEV76] studied a Pb–Cd eutectic in which the second component was practically a normal metal.[6] For a full discussion of the results of these experiments and their bearing on the Fulde–Moormann theory, the original articles [LEC72], [BEV76] should be consulted.

11.9 Proximity Effects in $\omega+\beta$-Phase Transition-Metal Alloys—A Case Study of Ti–Mo(10.3 at.%)

Alloys which exhibit fine-scale precipitates may provide further examples of the proximity effect in *in situ*-formed superconductive couples. By the usual casting and heat-treatment methods it is again possible to prepare samples sufficiently massive to enable the techniques of magnetic susceptibility and low-temperature specific heat to be applied. The influences of aging-modified ω-phase precipitation (Section 2.2) on the magnetic and calorimetric properties of several Ti–TM alloys have been considered in Sections 7.14 and 10.5, respectively. In what follows, evidence will be presented to support the conclusion that as-quenched and "saturation-aged" (i.e., aged to metastable equilibrium) Ti–Mo(10 at.%) alloys undergo superconducting transitions homogeneously throughout the entire two-phase sample volume at temperatures controlled by the operation of a Cooper-limited (i.e., "complete") proximity effect.

11.9.1 Experimental Results—General

Using low-temperature specific heat the influence of aging on the superconducting transitions of several Ti–TM alloys has been investigated. As mentioned in Section 10.5, alloys that have been studied include Ti–V(15 at.%) [COL75, COL76[b], COL82], Ti–Mo(10 at.%) [COL72[a], COL82], [HO73], and Ti–Fe(7.5 at.%) [HO74]. Throughout the entire precipitation process the calorimetric data, plotted in the format C/T *versus* T^2, showed only single superconductive transitions (Figure 10-10). Two classes of behavior were noted as functions of aging time, (i) a broadening of the transition accompanied by an increase in the threshold T_c (Ti–V) and (ii) a transition that remained relatively sharp during $\omega+\beta$-aging while T_c decreased (Ti–Mo and Ti–Fe). These tendencies have to do with the degree of completeness of the proximity effect; i.e., with the degree to which the Cooper limit, $D \ll \xi$, is applicable as gauged by the relationship between the coherence length, ξ, and the mean minor radii of the developing precipitate particles.

[6] $T_{c,\mathrm{Cd}} = 0.56$ K.

In what follows attention is focussed on Ti–Mo(10.3 at.%). The aging of this alloy for up to 880 h at 350°C has been monitored using both magnetic and calorimetric techniques. To the analysis of the magnetic susceptibility and electronic specific heat coefficient data the usual bulk-averaging techniques, using continuity equations such as (7-20), have been applied. As for the superconducting transition temperature itself, the results of the aging study have provided an opportunity for quantitatively evaluating the applicability of the Cooper–de Gennes complete-proximity-effect theory in the form of a modified version of (11-13).

11.9.2 Experimental Results—Ti–Mo(10.3 at.%)

As indicated in Figure 11-3, which summarizes the specific heat results taken at the beginning (a) and end (e) of the aging sequence (see Figure 10-10(b)), the calorimetrically determined superconducting transition in Ti–Mo(10 at.%) after aging for some 900 h at 350°C was just as sharp as in the as-quenched alloy. Also, according to Figure 11-4 (see also Figures 8-6 and 8-10), the relative height of the specific heat jump was well within the range expected for complete superconducting transitions in Ti–Mo alloys. It was only after an additional heat treatment of 300 h/500°C, which resulted in a decomposition of the alloy into macroscopic lamellae (repeat distance ~2000 Å) of alternate solute-lean α- and solute-rich β-phases

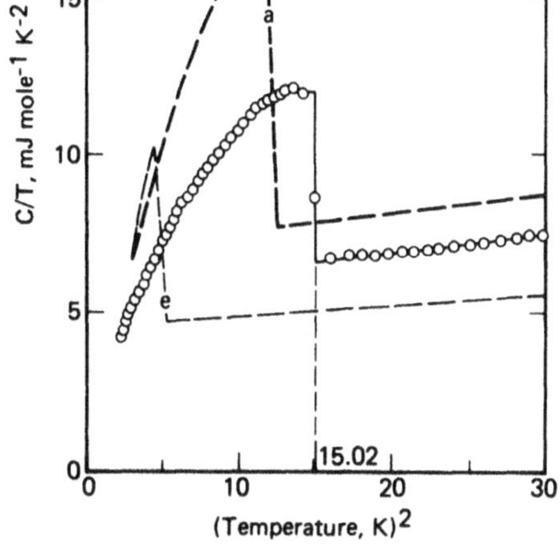

Figure 11-3. Low-temperature specific heat in the usual format C/T *versus* T^2 for Ti–Mo(10 at.%) in the as-quenched condition (curve a), after aging for some 900 h at 350°C (curve e), and after aging for 300 h at 550°C in order to equilibrate the $\alpha+\beta$-phase (data points)—see curves a and e, respectively, of Figure 10-10(b). Practically unrounded transitions are noted in all cases—after COLLINGS [COL84], see also HO and COLLINGS [HO73].

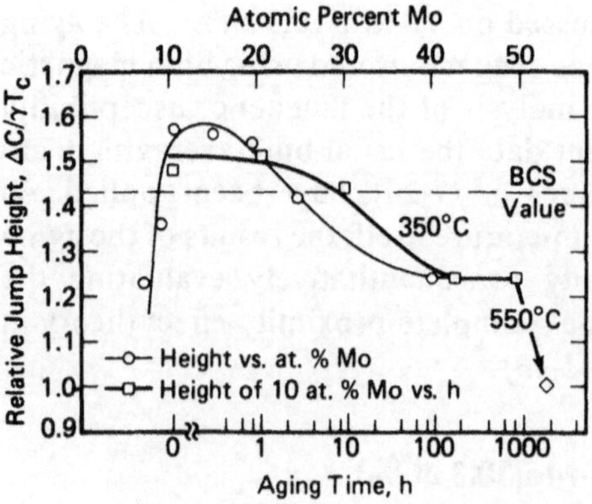

Figure 11-4. Relative height of the specific heat jump at T_c, $\Delta C/\gamma T_c$, *versus* aging time at 350°C in the meta-equilibrium $\omega+\beta$-field of the alloy Ti-Mo(10 at.%) (□) (cf. schedule, Figure 10-10(b)) followed by aging 300 h/550°C in order to equilibrate the $\alpha+\beta$-phase (◇). The $\omega+\beta$-phase in this case yields a complete proximity-effect-controlled transition; that of the $\alpha+\beta$-phase is incomplete. Repeated for comparison is the jump-height *versus* composition curve for the quenched Ti-Mo series (○) (cf. Figure 8-6) and the BCS-value of 1.43 (cf. (8-14))—after COLLINGS [COL84], see also HO and COLLINGS [HO73].

[BLA68], that the transition became incomplete (Figure 11-4) and, dominated by that of the β-phase component, increased in temperature (Figure 11-3).

The occurrence of ω-phase in Ti-TM alloys and its development during isothermal aging within the temperature range 300–450°C has been discussed by HICKMAN [HIC69[a]] (Ti-Fe, -Cr, -Nb, -Mo) and BLACKBURN and WILLIAMS [BLA68] (Ti-V, -Mo). Some results for Ti-Mo(6.2 at.%) from the work of the latter authors are presented in Table 11-5. Attention is drawn to the fact that a change of heat-treatment condition from 10 h/400°C to 960 h/350°C is accompanied by no more than a doubling of the mean particle diameter. Figure 11-5(a) is a dark-field TEM (cf. Figure 1-14) taken from a massive as-quenched Ti-Mo(10.3 at.%) alloy, simple measurement of which indicates that the minor diameters of the precipitate particles are

Table 11-5. Particle Sizes in Ti-Mo(6.2 at.%) Quenched from 950°C and Aged at Various Temperatures—After Blackburn and Williams [BLA68]

Aging Conditions		Particle Size[a], Å	
Temperature °C	Time h	Arithmetic Mean	Standard Deviation
350	960	270	100
400	10	145	55
	48	188	70
	115	240	110
	550	380	140
450	144	680	250

[a]Minor axis (axial ratio, 2.2:1).

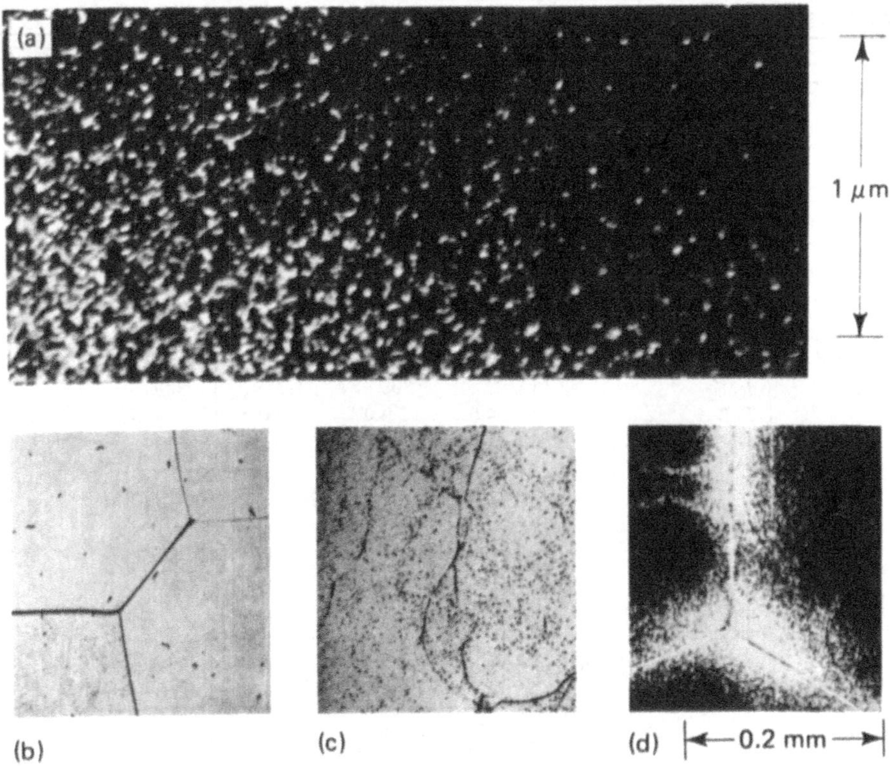

Figure 11-5. (a) Dark-field electron micrograph of a specimen taken from an ice-brine-quenched 30-g ingot of Ti–Mo(10 at.%). (b), (c), and (d) are optical micrographs depicting the further development of ω-phase precipitation during successive agings of 1 h/300°C, 8 h/350°C, and 150 h/350°C—after Ho and Collings [Ho72].

about 100 Å. The coherence lengths of several relevant β-phase Ti–Mo alloys are listed in Table 11-1. Taken together these data indicate that after quenching, and at all times during aging at 350°C, the ω-phase particles are sufficiently small compared to a coherence length as to be fully transparent to the pair wavefunctions, as a consequence of which Ti–Mo(10.3 at.%) is capable of exhibiting a complete proximity effect.

11.9.3 Data Analysis—Ti–Mo(10.3 at.%) Aged 880 h/350°C

(a) Procedures. It is assumed that after 880 h at 350°C, Ti–Mo(10 at.%) is aged to ω+β-phase (metastable) equilibrium in which state the ω-phase component has, according to an argument by Hickman [Hic69[a]] (see also Table 1-4), a composition of 4.3 ± 0.4 at.%. Next, in accordance with a suggested phase diagram for ω+β-phase Ti–Mo due to de Fontaine, Paton and Williams [Def71], upon which Figure 11-6 was

Table 11-6. Analysis of the Magnetic and Calorimetric Data for Aged (879 h/350°C) Ti–Mo(10 at.%)

		Block 1(a)			Block 1(b)
	Structure	β			Aged $\omega+\beta$
	[Mo] at.%	19	21	23	10.3
Experimentally Obtained Data	f	0.41	0.36	0.32	
	$\chi\ 10^{-6}\ \mathrm{cm^3\ mol^{-1}}$	203.6	202.6	201.5	167.9
	$\gamma\ \mathrm{mJ\ mol^{-1}\ K^{-2}}$	7.22	7.16	7.10	4.5
	θ_D K	302	308	313	380
	T_c K	4.0	4.0	4.0	2.18

Continuity Relationship—Equation (11-35)

Block 2				
Structure ($[\mathrm{Mo}]_\omega$)	ω(4.3 at.% Mo)			
	$[\mathrm{Mo}]_\beta$	19	21	23
Derived Data	f_ω	0.59	0.64	0.68
	χ_ω	143.1	148.5	152.1
	γ_ω	2.61	3.00	3.29
	$\theta_{D,\omega}$	434	420	411

Block 3		Figure 11-7		
Input	$1/0.212\gamma_\omega$ [a]	1.81	1.57	1.43
Output	$10^3(T_c/\theta_D)_\omega$	0.14	0.36_5	0.63
	$T_{c,\omega}$	0.06	0.15	0.26

Block 4

	Equation (11-36)			ω and β Data → / Figure 11-7
$\gamma_{\mathrm{eff},\,(\omega+\beta)}$ [a]	5.48	5.23	5.06	5.42

[a] Note: γ is here in $\mathrm{mJ\ mol^{-1}\ K^{-2}}$.

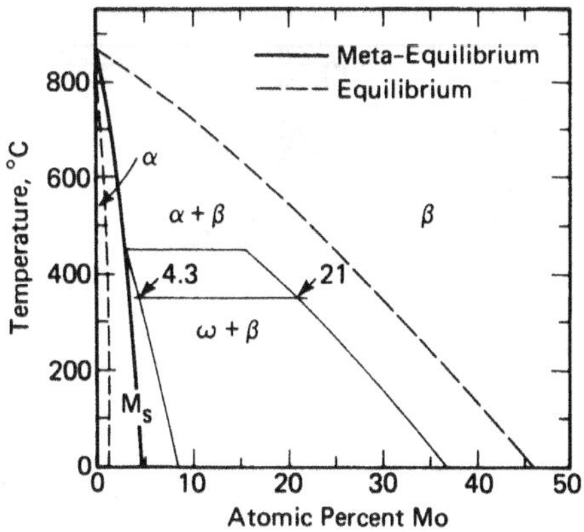

Figure 11-6. Metaequilibrium phase diagram for Ti-Mo indicating the $\omega/(\omega+\beta)$- and $(\alpha+\beta)/\beta$-phase boundaries (——). The 350°C isotherm is shown intersecting the transi at 4.3 and 21 at.% Mo. Also indicated, with dashed lines, is a standard equilibrium phase diagram—based on DE FONTAINE, PATON and WILLIAMS [DEF71].

based, the Mo concentration in the β-phase component of the metaequilibrated (at 350°C) alloy was about 22 at.%. The *mole fraction* of each component is given by substituting in the appropriate version of the general conservation equation:

$$\langle P\rangle = f_\omega P_\omega + f_\beta P_\beta \tag{11-35}$$

(cf. (7-20)) where P represents a molar physical property or some multiple of it (in this case the Mo concentration in at.%) and f represents mole fraction. Assigning to $\langle P\rangle$ and P_ω the values 10.3 and 4.3 at.%, respectively, and to P_β the values 19, 21, and 23 at.%, a corresponding set of f values is derived for use in further computations.

The goal of the analysis is twofold: (i) to determine a set of physical property data for ω-Ti-Mo(4.3 at.%), and (ii) to verify the de Gennes formula for Cooper-limited proximity effect, (11-13). Table 11-6 indicates the steps in the analysis and the interim results obtained. The "input data"—experimental results such as those depicted in Figure 10-8—are listed in Blocks 1(a) and 1(b) of the table. In support of the first goal, use is made of (11-35) with P replaced in turn by χ, γ, and θ_D. The results are presented in Block 2 of Table 11-6. $T_{c,\omega}$ is computed by substituting γ_ω into the BCS-Morel-Morin-Maita formula for $(T_c/\theta_D)_\omega$, (8-34), recognizing that $n(E_F)V = 212\gamma\, V_{\text{app}}$ (from (8-36)), with γ in J mol^{-1} K^{-2} and in which $V_{\text{app}} = 0.26$ eV as for the other members of the Ti-Mo family (Figure 8-12). In other words, the appropriate values of γ_ω from Block 2 of the table are inserted into Figure 11-7, an extension of the figure referred to earlier, after which corresponding values of $(T_c/\theta_D)_\omega$ are read from it and converted into $T_{c,\omega}$s with the aid of the $\theta_{D,\omega}$s already acquired as described in Section 10.4.2. The results are given in Block 3 of Table 11-6.

(b) Proximity Effect. The foregoing analysis has nothing to do with proximity effect, the continuity relationship, (11-35), having been applied

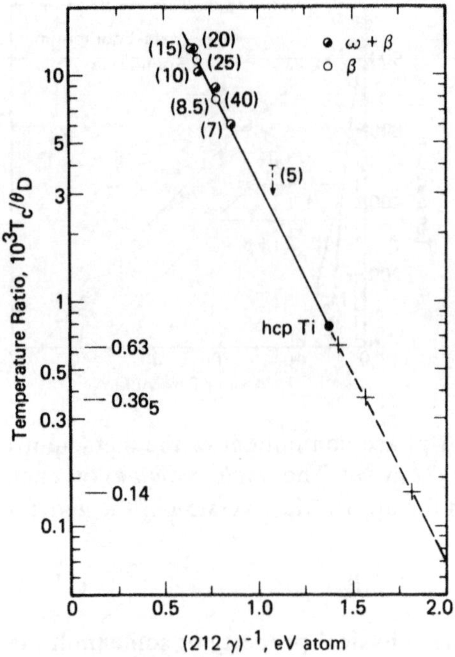

Figure 11-7. Semilog plot based on the relationship $\log_{10}(T_c/\theta_D) \propto (212\gamma V_{app})^{-1}$, with γ in J mol^{-1} K^{-2} and V_{app} in eV, for use in the procedure outlined in Table 11-6 for estimating the superconducting transition temperature of a metaequilibrium ω-phase Ti–Mo alloy—after COLLINGS [COL72[a], COL84].

to the evaluation of γ_ω and $\theta_{D,\omega}$ in order to determine $T_{c,\omega}$ within the BCS weak-coupling approximation assuming an apparent pairing potential, V_{app}, of 0.26 eV. In what follows, however, a test will be made of the validity of the de Gennes-modified Cooper-limit proximity relationship, (11-17), which after making the substitutions prescribed by (8-3) and (8-36) becomes, under the assumption that V_{app} is constant for all members of the Ti–Mo family,

$$\gamma_{eff} = \frac{\dfrac{\gamma_\omega^2 f_\omega}{1+\lambda_\omega} + \dfrac{\gamma_\beta^2 f_\beta}{1+\lambda_\beta}}{\dfrac{\gamma_\omega f_\omega}{1+\lambda_\omega} + \dfrac{\gamma_\beta f_\beta}{1+\lambda_\beta}} \tag{11-36}$$

(i) *Evaluation of the Left-Hand Side of (11-36).* Insertion of the value of $(T_c/\theta_D)_\omega$ from Block 1(b) of Table 11-6 into Figure 11-7 yields immediately $\gamma_{eff} = 5.42$, a value against which the right-hand side of (11-36) is to be compared.

(ii) *Evaluation of the Right-Hand Side of (11-36).* For estimates of the electron-phonon coupling constants, λ, we refer to Table 8-6, which suggests that for ω-Ti–Mo(4.3 at.%) and β-Ti–Mo(15 ~ 35 at.%), the values $\lambda_\omega = 0.2$ and $\lambda_\beta = 0.4$ would be reasonable. Substituting these, and data from Block 1(a) (f_β, γ_β) and Block 2 (f_ω, γ_ω) into the right-hand side of (11-36) yields 5.48, 5.23, or 5.06 depending on whether $[\text{Mo}]_\beta$ is taken to be 19, 21, or 23 at.%. Clearly the selection of $[\text{Mo}]_\beta \cong 20$ at.% will equalize the left-hand and right-hand sides of (11-36) in a demonstration, in the special case of aged Ti–Mo(10.3 at.%), of the operation of a complete de Gennes-modified Cooper-limit proximity effect between the ω-phase precipitates and the β-matrix.

11.9.4 Conclusion

According to the above analysis, satisfactory agreement with the predictions of (11-17) would be obtained with $[\text{Mo}]_\beta \cong 20$ at.%, the agreement being rather insensitive to the actual choice of $[\text{Mo}]_\beta$. The phase diagram of DE FONTAINE *et al.* [DEF71] actually indicated a saturation Mo concentration in the β-phase of 22 at.%. A compromise of $[\text{Mo}]_\beta = 21$ at.% is suggested (and was used in the construction of Figure 11-6 [Ho73]), enabling the final estimates of the physical properties of ω-Ti–Mo(4.3 at.%) to be accepted as those values listed down the center column of Table 11-6. In conclusion we note that the values of χ_ω, γ_ω, $T_{c,\omega}$, and $\theta_{D,\omega}$ so derived are able to nicely terminate the corresponding experimental curves for $\omega+\beta$ Ti–Mo presented in Figure 10-8 and which are obtained from the results of magnetic and calorimetric measurements on the quenched alloys.

Chapter 12

The Superconducting Transition Temperature

After the presentation of a brief survey of the principal formulae of weak-coupling and strong-coupling superconductivity, followed by discussion of the priorities under which the transition temperature is governed by atomic, band-electronic, and lattice-dynamical effects, the compositional systematics of T_c in amorphous and crystalline transition-metal-alloy superconductors are reviewed. The chapter concludes with a documented presentation, in numerical and graphical form, of the transition temperatures of representative and important Ti-base alloy systems.

As implied by the dual numbering scheme, some of the following equations have already been introduced in Chapter 8 during a discussion of calorimetrically determinable measures of superconductive coupling strength.

12.1 The BCS Weak-Coupling Result

The BARDEEN, COOPER, and SCHRIEFFER (BCS) [BAR57] weak-coupling theory of superconductivity led to the following expression for the transition temperature:

$$k_B T_c = 1.14 h\nu \exp[-1/n(E_F)V] \qquad [8\text{-}15a](12\text{-}1)$$

where ν is a characteristic phonon frequency and V represents a net electron–phonon pairing potential, which takes into account both electron-phonon and screened-Coulomb interactions. Equation (12-1) has also been written in the form

$$T_c = 1.14\langle\omega\rangle \exp[-1/n(E_F)V] \qquad [8\text{-}15b](12\text{-}2)$$

where $\langle\omega\rangle$ is the mean "phonon frequency," which, defined according to $\langle\omega\rangle \equiv h\nu/k_B$, has the dimensions of temperature (e.g., [McM68]), and also written in the form due to MOREL [MOR59], in which $h\nu$ is replaced by $C_0 k_B \theta_D$ where $C_0 = 3/4$ on the bases of which

$$T_c = 0.855\theta_D \exp[-1/n(E_F)V] \tag{12-3}$$

(see (8-33) and (8-34)). It may be recalled, however, that according to MORIN and MAITA [MOR63], who showed experimentally that C_0 could range between 0.08 and 0.44 (see Section 8.7.2), the adherence to a fixed 0.75 is not a good assumption.

12.2 Strong-Coupling Theory

The modern theory of transition temperature in strong-coupling superconductors commenced with MCMILLAN's solution of the Eliashberg equations [McM68]. The general form of the transition temperature that emerged from McMillan's theory is

$$T_c = \omega_0 \exp\left[\frac{-(1+\lambda)}{\lambda - \mu^* - (\langle\omega\rangle/\omega_0)\lambda\mu^*}\right] \qquad [8\text{-}16](12\text{-}4)$$

where ω_0 is the maximum phonon "frequency" in the form of a temperature, $\langle\omega\rangle \sim 0.5\omega_0$ is an average phonon frequency, μ^* is a Coulomb pseudopotential for electron–electron interaction, and λ is the electron–phonon coupling constant.

The quantity λ and the moments $\langle\omega^n\rangle$ of the phonon spectrum (the first of which appears in (12-4)) can be expressed in terms of $F(\omega)$, the phonon density-of-states, and $\alpha^2(\omega)$, the average electron–phonon interaction for phonons of frequency ω [McM68] (such that $\alpha^2(\omega)F(\omega)$ is the electron–phonon coupling function or spectral function [DYN76]), in the following way [BUT77]:

$$\lambda = 2\int d\omega\, \alpha^2(\omega)F(\omega)/\omega \tag{12-5}$$

$$\langle\omega^n\rangle \equiv \frac{\int d\omega\, \omega^n \alpha^2(\omega)F(\omega)/\omega}{\int d\omega\, \alpha^2(\omega)F(\omega)/\omega} \tag{12-6}$$

Finally, for subsequent use, a quantity $\omega_{\log}$ is defined as [DYN76]

$$\omega_{\log} \equiv \frac{\int d\omega \log \omega\alpha^2(\omega)F(\omega)/\omega}{\int d\omega\, \alpha^2(\omega)F(\omega)/\omega} \tag{12-7a}$$

such that

$$\omega_{\log} \cong 2\langle\omega\rangle - \langle\omega^2\rangle^{1/2} \tag{12-7b}$$

Although the integrals are often taken over the range $0 \to \infty$ [ALL75], [DYN76], it must be remembered that it is usual to assume that the phonons have a high-frequency cutoff as represented above by ω_0 [MCM68].

In a weak-coupling limit (i.e., $\lambda \to 0$) it is reassuring to note that (12-4) reduces to an equation similar to (12-2), with $\lambda - \mu^*$ playing the role of $n(E_F)V$.

A numerical solution of the Eliashberg gap equations next resulted in a refinement of (12-4) into a more accurate form suitable for experimental evaluation. In taking this step, MCMILLAN [MCM68] identified $F(\omega)$ with the phonon density-of-states for Nb as obtained from the results of neutron-diffraction measurements; he also chose α^2 to be independent of ω. The outcome of the calculation was the well-known semiempirical equation

$$T_c = \frac{\theta}{1.45} \exp\left[\frac{-1.04(1+\lambda)}{\lambda - \mu^*(1+0.62\lambda)}\right] \tag{12-8}$$

in which θ replaces the characteristic phonon frequency and is identified as the Debye temperature, θ_D. Subsequently, DYNES [DYN72] replaced the prefactor in (12-8) with $\langle\omega\rangle/1.20$ [ALL75].

The central parameter of (12-8) is, of course, the electron–phonon coupling constant, λ. After setting out to express this in terms of fundamental microscopic physical quantities involving electronic and lattice-dynamical parameters, MCMILLAN succeeded in rigorously showing that

$$\lambda = \frac{n(E_F)\langle I^2\rangle}{M\langle\omega^2\rangle} \qquad [8\text{-}19](12\text{-}9)$$

in which $\langle I^2\rangle$ is the average over the Fermi surface of the squared electron-ion (phonon) matrix elements, and M is the atomic mass. This equation defines $\langle\omega^2\rangle$, which, in favorable cases, can be estimated independently [ALL82]. Numerous authors have attempted to predict or justify the manner in which λ varies from one material to another and, in particular, along the transition-metal rows of the periodic table. In a fairly recent paper, in which much of the earlier work in the field was reviewed, BUTLER [BUT77] calculated the electronic components of λ (as embodied in the numerator of the right-hand side of (12-9)) relying on experiment for the phonon contribution, $M\langle\omega^2\rangle$. The results were discussed in terms of the various electronic and vibrational factors which compete with each other to limit the attainment of very high transition temperatures.

It has always been believed that in order to attain the desirable goal of a high transition temperature, λ should be as large as possible. Materials "design" or selection to achieve this end is not a trivial matter, since any modification that increases the numerator in (12-9) usually increases the denominator as well [BUT77]. For example, the full effect on λ of an increase in $\langle I^2\rangle$ brought about by strong interatomic-bond selection is offset by the high values of $\langle\omega^2\rangle$ that this implies. Furthermore, strong bonding in transition metals also tends to reduce $n(E_F)$ by spreading out the d band. By careful choice of system it may, in principle at least, be possible to avoid some of these difficulties. Thus, BUTLER speculated upon the possibility, in rare earth and actinide systems, of obtaining a high $\langle I^2\rangle$ whose effect would at the same time be undiminished by a concomitant high $\langle\omega^2\rangle$ [BUT77].

The range of validity of McMillan theory had been confined by the analysis to $\lambda < 1.5$, and consequently was unsuitable for predicting the course of T_c should λ be indefinitely increased. ALLEN and DYNES [ALL75], [DYN76] attended to this deficiency by developing an extension to McMillan theory that is valid not only for large values of λ but also unconstrained by any particular choice of phonon spectral function, $\alpha^2(\omega)F(\omega)$.[1] The strong-coupling expression for T_c obtained by those authors was

$$T_c = f_1 f_2 \frac{\omega_{\log}}{1.2} \exp\left[\frac{-1.04(1+\lambda)}{\lambda - \mu^*(1+0.62\lambda)}\right] \qquad [8\text{-}17](12\text{-}10)$$

a form that reverts to the McMillan equation[2] as f_1 and f_2 tend to unity and $\omega_{\log}$ tends towards $\langle\omega\rangle$ at small λ. When λ is large, $f_1(\lambda, \mu^*)$, the "strong-coupling correction" factor, ensures that $T_c \to \lambda^{1/2}$, while $f_2(\lambda, \mu^*, \omega)$, the "shape-correction" factor, permits $T_c \propto \langle\omega^2\rangle^{1/2}$. Together they ensure that T_c increases monotonically with λ according to $\sqrt{\lambda\langle\omega^2\rangle}$ at large λ.

The importance of this modification is that it predicts a T_c that, in terms of its variation with λ, is unbounded. This is in contrast to the pessimistic conclusion of McMillan theory, which had $T_c(\lambda)$ tending

[1]By this statement is meant that the Dynes-Allen treatment yielded comparable results using the spectral functions of "Einstein," Pb, or Hg.

[2]The prefactors are, of course, different, and $\omega_{\log}/1.2$ is generally less than $\theta_D/1.45$ [BUT77].

towards a limit or a maximum as λ increased beyond 2 (a value which was actually outside the range of the original theory itself).

12.3 The Cardinal Determiner of the Transition Temperature

In the meantime, between the formulation of (12-8) and its recasting in generalized-strong-coupling form by ALLEN and DYNES [ALL75], [DYN76], numerous authors, beginning with MCMILLAN himself, had been at work on the task of determining which of the physical parameters embodied in (12-9) dominated λ and in particular, for a given series of bcc transition-metal alloys, controlled the manner in which T_c varied as function of, say, e/a ratio.

MCMILLAN pointed out in a survey of various bcc transition elements, that although $n(E_F)$ and $\langle I^2\rangle$ individually varied by factors of 12, their product $n(E_F)\langle I^2\rangle$ was practically constant.[3] With

$$n(E_F)\langle I^2\rangle \cong \text{constant, say } C_M \tag{12-11}$$

it then seemed reasonable to claim that λ, and hence T_c, was governed by the phonon factor $M\langle\omega^2\rangle$—i.e., by the stiffness of the lattice. This was in marked contrast to the previously often-quoted result from BCS theory, that λ was governed by $n(E_F)\,V$, and hence by $n(E_F)$ since V was practically constant for an alloy series. MCMILLAN did, however, recognize some density-of-states involvement by admitting that $n(E_F)$ was inversely correlated with elastic stiffness.

HOPFIELD [HOP71] then went on to make the suggestion that λ was controlled by a combination of $M\langle\omega^2\rangle$ and a parameter, η, a property not of the band or crystal structure, but of the individual atoms. Thus, according to HOPFIELD,

$$\lambda_H = \frac{\eta}{M\langle\omega^2\rangle} \qquad [8\text{-}19](12\text{-}12)$$

in which η, which replaces the $n(E_F)\langle I^2\rangle$ of the previous theory, is no longer constant but varies strongly but smoothly along the rows of transition elements; i.e., it varies with, say, group number or e/a ratio. The ability of Hopfield theory to cope with transition-metal-binary alloys has been discussed by CAPPELLETTI and coworkers [WAN78].

Subsequently, BARIŠIĆ *et al.* [BAR70[b]] and BENNEMANN and GARLAND [BEN72] also produced expressions for λ in terms of atomic parameters

[3]As discussed by BUTLER [BUT77], subsequent work (see below) has revealed that this product, also represented by η, is not generally constant.

and phonon frequencies. The Bennemann–Garland parameter, η_0, was analogous to the η of (12-12) and varied with e/a ratio just about as strongly, but passed through a maximum towards the ends of the transition-metals rows. The authors concluded that λ (and hence T_c) depended on a combination of their η_0 (which was a local property, independent of $n(E_F)$) and $M\langle\omega^2\rangle$, which they claimed was not primarily determined by $n(E_F)$. An excellent review and summary of further work in this field has been provided by BUTLER [BUT77], whose discussion of the electronic factors influencing the electron–phonon coupling constant was referred to in the previous section.

A convincing picture of the hierarchial arrangement of the factors which control λ, hence T_c, has been offered by VARMA and DYNES [VAR76]. By recognizing that plots of λ *versus* $n(E_F)$ fell into groups of straight lines, one for each "subrow"[4] of transition elements and alloys in the manner first pointed out by HULM and BLAUGHER [HUL72], VARMA and DYNES directed attention towards the experimental observation that for groups or classes of alloys, and even A15 compounds,

$$\frac{\lambda}{n(E_F)} = \text{constant, say } C_2 \tag{12-13}$$

It follows from (12-9) that

$$\frac{\langle I^2\rangle}{M\langle\omega^2\rangle} = C_2 \tag{12-14}$$

Then to the extent that $n(E_F)\langle I^2\rangle$ may be regarded as a constant, say C_M, (but see the above note on the constancy of η) it may be deduced that

$$n(E_F) = (C_M/C_2)(M\langle\omega^2\rangle)^{-1} \tag{12-15}$$

which is an expression, in algebraic form, of one of the statements made at the very beginning by MCMILLAN. There is, of course, ample calorimetric evidence in the form of plots of γ and θ_D *versus* composition or e/a ratio for an inverse correlation between $n(E_F)$ and θ_D (e.g., Figure 8-1), and hence some simple function of ω.

These results demand that (12-9) now be viewed in a slightly different light; *viz.*, that it is no longer sufficient to ascribe the variation of T_c among members of a class of alloys to variaton solely of $n(E_F)$ (as in BCS) or $M\langle\omega^2\rangle$ (following MCMILLAN) but that all three of the quantities $n(E_F)$,

[4]Each row is subdivided into two portions, one above and one below the density-of-states minimum.

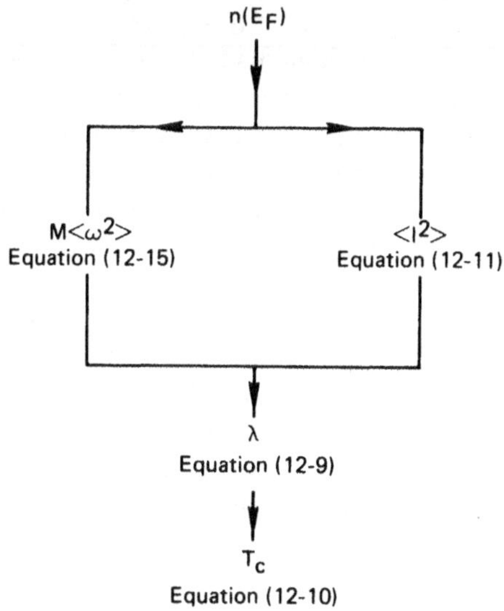

Figure 12-1. Status of the interrelationship between $n(E_F)$, $M\langle\omega^2\rangle$, $\langle I^2\rangle$, and consequently λ and T_c, according to VARMA and DYNES [VAR76].

$M\langle\omega^2\rangle$, and $\langle I^2\rangle$ must be considered in that they are interrelated (see for example [BUT77]). With this in mind, VARMA and DYNES [VAR76] asked which was the dominant or motivating parameter. They reached the conclusion that $n(E_F)$, as an electronic property of the equilibrium zero-K lattice, was the independent variable while the other quantities, being associated with *perturbations* of that lattice, were derivatives of the electronic structure. In other words, in terms of the equations cited above the priorities governing the variation of $n(E_F)$, $M\langle\omega^2\rangle$, and $\langle I^2\rangle$, and hence T_c, can be represented as in Figure 12-1.

12.4 Transition Temperature Systematics in Crystalline and Amorphous Transition-Metal Alloys

12.4.1 Transition Temperatures of bcc Transition-Metal Alloys

MATTHIAS was the first to draw attention to the oscillatory manner in which the transition temperatures of the elements seemed to vary with e/a ratio and, in particular, to the fact that in the long or transition-metal rows of the periodic table T_c possessed "maxima near three, five, and seven valence electrons per atom" [MAT57]. A similar variation noted by DAUNT [DAU55] in the electronic specific heat coefficients of the transition elements and by KRIESSMAN and CALLEN [KRI54] in their paramagnetic susceptibilities set the stage for T_c to be correlated with $n(E_F)$ using e/a ratio as a display parameter. The fact that a smooth curve connecting the "$n(E_F)$"s of the transition elements paralleled a calculated density-of-states curve for

Cu (e.g., [DAU55]) suggested the existence of common d bands for the transition elements of the three respective long periods, which, with increasing group number, were amenable to being filled to successively higher levels by their $s + d$ valence electrons. Whether or not this so-called "rigid-band-model" has any physical validity is not a part of the present discussion; what are important are the properties *at the Fermi levels* of the *individual* elements and the manner in which their T_cs and $n(E_F)$s appeared to vary in synchronism.

Subsequent workers using bcc solid solutions (and the results of magnetic susceptibility and low-temperature specific heat measurements as a substitute for $n(E_F)$ data) were able to fill in the spaces between the pure elements, in particular: (i) Relevant magnetic susceptibility data were acquired by TANIGUCHI *et al.* (e.g., [TAN62]) and others. (ii) Low-temperature specific heat measurements of transition-metal alloys, in support of both the rigid-band model and contemporary superconductivity studies (see below), were conducted by BECK and colleagues (e.g., [CHE60]

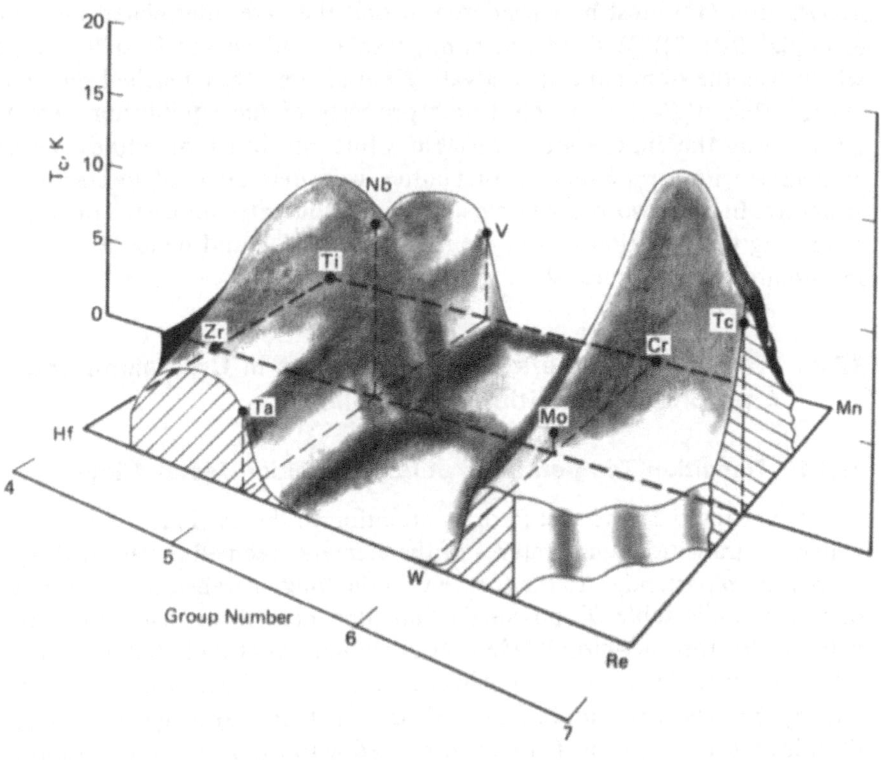

Figure 12-2. Transition-temperature/period/group-number surface for bcc transition elements and alloys formed from them—after HULM and BLAUGHER [HUL61, HUL72].

and CHE62], respectively) and others. The low-temperature specific heat of transition-metal alloys has been fully discussed by HEINIGER, BUCHER, and MÜLLER [HEI66] in a review which still remains the standard critical work on the subject. (iii) The superconducting transition temperatures of bcc transition-metal binary solid solutions were measured in a classic series of experiments by HULM and BLAUGHER [HUL61]. Their results are summarized schematically in Figure 12-2 which is based on a similar diagram that appeared in a recent review of the subject by those authors [HUL72]. The double-humped T_c *versus* e/a behavior predicted by the lines connecting the T_cs of the pure elements was substantiated by the alloys. In fact the peaks in Figure 12-2 are so pronounced that the two-dimensional T_c *versus* e/a profiles that appear on slices of the solid taken parallel to the $(T_c, e/a)$ plane (cf. the T_c *versus* e/a curves for $3d$–$3d$, $4d$–$4d$, and $5d$–$5d$ alloys due to MATTHIAS [MAT70]), or on the diagonal, all exhibit the characteristic double humps.

12.4.2 Superconductivity in Crystalline and Amorphous Transition-Metal Alloys

There is ample evidence, in the form of curves of χ, γ, and T_c *versus* e/a, that T_c parallels the variation of $n(E_F)$. As far as the bcc alloys themselves are concerned it is useful to note that although, as pointed out by HULM and BLAUGHER [HUL72], binary TM alloys form an unusually broad range of bcc solid solutions, their range in terms of e/a does not extend beyond the peaks of the humps in either direction; the bcc field is in fact delimited on the low-e/a side by ω-phase precipitation and on the high-e/a side by transformations to σ-phase or other structures. The minimum at group-number VI, in the middle of what is now a U-shaped $n(E_F)$ curve for bcc alloys, represents the Fermi state density of the half filled d band of the corresponding crystalline transition element.

It is interesting to inquire into the form of the $n(E_F)$ *versus* e/a functions for transition-metal alloys in the amorphous state. In the absence of long-range order it might be expected that the Fermi level in an amorphous group-VI element would *not* be found within a gap between a pair of conducting subbands as is the case with its crystalline counterpart. An analogy could be drawn with liquid Ge, which is metallic and in which structural disorder has "washed out" the gap between the erstwhile valence and conduction bands. Thus, as pointed out by JOHNSON [JOH79], amorphous transition metals should be in possession of density-of-states *versus* energy bands that, although comparable in overall width to those of the corresponding crystalline metals, are devoid of fine structure and rise to single, broad, more or less centrally located maxima. By analogy with the crystalline case, it is then not unreasonable to suppose that $n(E_F)$ *versus*

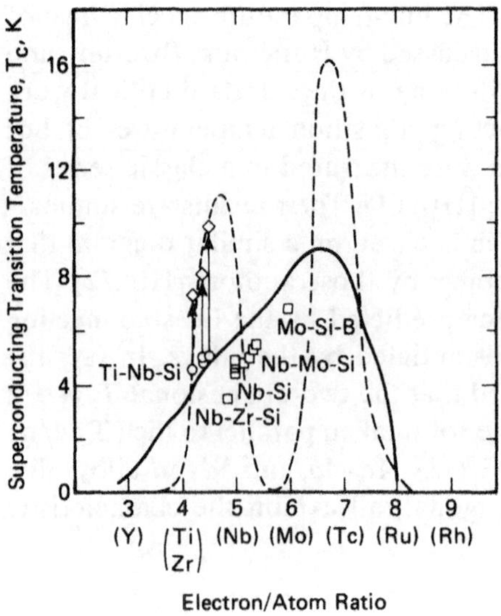

Figure 12-3. Superconducting transition temperatures plotted as a function of average group number or e/a ratio for amorphous vapor-cryodeposited (full line) and crystalline (chain line) 4d-transition-metal alloys (COLLVER and HAMMOND [COL73[c]], JOHNSON [JOH79]). Also shown are those of metallic-glass Nb-base and Mo-base alloys (□) (after MASUMOTO *et al.* [MAS80]), and those of Ti-Nb-Si metallic-glass ribbons before (○) and after (◇) recrystallization (after INOUE *et al.* [INO80]).

e/a for a *series* of amorphous transition-metal alloys, and indeed $T_{c,\text{amorph}}$ itself which is expected to scale with $n(E_F)_{\text{amorph}}$, should also possess such a maximum. Indeed according to the by now classic experiments of COLLVER and HAMMOND [COL73[c]], who studied the T_cs of vapor-cryodeposited films of 4d- and 5d-binary-TM alloys, $T_{c,\text{amorph}}$ *versus* e/a increased and decreased again in a rather structureless way with maxima situated at $e/a(4d) = 6.4$ and $e/a(5d) = 6.8$.[5]

The 4d results are indicated schematically in Figure 12-3 (full line). The Collver-Hammond curves, as they might be termed, have established a kind of standard or bench-mark for amorphous binary alloy transition temperatures against which those of the crystalline alloys, and other amorphous, and amorphous-plus-recrystallized alloys may be referred. Following the pioneering studies of COLLVER and HAMMOND, other workers have measured the transition temperatures of amorphous glassy alloys (metallic glasses), which are obtainable in splat or ribbon form and have the distinct advantage of being relatively stable at room temperature. The first of such results were reported in 1975 by JOHNSON [JOH75], who recently authored a comprehensive review of the subject [JOH79].

In Figure 12-3 the 4d-amorphous-alloy T_c is juxtaposed against a curve representing those of the corresponding crystalline alloys—the medial section of Figure 12-2. A comparison of the two curves clearly indicates that,

[5]This type of behavior was originally taken as evidence that the amorphous alloys were providing an opportunity for T_c to be described by (12-12) and, therefore, controlled in its variation with e/a by the atomic-like parameter, η [COL73[c]]. Currently it is believed that T_c is guided by the variation of $n(E_F)$ along the series just as it is in the crystalline alloys [JOH79].

although the prognosis for Nb-base alloys is not good, amorphous alloys based on Mo have much higher T_cs than their crystalline counterparts. The data points in Figure 12-3, which refer to the transition temperatures of the melt-spun metallic-glass alloy ribbons produced by MASUMOTO and colleagues (e.g., see [MAS80], for Nb- and Mo-alloys; [INO80], for Ti-alloys), closely follow the Collver-Hammond curve. The slight departures from it that do occur (positive for the Ti-base alloys and negative for the Nb-base ones) were tentatively attributed by INOUE *et al.* [INO80] to the presence of the metalloid atoms, the combining of transition elements from different rows of the periodic table, and/or to other artifacts. The results of the amorphous Ti-base alloy study, although disappointing from a practical transition-temperature standpoint, provided strong verification of the validities of the two curves in Figure 12-3 in that, although the amorphous Ti–Nb–Si data points lay close to the Collver-Hammond curve, they rose to the Matthias–Hulm–Blaugher curve upon recrystallization.

12.5 Transition Temperatures of Titanium–Niobium-Base Alloy Superconductors—Some Experimental Results

Comparative studies of the superconducting properties of Zr–Nb and Ti–Nb alloys were undertaken in the early 1960s by RALLS [RAL64] under the supervision of J. Wulff at the Massachusetts Institute of Technology. The report based on this work clearly demonstrated that although the maximum T_c in the Zr–Nb system (about 11 K, 80 at.% Nb) is higher than that in Ti–Nb (about 10 K, 60–70 at.% Nb) this advantage is offset by a significantly smaller H_{c2}. For Zr–Nb, $H_{c2,\max} \cong 91$ kOe, compared to the $H_{c2,\max}$ value of about 115 kOe for Ti–Nb. This fact, plus the better workability of Ti–Nb (Zr–Nb(50 at.%), for example, becomes brittle below about 100°C) led to its replacing Zr–Nb as a candidate for further technical development. Figure 3-37 has demonstrated how a search for improvements to the *workability* of Zr–Nb might be regarded as having led not only to Ti–Nb but also to Ti–Nb–Ta and Ti–Zr–Nb–Ta. The influences of ternary and quaternary additions to Ti–Nb were also investigated in the search for improved *superconductive* properties. It is impractical to deal adequately with such a topic in this space; however, for the remainder of the chapter attention will be given to the transition temperatures of the more important members of the Ti–Nb-family of alloys.

12.5.1 Transition Temperature of Ti–Nb

As part of their comprehensive study of the superconducting transition temperatures of transition-metal-binary alloys, HULM and BLAUGHER

[HUL61] investigated a series of homogenized Ti–Nb(2–97 at.%) alloys. Samples were measured magnetically in both the as-quenched and slow-cooled conditions, and the responses of T_c to changes of compositional and thermal variables were interpreted in terms of microstructural condition as estimated from the equilibrium phase diagram of HANSEN *et al.* [HAN51] in conjunction with a martensitic transformation curve due to DUWEZ [DUW53]. The possible existence of ω-phase, which was to play an important role in interpreting both the decrease in T_c generally noted in quenched Ti–TM alloys for which $e/a \lesssim 4.3$ as well as the effects of aging on those alloys, was not recognized at the time. A T_c *versus* composition diagram, assembled from the results of HULM and BLAUGHER [HUL61], BUCHER and MÜLLER [BUC61], and DESORBO [DES63] is presented as Figure 12-4. In agreement with the expected scaling of T_c with the $n(E_F)$-controlled λ (Figure 12-1 and (12-10)), and hence the electronic specific heat coefficient, γ (Figure 8-4), the transition temperature rises to a maximum at a Nb concentration of 60–65 at.%. To be seen in Figure 12-4 is a curious local minimum near 85 at.% Nb, to some extent paralleling a minimum which is also discernible in the γ *versus* e/a curve of Figure 8-4. If the subsequently obtained data of BELLIN *et al.* [BER69] and SUENAGA and RALLS [SUE69] (who obtained a maximum T_c for the system of 10.1 K at 70 at.% Nb) are included in the figure, the local minimum disappears (see LARBALESTIER [LAR81]) only to reappear again if weight is given to the calorimetric results of SUKHAREVSKII *et al.* [SUK71], as in Figure 10-1.

The technically important Ti–Nb alloys with compositions of about 34 at.% Nb (50 wt.%) selected for their high upper critical fields in combination with an amenability towards high critical current density development, possess zero-field transition temperatures centered on about 8 K (Figure

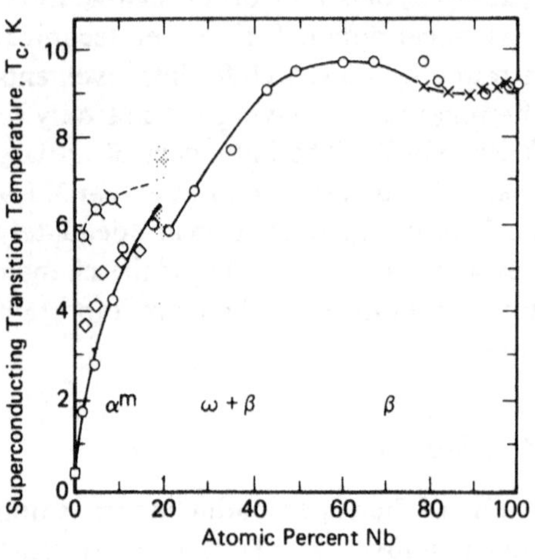

Figure 12-4. Superconducting transition temperature as a function of composition for Ti–Nb alloys. Data sources are: HULM and BLAUGHER [HUL61], water-quenched (○) and slow-cooled (ᾶ) samples; BUCHER and MÜLLER [BUC61] (◇); DESORBO [DES63] (×). The pure-Ti value, 0.36 K, is a critical value taken from several sources.

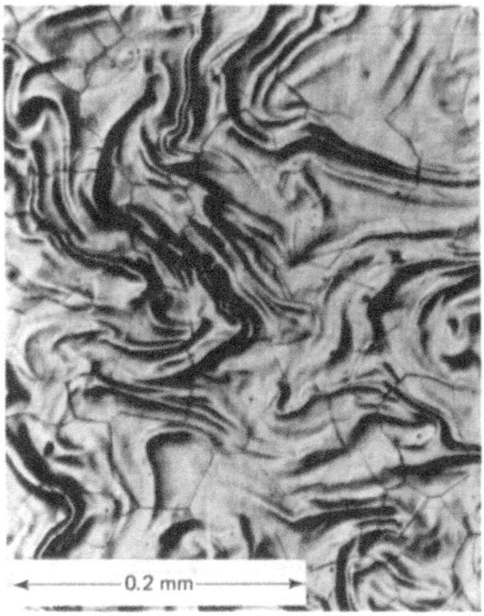

Figure 12-5. Cross section of an extruded-and-drawn 1-cm$^{\varnothing}$ rod of Ti–45Nb, recrystallized at 800°C and water quenched, and polished and etched with $HNO_3/HF/H_2O$. The typical swirled pattern arises from dendritic compositional segregation (coring) always present in the starting ingot—after READ [REA78]; photograph (original magnification, 350×) courtesy of D. T. Read. Copyright © 1978, IPC Science and Technology Press, reproduced with permission.

12-4). In this regime T_c exhibits its strongest composition dependence; the slope of the curve in the vicinity of the 8 K point is about 0.14 K per at.% Nb.

A glance at the equilibrium phase diagram for Ti–Nb (Figure 1-7) reveals a very broad liquid + solid zone. The diagram shows in fact that during slow solidification an opportunity exists for compositional modulation, with an amplitude as large as 35 at.% Nb, to take place. This does not happen in practice; nevertheless as pointed out by CORDIER and MCDONALD [COR75] microscopic solidification-modulation (so-called "coring"), in which the Nb content of a massive as-cast Ti–55 Nb ingot varies from 37.6 at.% Nb to 42.9 at.% Nb with a half-wavelength of some 0.1 mm, can be anticipated. The width of the superconducting transition temperature in a sample cut from such an ingot would be about ±0.4 K. During the subsequent metalworking processes, the regular dendritic coring structure becomes distorted and replaced by what, after etching, is revealed to be an intricate pattern of swirls, Figure 12-5. Finally, during wire drawing it can be expected that the original cored structure will become translated into a compositional modulation along the length of the conductor. This could have serious consequences for the stability of a monofilament [MOO68]. In a multifilamentary conductor, fabricated from many rods of cast-and-drawn Ti–Nb, although the *average* cross-sectional properties are not likely to vary with distance along the conductor, the effective critical current density may be lower than expected.

12.5.2 Simple Metal Additions to Ti–Nb

Although the variation of T_c with composition in Ti–Nb alloys is traceable to a corresponding variation of $n(E_F)$, the influences of simple-metal additions to the binary Ti–Nb base tend to be dominated by metallurgical rather than electronic considerations. Thus, solutes like Sn, for example, act to stabilize the bcc structure when dissolved in it, although not as "β-stabilizers" in the conventional sense but as strengtheners of the bcc lattice against the ω-instability (see Section 1.8). With reference to Figure 2-4, then, the addition of such solutes may cause the $\omega+\beta$-region to shrink in width such that a given host alloy previously within it may find itself within the $\beta'+\beta$-phase-separated zone (Section 2.3) and prone to α-phase precipitation at the β'-sites (Section 2.4.2).

Although the influences of the nontransition metals Al, Si, Ga, Y, Ag, In, Sn, Sb, Au, Pb, and U (listed in order of increasing atomic number) on the superconducting properties of some low-concentration Ti–Nb alloys have been considered (e.g., by BACHMANN *et al.* [BAC68]), the greatest attention has been devoted to the effects of Cu and Ge additions to Ti–Nb alloys with compositions near the technically important 50 wt.%.

(a) Transition Temperatures of Ti–Nb–Cu Alloys. The superconducting properties of Ti–Nb–Cu have been investigated extensively by ZWICKER *et al.* [BAC68], [ZWI70], [HEL71[a], HEL72], [LOH71, LOH73], ARNDT *et al.* [ARN74], [WIL75[a]], and WADA *et al.* [WAD80]. In several of these investigations T_c was measured as a function of aging time, during which it was generally noted that T_c eventually rose to values higher than those of the binary Ti–Nb host. The following types of experiments were conducted: (i) low-concentration Ti–Nb alloys (15–20 at.% Nb) doped with about 5 at.% Cu were homogenized and aged for various times and temperatures [WAD80]. In one such experiment the T_cs of Ti–Nb_{15}–Cu and Ti–Nb_{20}–Cu,[6] during aging at 500°C, rose to about 8.5 K, the T_c of binary Ti–Nb(35 at.%). (ii) Intermediate-concentration Ti–Nb alloys (e.g., Ti–Nb(33 at.%)) doped with 3 and 5 at.% Cu were aged after (a) homogenization (3 h/1000°C) followed by quenching, and (b) casting followed by cold deformation (97.2%) [LOH71]. In either case T_c usually passed through a minimum before steadily increasing with aging time at 400°C, presumably due to the partitioning of Nb to the parent β-phase.

(b) Transition Temperatures of Ti–Nb–Ge Alloys. The transition temperatures of Ti–Nb–Ge alloys have been extensively investigated by ZWICKER *et al.* [BAC68], [ZWI70], [HEL71] and by KITADA [KIT73]. Again, the alloying and aging results were usually discussed in terms of the

[6] In multicomponent alloys, subscripts are used to indicate composition in atomic percent.

partitioning of Nb and Ge between the host alloy and the precipitated phases. For example, the behavior of homogenized-and-aged Ti–Nb_{33}–$Ge_{0.5-4}$ was discussed by HELLER [HEL71] in terms of the dissolving of Ge in the β-phase during homogenization at 1080°C (which lowered the T_c) followed by the precipitation of Ti-rich Ti_5Ge_3 and α-phases accompanied by Nb-enrichment of the β-matrix during aging at 400°C, which had the effect of raising the inductively measured T_c to a value higher than that of binary Ti–Nb(33 at.%) itself.

12.5.3 Substitutes for Titanium in Ti–Nb Alloys

The two transition elements of group IV other than Ti in the periodic table are Zr and Hf. The influences on the superconducting properties of Ti–Nb after partial substitutions for Ti of either of these elements have been studied in several laboratories worldwide.

(a) Transition Temperatures of Ti–Zr–Nb Alloys. The transition temperatures of large groups of Ti-Zr-Nb alloys have been studied and documented by: (i) LAZAREV *et al.* [LAZ68], with thirteen ternary alloys in the as-cast condition and after annealing 24 h/520°C and 120 h/560°C; (ii) ZWICKER *et al.* [ZWI68], with five ternary alloys in both the cast-plus-deformed and the heat-treated conditions; (iii) SAVITSKII *et al.* [SAV73, p. 312], with eleven ternary single-phase-bcc alloys; (iv) DOI [DOI66], with twenty-four ternary alloys. The results obtained by the latter group, augmented by those for five Ti–Nb and five Zr–Nb alloys obtained by other workers, are presented on a composition triangle in Figure 12-6. In agreement with SAVITSKII *et al.*, whose results were also displayed on a composition triangle, nowhere within the diagram did the ternary-alloy T_cs exceed that of the best Zr–Nb alloy, *viz.*, 10.9 K for Zr–Nb(80 at.%).

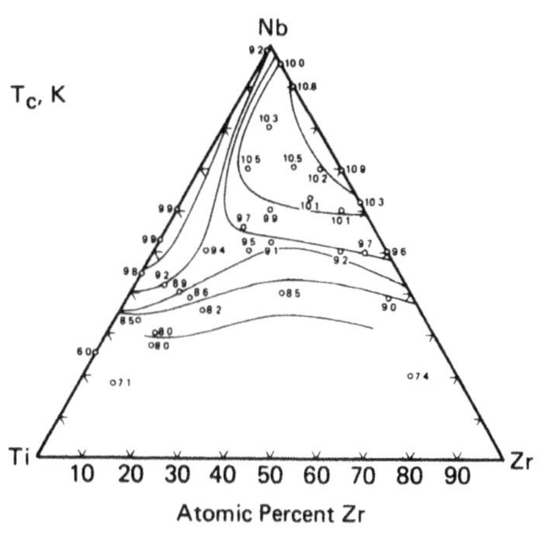

Figure 12-6. Ternary diagram (at.% linear) displaying the superconducting transition temperatures of the twenty-four Ti-Zr-Nb alloys measured by DOI *et al.* [DOI66] augmented by those of the Ti-Nb and Zr-Nb alloys measured by HULM and BLAUGHER [HUL61]—after DOI *et al.* [DOI66].

(b) Transition Temperatures of Ti–Hf–Nb Alloys. In a study conducted in the mid 1960s, BYCHKOVA *et al.* [BYC70[a]] measured the transition temperatures and other properties of homogenized and cold-worked (99.9%) Ti-Nb(25 at.%), Ti_{65}-Hf_{10}-Nb_{25}, and Ti_{50}-Hf_{25}-Nb_{25}, noting that T_c first dropped very slightly from an initial value of 7.1 K and then remained fairly constant at about 6.8 K with increasing Hf content. Then quite recently (~1979), in response to the generally recognized necessity of improving the properties of Ti-alloy superconductors, HAWKSWORTH and LARBALESTIER [HAW80] explored the effects of including 2.5, 5, and 10 at.% Hf in a series of Ti-Nb alloys whose compositions (25-38 at.% Nb) embraced those of technical interest. In the systems studied, viz., Ti-$Hf_{2.5}$-$Nb_{27.5-37.5}$, Ti-Hf_5-$Nb_{27.5-42.5}$, and Ti-Hf_{10}-Nb_{25-35}, it was noted that substitutions of Hf for Ti decreased T_c at rates of about 0.08 K per at.% Hf, a conclusion which tended to substantiate the results of the more limited earlier experiments of BYCHKOVA *et al.* [BYC70[a]].

12.5.4 Substitutes for Niobium in Ti–Nb Alloys

The two transition elements of group V other than Nb in the periodic table are V and Ta. Research on the effects of additions of these elements to the basic Ti-Nb alloy was motivated by a desire to increase the upper critical field through hoped-for reductions in Pauli paramagnetic limitation [SUE69], [HAW80], [HOR73]. It can be stated at the outset that as far as transition temperature is concerned both V and Ta reduced the T_c of the host alloy.

(a) Transition Temperatures of Ti–Nb–V Alloys. The T_cs of Ti-Nb-V alloys have been measured by SAVITSKII *et al.* [SAV71, SAV73, p. 326] and BELLIN *et al.* [BEL69]. As an example of the detrimental influence of V additions we note, from the results of the latter group, that at a constant 50 at.% Ti the T_cs of Ti-Nb-V corresponding to 0, 10, 20, 30, and 40 at.% V were 9.9, 8.8, 8.3, 7.8_3, and 7.8 K, respectively.

(b) Transition Temperatures of Ti–Nb–Ta Alloys. On account of its promising current-carrying capacity, particularly at high fields and low temperatures, Ti-Nb-Ta has begun to assume considerable importance as a technical superconducting alloy [STE78], [SEG79, SEG81], and consequently the influence of Ta on the transition temperature of Ti-Nb is of special interest. Major studies of Ti-Nb-Ta were conducted by BARANOV and coworkers [BAR70[a]] (see also SAVITSKII *et al.* [SAV73, p. 329] and SUENAGA and RALLS [SUE69]). Subsequently, the Soviet work was the subject of multiple-regression analysis to establish a T_c *versus* composition relationship. The results of these investigations were in reasonable agree-

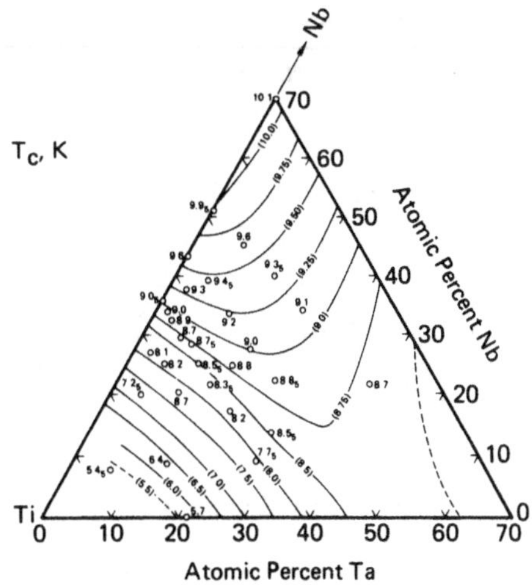

Figure 12-7. Superconducting transition temperatures of Ti–Nb–Ta alloys. The isothermal temperatures are in parentheses. The data indicated are the magnetic half-transition points—after SUENAGA and RALLS [SUE69].

ment and showed that the superconducting transition temperatures of all isoelectronic quenched Ti–Nb–Ta alloys decreased as Ta was substituted for Nb, both in the single-phase bcc regime and in the Ti-rich corner of the ternary diagram where the alloys contained some α-phase precipitates.

The composition dependence of T_c in the Ti–Nb–Ta system, from the work of SUENAGA and RALLS [SUE69], is displayed on a composition triangle in Figure 12-7. As an example of the rate of decrease of T_c with increasing Ta content, we note that at 50 at.% Ti the magnetically determined transition temperatures corresponding to 0, 10, 20, 30, and 40 at.% Ta were, respectively, about 9.9, 9.4, 9.1, 8.9, and 8.7 K, which is not quite as steep a rate of decrease as that found in response to V additions under otherwise similar conditions. In some recent studies of Ti–Nb–Ta by HAWKSWORTH and LARBALESTIER [HAW80] it was shown that the substitution of 5, 10, 15, and 20 at.% Ta for Nb in Ti–Nb(35 at.%) decreased T_c at the rate of about 0.04 K per at.% Ta.

12.5.5 Substitutes for both Ti and Nb in Ti–Nb Alloys

Quaternary alloys included under this heading might be represented by the general formula

$$\underbrace{\text{Ti-(Zr-Hf)}}_{\text{group IV}}\text{-}\underbrace{\text{(V-Ta)-Nb}}_{\text{group V}}$$

in which, within each triad, the elements are arranged in order of increasing atomic number. The transition temperatures of quaternary alloys based on this scheme have been measured by: (i) RAYEVSKII *et al.* [RAY71], who has

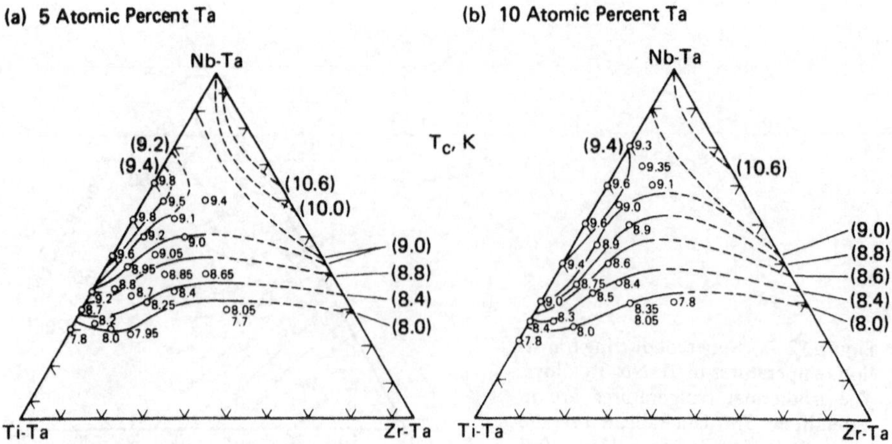

Figure 12-8. Pseudoternary diagrams (at.% linear) indicating the superconducting transition temperatures of quaternary Ti-Zr-Nb-Ta alloys. Two series of alloys are represented: (a) 5 at.% Ta = constant; (b) 10 at.% Ta = constant. The isothermal temperatures are in parentheses. The dashed lines which are based on binary-alloy data suggest the approximate positions of the unmeasured isothermals. The full curves (experimentally determined isothermals) are from HORIUCHI *et al.* [HOR73].

taken the basic Ti-Nb(30 at.%) alloy and substituted into it varying amounts of Zr and/or Hf for some of the Ti; (ii) KITADA and DOI [KIT70[d]], who took Ti_{10}-Zr_{40}-Nb_{50} and made small substitutions of V, Ta, and Mo for Nb; (iii) HORIUCHI *et al.* [HOR73], who examined a wide range of Ti-Zr-Nb alloys with 5 at.% and 10 at.% substitutions of Ta. Of all these systems, the most important from a technical standpoint is Ti-Zr-Nb-Ta, and, in particular, the alloy Ti_{61}-Zr_6-Nb_{27}-Ta_6, which, under the trade-name "Cryozitt," is being manufactured in stabilized multifilamentary form by Kobe Steel Ltd. The remainder of this section is devoted to a description of the transition temperature of the system of which this alloy is a member.

In the ternary Ti-Zr-Nb system, according to Figure 12-6, the highest T_cs lay along the Zr-Nb and Ti-Nb edges of the composition triangle, the maxima occurring near 20 at.% Zr ($T_c \cong 11$ K) and 30-40 at.% Ti ($T_c \cong 9.6$ K), respectively. The T_c surface itself is a distorted trough, narrow near the Nb corner and widening out towards the Ti-Zr base. With regard to the quaternary Ti-Zr-Nb-Ta system, similar features are suggested by the (incomplete) data of HORIUCHI *et al.* [HOR73], excepting that with the addition of first 5 at.% and then of 10 at.% Ta the entire Ti-Zr-Nb surface (for 2 h/1000°C-annealed alloys) appears to drop uniformly in two steps of up to about 0.5 K each. The superconducting transition temperatures of Ti-Zr-Nb-$Ta_{5,10}$, from the work of HORIUCHI *et al.* [HOR73], are presented in Figure 12-8.

IV
The Mixed State

Chapter 13

Magnetic Properties of Superconductors

13.1 Development of the Classical Models

Although initially discovered and investigated by ONNES (e.g., [ONN11]) in the period between 1911 and 1913 as a zero-resistivity phenomenon (and ultimately exploited as such), it was soon recognized that the essential physics of superconductivity was couched in terms of electromagnetism and, in particular, in terms of the responses of superconductors to applied magnetic fields. In 1950 GINZBURG and LANDAU [GIN50] produced a phenomenological theory of superconductivity, central to which was a superconducting order parameter cast in the form of a wavefunction Ψ, such that $|\Psi|^2 = n_s$, the number density of superelectrons. In 1952 ABRIKOSOV [ABR52] noted that this theory led to the existence of two types of superconductor distinguishable by the magnitude of a parameter κ_{GL} of the Ginzburg-Landau (G-L) theory and, in particular, by whether or not that parameter was greater than $1/\sqrt{2}$. It was noted at the time that $\kappa_{GL} < 1/\sqrt{2}$ implied a positive s/n interface surface energy and that $\kappa_{GL} > 1/\sqrt{2}$ implied a negative one. In a subsequent paper, while investigating in detail the magnetic properties of a superconductor belonging to this "second group" (i.e., that for which $\kappa_{GL} > 1/\sqrt{2}$), ABRIKOSOV [ABR57] predicted that within a suitable magnetic field range (the "unstable" region of G-L theory [GIN50]) the lines of magnetic flux penetrating the sample, now in the "mixed state," would arrange themselves in the form of a long-range-ordered "lattice." The calculations yielded a square lattice, and the first picture of such an arrangement appeared in ABRIKOSOV's paper. Following some preliminary calculations published in 1958 [GOR58], GOR'KOV [GOR59] explicitly showed that the Ginzburg-Landau equations, with Ψ representing a wavefunction for Cooper pairs, could be recovered from the

BARDEEN-COOPER-SCHRIEFFER (BCS) [BAR57] and BOGOLIUBOV [BOG58] microscopic theories of superconductivity.

In a subsequent series of papers, GOODMAN showed that κ_{GL}-related energy considerations, applied to the interface between the normal and superconducting regions, could lead to a laminar model [GOO61] for what was later to be referred to as the mixed state. Also, in discussing the combined theories of GINZBURG, LANDAU, ABRIKOSOV, and GOR'KOV (so-called "GLAG theory"), GOODMAN produced some useful expressions for κ_{GL} and a so-called "coherence length" ξ_{GL} in terms of atomic constants and normal-state measurables [GOO62]. Since the G-L equations were valid only in a narrow temperature interval below T_c, theorists following GOR'KOV attempted to extend the temperature range of the microscopic theory and to generalize it further in terms of microscopic materials properties such as the Fermi velocity, v_F, the Fermi density-of-states, $n(E_F)$, and the electronic mean-free-path length, l. Also taken into consideration has been the effect of a wide range of electronic mean-free-path lengths, from the "clean" ($l > \xi_0$) to "dirty" ($l \ll \xi_0$) limits, where ξ_0 represents the BCS pure solvent coherence length. As pointed out by FIETZ [FIE67[a]], most of the theoretical results pertain to the region of the *second-order transition* near H_{c2}, where the gap function Δ_{GL} is small and forms a convenient expansion parameter. The theoretical extensions to GLAG theory resulted in the definition of three generalized temperature-dependent "G-L" parameters—$\kappa_1(t)$, $\kappa_2(t)$, and $\kappa_3(t)$—all of which reduce to the usual G-L parameter κ_{GL} at T_c. The purpose of critical-field temperature-dependence studies was to put these improvements to the test.

With the upsurge of interest in GLAG theory beginning at about 1961 it was recognized that both normal-state and mixed-state Pauli paramagnetic susceptibilities would play important roles in adjusting the energy balance between the superconducting and normal states. Theoretical studies of the effects of Pauli paramagnetism on the upper critical field were initiated independently by CLOGSTON and CHANDRASEKHAR and extended by MAKI and by WERTHAMER and colleagues, as indicated in Tables 13-1 and 13-2. An overview of the historical development of superconductivity, which places in perspective the contributions made by CLOGSTON, the LONDONS, GORTER and CASIMIR, PIPPARD, FRÖLICH, SCHAFROTH and colleagues, the Soviet scientists GINZBURG, LANDAU, ABRIKOSOV and GOR'KOV, and finally, BARDEEN, COOPER, and SCHRIEFFER, has been provided by RICKAYZEN [RIC65], see also Tables 13-1 and 13-2 (Section A in each case). The more recent literature, which deals with improvements to GLAG theory through the introduction of normal- and mixed-state paramagnetisms, and electron-phonon and electron-electron interactions, is summarized by Section B of those tables. Stemming from the London electromagnetic equations by way of MENDELSSOHN's filamentary-mesh hypothesis was the

Table 13-1. Selected Chronology in the Development of Microscopic and Electromagnetic Theories of Superconductivity

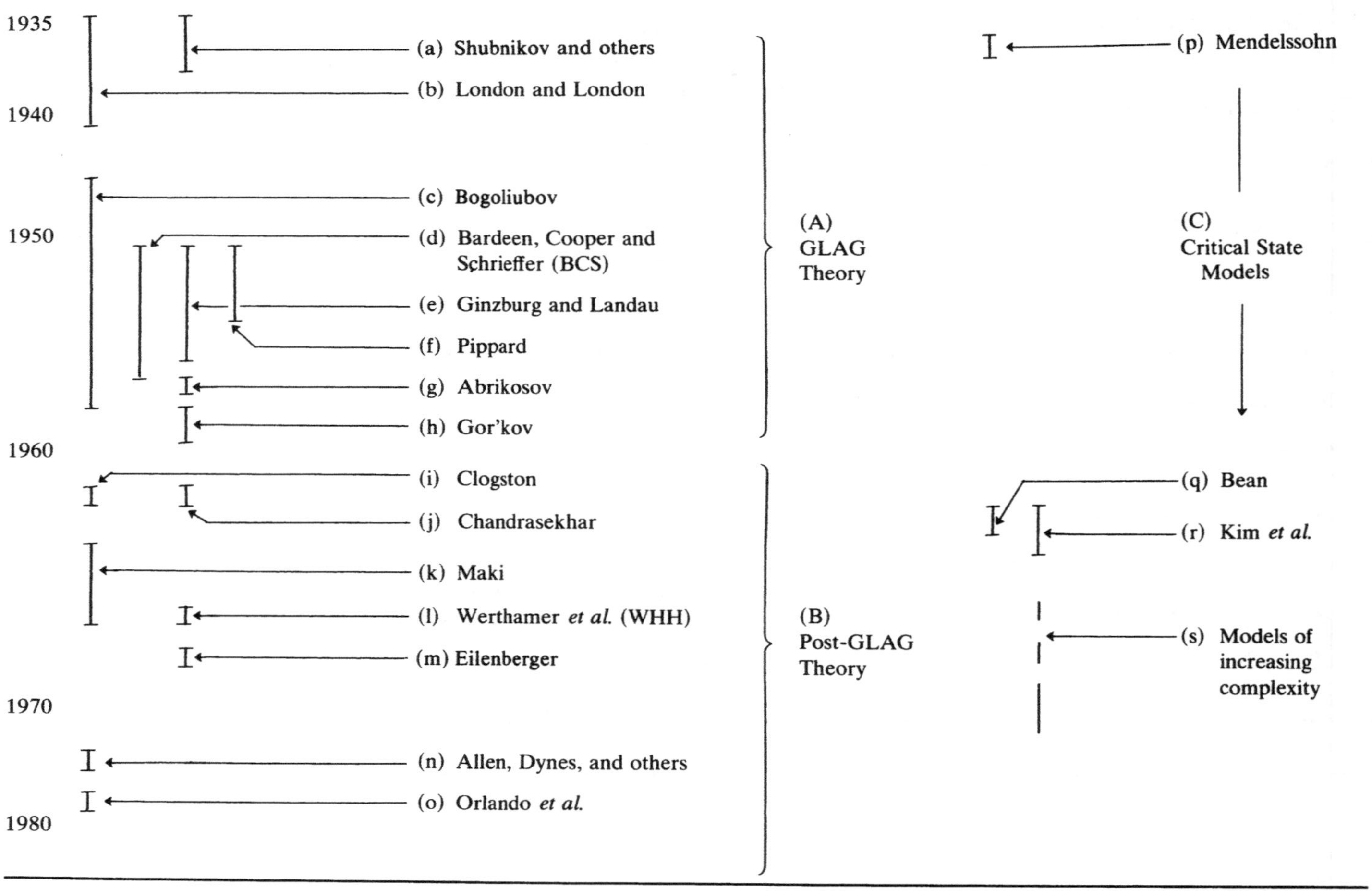

Table 13-2. Index to Table 13-1—Chronological Development of Theories of the Mixed State

Table Entry	Indication of Contribution	Literature
	A. *GLAG Theory—Chronological*	
Shubnikov and others	Experimental evidence for flux penetration at fields between what were to be known as H_{c1} and H_{c2}.	K. Mendelssohn *et al.*, *Nature* (London) **135**, 826 (1935) W. J. de Haas *et al.*, *Nature* (London) **135**, 30 (1935) L. W. Shubnikov *et al.*, *Zh. Eksp. Teor. Fiz.* **7**, 221 (1937)
London and London	Phenomenological electromagnetic equations and introduction of flux penetration (λ)	F. London and H. London, *Proc. R. Soc. London A* **149**, 71 (1935) H. London, *Proc. R. Soc. London A* **176**, 522 (1940)
Bogoliubov	Extensive studies of He II and superconductivity and eventual development of theory closely related to BCS.	See Rickayzen, *Theory of Superconductivity*, Wiley Interscience, New York (1965)
Bardeen, Cooper, and Schrieffer (BCS)	Development of the phonon-moderated-electron/electron quasiparticle leading to the BCS theory.	See above; also J. Bardeen *et al.*, *Phys. Rev.* **108**, 1175 (1957)
Ginzburg and Landau	Macroscopic approach leading to the selection and solution of a wave-like equation in Ψ, an order parameter, selected such that $\|\Psi\|^2 = n_s$, the number density of superconducting electrons. Introduced the parameter κ_{GL} and considered $\kappa_{GL} \ll 1$.	V. L. Ginzburg *et al.*, *Zh. Eksp. Teor. Fiz.* **20**, 1064 (1950) V. L. Ginzburg, *Nuovo Cimento* **2**, 1234 (1955)
Pippard	Introduced the concept of coherence distance (ξ).	A. B. Pippard, *Proc. R. Soc. London A* **216**, 547 (1953)

Abrikosov	Solved the Ginzburg–Landau equation for $\kappa_{GL} > 1/\sqrt{2}$ and computed the existence of a flux lattice.	A. A. Abrikosov, *Sov. Phys. JETP* **5**, 1174 (1957); [transl. *Zh. Eksp. Teor. Fiz.* **32**, 1442 (1957)]
Gor'kov	Showed that the phenomenological Ginzburg–Landau equations followed from the BCS/Bogoliubov microscopic (electron/phonon) theories.	L. P. Gor'kov, *Sov. Phys. JETP* **9**, 1364 (1959); **10**, 593, 998 (1960); [transl. *Zh. Eksp. Teor. Fiz.* **36**, 1918; **37**, 833, 1407 (1959)]
	B. *Post-GLAG Theory—Chronological*	
Clogston	Showed by means of a free-energy argument, that the magnetic-field reduction of *normal-state* free energy led to a paramagnetically induced lowering of H_{c2}. Also suggested that electron–phonon renormalization could reduce the Pauli limitation.	A. M. Clogston, *Phys. Rev. Lett.* **9**, 266 (1962)
Chandrasekhar	The same conclusion independently drawn using an argument based on $2\mu_B H \equiv$ (zero K BCS energy gap).	B. S. Chandrasekhar, *Appl. Phys. Lett.* **1**, 7 (1962)
Maki	Developed a generalized Ginzburg–Landau equation in the dirty limit valid for all temperatures such that the temperature dependences of H_{c2}, $(dM/dH)(H = H_{c2})$, and H_{c1} can be expressed in terms of $\kappa_1(t)$, $\kappa_2(t)$, and $\kappa_3(t)$.	K. Maki, *Physics* **1**, 21, 127 (1964); see also, C. Caroli *et al.*, *Solid State Commun.* **4**, 17 (1966)
	Extended the above analysis to deal with the effect of Pauli paramagnetism on terms parameterized by $\kappa_1(t)$ and $\kappa_2(t)$.	K. Maki, *Physics* **1**, 127 (1964)
	Analyzed the effect of Pauli paramagnetism on superconductivity in the mixed state and showed that in its presence the s/n transition changes from first order to second order as the temperature increases. Went on to study, for the first time, the effect of finite mean-free-path including the spin-orbit part of the scattering.	K. Maki and T. Tsuneto, *Prog. Theor. Phys.* (Kyoto) **31**, 945 (1964)

Continued

Table 13-2. *(cont.)*

Table Entry	Indication of Contribution	Literature
	Studied the magnetic properties of the mixed state in the *clean limit* in the immediate subcritical region and again calculated the parameters $\kappa_1(t)$ and $\kappa_2(t)$, which as before coincide with κ_{GL} at $t = 1$.	K. Maki and T. Tsuzuki, *Phys. Rev. A* **139**, 868 (1965)
	Derived expressions to take into account the paramagnetic effect (parameterized by α) in the dirty limit and its reduction by spin-orbit scattering (parameterized by λ_{so}) and introduced for convenience an additional parameter $\beta_M(\alpha, \lambda_{so})$ given by $\beta_M^2 \propto \alpha^2/\lambda_{so}$.	K. Maki, *Phys. Rev.* **148**, 362 (1966)
Werthamer, Helfand, and Hohenberg (WHH)	Developed solutions of the linearized Gor'kov equations yielding the temperature dependence of H_{c2} for all values of the electronic mean-free-path (l). Computed results were plotted in the format $\kappa_{red} \equiv \kappa_1(t, \lambda)/\kappa_1(1, \lambda)$ *versus* t where $\kappa_1(t, \lambda) = H_{c2}(t, \lambda)/\sqrt{2}H_c(t)$, and λ, the impurity parameter ($\equiv$Gor'kov's ρ_G), $= 0.882\ \xi_0/l$.	E. Helfand and N. R. Werthamer, *Phys. Rev. Lett.* **13**, 686 (1964)
	The same, but presented solutions in the format κ_{red} *versus* t for various λ as before, and in addition, h^* *versus* t where $h^*(t, \lambda) \equiv H_{c2}(t, \lambda)/(-dH_{c2}/dt)_{t=1}$.	E. Helfand and N. R. Werthamer, *Phys. Rev.* **147**, 288 (1966)
	Developed a theory of H_{c2} in the general (unlimited) case (taking into account only s-wave scattering by impurities). Considered both Pauli limitation and spin-orbit scattering, and derived an expression suitable for computation parameterized by α and λ_{so}, enabling a function $h^*(t, \alpha, \lambda_{so})$, related to H_{c2}, to be compared with experiment.	N. R. Werthamer, E. Helfand and P. C. Hohenberg, *Phys. Rev.* **147**, 195 (1966)

Eilenberger	In what was essentially an extension, to *s*- and *p*-wave (anisotropic) electron scattering parameterized by $\lambda_{tr} = \xi_0/l_{tr}$, of References [MAK64, MAK64[a]] and [HEL64, HEL66] above, EILENBERGER calculated and plotted *versus* t the parameters κ_1 and κ_2 which are, in this work, functions of t, l_{tr}/l, and ξ_0/l_{tr}.	G. Eilenberger, *Phys. Rev.* **153**, 584 (1967)
Allen, Dynes, and others	Studied corrections to the G–L parameters due to strong electron–phonon interaction.	P. B. Allen *et al.*, *Phys. Rev. B* **12**, 905 (1975)
Orlando *et al.*	Studied influences of electron–phonon renormalization of normal-state parameters on the Pauli-limited H_{c2} and the magnitude of adjustable spin-orbit scattering requirements.	T. P. Orlando *et al.*, *Phys. Rev. B* **19**, 4545 (1979)
	C. *Critical State Models—Chronological*	
Mendelssohn	Introduction of the "sponge" or "filamentary" model of magnetic irreversibility (flux trapping).	K. Mendelssohn *et al.*, *Nature* (London) **135**, 826 (1935) (see also, *Rev. Mod. Phys.* **36**, 50 (1964))
Bean	Application of London theory within the framework of the filamentary model to couple current with field in irreversible type-II superconductors.	C. P. Bean, *Phys. Rev. Lett.* **8**, 250 (1962)
Kim *et al.*	Reinterpretation of the critical state concept in terms of the balancing of the Lorentz force on Abrikosov flux bundles against a pinning force.	Y. B. Kim *et al.*, *Phys. Rev. Lett.* **9**, 306 (1962)
Increasing Complexity	Introduction of the idea of a pinning force to be expressed in terms of $F_p(T, H)$, the bulk pinning force, paved the way for the development of more and more powerful critical state models as flux-pinning theory eventually led to relationships of the form $F_p = \alpha(T)H_{c2}^{l}h^{m}(1-h)^{n}$ at high fields.	W. A. Fietz *et al.*, *Phys. Rev. A* **136**, 335 (1964) M. A. R. LeBlanc *et al.*, *Phys. Rev. Lett.* **14**, 704 (1965) K. Yasukōchi *et al.*, *J. Phys. Soc. Jpn* **21**, 89 (1966) B. C. Belanger, *Dissertation*, University Microfilms International, Ann Arbor, Michigan (1968) R. Gauthier *et al.*, *Low Temperature Physics–LT13*, Plenum Press, New York, 1974, pp. 241–246 —etc.

phenomenological theory of Bean, which described the response of a superconductor to an applied magnetic field in terms of the flow of critically dense current. The recognition of the Abrikosov flux lattice next required the reinterpretation of the Bean model in terms of flux pinning, heralding the development of a family of critical-state models generated by variation of the functional form of the bulk pinning force F_p. This branch of mixed-state theory is indicated in Tables 13-1 (Section C) and 13-2 (Section C).

13.2 Type-I and Type-II Superconductors

Type-I and type-II superconductors are experimentally distinguishable by their responses to an applied magnetic field, H_a. The phase diagram of Figure 13-1 serves to define some of the essential parameters and properties [Kim64], [Cat66], [Fer70]. In studying the magnetization of superconductors it is usual to deal, both theoretically and experimentally, with long cylinders in axially directed magnetic fields in order to avoid the complication of having to account for demagnetization effects. During the increase of an applied field up to the critical value, H_c, (or H_{c0} at zero K) a type-I superconductor remains ideally diamagnetic. The applied field is excluded from the interior of the sample by supercurrents that circulate (hence the London relationship $\mathbf{\nabla} \times \mathbf{J} \propto \mathbf{H}$ [Lyn64, p. 29]) within a thin surface layer, which is typically about 500 Å in thickness (almost exactly that value in the case of Sn and Al [Lyn64, p. 38]). This is known as the Meissner effect. Further increase of the field results in an immediate transformation to the normal state, Figure 13-1(a). In the type-II superconductor (a category to which most superconducting materials belong) Meissner flux exclusion, which takes place immediately upon application of a field, gives way to flux penetration, which commences as soon as H_a exceeds a "lower critical field" H_{c1}, Figure 13-1(b). The Abrikosov "mixed state" characterized by the existence of a regular array of magnetic fluxoids aligned parallel to the

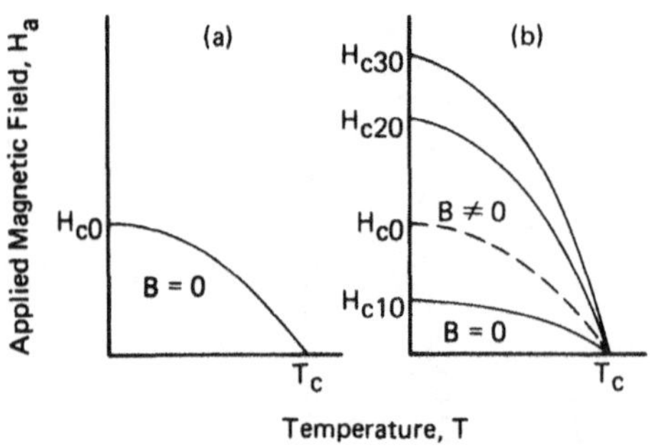

Figure 13-1. Schematic representation in field-temperature space of (a) type-I, and (b) type-II superconductivity. The regions designated $B = 0$ represents the Meissner flux-exclusion state; that designated $B \neq 0$ represents the mixed state (or Shubnikov phase). Surface superconductivity exists at applied fields, H_a, between H_{c2} and H_{c3}—after standard sources (see also Hake [Hak69]).

applied field, H_a, persists until that field is increased beyond H_{c2}, its "upper critical" value, at which point the bulk of the superconductor undergoes a second-order transition to the normal state. In type-II superconductors two additional effects must be taken into consideration as (i) the applied field is increased beyond H_{c2}, and (ii) the temperature (at, say, zero applied field) is raised above T_{c0}. In the first case, with $H_{c2} < H_a < H_{c3}$, superconductivity is confined to a surface sheath. Surface superconductivity which is due to the influence of a surface boundary on the mixed state has been discussed at length by SAINT-JAMES and SARMA [SAI69, pp. 81ff]. According to SAINT-JAMES and DE GENNES [SAI63], the surface sheath critical field, H_{c3}, has the value 1.695 H_{c2} or 1.0 H_{c2} depending on whether the applied field is parallel or perpendicular to the surface. They also suggested that formation of a surface sheath at $H_{c3} > H_a > H_{c2}$ in parallel-field geometry may explain some apparent discrepancies that had been noted between the magnitudes of the magnetically measured (on cylinders in parallel fields) and resistively measured (on wires in transverse fields) upper critical fields. Further discussion of this topic is deferred to Section 13.9.

Another interesting "supercritical" phenomenon has been observed as the temperature is raised above T_c, *viz.*, "fluctuation superconductivity" [HAK69[a]], a critical fluctuation encountered just outside the mixed-state/normal-state second-order phase boundary. According to HAKE [HAK69[a], HAK70, HAK71, HAK74], who has made an experimental study of the effect, fluctuation superconductivity has been found at temperatures as high as $2T_c$.

Typical magnetization curves for type-I and type-II superconductors are schematically depicted in Figure 13-2. Whether a material is type-I or type-II (or has been converted from one to the other by alloying) depends on the magnitude of the parameter $\kappa_{GL} = \lambda/\xi$ (to be considered in detail below) and its relationship to $1/\sqrt{2}$. In particular: $\kappa_{GL} < 1/\sqrt{2}$—type-I; $\kappa_{GL} > 1/\sqrt{2}$—type-II. The origin and development of the concepts of "penetration depth," λ, and "coherence length," ξ, are outlined below.

13.3 The London Penetration Depth, λ_L

The Meissner flux-exclusion effect requires the termination of magnetic flux at the surface of the sample. Since the circulating current needed to accomplish this requires a surface layer of characteristic thickness (say λ_L) in which to flow, it follows according to the LONDONS (e.g., Tables 13-1 and 13-2) that

$$\mathbf{H} = -\left(\frac{4\pi}{10}\lambda_L^2\right)\nabla \times \mathbf{J} \qquad (13\text{-}1)$$

Combining this with the Maxwell equation,

$$\nabla \times \mathbf{H} = \left(\frac{4\pi}{10}\right)\mathbf{J} \tag{13-2}$$

leads directly to

$$\nabla^2\mathbf{H} = \lambda_L^{-2}\mathbf{H} \tag{13-3}$$

A corresponding relationship in terms of **J**, *viz.*,

$$\nabla^2\mathbf{J} = \lambda_L^{-2}\mathbf{J} \tag{13-4}$$

can also be shown to exist. Taken together, (13-3) and (13-4) portray both field and current decaying exponentially towards the interior of the sample. According to the London theory the zero K penetration depth, λ_{L0}, was given by

$$\lambda_{L0} = (mc^2/4\pi n_s e^2)^{1/2} \tag{13-5}$$

where m is the electronic mass, c is the velocity of light, e is the electronic charge in e.s.u., and n_s is the number density of superconducting electrons (a typical value for which might be 3×10^{23} cm^{-3} [HAK67[b]]). The introduction of a temperature dependence factor taken from the Gorter–Casimir two-fluid model [LYN64, p. 19] results in a temperature-dependent London penetration depth of

$$\lambda_L = \lambda_{L0}(1 - t^4)^{-1/2} \tag{13-6a}$$

$$\cong (\lambda_{L0}/2)(1 - t)^{-1/2} \tag{13-6b}$$

provided that $t(\equiv T/T_c)$ is close to unity.

13.4 Extension of London Theory

13.4.1 The Coherence Length, ξ

The basic London equations led to

$$\mathbf{J}(r) = -\left[\frac{c}{4\pi\lambda_L^2}\right]\mathbf{A}(r) \tag{13-7}$$

(where,. of course, $\mathbf{H} = \nabla \times \mathbf{A}$).

Thus,

$$\mathbf{J}(r) = -\left(\frac{n_s e^2}{m}\right)\mathbf{A}(r) \tag{13-8}$$

which may be compared with Ohm's law

$$\mathbf{J}(r) = \sigma(l)\mathbf{E}(r) \tag{13-9}$$

where σ is the electrical conductivity. But just as the electronic mean-free-path, l, provides a characteristic length for normal-state electrical conduction, superconductivity, according to PIPPARD (see Tables 13-1 and 13-2), should also be associated with a characteristic length—one which was to become known, for the following reason, as the "electromagnetic coherence length."

The nonlocal relationship between current and field within the context of normal-state conductivity leads to the anomalous skin effect expressible in the form

$$\mathbf{J}(r) = \left(\frac{3}{4\pi}\right)\frac{\sigma}{l}\int\frac{\mathbf{r}(\mathbf{r}\cdot\mathbf{E})\,e^{-r/l}\,d\tau}{r^4} \tag{13-10}$$

where the integration is over a volume. Corresponding to this is the following relationship due to PIPPARD [LYN64, p. 34] for the electromagnetic response of a clean superconductor:

$$\mathbf{J}(r) = \left(\frac{3}{4\pi}\right)\frac{n_s e^2}{m\xi_{0P}}\int\frac{\mathbf{r}(\mathbf{r}\cdot\mathbf{A})\,e^{-r/\xi}\,d\tau}{r^4} \tag{13-11}$$

Corresponding to the normal-state mean-free-path, l, is the electromagnetic coherence length designated ξ_{0P}, in which the additional subscript P is used to distinguish it from the BCS counterpart. At the time in which these equations were developed, the term "coherence" made reference to the range of coherence of the superconductive order parameter of the Gorter–Casimir model. The Pippard parameter does, however, relate to the Ginzburg–Landau coherence length ξ_{GL} that characterizes the distance over which the pseudowavefunction or order parameter $\Psi(\mathbf{r})$ may vary while causing little increase in energy (but see [TIN75, p. 111]).

Using uncertainty principle arguments, PIPPARD concluded that

$$\xi_{0P} = a\frac{\hbar v_F}{k_B T_c} \tag{13-12}$$

where v_F is the Fermi velocity, and a is an adjustable constant. BARDEEN, COOPER, and SCHREIFFER [BAR57, p. 1196] subsequently developed a related coherence length given by

$$\xi_0 = \frac{e^\gamma}{\pi^2} \frac{\hbar v_F}{k_B T_c} = 0.18 \frac{\hbar v_F}{k_B T_c} \qquad [11\text{-}3](13\text{-}13)$$

13.4.2 The Penetration Depth, λ

In 1953, PIPPARD demonstrated [LYN64, p. 41], as a result of measurements on alloys, that λ_L was mean-free-path length dependent (a feature not present in the original London parameter), and went on to show that the influence of mean-free-path length on penetration could be taken into account with the aid of a factor $\xi_{0P}/\xi(l)$ where the new electromagnetic coherence length was given by

$$\frac{1}{\xi(l)} = \frac{1}{\xi_{0P}} + \frac{1}{\alpha l} \qquad (13\text{-}14)$$

with $\alpha = 0.8$ (see GOODMAN [GOO62]), hence approximately equal to 1.0, such that

$$\lambda = (\xi_{0P}/\xi(l))^{1/2} \lambda_L \qquad (13\text{-}15)$$

Inspection of (13-14) shows that when $l \ll \xi_{0P}$, coherence is controlled by the normal-state mean-free-path length, and that under such conditions, and with ξ_{0P} henceforth replaced by ξ_0, the BCS coherence length (13-13) becomes

$$\lambda = (\xi_0/l)^{1/2} \lambda_L \qquad (13\text{-}16)$$

It is interesting to note in conclusion that the neglect of the coherence idea in the original London formulation was equivalent to assuming $\xi \to 0$, and in particular that $\xi \ll \lambda_L$. In current terminology this is an expression of the dirty limit, to which the London theory was not intended to be confined. Nevertheless, it is now recognized that London theory is valid only in this limit, and then only at low fields. What has been referred to as the penetration depth in the Pippard limit, λ_∞, [LYN64, pp. 41ff] is given by a more complicated expression than (13–15) above. To summarize, the two limiting-case penetration depths within the present context are given by

$$\lambda_\infty = \left[\frac{\sqrt{3}}{2\pi} \xi_0 \lambda_L^2\right]^{1/3} \quad \begin{cases} \xi \gg \lambda \\ \text{“Pippard limit”} \\ \text{Type-I} \end{cases} \qquad (13\text{-}17)$$

and

$$\lambda = [\xi_0/\xi(l)]^{1/2}\lambda_L \qquad \begin{cases} \xi \ll \lambda \\ \text{"London limit"} \\ \text{type-II} \end{cases} \tag{13-18}$$

in which $\xi(l) \cong l$ can be computed using

$$l = \frac{\pi^2 k_B^2}{e^2 \gamma_v v_F \rho_n} \quad \text{(cm)} \tag{13-19}$$

where all the quantities are in c.g.s. units, including the resistivity ρ_n.[1]

13.5 Parameters of the Ginzburg–Landau (G–L) Theory

13.5.1 Penetration Depths and Coherence Lengths

As indicated in Table 13-2, GOR'KOV was able to show that the Ginzburg-Landau equations, valid for both types of superconductivity, could be recovered from BCS theory. For example, the G–L order parameter, $\Psi(\mathbf{r})$, may be related to the BCS energy gap, $\Delta_0(\mathbf{r})$, through

$$\Psi(\mathbf{r}) = \left[\frac{8.4 n_e}{4\pi^2 T_c}\right]^{1/2} \Delta_0(\mathbf{r}) \tag{13-20}$$

where n_e is the number density of the conduction electrons [GRA75, p. 55]. Other microscopic interpretations of the G–L results led to a new representation of the penetration depth and coherence length. Based on a full set of equations to be presented in Table 14-1 are the following simplified expressions, in terms of λ_{L0}, l, and ξ_0, for the clean- and dirty-limit parameters [ORL79].

(a) The Penetration Depths. For the clean limit

$$\lambda_{GL}^c = (1/\sqrt{2})\lambda_{L0}(1-t)^{-1/2} \tag{13-21}$$

while for the dirty limit

$$\lambda_{GL}^d = (1/\sqrt{2})\lambda_{L0}(\xi_0/1.33l)^{1/2}(1-t)^{-1/2} \tag{13-22}$$

[1] If ρ_n were to be in Ω cm and all other quantities to remain the same, a prefactor of 9×10^{11} would be needed.

where λ_{L0} has been given in (13-5). It is interesting to note that λ^c_{GL} (clean limit) is given approximately by the λ_L of the original London theory, which we know recognize as a dirty-limit ($\xi \to 0$) approximation. The complementary term, λ^d_{GL}(dirty limit) is comparable to the Pippard-modified London expression, (13-16).

(b) The Coherence Lengths. For the clean limit

$$\xi^c_{\mathrm{GL}} = 0.739\xi_0(1 - t)^{-1/2} \tag{13-23}$$

while for the dirty limit

$$\xi^d_{\mathrm{GL}} = 0.852(\xi_0 l)^{1/2}(1 - t)^{-1/2} \tag{13-24}$$

Equation (13-24) shows ξ^d_{GL} to be approximately the geometrical mean of ξ_0 and the mean-free-path limited coherence length.

In studies of the superconductive proximity effect (see Chapter 11), DE GENNES and GUYON [DEG63], and subsequently WERTHAMER [WER63] (see also [DEU69, p. 1006]), have defined coherence distances ξ_n and ξ_s that are characteristic of either the normal (n) or superconductive (s) components of a bimetallic layer according to

$$\xi = (\hbar v_F l/6\pi k_B T_C)^{1/2} \tag{11-4}$$

which at $T = T_c$ is related to the BCS coherence length ξ_0, (13-13), according to:

$$\xi = (\xi_0 l/3.4)^{1/2} \cong (\xi_0 l/3)^{1/2} \tag{11-7}$$

13.5.2 The Ginzburg–Landau Parameter, κ_{GL}

Central to the theory of type-II superconductivity is the Ginzburg–Landau parameter

$$\kappa_{\mathrm{GL}}(T) = \frac{\lambda_e(T)}{\xi_{\mathrm{GL}}(T)} \tag{13-25}$$

where λ_e is an "effective" or measured penetration depth in the weak-field London limit [GOO62], [TIN75, p. 113]. It may be expressed in the following useful form:

$$\kappa_{\mathrm{GL}}(T) = \frac{2\sqrt{2}\pi H_c(T)\lambda_e^2(T)}{\Phi_0} \tag{13-26}$$

where Φ_0 is the flux quantum and H_c is the "thermodynamic" critical field. Next, since

$$H_c \cong H_{c0}(1 - t^2) \tag{8-7}$$

and

$$\lambda_L \cong \lambda_{L0}(1 - t^4)^{-1/2} \tag{13-6a}$$

it follows that $\kappa_{GL}(t) \propto (1 + t^2)^{-1}$ near $t = 1$, the temperature regime to which G–L theory is restricted, or

$$\kappa_{GL}(t) = 2\kappa_{GL1}(1 + t^2)^{-1} \tag{13-27}$$

Since, in the vicinity of $t = 1$, this implies only a weak temperature dependence, in what follows κ_{GL} will be taken as independent of temperature.

By making appropriate substitutions it can be shown [TIN75, p. 114], [ORL79] that, for the clean limit,

$$\kappa_{GL}^c = 0.957\lambda_{L0}/\xi_0 \tag{13-28}$$

and for the dirty limit,

$$\kappa_{GL}^d = 0.720\lambda_{L0}/l \tag{13-29}$$

A comparable result can be deduced using the following approach based on GOR'KOV's χ_G-function [GOR60^a], [GOO62]: The Ginzburg–Landau parameter, according to GOR'KOV, is defined as

$$\kappa_{GL} = \kappa_{GL}^c/\chi_G(\rho_G) \tag{13-30}$$

where, as pointed out in WERTHAMER's comprehensive treatment of G–L theory [WER69], the function χ_G is such that

$$\chi_G^c(\rho_G) = (1 + \rho_G)^{-1} \cong 1 \qquad (\text{"small-}\rho_G\text{," clean limit}) \tag{13-31}$$

$$\chi_G^d(\rho_G) = 1.17(1 + \rho_G)^{-1} \cong 1.17/\rho_G \qquad (\text{"large-}\rho_G\text{," dirty limit}) \tag{13-32}$$

while ρ_G itself is defined by

$$\rho_G \equiv \frac{\hbar}{2\pi k_B T_c \tau} = \frac{\hbar v_F}{2\pi k_B T_c l} \tag{13-33}$$

where τ is the transport scattering relaxation time and is equal to l/v_F.

Next, after substituting ξ_0 from (13-13) into (13-33)

$$\rho_G = \frac{\xi_0}{0.18} \cdot \frac{1}{2\pi l} = 0.884 \frac{\xi_0}{l} \tag{13-34}$$

Finally by equating κ_{GL} to the sum of the clean limit given by (13-28) and the "large-ρ_G" dirty limit given by (13-30), and after replacing ρ_G with the aid of (13-34) we find, for comparison with (13-28) and (13-29),

$$\kappa_{GL} = \kappa_{GL}^c + \kappa_{GL}^d \tag{13-35}$$

$$= \kappa_{GL}^c + \kappa_{GL}^c \chi^{-1}(\rho_G^d)$$

that is,

$$\kappa_{GL} = 0.957 \left(\frac{\lambda_{LO}}{\xi_0}\right) + 0.957 \left(\frac{\lambda_{LO}}{1.32l}\right) \tag{13-36}$$

The two components of κ_{GL} as defined in (13-36) or its predecessor have been referred to by HAKE [HAK67[b]] as the "intrinsic" and "extrinsic" G–L parameters, respectively. The relationship as stated can also be derived by taking the clean-limit parameter, (13-28), and replacing $1/\xi_0$ with $(1/\xi_0) + (1/\alpha' l)$, which is reminiscent of a coherence-length transformation proposed earlier by PIPPARD and presented here in (13-14).

13.5.3 Clean and Dirty Limits of κ_{GL} in Type-II Superconductors

In type-II superconductors, κ_{GL} is lower bounded by $1/\sqrt{2}$ but has no upper limit. As Ti is added to pure Nb (an elemental type-II superconductor), for example, κ_{GL} increases from about 1 to about 80 at 68 at.% Ti. In such systems then, (13-35) reduces from

$$\kappa_{GL} = \kappa_{GL}^c + \kappa_{GL}^d \tag{13-35}$$

to

$$\kappa_{GL} = 1 + \kappa_{GL}^d \tag{13-37}$$

Four conclusions can be drawn from these simple observations: (i) Since $\kappa_{GL}^c \cong 1$ it follows that $\lambda_{LO} \cong \xi_0$, which, since neither of these parameters contains l, should remain valid under all departures from purity. (ii) With reference to (13-29), $\kappa_{GL}^d = 0.720\lambda_{LO}/l \cong 0.720\xi_0/l$. (iii) Since the clean limit is dominated by κ_{GL}^c the second component in that limit must be small compared to unity, from which we deduce that both λ_{LO}/l and ξ_0/l are significantly less than unity in that limit. (iv) Conversely, since the dirty limit is dominated by κ_{GL}^d it follows, in that limit, that λ_{LO}/l, $\xi_0/l \gg 1$. These conclusions may be summarized in the following way:

$$\text{clean limit} \quad \begin{cases} \kappa_{GL} \cong \kappa_{GL}^{c} = 0.957\left(\dfrac{\lambda_{L0}}{\xi_0}\right) \cong 1 \\ \left.\begin{array}{l} \rho_G = 0.884\left(\dfrac{\xi_0}{l}\right) \\ \dfrac{\lambda_{L0}}{l}, \dfrac{\xi_0}{l} \end{array}\right\} < 1 \end{cases} \tag{13-38}$$

$$\text{dirty limit} \quad \begin{cases} \left.\begin{array}{l} \kappa_{GL} \cong \kappa_{GL}^{d} = 0.720\left(\dfrac{\lambda_{L0}}{l}\right) \cong \left(\dfrac{\xi_0}{l}\right) \\ \rho_G = 0.884\left(\dfrac{\xi_0}{l}\right) \\ \dfrac{\lambda_{L0}}{l}, \dfrac{\xi_0}{l} \end{array}\right\} \gg 1 \end{cases} \tag{13-39}$$

Verification of these general statements is provided by Table 13-3, which lists specific examples of the numerical values assumed by the key parameters l, ξ_0, λ_{L0}, κ_{GL}^{c}, and κ_{GL}^{d} for two representative groupings of Ti-alloy superconductors.

In the present context, the characterization of superconductors according to $l > \xi_0$ or $l \ll \xi_0$ would suggest that an almost even subdivision of superconductors into one or other of the two types is possible. Such is not the case. Titanium alloys, for example, are always more or less "dirty." Thus, with l of the order 10 Å and ξ_0 typically several hundred Å, the characteristic quotient ξ_0/l is usually much greater than unity. The situation with these alloys is best presented in terms of κ_{GL}s being either $\cong 1$ or $\gg 1$, which quite correctly implies that the clean limit is a very special case.

The GOR'KOV–BCS microscopic reinterpretation of Ginzburg–Landau theory naturally leads to the use of ξ_0/l as a purity parameter. The significance of this quantity is referred to in the next chapter, which deals further with the basic theories of the mixed state.

13.6 The Thermodynamic Critical Field, H_c

13.6.1 Thermodynamic Relationships

The thermodynamic critical field, H_c, which is a directly measurable quantity in type-I superconductors (see Figure 13-2(a)), is definable in terms

Table 13-3. Characteristic Lengths and Ginzburg–Landau

Nb Concentration at.%	Residual Resistivity ρ_n $10^{-6}\ \Omega$ cm	Electronic Mean-Free-Path l Å	Transition Temperature T_c K	Electronic Specific Heat Coefficient γ_v 10^3 erg cm^{-3} K^{-2}
32	102	4.1	8.1_4	10.0
55	60	6.8	9.7_2	10.8
62	40	10	9.6_7	10.8
83	24	16	9.2_7	10.0
91	8.9	42	9.2	7.9
100	0.6	763	9.1	7.0
V Concentration at.%				
22.5	146	3.6	4.7	8.2
25	132	4.0	5.3	9.3
36	143	3.1	6.6_8	10.6
70	47	8.1	7.6_5	11.5
82	40	9.2	7.5_4	11.4

Formulae

$\kappa^d_{GL} = 7.49 \times 10^3 \rho_n \gamma_v^{1/2}$

$\xi^d_{GL0} = 8.57 \times 10^{-7}(\rho_n \gamma_v T_c)^{-1/2}$ cm

$\lambda^d_{GL0} = 6.42 \times 10^{-3}(\rho_n / T_c)^{1/2}$ cm

After Table 14-1, with ρ_n in Ω cm and γ_v in erg cm^{-3} K^{-2}.

of the difference between the free energies[2] of the superconductive and normal states. Starting with

$$dG = V\,dP - S\,dT - I\,dH_a \tag{13-40}$$

in which H_a is the applied field, I is the magnetic moment, V is the volume of the sample, and the other quantities have their usual thermodynamic meanings [HAK69], and then introducing the Meissner effect in the form

$$0 = B = H_a + 4\pi M \tag{13-41}$$

where B is the induction within the specimen and M is its magnetization,

[2]As pointed out by HAKE [HAK69] in a valuable discussion of the thermodynamics of superconductors, if $G = U + PV - HI - TS$ then since $dU = T\,dS - P\,dV + H\,dI$, it follows that $dG = dU + P\,dV + V\,dP - H\,dI - I\,dH - T\,dS - S\,dT = V\,dP - S\,dT - I\,dH$.

Parameters for Two Series of Titanium-Alloy Superconductors

BCS Coherence Length ξ_0 Å	Zero-K London Penetration Depth λ_{L0} Å	Ginzburg-Landau Parameters: Clean κ_{GL}^c	Dirty κ_{GL}^d	Full κ_{GL}	Zero-K Coherence Length ξ_{GL0}^d Å	Zero-K Penetration Depth λ_{GL0}^d Å
296	438	1.4	76.4	77.8	29.7	2273
238	440	1.8	46.7	48.5	34.1	1595
240	435	1.7	31.1	32.8	41.9	1306
280	406	1.4	18.0	19.4	57.5	1033
357	—	1.0	5.9	6.9	107	631
407	—	0.8	0.4	1.2	438	165
496	499	1.0	99.0	100	36.1	3578
391	527	1.3	95.3	96.6	33.6	3204
318	485	1.5	110	112	26.9	2970
300	430	1.4	37.8	39.2	42.1	1591
314	417	1.3	32.0	33.3	46.2	1479

Data Sources
Ti-Nb(32–83 at.%), cold-worked; Ti-V(36–82 at.%): BRAND [BRA72]
Ti-Nb(91, 100 at.%), cold-worked: FIETZ [FIE67]
Ti-V(22.5 at.%), as-cast; Ti-V(25 at.%), cold worked: HAKE [HAK67[b]]

and considering *unit volume* at constant pressure and temperature, it follows that

$$g = -\int M\,dH_a + \text{constant} = \frac{1}{4\pi}\int H_a\,dH_a + \text{constant} \qquad (13\text{-}42)$$

Next, integrating from H_c to H_a, and recognizing that $g_s(H_c)$ and $g_n(H_c)$

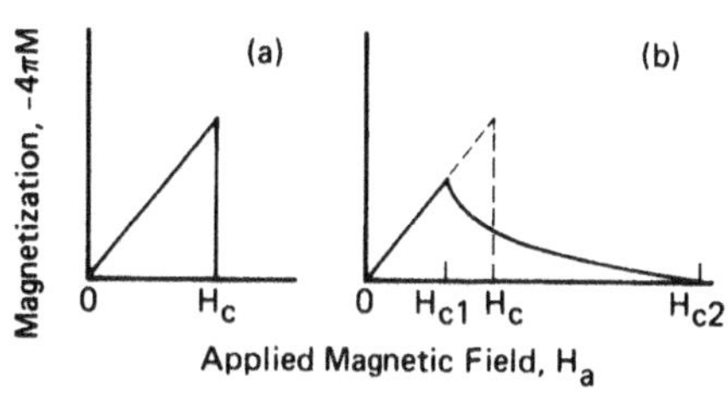

Figure 13-2. Magnetization diagrams representative of (a) type-I, and (b) reversible type-II superconductivity in which the mixed-state diamagnetism decreases monotonically from the Meissner value as the applied field, H_a, increases between H_{c1} and H_{c2}. The absence of the subscript zero (cf. Figure 13-1) implies an arbitrary temperature $0 < T < T_c$.

must be equal,

$$g_s(H_a) = g_n(H_c) - \frac{H_c^2 - H_a^2}{8\pi} \tag{13-43}$$

Assuming the normal state to have a vanishingly small susceptibility, such that $g_n(H_c) \equiv g_n(H_a)$ for all H_a, including $H_a = 0$, we have from (13-43) that

$$g_n(H_a) - g_s(H_a) = \frac{H_c^2 - H_a^2}{8\pi} \tag{13-44}$$

and

$$g_n(0) - g_s(0) = \frac{H_c^2}{8\pi} \tag{13-45}$$

It follows, with the aid (13-42), that

$$\frac{H_c^2}{2} = \int_0^{H_c} (-4\pi M)\, dH_a \tag{13-46}$$

which is, of course, the area under the type-I magnetization curve, Figure 13-2a.

A hypothetical "thermodynamic critical field" for type-II superconductors is similarly defined, with reference to Figure (13-2b), by

$$\int_0^{H_c} (-4\pi M_I)\, dH_a = \frac{H_c^2}{2} = \int_0^{H_{c2}} (-4\pi M_{II})\, dH_a \tag{13-47}$$

indicating equality in the areas under by the two reversible magnetization curves.

13.6.2 The BCS "Thermodynamic" Critical Field

Two useful results from BCS theory [BAR57] are the following expressions for the thermodynamic critical field at and near the absolute zero of temperature: for the zero K thermodynamic critical field

$$H_{c0} = 2.43\, \gamma_v^{1/2} T_c \qquad \text{(Oe)} \tag{13-48}$$

and for the limiting slope at T_c of the critical field temperature dependence

$$\left(\frac{dH_c}{dT}\right)_{T_c} = (18.0\gamma_v)^{1/2} \tag{13-49}$$

with H in Oe and γ_v in erg cm^{-3} K^{-2}.

13.7 The Lower Critical Field, H_{c1}—Onset of the Mixed State

In an early review of type-II superconductivity, GOODMAN [GOO64] pointed out that field penetration would occur and his laminar model of the mixed state would become established when the applied field exceeded a value H_{c1} given by

$$H_{c1} = \frac{H_c}{\sqrt{\kappa_{GL}}} \tag{13-50}$$

(followed, with increasing field according to GORTER and VAN BEELEN [GOR64], by a second-order transition to the normal state at some higher field, H_{c2}). GOODMAN [GOO64] then went on to demonstrate that the use of a flux-line model led to

$$H_{c1} = \frac{H_c}{\kappa_{GL}}(\ln \kappa_{GL} + 0.24) \tag{13-51a}$$

in case $\kappa_{GL} \gg 1$. According to ABRIKOSOV [ABR57],

$$H_{c1} = \frac{H_c}{2\kappa_{GL}}(\ln \kappa_{GL} + 0.08) \tag{13-51b}$$

for $\kappa_{GL} \gg 1$; but the GLAG expression for H_{c1} frequently cited (e.g., [BLA65], [CAT66]) and usually attributed to ABRIKOSOV is

$$H_{c1} = \frac{H_c}{\sqrt{2}\kappa_{GL}}(\ln \kappa_{GL} + 0.08) \tag{13-51c}$$

Since the region of validity was claimed to be $\kappa_{GL} \gg 1$ (and, of course, the formula fails for $\kappa_{GL} \cong 1/\sqrt{2}$) the numerical term in the bracket serves little purpose and is now dropped, resulting in

$$H_{c1} = \frac{H_c}{\sqrt{2}\kappa_{GL}} \ln \kappa_{GL} \tag{13-52}$$

As suggested by (13-51), the early formulations of GLAG theory were prone to be in error by factors of about $\sqrt{2}$. Equation (13-52), the currently accepted form, is correct as are the other expressions for the critical fields presented below.

Reference to the development of theories of the lower critical field is not complete without mention of the work of HARDEN and ARP [HAR63], who, commencing with Ginzburg–Landau–Abrikosov (GLA) theory, set up for numerical solution a relationship between H_{c1}/H_c and all values of κ_{GL} greater than $1/\sqrt{2}$. When logarithmically plotted, HARDEN and ARP'S H_{c1}/H_c function was asymptotic to ABRIKOSOV'S approximation ((13-51c)) for $\kappa_{GL} \gtrsim 30$, but deviated quite strongly from it as κ_{GL} dropped below 10. Stemming as it did from GLA theory, the result was believed to be valid only at temperatures near T_c.

Further discussion of the lower critical field, with particular reference to Maki theory, is postponed until Chapter 14.

13.8 The Upper Critical Field, H_{c2}—Onset of the Normal State

13.8.1 Microscopy Theory

With the aid of the Ginzburg–Landau–Gor'kov microscopic theory it is possible to trace a direct relationship between H_{c2} and H_c [Goo64], according to which

$$H_{c2} = \sqrt{2}\kappa_{GL} H_c \tag{13-53}$$

Next by invoking the BCS expression $H_{c0} = 2.43\gamma_v^{1/2} T_c$, it is possible to write down the following useful relationships between the zero K upper critical field and the transition temperature:

$$H_{c20} = 3.44\kappa_{GL}\gamma_v^{1/2} T_c \tag{13-54}$$

which, with the aid of the G–L expression for κ_{GL} in the dirty limit (see(10-26)), becomes[3]

$$H_{c20} = 2.57 \times 10^4 \rho_n \gamma_v T_c \qquad \text{(Oe)} \tag{13-55}$$

with ρ_n in Ω cm and γ_v in erg cm^{-3} K^{-2}.

Further representations of the upper critical field in terms of both superconductive-state and normal-state parameters, with particular reference to the theories of MAKI, WERTHAMER and HOHENBERG, and EILENBERGER, are given in Chapter 14.

[3]The nonparamagnetically limited version of the Maki dirty-limit theory will yield, for the same upper critical field, the value $3.06 \times 10^4\ \rho_n \gamma_v T_c$ (Chapter 14).

13.8.2 Thermodynamic Relationships

Using a straightforward argument based on GLAG theory, TINKHAM [TIN75, p. 155] demonstrated that

$$B = H_a + 4\pi M = H_a - \frac{H_{c2} - H_a}{(2\kappa_{GL}^2 - 1)\beta_A} \tag{13-56}$$

where β_A, a constant due to ABRIKOSOV, takes the value 1.16 for the usual triangular flux lattice. Next, using the same arguments that produced (13-42) to (13-44) but with M replaced by the expression implied in (13-56), it follows that

$$g_n(H_a) - g_s(H_a) = \frac{(H_{c2} - H_a)^2}{8\pi(2\kappa_{GL}^2 - 1)\beta_A} \tag{13-57a}$$

$$= \frac{H_{c2}^2}{8\pi} \frac{(1-h)^2}{(2\kappa_{GL}^2 - 1)\beta_A} \tag{13-57b}$$

where h is the reduced applied field, H_a/H_{c2}.[4] This is a very important result in that it is responsible for the $(1 - h)$ factor that appears in numerous expressions for the local pinning site interaction, f_p, and consequently in the appropriately-summed bulk pinning force $F_p = \sum f_p$. As such it is an "existence" term—i.e., it describes the field-range of existence of the Abrikosov flux lattice. If individual pinning forces are summed within LABUSCH's elastic-lattice model [LAB69], such that for example $F_p \propto N_p f_p^2/\sqrt{C_{66}C_{44}}$ where N_p is the effective pinning-site density and the C_{ij}s are the elastic stiffness moduli of the flux lattice (see Sections 21.4ff), the $(1 - h)$ again appears as a factor of the modulus C_{66}[5] and expresses the softening of the lattice with increasing field and its eventual "melting" at H_{c2}. Thus the "resonance shape" of the pinning function—$h(1 - h)$ in its simplest form—is not mechanistic in origin, but arises out of the thermodynamic treatment of the Abrikosov lattice.

13.9 The Surface Sheath Critical Field, H_{c3}

It was theoretically predicted by SAINT-JAMES and DE GENNES [SAI63] that, with an applied field parallel to the surface of an ideal sample,

[4]A comparable calculation of the free energy of the *flux lattice* (TINKHAM [TIN75, p. 155]) leads to an expression similar to (13-57) but with an extra term, $-(H_{c2}^2 - H_a^2)/8\pi$.

[5]Anticipating a result to be discussed in Section 21.5 and combining it with (13-57), remembering that g as usual represents free energy per unit volume: $C_{66}(H_a) = 0.41[g_n(H_a) - g_s(H_a)]$.

superconductivity would persist up to (or conversely, nucleate at) a field H_{c3} given by

$$H_{c3} = 2.397\ \kappa_{GL} H_c \tag{13-58}$$

The results of ROSENBLUM *et al.* [ROS64] and STRONGIN *et al.* [STR64] on the dilute type-I alloys Pb–Tl ($H_{c3}/\kappa_{GL}H_c = 2.69$) and Pb–Bi ($H_{c3}/\kappa_{GL}H_c = 2.63$) were in reasonable accord with this prediction. For type-II superconductors, since $H_{c2} = \sqrt{2}\ \kappa_{GL}H_c$, one expects that

$$H_{c3} = 1.695\ H_{c2} \tag{13-59}$$

Superconductivity in fields between H_c or H_{c2} and H_{c3} is attributed to the existence of a surface film about one coherence length in thickness. As indicated by SAINT-JAMES, SARMA, and THOMAS, who devoted an entire chapter in their book [SAI69] to this subject, the results of some 12 measurements of the ratio H_{c3}/H_{c2} in simple-metal type-II superconductors have yielded an average value of 1.64 ± 0.25, in very good agreement with (13-59). Indeed in experimental studies of type-II Sn–In and Pb–Tl alloys, coefficients of 1.67 and 1.69 have been obtained.

The existence of a surface sheath is well-established both theoretically and experimentally [GOO66], [SAI69]. It may be thought of as arising through the imposition of a surface boundary condition upon the Ginzburg–Landau equations (e.g., [FIN69], [SHM70]). Using a simple variational approach, TINKHAM [TIN75, p. 130] has provided a useful derivation, within the G–L framework, of (13-59)

The surface critical field is strongly dependent on the angle, β, between the applied field and the sample surface; as H_a is rotated, the critical field, equal to H_{c2} with $\beta = 90°$ (the normal orientation), increases and passes through a cusp of magnitude H_{c3}, given by (13-59), at $\beta = 0°$. Early angular dependence studies were cited by GOODMAN [GOO66] in a review of the experimental foundations (1963–1965) of surface superconductivity; subsequently, KARASIK *et al.* [KAR72] studied an anomalous angular dependence of H_{c3}/H_{c2} in a series of variously heat-treated Ti–Nb(22 at.%) alloys.

As pointed out by SAINT-JAMES *et al.* [SAI69], the existence of the surface superconductive sheath explains the following effects which might otherwise be regarded as metallurgically produced artifacts: (i) a critical field which, as indicated in Section 13.2, depends on whether its measurement was made magnetically in parallel fields or resistively in transverse fields; (ii) broad and/or measuring-current-sensitive resistive critical field (H_r) transitions; (iii) the observed β-dependence of H_r. The first two points are illustrated in Figure 13-3, from the work of KARASIK *et al.* [KAR72] with Ti–Nb(22 at.%) heat treated 1 h/800°C/WQ plus 1 h/390°C. The figure

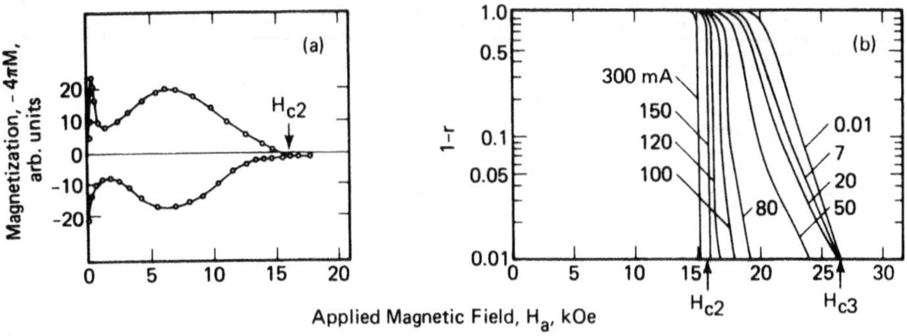

Figure 13-3. Determination of the critical fields H_{c2} and H_{c3} of a sample of Ti–Nb(22 at.%) (condition: recrystallized 1 h/800°C, aged 1 h/390°C) at the reduced temperature $t = 0.88$. (a) Magnetization as measured ballistically; (b) resistance function $1 - (R/R_n)$ plotted *versus* H_a oriented parallel to the transport current—after KARASIK *et al.* [KAR72].

depicts a magnetically determined upper critical field ($T = 0.88\ T_c$) of 16 kOe, but a resistive transition whose tail extends to fields as high as 26.5 kOe. Assuming the cutoff to be H_{c3}, the critical-field ratio in this case is 1.67, in excellent agreement with prediction (13-2).

KARASIK *et al.* [KAR72] also undertook a comprehensive study of surface critical field and current in Ti–Nb(22 at.%) recrystallized and heat-treated in such a way as to produce ω-phase (1–10 h/390°C) or α-phase (1–56 h/500°C) precipitation. Some representative results are reproduced in Table 13-4, which shows that as H_{c2} increases in response to the metallurgical effects of aging at 390°C (i.e., a Nb enrichment of the β-phase) the ratio H_{c3}/H_{c2} remains, to a first approximation, constant and equal to 2.1 ± 0.2. The critical-field ratio was unexpectedly high; attention was drawn to this and to the additional observation that H_{c3}/H_{c2} increased rapidly as the temperature was raised from 0.8 T_c to about T_c. It was suggested, with reference to work to be considered below, that the underlying

Table 13-4. Critical Fields of Ti–Nb(22 at.%) at $T = 0.9\ T_c$—After Karasik *et al.* [KAR72]

Condition	Phases Present	Critical Fields, kOe		
		H_{c2}	H_{c3}	H_{c3}/H_{c2}
1 h/800°C/WQ	$\omega+\beta$	8.2	18.6	2.27
1 h/800°C/WQ + 1 h/390°C	$\omega+\beta$	12.8	22.9	1.79
1 h/800°C/WQ + 3 h/390°C	$\omega+\beta$	13.8	30	2.17
1 h/800°C/WQ + 10 h/390°C	$\omega+\beta$	17.0	37	2.18
				$[H_{c3}/H_{c2}]_{av} = 2.1 \pm 0.2$

cause of these effects was an enhanced *transition temperature* associated with the surface layer.

SHMIDT [SHM70] and others [GOO66, p. 474] have observed that H_{c3}/H_{c2} depended strongly on the condition of the sample surface; pronounced changes in the field ratio could be induced by electropolishing, etching, and annealing [GYG64], by rubbing with fine abrasive paper [FIN69], or by electroplating which forces H_{c3} towards H_{c2} as the superconductor interfaces with a normal metal rather than a dielectric. Noting that the feature associated with most departures from ideal surface preparation was a reduced electron mean-free-path, l, SHMIDT [SHM70] suggested that the variation in H_{c3}/H_{c2} with surface treatment could be modelled in terms of a variation in κ_{GL}, whose dirty limit, according to (10-26), is directly proportional to ρ_n. Calculations within the framework of the G–L model demonstrated that a reduction in l, such as would result from some form of mechanical deformation of the surface (as in machining, etc.), would lead to an increase in H_{c3}. The same conclusion was reached by FINK and JOINER [FIN69], who, from a different standpoint but again within the context of G–L theory, suggested that an enhanced H_{c3} should result from an increase in the magnitude of the order parameter as it approaches the surface, a situation which would also be accompanied by an enhanced surface T_c. Such a prediction is open to confirmation by measurement; indeed the surface roughening of In–Bi foils was found to be accompanied by conjoint increases in both H_{c3} *and* T_c [FIN69]. The T_c enhancement itself is also reminiscent of similar effects that have been found to accompany granularity in Al and stress-induced structural disordering in Ti-Mo (Section 10.3.1). The measurement and properties of surface-sheath critical currents are discussed in Section 22.5.

Chapter 14

The Mixed State

The GLAG theory in its original form was restricted in its range of validity to temperatures near T_c. Further developments of the theory were therefore directed towards extending its temperature range and, in particular, to providing descriptions of the temperature dependences of the critical fields. At first the effects of the mixed-state and normal-state paramagnetisms were neglected, and later they were fully taken into account.

This chapter is confined to *nonparamagnetic* theories of the mixed state. The first few sections, which deal with properties of, or related to, the GLAG theory itself, conclude with a presentation of the most important GLAG–BCS equations taken from a collection recently compiled by Beasley, his student, and colleagues [Orl79]. Next to be reviewed in this chapter are the nonparamagnetic "post-GLAG" theories of the mixed state, as discussed in the early papers of Werthamer *et al.* and Maki *et al.*, and subsequently by Eilenberger and by Fietz and Webb, in the course of which some equations for the relative critical field temperature dependences are presented. Still within the nonparamagnetic post-GLAG framework, the chapter goes on to deal with some expressions for the upper critical field in terms both of normal-state parameters (i.e., ρ_n and γ_v) and superconductive-state parameters (i.e., $(dH_{c2}/dT)_{T_c}$). Finally, with reference to the results of two important case studies of Ti–Nb alloys, the chapter concludes by describing some experimental methods for evaluating the thermodynamic and lower critical fields.

14.1 Temperature Dependences of the Critical Fields

The temperature dependences of the critical fields have been studied and reviewed, within the framework of GLAG theory, by Kinsel, Lynton, and Serin [Kin64] and others whose work is considered below. Our full current understanding of the mixed state is based on this work and on those improvements to the GLAG model to be considered in the next chapter, which have taken into account the influences of normal- and superconductive-state paramagnetisms and of electron-scattering effects of various kinds. A useful reference work dealing with both GLAG theory and these improvements to it has been authored by SAINT-JAMES, THOMAS, and SARMA [SAI69].

14.1.1 Early Experimental Studies of $H_{c1}(T)$

In reversible or "ideal" nonparamagnetic type-II superconductors, as the applied field passes across the threshold, H_{c1}, of the mixed state the (diamagnetic) magnetization, $-4\pi M$, previously increasing with increasing field (Figure 13-2), passes through a cusp before dropping to zero at H_{c2}. In irreversible type-II superconductors, on the other hand, in which flux pinning impedes both the ingress and egress of flux, the peak in the magnetization curve becomes rounded off and moves to higher values, say H_{c1}^* [BLA65], of the applied field. Curves illustrating this effect are shown in Chapter 16 of this book where the so-called "critical state," which relies entirely on the principle of irreversibility, is fully discussed.

In an early investigation, JOINER and BLAUGHER [JOI64] measured the lower critical field temperature dependences of well-annealed samples of Mo–Re(15 and 25 at.%). With Ginzburg–Landau parameters of 2.62 and 3.68, respectively, these samples were practically "ideal" and yielded values of H_{c1} directly from the magnetization curves. This work was followed by a comparable investigation by BLAUGHER [BLA65] of H_{c1}^* in the highly irreversible alloys Ti–V(38.5, 48.4, and 58.5 at.%). Although vacuum-annealed, the presence of flux-pinning impurities had raised the Ginzburg–Landau parameters of these samples to 74, 60, and 49, respectively. In both the Mo–Re and Ti–V studies, a $(1 - t^2)$ type of temperature dependence was noted.

14.1.2 Early Semiempirical Studies of $H_{c2}(T)$

Several semiquantitative evaluations of H_{c2} *versus* T can be made by investigating the properties of G–L theory, albeit outside of its regime of validity (*viz.*, $t \cong 1$). JONES *et al.* [JON64] compared the validities of three expressions for $H_{c2}(T)$ derived from: (i) G–L theory; (ii) calculations by GOR'KOV, (iii) calculations by SHAPOVAL.

(i) The "Abrikosov-Ginzburg" approach is based on (13-53) for H_{c2}, but assigns some of the temperature dependence to κ_{GL} according to (13-27). The coupling of this pair of equations, together with the use of the empirical

$(1 - t^2)$ temperature dependence for H_c itself results in

$$H_{c2}(t) = 2\sqrt{2}\kappa_{GL1}H_{c0}(1 - t^2)/(1 + t^2) \tag{14-1}$$

where κ_{GL1} is the value of κ_{GL} at $t = 1$. At $t = 0$ this reduces to

$$H_{c20} = 2\sqrt{2}\kappa_{GL1}H_{c0} \tag{14-2}$$

(ii) GOR'KOV [GOR60, GOR60[a]] showed from microscopic theory that near $t = 0$

$$H_{c20} = 1.77\kappa_{GL1}H_{c0} \tag{14-3}$$

and for the temperature dependence he suggested a premultiplier $(1.77 - 0.43t^2 + 0.07t^4)$ leading to the following expression for the full temperature dependence

$$H_{c2}(t) = \kappa_{GL1}H_{c0}(1.77 - 0.43t^2 + 0.07t^4)(1 - t^2) \tag{14-4a}$$

which of course tends to

$$H_{c2}(t) = \sqrt{2}\kappa_{GL1}H_{c0}(1 - t^2) \tag{14-4b}$$

that is, to (14-1), as $t \to 1$.

(iii) SHAPOVAL [SHA62], calculating H_{c20} in the dirty limit using microscopic theory, obtained [JON64], [ELB63[a], ELB64]

$$H_{c20} = 3.03\kappa_{GL1}H_{c0} \tag{14-5}$$

and for the temperature dependence concluded that [ELB64]

$$H_{c2}(t) = \begin{cases} H_{c20}(1 - 2t\ln 2) & t \ll 1 \\ H_{c20}(1 - t)8/\pi^2 & t \cong 1 \end{cases} \tag{14-6}$$

The latter expression, in conjunction with (14-5), is, of course, equivalent to the by-now familiar

$$H_{c2}(t) = \sqrt{2}\kappa_{GL1}H_{c0} \tag{13-52}$$

at an intermediate temperature (in particular $t \cong 0.42$).

At this stage in the development of upper critical field "theory," experimental research concerned itself with evaluating the degree to which the actual data fitted the normalized temperature functions

$$\frac{H_{c2}(t)}{H_{c20}} = \begin{cases} \dfrac{1-t^2}{1+t^2} & \text{(Abrikosov-Ginzburg)} \quad (14\text{-}1,2) \\[2ex] \dfrac{(1.77 - 0.43t^2 + 0.07t^4)(1-t^2)}{1.77} & \text{(Gor'kov)} \quad (14\text{-}3,4) \\[2ex] \left.\begin{matrix} (1 - 2t \ln 2) & t \ll 1 \\ (1-t)8/\pi^2 & t \cong 1 \end{matrix}\right\} & \text{(Shapoval)} \quad (14\text{-}6) \end{cases}$$

As a result of this testing [JON64] it was possible to conclude that the Shapoval function did not fit the experimental results; both Abrikosov-Ginzburg and Gor'kov did, however, fit the data, with the points for a series of alloys shifting from the Abrikosov-Ginzburg to the Gor'kov curve with increasing κ_{GL1}. Further contributions to this subject have been made by HULBERT [HUL64].

Subsequent extensions of GLAG theory leading to a series of new formulations of the upper critical field temperature dependence are reviewed below. These advances, together with those resulting from the introduction of Pauli paramagnetic effects to be considered in Chapter 15, are appropriately referred to as "neo-GLAG"- [HAK67[b]] or "post-GLAG" theory.

14.2 Foundations of the Ginzburg–Landau–Abrikosov–Gor'kov (GLAG) Theory of the Mixed State

14.2.1 The Ginzburg–Landau Parameter and Its Response to Alloying

As pointed out by ABRIKOSOV [ABR57] the theory of GINZBURG and LANDAU [GIN50] led to the conclusion that the two classes of superconductor, later to be known as "type-I" and "type-II," were distinguishable in terms of the magnitude of $\kappa_{GL} \cong \lambda_{L0}/\xi_0$ and, in particular, whether this parameter was less than or greater than $1/\sqrt{2}(= 0.71)$, respectively. These two classes of superconductor, and the manner in which they are penetrated by an applied mangetic field, are illustrated schematically in Figure 14-1. Just as the conversion from type-I to type-II could be achieved by alloying so as to increase κ_{GL} through the value 0.71 [LIV63, p. 204], the properties of type-II superconductors themselves can be "tuned" from the clean ($\kappa_{GL} \cong 1$) to the dirty ($\kappa_{GL} \gg 1$) limits (Section 13.5.3). For example, pure annealed Nb is a type-II superconductor with a κ_{GL} of 0.9 [FIE67[a]]. By alloying Nb with Ti, and with some slight assistance from cold work, κ_{GL}

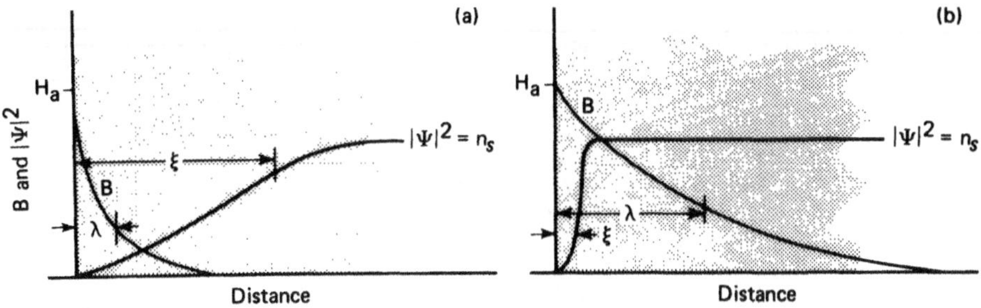

Figure 14-1. Representations of the variation of B and the corresponding variation of $|\Psi|^2 \equiv n_s$ (the fraction of superelectrons) with distance into the superconducting part of a normal-superconducting pair in the presence of an applied field, H_a. In (a) the superconductor is type-I ($\lambda/\xi \gtrsim 0.7$); in (b) it is type-II ($\lambda/\xi \lesssim 0.7$)—after standard sources, see in particular KIM [KIM64].

can be continuously increased to values as high as 78 (68 at.% Ti) [BRA72], Figure 14-2, while in Ti–V(22.5–25 at.%) even higher values of κ_{GL} (>100) have been achieved [HAK67[b]], [BRA72], Table 13-3. Such materials, which have been referred to as "extreme" type-II superconductors [HAK67[b]], exist in the dirty limit referred to above.

GOR'KOV [GOR59], in one of a series of papers, showed that the equations of the phenomenological (or "quasimicroscopic" in the words of ABRIKOSOV [ABR57]) theory of GINZBURG and LANDAU could be deduced by application, near T_c, of the theory of BARDEEN, COOPER, and SCHRIEFFER (BCS) [BAR57], and in so doing showed that the e^* (electronic charge) of G–L theory was actually $2e$—consistent with the Cooper-pair (i.e., quasiparticle) idea. In the meantime (in the same year as BCS, 1957) ABRIKOSOV [ABR57] had solved the G–L equation (hence T close to T_c)

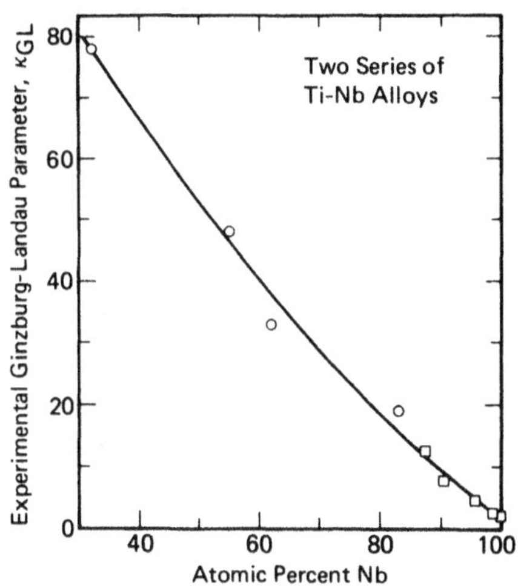

Figure 14-2. Experimental values of the Ginzburg–Landau parameter κ_{GL} for two series of cold-worked Ti–Nb alloys. Data for Ti–Nb(19, 33, 48, 78 at.%) from BRAND [BRA72]; data for Ti–Nb(87.5, 91, 95.5, 98.5, 100 at.%) from FIETZ and WEBB [FIE67[a]].

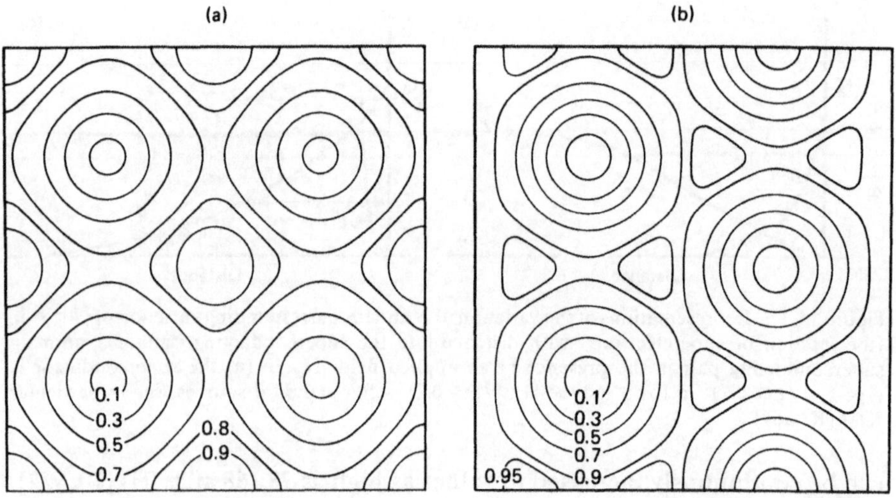

Figure 14-3. Normalized $|\Psi|^2$ contours representing a regular array of fluxoids penetrating a type-II superconductor in the mixed state. (a) The lowest-energy solution to the Ginzburg-Landau equations, according to ABRIKOSOV's original article [ABR57], was a square lattice. (b) On the other hand, according to KLEINER *et al.* [KLE64[a]] a triangular lattice is the stable structure just below H_{c2}; but also see text.

and studied the magnetic properties of bulk superconductors in the regime $\kappa_{GL} > 1/\sqrt{2}$. The result of so doing was the prediction of what was to become known as the Abrikosov mixed state in which the applied field, H_a, when greater than the lower critical field, H_{c1}, is able to cause flux penetration of the superconductor and produce within it a regularly arranged (in two dimensions) flux lattice. In the light of the earlier experimental results of SHUBNIKOV *et al.* [SHU37] and the studies of vortices in He-II by ONSAGER [ONS49] and FEYNMANN [FEY55], a very complete picture of the mixed state and the fluxoids from which it is constructed emerged. A diagram of ABRIKOSOV's vortex lattice, in Ψ-space, is reproduced in Figure 14-3(a). At the time a square lattice was predicted; today it is known that both square and triangular lattices are possible depending on the details of the interaction between the fluxoids [ORL79], but that the triangular lattice is by far the more common [BEA82[a]].

14.2.2 Structure of the Flux Lattice

The structure of the individual fluxoid [DEG64], [KIM64] is depicted in Figure 14-4. Each fluxoid, or vortex, is associated with a single quantum of flux represented by the symbol Φ_0, of value

$$\Phi_0 = \frac{hc}{2e} = 2.07 \times 10^{-7} \qquad (\text{G cm}^2) \tag{14-7}$$

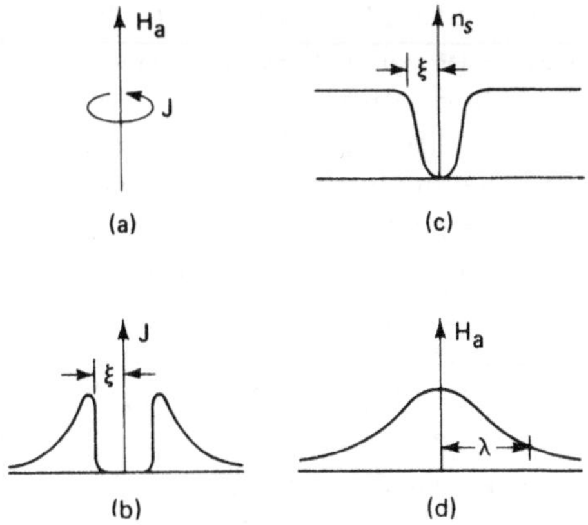

Figure 14-4. Structure of a fluxoid. It is supported by a circulating supercurrent of density J, (a), which occupies a cylinder of characteristics radius ξ, (b), within which the density of superelectrons drops to zero, (c), and whose field extends a characteristic distance λ into the surroundings, (d)—after KIM [KIM64].

and is surrounded by a circulating supercurrent, J, of spatial extent, λ. The "core" of the fluxoid is normal and is defined by the region of radius equal to a coherence length, ξ_{GL}, within which $|\Psi|^2 = n_s \to 0$.

As the applied field increases, the fluxoids begin to interact and as a consequence assemble themselves into a lattice. A simple geometrical argument for the spacing, d, of a triangular lattice then gives

$$d^2 = \frac{2}{\sqrt{3}} \frac{\Phi_0}{B} \tag{14-8}$$

where B is the induction. Under increase of the applied field, superconductivity could be imagined to vanish with either (i) the touching of the normal cores of the triangular lattice or (ii) the consuming of the entire sample by normal (core-like) material. Approximate expressions for the upper critical field, H_{c2}, in terms of the induction B_{c2} at the bulk transition to the normal state are therefore given, in c.g.s. units, by

$$\xi_{GL}^2 \cong d^2 = \frac{2}{\sqrt{3}} \frac{\Phi_0}{B_{c2}}$$

that is,

$$B_{c2} = \frac{2}{\sqrt{3}} \frac{\Phi_0}{\xi_{GL}^2} \qquad \text{(cores touching)} \tag{14-9a}$$

or

$$B_{c2} = \frac{1}{2\pi} \frac{\Phi_0}{\xi_{GL}^2} \qquad \text{(area consumed)} \tag{14-9b}$$

In the microscopic superconducting region outside the core, where $|\Psi| =$ constant and λ is the only variable parameter, the London equations (e.g.,

(13-3)) are valid. This will still be approximately true for the bulk superconductor provided that the cores take up little volume, i.e., in the extreme or dirty limit, provided that ξ is relatively small (see [HAK67[b]]).

As a result of his initial calculations, ABRIKOSOV [ABR57] anticipated that the flux lattice, initially triangular in a plane perpendicular to the field, would transform to square at fields not far above H_{c1}, Figure 14-3(a); but a reinvestigation of the vortex structure by KLEINER *et al.* [KLE64[a]] demonstrated that triangular was the higher-field stable structure, Figure 14-3(b). Direct observations of the flux-lattice structure were made by LAZAREV *et al.* [LAZ73] and OBST and BRANDT [OBS78] using the ferromagnetic-particle decoration technique first developed by TRÄUBLE and ESSMANN [TRA66, TRA67]. In their experiments on a low-κ_{GL} ($\cong 0.72$) Pb–Tl crystal, OBST and BRANDT [OBS78] noted a gradual transition from the triangular to the square flux-line lattice with increasing applied field.

14.3 Dirtiness and Irreversibility in Type-II Superconductors

14.3.1 The Ginzburg–Landau–Gor'kov Impurity Parameter, ξ_0/l

The Ginzburg-Landau parameter, κ_{GL}, has been defined in the clean and dirty limits with the aid of Gor'kov parameter, $\rho_G \equiv 0.882\ \xi_0/l$, (13-34). Precisely the same quantity, represented by λ, was later introduced by HELFAND and WERTHAMER [HEL64, HEL66] to define their "reduced mean collision frequency," $\lambda = \hbar/2\pi k_B T_c \tau$ $(= 0.882\ \xi_0/l)$, where $\tau = l/v_F$ is a transport scattering relaxation time. Finally, as will be explained below, EILENBERGER [EIL67] calculated the full temperature dependences of the reduced quantities κ_1/κ_{GL} and κ_2/κ_{GL} in terms of an impurity parameter $\lambda_{tr} = \xi_0/l_{tr}$ where the l_{tr} is a mean-free-path length for s-wave *and* p-wave scattering. This reduces to the earlier l when only the s scattering is considered.[1]

14.3.2 Irreversible Alloy Superconductors

As implied by the results of Section 13.5, a superconducting metal may be converted from type-I to type-II by the addition of solute to reduce the electronic mean-free-path; and as directly pointed out there, the

[1]The three important λs of mixed-state theory then become [BEA82[a]]: λ = penetration depth; $\lambda_{tr} = \hbar/2\pi k_B T_c \tau$, the transport collision frequency; $\lambda_{so} = 2\hbar/3\pi k_B T_c \tau_{so}$, the spin-orbit-scattering frequency (see Chapter 15). In Table 14-2 the subscript tr is included in recognition of the existence of anisotropic scattering, which is, of course, a part of "post-GLAG" theory as outlined in Section 14.5, and also in order to avert possible confusion with the penetration depth λ.

Ginzburg–Landau parameter of the resulting type-II alloy, hence its dirtiness, may be still further increased by continued alloying. In this way several concentrated (so-called "extreme", in the κ_{GL} sense [HAK67[b]]) type-II Ti–TM alloys have been prepared for study.

Such alloys may still be reversible. However, if contaminated by interstitial impurities during preparation, or if suitably heat treated, the precipitates that may form in concentrated alloys are able to pin or impede the motion of the mixed-state fluxoids (see Chapter 21), thereby rendering them magnetically irreversible. Clearly, all densely precipitated alloys are both dirty and irreversible, but not all dirty alloys are irreversible.

14.4 The Full Ginzburg–Landau–Gor'kov–Bardeen–Cooper–Schrieffer Relationships

To conclude this summary of the GLAG model of the superconducting mixed state is a collection of identities, presented in Table 14-1, describing the GLAG–BCS quantities ξ_0, ξ_{GL}, λ_{GL}, and κ_{GL}, in terms of the normal-state parameters ρ_n, γ_v, and S (the Fermi-surface area), and T_c. The table in its present form was first published by ORLANDO, MCNIFF, FONER, and BEASLEY [ORL79] as an improvement upon an earlier such collection of equations by HAKE [HAK67[b]].

14.5 Evolution of Nonparamagnetic Post-GLAG Theories of the Upper Critical Field Temperature Dependences

14.5.1 Symbols for the Upper Critical Fields

In preparation for the following discussions, and for those of Chapter 15, it is appropriate to mention here that, wherever it has been found useful for the distinctions to be preserved, an attempt has been made to adhere to the critical field formalisms suggested by HAKE (e.g., [HAK67[b]]), *viz.*,

- H_p : Clogston paramagnetic limit
- H^*_{c2} : GLAG-calculated or nonparamagnetically limited ($\alpha \to 0$ or $\lambda_{so} \to \infty$, Section 15.4) post-GLAG-calculated upper critical field
- H_{c2} : fully corrected theoretical upper critical field, or simply as an abbreviation for the words "upper critical field"
- H_r : resistively *measured* upper critical field
- H_u : magnetically or otherwise *measured* upper critical field

To each of these symbols a further subscript 0 may be attached to indicate a zero-K property.

Table 14-1. Descriptions in Microscopic Terms of the GLAG–BCS Properties of the Mixed State—After Orlando *et al.* [ORL79].[a] The following quantities (and units) are employed: e, the electronic charge (e.s.u.); γ_v, the unit-volume normal-state electronic specific heat coefficient (erg cm^{-3} K^{-2}); $\hbar$, Planck's constant/2π (erg s); k_B, Boltzmann's constant (erg K^{-1}); n_e, the conduction electron density (cm^{-3}); $n_v(E_F)$, the Fermi density-of-states per unit volume for one spin direction (erg^{-1} cm^{-3} spin^{-1}); ρ_n, the normal-state residual electrical resistivity (Ω cm); S/S_F, the ratio of the Fermi surface area to that of a free-electron gas of density, n_e; v_F, the Fermi velocity (cm sec^{-1}); $\zeta(3) \cong 1.202$; $e^\gamma \cong 1.781$.

1. Average Fermi velocity, $\langle v_F \rangle$, cm sec^{-1}:

$$\langle v_F \rangle = k_B^2 \hbar^{-1}(\pi^4/3)^{1/3}(n_e^{2/3} S/S_F)\gamma_v^{-1}$$
$$= 5.77 \times 10^{-5}(n_e^{2/3} S/S_F)\gamma_v^{-1}$$

2. Electronic mean-free path, l_{tr}, cm:

$$l_{tr} = 9 \times 10^{11}\hbar(3\pi^2)^{1/3}[e^2\rho_n(n^{2/3} S/S_F)]^{-1}$$
$$= 1.27 \times 10^4[\rho_n(n_e^{2/3} S/S_F)]^{-1}$$

3. Density-of-states of one spin direction, $n_v(E_F)$, erg^{-1} cm^{-3} spin^{-1}:

$$n_v(E_F) = (2\pi^2 k_B^2/3)^{-1}\gamma_v$$
$$= 7.97 \times 10^{30}\gamma_v, \text{ which is of course equivalent to (7-4b)}$$

4. BCS coherence length, ξ_0, cm:

$$\xi_0 = \hbar\langle v_F \rangle(\pi^2 e^{-\gamma} k_B T_c)^{-1}$$
$$= e^\gamma k_B(3\pi^2)^{-1/3}(n_e^{2/3} S/S_F)(\gamma_v T_c)^{-1}$$
$$= 7.95 \times 10^{-17}(n_e^{2/3} S/S_F)(\gamma_v T_c)^{-1}$$

5. London penetration depth at zero K, λ_{L0}, cm:

$$\lambda_{L0} = 3\hbar c \pi^{1/2}[2(3\pi^2)^{2/3} k_B e]^{-1}\gamma_v^{1/2}(n_e^{2/3} S/S_F)^{-1}$$
$$= 1.33 \times 10^8 \gamma_v^{1/2}(n_e^{2/3} S/S_F)^{-1}$$

6. Gor'kov's χ function (of the impurity parameter, λ or λ_{tr}), Section 14.5:

a. $\chi(\lambda_{tr}) = (1 + \lambda_{tr})^{-1} R(\lambda_{tr})$

where $R(0) = 1$ and $R(\infty) = \pi^2[7\zeta(3)]^{-1} = 1.17$ (see [WER69])

b. $\lambda_{tr} = \pi e^{-\gamma}\xi_0(2l_{tr})^{-1} = 0.882\xi_0/l_{tr}$

$$= k_B e^2 \rho_n(3\pi^2)^{1/3}(n_e^{2/3} S/S_F)^2(54 \times 10^{11}\pi\hbar\gamma_v T_c)^{-1}$$
$$= 5.51 \times 10^{-21}\rho_n(n_e^{2/3} S/S_F)^2(\gamma_v T_c)^{-1}$$

Table 14.1. (*cont.*)

7. Ginzburg–Landau coherence length, ξ_{GL}, cm:
   a. Full value (λ_{tr} arbitrary):

$$\xi_{GL} = \pi e^{-\gamma}[7\zeta(3)/48]^{1/2}\xi_0[\chi(\lambda_{tr})]^{1/2}(1-t)^{-1/2}$$
$$= 0.739[\xi_0^{-2} + 0.882(\xi_0 l_{tr})^{-1}]^{-1/2}[R(\lambda_{tr})]^{1/2}(1-t)^{-1/2}$$
$$= \{144T_c^2\gamma_v^2[7\zeta(3)k_B^2(3\pi^2)^{1/3}(n_e^{2/3}S/S_F)^2]^{-1}$$
$$+ 24e^2\rho_n\gamma_v T_c[(63\times 10^{11})\zeta(3)\pi k_B\hbar]^{-1}\}^{-1/2}R^{1/2}(\lambda_{tr})(1-t)^{-1/2}$$
$$= [2.90\times 10^{32}T_c^2\gamma_v^2(n_e^{2/3}S/S_F)^{-2}$$
$$+ 1.60\times 10^{12}\rho_n\gamma_v T_c]^{-1/2}[R(\lambda_{tr})]^{1/2}(1-t)^{-1/2}$$

   b. Clean limit ($\lambda_{tr} \ll 1$):

$$\xi_{GL}^c = \pi e^{-\gamma}[7\zeta(3)/48]^{1/2}\xi_0(1-t)^{-1/2} = 0.739\xi_0(1-t)^{-1/2}$$
$$= k_B(3\pi^2)^{2/3}[7\zeta(3)/3]^{1/2}(n_e^{2/3}S/S_F)(12\pi\gamma_v T_c)^{-1}$$
$$= 5.87\times 10^{-17}(n_e^{2/3}S/S_F)(\gamma_v T_c)^{-1}$$

   c. Dirty limit ($\lambda_{tr} \gg 1$):

$$\xi_{GL}^d = (\pi^3 e^{-\gamma}/24)^{1/2}(\xi_0 l_{tr})^{1/2}(1-t)^{-1/2} = 0.852(\xi_0 l_{tr})^{1/2}(1-t)^{-1/2}$$
$$= (9\times 10^{11}\pi^3 k_B\hbar)^{1/2}(24e^2\rho_n\gamma_v T_c)^{-1/2}(1-t)^{-1/2}$$
$$= 8.57\times 10^{-7}(\rho_n\gamma_v T_c)^{-1/2}(1-t)^{-1/2}$$

8. Ginzburg–Landau penetration depth, λ_{GL}, cm:
   a. Full value:

$$\lambda_{GL} = 2^{-1/2}\lambda_{L0}(1 + \pi e^{-\gamma}\xi_0/2l_{tr})^{1/2}[R(\lambda_{tr})]^{-1/2}(1-t)^{-1/2}$$
$$= 2^{-1/2}\lambda_{L0}(1 + 0.882\xi_0/l_{tr})^{1/2}[R(\lambda_{tr})]^{-1/2}(1-t)^{-1/2}$$
$$= \{3c^2\hbar^2\gamma_v[8\pi e^2k_B^2(3\pi^2)^{1/3}(n_e^{2/3}S/S_F)^2]^{-1}$$
$$+ c^2\hbar\rho_n(144\times 10^{11}\pi^2k_BT_c)^{-1}\}^{1/2}[R(\lambda_{tr})]^{-1/2}(1-t)^{-1/2}$$
$$= [8.77\times 10^{15}\gamma_v(n_e^{2/3}S/S_F)^{-2}$$
$$+ 4.83\times 10^{-5}\rho_n T_c^{-1}]^{1/2}[R(\lambda_{tr})]^{-1/2}(1-t)^{-1/2}$$

   b. Clean limit:

$$\lambda_{GL}^c = 2^{-1/2}\lambda_{L0}(1-t)^{-1/2}$$
$$= 3(2\pi)^{1/2}c\hbar\gamma_v^{1/2}[4ek_B(3\pi^2)^{2/3}(n_e^{2/3}S/S_F)]^{-1}(1-t)^{-1/2}$$
$$= 9.37\times 10^7\gamma_v^{1/2}(n_e^{2/3}S/S_F)^{-1}(1-t)^{-1/2}$$

Continued

Table 14-1. (*cont.*)

c. Dirty limit:

$$\lambda_{GL}^{d} = 2^{-1/2}\lambda_{L0}[7e^{-\gamma}\zeta(3)\xi_0/2\pi l_{tr}]^{1/2}(1-t)^{-1/2}$$
$$= 2^{-1/2}\lambda_{L0}(\xi_0/1.33l_{tr})^{1/2}(1-t)^{-1/2}$$
$$= [7c^2\hbar\zeta(3)\rho_n]^{1/2}(144\times 10^{11}\pi^4 k_B T_c)^{-1/2}(1-t)^{-1/2}$$
$$= 6.42\times 10^{-3}(\rho_n/T_c)^{1/2}(1-t)^{-1/2}$$

9. Ginzburg-Landau Parameter, κ_{GL}:
 a. Full value:

$$\kappa_{GL} = e^{\gamma}\lambda_{L0}(\xi_0\pi)^{-1}[24/7\zeta(3)]^{1/2}(1+\pi e^{-\gamma}\xi_0/2l_{tr})[R(\lambda_{tr})]^{-1}$$
$$= 0.957\lambda_{L0}\xi_0^{-1}(1+0.882\xi_0/l_{tr})[R(\lambda_{tr})]^{-1}$$
$$= \{3c\hbar T_c\gamma_v^{3/2}[6/7\pi\zeta(3)]^{1/2}[ek_B^2(3\pi^2)^{1/3}(n_e^{2/3}S/S_F)^2]^{-1}$$
$$+ ce\gamma_v^{1/2}\rho_n[6/7\pi^3\zeta(3)]^{1/2}(18\times 10^{11}k_B)^{-1}\}[R(\lambda_{tr})]^{-1}$$
$$= [1.60\times 10^{24}T_c\gamma_v^{3/2}(n_e^{2/3}S/S_F)^{-2} + 8.78\times 10^3\rho_n\gamma_v^{1/2}][R(\lambda_{tr})]^{-1}$$

 b. Clean limit:

$$\kappa_{GL}^{c} = e^{\gamma}\lambda_{L0}(\xi_0\pi)^{-1}[24/7\zeta(3)]^{1/2} = 0.957\lambda_{L0}/\xi_0$$
$$= 3c\hbar T_c\gamma_v^{3/2}[6/7\pi\zeta(3)]^{1/2}[ek_B^2(3\pi^2)^{1/3}(n_e^{2/3}S/S_F)^2]^{-1}$$
$$= 1.60\times 10^{24}T_c\gamma_v^{3/2}(n_e^{2/3}S/S_F)^{-2}$$

 c. Dirty limit:

$$\kappa_{GL}^{d} = \lambda_{L0}l_{tr}^{-1}[42\zeta(3)/\pi^4]^{1/2} = 0.720\lambda_{L0}/l_{tr}$$
$$= ce\rho_n\gamma_v^{1/2}[42\zeta(3)/\pi^7]^{1/2}(18\times 10^{11}k_B)^{-1}$$
$$= 7.49\times 10^3\rho_n\gamma_v^{1/2}$$

[a] Reproduced by permission of The American Physical Society and M. R. Beasley.

14.5.2 Development of the Maki Dirty-Limit Equations

Efforts to improve upon the original GLAG theory of the mixed state, which at first were focused solely upon extending its temperature range of validity, in due course became directed towards incorporating into microscopic theory the effects[2] of the interactions between the conduction electrons and the applied field (Pauli paramagnetism), the lattice (spin-orbit scattering), the phonons (electron-phonon interaction), and each other

[2] ORLANDO *et al.* [ORL79] presented a useful discussion of these effects within the framework of A15-compound superconductivity.

(electron–electron interaction). The early stages of this evolutionary process, prior to the introduction of paramagnetic effects, are considered below.

As indicated in Table 14-2 (which is intended to be associated with Table 15-1 since the work referred to therein provides a necessary background to the studies of the paramagnetic mixed state discussed in Chapter 15), developments leading towards the present method of dealing with upper critical field temperature dependence commenced in 1964 with a pair of papers by MAKI [MAK64, MAK64ª], who derived, using microscopic theory, a general Ginzburg–Landau equation, valid in the dirty limit ($\xi_0/l \gg 1$) for all temperatures. In MAKI's formulation, the required temperature dependence was assigned to three generalized Ginzburg–Landau parameters—$\kappa_1(t)$, $\kappa_2(t)$ [MAK64], and $\kappa_3(t)$ [MAK64ª]—all of which reduced to κ_{GL}^d at the critical temperature, $t = 1$. The critical fields, and the limiting

Table 14-2. Evolution of Contemporary Theories of the Mixed State—The Introduction of Temperature, and Conduction-Electron Interactions with: The Applied Field (Pauli Paramagnetism), the Lattice (Spin-Orbit Scattering), the Phonons (Electron–Phonon Interaction) and Each Other (Electron–Electron Interaction)—Part I[a]

Evolution of Nonparamagnetic Theories	Literature
H_{c1} and H_{c2} and $4\pi M$ as functions of t in terms of $\kappa_1(t)$, $\kappa_2(t)$ and $\kappa_3(t)$—*Dirty limit*	Maki [MAK64, MAK64ª]
H_{c2} and $4\pi M$ as functions of t in terms of $\kappa_1(t)$ and $\kappa_2(t)$—*Clean limit*	Maki and Tsuzuki [MAK65ª]
H_{c2} as function of t, and electron-lattice scattering parameterized by λ, in terms of $\kappa_1(t, \lambda)$—*Unrestricted* mean-free-path	Helfand and Werthamer (H–W) [HEL64]
Same as above, but in addition the representation of $H_{c2}(t)$ in terms of a reduced parameter $h^*(t)$—*Unrestricted* mean-free-path	Helfand and Werthamer (H–W) [HEL66]
H_{c2} as function of t and electron-lattice scattering, taking into account both s- and p-wave (anisotropic) scattering parameterized by λ_{tr}—*Unrestricted* mean-free-path	Eilenberger [EIL67]
Experimental evaluation of the predictions of H–W—expressed in terms of $h^*(t, \lambda)$; and Eilenberger—expressed in terms of $\kappa_i[t, (l_{tr}/l), (\xi_0/l_{tr})]/\kappa_{GL}$—*Case study* of Nb–Ti (0–12.5 at.%)	Fietz and Webb [FIE67ª]

[a] Part II, *Paramagnetic Theories*, is Table 15-1

slope of the magnetization curve, $d(4\pi M)/dH|_{H_{c2}}$, were given by

$$H_{c2}(t) = \sqrt{2}\kappa_1(t)H_c(t) \tag{14-10}$$

$$-4\pi M|_{H_{c2}} = (H_{c2} - H_a)/(2\kappa_2^2(t) - 1)\beta_A \tag{14-11a}$$

$$4\pi(dM/dH)_{H_{c2}} = 1/(2\kappa_2^2(t) - 1)\beta_A \tag{14-11b}$$

$$H_{c1}(t) = H_c(t)\frac{\ln \kappa_3(t)}{\sqrt{2}\kappa_3(t)} \tag{14-12}$$

where H_a is the applied field, H_c the thermodynamic critical field, and β_A is the Abrikosov flux-lattice factor (= 1.16 for the usual triangular lattice).

The temperature dependences of $\kappa_1(t)$ and $\kappa_2(t)$ in the form in which they were originally calculated by MAKI [MAK64] are shown in the inset to Figure 14-5(a). Also shown in that figure are the results of a recalculation by CAROLI *et al.* [CAR66], who demonstrated that in the dirty limit κ_1/κ_{GL}^d and κ_2/κ_{GL}^d in fact differed by only a few percent over the entire temperature range. Expressions for the reduced Maki parameters [MAK64, MAK66], obtained by taking the nonparamagnetic ($\alpha \to 0$) limits of the corrected [MAK66] paramagnetic forms[3], are given by (14-13) and (14-14).

$$\kappa_1(t)/\kappa_{GL}^d = \begin{cases} 1.20(1 - 1.06t^2) & t \ll 1 \\ 1 + 0.119(1 - t) & t \cong 1 \end{cases} \tag{14-13}$$

$$\kappa_2(t)/\kappa_{GL}^d = \begin{cases} 1.20(1 - 2.53t^2) & t \ll 1 \\ 1 + 0.105(1 - t) & t \cong 1 \end{cases} \tag{14-14}$$

The Maki-calculated $\kappa_3(t)/\kappa_{GL}^d$, whose endpoints are

$$\kappa_{30}/\kappa_{GL}^d = 1.53 \tag{14-15a}$$

and

$$\kappa_{31}/\kappa_{GL}^d = 1.0 \tag{14-15b}$$

is depicted in Figure 14-5(b).

A comparison of (14-13), (14-14), and (14-15a) yields the following relationships among the parameters:

$$\kappa_{10} = \kappa_{20} = 0.784\kappa_{30} = 1.20\kappa_{GL}^d \tag{14-16}$$

[3]Corresponding expressions for the paramagnetically limited parameters $\kappa_1'(t, \alpha)$ and the spin-orbit-scattering-modified paramagnetically limited parameters $\kappa_1(t, \beta)$ are given in Section 15.5.

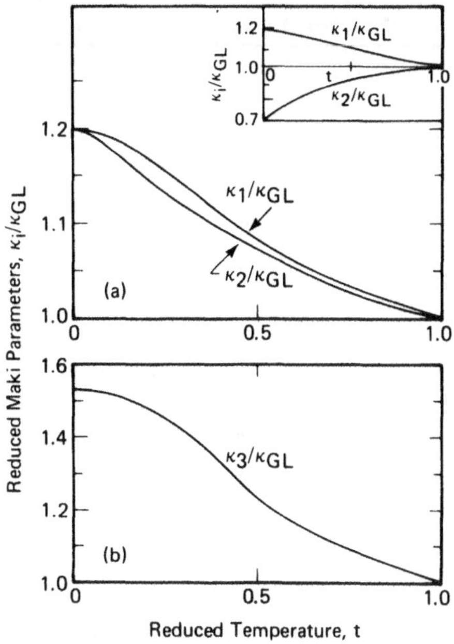

Figure 14-5. Temperature dependences of the reduced Maki parameters: (a) κ_1/κ_{GL} and κ_2/κ_{GL}; (b) κ_3/κ_{GL}. The $\kappa_1(t)$ and $\kappa_3(t)$ are due to MAKI ([MAK64] and [MAK64[a]], respectively). The corrected $\kappa_2(t)$ is due to CAROLI *et al.* [CAR66], the results of the original Maki calculation [MAK64] being portrayed in the inset. According to CAROLI [CAR66] the difference between $\kappa_1(t)$ and $\kappa_2(t)$ is extremely small—always less than 0.02 κ_{GL}.

where κ_{GL}^d, according to Table 14-1, is given by

$$\kappa_{GL}^d = 7.49 \times 10^3 \rho_n \gamma_v^{1/2} \tag{14-17}$$

These equations enable the Maki parameters (and hence the critical field temperature dependences) to be obtained from the results of resistive and calorimetric measurements. An alternative approach to the determination of experimental values of κ_1 and κ_2 was suggested by BARNES and HAKE [BAR66, BAR67] in connection with their studies of the paramagnetically limited superconductors to be considered in Chapter 15. They have shown how experimental values of the Maki parameters may also be obtained by combining the results of magnetic and calorimetric measurements of the mixed state.

14.5.3 Magnetic and Calorimetric Determinations of $\kappa_1(t)$

The parameter $\kappa_1(t)$ may, of course, be calculated with the aid of (14-10) in the form

$$\kappa_1(t) = H_u(t)/\sqrt{2}H_c(t) \tag{14-10}$$

after H_c has first been determined magnetically or, following BARNES and HAKE [BAR67], by double integration of the zero-field specific heat functions C_s/T and C_n/T. Then as a check on the result a special value, $\kappa_1(T_c)$, can be obtained by combining (14-10) in the form

$$\kappa_1(t) = (\partial H_u/\partial t)/\sqrt{2}(\partial H_c/\partial t) \tag{14-18}$$

with the Rutgers relation (e.g., [LYN64, p. 15])

$$\Delta C(T_c) \equiv (C_s - C_n)|_{T=T_c} = (VT/4\pi)(\partial H_c/\partial T)^2|_{T=T_c} \tag{14-19}$$

where V is the volume. The result of so doing is

$$\kappa_1(T_c) = (\partial H_u/\partial T)(VT/8\pi\Delta C)^{1/2}|_{T=T_c} \tag{14-20}$$

which should be found to agree with the Gor'kov-Goodman-calculated κ^d_{GL}, (14-17).

14.5.4 Magnetic and Calorimetric Determinations of $\kappa_2(t)$

As indicated above, the parameter $\kappa_2(t)$ may be extracted from the slope, S_M, of the magnetization curve at the upper critical field, H_u, using

$$S_M \equiv 4\pi(dM/dH)|_{H_u} = 1/(2\kappa_2^2(t) - 1)\beta_{\mathrm{A}} \tag{14-11b}$$

Again following BARNES and HAKE [BAR67], a value of $\kappa_2(t)$ can also be obtained by combining (14-11b) in the very dirty limit ($\kappa_2 \gg 1$) with the Ehrenfest function [LYN64, p. 15] in the form

$$\Delta C = (VTS_M/4\pi)(\partial H_u/\partial T)^2 \tag{14-21}$$

evaluated at $T = T_s(H_a)$.[4] The result of so doing is

$$\kappa_2(T_s) = (\partial H_u/\partial T)(VT/8\pi\beta_{\mathrm{A}}\Delta C)^{1/2}|_{T=T_s} \tag{14-22}$$

It is important to recognize that the desired temperature variation, usually written $\kappa_2(t)$, can be achieved through variation of $T_s(H_a)$ by exposing the sample to a variable magnetic field.

14.5.5 Final Developments in Nonparamagnetic Mixed-State Theory

At about the same time as MAKI's dirty-limit theories were making their appearance, MAKI and TSUZUKI [MAK65[a]] published a theory of the

[4]When exposure to a magnetic field is to be considered, it is no longer sufficient to represent the transition temperature simply in terms of $t = 1$. The zero-field transition temperature is represented by T_c, while the transition temperature in some applied field, $H_a < H_{c2}$, is represented by T_s.

mixed state and produced expressions for the parameters $\kappa_1(t)$ and $\kappa_2(t)$ that were valid in the clean limit of infinite electronic mean-free-path. HELFAND and WERTHAMER (H–W), in a pair of papers [HEL64, HEL66], then went on to develop solutions of the microscopic (Gor'kov) Ginzburg–Landau equations valid for *all values* of the electronic mean-free-path, l, and analyzed the joint temperature- and l-dependences of H_{c2}. The effect of alloying on l was taken into account through the "impurity parameter," λ, mentioned above, which, defined in terms of transport collision frequency by

$$\lambda \equiv \hbar/2\pi k_B T_c \tau \equiv 0.882\xi_0/l \tag{14-23}$$

was identical to GOR'KOV's ρ_G parameter ((13-34)). The t- and l-dependences of H_{c2}, defined in terms of a generalized form of (14-10), *viz.*,

$$H_{c2}(t, \lambda) = \sqrt{2}\kappa_1(t, \lambda)H_c(t) \tag{14-24}$$

could be deduced with the aid of computed plots of $\kappa_1(t, \lambda)/\kappa_1(1, \lambda)$ *versus* t for values of λ between 0 and ∞ (Figure 14-6(a)). In the companion paper [HEL66], as well as reiterating the above analysis HELFAND and WERTHAMER also described, as an alternative to the κ-parameter representation, procedures for expressing the t- and λ-dependences of H_{c2} in terms of a reduced parameter $h^*(t)$, defined by

$$h^*(t, \lambda) \equiv H_{c2}/(-dH_{c2}/dt)_{t=1} = h/(-dh/dt)_{t=1} \tag{14-25}$$

where $h^*(t)$ and h are further defined in Table 15-3. A computed tabulation of the λ-dependence of $(-dh/dt)_{t=1}$, for $0 < \lambda < 10$, was given together with plots of $h^*(t)$ *versus* t for all λ from 0 to ∞, Figure 14-6(b), enabling the results of critical field experiments to be conveniently compared with the theory.

In 1967 EILENBERGER [EIL67] returned to the κ-representational form of H_{c2} temperature dependence in developing what was essentially an extension of the H–W theory [HEL64, HEL66] for the temperature- and mean-free-path dependences of H_{c2}. The impurity parameter in this case was designated by $\lambda_{tr} = \xi_0/l_{tr}$ where the electronic mean-free-path length, l_{tr}, took into account both s-wave (represented by the λ and l of H–W theory) and p-wave (i.e., anisotropic) electron scattering. The results of the theory, which covered the entire range of mean-free-path lengths, was expressed in terms of a generalization of the normalized Ginzburg–Landau parameters κ_1/κ_{GL} and κ_2/κ_{GL}. The κ_i, which are functions of t, ξ_0/l_{tr}, and l_{tr}/l, were represented in the form of calculated plots of the triplets $\kappa_i(t, l_{tr}/l = 1, 1.5, 2)$ *versus* t for $\xi_0/l_{tr} = 0$–50. Figure 14-7, for $\xi_0/l_{tr} = 6.0$, is a typical example. It is reassuring to note that near the dirty limit (in this case for $\xi_0/l_{tr} = 50$) the triplets corresponding to $l_{tr}/l = 1, 1.5$, and 2 coalesce

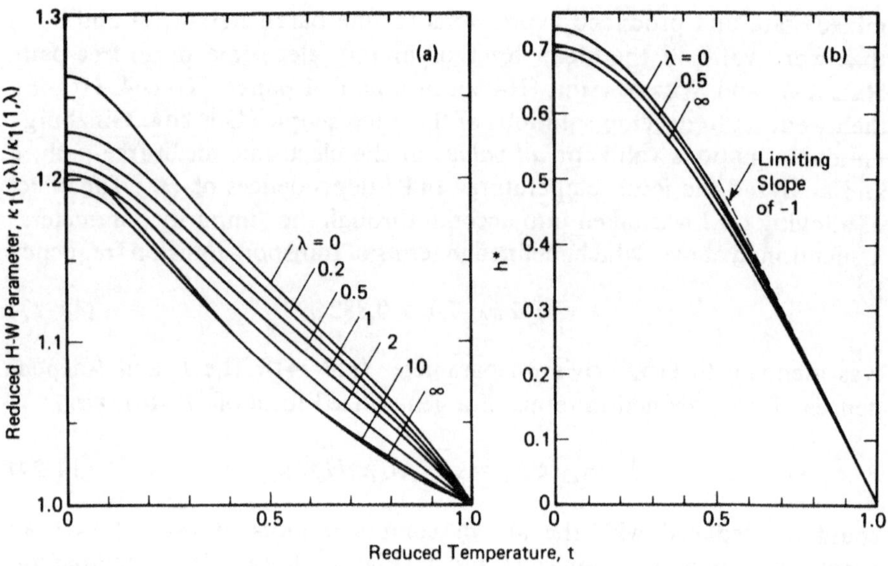

Figure 14-6. Temperature dependences of (a) the HELFAND–WERTHAMER [HEL64, HEL66] reduced Ginzburg–Landau parameter, $\kappa_1(t,\lambda)/\kappa_1(1,\lambda)$, (i.e., $\kappa_1(t,\lambda)$ normalized to its value at $t = 1$) for various values of the "impurity parameter" (related to s-electron scattering and defined by $\lambda \equiv 0.882\ \xi_0/l$); and (b) the HELFAND–WERTHAMER [HEL66] reduced upper critical field parameter

$$h^*(t,\lambda) \equiv \frac{H_{c2}}{(-dH_{c2}/dt)_{t=1}} \equiv \frac{h}{(-dh/dt)_{t=1}}$$

(see Table 15-3), also for various values of λ ($\lambda = 0$ is clean, $\lambda = \infty$ is dirty). Both figures are due to HELFAND and WERTHAMER [HEL66].

at all except the lowest temperatures, and that $\kappa_1/\kappa_{GL} \cong \kappa_2/\kappa_{GL}$ in agreement with Figure 14-5(a).

In an experimental study planned to test these theories, FIETZ and WEBB [FIE67[a]] measured the magnetizations of a series of Nb–Ti(0 to 15 at.%) alloys selected such that although their basic superconductive parameters, as represented by their transition temperatures and thermodynamic critical fields, remained nearly constant (at ~9 K and ~2 kOe, respectively), their electronic mean-free-path ranged from about 0.1 ξ_0 to 15 ξ_0 and could be regarded as an almost independent variable. The experiment of FIETZ and WEBB was designed to test the basic assumptions of the Maki and H–W theories, and the effect of Eilenberger p-wave scattering. Since H_{c2} was relatively low ($\lesssim$40 kOe) in most cases, paramagnetic effects (to be considered in Chapter 15) were neglected except for in the highest concentration (dirty-limit) alloy. The results turned out to be in poor agreement with the predictions of EILENBERGER (in terms of κ_1 and κ_2) except in the dirty limit [FIE67[a]]. In a comparison of the experimental results with those of H–W theory (in terms of h^*) agreement within experi-

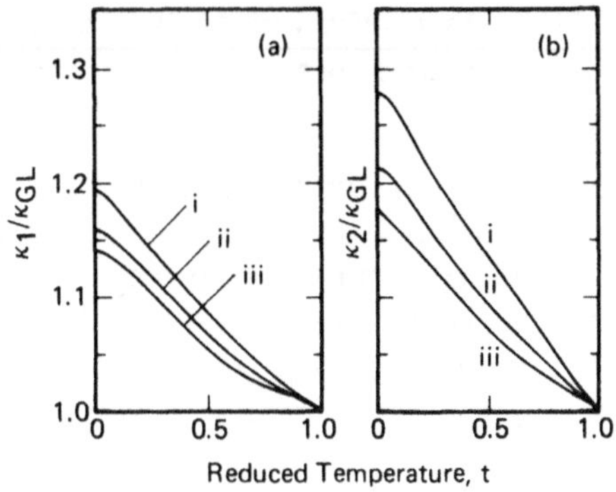

Figure 14-7. Temperature dependences of the Eilenberger reduced parameters (a) $\kappa_1(t, \lambda_{tr})/\kappa_{GL}$ and (b) $\kappa_2(t, \lambda_{tr})/\kappa_{GL}$ for a selected value of the impurity parameter $\lambda_{tr} \equiv \xi_0/l_{tr} =$ 6.0 (where the subscript *tr* indicates both *s*- and *p*-wave scattering) and for three values of the ratio l_{tr}/l, *viz.*, (i) 1.0, (ii) 1.5, and (iii) 2.0—after EILENBERGER [EIL67].

mental error was claimed for all except the highest concentration alloy, which was expected to exhibit paramagnetic lowering of the upper critical field.

Further advances in the theory of upper critical field temperature dependence, which require the introduction of the effects of both normal-state and mixed-state paramagnetism and of electron–phonon interaction, are considered in Chapter 15.

14.6 Evaluation of the Nonparamagnetic Upper Critical Field

14.6.1 Evaluation of H_{c2} in Terms of Normal-State Properties

The nonparamagnetic upper critical field may be evaluated by combining

$$H^*_{c20} = \sqrt{2}\kappa_{10}H_{c0} \qquad [14\text{-}10](14\text{-}26)$$

$$H_{c0} = 2.43\gamma_v^{1/2}T_c \qquad (13\text{-}48)$$

and

$$\kappa^d_{GL} = 7.49 \times 10^3 \rho_n \gamma_v^{1/2} \qquad (14\text{-}17)$$

with a suitable value of $\kappa_{10}/\kappa^d_{GL}$ selected from among the values 1.20 (Figure 14-5(a)), 1.195 (Figure 14-6(a)), or about 1.193 (Figure 29 of [EIL67]). Taking the value 1.19 and performing the substitutions suggested by the above group of equations yields

$$H^*_{c20}\,(\text{dirty}) = 3.06 \times 10^4 \rho_n \gamma_v T_c \qquad (\text{Oe}) \qquad [5\text{-}1](14\text{-}27)$$

where the asterisk indicates that H_{c20} has been taken to the nonparamagnetic (and in this case, dirty) limit. The normal-state quantities ρ_n and γ_v are in Ω cm and erg cm^{-3} K^{-2}, respectively.

14.6.2 Evaluation of H_{c2} in Terms of Superconductive-State Properties

(a) First Approach. From the analysis of HELFAND and WERTHAMER, expressed in terms of a plot of the reduced parameter[5] $h^*(t, \lambda) \equiv H^*_{c2}/(-dH_u/dt)_{t=1}$ *versus* t as depicted in Figure 14-6(b), it turns out that in the dirty limit ($\lambda = \infty$) the intercept at $t = 0$ is 0.68. This suggests that a value for H^*_{c20} can be extracted from the experimentally determined value of the limiting slope at $t = 1$ of the measured temperature dependence of the upper critical field, H_u, with the aid of

$$h^*(0, \infty) \equiv H^*_{c20}/(-dH_u/dt)_{t=1} = 0.68 \qquad [14\text{-}20](14\text{-}28)$$

that is,

$$H^*_{c20}\,(\text{dirty}) = 0.68(-dH_u/dt)_{t=1} \qquad (14\text{-}29a)$$

$$= 0.68\,T_c(-dH_u/dT)_{T_c} \qquad (14\text{-}29b)^6$$

(b) Second Approach. Alternatively, by taking the derivative with respect to T, again at $T = T_c$, of the Maki equation (14-10) in the form

$$-(dH_{c2}/dT)_{T_c} = \sqrt{2}\kappa_{11}(dH_c/dT)_{T_c} \qquad (14\text{-}30)$$

with $\kappa_{11} \equiv \kappa^d_{GL}$, and combining it with (14-17) for κ^d_{GL}, (13-49) for $(dH_c/dT)_{T_c}$, and (14-27) for H^*_{c20}, we find that

$$-(dH_{c2}/dT)_{T_c} = 4.494 \times 10^4 \rho_n \gamma_v \qquad (14\text{-}31)$$

$$= (0.681\,T_c)^{-1} H^*_{c20} \qquad (14\text{-}29c)$$

thus confirming the above result.

(c) Third Approach. As pointed out in Section 14.2.2, expressions for the upper critical field in terms of Φ_0, the flux quantum, and ξ, the coherence length, are obtainable by defining H_{c2} as that field at which the cores of the flux lattice either touch ((14-9a)) or consume the entire sample volume ((14-9b)).

(d) Fourth Approach. Finally, attention is drawn to a result based on Bardeen–Stephen theory as discussed in Section 23.8, which shows that the

[5] An index of reduced parameters is assembled in Table 15-3.
[6] In the clean limit ($l > \xi_0$): H^*_{c20} (clean) $= 0.72\,T_c(-dH_u/dT)_{T_c}$.

nonparamagnetically limited upper critical field is proportional to the ratio of the flux-flow- and normal-state resistivities, ρ_f and ρ_n, respectively, according to

$$H_a/H^*_{c20} = \rho_f/\rho_n \tag{14-32}$$

14.7 Evaluation of the Thermodynamic Critical Field

14.7.1 H_{c0} in Terms of Measurable Parameters

A useful expression enabling the thermodynamic critical field to be evaluated in terms of mixed-state measurables can be developed in the following way. Eliminating γ_v between (13-48) and (14-27) yields

$$H_{c0} = 1.39 \times 10^{-2}(H^*_{c20}T_c/\rho_n)^{1/2} \tag{14-33}$$

This, with the aid of (14-32) connecting H^*_{c20} with the flux-flow resistivity, $\rho_f(H_a)$, reduces to

$$H_{c0} = 1.39 \times 10^{-2}(H_aT_c/\rho_f)^{1/2} \tag{14-34}$$

with the fields in Oe and ρ_f in Ω cm. The thermodynamic critical field may also be calorimetrically determined as indicated in a work by Barnes and Hake [Bar67] and in Section 14.5.3.

14.7.2 A Case Study with Ti–Nb

The upper critical field can be obtained from the results of magnetic measurements on either reversible ("ideal") or irreversible type-II superconductors. Although the lower critical field can also occasionally be determined directly in this way, the extraction of an H_{c1} from highly hysteretic magnetization loops is generally not possible with any reasonable degree of accuracy (Section 14.1.1). Finally, for the direct determination of the thermodynamic critical field, reversible material is mandatory. For these reasons, in their experiments on Nb-Ti alloys, Fietz and Webb [Fie67, Fie67[a]] prepared pairs of compositionally similar alloys, administered cold work to one member and annealed the other in preparation for H_c determination.

As pointed out in Section 13.6, the thermodynamic critical field is definable in terms of the area, say $A_{II} = H_c^2/2$ ((13-47)), under the reversible magnetization curve. In studies of Nb-Ti(0, 0.5, 1.5, 4.5, and 9.0 at.%—annealed), Fietz measured this area following a half-cycle of magnetization. In practice some flux trapping was encountered, leading to hysteresis corresponding to some 5 to 20% of the area under the increasing-field magnetization curve. Analysis suggested that rather than considering an average based on field-increasing and field-decreasing data, it was better to take for $H_c^2/2$ the area under the increasing-field curve traced out from an initial state of zero magnetization (i.e., no trapped flux).

14.8 Evaluation of the Maki Lower Critical Field

14.8.1 H_{c10} in Terms of Measurable Parameters

(a) The Method of Fietz and Webb for Irreversible Superconductors. The method of FIETZ and WEBB [FIE67, FIE67[a]] provides a means of determining, indirectly from the results of magnetization measurements, the H_{c1}s of a series of samples of the same alloy, all of which are irreversible as a result of flux-pinning precipitation except one which has been suitably annealed and whose H_{c1} could be measured directly if needed (as in Section 14.1.1) and whose H_c can be obtained in the manner of the previous subsection.[7] The dirty-limit Maki representations of Section 14.5 are assumed to apply to the alloys of interest.

A zero-K version of (14-12) is

$$H_{c10} = H_{c0}\frac{\ln \kappa_{30}}{\sqrt{2}\kappa_{30}} \tag{14-35}$$

Next, by transforming κ_{30} to κ_{10} with the aid of (14-16) and substituting for κ_{10} its value as prescribed by (14-10), it follows that

$$H_{c10} = \frac{\ln(0.902 H_{c20}^*/H_{c0})}{1.276 H_{c20}^*/H_{c0}} \tag{14-36}$$

Thus, the H_{c1}s of the irreversible members of the series referred to above may be extracted from their H_{c20}^*s, as determined from the temperature dependences of their upper critical fields as in Section 14.6.3, using the measured H_{c0} of one well-annealed reversible representative of the set.

(b) The Method of Lubell and Kernohan. The starting point of the method of LUBELL and KERNOHAN [LUB71] is again (14-35), but this time κ_{30} is replaced by κ_{GL}^d from the second part of (14-16) and κ_{GL}^d is in turn replaced by its value in terms of ρ_n and γ_v using (14-17), following which (14-27) is applied to eliminate the γ_v. The result is an expression for H_{c10} (Oe) in terms of the measurables ρ_n (Ω cm), T_c (K), and H_{c20}^* (Oe) in the

[7] A pair of magnetization loops representing the situation just described is given in Figure 16-1.

form

$$H_{c10} = 7.50 \times 10^{-5}(T_c/\rho_n)\ln(4.30 \times 10^3 H^*_{c20}\rho_n/T_c) \qquad (14\text{-}37)$$

Since, according to (14-32), H^*_{c20} is itself available from flux-flow resistivity data, all the unknowns in (14-37) can be obtained by performing an appropriate set of four-terminal potentiometric measurements.

14.8.2 Validity of Nonparamagnetic Maki Theory as a Descripter of H_{c1} in Intermediate-κ_{GL} Alloys—A Case Study of Ti-Doped Nb

LUBELL and KERNOHAN [LUB71] investigated the validity of (14-37) as a descripter of the zero-K lower critical field of a series of Ti–Nb(75–95 at.%). Unfortunately the "extrinsic" κ_{GL}-values of these alloys were sufficiently low ($5 < \kappa^d_{GL} < 24$) as to place them below the dirty limit envisioned in MAKI's original theory. All the relevant materials properties listed in Table 14-3 were individually measured enabling several estimates of H_{c10} to be made using (i) the method of HARDON and ARP [HAR63] (Section 13.6), (ii) the method of LUBELL and KERNOHAN [LUB71] ((14-32)), and (iii) the method of FIETZ and WEBB [FIE67[a]] (Section 14.8.1) for comparison with the results of direct magnetic determination. For use in the evaluation, T_c was magnetically obtained, ρ_n and H^*_{c20} were obtained by normal- and flux-flow four-terminal resistometry, respectively, and H_{c10} (in the form of $H_{c1}(t \to 0)$) was determined using a magnetization technique which sought the lowest applied field that was associated with the absence of trapped flux (residual moment) following its removal.

The results of these investigations, as summarized in Table 14-3, reveal a considerable difference between the measured values of H_{c10} and those calculated using either the method of HARDON and ARP [HAR63] or the Maki approach via (14-37). In the discussion that followed the presentation of the results, inadequacy of the Maki theory (augmented by other small contributions, such as uncertainty in the estimation of H_{c0}) was identified as the most likely cause of the discrepancy. The principal factors leading to this conclusion were as follows:

1. The inapplicability of Maki theory to alloys with κ^d_{GL}s less than 24 would be the most obvious possible cause of the discrepancy between theory and experiment. But militating against this interpretation was the observation by DECKER and LACQUER [DEC69] that Maki theory did not seem to agree with experiment even for high-κ_{GL} (i.e., $\kappa_{GL} \cong 60$) material, coupled with the observation from Table 14-3 that the theoretical–experimental discrepancy actually became worse as κ_{GL} *increased.*

Table 14-3. Evaluation of the Zero-K Lower Critical Field—A Case Study by Lubell and Kernohan [LUB71, LUB71[a]]

Experimental Data					Calculated Data						
Ti–Nb Nominal Composition at.% Nb	T_c K	ρ_n $10^{-6}\ \Omega$ cm	H^*_{c20} kOe	H_{c10} Oe	H_{c0} kOe (a)	κ^d_{GL} (b)	κ_{10} (c)	κ_{30} (d)	H_{c10} Oe (e)	H_{c10} Oe (f)	H_{c10} Oe (g)
95	9.41	6.4_2	18	675	2.24	4.7_4	5.7	7.3	556	433	436 (95.5 at.%)
90	9.61	12.1	35	500	2.30	9.0	11	14	361	310	314 (91 at.%)
75	9.93	35.5	$90._5$	350	2.20	24.3	29	37	174	151	—

(a) Eq. (14-33).
(b) By eliminating γ_v between Eqs. (14-17) and (14-27).
(c) Above plus use of Eq. (14-16).
(d) From $\kappa_{30} = 1.275\ \kappa_{10} = 1.53\ \kappa^d_{GL}$, i.e., Eq. (14-16).
(e) Harden and Arp [HAR63].
(f) Eq. (14-37).
(g) Via Eq. (14-36) using zero-K-extrapolated data of Fietz and Webb [FIE67[a]] for alloys of the compositions indicated.

2. A rigorous test of the theory was subverted by uncertainty associated with the calculation of H_{c0} via (13-48) with the aid of estimated rather than accurately measured values of γ_v.

3. In spite of the discrepancy between experiment and calculation, good agreement existed among the theoretical values themselves, especially between those obtained using (14-37) (column (f)), and those calculated via (14-36) using the $H_u(t \to 0)$ and $\kappa_1(t \to 0)$ data obtained by FIETZ and WEBB [FIE67[a]] from magnetic measurements of Ti–Nb(91 and 95.5 at.%). The mutual accord among the calculated results lent credence to their correctness and suggested that the discrepancy lay with either the Maki model or its regime of applicability rather than with numerical error.

4. LUBELL and KERNOHAN [LUB71] suggested that in order to obtain better agreement between the experimental and calculated H_{c10} the theory may be in need of further refinement such as the taking into account of Fermi-surface anisotropy, multiple-band effects (e.g., *p*- as well as *s*-wave scattering), and the influence of normal-state Pauli paramagnetism on the *lower* critical field. It seems that these questions, and possibly others associated with lower-critical-field theory, still remain to be addressed.

Chapter 15

The Paramagnetic Mixed State

It will be recalled that (13-57) and its predecessor (13-44) relied on the validity of replacing $g_n(H_c)$ with $g_n(H_a)$ (cf. (13-43) and (13-44)). Obviously if g is field-dependent to an extent beyond that already implied in the derivations of these relationships, as it is when the normal-state and mixed-state paramagnetisms must be taken into consideration, the original energy-balance is upset and the equilibrium has to be reevaluated. The goals of the various paramagnetic theories of the mixed state have been to take into account the effects of these additional paramagnetic energy terms while developing expressions for the temperature dependence of the upper critical field (now reduced from H_{c2}^* to H_{c2} through the intervention of a paramagnetically limited upper critical field, H_p), and, peripherally, to provide guidance to the designers of high-H_{c2} superconductors. The theoretical burden of this development was initially carried by MAKI and by WERTHAMER, HELFAND, and HOHENBERG (WHH), while early comparisons of theory with experiment were conducted by NEURINGER and SHAPIRA, STRNAD and KIM, and by HAKE, who went on to carry out some very detailed evaluations using Ti–TM alloys as test materials. All this work was conducted during the mid-1960s. More than a decade later the paramagnetic mixed state began to undergo close reevaluation both experimentally and theoretically. This time the focus of experimental attention tended to be upon the A15 superconductors. Nevertheless, the suggested improvements to paramagnetic mixed-state theory which are coming out of this work, being generally applicable, are considered in this chapter.

In describing the background and present status of the subject, this chapter commences with a description of the Pauli paramagnetic limitation (PPL) idea as first proposed by CLOGSTON and CHANDRASEKHAR and its development in the hands of MAKI. The review then goes on to deal, both separately and conjointly, with the theories of MAKI and WHH as they relate to the partial offsetting, through spin-orbit-scattering (SOS) of the Cooper pairs, of the effects of normal-state paramagnetism, and concludes with a discussion of the limitations of the merged Maki–WHH approach and the manner in which it can be improved.

The reduction of the upper critical field below its ideal value H_{c2}^* as a consequence of the presence of a finite H_p has been described by the Maki parameter $\alpha \equiv \sqrt{2}H_{c20}^*/H_{p0}$. The removal of PPL thus corresponds to $\alpha \to 0$. The increase of the actual H_{c2} to some higher field closer to H_{c2}^* due to the intervention of *mixed-state* paramagnetism via the SOS of superelectron pairs is parameterized in WHH theory by an SOS frequency parameter, λ_{so}, and in Maki theory by a combined parameter $\beta_M^2 \equiv \alpha^2(3\Delta_{00}/\hbar)\tau_{so}$. As $\lambda_{so} \to \infty$, $H_{c2} \to H_{c2}^*$ with the vanishing of the paramagnetic difference between the mixed and normal states. Clearly the condition $\alpha \to 0$ is operationally equivalent to, although physically different from, this $\lambda_{so} \to \infty$ condition. As it then stood, Maki–WHH theory predicted that heavy-element doping, by increasing the frequency of spin-orbit scattering, should increase H_{c2}. Treated as an engineering fact rather than as the scientific suggestion that it was, the concept of heavy-element alloying then continued for fifteen years to guide numerous subsequent attempts to raise the H_{c2} of Ti–Nb-base superconductors. But as a justification of the simple SOS mechanism for a relief of PPL, one of the early experimental studies [NEU66] contained a serious flaw, the significance of which had passed more or less unnoticed until recently—*viz.*, that the *experimental* critical field of a Ti–Ta alloy exceeded theoretical expectation even with λ_{so} set equal to ∞ (and incidentally then went beyond the range of WHH theory itself, which required $\tau_{so} \gg \tau_{tr}$).

HAKE was able to show that, although effects qualitatively interpretable in terms of Maki–WHH theory were exhibited by the alloys that he was studying, the finer details of the temperature dependences could not be duplicated by theory [HAK67[b]]. Much more serious, however, was the discrepancy referred to above between the theoretical and experimental values of the upper critical fields themselves, a result that implied that factors other than (or in addition to) SOS were at work to elevate H_{c2}. The possibility of an electron–phonon-renormalized density-of-states increase of the Pauli limiting field, H_p, had already been considered by CLOGSTON in his original paper [CLO62[a]] and subsequently by HANCOX [HAN66] only to be forgotten until very recently. The roles played by this and other factors in reducing the need for unrealistically rapid SOS rates and in influencing the design requirements for high-H_{c2} alloy superconductors, which have been the subjects of the most recent investigations of this topic, are discussed in the concluding sections of this chapter and again in Chapter 17.

15.1 Pauli Paramagnetic Limitation

According to (13-45) and (13-57) the free-energy difference between the superconductive and normal phases, equal to $H_c^2/8\pi$ at zero field, vanishes at $H_a = H_{c2}^*$ as a result of a quadratic field-induced *increase* in $g_s(H_a)$ (Figure 15-1). H_{c2}^* is calculable from fundamental measurables with the aid of (14-27) or (14-29), enabling theory and experiment to be compared. In a classic study of high-field superconductivity in a series of binary transition metal alloys such as Ti–V, –Nb, –Ta, etc., BERLINCOURT and HAKE [BER63] noted that in most cases the resistively measured upper critical field, H_r, failed to reach its theoretical expectation. This result had been anticipated by CHANDRASEKHAR [CHA62] and CLOGSTON [CLO62ᵃ] who, working independently, had each proposed mechanisms whereby the upper critical field could be lowered by normal-state paramagnetism.

In 1958 YOSIDA [YOS58] demonstrated from BCS theory that the superconductive state at zero K, with unbroken spin-up/spin-down pairing, should have zero Pauli spin susceptibility, χ_p. Thus, the spin-magnetic energies of the superconductive and normal states should differ by the full Pauli spin energy. With reference to this work, to the results of BERLINCOURT and HAKE [BER63], and the filamentary-mesh model (Section 16.1), CHANDRASEKHAR [CHA62], suggested that an upper limit, H_{p0}, to H_{c20} would correspond to the establishment of an electron paramagnetic energy about equal in magnitude to the zero-K BCS energy gap, $2\Delta_{00}$ ($=3.5\ k_BT_c$), i.e., to

$$2\mu_B H_{p0} \cong 2\Delta_{00} \tag{15-1}$$

CLOGSTON [CLO62ᵃ], on the other hand, cast the normal-state paramagnetic effect into a different form by suggesting that an extreme zero-K upper limit, H_{p0}, to H_{c20} would correspond to the reduction of the normal-state paramagnetic energy by an amount, $\frac{1}{2}\chi_p H_{p0}^2 \equiv \mu_B^2 n(E_F) H_{p0}^2$, sufficient to make up the difference between a field-independent superconductive free energy, $g_s(H_a) = g_s(0)$, and the normal-state free energy $g_n(0)$—i.e., to

$$\mu_B^2 n(E_F) H_{p0}^2 = \{g_n(0) - g_s(0)\}|_{0=\text{zero K}} \qquad (=H_{c0}^2/8\pi) \tag{15-2}$$

If the right-hand side of (15-2) is then expressed in terms of the gap energy, $2\Delta_{00}$, a thermodynamic expression is transformed into a microscopic one and a new critical equation emerges, *viz.*,

$$(\mu_B H_{p0})^2 n(E_F) = \tfrac{1}{2}\Delta_{00}^2 n(E_F) \tag{15-3a}$$

that is,

$$\sqrt{2}\mu_B H_{p0} = \Delta_{00} \tag{15-3b}$$

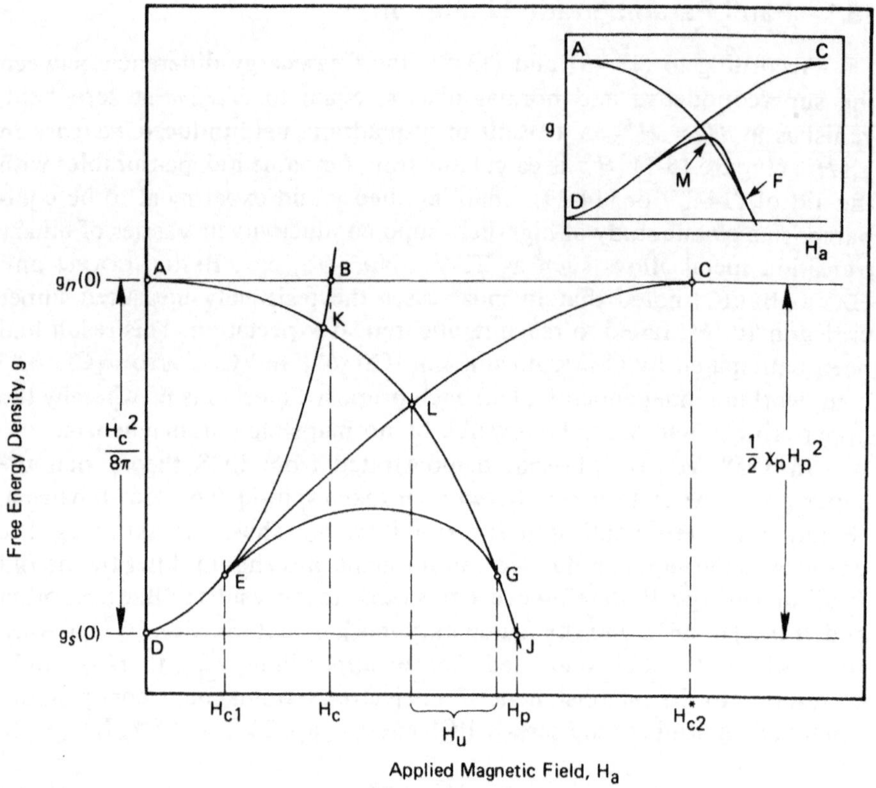

Figure 15-1. Schematic diagram of normal-state and superconducting-state free energies *versus* applied magnetic field, H_a. An arbitrary temperature $0 < T < T_c$ is assumed. The normal state in the absence of Pauli spin paramagnetism is represented by AC and its continuation; that in the presence of spin paramagnetism, of susceptibility χ_p, by the parabola AJ. The field-ignoring superconducting state is represented by DJ and its continuation; the point J defines the Clogston–Chandrasekhar paramagnetically limited first-order critical field, H_p. The parabola DB represents the response of a type-I superconductor to H_a, with point B itself at the thermodynamic critical field, H_c. The magnetization of a GLAG (nonparamagnetic) type-II superconductor is represented by DEC with lower critical field, H_{c1}, and upper critical field, H_{c2}^*, at E and C, respectively. The inclusion of spin in the normal state leads to a first-order transition at an upper critical field corresponding to L. The further inclusion of spin in the superconducting state can lead to a second-order transition at an upper critical field corresponding to G. Both these upper critical fields, designated H_u in the figure, are lower than H_{c2}^*. The inset suggests that within the context of paramagnetic theory it is also possible to have a first-order s/n transition (at M) followed by a second-order transition at F with a metastable state lying between the two—after WERTHAMER *et al.* [WER66].

This differs from the Chandrasekhar equation by a factor of only $\sqrt{2}$. The Clogston result,

$$H_{p0} = 1.84 \times 10^4 T_c \qquad \text{(Oe)} \tag{15-4}$$

is generally accepted as the pure "paramagnetic limit" (pure in the sense that it determines critical field in the absence of any other effect [BEA82[a]]). Modifications to the $n(E_F)$s of (15-3a), which must be carried out differently on each side of the equation, severely upset the Clogston–BCS energy balance. They led to the improvements to classic paramagnetic mixed state theory discussed in Section 15.8.

The paramagnetic effect is illustrated in Figure 15-1. Clearly, in the presence of a finite H_p the experimental critical field is reduced below its "nonparamagnetically limited" value. Although an effect is to be expected *whatever* the value of H_p, it will be strongest when $H_p < H_{c2}^*$.

H_{p0} is quite a large field, equal to 150 kOe in Ti–Nb(32 at.%), for example. But as predicted in Figure 15-1, which shows the superconductive-state free-energy curve DEC intersecting the normal-state paramagnetic free-energy curve AJ in a *first-order* superconductive/normal transition at L, the experimentally-measured value, H_u, must always lie beneath the lower of H_{c2}^* and H_p. The same is true for the curve DEG representing MAKI's (1964) paramagnetic theory [MAK64[a]], which predicted a *first- or second-order* transition to the normal state depending on whether the parameter α was greater than or less than unity, respectively [MAK66], except in the presence of strong SOS when the transition was always second-order [MAK66]. Before proceeding it is appropriate to mention here, with reference to Figure 15-1 (inset), that a third kind of high-field behavior is possible in the paramagnetic model: the $g_s(H_a)$ curve may intersect $g_n(H_a)$ in a first-order transition at M then go on to merge smoothly into it at F between which two points the superconductivity may exist in a metastable super-cooled state. The extent of the deviation of H_u from H_p and H_{c2}^* has been experimentally investigated in an important case study of high-field superconductivity in Ti–TM and other binary transition-metal alloys by BERLINCOURT and HAKE [BER63]. A typical set of results is summarized in Figure 15-2.

In attempting to quantify the relationships between these three upper critical fields HANCOX [HAN66], using a simple functional form for the magnetization curve, $M(H_a)$, found that

$$H_u = H_{c2}^* H_p/(H_{c2}^* + H_p) \tag{15-5a}$$

based on the assumption that all the electrons in the superconductor were paired. Next, by recognizing the fact that the superconducting state was in fact mixed, i.e., by taking into account the effect of the normal electrons in the superconductor, HANCOX succeeded in increasing the estimated experimental upper critical field to

$$H_u = H_p(\sqrt{H_p^2 + 4H_{c2}^{*2}} - H_p)/2H_{c2}^* \tag{15-5b}$$

in agreement with a result derived by MAKI and TSUNETO [MAK64[b]] using

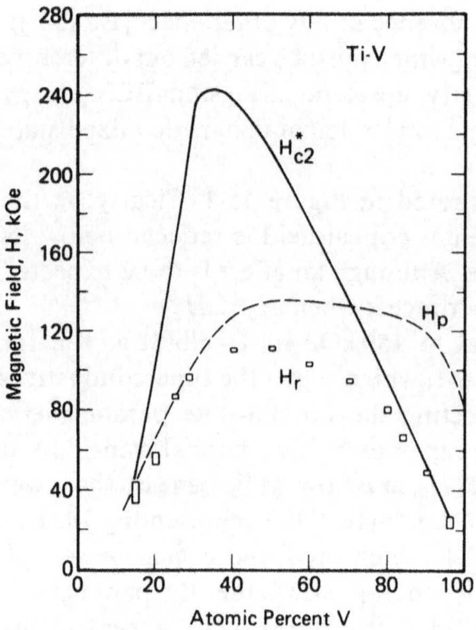

Figure 15-2. Upper critical fields at 1.2 K for Ti–V alloys. A comparison of the resistively measured values, H_r (entire transition range depicted, measured at 10 A cm^{-2}), with the Clogston paramagnetic limit, H_p, and the GLAG-calculated upper critical field, H^*_{c2}—after BERLINCOURT and HAKE [BER63].

the generalized Gor'kov equations. Both (15-5a) and (15-5b) provide for an experimental upper critical field that lies below the smaller of H^*_{c2} and H_p as indicated previously. MAKI [MAK64[a]] and subsequently others (see for example [SAI69, p. 169ff]), using a microscopic approach based on the Ginzburg–Landau equations, showed that the minimum allowed upper critical field at zero K is given by

$$H_{u0} = H^*_{c20} H_{p0} / \sqrt{2H^{*2}_{c20} + H^2_{p0}} \tag{15-6}$$

which again, of course, falls below H^*_{c20} and H_{p0}.

15.2 Mechanisms for the Relief of Pauli Paramagnetic Limitation

15.2.1 Early Observations

In comparing the (15-5a)- and (15-5b)-calculated paramagnetically limited minimum upper critical fields with experiment (for example the Ti–V data of BERLINCOURT and HAKE [BER63]) HANCOX [HAN66] found that the critical field depression was in both cases overestimated (Figure 15-3). Clearly the influence of normal-state Pauli paramagnetism was being opposed by other effects yet to be accounted for. By way of explanation, several mechanisms were postulated, two of which called for increases in

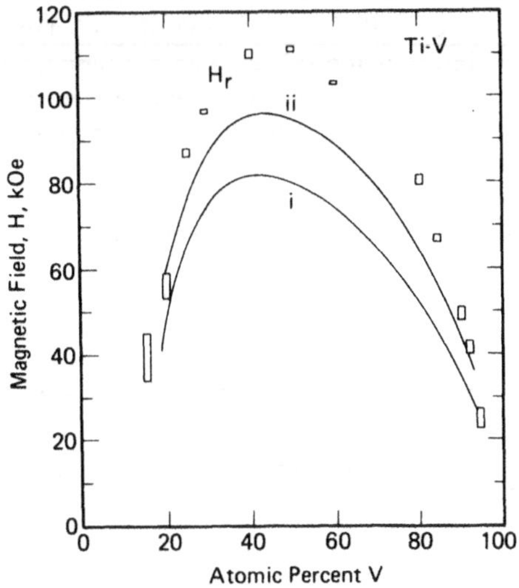

Figure 15-3. Upper critical fields of Ti–V alloys. A comparison of the 1.2 K experimental data of BERLINCOURT and HAKE, H_r, with a pair of suggested (and temperature-ignoring) calculated results: curve (i) represents (15-5a), and curve (ii) represents (15-5b)—after HANCOX [HAN66].

H_p itself: (i) It was suggested by CLOGSTON [CLO62[a]], and subsequently overlooked (to be rediscovered in more satisfactory form in the 1970s), that the taking into account of the electron–phonon interaction would cause the normal-state χ_p to be "reduced by a factor $1 + n(E_F)V$ where V is the average matrix element used by BCS" (see (8-34) through 8-36)) and by so doing would increase H_p.[1] (ii) As pointed out in a discussion by HANCOX [HAN66], strong-coupling corrections to Δ_{00} could, via (15-3b), also significantly increase H_p—by up to 30–40% in fact. (iii) Finally, the idea that attracted most attention in the mid-1960s had to do with the development, as a result of SOS, of a *superconductive-state* paramagnetism, a property which would then offset the effect of paramagnetism in the normal state. As the superconductive state paramagnetism approached that of the normal state, with increasing λ_{so}, the paramagnetic energy difference between those states would become smaller and H_{c2} would approach the GLAG-calculated upper critical field designated H^*_{c2} in this context.

15.2.2 Thermodynamic Model for the SOS Relief of PPL

The contributions of MAKI and WERTHAMER *et al.* (WHH), together with those of the authors of several important experimental case studies, to the evolution of the paramagnetic theories of the mixed state are indicated

[1]This statement was actually implying that $\chi_{\text{spin}} = \chi_{\text{specific heat}}[1 + n(E_F)V]^{-1}$ [CLO62]. In present terminology this is stating that the density-of-states entering γ_v (or what is the same thing, the condensation energy on the right-hand side of (15-3)) is being renormalized by the factor $(1 + \lambda)$. The effect of this is to increase H_{p0} by the factor $(1 + \lambda)^{1/2}$[ORL79], see Section 15.8.

Table 15-1. Evolution of Contemporary Theories of the Mixed State—The Introduction of Temperature, and Conduction-Electron Interactions with: The Applied Field (Pauli Paramagnetism), the Lattice (Spin-Orbit Scattering), the Phonons (Electron-Phonon Interaction) and Each Other (Electron-Electron Interaction)—Part II[a]

Evolution of Paramagnetic Theories	Literature
A. Introduction of Pauli Paramagnetism	
Limitation on H_{c2} imposed by normal-state Pauli paramagnetism.	Chandrasekhar [CHA62] Clogston [CLO62[a]]
Composition dependence of $H_u(1.2\text{ K})$ compared to the GLAG-calculated H_{c2} and the Clogston-calculated H_p.	Berlincourt and Hake [BER63]
Case studies of Ti–V, -Nb, -Ta and other binary Ti-TM alloys.	
Introduction for the first time, within the context of Pauli limitation, of the effect of finite mean-free-path. Showed that the paramagnetic reduction could be expressed in terms of $H_{c20}/H^*_{c20} = (\sqrt{1+2\alpha^2}-1)/\alpha^2$ if α, as subsequently defined, is $\sqrt{2}H_{c20}/H_{p0}$, the Pauli limitation parameter.	Maki and Tsuneto [MAK64[b]]
H_{c2} and $4\pi M$ as functions of t and the effect of normal-state Pauli paramagnetism in terms of $\kappa_1(t)$ and $\kappa_2(t)$ and parameterized by $\alpha \equiv \sqrt{2}H_{c20}/H_{p0}$ such that $H_{c20}/H^*_{c20} = (1+\alpha^2)^{-1/2}$. —*Dirty limit*	Maki [MAK64[a]]
B. Introduction of Spin-Orbit Scattering and Other Effects	
Influence of Pauli paramagnetism in conjunction with strong spin-orbit scattering on H_{c2} and $4\pi M$, expressed in terms of $\kappa_1(t)$ and $\kappa_2(t)$ and parameterized by α and β_M which is itself a function of α and τ_{so}. —*Dirty limit*	Maki [MAK66]
Influence of normal-state Pauli paramagnetism and its modification due to spin-orbit scattering on the temperature dependence of H_{c2} described in terms of an $h^*(t)$ and parameterized by α and λ_{so}. —*Initially unrestricted* electron mean-free-path, then specialized to the dirty limit.	Werthamer, Helfand and Hohenberg (WHH) [WER66]

Table 15-1. (*cont.*)

Evolution of Paramagnetic Theories	Literature
Temperature dependence of H_{c2} with reference to Pauli limitation and spin-orbit scattering theories.	
Case study of Ti–V	Kim and Strnad [KIM66]
Case study of Ti–Nb	Shapira and Neuringer [SHA65]
Case study of Ti–Mo	Cape[CAP66]
Case studies of Ti–V, -Nb, -Ta	Neuringer and Shapira [NEU66]
Case studies of Ti–V, Ti–V–Cr, Ti–Mo	Hake [HAK67b]
Case studies of Ti–V and Ti–Nb	Brand [BRA72]
Temperature dependence of H_{c2} with reference to Pauli limitation, spin-orbit scattering, electron-electron interaction and electron–phonon interaction.	
Case study of Nb_3Sn and V_3Si	Orlando, McNiff, Foner and Beasley [ORL79]

[a] Part I, *Nonparamagnetic Theories*, is Table 14-2.

in Table 15-1. A thermodynamic model of the SOS–PPL effects (which were also mechanistically interpreted by the final versions of the theories of MAKI [MAK66] and WHH [WER66]) is embodied in Figure 15-1. As indicated in that figure and elsewhere [WER66], [HAK67b], the curve DEGJ representing superconductive free energy, which initially increases with H_a and must eventually decrease to merge into that of the paramagnetic normal state at

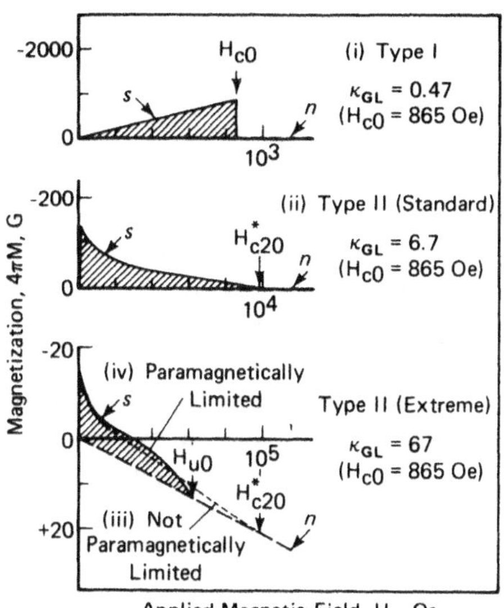

Figure 15-4. Schematic representation of the zero-K, reversible magnetization *versus* field diagrams for four classes of superconductor: type-I, "standard" type-II, "extreme" type-II in the absence and presence, respectively, of paramagnetic limitation. The symbols *n* and *s* indicate the normal-state and superconducting-state magnetization curves, respectively. Details of the materials parameters represented are given in [HAK67b]. Parameters selected result in an equality among the $\int MdH_a$ condensation energies; since the $4\pi M$ and H_a axes have been oppositely scaled by successive factors of ten, the cross-hatched areas are actually equal—after HAKE [HAK67b].

G, passes through a maximum at some intermediate field which (since $M = -dg/dH_a$, (13-40)) corresponds to a point of zero net magnetization (hence $M_m = M_n$, where "m" and "n" stand for "mixed" and "normal," respectively). The region above this point has been referred to by HAKE [HAK67[b]] as the "paramagnetic mixed state." Since $M_n(H_a)$ is now no longer implicitly zero, $\Delta M = M_m - M_n$ replaces the M in (14-11) which defines the parameter $\kappa_2(t)$ in this state.

The magnetic behaviors of four classes of type-I and type-II superconductors are schematically depicted in Figure 15-4: (i) type-I, (ii) "standard" (i.e., nonmagnetic normal state) type-II, (iii) paramagnetic normal-state but not paramagnetically limited type-II, (iv) paramagnetically limited type-II. Real materials falling into these categories according to HAKE [HAK67[b]] would be, respectively, (i) Ta, (ii) V–Ta(5 at.%), and (omitting (iii)), (iv) Ti–Mo(16 at.%) As with the alloys prepared by FIETZ and WEBB for their study (Section 14.7.2), these three materials possess similar electronic properties but differ in their resistivities which, being approximately in the ratio $1:10^2:10^3$, are responsible for $\kappa_{\rm GL}$ values of 0.47, 6.7, and 67, respectively. The figure is constructed for materials that have the same T_c and zero-K superconductive condensation energies such that (cf. (13-45) to (13-47)):

$$\left.\begin{aligned} H_{c0}^2/8\pi &= \Delta g_{n,s}(0)|_{\text{zero K}} \\ &= \{g_n(0) - g_s(0)\}|_{\text{zero K}} \\ &= \int_0^{H_{u0}} M_n(H_a)\,dH_a - \int_0^{H_{u0}} M_s(H_a)\,dH_a \\ &= \oint M(H_a)\,dH_a \qquad \text{(represented by the shaded areas between } M_s \text{ and } M_n\text{)} \end{aligned}\right\} \tag{15-7}$$

It is interesting to note that, as a result of the choice of compensatory scale factors for the vertical and horizontal axes of Figure 15-4, the shaded areas are in actuality all equal.

The properties of the paramagnetically limited, and paramagnetic mixed states were thoroughly explored both magnetically and calorimetrically by HAKE and colleagues at the then North American Aviation Science Center. Figure 15-5, from the work by CAPE [CAP66] and HAKE [HAK67[b]], presents some magnetization results for Ti–Mo(16 at.%), a good example of an ideal, extremely-dirty ($\kappa_{\rm GL} = 67.6$) paramagnetically limited ($\chi_{p,n} = +10.3 \times 10^{-6}$)[2] type-II superconductor. The figure exhibits the following features: (i) a reversible high-field magnetization, (ii) a transition (as measured point by point with increasing applied field) from a *diamagnetic*

[2]The units are e.m.u. volume units ($\text{cm}^3\,\text{cm}^{-3}$); χ in these units is obviously dimensionless.

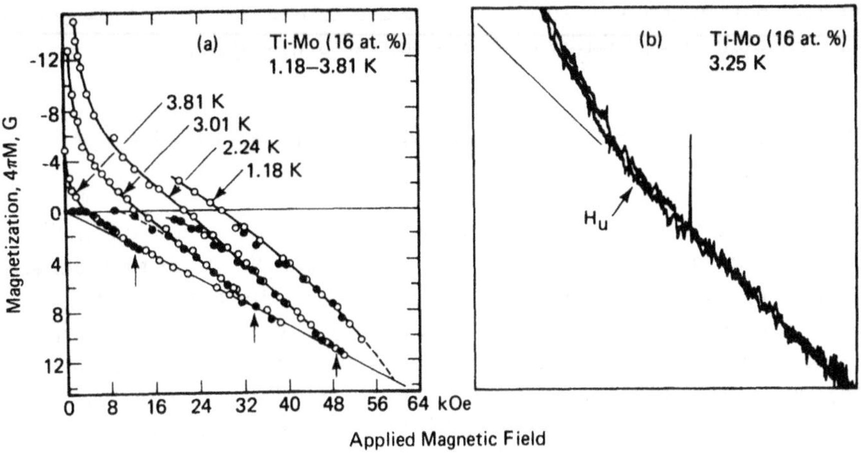

Figure 15-5. (a) Isothermal magnetization, $4\pi M$, *versus* applied magnetic field, H_a, for a Ti-Mo(16 at.%) alloy ($T_c = 4.18$ K, $\kappa_{GL} = 68$). For clarity not all the data points are included. Arrows indicate the s/n transitions. The data taken with field increasing (open circles) followed by field decreasing (closed circles) exhibit *complete reversibility* in a high-field *paramagnetic* mixed state—after HAKE [HAK67[b]]. (b) Continuous record of the reversible magnetization change with increasing and decreasing applied field across the second-order boundary of the mixed and normal states at H_u as measured using a vibrating sample magnetometer—after CAPE [CAP66].

to a *paramagnetic* mixed superconductive state, (iii) as depicted both in Part (a) and in Part (b) (a continuous record of the magnetization change), a *second-order* transition from a paramagnetic mixed superconductive state to a paramagnetic normal state.

15.2.3 Mechanistic Interpretation of the SOS/PPL Effects

In the absence of SOS, each electronic component of a quasibound Cooper pair, or quasiparticle, in a dirty superconductor is formed from wavefunctions of a single spin direction. The result is a spin-up/spin-down quasiparticle. A magnetic field applied to a system of such particles can produce polarization (Zeeman splitting, in quantum terms) only at the expense of pair breaking. According to (15-3b), the requisite energy for so doing can be acquired only when H_a reaches H_{p0}—which is the condition for normal-state PPL. Under the condition of random SOS, spin is no longer a good quantum number [BEA82[a]]. The quasiparticle components are formed from equal numbers of spin-up and spin-down wavefunctions; the particle still has a net spin of zero but can no longer be treated as an oppositely polarized pair.[3] In the presence of an applied field, Zeeman splitting of a population of such particles into spin-up and spin-down states is possible

[3]This entire process has been referred to, on good authority, by authors such as HAKE [HAK67[b]] and BRAND [BRA72] as "pair-spin decoupling."

without pair breaking as soon as the field is turned on. Thus, SOS confers paramagnetic susceptibility on the superconducting state.

15.3 Calorimetric Evidence for the Paramagnetic Mixed State

The results of magnetization studies such as those depicted in Figure 15-5 provide ample evidence for a smooth magnetic transition between diamagnetic and paramagnetic states of the alloy. Furthermore, the results of resistive measurements such as those reported by HAKE in several articles [HAK65, HAK67] indicate resistive transitions to the normal state at fields, H_r, well into the paramagnetic regime. But one of the most conclusive tests of a bulk superconductive transition is low-temperature specific heat. BARNES and HAKE [BAR66, BAR67], by coupling magnetic with low-temperature calorimetric measurements in a study of Ti–Mo(16 at.%), determined that the paramagnetic mixed state, as defined by the resistive and magnetic measurements, (i) does indeed exist as a single thermodynamic phase, and (ii) undergoes a second-order transition into the normal state at a sufficiently high field $H_a = H_u$.

The low-temperature specific heat measurement was shown by BARNES and HAKE [BAR66, BAR67] to be an important source of data descriptive of the electronic properties of the normal and superconducting states near the transition temperature. Furthermore, as indicated by MAKI [MAK66] and pointed out by BARNES and HAKE, the results of magnetic and calorimetric measurements when suitably combined could be made to yield experimental values of H_c and the GLAG–Maki parameters $\kappa_1(t)$ and $\kappa_2(t)$ (see Sections 14.5.2 to 14.5.4). As indicated in the sections just mentioned, the temperature dependences of the Maki parameters may be described in terms of

$$\kappa_1(t) = H_u(t)/\sqrt{2}H_c(t) \tag{14-10}$$

with

$$\kappa_1(T_c) = (\partial H_u/\partial T)(VT/8\pi\Delta C)^{1/2}|_{T=T_c} \tag{14-20}$$

and

$$\kappa_2(T_s) = (\partial H_u/\partial T)(VT/8\pi\beta_A\Delta C)^{1/2}|_{T=T_s} \tag{14-22}$$

with

$$T_s = T_c(H_a)$$

Using an annealed sample of Ti–Mo(16 at.%) ($\kappa_{GL} = 67.6$) as a working substance, BARNES and HAKE [BAR66, BAR67] evaluated the temperature dependence of $\kappa_1(t)$ using (14-10) and that of $\kappa_2(t)$ using (14-22), controlling t (i.e., T_s) in the latter case by varying the strength of a magnetic field being applied to the calorimetric sample *via* a superconducting Helmholtz pair. The results, to be considered in more detail below (see Figure 15-6), showed $\kappa_1(t)$ and $\kappa_2(t)$ *decreasing* with decreasing t, in sharp contrast to the behaviors of the nonparamagnetic parameters, Figure 14-5.

15.4 The Spin-Paramagnetic Theories of MAKI and of WERTHAMER, HELFAND, and HOHENBERG

15.4.1 The Theories of Maki

In a pioneering paper, MAKI [MAK64] derived a general Ginzburg-Landau-Gor'kov equation valid in the dirty limit ($\xi_0/l \gg 1$) but for all temperatures, and went on to develop a set of equations descriptive of the temperature dependences of the magnetic properties. In his second 1964 paper, MAKI [MAK64[a]] introduced normal-state Pauli paramagnetism and the quantity $\alpha \equiv \sqrt{2}H^*_{c20}/H_{p0}$, which expressed the degree to which this paramagnetism lowered the upper critical field. Furthermore, a zero-K reduced upper critical field, $h^*(0)$, was shown to be lowered by Pauli paramagnetism according to

$$h^*(0) \equiv H_{c20}/(-dH_{c2}/dt)_{t=1} \tag{15-8}$$

or

$$h^*(0) = 0.68(1+\alpha^2)^{-1/2} \tag{15-9}$$

(the latter of course reduces to (14-28) in the limit $\alpha = 0$). Next, recognizing that the measured upper critical field, H_u, was always greater than and its temperature dependence always less than the value predicted by the original paramagnetic theory, several extensions to the theory were developed (see Table 15-1). This work culminated in the classic 1966 paper [MAK66] which considered not only normal-state Pauli paramagnetism but also the effect on the mixed-state properties of SOS. For comparison with experiment, the upper critical field and magnetization temperature dependences were described in terms of the parameters $\kappa_1(t, \alpha, \beta_M)$ and $\kappa_2(t, \alpha, \beta_M)$, respectively, in which β_M was a function of both α, the dimensionless paramagnetic-limitation parameter, and τ_{so}, the SOS relaxation time. The upper critical field, $H_{c2}(t)$, was also explicitly presented as a function of μ_B and τ_{so}/τ_{tr}, where τ_{tr} is the transport relaxation time. The resulting expression was transcribed by HAKE [HAK67[b]] into the reduced-functional form, $h^*(t)$, considered below.

15.4.2 The Theory of Werthamer, Helfand, and Hohenberg (WHH)

The nonparamagnetic theories of HELFAND and WERTHAMER described in Chapter 14 (see Table 14-2) culminated in the theory of WERTHAMER, HELFAND, and HOHENBERG (WHH), who in 1966 published a gauge-invariant solution of the linearized Gor'kov equations, neglecting strong coupling and Fermi-surface anisotropy effects as in earlier calculations (i.e., neglecting *p*-wave scattering—cf. EILENBERGER'S [EIL67] nonparamagnetic theory) but including both Pauli paramagnetism and SOS [WER66]. The general WHH theory was developed without restriction on the electronic mean-free-path, but prior to implementing it in experimental situations it was deemed advisable to specialize it to the dirty limit. At this point the dimensionless variables α (as before) and λ_{so}, the SOS frequency parameter, were introduced, and codes were developed for the calculation of a reduced field, $\bar{h}$ (numerically related to h^*), as a function of t and parameterized by α and λ_{so}.

15.4.3 The Conjoint Theories of Maki and WHH

The several case studies which have been offered as experimental tests of the Maki and WHH theories are listed in Table 15-2.

Most discussions of the SOS-influenced paramagnetic mixed state combine the Maki and WHH results into a single mutually consistent model (Section 15.7). But before proceeding to do so here, the salient features of the individual Maki and WHH theories will be treated separately in the following two sections.

The student of Maki–WHH theory and its experimental evaluation is inevitably confronted by a confusion of reduced magnetic fields, represented in the various works by the symbols h, h_{c2}, h^*, and $\bar{h}$. Table 15-3 is a catalog or index of such reduced-field notations. Reduced or normalized quantities have also been designated by some authors through the use of asterisks, as have special quantities such as the nonparamagnetically limited upper critical field, H^*_{c2}. A relevant note on the use of asterisk notation is presented as a supplement to Table 15-3.

15.5 The Maki Result

15.5.1 Pauli Paramagnetic Limitation (PPL)

PPL in MAKI'S theory is introduced through the parameter

$$\alpha \equiv \sqrt{2}H^*_{c20}/H_{p0} \qquad (15\text{-}10)$$

which may be evaluated using either superconductive-state (hence α_s) or

Table 15-2. Experimental Verification of the Maki and WHH Theories of Paramagnetic Mixed State Upper Critical Field Temperature Dependence—Some Case Studies of Titanium-Alloy Superconductors

Alloy System	Representation[a]	Literature
Ti–V(65 at.%)	h^* *versus* t	Kim and Strnad [KIM66]
Ti–V(58 at.%)	h^* *versus* t	Neuringer and Shapira [NEU66]
Ti–V(22.5, 25 at.%)	h^* *versus* t	Hake [HAK67
	$\kappa_i^*(t) \equiv \kappa_i(t)/\kappa_i(t=1)$ *versus* t $(i = 1, 2)$	
Ti–V(36, 70, 82 at.%)	H_{c2} *versus* T	Brand [BRA72]
Ti_{30}–V_{60}–Cr_{10}	h^* *versus* t	
	$\kappa_i^*(t)$ *versus* t $(i = 1, 2)$	Hake [HAK67]
Ti–Nb(44, 63 at.%)	H_{c2} *versus* T	Shapira and Neuringer [SHA65]
Ti–Nb(44 at.%)	h^* *versus* t	Neuringer and Shapira [NEU66]
Ti–Nb(32, 55, 62, 83 at.%)	H_{c2} *versus* T	Brand [BRA72]
Ti–Nb(11 at.%)	h^* *versus* t	Missell *et al.* [MIS75]
Ti–Ta(52 at.%)	h^* *versus* t	Neuringer and Shapira [NEU66]
	H_{c2} *versus* T^2	Cape [CAP66]
	H_{c2} *versus* T^2 }	
Ti–Mo(16 at.%)	h^* *versus* t }	Hake [HAK67[b]]
	$\kappa_i^*(t)$ *versus* t $(i = 1, 2)$ }	
	h^* *versus* t	Hake [HAK67]

[a] $h^*(t) \equiv H_{c2}(t)/(-\partial H_{c2}/\partial t)_{t=1}$
$= 0.693\, H_{c2}(t)/H^*_{c20}$ (Hake) or $0.680\, H_{c2}(t)/H^*_{c20}$ (Brand)—representing the clean and dirty limits, respectively, see footnote 7, p. 544.
$\kappa_1(t)$ and $\kappa_2(t)$ are as defined in Section 14.5; Hake represented the normalized quantities, $\kappa_i(t)/\kappa_i(t=1)$, by $\kappa_i^*(t)$. See appendix to Table 15-3 for a note on asterisk notation.

Table 15-3. Reduced-Field Representations with Particular Reference to the H–W, Maki, and WHH Theories

Relationship	Origin	Literature
$t = T/T_c$	General	
$h = H/H_{c2}$	General	
$b = B/H_{c2}$	General	
$h^*(t,\lambda) \equiv \dfrac{H_{c2}}{(-dH_{c2}/dt)_{t=1}}$ $\equiv \dfrac{h}{(-dh/dt)_{t=1}}$ $h \equiv \dfrac{0.5807\, H_{c2}}{\kappa_{GL}^c(t=1)H_{c0}}$	H–W	Helfand and Werthamer [HEL66]
$0.680 = \dfrac{H^*_{c20}}{(-dH_{c2}/dt)_{t=1}}$	General	Helfand and Werthamer [HEL66]
$h^*(0,\infty) = 0.680$	H–W (special case)	Helfand and Werthamer [HEL66]
$h^*(t,\beta_M) \equiv \dfrac{H_{c2}}{(-dH_{c2}/dt)_{t=1}}$ $= \dfrac{1.36\, h_{c2}}{1+(1+\beta_M^2 h_{c2})^{1/2}}$ $h_{c2} = H^*_{c2}/H^*_{c20}$	Maki-derived	Hake [HAK67[b]]
$\ln\dfrac{1}{t} = f(\bar{h},\alpha,\lambda_{so})$ $\bar{h} \equiv 2(e/c)(v_F^2\tau_{tr}/6\pi k_B T_c)H_{c2}$ $h^*(t,\alpha,\lambda_{so}) \equiv \dfrac{H_{c2}}{(-dH_{c2}/dt)_{t=1}}$ $= \dfrac{\bar{h}}{(-d\bar{h}/dt)_{t=1}}$ $= (\pi^2/4)\bar{h}$	WHH	Werthamer, Helfand and Hohenberg [WER66]

Appendix to Table 15-3. Asterisk Notation and Reduced Quantities.

Some authors have assigned an asterisk to the symbol for a "reduced" (i.e., normalized) quantity; for example Hake [HAK67[b]] defines a reduced Ginzburg-Landau parameter according to: $\kappa_i^*(t) \equiv \kappa_i(t)/\kappa_i(t=1) = \kappa_i(t)/\kappa_{GL}$. An asterisk has also been used to represent the nonparamagnetic limit (i.e., $\alpha \to 0$, $\lambda_{so} \to \infty$) of a fully dressed quantity which itself carries a bare symbol (whether or not the paramagnetic limitation is influenced by spin-orbit scattering). Confusion has thereby been created when the reduced form of the nonparamagnetically limited quantity is needed. Hake [HAK67[b]] dealt with this by *omitting* the asterisk, as in $h_{c2} \equiv H_{c2}^*/H_{c20}^*$. Brand [BRA72, BRA75] has assigned an asterisk to the nonparamagnetic limit of any quantity (e.g., H_{c2}^*, $\kappa_1^*(t)$) and occasionally [BRA75] used a lowercase symbol to represent a reduced quantity; thus $h^* \equiv H/H_{c2}^*$, $h \equiv H/H_{c2}$, $k \equiv \kappa/\kappa_{GL}$, etc. Suenaga and Ralls [SUE69] have used a superscript " + " to signify the inclusion of only Maki [MAK64, MAK66] paramagnetic limitation. Saint-James *et al.* [SAI69] alone adopt what is perhaps the most rational approach of assigning no asterisk to the nonparamagnetic GLAG-type quantity, one asterisk to a paramagnetically-limited quantity and a double asterisk when, in addition to paramagnetic effects, spin-orbit scattering is operative. No symbol is assigned to a reduced quantity, which is simply expressed as a quotient. Tempting as it is to follow this lead, it was deemed appropriate that this review should adhere as closely as possible to the literature consensus, while at the same time avoiding one of the pitfalls referred to above. The scheme adopted is as follows:

(a) *No Asterisk*: Pauli paramagnetic limited with or without spin-orbit scattering.
(b) *Asterisk*: Nonparamagnetic limit of the above, e.g.,

$$\kappa^*(t) \equiv \begin{cases} \kappa(t, \beta = 0) & \text{Maki} \\ \kappa(t, \alpha = 0, \lambda_{so}) & \\ \kappa(t, \alpha, \lambda_{so} \to \infty) & \text{WHH} \end{cases}$$

(c) *Reduced Quantities*: These are designated by lowercase symbols, asterisked or otherwise according to (a) and (b) above, e.g.,

$$h \equiv H/H_{c2}, \qquad h^* \equiv H/H_{c2}^*, \qquad k^* \equiv \kappa^*(t)/\kappa_{GL}$$

except that the Maki-WHH definition

$$\left.\begin{matrix} h^*(t, \lambda) \\ h^*(t, \beta) \\ h^*(t, \alpha, \lambda) \end{matrix}\right\} \equiv H_{c2}/(-dH_{c2}/dt)_{t=1} \quad \text{(see above table)}$$

is retained.

normal-state (hence α_n) quantities: (i) The superconductive-state representation is obtained by combining the above definition with (14-29b) for H_{c20}^* and (15-4) for H_{p0}, leading to

$$\alpha_s = 5.23 \times 10^{-5}(-dH_{c2}/dT)_{T_c} \tag{15-11}$$

(ii) For the normal-state representation three alternative forms in terms of the usual atomic constants and physical measurables are possible, *viz.*,

$$\alpha_n = 3\mu_B c/ev_F^2\tau_{tr} \tag{15-12^4}$$

$$\alpha_n = 3\hbar/2mv_F^2\tau_{tr} \tag{15-13}$$

$$\alpha_n = 2.37\rho_n\gamma_v \tag{15-14}$$

[4]In defining α_n as stated, MAKI [MAK66] had adopted the $c = 1$ convention.

MAKI's relationships for the temperature-dependent paramagnetically limited Ginzburg–Landau parameters in the dirty limit, $\kappa_1(t, \alpha)$, presented initially in [MAK64[a]] and in their final corrected form in [MAK66], are

$$\kappa_1(t,\alpha)/\kappa_{GL}^d = \begin{cases} 1.20\dfrac{1}{\sqrt{1+\alpha^2}}\left[1-1.06t^2\left(\dfrac{1-2\alpha^2}{1+\alpha^2}\right)\right] & t \ll 1 \quad (15\text{-}15a) \\ 1+(0.119-0.346\alpha^2)(1-t) & t \cong 1 \quad (15\text{-}15b) \end{cases}$$

$$\kappa_2(t,\alpha)/\kappa_{GL}^d = \begin{cases} 1.20\dfrac{\sqrt{1-\alpha^2}}{1+\alpha^2}\left[1-2.11t^2\left(\dfrac{1-14\alpha^2+\alpha^4}{1-\alpha^4}\right)\right] & t \ll 1 \quad (15\text{-}16a) \\ 1+(0.105-0.691\alpha^2)(1-t) & t \cong 1 \quad (15\text{-}16b) \end{cases}$$

It is interesting to compare the forms of these equations with those of (14-13) and (14-14) for the nonparamagnetic $\kappa_1(t)/\kappa_{GL}^d$ and $\kappa_2(t)/\kappa_{GL}^d$, respectively, whose temperature-dependence curves were presented in Figure 14-5(a). A comparable set of curves for MAKI's simple paramagnetically limited $\kappa_i(\alpha, t)/\kappa_{GL}^d$s is not provided here.

15.5.2 Spin-Orbit Scattering (SOS)

In the Maki model SOS is introduced *via* a combined parameter

$$\beta_M^2 \equiv \alpha^2(3\Delta_{00}/\hbar)\tau_{so} \qquad (15\text{-}17)^5$$

where Δ_{00} is the BCS half-energy-gap and τ_{so} is the spin-orbit relaxation time. This, by applying (15-10) to one of the αs and (15-12) to the other, and after expressing Δ_{00} in terms of the Clogston relationship, (15-3b), can be arranged into

$$\beta_M^2 = \frac{18\mu_B^2 c}{v_F^2 e\hbar}\left(\frac{\tau_{so}}{\tau_{tr}}\right)H_{c20}^* \qquad (15\text{-}18)$$

a convenient form for comparing the spin-orbit and transport relaxation times.

[5] In defining β_M^2 as stated, MAKI [MAK66] had adopted the $\hbar = 1$ convention.

15.5.3 The Maki Mixed State $\kappa_i(t)$ and $H_{c2}(t)$ Relationships

(a) The Full Maki Mixed-State κ_i-Parameter. Expressions for the Ginzburg–Landau–Maki parameters descriptive of the effects of both PPL (via α) and its relief through the medium of mixed-state SOS (characterized by τ_{so}) were developed by MAKI. The asymptotic forms of these new relationships, as presented by MAKI [MAK66] for the temperature limits $t \ll 1$ and $t \cong 1$, are

$$\kappa_1(t, \beta)/\kappa_{GL}^d = \begin{cases} 1.20 \dfrac{2}{1+\sqrt{1+\beta^2}}\left(1 - 1.06\dfrac{t^2}{\sqrt{1+\beta^2}}\right) & t \ll 1 \quad (15\text{-}19a) \\ 1 + (0.119 - 0.361\beta^2)(1-t) & t \cong 1 \quad (15\text{-}19b) \end{cases}$$

$$\kappa_2(t, \beta)/\kappa_{GL}^d = \begin{cases} \dfrac{1.20(2 + 2\sqrt{1+\beta^2} + \frac{1}{3}\beta^2)^{1/2}}{\sqrt{1+\beta^2}(1+\sqrt{1+\beta^2})}(1 - \mathrm{O}t^2) & t \ll 1 \quad (15\text{-}20a) \\ 1 + (0.105 - 0.722\beta^2)(1-t) & t \cong 1 \quad (15\text{-}20b) \end{cases}$$

where $\beta^2 \equiv \beta_M^2$.

Sets of curves representing the normalized full Maki parameters have been computed and plotted, as in Figures 15-6(a) and (b), enabling theory to be conveniently compared with the results of experiment; as expected, the curves for $\beta_M = 0$ agree with those depicted in Figure 14-5(a), the nonparamagnetic case. A glance at Figure 15-6(a) shows the manner in which $\kappa_1(t, \beta_M)$, and consequently H_{c2} (*via* a generalization of (14-10)), drops away from the nonparamagnetic curve with increasing α (i.e., increasing β_M), and conversely how a reduction of the SOS relaxation time, τ_{so}, which in turn reduces β_M, permits κ_1 to recover.

(b) The Maki Upper Critical Field. MAKI has also explicitly expressed H_{c2} in terms of H_{c2}^*, μ_B^2, and τ_{so}/τ_{tr}. His equation for the upper critical field is

$$H_{c2}(t) = \frac{2H_{c2}^*(t)}{1 + [1 + (18\mu_B^2 c\tau_{so}/\tau_{tr}v_F^2 e\hbar)H_{c2}^*(t)]^{1/2}} \qquad (15\text{-}21)^6$$

a relationship which, with the aid of the reduced parameters (see Table 15-3)

$$h^*(t) \equiv 0.68 H_{c2}/H_{c20}^* \qquad (15\text{-}22)$$

and

$$h_{c2}(t) \equiv H_{c2}^*/H_{c20}^* \qquad (15\text{-}23)$$

[6]In presenting this equation MAKI [MAK66] had adopted the $c = \hbar = 1$ convention.

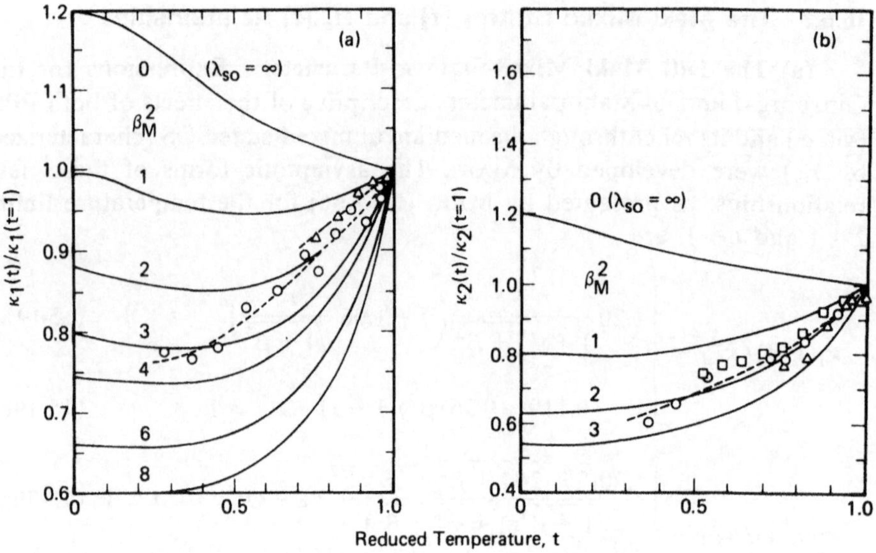

Figure 15-6. Temperature dependences of the Maki paramagnetic spin-orbit parameters $\kappa_1(t, \beta_M)$ and $\kappa_2(t, \beta_M)$ plotted in the formats: (a) $\kappa_1(t)/\kappa_{GL}$ *versus* t, and (b) $\kappa_2(t)/\kappa_{GL}$ *versus* t for various values of the Maki β-parameter defined in this diagram as $\beta_M^2 \equiv \alpha^2/1.78\lambda_{so}$ (but see Section 15.7.1 and footnote). Comparison is made with some experimental results for Ti-Mo(16 at.%). The data portrayed are: (i) the magnetic results of HAKE [HAK67b] (○); (ii) the magnetic results of CAPE [CAP66] (□); (iii) the calorimetric results of BARNES and HAKE [BAR66, BAR67] (△)—after HAKE [HAK67b].

and with the assistance of (15-18), was converted by HAKE [HAK67b] into the following form:

$$h^*(t) = \frac{1.36h_{c2}}{1 + (1 + \beta_M^2 h_{c2})^{1/2}} \qquad (15\text{-}24)^7$$

This equation provides an alternative format—*viz.*, h^* *versus* t parameterized by α and τ_{so}—for the representation of the upper critical field temperature dependence.

The $h^*(t) = 0.680H_{c2}/H_{c20}^*$ conversion used here, combined with the general result

$$H_{c20}^*(\text{dirty}) = 0.68(-dH_{c2}/dt)_{t=1} \qquad (14\text{-}29a)$$

is equivalent to

$$h^*(t) \equiv H_{c2}/(-dH_{c2}/dt)_{t=1} \qquad [15\text{-}8](14\text{-}25)$$

[7] The numerical coefficient in HAKE's version of (15-24) was 1.39, based upon the use of the *clean-limit* rather than the *dirty-limit* expression for H_{c20}^* (i.e., with reference to (14-29a), based upon the use of a prefactor 0.693 instead of 0.68).

which is the definition of $h^*(t)$ adopted by H–W [HEL66, Eq. (27)] and WHH [WER66, Eq. (33)]; it thereby enables the H–W (nonparamagnetic, s-wave scattering), Maki (paramagnetic, spin-orbit-scattering, dirty limit), and WHH (the same) results to be intercompared.

15.5.4 An Application of Maki Theory

CAPE [CAP66], using vibrating sample magnetometry, and BARNES and HAKE [BAR66, BAR67], using low temperature calorimetry in an applied magnetic field, have accumulated data from which the paramagnetic mixed-state properties of Ti–Mo(16 at.%) can be deduced for comparison with the predictions of Maki theory (and, of course, with WHH theory as discussed below). The comparison was carried out by obtaining experimental values of $\kappa_1(t)$ and $\kappa_2(t)$ and plotting them on Maki diagrams as in Figure 15-6. In doing so, the following approaches were used to convert the experimental data into κ-values:

(a) At $t = 1$, the κs are all equal to κ_{GL}^d which, according to (14-17), is $7.49 \times 10^3 \rho_n \gamma_v^{1/2}$. κ_{11} has also been derived from the calorimetric data [BAR66] using (14-20).

(b) The evaluation of $\kappa_1(t)$, according to the defining equation $\kappa_1(t) = H_u(t)/\sqrt{2}H_c(t)$, requires: (i) a knowledge of $H_c(t)$, obtainable either from BCS in the form of (13-48) aided by (8-7) for the temperature dependence, or from calorimetric data if these are available (see Section 14.5.3); and (ii) either calorimetrically or magnetically measured values of $H_u(t)$ (see BARNES and HAKE [BAR66, BAR67] and Section 14.5.3).

(c) $\kappa_2(t)$ is obtainable with the aid of (14-22) if the appropriate calorimetric data are available (see Section 14.5.4), otherwise, or in addition, it may be calculated from the definition (14-11) using the results of a magnetization study.

The experimentally deduced $\kappa_i(t)$ are then plotted against a background of Maki-computed curves, as in Figure 15-6. The parameter β_M for a selected temperature range is obtainable by inspection; a value for the SOS relaxation time may then be extracted using (15-17), after rearrangement of that equation into one or other of the following forms for use with the appropriate experimental data:

$$\tau_{so} = 3.39 \times 10^{-2} \hbar (\beta_M / \rho_n \gamma_v)^2 / k_B T_c \qquad (15\text{-}25)$$

in terms of the fundamental "electronic" measurables, or

$$\tau_{so} = 2.17 \times 10^3 (\beta_M / H_{c20}^*)^2 (\hbar T_c / \mu_B) \qquad (15\text{-}26)$$

that is,

$$\tau_{so} = 4.68 \times 10^3 \beta_M^2 \hbar / [(\partial H_u/\partial T)^2 \mu_B T]_{T=T_c} \tag{15-27}$$

in terms of magnetic quantities. Finally, substitution of the experimental results into (15-18) would yield the ratio of the spin-orbit and transport relaxation times. τ_{so} is not amenable to prior calculation, and has been generally regarded as an experimental result or fitting parameter. Even so, having been obtained by matching experimental results to the family of Maki curves over a limited temperature range, a knowledge of τ_{so} may be put to further use in extrapolating the data. The technique of spin-polarized tunneling, developed during the 1970s by TEDROW and MESERVEY, has recently been applied to a direct determination of τ_{so} in thin films of Al and Ti–V [TED79]. Although the results so far have not agreed well with Maki theory, the scattering properties of many of the important high-field superconductors await examination by the new method.[8]

The resistometric, magnetic, and calorimetric investigations of Ti-Mo(16 at.%) by CAPE [CAP66], HAKE [HAK67^b], and BARNES and HAKE [BAR66, BAR67], whose results in terms of $\kappa_1(t)$ and $\kappa_2(t)$ are plotted in Figure 15-6, also yielded the following data [BAR66], [HAK67^b]:

β_M^2 (from Figure 15-6)	$= 2$
ρ_n	$= 1.03 \times 10^{-4}\ \Omega$ cm
γ_v	$= 7.49 \times 10^3$ erg cm^{-3} K^{-2}
T_{c0} (magnetization)[9]	$= 4.18$ K
T_{c0} (calorimetry)	$= 4.25$ K
H_{c20}^*	$\cong 1 \times 10^5$ Oe
$(\partial H_u/\partial T)_{T_{c0}}$ (calorimetry)	$= -3.41 \times 10^4$ Oe K^{-1}

These, when substituted into (15-25), (15-26), and (15-27) in association with the $T_{c0} = 4.18$ K option, yield, respectively,

$$\tau_{so} = \begin{cases} 2.08 \times 10^{-13} \text{ using (15-25)} \\ 2.06 \times 10^{-13} \text{ using (15-26)} \\ 2.19 \times 10^{-13} \text{ using (15-27)} \end{cases}$$

and when substituted into (15-18) they yield

$$(\tau_{so}/\tau_{tr}) = 79$$

[8] I am grateful to S. Foner for drawing my attention to these studies.
[9] The subscript 0 is used here to indicate *zero-field* quantities.

15.6 The Werthamer, Helfand, and Hohenberg (WHH) Result

15.6.1 Essential Parameters and Formalisms of WHH Theory

Unlike Maki theory, which provided expressions for both $\kappa_i(t)$ and $h^*(t)$, WHH theory concentrated attention on the upper critical field expressed in terms of a reduced field, $\bar{h}$, such that $(\pi^2/4)\bar{h}$ was exactly equal to the $h^*(t) \equiv H_{c2}/(-dH_{c2}/dt)_{t=1}$ common to all the above-mentioned models (see Table 15-3). The approach adopted by WHH was to solve the linearized Gor'kov equations corresponding to H_{c2}: *assuming* (i) that the transition was of second order, and (ii) that the spin-orbit-scattering frequency (parameterized by $\lambda_{so} \propto 1/\tau_{so}$) was much less than the normal scattering frequency;[10] *neglecting* (i) strong coupling and (ii) Fermi-surface anisotropy (i.e., Eilenberger p-wave scattering); *considering* (i) normal-state spin (Pauli) paramagnetism and (ii) mixed-state SOS. An implicit expression for H_{c2} is imbedded in a series of complicated equations, tractable solutions to which become available in the dirty limit, where a relationship between $\bar{h}$ and t, α, and λ_{so} emerges in the following form suitable for numerical solution:

$$\ln\left(\frac{1}{t}\right) = \sum_{n=-\infty}^{n=+\infty}\left[\frac{1}{|2n+1|} - \frac{1}{|2n+1| + \bar{h}/t + \dfrac{(\alpha\bar{h}/t)^2}{|2n+1| + (\bar{h}+\lambda_{so})/t}}\right] \tag{15-28}$$

(15-28)[11]

A glance at the last term of (15-28) reveals the interesting result that, from the standpoint of the algebraic relationship between $\bar{h}$ and t, $\alpha \to 0$ (the normal-state property) yields the same result as $\alpha \neq 0$, $\lambda_{so} \to \infty$ (which affects the mixed state), as expected from the physics of the problem. In other words, in the presence of nonzero α (Pauli limitation) SOS raises the paramagnetic susceptibility of the mixed state, bringing it closer to that of the normal state and offsetting the effect of the limitation. A reduction of normal-state Pauli paramagnetism ($\alpha \to 0$) would, in principle, achieve the same result. The PPL parameter, α, of WHH theory is identical to that of MAKI; the SOS frequency parameter, λ_{so}, is inversely related to the relaxation time, τ_{so}, of both Maki and WHH theory. In short, the essential

[10] The restriction $\lambda_{so} \ll \lambda$ has recently been lifted in a reevaluation of WHH theory by SCHOPOHL and SCHARNBERG [SCH81].

[11] As a result of their recalculation, SCHOPOHL and SCHARNBERG [SCH81] presented a modified version of (15-28).

parameters of WHH theory[12] are (i) the paramagnetic limitation parameter

$$\alpha = 3\hbar/2mv_F^2\tau_{tr} \tag{15-13}$$

(ii) the spin-orbit scattering frequency parameter

$$\lambda_{so} = 2\hbar/3\pi k_B T_c \tau_{so} \qquad (15\text{-}29)^{13}$$

which, together with (15-13), yield

$$(\tau_{so}/\tau_{tr}) = (4mv_F^2/9\pi k_B T_c)(\alpha/\lambda_{so}) \tag{15-30}$$

and (iii) the reduced upper critical field

$$\bar{h}(t) = 2(e/c)(v_F^2\tau_{tr}/6\pi k_B T_c)H_{c2} \qquad (15\text{-}31)^{14}$$

Finally, WERTHAMER, HELFAND, and HOHENBERG elected to plot reduced-field results not in terms of $\bar{h}(t)$ but rather in terms of the function of it introduced above (see (14-25)), *viz.*,

$$h^*(t) \equiv \frac{H_{c2}}{(-dH_{c2}/dt)_{t=1}} \equiv \frac{\bar{h}}{(-d\bar{h}/dt)_{t=1}} = (\pi^2/4)\bar{h} \qquad (15\text{-}32)^{15}$$

15.6.2 The Order of the Transition at H_{c2}

WERTHAMER, HELFAND, and HOHENBERG briefly discussed the thermodynamic order of the transition at H_{c2}. As illustrated in Figure 15-1

[12]WHH [WER66], in presenting equations (15-13) to (15-32), adopted the $\hbar = c = k_B = 1$ convention.

[13]As pointed out by FETTER and HOHENBERG [FET69] the prefactor 2 was omitted in the original WHH [WER66] expression for λ_{so}. I am grateful to Dr. Hohenberg for drawing my attention to this error which has propagated into HAKE [HAK67[b]] (and was subsequently corrected by him (footnote in FETTER *et al.*, above)), SAINT-JAMES *et al.* [SAI69, p. 175], BRECHNA [BRE73, p. 183] and, presumably, elsewhere.

[14]SAINT-JAMES *et al.* [SAI69, p. 175] in their rendering of (15-31) have dropped the prefactor 2.

[15]Substitution of experimental data, e.g., for Ti–Mo(16 at.%), into (15-31) and (15-32) as presented above will be found to yield $h^*(0) = 0.46$. Thus, with reference to Figure 14 of HAKE [HAK67[b]] these equations and the experimental literature are in mutual accord. This agreement would not be sustained in the absence of the "2" prefactor in (15-31). The denominator of (15-32) is evaluated by forming: $(-d\bar{h}/dt)_{t=1} = T_c(-d\bar{h}/dT) = (2ev_F^2\tau_{tr}/6\pi k_B c)(-dH_{c2}/dT)_{T=T_c}$, and inserting $(-dH_{c2}/dT)_{T_c}^d = 12ec\rho_n\gamma_v(9\times10^{11}\pi^3 k_B)^{-1}$, [ORL79] upon which $(-d\bar{h}/dt)_{t=1} = 4/\pi^2$, exactly.

Table 15-4. Parameters of the First Application of WHH Theory [WER66]

Alloy	Paramagnetic Limitation α	Spin-Orbit-Scattering λ_{so}
Ti–V(65 at.%)	1.37	0.75
Ti–Nb(44 at.%)	1.22	1.50

(inset), the transition is first-order if the normal state is sufficiently paramagnetic, but as α decreases it eventually becomes second-order below a critical value [WER66], α_c, given by

$$\alpha_c = (\bar{h} + \lambda_{so})/\bar{h} \tag{15-33}$$

In the absence of SOS, $\alpha_c = 1$ in agreement with MAKI [MAK64[a]]. On the other hand, no matter how strong the paramagnetism α, the transition is *always* second-order [WER66] provided

$$\lambda_{so} > \lambda_{so}^{c} = 0.5139 \tag{15-34}$$

which, as the fitting of WHH theory to the experimental results shows, is usually the case (see Tables 15-4 and 15-5).

15.6.3 Early Applications of WHH Theory

Just prior to the appearance of the Maki and WHH theories, COLLING, RALLS, and WULFF [COL66] studied the resistive upper critical fields, H_r, of a series of Ti–Ta alloys, comparing the results with the calculated[16] H_p and H_{c2}^* adjusted to 4.2 K using a $(1 - t^2)$ factor. As a result of this work it was noted that the mixed states of alloys with more than about 70 at.% Ta were not paramagnetically limited—i.e., were limited only by H_{c2}^* itself—a fact that was attributed to the possible operation of the new SOS mechanism yet to be published. Subsequently, in the first direct test of WHH theory, its authors [WER66] compared their predictions with the experimentally measured upper critical field temperature dependences of Ti–V(65 at.%) and Ti–Nb(44 at.%) as measured by KIM and STRNAD [KIM66] and SHAPIRA and NEURINGER [SHA65], respectively. The data were plotted in the format $h^*(t)$ *versus* t, with h^* defined as in (15-32) but with H_{c2} replaced by H_u, the experimental value. The PPL parameter was taken as fixed at $\alpha = 2.37\,\rho_n\gamma_v$, (15-14), while λ_{so} was regarded as an adjustable parameter

[16]According to COLLING *et al.* (with reference to RALLS *et al.* [RAL65]), $H_{c20}^* = 3.24 \times 10^4\,\rho_n\gamma_v T_c$ instead of the $3.06 \times 10^4 \rho_n\gamma_v T_c$ of (14-27).

Table 15-5. Application of WHH Theory to Ti-V, Ti-Nb, and Ti-Ta Alloys [NEU66]

Alloy	Mean Atomic No. $\langle Z \rangle$	T_c K	ρ_n $\mu\Omega$ cm	$(-dH_u/dt)_{t=1}$ kOe	H^*_{c20} (dirty)[a] kOe	H_{u0}[b] kOe	α	λ_{so}	10^{-5} $\langle Z \rangle^4$	τ_{so}[c] 10^{-13} s	$H^*_{c20} - H_{u0}$ kOe
Ti-V(58 at.%)	22.6	7.52	59	221 ± 8	150 ± 5	107	1.56	0.7	3	3	43
Ti-Nb(44 at.%)	30.4	8.99	53	227 ± 16	154 ± 11	143	1.34	4.5	9	0.4	11
Ti-Ta(52 at.%)	48.5	7.86	51	194 ± 5	132 ± 3	138	1.31	∞	55	0	−6

[a] From Eq. (14-29a).
[b] From published curve.
[c] ×2 correction (see Eq. (15-29) and associated footnote 13) has been applied to the Neuringer-Shapira data [NEU66].

chosen to yield what seemed to be the best fit to the data. The results of this procedure are summarized in Table 15-4.

In a second test of the theory, intended to examine the effect of atomic number Z on the spin-orbit scattering rate, NEURINGER and SHAPIRA [NEU66] investigated the H_u temperature dependences of alloys of Ti with the elements $^{51}_{23}V$, $^{93}_{41}Nb$, and $^{181}_{73}Ta$. The alloys studied were Ti–V(58 at.%), Ti–Nb(44 at.%), and Ti–Ta(52 at.%); the determination of α and λ_{so} proceeded as before and yielded the results summarized in Table 15-5.

15.6.4 Experimental Spin-Orbit Relaxation Time, τ_{so}

Assuming a normal-scattering-relaxation time for alloys such as the above of about 10^{-15} s (see, e.g., [HAK67[b]]) relaxation-time ratios, τ_{so}/τ_{tr}, when obtainable are typically about 100. This is a reasonable value, and one that is in agreement both with the Maki-generated result of Section 15.5.4 and with the comments of WHH concerning the results of their analyses of the KIM–STRNAD [KIM66] and SHAPIRA–NEURINGER [SHA65] data.

15.6.5 Influence of Atomic Number on the Spin-Orbit-Scattering Frequency, τ_{so}

Table 15-5, which summarizes the results of NEURINGER and SHAPIRA [NEU66], shows that for the three alloys studied, which had comparable values (within ~15%) of T_c, ρ_n, H^*_{c20}, and α, a rapid variation of λ_{so} with $\langle Z \rangle$ was encountered in accord with an earlier prediction by ABRIKOSOV and GOR'KOV [ABR62]. A Z^4 proportionality had been suggested by NEURINGER and SHAPIRA [NEU66] and for this reason λ_{so} and $\langle Z \rangle^4$ are juxtaposed in Table 15-5, the final two columns of which compare the λ_{so}-calculated spin-orbit relaxation times with a zero-K Pauli-limitation term.

SUENAGA and RALLS [SUE69] have, in a sense, extended the work of NEURINGER and SHAPIRA [NEU66] by studying the influence of Ta additions on the upper critical fields of a wide range of Ti–Nb alloys and discussing the results in terms of the influence of Ta on the supposed $\langle Z \rangle^4$-dependent SOS frequency parameter, λ_{so}. Particular attention was given to the 64 at.% Ti series, which, with 36 at.% total group-V solute content, lies within the technically-important composition range. It was noted that Ti(64 at.%)–Nb–Ta alloys with more than 12 at.% Ta were not Pauli limited, suggesting at the time that, with the increased $\langle Z \rangle$, λ_{so} was large enough to produce compensation in the mixed state for the normal-state paramagnetic susceptibility. On the other hand, as [Ta] decreased below about 12 at.%, λ_{so} dropped from 8 to 0.6 suggesting that with reduction in $\langle Z \rangle$ the SOS was becoming less and less effective.

Referring again to Table 15-5, we note that Ti–Ta(52 at.%) experienced no PPL, thus requiring that $\lambda_{so} \cong \infty$ in order to fit the data in the presence of a nonadjustable $\alpha = 1.31$. As indicated in Section 15.6.1, such a high SOS rate would be outside the approximation of WHH, who under similar circumstances would have regarded α as adjustable and put it equal to zero. In the absence of additional confirmation, either theoretical or experimental, it could well be that the correlations demonstrated by Table 15-5, and referred to in the discussion of SUENAGA's study of Ti–Nb–Ta, are fortuitous and that additional or alternative mechanisms may be needed to explain the reduction of PPL with change of solute species and, in particular, its complete absence in Ti–Ta [NEU66] and Ti-rich Ti–Nb–Ta [SUE69]. For further treatment of this topic, with particular reference to an absence of PPL in Nb_3Sn attributable to electron-phonon interaction, the work of ORLANDO *et al.* [ORL79, ORL81], [BEA82] (see below) is recommended. It has been pointed out by those authors that with the introduction of electron-phonon interaction, which increases the Clogston limiting field H_{p0} above its BCS value and thereby raises H_{u0} (see Figure 15.1), less spin-orbit scattering is needed to account for what then becomes a smaller difference between H^*_{c20} and H_{u0}.

15.7 Application of the Coupled Results of MAKI and WHH

15.7.1 Interrelationships Between the Maki and WHH Theories

The Maki and WHH theories are coupled through (i) identical values for the PPL parameter, α, and (ii) the inversely proportional SOS terms,

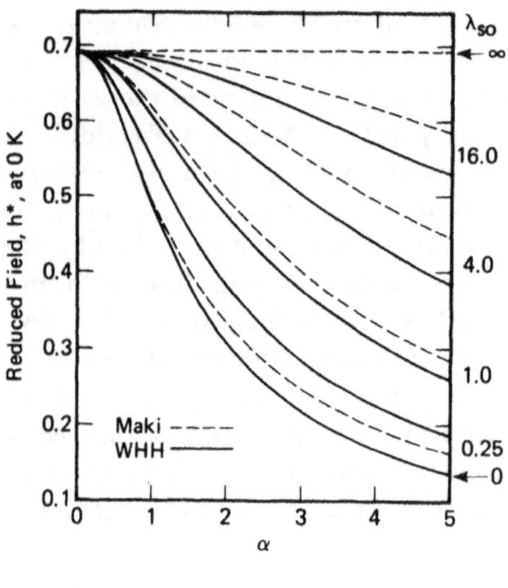

Figure 15-7. Reduced upper critical field at zero K in the dirty limit *versus* the Maki paramagnetic limitation parameter, α (calculated according to (15-14)), for various values of the spin-orbit-scattering frequency parameter λ_{so} according to the theories of Maki and WHH—after HAKE [HAK67b].

τ_{so} (MAKI and WHH) and λ_{so} (WHH), (15-29). On the other hand there is little formal similarity between the expressions offered by those authors for the dirty-limit upper critical fields, (15-21, -24) (MAKI) and (15-31, -32) (WHH). Figure 15-7, from the work of HAKE [HAK67[b]] (see Figure 1 of [HAK67[a]] for a wider range of α values), compares the responses of the reduced critical fields, at zero K, to variations of α and λ_{so}. In addition to revealing excellent numerical agreement between the two theories, the figure graphically demonstrates the manner in which the rapid decrease in $h^*(0)$ with increasing α is counteracted by an increasing λ_{so}.

Further coupling between the theories is provided by the Maki combined parameter $\beta_M^2 = \alpha^2(3\Delta_{00}/\hbar)\tau_{so}$, which, when inserted into (15-29) for λ_{so}, yields[17]

$$\beta_M^2 = 1.11\frac{\alpha^2}{\lambda_{so}} \tag{15-35}$$

WHH theory does not provide expressions for the $\kappa_i(t)$ needed for a full description of the mixed state. But these parameters are available from Maki theory in terms of β_M (see e.g., Figure 15-6), which as (15-35) shows is directly proportional to α^2/λ_{so}, the ratio of the nonadjustable and adjustable parameters of WHH theory. The parameters of the two theories are

MAKI	WHH
$h^*(t, \beta_M)$	$h^*(t, \alpha, \lambda_{so})$
$\kappa_i(t, \beta_M)/\kappa_{GL}$	

where $\beta_M = \beta_M(\alpha, \tau_{so})$, and their region of validity is defined by $\xi_0/l \gg 1$ (dirty limit) and $\tau_{tr} \ll \tau_{so} \ll \hbar/\Delta_{00}$, or in terms of λ_{so} (reciprocally related to τ_{so} as in (15-29)) by the inequality

$$7.3 \times 10^{-21}[\rho_n n_e^{4/3}(S/S_F)^2](\gamma_v T_c)^{-1} \gg \lambda_{so} \gg 0.4 \tag{15-36}$$

This is not a severe restriction and is one which, moreover, practically ensures a second-order transition at H_{c2}.

In utilizing MAKI–WHH theory [BRA72], the limit of small fields is described in terms of $\beta_M^2 = 0$ (MAKI) and $\alpha = 0$ or $\lambda_{so} \to \infty$ (algebraically, WHH); while in the paramagnetic region near H_{c2} ($h \to 1$), the mixed state is described in terms of $H_{c2}(t, \beta_M^2)$, $\kappa_i(t, \beta_M^2)$ (MAKI) and $H_{c2}(t, \alpha, \lambda_{so})$ (WHH). An example of this approach is given in Section 21.8 in connection with a discussion of pinning-force scaling in the paramagnetic mixed state.

[17] CODY (in BRECHNA [BRE73, p. 183]) finds $\beta^2 = \alpha^2/1.78\,\lambda_{so}$, from the use of $\Delta_{00} = 1.76\,k_BT_c$, instead of the 1.75 k_BT_c used here, in association with the use of $\lambda_{so} = \hbar/3\pi k_BT_c\tau_{so}$ instead of twice that value (see (15-29) and also the associated footnote 13).

15.7.2 Applications of the Coupled Maki–WHH Theories

Both the Maki and WHH theories are clearly dealing with the same set of physical phenomena. WHH confined their calculations to the reduced critical field $h^*(t, \alpha, \lambda_{so})$; in that MAKI extended his to the development of an expression for a similar reduced field, symbolized by $h^*(t, \beta_M)$, a common ground has been established upon which the two theories can be intercompared. HAKE [HAK67[b]] has presented such an intercomparison in the form of a plot of $h^*(t = 0)$ *versus* α as in Figure 15-7. Reasonably close agreement between the results of the two approaches is evident. It is, therefore, adequate in an experimental test of the now "Maki-WHH" theory of the paramagnetic mixed state to compare the experimental with the computed temperature dependences of *either* the κ_i *or* the h^* parameters.

(a) The Experiments of Barnes and Hake. Using both magnetic [CAP66], [HAK67[b]] and calorimetric [BAR67] techniques, HAKE has measured the mixed-state properties of Ti-V(22.5 and 25 at.%), Ti-Mo(16 at.%), and Ti_{30}-V_{60}-Cr_{10}, and superposed the results on plots of $\kappa_i(t)/\kappa^d_{GL}$ *versus* t and h^* *versus* t, thereby accumulating evidence for spin-orbit effects and acquiring numerical values for the parameters. Some

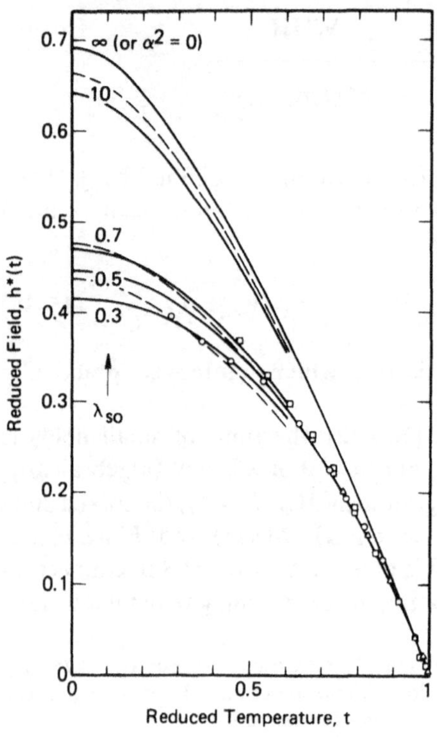

Figure 15-8. Temperature dependences of the reduced upper critical field $h^*(t) = H_u(t)/(-dH_u/dt)_{t=1}$ according to Maki theory (——) and WHH theory (- - -) for $\alpha^2 = 3.3$ and various values (as indicated) of λ_{so}, the spin-orbit-scattering frequency parameter. ($\lambda_{so} = \infty$, $\alpha^2 = 3.3$) is of course the "same" as (λ_{so} = finite, $\alpha^2 = 0$). In the region where dh^*/dt is positive, $h^*(t)$ is a supercooling field. Comparison is made with some experimental results for Ti-Mo(16 at.%). The data portrayed are: (i) the magnetic results of HAKE [HAK67[b]] (○); the magnetic results of CAPE [CAP66] (□); the calorimetric results of BARNES and HAKE [BAR66, BAR67] (△)—after HAKE [HAK67[b]].

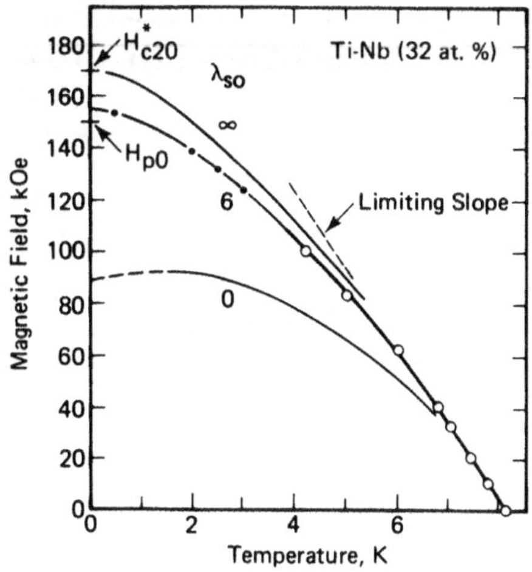

Figure 15-9. Measured and calculated upper critical fields for Ti-Nb(32 at.%). The experimental results are indicated by the open circles. The line through the closed circles, calculated from WHH theory with $\alpha = 1.56$ and $\lambda_{so} = 6$ extrapolates the experimental results into an otherwise inaccessible regime. H_{p0} is the zero-temperature Clogston field; H^*_{c20}, the GLAG-calculated nonparamagnetically limited field, terminates the WHH ($\lambda_{so} \to \infty$) curve; the lowest curve, also calculated using WHH, represents the effect of full paramagnetic limitation ($\alpha = 1.56$, $\lambda_{so} = 0$) —after BRAND [BRA72].

typical results, those for Ti-Mo(16 at.%), are depicted in Figures 15-6 and 15-8.

(b) The Experiments of Brand. BRAND [BRA72] has carried out a Maki-WHH analysis of the results of a critical field temperature dependence study of the alloy systems Ti-V(36, 70, and 82 at.%) and Ti-Nb(32, 55, 62, and 83 at.%) in order to acquire values for the temperature dependences of the essential mixed-state quantities H_r (measured), H^*_{c2} (calculated using WHH with $\lambda_{so} \to \infty$), and $\kappa_2(t)$ and $\kappa^*_2(t)$ (deduced from MAKI with input from WHH),[18] for subsequent use in examining the applicability of several variants of elastic pinning force theory to the paramagnetic mixed state (to be discussed in Section 21.8). The procedure used is outlined below with reference to the data for Ti-Nb(32 at.%), an alloy which is not only of technical importance but also one whose upper critical field temperature dependence provides an excellent example of the essential features of SOS-relieved PPL. Figure 15-9 illustrates the critical field behavior, and Table 15-6 lists some of the essential physical properties.

The first step in analyzing the experimental results is to compute H_{p0}, H^*_{c20}, and α from the superconductive- or normal-state results. After inserting the computed $\alpha_s(= 1.56)$ into the WHH expression (15-28), BRAND varied λ_{so} until the best computed fit to the H_u *versus* T data was obtained (with $\lambda_{so} = 6$) [BRA72]. It was then possible, as in Figure 15-9, to project $H_r(t, \alpha, \lambda_{so}) = H_r(t, 1.56, 6)$ into a range of temperatures lower than those available at the time of the experiment and, in particular, to obtain a zero-K value for the upper critical field. For this particular alloy it turned out

[18] $\kappa^*_2(t) \equiv \kappa_2(t, \beta^2_M)$ with $\beta^2_M = 0$, see Table 15-3.

Table 15-6. Some Essential Experimental and Deduced Properties of the Paramagnetically-Limited Type-II Superconductor, Ti-Nb(32 at.%)—After Brand [BRA72]

Property	Equation No.	Symbol	Unit	Value
A. *Experimental*				
Normal-state residual resistivity		ρ_n	$\mu\Omega$ cm	102
Superconducting transition temperature		T_c	K	8.14
Limiting upper critical field temperature dependence		$(-dH_u/dT)_{T_c}$	kOe K^{-1}	30.1
Electronic specific heat coefficient		γ_v	10^4 erg cm^{-3} K^{-2}	1.0
B. *Deduced*				
Clogston zero-K paramagnetic limit	(15-4)	H_{p0}	kOe	150
Nonparamagnetic dirty-limit upper critical field (normal-state calculated)	(14-27)	H^{*n}_{c20} (dirty)	kOe	254
The same (superconductive-state calculated)	(14-29a)	H^{*s}_{c20} (dirty)	kOe	166
Paramagnetic limitation parameter (normal-state calculated)	(15-14)	α_n	—	2.42
The same (superconductive-state calculated)	(15-10)	α_s	—	1.57

that H_{r0} was not lower than both H_{p0} and H^*_{c20} (cf. Figures 15-1, 15-2, and 17-2). To show the extent of the PPL–SOS effects it is customary, as in Figure 15-9, to present, on the same plot as the experimental and fitted results, curves of $H_{c2}(t, \alpha, \lambda_{so} = 0)$ and $H_{c2}(t, \alpha, \lambda_{so} = \infty) \equiv H^*_{c2}(t)$ which represent, respectively, the fully limited and nonparamagnetically limited[19] upper critical fields. $H^*_{c2}(t)$ was needed to characterize the nonparamagnetic mixed-state behavior. For complete characterization of the superconductive properties $\kappa^*_2(t)$ and $\kappa_2(t)$ are required. These are obtained from the Maki-calculated functions (Figure 15-6(b)) by setting $\beta^2_M = 0$ and $\beta^2_M = 1.11\,\alpha^2_s/\lambda_{so} = 0.45$, respectively, with the λ_{so} obtained from the above fitting procedure.

The paramagnetic mixed state is, thus, fully characterized within the framework of the PPL–SOS model. With the aid of the deduced parameters,

[19]Of course $\lambda_{so} = \infty$ violates the restrictions placed on WHH. Although $\lambda_{so} = \infty$ is algebraically equivalent to $\alpha = 0$, it is physically preferable to use the latter form and to bracket the fitted curve between those designated $H_{c2}(t, \alpha = 0, \lambda_{so} = 0)$ and $H_{c2}(t, \alpha, \lambda_{so} = 0)$ as in the original WHH treatment of the Ti–V(65 at.%) and Ti–Nb(44 at.%) data referred to above (see Figures 3 and 4 of reference [WER66]).

BRAND [BRA72] was able to demonstrate the validity of a pair of relationships for the bulk pinning force, F_p, valid for (i) low fields and (ii) fields near H_{c2} based on, respectively,

$$F_p \propto H_{c2}^{5/2} h^{2/3}(1-h)/\kappa_2^3(t) \qquad h \cong 1 \qquad (15\text{-}37)$$

$$F_p \propto H_{c2}^{*5/2} h^{*2/3}(1-h^*)/\kappa_2^{*3}(t) \qquad h^* \ll 1 \qquad (15\text{-}38)$$

in which H_{c2} is the actual measured H_u and the H_{c2}^*, $\kappa_2(t)$, and $\kappa_2^*(t)$ have been deduced from it in the manner just described.

15.8 The Breakdown of Simple WHH Theory—Consideration of Many-Body Effects and Spin-Orbit-Scattering Frequency

As a result of PPL, the "measured" zero-K upper critical field is expected to fall below the lower of H_{p0} and H_{c20}^* ((15-6)), after which the effect of SOS may be strong enough to place H_{u0} between these two fields. This was shown to be the case for Ti–Nb(32 at.%), whose $H_u(t)$ was characterized by $\alpha = 1.56$, $\lambda_{so} = 6$ (Figure 15-9).

But in the experiments on Ti–Ta(52 at.%) referred to above, NEURINGER and SHAPIRA [NEU66] found that the measured data actually lay *above* the $H_{c2}(t)$ curve for $\alpha = 0$, i.e., above $H_{c2}^*(t)$. In other words that alloy, although known to be paramagnetic in the normal state, was not paramagnetically limited. Subsequently, comparable results were obtained by MISSELL, OLIVEIRA, and SHAPIRA [MIS75] in studies of arc-melted and vacuum anealed Ti–Nb(11 at.%), whose reduced critical field temperature dependence was found to lie slightly above the WHH $h^*(t, \alpha = 0, \lambda_{so} = 0)$. As before, for finite α it was impossible to fit the data using any value of λ_{so}, although the closest fit was with $\lambda_{so} = \infty$, in spite of the fact that the dirty-limit requirement of the final WHH relationship should have been well satisfied. Finally, in a study of A15 superconductors, ORLANDO *et al.* [ORL79] noted the complete absence of PPL in Nb_3Sn. The full cancellation of the limiting effect of Pauli paramagnetism in this material again would have required the operation of an "infinite" SOS frequency were it not for the introduction of electron–phonon density-of-states renormalization, to be discussed below.

Before proceeding with the central theme of this section, the reader is reminded that nothing in the foregoing discussion has suggested that H_u might ever *exceed* H_{c2}^*. Consequently, provided that the observation of $H_u > H_{c2}^*$ referred to above was not just the result of an inappropriately chosen transition criterion [NEU66], an additional mechanism other than simple SOS must be operating in some alloy systems. Possible explanations for superenhanced H_us have been advanced from time to time in terms of the effects on the upper critical field of: (i) strong-coupling corrections [HAN66], [NEU66], [RAI73], [DOM78]; (ii) impurity scattering and short mean-free-paths [NEU66], [DOM78]; (iii) anisotropy of the electron–phonon interaction [DOM78]; (iv) microinhomogeneities and anomalous flux-pinning effects, especially in amorphous alloys [POO83[a]]. The discussion to follow ignores all these possibilities and assumes that H_u has an absolute upper limit of $H_{c2}(\lambda_{so} = \infty)$, i.e., H_{c2}^*.

15.8.1 The Influence of Many-Body Interaction on the Clogston Limiting Field

The manner in which the effect of many-body interactions upsets the simple Clogston–BCS energy balance, and with it the Pauli paramagnetically limiting upper critical field, H_{p0}, ((15-2) and (15-3a)) has been considered by ORLANDO and BEASLEY [ORL79, ORL81], [BEA82] following and improving on an early but long-neglected suggestion by CLOGSTON himself [CLO62[a]]. The two types of many-body effect to be considered are: (i) the electron–electron interaction, $n(E_F)\phi$, parameterized by α^e (see Section 7.3.1) and the electron–phonon interaction $n(E_F)V$, parameterized by λ (see Section 7.3.2).

It is well known that although λ does not enter into the density-of-states renormalization for Pauli paramagnetic susceptibility, α^e, the electron–electron interaction does so according to (7-3), yielding what might be termed a "susceptibility density-of-states", $n^\chi(E_F)$, given by

$$n^\chi(E_F) = n(E_F)/(1 - \alpha^e) \tag{15-39}$$

where the denominator is known as the "Stoner factor." As indicated in Section 7.3.1, this correction may be negligible for early transition elements whose densities-of-states lie in the shallow regions of the universal $n(E_F)$ *versus* e/a curve, but may be quite significant for others, especially towards the Pd end of the transition-metal block.

Of general importance as a correction to the electronic specific heat coefficient is the electron–phonon interaction parameter, λ. Following (7-5), the factor $(1 + \lambda)$ renormalizes the density-of-states entering the right-hand side of (15-2) (which contains H_c, hence γ—(13-48)) according to

$$n^\gamma(E_F) = n(E_F)(1 + \lambda) \tag{15-40a}$$

or

$$n^\gamma(E_F) = n(E_F)(1 + \lambda + \lambda_s) \tag{15-40b}$$

when spin–spin interaction is sufficiently strong that the effects of spin fluctuations, parameterized by λ_s, cannot be neglected. Also as a result of the electron–phonon interaction the Δ_{00} of (15-3a) is enhanced by a "strong-coupling" factor (see Section 8.5.1, in particular (8-21)), designated η by RAINER *et al.* [RAI73] and η_{H_c} by ORLANDO *et al.* [ORL79].

Bringing together all these considerations, (15-2) or (15-3a) can be recast in the form

$$(\mu_B H_{p0})^2 n(E_F)/(1-\alpha^e) = \tfrac{1}{2}\Delta_{00}^2 \eta_{H_c}^2 n(E_F)(1+\lambda+\lambda_s) \qquad (15\text{-}41)$$

leading to the following correction to the Clogston–BCS limiting field, now represented by H_{p0}^{CL}:

$$H_{p0} = \eta_{H_c}(1-\alpha^e)^{1/2}(1+\lambda+\lambda_s)^{1/2} H_{p0}^{\mathrm{CL}} \qquad \text{(first order)} \qquad (15\text{-}42)$$

The application of this result to Nb_3Sn by ORLANDO *et al.* [ORL79] was found to raise H_{p0} sufficiently to allow H_{c2}^{WHH} to equal the experimental result for $\tau_{so} \cong \tau$, a considerable improvement over the $\tau_{so} = 0$ that would have been needed to explain the apparent absence of paramagnetic limitation in the face of the previously much lower value of H_{p0}, *viz.*, H_{p0}^{CL}.

Subsequently, BEASLEY pointed out [ORL81], [BEA82] that (15-42), which is based on thermodynamic arguments, only applies when the s/n transition is of *first order*. In practical materials, the transition is always *second order* (see Section 15.6.2), in which case the appropriate theory yields

$$H_{p0} = (1-\alpha^e)(1+\lambda+\lambda_s) H_{p0}^{\mathrm{CL}} \qquad \text{(second order)} \qquad (15\text{-}43)$$

The form that η_{H_c} takes in the second-order expression has not been worked out, but being a small correction in some cases (see Section 15.2.1) it has been dropped from (15-43). The influences of strong electron–electron interaction and the accompanying spin fluctuations on the superconductivity of several elements and compounds are discussed by ORLANDO and BEASLEY [ORL81].

For transition-metal alloy superconductors, the effect of electron–phonon renormalization is generally very significant. Semiempirical and empirical methods of acquiring values of λ have been considered in Sections 7.3 and 8.7, and data for several unalloyed transition metals and members of a model Ti–TM alloy series have been provided in Tables 7-2 and 8-6, respectively. Some typical values of λ are: 0.38 (Ti), 0.82 (Nb), and 0.39 (Ti–Mo(25 at.%)). The effect of electron–electron interaction, parameterized by $\alpha^e = n(E_F)\phi$, with $0.03 < \phi < 0.30$ eV for transition elements, has been discussed in Section 7.3.1. Consider for example Nb, for which a semiempirical value for the electron–electron interaction parameter, *viz.*,

$\alpha^e = 0.2\text{–}0.3$, is available from [ORL81].[20] The presence of a term as large as this in the Stoner factor significantly offsets the effect of the $(1+\lambda)$ renormalization. Nevertheless, the product $(1-\alpha^e)(1+\lambda)|_{Nb} \cong 1.3$ still provides a 30% increase in H_{p0}, thereby elevating H_u and considerably reducing the difference between it and H^*_{c2}. In some materials, evidently, the effect is sufficient to reduce this difference to zero.

15.8.2 Spin-Orbit Scattering Frequency

The requirement $\tau_{so} \cong \tau$, adequate to justify the absence of paramagnetic limitation in Nb_3Sn even in the presence of strong electron-phonon coupling, still exceeded the range of validity of the original WHH theory which required $\tau_{so} \gg \tau$ (Section 15.7.1). In response to this problem, SCHOPOHL and SCHARNBERG (SS) reexamined WHH theory, derived an improved expression for the upper critical field, and as a result came to two important conclusions [SCH81]—*viz.*, (i) for a given SOS frequency, $H^{SS}_{c2} > H^{WHH}_{c2}$, thus bringing the upper critical field closer to H_p with all other factors remaining unchanged; (ii) the recalculated theory would retain its validity even when spin-orbit was the only scattering mechanism, for arbitrary values of τ_{so}/τ, and certainly for $\tau_{so} \cong \tau$, thereby providing the final justification for the Orlando-Beasley approaches [ORL79, ORL81].

15.9 Conclusion—Summary of Essential Factors Controlling the Magnitude of the Upper Critical Field

The upper critical field is controlled by the following relationships and considerations:

(a) $H^*_{c20}\,(\text{dirty}) = 3.06 \times 10^4 \rho_n \gamma_v T_c$ (Oe) (14-27)

(b) $H^{CL}_{p0} = 1.84 \times 10^4 T_c$ (Oe) (15-4)

(c) $H^{min}_{c20}(\text{PPL}) = \dfrac{H^*_{c20} H_{p0}}{\sqrt{2H^{*2}_{c20} + H^2_{p0}}}$ (Oe) (15-6)

In other words, H_{c20} will lie beneath the lower of H_{p0} and H^*_{c20} unless relieved by the effect of SOS, in which case it is possible for $H_{p0} < H_{c20}(\alpha, \lambda_{so}) < H^*_{c20}$, for example—the final magnitude of H_{c20} depending on the value of λ_{so}. In the original WHH theory the absence of PPL implied $\lambda_{so} = \infty$; this is not so if the electron-phonon-modified approach is adopted.

[20] Alternatively, with $\phi = 0.05\text{–}0.1$ eV for Nb (see Section 7.3.1) and $n(E_F) = 1.8$ eV atom per spin [ORL81], $\alpha^e = 2n(E_F)\phi = 0.18\text{–}0.36$.

(d) Under electron–phonon (offset by electron–electron) renormalization of the densities-of-states, the paramagnetically limiting field is raised by a factor $(1 - \alpha^e)(1 + \lambda) > 1$, reducing to an appreciable extent the spin-orbit scattering rate required to match theory with experiment in some cases.

(e) A recent recalculation of WHH theory extends the SOS frequency to $\lambda_{so} \cong \lambda_{tr}$ and beyond, thereby allowing full advantage to be taken of the results of paragraph (d).

In the light of these considerations the upper critical field will increase in response to

(1) increase in ρ_n
(2) increase in γ_v, and/or T_c
(3) increase in λ_{so}, the spin-orbit scattering frequency
(4) increase in λ, the electron–phonon interaction parameter
(5) decrease in α^e, the electron–electron interaction parameter

The possibilities for achieving actual increases in H_{c2} through the manipulation of these quantities are discussed in Section 17.3.

Chapter 16

The Critical State

16.1 Reversible and Irreversible Type-II Superconductors

The first experimental observations of the existence of what is now known as the mixed state of type-II superconductors took place almost simultaneously (1934–1935) in Oxford, Leiden, and Kharkov [MEN64]. The observed magnetic irreversibility of these materials precluded at the time a thermodynamic description of their properties, and they were known as "non-ideal" superconductors. The residual magnetization represented by the magnetic irreversibility was reminiscent of magnetic coercivity and led to their being characterized as "hard" superconductors. Workers at the time recognized that the magnetic irreversibility represented the trapping of flux within the bulk of the specimen during field removal, to account for which MENDELSSOHN [MEN35] proposed the existence of a multiply connected net of superconductive filaments. In recent years, artificial sponges exhibiting magnetic hysteresis have been prepared by the impregnation of porous glass [BEA64]. In spite of the fact that MENDELSSOHN very quickly recognized that the flux trapping had more to do more with the metallurgical condition of the superconductor than with some intrinsic property [MEN36], the seeds of confusion between irreversibility and dirtiness in type-II superconductivity had been planted by the early observations.[1] In studies conducted in or about 1956, MENDELSSOHN and colleagues [CAL61] showed that although zone-refined Ta was a type-I superconductor, zone-refined Nb was both type-II and magnetically reversible. Today it is recognized that flux trapping, which gives rise to the properties associated with "irreversible," "non-ideal," or "hard" superconductors, occurs as a result of the

[1] Indeed in 1955 PIPPARD, after completing a series of flux-trapping measurements on Sn–In alloys, concluded that the magnetic irreversibility encountered was a consequence of some fundamental property of the short mean-free-path material rather than metallurgical inhomogeneity [PIP55].

presence of flux-pinning metallurgical artifacts. For the same reason such superconductors were also found to be able to support large transport current densities in strong magnetic fields. In early studies of current-carrying superconductivity in ductile alloys, the existence of a fibrous cold-worked structure lent strong support to the filamentary model [BER62, BER63[a]]. Upon such a foundation was also constructed the "critical state" model of BEAN [BEA62] and LONDON [LON63], in which the entry of flux was supposed to be accompanied by the flow of critically dense current. Such a concept is untenable in a homogeneous solid, the filaments of the sponge model being needed to define current paths within which current can flow at its critical density. But once having become established in this way, the critical state model in one or another of its more recent variants is still extensively employed in describing the responses of superconductors to applied DC and AC fields.

Alloying is a useful technique for raising the upper critical field, which, when expressed in the form $H^*_{c20}(\text{dirty}) = 3.06 \times 10^4\ \rho_n \gamma_v T_c$ ((14-27)), can be seen to respond to the addition of solutes that increase the residual resistivity, ρ_n and, hopefully, both γ_v and T_c. Annealed solid-solution alloys based on Nb may be reversible type-II superconductors. But if the metallurgical consequences of alloying and possibly some subsequent heat treatment include precipitation, or if the sample has been cold worked so as to develop

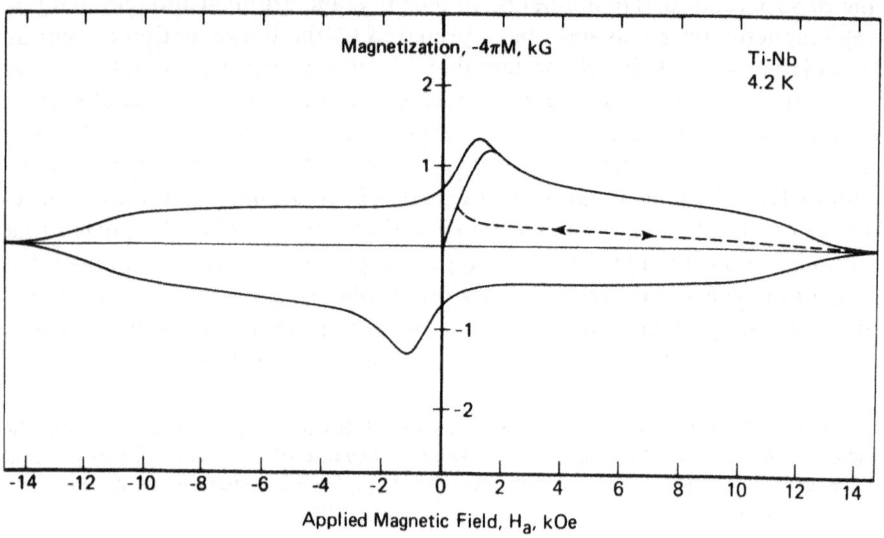

Figure 16-1. Irreversible magnetization loop for a heavily cold-worked type-II superconductor (——) compared with the reversible magnetization curve (- ← → -) obtained from an annealed specimen of the same composition. Measurements performed on Ti–Nb(96 at.%) at 4.2 K—after FIETZ [FIE67].

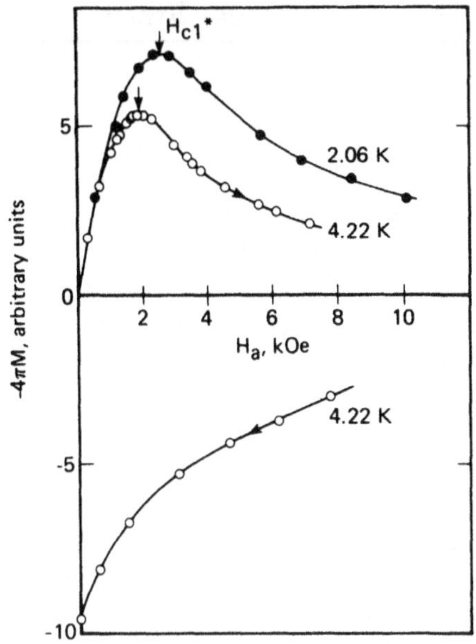

Figure 16-2. Low-field detail of the hysteretic magnetization loop of a sample of Ti–V(58.5 at.%) prepared by alternate cold rolling and vacuum annealing and measured at the two temperatures indicated. The relationship between the "field-at-maximum," H_{c1}^*, and the lower critical field, H_{c1}, was discussed in the original article—after BLAUGHER [BLA65].

a submicroscopic deformation structure, the resulting superconductor is likely to exhibit magnetic hysteresis, i.e., irreversibility.

FIETZ [FIE67] has studied a series of Nb-rich Ti–Nb alloys that are reversible when annealed but irreversible when in the cold-worked condition. Figure 16-1, for Ti–Nb(96 at.%), is an example of this. In both the annealed and the cold worked states the alloy has similar κ_{GL}s (4.2 and 4.4, respectively) and residual resistivities ($\rho_n = 3.5$ and $4.6 \times 10^{-6}\ \Omega$ cm, respectively), but in the cold-worked alloy the presence of lattice defects are responsible for the magnetic hysteresis. Figure 16-2, for some alternately cold-rolled and annealed Ti–V alloys [BLA65], presents a detail of the low-field portion of the magnetic-hysteresis half-cycle.

The influence of precipitates on magnetic irreversibility was noted by HAKE [HAK67^b] in a study of Ti–V(25 at.%), whose magnetization curve passed from reversible to irreversible in response to: (i) the cold rolling and (ii) the precipitation heat treatment of a previously as-cast specimen. Concentrated solid solution alloying itself need not result in magnetic irreversibility: in the same study, HAKE [HAK67^b] demonstrated the existence of complete reversibility in alloys such as annealed Ti–Mo(16 at.%) with a κ_{GL} of 68 and as-cast Ti–V(22.5 at.%) with a κ_{GL} of 100, both of which were referred to as "extreme" (i.e., extremely dirty) type-II superconductors.

16.2 The Critical State

16.2.1 Introduction

Magnetic irreversibility is a result of flux pinning. The role of flux pinning in assuring the existence of a transport current in the presence of an applied transverse magnetic field is the burden of Chapter 21. In the present chapter transport current is not the primary concern; instead attention is focused on the pinning-center-stabilized gradient of the magnetic induction. As the external field is increased, flux flows into the superconductor (then in its "shielding mode") by the overcoming of pinning forces; as it is decreased, flux leaves the interior of the superconductor (then in its "trapping mode") by the same "flux-flow" mechanism—to be discussed in Chapter 18. Thus, if the external field is held steady at some particular value, the flux-density gradient is such that the flux-lattice *remains poised at the point of breaking away* from the restraining forces of the pinning centers usually represented by a bulk pinning force, F_p; consequently it is said to be in the *critical state.*

The critical state is usually discussed in terms of a superconducting cylinder immersed in an applied axial (z direction) magnetic field, H_a. If the material from which the cylinder is made is "reversible," i.e., if its *magnetization* is reversible as exemplified by the dashed curve of Figure 16-1, the field will produce within it a uniform equilibrium value of the induction B_z given by

$$B_z = H_a + 4\pi M_z \tag{16-1}$$

where $M_z \equiv M_{eq}$ is a uniform equilibrium magnetization. On the other hand, if the cylinder has been machined from "irreversible" material, such that its magnetization follows the full curve of Figure 16-1, the induction within it is a function of radial (r) position (although independent of θ, the angle in the transverse plane) and the induction is critically supported by the pinning centers. When dealing with magnetic driving forces in irreversible material, which involve the gradient of the local induction, it is customary to relate, differentially, the local induction, $B(r)$, to some magnetic field, $H(B(r))$, that would uniformly and reversibly provide such an induction in an unpinned sample of the same material. This relationship is provided by a differential equilibrium permeability, μ_{eq}, defined as

$$\mu_{eq} \equiv dB/dH(B) \tag{16-2}$$

The purpose of introducing such a parameter is to enable a local field gradient—the immediate result of the basic force-balance calculation, and to which the Maxwell $\nabla \times \mathbf{H}$ equation can be applied—to be converted into an induction gradient, which experiment has shown exists within the pinned specimen.

The present treatment of critical-state-related properties tends to follow chronological order, although, in anticipation of more recent developments, provision has been made throughout for the inclusion of a surface-boundary condition (see Section 16.7) not present in the early analyses.

16.2.2 Thermodynamic Equilibrium in the Critical State

According to CAMPBELL and EVETTS [CAM72, p. 14] the most satisfactory starting point for a discussion of thermodynamic equilibrium in the mixed state is by way of the Helmholtz free-energy function[2] $F^H = U - TS$ (or $f^H = u - Ts$ on a per-unit-volume basis). Next, returning to the superconducting cylinder immersed in an axially directed applied field, H_a, suppose the induction in it changes by a small amount δB_z, then the change in specific free energy of the cylinder which accompanies this is $\delta f^H = H_a \delta M_z = (1/4\pi) H_a \delta B_z$. It follows that the *equilibrium* value of f^H is such that

$$\frac{df^H}{dB_z} \equiv \frac{H_a}{4\pi} \tag{16-3}$$

A bundle of magnetic flux has a tendency for mutual repulsion, i.e., to expand laterally, a tendency which, incidentally, is responsible for the free-energy term $H^2/8\pi$ already encountered in Sections 13.6 and 15.2. A gradient in the flux density biases this mutual repulsion giving rise to a force in the direction of the gradient. It is the purpose of this section to calculate the force. But before proceeding it is necessary to introduce one other identity connecting the field produced by a fixed number of fluxoids to the cross-sectional area, A, of the bundle, *viz.*,

$$\frac{dB_z}{B_z} \equiv -\frac{dA}{A} \tag{16-4}$$

Next, consider a bundle of unit-length fluxoids; the repulsive pressure

[2]CAMPBELL and EVETTS discussed the applicabilities of the Helmholtz and Gibbs free-energy functions [CAM72]. The Gibbs function, used elsewhere in this book (e.g., Sections 1.7, 8.4, 13.6, 13.8, 15.2) may be used in the present context to describe the thermodynamics of long cylinders in parallel fields with straight flux lines and no transport currents [CAM72, p. 16].

(≡energy/volume) referred to above is then

$$P = -\frac{dF^H}{dA} = -\frac{dAf^H}{dA}$$

$$= -f^H - A\frac{df^H}{dA}$$

which after an application of (16-4) becomes

$$P = -f^H + B_z\frac{df^H}{dB_z} \tag{16-5}$$

The force per unit area, F, on the vortices within the bundle is the negative gradient of this pressure, $-dP/dr$. To obtain it, (16-5) is differentiated and (16-3) is applied to the result [CAM72, p. 18] yielding

$$F = -\frac{dP}{dr}$$

$$= -B_z\frac{1}{4\pi}\frac{dH(B_z)}{dr} \tag{16-6}$$

With the aid of (16-2) this then becomes

$$F = -\frac{1}{4\pi}B_z\frac{1}{\mu_{eq}}\frac{dB_z}{dr} \tag{16-7}$$

In its critical state the flux lattice is just on the point of breaking away from its pins, a condition that is expressed by

$$F_c + F_p = 0 \tag{16-8}$$

which defines a critical force, F_c, and bulk pinning force, F_p, given by

$$-F_c = F_p = \frac{1}{4\pi}B_z\frac{1}{\mu_{eq}}\frac{dB_z}{dr}\bigg|_{\text{critical}} \tag{16-9}$$

16.2.3 The Elementary Pinning Force, f_p

Using an argument similar to that presented above, FRIEDEL *et al.* [FRI63] have shown that the force per unit length acting in the r-direction on a single vortex (or fluxoid, Φ_0) in a nonuniform distribution of vortices is

$$f_\phi = -\frac{\Phi_0}{4\pi}\frac{dH(B_z)}{dr} = -\frac{\Phi_0}{4\pi}\frac{1}{\mu_{eq}}\frac{dB_z}{dr} \tag{16-10}$$

which in the critical state is restrained by an equal and opposite elementary pinning force, f_p, given by

$$f_p = \frac{\Phi_0}{4\pi}\frac{1}{\mu_{eq}}\frac{dB_z}{dr}\bigg|_{\text{critical}} \tag{16-11}$$

This, of course, would revert to (16-9) if a linear summation of elementary pinning forces were permissible. The summation of elementary pinning forces is introduced in Section 21.2.

16.2.4 Introduction of the Maxwellian Supercurrent

The Maxwell curl equation in cylindrical coordinates and c.g.s. units is

$$\nabla \times \mathbf{H} \equiv \begin{vmatrix} \frac{1}{r}\hat{r} & \hat{\theta} & \frac{1}{r}\hat{k} \\ \frac{\partial}{\partial r} & \frac{\partial}{\partial \theta} & \frac{\partial}{\partial z} \\ H_r & rH_\theta & H_z \end{vmatrix} = \left(\frac{4\pi}{10}\right)\mathbf{J} \tag{16-12a}$$

which, for field in the z (i.e., $\hat{k}$) direction, reduces to

$$\frac{dH_z}{dr} = -\left(\frac{4\pi}{10}\right)J_\theta \tag{16-12b}$$

which, in case H_a is constant and the gradient is associated with B_z, can be interpreted to mean that

$$\frac{1}{\mu_{eq}}\frac{dB_z}{dr} = -\left(\frac{4\pi}{10}\right)J_\theta \tag{16-13}$$

where J_θ is a current density. Comparing this with (16-7) and (16-9) in turn, it is possible to imagine that the flux lattice is being driven by a force

$$F = -\frac{dP}{dr} = \frac{B_z J_\theta}{10} \tag{16-14}$$

and restrained under critical conditions by a bulk pinning force

$$F_p = \frac{-B_z J_{\theta c}}{10} \tag{16-15}$$

where $J_{\theta c}$ is a critical current density. With B_z in gauss and $J_{\theta c}$ in A cm^{-2}, the force unit[3] is dyn cm^{-3}. It is important to note at this juncture that the right-hand sides of (16-14) and (16-15) are in the form of Lorentz forces.

Equation (16-19) and (16-15) taken together imply the equivalence of critical flux density gradient and critical current density. Admittedly these equations are equivalent statements of the same phenomenon, and $J_{\theta c}$ is none other than the critical circulating current, present in response to B_z. This same pair of equations also emphasizes the equivalence of flux density gradient and Lorentz force as fluxoid driving mechanisms, thereby establishing a connection between Chapter 18, in which the driving force on the flux lattice is supplied by the flux density gradient, and Chapter 23, in which a *uniform field* is set in motion by a transport-current-controlled Lorentz force.

16.3 Critical State Models

The various models of the critical state are macroscopic descriptions of the electromagnetic properties of superconductors constructed from the Maxwell curl equation, some information or assumptions concerning the field dependences of J_c, and one or two assumptions regarding the microscopic mechanism of magnetic irreversibility. An excellent comparative discussion of magnetic hysteresis within the models of BEAN [BEA64], KIM *et al.* [KIM62, KIM63], DE JONG and BLAISSE [DEJ65], GOEDEMOED *et al.* [GOE67], FIETZ and WEBB [FIE69], CAMPBELL *et al.* [CAM64], SILCOX and ROLLINS [SIL63], and YASUKŌCHI and IRIE [YAS64[a]], [IRI67], has been presented by INHABER [INH73]. The forerunner of all critical state models is that devised in the early 1960s, independently it seems, by BEAN [BEA62] and LONDON [LON63].[4] The current-support mechanism was modelled on

[3]With B_z in gauss and $J_{\theta c}$ in A cm^{-2}, the bare product $B_z J_{\theta c}$ is numerically the same as if it were in N m^{-3}.

[4]As far as timing is concerned, BEAN's paper on magnetization received in October, 1961, clearly preceded LONDON's AC-loss paper which was received in August, 1963.

MENDELSSOHN's earlier sponge or filamentary model of high-field superconductivity [MEN64], in which the magnetic irreversibility was regarded as being a result of flux-trapping within the cells of a sponge-like body, and current was thought to flow along the multiply-connected filamentary paths associated with such a structure. Indeed, experiments intended to test the early theory were performed on sintered powders and Pb-impregnated porous glass [BEA62, BEA64]. As a consequence of the "filamentary" assumption, since no additional statement was necessary with regard to flux pinning, the bulk pinning force function $F_p(t, H_a, H_{c2})$, which plays an essential role in recent critical state formulations, was completely absent from the model. Embellishments of the Bean model that took place during the 1960s and early 1970s incorporated the results of contemporary flux pinning theory as they appeared from time to time. The result was a continuous improvement in the agreement between theory and the results of magnetization studies of various kinds. But the essential macroscopic physical features of all the newer approaches are contained within the Bean model. Because of this, and its simplicity, the Bean model continues to be in popular use even today.

16.4 The Bean Model of the Critical State

16.4.1 Basic Phenomenological Equations of the Model

The Bean model postulates the existence of a critical supercurrent of density J_c whenever flux motion, however small, has taken place, and that three states of such current exist: $+J_c$, 0, $-J_c$, depending on the sense of the e.m.f. that accompanied the last local change of field. The model as proposed was essentially macroscopic and, although it did not dwell on the detailed distribution or origin of the current, admitted that it could be treated either as a property of the walls of the sponge or a consequence of the gradient of Abrikosov flux [BEA64]. Having postulated the existence of current according to this prescription its connection with flux gradient was made via the Maxwell $\nabla \times \mathbf{H}$ equation. The Bean model is thus defined in terms of

$$\left.\begin{aligned} &\nabla \times \mathbf{H} = (4\pi/10)\mathbf{J} \\ \text{within which} \quad &\mathbf{J} = \mathbf{J}_c = \text{constant} \end{aligned}\right\} \tag{16-16}$$

The term "critical state" can be thought of as implying the existence of Bean current-filaments of critical density, or alternatively, of a critical flux-density-gradient, the magnetic force associated with which is exactly

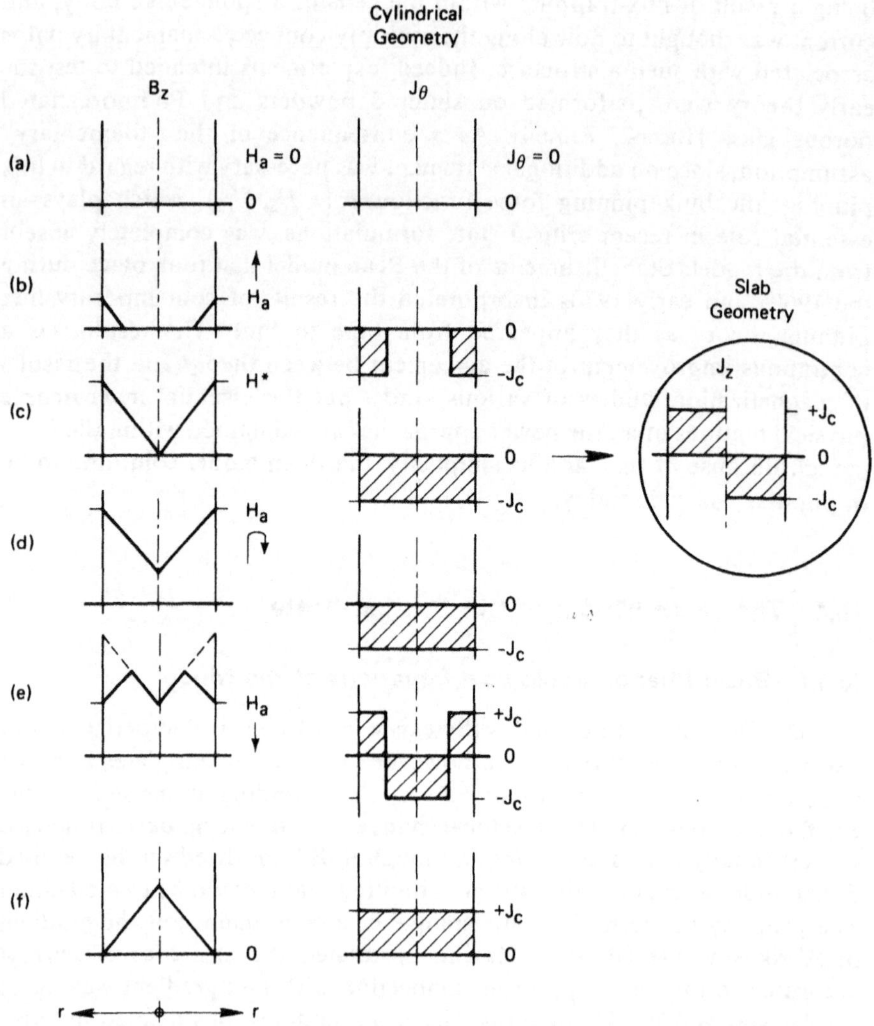

Figure 16-3. Magnetization of an infinite cylinder in a variable axial applied magnetic field, H_a, treated in the Bean approximation, $dB_z/dr = -(4\pi/10)J_{\theta c}$. $J_{\theta c}$ is a diamagnetic circulating current; within the model it is either zero or equal to J_c, the critical current density. In the figure, the radial direction is regarded as positive, thus in (c), for example, $J_{\theta c}$ is uniformly negative across the cylinder. If, as in some such diagrams, the edge of an infinite slab were being portrayed (e.g., [BRE73, p. 237]), it would be necessary to invert one side of the, then, J_z diagram as in the inset, and so on for the others (an unnecessary complication)—after standard sources, see also BELANGER [BEL68, p. 24].

counterbalanced by the bulk pinning force (i.e., the maximum force available to resist further ingress or egress of flux).

As an example of the application of the model, consider the magnetization of an infinite cylinder in an axial applied field, H_a. The Maxwell equation in cylindrical coordinates, (16-12a), has been shown to reduce in the general case to

$$\frac{1}{\mu_{\text{eq}}}\frac{dB_z}{dr} = -\left(\frac{4\pi}{10}\right)J_\theta \tag{16-13}$$

for field and flux in the z direction, and furthermore to lead to

$$\left.\frac{1}{\mu_{\text{eq}}}\frac{dB_z}{dr}\right|_{\text{critical}} = -\left(\frac{4\pi}{10}\right)J_{\theta c} \qquad \text{(shielding)} \tag{16-17}$$

upon the establishment of the shielding critical state. In other words, an induction linearly decreasing at a rate of dB_z/dr from its value at the surface, B_0, is supported by a circulating supercurrent, $J_{\theta c} \equiv J_c$ (for the material) that is fixed in magnitude and whose direction depends on the sign of the flux gradient. This gives rise to the magnetization cycle presented, without further explanation, in Figure 16-3.

16.4.2 Cylinder Magnetization in the Bean Model

The magnetization of a superconducting cylinder in an *increasing* longitudinal field (shielding mode) will be considered in the Bean approximation. The problem is treated with reference to Figure 16-4, and in order to satisfy (16-17) a counterclockwise (shielding) current is defined as negative. By geometry, the local induction is given by

$$B_z(r) = B_0 - \left(\frac{dB_z}{dr}\right)(R_0 - r) \tag{16-18}$$

The next step is taken by comparing this with (16-1) in the form

$$B_z(r) = H_a + 4\pi M_z(r) \tag{16-1}$$

and making the assumption that $H_a = B_0$ at the surface. It can then be concluded, by inspection, that

$$4\pi M_z(r) = -\left(\frac{dB_z}{dr}\right)(R_0 - r) \tag{16-19}$$

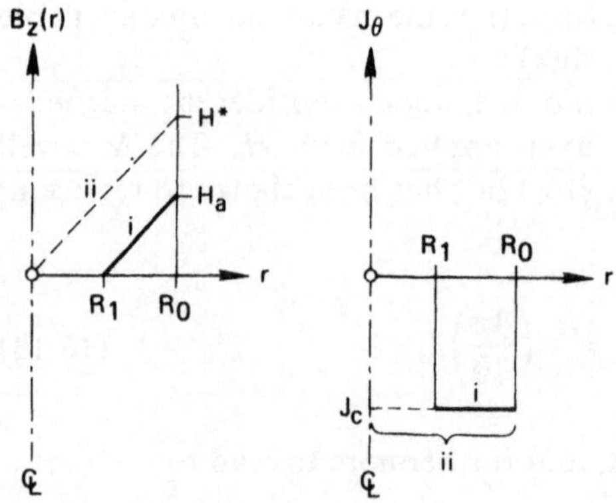

Figure 16-4. Magnetization of an infinite cylinder in a variable axial applied field, H_a, treated in the Bean approximation. The two cases depicted are: (i) the *general* case of a small increasing applied field, and (ii) the *special* case, in which the induction has just penetrated to the center of the cylinder—detail from Figure 16-3.

Substituting the Maxwell expression for (dB_z/dr) from (16-17) yields

$$4\pi M_z(r) = \mu_{eq}\left(\frac{4\pi}{10}\right) J_{\theta c}(R_0 - r) \tag{16-20}$$

a relationship which shows that within the Bean approximation (in particular, the $J_c = 0$ or constant assumption) the local magnetization is either zero or some constant value independent of the magnetizing field. This condition eventually leads to magnetization saturation at the special value of the surface induction, B^*, at which $B_z(r)$ has just "reached" the axis[5] of the cylinder (i.e., the field at the axis is on the point of increasing from zero). Returning to (16-20), and recalling (16-17) in the form

$$\frac{dB_z}{dr} = \frac{B^*}{R_0} = -\mu_{eq}\left(\frac{4\pi}{10}\right) J_{\theta c} \tag{16-17}$$

we are now able to express the local magnetization in the convenient form

$$4\pi M_z(r) = -\frac{B^*}{R_0}(R_0 - r) \tag{16-21}$$

clearly diamagnetic in this case. The mean magnetization is then obtained by averaging over the entire radius of the cylinder, but recognizing that J is zero between $r = 0$ and R_1 (see Figure 16-4). The result of the integration

$$4\pi\langle M_z(r)\rangle = \int_{R_1}^{R_0} 4\pi M_z(r) 2\pi r\, dr \Big/ \int_0^{R_0} 2\pi r\, dr \tag{16-22}$$

[5]Other geometries would specify the "inside wall" in the case of a tube, or the center plane in the case of a two-sided slab.

after it has been assumed for simplicity that at the surface $B_0 = H_a$, in general, such that B^* can be replaced by H^* and R_1/R_0 by $1 - h$ (where, in this context, $h \equiv H/H^*$), leads directly via

$$4\pi\langle M_z(r)\rangle = -\frac{2H^*}{R_0^3}\int_{R_1}^{R_0}(R_0 - r)r\,dr \tag{16-23}$$

to

$$4\pi\langle M_z\rangle = -2H^*(\tfrac{1}{2}h^2 - \tfrac{1}{3}h^3) \tag{16-24}$$ [6]

This analysis leads to several important conclusions:

1. For a solid cylinder ($R_1 = 0$) immersed in an external field H^*, and consequently for $H \gtrless H^*$,

$$4\pi\langle M\rangle_{\text{sat}} = -H^*/3 \tag{16-25}$$

which, with the aid of the Maxwell relation (say (16-20)), becomes

$$4\pi\langle M\rangle_{\text{sat}} = \left(\frac{4\pi}{10}\right)\left(\frac{R_0}{3}\right)J_{\theta c} \tag{16-26}$$

a diamagnetic saturated magnetization (for $H \gtrless H^*$) proportional only to R_0 and $J_{\theta c}$. This is a very useful result in that it points the way towards the design of contactless magnetization techniques for the measurement of critical current density.

2. By extending the above calculations in such a way as to complete the entire magnetization–demagnetization cycle, BEAN was able to quantitatively account for the unsaturated hysteresis loop ($4\pi M$ *versus* H_a) obtained from a synthetic superconductive sponge (Pb in porous glass at 4.2 K) [BEA64] in an impressive demonstration of the power of the simple model. Subsequently FIETZ *et al.* [FIE64], in an important case study of the applicability of the critical state idea to the cylindrical magnetization of commercial hard superconductors, investigated the hysteretic magnetization of a sample of Zr–Nb(75 at.%), analyzing the results with the aid of a truncated KIM-type [KIM62] critical-pinning-force equation (see Table 16-1).

3. Equation (16-24) suggests that during the magnetic cycling of a cylinder in an oscillating magnetic field of amplitude $H_m < H^*$, the energy

[6] BEAN [BEA64], using a definition for magnetic field different from that implied throughout this work obtained

$$4\pi\langle M_z\rangle = -H^*(h - h^2 + \tfrac{1}{3}h^3)$$

which agrees with (16-24) only at $h = 0$ and $h = 1$.

Table 16-1. Early Pinning-Force or Critical-State Functions

Critical-State Function Based on $-\mathbf{J}_c \times \mathbf{B} = \mathbf{F}_p$ [a] or J_c-Function, when preferable	Literature
$J_cB = \alpha(T)B^b$	Bean [BEA64]
$J_c = \alpha(T)/(b_0 + B + a_2B^2 + a_3B^3 + \cdots)$	Kim *et al.* [KIM62]
$\rightarrow \alpha(T)/(b_0 + B)$	Kim *et al.* [KIM62, KIM64[a]]
$\rightarrow \alpha(T)/B \quad (b_0 \ll B)$	Anderson and Kim [AND64]
$J_cB = \alpha(T)B^{1/2}$	Yasukōchi *et al.* [YAS64, YAS66]
$J_c = (a_0/b_0)\exp(-B/b_0) + c_0$	Fietz *et al.* [FIE64]

[a] $\mathbf{F}_p = \boldsymbol{\beta}(T, B) = \alpha(T)f(\mathbf{B})$. Depending on application, the force-balance may be taken either as "bulk" as shown, or local—in which case $-\mathbf{F}_p(r, B, T) = \mathbf{J}_c(r, B, T) \times \mathbf{B}(r)$. If J_c is in A cm^{-2}, and B is in gauss, then $F_p/10$ would be in dyn cm^{-3}; in this context, however, the units are often selected so that the "10" may be dropped, in order to simplify the interchange of F_p, J_cB, and $\alpha(T)f(B)$.

[b] The α's being in general all different, they may be distinguished by subscripts to match the index of the associated B^n.

dissipated per cycle, $\oint MdH$, can be expected to be of the form $aH_m^3 + bH_m^4$. The actual expression developed by BEAN [BEA64] for the hysteretic energy dissipated per cycle in a cylinder was

$$Q_h = H_m^3 V/3\pi H^* - 5H_m^4 V/16\pi H^{*2} \qquad H_m \lesseqgtr H^* \tag{16-27}$$

For a slab BEAN obtained

$$Q_h = H_m^3 V/6\pi H^* \qquad (\text{erg cycle}^{-1}) \qquad H_m \lesseqgtr H^* \tag{16-28}$$

He also went on to show that when $H_m \ll H^*$ the losses take place in a surface layer (not to be confused with the "surface sheath" of Section 13.9), under which condition the cyclical hysteretic loss per unit surface area of both slabs and cylinders becomes

$$Q_{hs} = 5H_m^3/12\pi^2 J_c \qquad (\text{erg cm}^{-2}\,\text{cycle}^{-1}) \tag{16-29}$$

While briefly touching on the subject of hysteretic AC loss, which is to be considered in detail in Chapter 26, it is useful to note that, for complete field-penetration ($H_m > H^*$) of a two-sided slab of thickness w, LONDON [LON63] showed that the hysteretic cyclical loss was

$$Q_h = (J_c w V H_m/10)(1 - 2H^*/3H_m) \qquad H_m > H^* \tag{16-30}$$

$$\cong J_c w V H_m/10 \qquad (\text{erg cycle}^{-1}) \qquad H_m \gg H^* \tag{16-31}$$

A comparison of (16-28) and (16-31)[7] leads to the very important result, to be encountered again in Section 26.2, that whereas with incomplete penetration the loss is characterized by an H^3 field dependence, with full penetration the field dependence becomes linear.

For a recent discussion of AC loss within the framework of critical-mixed-state models, the work of OHMER, WOLLEN, and LAWSON [OHM75] is recommended. To be discussed in detail in Chapter 26 are the mechanisms of AC loss in composite conductors, *viz.*, the hysteretic component (as considered here) and the eddy-current component (which arises in the Cu matrix).

As indicated above in connection with the presentation of (16-26), a critical current density, albeit model dependent, can be computed from the results of saturated magnetization measurement. This is to be dealt with in Section 19.3. In one of his early papers, BEAN also established the AC-modulation technique as a viable method of contactless J_c measurement (see Section 19.4). Again starting with the simple model where J_c = constant, BEAN showed that the output from a superconducting transformer (secondary: N turns, radius R_0; wound on the superconducting cylinder as core) immersed in a steady bias field, and supplied with a relatively small oscillating current whose field is $h = h_0 \cos \omega t$ could be expressed as a series of odd harmonics of amplitude V_n given by

$$\left.\begin{aligned} V_1 &= 1.088\ V_3 \\ V_2 &= -(2h_0^2\omega N R_0/\pi J_c) \times 10^{-8} \qquad \text{(volts)} \\ V_5 &= 0.238\ V_3 \\ V_7 &= 0.111\ V_3 \\ V_n &= 5\ V_3/(n-2)(n+2) \end{aligned}\right\} \qquad (16\text{-}32)$$

Results such as those represented by (16-26) and (16-32) demonstrate the utility of the hysteretic magnetization technique for determining the critical current density.

The method of harmonic analysis, first devised by BEAN [BEA64], was subsequently used by LOVE [LOV66], ULLMAIER [ULL66], KROEGER *et al.* [KRO73, KRO75], and ALAIS and SIMON [ALA67] to measure surface and bulk critical currents, and by NAKAYAMA *et al.* [NAK74] in the investigation of longitudinal-field AC losses in Ti–Nb wires and cylinders. It is considered further in Section 19.4.

[7]When $H_m = H^*$ (16-28) and (16-31) are identical.

16.5 Models for the Pinned Critical State

BEAN's model makes no specific claims with regard to the microscopic mechanism governing the trapping of the transport or circulating current. Adopting an Abrikosov model for the mixed state, if the flux lattice is thought of as being restrained by pinning centers the critical-current prerequisite of the Bean model, *viz.*,

$$J_c = \text{constant}|_T = \text{say, } \alpha(T) \tag{16-33a}$$

immediately proceeds to

$$-\mathbf{F}_p = \mathbf{J}_c \times \mathbf{B} = \alpha(T)\mathbf{B} \tag{16-33b}$$

Thus, the Bean model specifies, by implication, a bulk pinning force proportional to $B(r)$. It then requires only a small conceptual step by way of, say,

$$-\mathbf{F}_p = \mathbf{J}_c \times \mathbf{B} = \alpha(T)f(\mathbf{B}) \tag{16-34a}$$

$$\text{or} \quad = \alpha(T)f\{\mathbf{H}(B)\} \tag{16-34b}$$

where as before **H** is understood to represent the external field that under equilibrium conditions would provide the induction, **B**, and where $f(\mathbf{B})$ is a "bulk pinning function," to move from the Bean approximation into the realm of modern flux-pinning theory. The essential ingredients of the transition from the low-field to high-field critical state are delineated in Section 20.6. In the present context the right-hand side of the pinning force equation, (16-34), is referred to as a "critical state function." A list of early (pre-1966) such functions is presented in Table 16-1. As a result of subsequent developments in flux-pinning theory, primarily within the context of the current-carrying mixed state, it would be possible to make substantial additions to this list in the form of generalized expressions for the bulk pinning force. This matter is fully discussed, with particular reference to elastic-lattice pinning-force summation, in Chapter 21.

According to BEAN, J_c = constant; in practice, of course, the mean critical current density always decreases with increasing field, and it is suspected that the local value $J(r)$ should respond in a like manner to the local induction, $B(r)$. The YASUKŌCHI [YAS64, YAS64[a], YAS66] formula

$$J_c = \alpha(T)/B^{1/2} \tag{16-35}$$

is a simple way of expressing this. As a consequence of the B-dependence

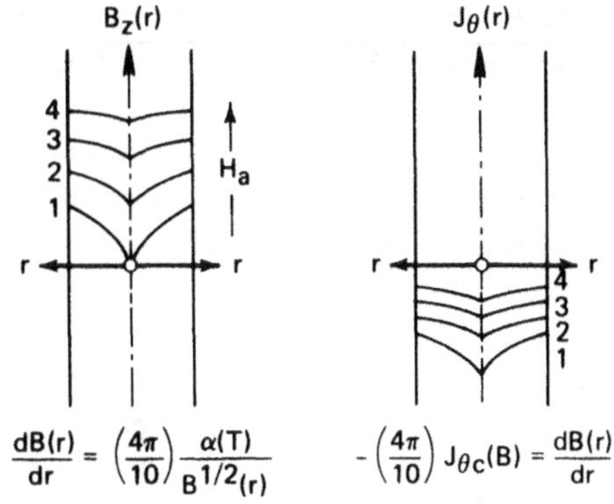

Figure 16-5. Magnetization of an infinite cylinder in a variable axial applied field. In this generalization of the Bean model (cf. Figure 16-3(c) and (d)), $J_{\theta c}$ is a function of B_z which in turn is no longer a linear function of r. In the figure, the Yasukōchi formula is depicted. H_a increases in the direction of the arrow; $|J_{\theta c}|$ undergoes a corresponding decrease.

of J_c, the induction is no longer a linear function of r as in Figures 16-3 and 16-4. Instead, the relationship

$$\frac{1}{\mu_{eq}} \frac{dB_z(r)}{dr}\bigg|_{\text{critical}} = -\left(\frac{4\pi}{10}\right) J_{\theta c}(B) \qquad \text{(shielding)} \qquad (16\text{-}17)$$

after substitution of the Yasukōchi formula, leads to

$$\frac{1}{\mu_{eq}} \frac{dB}{dr} = -\left(\frac{4\pi}{10}\right) \frac{\alpha(T)}{B^{1/2}(r)} \qquad (16\text{-}36)$$

and to curvilinear flux- and critical-current profiles of the type depicted in Figure 16-5.

16.6 Applications of the Critical State Models to Tube and Coil Magnetization

16.6.1 Tube and Coil Magnetization Studies—A General Introduction

Applications of simple critical state models of the kind presented in Table 16-1 to the low-field magnetization of hard type-II superconductors have been reasonably successful. Chang [Cha68] and Fietz *et al.* [Fie64] studied the magnetizations of cylindrical samples fabricated from the alloys Ti–Nb(36 at.%) [Cha68] and Zr–Nb(75 at.%) [Fie64], [Cha68], comparing the results with computed plots of $4\pi\langle M_z\rangle$ *versus* H_a. The mean magnetization was obtained by averaging $M_z(r)$ over the cylinder (or solenoid) cross section; i.e., by computing

$$4\pi\langle M_z\rangle = \frac{1}{\pi R_0^2}\int_0^{R_0} 2\pi r\, dr[B_z(r) - H_a] \qquad (16\text{-}37)$$

where $B_z(r)$ is obtained by suitably integrating the pinning-function-substituted Maxwell relation, (16-17), with $J_{\theta c}$ replaced by $\alpha(t)f(B)/B$ as in (16-34a).

For the successful accomplishment of the tests it is not necessary for the sample to be in the form of a monolithic machined solid or hollow cylinder. Moreover, since material is more readily available as wire than in ingot form it is not unusual to conduct the experiments with either single-layer coils [CHA64], bundles of straight insulated wires [CHA64], or open-circuited multilayer solenoids [FIE64]. CHANG [CHA64] examined the applicability of $J_cB = \alpha(T)B^n$ with $n = 0$, 0.75, and 1.0 to Ti–Nb(36 at.%) in the field range −5 to +15 kOe, and with $n = 0$ and 0.5 to Zr–Nb(75 at.%) in the field range 0 to 7 kOe. A typical set of semiloops for the latter material is shown in Figure 16-6. FIETZ *et al.* [FIE64] developed a set of analytical expressions based on (16-37) and the Kim critical state equation $J_c = \alpha(T)/(b_0 + B)$ to describe the complete hysteresis loop (although truncated at very high positive and negative fields) for comparison with experimental results for two sample coils of Zr–Nb(75 at.%). Data for the two coils, which were wound to different specifications from different batches of wire, when fitted to the Kim equation within the applied field range of 3–15 kOe yielded the following parameters:

$$J_c = \frac{1.9 \times 10^6 \text{ kG A cm}^{-2}}{(2.6 + B) \text{ kG}} \qquad \text{(coil A)}$$

$$J_c = \frac{0.95 \times 10^6 \text{ kG A cm}^{-2}}{(1.3 + B) \text{ kG}} \qquad \text{(coil B)}$$

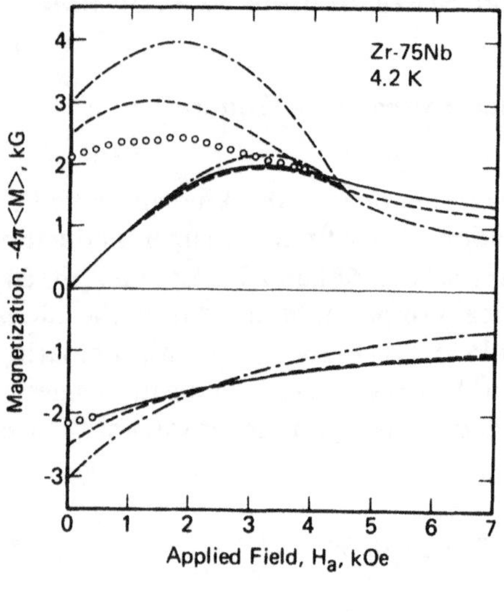

Figure 16-6. Experimental magnetization half-loop for a sample of Zr–Nb(75 at.%) (solid curves and circles) compared with predictions based on the critical state relationship $J_cB = \alpha(T)B^n$ for $n = 0$ (-·-·-) and $n = \frac{1}{2}$(- - - -). The sample was a simulated infinite slab formed from a closely wound coil of cold-rolled ribbon; measurements were made at 4.2 K—after CHANG [CHA68, p. 36].

It was noted that the fit was not completely satisfactory, especially at high fields where use of the Kim equation required the magnetization to drop to zero in contrast to the observed high-field saturation (cf. (16-26)). FIETZ *et al.* [FIE64] pointed out that a remarkably good fit to the data could be obtained using the expression

$$J_c = (a_0/b_0)\exp(-B/b_0) + c_0 \tag{16-38}$$

(where the a_0, b_0, and c_0 are fitting parameters), which, although apparently arbitrary, is actually a form of the KIM *et al.* function [KIM62] $J_c = \alpha(T)/\sum a_n B^n$ (Table 16-1). Values assumed by the Fietz parameters for the same two coils referred to above were

	a_0/b_0, A cm^{-2}	b_0, kG	c_0, A cm^{-2}
Coil A:	2.4×10^5	3.6	0.79×10^5
Coil B:	2.3×10^5	4.6	0.48×10^5

A first integration of J_c, in the form in which it is given in (16-38), yielded a local induction $B(r)$ in closed form, and the second integration yielded a convergent series for $4\pi\langle M\rangle$.

Frequently noted in data presentations of the type shown in Figure 16-6, and to be discussed in more detail below, are two very important dynamical properties of the flux lattice—flux flow and flux jumping. The smooth transitions between successive critical states during magnetization and demagnetization take place by the flowing of "supercritical" flux. In heavily pinned material, flow is impeded until sufficient energy is stored in the lattice to enable it to collectively break away from the pinning centers in the form of a flux jump. Although both of these phenomena will themselves be dealt with in Chapter 18, it is appropriate to briefly consider an important experimental technique used in their study—tube magnetization—within the present context.

Tube magnetization measurements of the kind conducted by HECHT and HARPER [HEC65], KROEGER [KRO66], and numerous others following the original studies of KIM *et al.* [KIM62] anticipated the more sophisticated "field-replication experiments" [FRA77] of CAMPBELL and subsequent workers to be discussed in Section 16.9. In a typical tube-magnetization experiment, the core induction, B', of a long hollow superconducting cylinder is measured as a function of the external longitudinal applied field, H_a. The B–H curves (see Figure 16-7) exhibit several interesting features: (i) with H_a increasing, B' is zero until H^* is exceeded; (ii) with H_a decreasing, a core flux density, say B^*, remains after H_a has returned to zero; (iii) as H_a becomes very large, $B' \to H_a$, justifying the ignoring of $4\pi\langle M\rangle$ and the use of the approximation $B \equiv H$ when dealing with very high fields as in practical studies of the current-carrying mixed state.

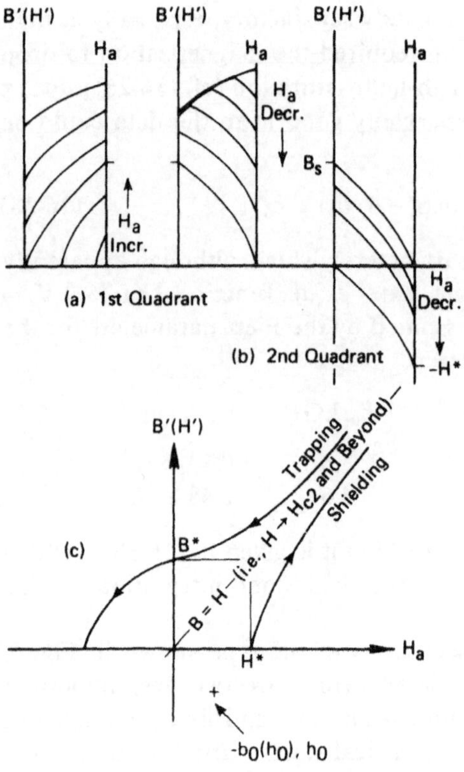

Figure 16-7. Magnetization of a hollow cylinder in the first and second quadrants. Indicated is the variation of induction (and in particular the core or inner-surface induction B') with applied longitudinal field, H_a; (a) H_a first increasing then decreasing back to zero; (b) H_a decreasing below zero. Also shown (c) is the corresponding B' *versus* H_a half-loop.

16.6.2 The Tube-Magnetization Experiments of Kim *et al.*

Important pioneering experiments on the magnetization of tubular samples of type-II superconducting alloys were described within the framework of a flux-pinning model by KIM, HEMPSTEAD, and STRNAD, in a series of three classic papers [KIM62, KIM63, KIM63[a]]. But before proceeding with an analysis of the static tube-magnetization experiment, it is necessary to recall, following ROLLINS [ROL74], that a steady field applied to the surface of a type-II superconductor will produce an induction B_0 just below the surface related to H_a by the equilibrium equation

$$B_0 = H_a + 4\pi M_{eq} \qquad \text{(cf. (16-1))} \tag{16-39}$$

that is, by

$$B_0 = \mu_{eq0} H_a \qquad \text{(cf. (16-2))} \tag{16-40}$$

where $\mu_{eq0} = B_0/H_a$ is the static equilibrium permeability. By the same token, the resulting induction, B', at the interior wall of the tube will be equivalent to a field $\mu_{eq0}H'$ within the bore.

In analyzing the experimental observations, the Kim approximation, which is usually understood to represent the local condition

$$J_c(r) \cong \alpha(T)/(b_0 + B(r)) \tag{16-41}$$

where b_0 is a constant, was restated in the form [KIM62]

$$\langle J_c \rangle \cong \alpha(T)/(b_0 + \langle B \rangle) \tag{16-42}$$

and coupled with the Maxwell relation

$$\frac{H' - H_a}{w} = \pm k|\langle J_c \rangle| \tag{16-43}$$

where $k \equiv 4\pi/10$, the " $-$ " sign is used for $H_a > H' > 0$, and the " $+$ " sign for $H' > H_a > 0$ (Figures 16-7(a) and (b)). Then setting $\langle B \rangle = (B' + B_0)/2$, the solution of (16-42) and (16-43) in the first quadrant could be written in either of the forms

$$b_0 + B = \pm \frac{k\alpha w}{H' - H_a} \tag{16-44}$$

or

$$\left.\begin{aligned} (B' + b_0)^2 - (B_0 + b_0)^2 &= \pm 2\mu_{eq0} k\alpha w \\ (H' + h_0)^2 - (H_a + h_0)^2 &= \pm 2k\alpha w/\mu_{eq0} \end{aligned}\right\} \tag{16-45}$$

Equation (16-44) shows that the parameters of the Kim critical state function, (16-42), may be derived from the slope and intercept of plots of $\langle B \rangle$, as defined above, *versus* $(H' - H_a)^{-1}$; while (16-45) states that H' *versus* H_a describes a pair of hyperbolas, as in Figure 16-7(c).

In the second-quadrant diagram (Figure 16-7(b)), which describes the magnetization of the tube as the induction profiles cross the axis, $H' > 0 > H_a$, and the solution of (16-42) and (16-43) has been shown by KIM *et al.* [KIM63] to be (in H-notation) a circle

$$(H' + h_0)^2 + (H_a - h_0)^2 = 2(k\alpha w/\mu_{eq0} + h_0^2) \tag{16-46}$$

with center at $(-b_0, h_0)$. The new value of $\langle B \rangle$ (see (16-42)) under this condition is: $\langle B \rangle_{new} = \langle B \rangle (H' + H_a)/(H' - H_a)$.

FIETZ *et al.* [FIE64] provided a set of analytic expressions representing the mean magnetization $4\pi\langle M \rangle = \int dx[B(x) - H_a]$ of a superconducting plate, in the Kim approximation, valid in each of seven selected regions of the full magnetization cycle. For insights into further applications of the tube-magnetization experiment with particular reference to the dynamic magnetization process (particularly flux flow and flux jumping) the studies of KIM *et al.* [KIM63], HECHT and HARPER [HEC65], and KROEGER [KRO66] are also recommended.

16.6.3 Relationship Between the $4\pi M$ *versus* H and B *versus* H Diagrams

The $B(H)$ trace of Figure 16-7(c) bears a superficial resemblance to the $4\pi\langle M\rangle$ *versus* H curve of Figure 18-6 rotated through $\pi/4$ [KIM63]. The relationship between the curves is simply demonstrable with the aid of the usual formulae for rotation of the cartesian axes from (x, y) to (x′, y′). For an axis rotation of $\pi/4$ any figure is invariant under a change of axial scale conforming to $\sqrt{2}x' = y + x$; $\sqrt{2}y' = y - x$. In the present case, therefore,

$$\text{with} \quad \left\{\begin{matrix} x = H \\ y = B(r) \end{matrix}\right\} \quad \text{then} \quad \left\{\begin{matrix} \sqrt{2}x' = B(r) + H \\ \sqrt{2}y' = B(r) - H = 4\pi M(r) \end{matrix}\right\} \tag{16-47}$$

This is again illustrated in Figure 16-8 which includes, for use in further discussion, a representation of the reversible ("equilibrium") magnetization curve for an unpinned sample of the same material.

16.7 Relationship Between Applied Field and Induction in Irreversible (i.e., Hard) Type-II Superconductors

This section reviews the dynamic and static situations, respectively, under which the permeabilities μ_{eq}, as defined in (16-2), and μ_{eq0}, as defined

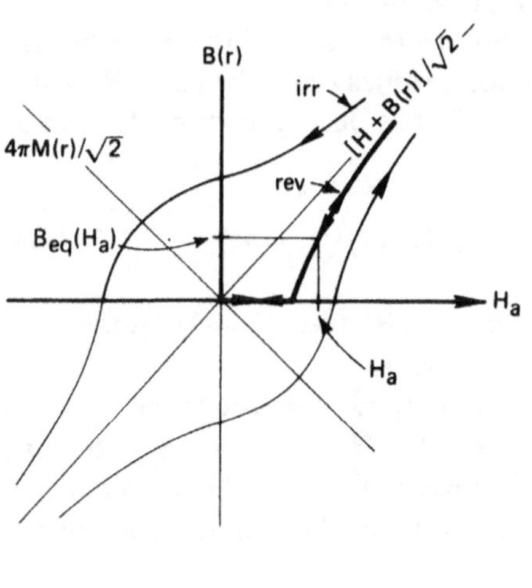

Figure 16-8. Complete (although truncated) B–H loop for an irreversible (irr) type-II superconductor (cf. previous figure) together with an indication of its relationship, via rotation and change of scale, to the magnetization loop of, for example, Figure 16-1. Included is the reversible (or "equilibrium") B–H curve (rev)—after KIM *et al.* [KIM63].

in (16-40), arise. In both cases (16-1), governing the equilibrium magnetization of an unpinned version of the material of interest, had been invoked. However, in dealing with flux-density gradients and forces the relationship $\mu_{eq} = dB/dH$ was needed, while under static conditions, the connection between B and H was achieved via $\mu_{eq0} = B/H$.

16.7.1 Relationship Between *B* and *H* at the Surface

A reversible (unpinned) type-II superconductor in the mixed state supports no flux-density gradient. For such a superconductor in thermodynamic equilibrium: $B(r) = \text{constant} = \langle B \rangle = B_0$, $M(r) = \text{constant} = \langle M \rangle = M_{eq}$ and B_0 is given by (16-39), thereby defining a parameter, $\mu_{eq0} = B/H$, as in (16-40). The early theories of the static critical state did not provide for the possibility that $\mu_{eq} \neq 1$, and in any event, unlike subsequent studies which included a calibration step, the early experiments would have been unable to deal with departures from $B_0 = H_a$ should they have been present.

Use has been made of these relationships by FIETZ *et al.* [FIE64, FIE67] and subsequently workers (e.g., ROLLINS *et al.* [ROL74]) in treating the boundary condition to be applied to the critical state of *pinned* superconductors. The assumption was that the surface exposed to the applied field could be treated reversibly. This implied that: (i) the boundary value of the flux, a function of r in such materials, was given by (16-40), and (ii) there was no surface hysteresis and B_0 was a single-valued function of H_a.

In analyzing the results of critical state magnetization experiments, (16-40) provides a convenient way of translating applied field into surface mixed-state flux density. Prior to performing the experiments, instrumental calibration factors are obtained by comparing magnetizations with the sample in the superconductive state and in the normal state above H_{c3}. To complete the calibration [ROL74] it is then only necessary to recognize that the ratio of surface flux densities for the normal and superconductive states is μ_{eq0}. This must be either (i) determined in a separate magnetization measurement on unpinned material otherwise similar to that of the sample under test, or (ii) set equal to unity if κ_{GL} is large and data are to be obtained in fields well removed from H_{c1} [ROL74]—conditions under which $\langle M \rangle$ is small (see Figure 20-3).

16.7.2 Relationship Between the Field Gradients in the Interior

The thermodynamics of the mixed state has been treated by numerous authors and for details the book by CAMPBELL and EVETTS [CAM72] is recommended. In developing equations such as (16-7) a relationship between $B(r)$ and an H was required. Similarly, in deriving expressions for the *local* Lorentz or pinning forces, the one-dimensional Maxwell curl

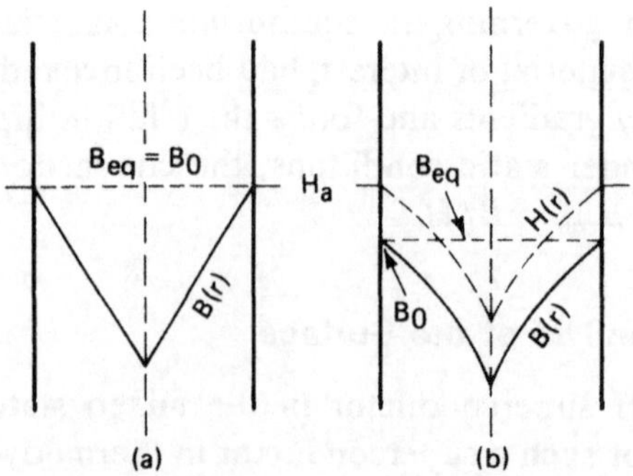

Figure 16-9. Relationships between fields in the critical mixed state. (a) Simple Bean model with no surface magnetization. (b) General critical state model showing the reversible surface induction $B_0 = B_{eq}(H_a) \equiv \mu_{eq0} H_a$ and the internal field $B(r)$ imagined to be in thermodynamic equilibrium everywhere with some $H(r)$ to which it is related differentially via $\mu_{eq} = dB(r)/dH(r)$.

equation $dH_z/dr = -(4\pi/10)J_c$, (16-12b), needed to be translated into an induction-gradient relationship. The device introduced by FRIEDEL *et al.* [FRI63] and followed by others [CAM68], [BEA69] was to equate f_Φ, the local fluxoid driving force, to $-(\Phi_0/4\pi)(dH(B_z)/dr)$, (16-10), where $H(B_z)$ was the magnetic field that would be in thermodynamic equilibrium with the magnetic induction at point r within the superconductor. After differentially connecting H and B as in (16-2), the induction gradient at point r then became $dB_z(r)/dr = -\mu_{eq}(4\pi/10)J_\theta$, (16-13).

16.7.3 Relationship Between μ_{eq0} and μ_{eq}

The relationship between applied field and magnetization, in the equilibrium and nonequilibrium situations, both within the Bean model with no surface magnetization and in the general state in which the surface boundary condition is properly taken into account, is depicted in Figure 16-9. The permeability, μ_{eq0}, is found associated with static $H \rightarrow B$ conversions; μ_{eq} is found associated with J_θ in equations such as (16-13), just mentioned. With reference to figures such as (16-8) and (18-11) to come, it can be seen that at high fields the approximation $\mu_{eq0} \equiv B/H \cong 1$ is good and $\mu_{eq} \equiv dB/dH \cong 1$ is even better. Obviously, caution must be exercised in assigning the appropriate permeability to the medium (μ_{eq} under dynamic conditions and μ_{eq0} under quasistatic) except in high-κ_{GL} samples of any given material when, provided H_a is not too small,

$$\mu_{eq} \cong \mu_{eq0} \cong 1 \tag{16-48}$$

16.8 The Role of H_{c1} in Critical State Theory

As depicted in the usual M–H or B–H diagrams (e.g., Figure 16-8) the Abrikosov mixed state exists only for fields above H_{c1}. It is, therefore, natural to inquire about the role of H_{c1} in magnetization experiments of the type considered in this chapter. The answer, of course, has to do with the irreversible properties of superconductors not represented in the equilibrium magnetization phase diagram, but depicted in the irreversible magnetization curves such as those of Figures 16-1 and 16-6, for example. In the present context, H_{c1} merely heralds the first entry into the mixed state during the magnetization of a virgin sample already cooled in zero field. Thereafter the critical mixed state is a permanent condition of the bulk material until such time as it is returned to the normal state by suitable increases in either the temperature or the applied magnetic field strength.

In a manner of speaking, it could also be said that H_{c1} influences the validity of (16-48) for a given alloy sample. With reference to Figure 16-8, it is noted that the smaller the value of H_{c1}, the closer the equilibrium magnetization curve approximates the line $B(r) = H_a$. This, of course, is consistent with the high-κ_{GL} requirement mentioned above. An increase in κ_{GL}, according to (14-10) and (14-12), increases H_{c2} and reduces H_{c1} thereby bringing the reversible magnetization curve nearer to the $B(r) = H_a$ diagonal at constant enclosed area.

16.9 Experimental Studies of Induction Profiles in the Critical State

Tube magnetization experiments of the type just discussed may in a sense be regarded as the logical forerunners (by way of dynamic magnetization experiments of the sort pioneered by FIETZ and colleagues [FIE64, FIE67]) of the modern techniques used in flux profile analysis. The manner in which critical state flux-density-related procedures can be pictured as having developed since the pioneering work of KIM *et al.* in 1962 is indicated in Figure 16-10 and the associated table, Table 16-2.

16.9.1 Induction Profile Scanning

The spatial distribution of magnetic induction adjacent to the surface of a superconductor can be directly measured using Hall-effect probes. Miniature thin-film sensors (0.005–0.01 cm$'$) were employed by COFFEY [COF67] in the first of such investigations. In a typical experiment, a motor-driven probe was drawn across the diameter of a narrow (0.5 mm) gap between a pair of 1 cm$^{\varnothing}$ × 6 cm^{l} superconducting cylinders exposed to an axial magnetic field which was varied up to some 40 kOe. Under such experimental conditions, the field measured by the probe was taken to be identical to the distribution of induction along the diameter of the monolithic specimen. Figure 16-11(a) is a tracing of the field distribution across a cylinder of Ti–Nb(40%). Important features recorded by the experiment

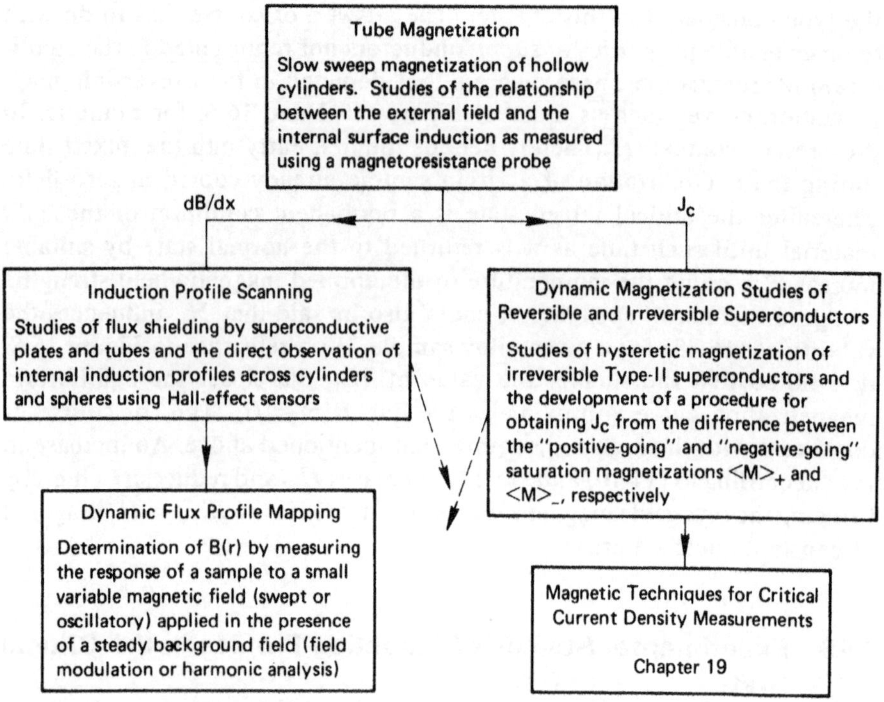

Figure 16-10. Developments of and interrelationships between studies of the magnetic structure of the mixed state. The literature cited is listed in Table 16-2.

were (i) a nonlinear $dB(r)/dr$, indicative of the B-dependence of J_c and/or a variation of pinning strength with r, and (ii) the occurrence of flux jumps that took place five times as the external field was raised to about 40 kOe. The figure presents direct evidence for (i) the breakdown of the critical state during jumping, accompanied by the uniform penetration of flux density into the center of the cylinder, and (ii) the subsequent reestablishment of a new critical state.

A flux-density gradient measurement having been made, it is only a short step to the determination of a critical current density appropriate to some fixed value of the applied field—not the bulk J_c such as is obtained from the usual four-terminal measurement, but rather a $J_c(r)$ that would respond to variation of pinning strength across the diameter of the sample. COFFEY [COF67] simply chose to drive the Hall probe at a constant speed, and performed a time differentiation on the output signal. In a like manner the J_c field dependence at some selected position could also be determined. Figure 16-11(b), which shows the results of such a measurement, taken about a point 1 mm in from the surface of the 1-cm$^{\varnothing}$ cylinder, indicates with vertical strokes the fields at which flux jumps took place. Other direct

Table 16-2. Experimental Studies of the Magnetic Structure of the Critical Mixed State

Tube Magnetization	
"Critical Persistent Currents in Hard Superconductors"	Kim, Hempstead, and Strnad [KIM62]
"Magnetization and Critical Supercurrents"	Kim, Hempstead, and Strnad [KIM63]
"Flux Creep in Hard Superconductors"	Kim, Hempstead, and Strnad [KIM63[a]]
Dynamic Magnetization Studies	
"Magnetization of Superconducting Nb-25% Zr Wire"	Fietz, Beasley, Silcox, and Webb [FIE64]
"Reversible and Irreversible Magnetic Properties of Some Type-II Alloy Superconductors"	Fietz [FIE67]
Induction Profile Scanning	
"Distribution of Magnetic Fields and Currents in Type-II Superconductors"	Coffey [COF67]
"Field Replication and Flux Shielding in Annular Superconductors"	Frankel and Garwin [FRA77]
Dynamic Flux Profile Mapping	
"The Response of Pinned Flux Vortices to Low-Frequency Fields"	Campbell [CAM69]
"Behaviour of Type-II Superconductors in Time-Dependent Magnetic Fields"	Fournet and Mailfert [FOU70]
"Experimental Study of Limited Instabilities of Hard Type-II Superconductors. Distribution of the Magnetic Induction Before and After a Flux Jump"	Noel, Mailfert, and Fournet [NOE71]
"Study of the Limit of Flux Jumps in Superconductors of the Second Kind"	Boyer, Fournet, Mailfert, and Noel [BOY71]
"Magnetic Field Profiles in Type-II Superconductors with Pinning Using a New AC Technique"	Rollins, Kuepfer, and Gey [ROL74]
"A Comparison of Methods for Measuring Flux Gradients in Type-II Superconductors"	Kroeger, Koch, and Charlesworth [KRO75]
"Experimental Investigation of the Flux Density Profile in Hard Superconductors"	Eckert and Handstein [ECK76]

microprobe measurements of internal fields have been undertaken by VOIGT [VOI68], SIKORA [SIK68], WEBER and RIEGLER [WEB73], and, most recently, FRANKEL and GARWIN [FRA77], who used a Hall-effect probe to investigate the field profiles (with particular reference to flux trapping and shielding) (i) within superconducting tubes, (ii) on the "shielded" sides of superconducting composite foils, and (iii) along the diameter of a superconducting Pb–Bi sphere.

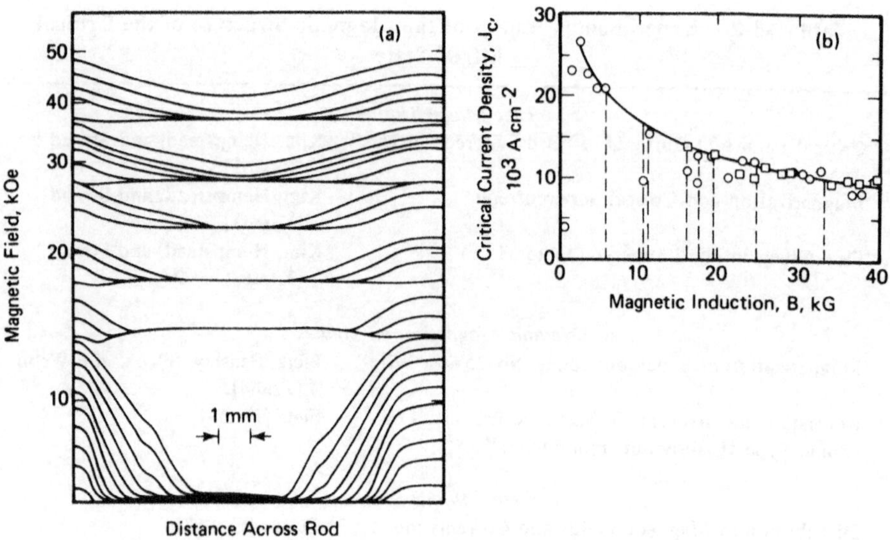

Figure 16-11. (a) Recorder tracings of the field distribution across a Ti-Nb(40%) cylinder obtained with the aid of a moving Hall probe. The applied field was increased at about 300 Oe min^{-1} between successive traces. Five flux jumps are indicated. Note the deeper penetration of the field after each flux jump, and flux flow after the last two jumps. (b) J_c *versus* B at a point 1 mm from the outside edge of the magnetized cylinder obtained from Hall-probe measurements of the field gradient at that point. The vertical lines indicate the occurrence of flux jumps. Data from two determinations are presented—after COFFEY [COF67].

16.9.2 Field Modulation Methods

Field modulation techniques have been devised and employed by several authors for the determination of critical current density profiles. The method used by FOURNET *et al.* [FOU70], [BOY71], [NOE71], which was developed independently of those of the other workers listed in Table 16-2, was also suitable for the study of flux jumping. In it, a steadily increasing field was applied to a sample in the presence of a bias field, and information obtained from the amplified output of a pickup coil made it possible to determine both J_c and the magnetic-induction distribution. The theory of the field modulation technique for field-profile mapping will be discussed with particular reference to CAMPBELL's method [CAM69] and its refinement in the hands of ROLLINS *et al.* [ROL74].

The modulation methods rely on the following known properties or assumptions:

(a) the sample is at all times in its critical state
(b) J_c depends on x, the distance measured from the surface
(c) J_c is independent of B within the amplitude of the modulation

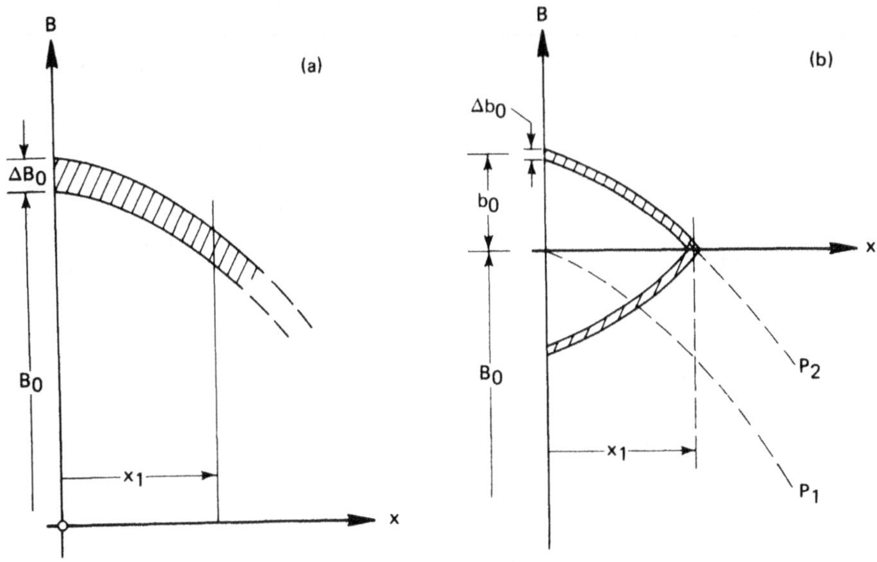

Figure 16-12. (a) Basic principle underlying the field-modulation techniques for the determination of induction profiles. Accompanying an increment, ΔB_0, in the surface induction, B_0, is a uniform displacement by that amount of the critical profile, whatever its shape. (b) *Campbell's method* (expressed in terms of surface and subsurface inductions), in which an oscillating applied field of amplitude h_0 is superposed upon a steady bias field, H_a (where $B_0 \equiv \mu_{eq0}H_a$, see Section 16.7).

With reference to Figure 16-12(a), these imply that if the surface induction, B_0, increases by a small amount, ΔB_0, the critical flux profile, whatever its shape, is displaced uniformly by that amount. It follows that the corresponding change of flux to depth x_1 is the area of the "bent parallelogram," $\Delta B_0 x_1$ multiplied by the width w (directed into the paper) of what is assumed here to be a one-sided slab.

CAMPBELL's method is illustrated in Figure 16-12(b). An oscillating field of amplitude h_0 is superposed upon a steady bias field H_a. With reference to the above note on surface boundary conditions for B (Section 16.7), the surface inductions in response to these fields are $b_0 = \mu_{eq0}h_0$ and $B_0 = \mu_{eq0}H_a$. Then recognizing that $dB/dx = \pm(4\pi/10)J_c$, the oscillating induction profiles are mirror symmetric about $B_0 = \text{constant}$. Under the conditions shown, the integrated output from a pickup coil is $S \propto \Phi = \int_0^{x_1} B(x)\,dx$, where Φ is the flux threading the coil. Suppose now the modulation amplitude to be changed by an amount Δh_0, then

that is

$$\left.\begin{aligned} \Delta S \propto \Delta\Phi &= \Delta b_0 w x_1 \\ \frac{\Delta S}{\Delta b_0} &= \text{constant} \cdot w x_1 \end{aligned}\right\} \qquad (16\text{-}49)$$

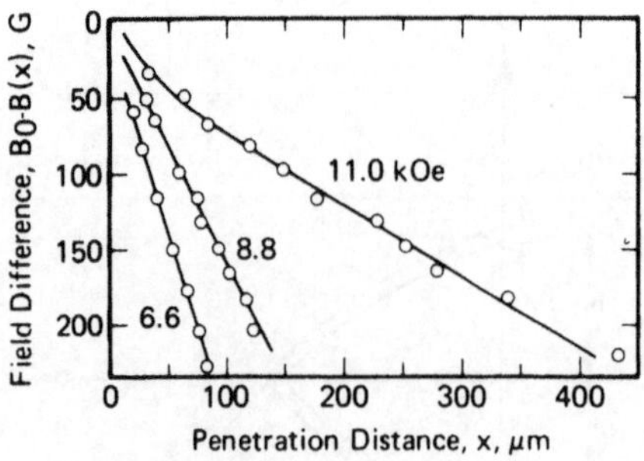

Figure 16-13. Induction profiles in a fairly strongly pinned sample (extruded Pb-Bi(33%)) for three values of the applied field. Obviously an H_a-increasing condition is implied. The results were obtained by CAMPBELL [CAM69] (cf. Figure 16-12(b), also cf. [CAM69, Figures 2 and 3] for a comparison of induction profiles in irreversible and reversible materials, respectively).

The instrumental constant is evaluated by putting the sample into the normal state above H_{c3} where

$$\left.\frac{\Delta S}{\Delta h_0}\right|_n = \text{constant} \cdot wd \tag{16-50}$$

in which d is the width of the slab. Equations (16-49) and (16-50) together yield

$$x_1 = d\frac{(\Delta S/\Delta h_0)}{(\Delta S/\Delta h_0)_n \mu_{\text{eq}0}} \tag{16-51a}$$

associated with which depth there is an induction $B(x_1)$ given simply by

$$B(x_1) = B_0 - b_0 = \mu_{\text{eq}0} Ha - \mu_{\text{eq}0} h_0 \cong H_a - h_0 \tag{16-51b}$$

upon the assumption that profile P_2 is just a vertical displacement of P_1, the profile under investigation. Each such experiment gives a single datum point (B, x_1) on the flux profile and must be repeated numerous times, each time with a different modulation amplitude, h_0, in order to map out the complete profile. A typical result is given in Figure 16-13.

In an interesting comparison of measurement techniques KROEGER *et al.* [KRO75] juxtaposed a pair of flux density profiles (H_a = 7.56 kOe) taken from annealed Ti-Nb(90 at.%) using CAMPBELL's method and the harmonic analysis method, respectively. Presented in Figure 16-14, the results clearly demonstrate the existence of a surface supercurrent that shields the interior of the sample from field excursions of half-amplitude $\Delta H_s/2$ (see Sections 19.4 and 22.5).

In the modification of CAMPBELL's method due to ROLLINS *et al.* [ROL74], the nonintegrated output resulting from low-frequency (3–300 Hz) field modulation of the general form $h_0 g(t)$, where $g(t)$ is a periodic function of unit amplitude, is manipulated so as to yield the entire $B(x)$ curve in

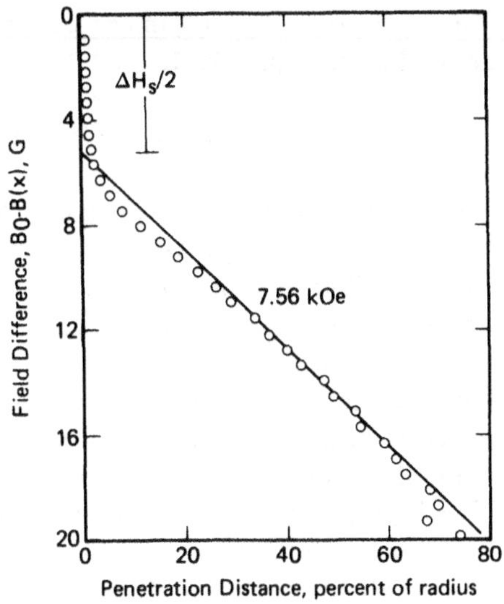

Figure 16-14. Induction profile for a sample of Ti–Nb(90 at.%) (annealed 1 h/1500°C) in an applied field of 7.56 kOe, as measured by the Campbell technique (○). The presence of a surface barrier is evident (see Sections 19.4 and 22.5.3). Shown for comparison is the field distribution derived from third-harmonic analysis (again see Section 19.4). The measurements were conducted by KROEGER *et al.* [KRO75].

one excursion of the total applied field. Modulation amplitude, h_0, then becomes a free variable which is available for use, along with frequency, as an internal check on the consistency of the results. ROLLINS' method is easily understood with the aid of Figure 16-15.

As the surface induction $B_0(t)$ is slowly swept between the limits $B_0 \pm b_0$ a pickup coil provides a signal output proportional to $d\Phi/dt$ such that, by an argument similar to that used in deriving (16-51a),

$$x_1(t) = \frac{(d\Phi/dt)}{(d\Phi/dt)_n \mu_{eq0}} \qquad (16\text{-}52a)$$

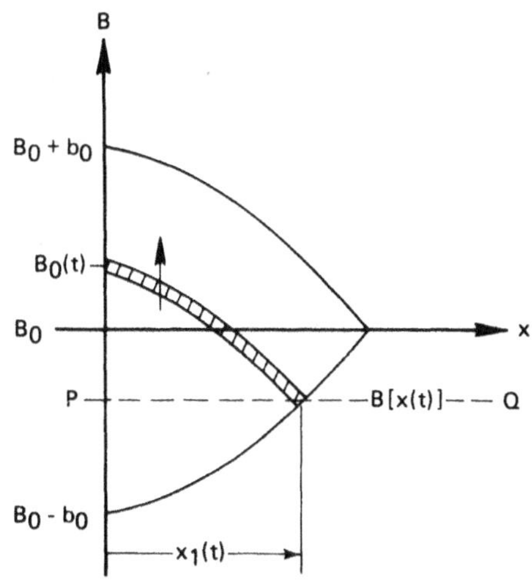

Figure 16-15. *Rollins' method* (expressed in terms of surface and subsurface induction; see caption to Figure 16-12) in which an applied field, H_a, is slowly swept between the limits $H_a \pm h_0$. The suitably manipulated output of a pickup coil yields the entire $B(x)$ curve in a single excursion of h_0— after ROLLINS *et al.* [ROL74].

and by a geometric argument based on the symmetry of the flux profiles about PQ in Figure 16-15

$$B[(x_1(t))] = (B_0 - b_0) + \tfrac{1}{2}\mu_{eq0}h_0[1 + g(t)] \qquad (16\text{-}52b)$$

The parametric pair (16-52a) and (16-52b) will *in principle* yield $B(x)$ as a function of x when provided with data taken point by point during a single sweep of $B_0(t)$. *In fact* this is done electronically as outlined in the original article by ROLLINS and coworkers [ROL74]. A typical set of results, taken from cold-worked Ti–Nb(95.5 at.%) in external bias fields of 1.8 through 9.0 kOe, is given in Figure 16-16. Although the above analysis was presented in terms of a field modulation of general wave form $g(t)$, in actuality the applied signal was sinusoidal. Subsequently ECKERT and HANDSTEIN [ECK76] employed triangular and trapezoidal field-modulation wave-forms, which they claim are simpler to process, in the study of flux-density profiles in V_3Si single crystals and cold-worked Ti–Zr–Nb alloys.

As indicated above, the slope of the flux-density profile is directly proportional to the critical current density. Thus, field-modulation experiments represent an alternative approach to the magnetic determination of J_c to be considered in more detail below. ROLLINS *et al.* [ROL74] extended their data analysis in this way and obtained values of J_c *versus* H_a for a pair of Ti-doped cold-worked Nb-base alloys that turned out to be in good agreement with the results obtained by FIETZ and WEBB [FIE69] who used

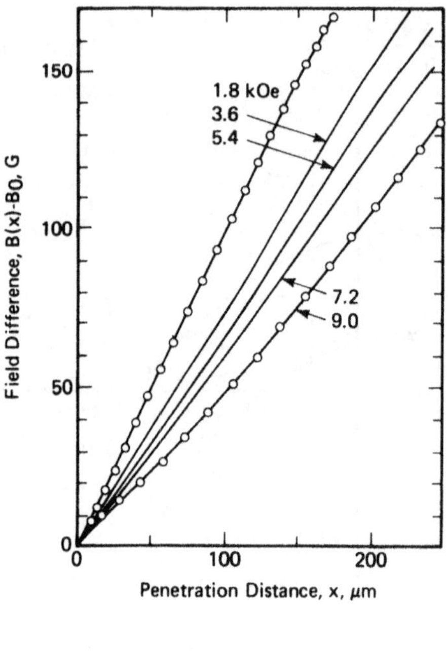

Figure 16-16. Induction profiles in cold-worked Ti–Nb(95.5 at.%) for several values of the applied field. The results shown were obtained after H_a had been *reduced* from $H_a = 3\ H_{c2}$ down to the desired values indicated—after ROLLINS *et al.* [ROL74].

a more conventional magnetization approach. As techniques for critical current density determination, the methods of FIETZ and WEBB [FIE64, FIE67, FIE69] and CAMPBELL [CAM69] have been compared by KROEGER *et al.* [KRO75] with the harmonic analysis technique introduced in Section 16.4.2 and with the results of direct four-probe resistometry. Samples of Ti–Nb(25 and 90 at.%) were used for the comparison and the relative merits of the techniques were critically reviewed. But the unique property of the field-modulation technique is its ability to provide information on the variation of pinning strength (or local $J_c(r)$) with depth below the surface. Its use in this regard has been an invaluable adjunct to the study of the influence of proton-radiation damage on flux pinning [ROL77], [ANJ79].

Chapter 17

The Upper Critical Field

17.1 The Nonparamagnetic Critical Fields

17.1.1 Temperature Dependences of the Critical Fields

According to GLAG theory, valid near $t = 1$, the lower and upper critical fields are given by $H_{c1} = H_c \ln \kappa_{GL}/\sqrt{2}\kappa_{GL}$, (13-52), and $H_{c2}^* = \sqrt{2}\kappa_{GL} H_c$, (18-53), where the asterisk notes that the effect of normal-state Pauli paramagnetism has not yet been taken into consideration.

The first successful attempt at calculating the temperature dependences of H_{c1} and H_{c2}^* resulted in the Maki (dirty-limit) equations

$$H_{c1}(t) = H_c(t)\frac{\ln \kappa_3(t)}{\sqrt{2}\kappa_3(t)} \tag{14-12}$$

$$H_{c2}^*(t) = \sqrt{2}\kappa_1(t)H_c(t) \tag{14-10}$$

in which contributions to the temperature dependences came from $H_c(t)$ as well as $\kappa_1(t)$ and $\kappa_3(t)$, both of which equal κ_{GL} at $t = 1$. The thermodynamic critical field, $H_c(t)$, is temperature dependent according to $(1 - t^2)$, (8-7). The Maki-calculated temperature dependences of $\kappa_1(t)$ and $\kappa_3(t)$ in the nonparamagnetic approximation have been presented in Figure 14-5. Obviously, as far as H_{c2}^* is concerned, since $\kappa_1(t)$ increases by only 20% between $t = 1$ and $t = 0$, the dominant temperature dependence under nonparamagnetic Maki theory should be the $(1 - t^2)$ contributed by $H_c(t)$ itself. Since paramagnetic improvements to Maki theory as they relate to H_{c2} can be transcribed in terms of modifications to the parameter $\kappa_1(t)$, according first of all to (15-15) (paramagnetic only) and then to (15-19) (paramagnetic plus spin-orbit scattering), the temperature dependence of H_{c2} would still be expected to be dominated by the same $(1 - t^2)$ factor.

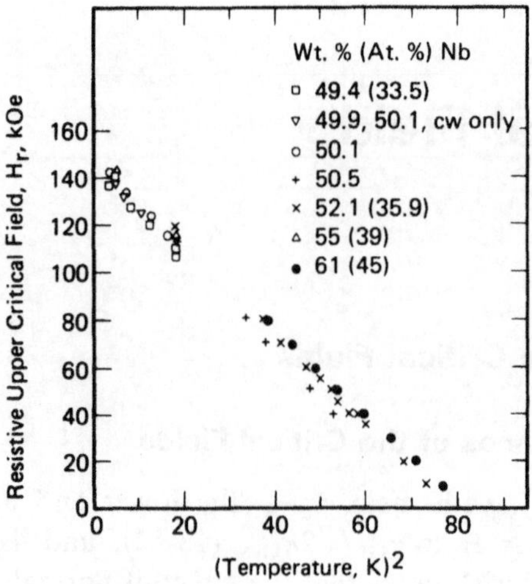

Figure 17-1. Temperature dependence of resistive upper critical field for seven technical monofilamentary or 60-filamentary Cu/Ti-Nb composite conductors. All seven were measured in the current-optimized condition; for comparison one wire (▽) was measured after cold work only. H_r can be seen to follow a $(1 - t^2)$ temperature dependence—after HILLMANN *et al.* [HIL79].

This does seem to be borne out in practice as indicated in Figure 17-1 for a series of technical Ti-Nb superconductors.

17.1.2 Evaluation of the Zero-K Upper Critical Field

A development from the Maki theory is the well-known expression for the GLAG-BCS-Goodman-Eilenberger zero-K nonparamagnetically limited upper critical field

$$H^*_{c20}\,(\text{dirty}) = 3.06 \times 10^4 \rho_n \gamma_v T_c \qquad (14\text{-}27)$$

that is easily derivable from (14-10) in the form $H^*_{c20} = \sqrt{2}\kappa_{10} H_{c0}$, (14-26), by way of $\kappa_{10}/\kappa^d_{GL} = 1.19$ (from Eilenberger theory [EIL67], also see Sections 14.5 and 14.6.1), $\kappa^d_{GL} = 7.49 \times 10^3\ \rho_n \gamma_v^{1/2}$, (14-17), (according to GOODMAN [GOO62]), and $H_{c0} = 2.43\gamma_v^{1/2} T_c$, (13-48), (according to BCS [BAR57]).

Another useful result from nonparamagnetic theory—an alternative expression for H^*_{c20} (dirty)—this time in terms of *superconductive-state* measurables, is

$$H^*_{c20}\,(\text{dirty}) = 0.68(-dH_{c2}/dt)_{t=1} \qquad (14\text{-}29)$$

which is derivable by way of the nonparamagnetic H-W alternative to early Maki theory (see Table 14-2). The two forms of H^*_{c20} may, when the distinction is useful, be referred to as H^{*n}_{c20} and H^{*s}_{c20}, respectively. In spite of the serious approximations involved, which restrict their validity as quantitative descriptions of the upper critical field, (14-27) and (14-29) continue to provide valuable guidelines for the design of high-field superconductors.

17.2 The Paramagnetically Limited Upper Critical Field and Its Temperature Dependence

17.2.1 Influence of Normal-State Pauli Paramagnetism

In considering a superconductor with a Pauli paramagnetic normal state, CLOGSTON [CLO62[a]] showed that quite apart from any field-dependent changes in the free-energy of the superconductive state, field-induced changes in the free-energy of the paramagnetic normal state would eventually destroy superconductivity at a zero-K field of

$$H_{p0} = 1.84 \times 10^4 T_c \qquad \text{(Oe)} \qquad (15\text{-}4)$$

In developing improved formulations for $H^*_{c2}(t)$ via (14-10), MAKI took this so-called Pauli paramagnetic limitation (PPL) effect into account by assigning, as before, part of the temperature dependence to what then became a new $\kappa_1(t, \alpha)$, in which α, defined by $\alpha \equiv \sqrt{2}H^*_{c20}/H_{p0}$, (15-10), was calculable using either superconductive-state (thence α_s) or normal-state quantities (thence α_n) using (15-11) to (15-14). Although initially intended to describe the temperature dependence of H_{c2} remote from the $t \cong 1$ constraint of the original GLAG theory, these new models and their subsequent further developments assumed much greater importance as quantitative descriptors of the *magnitude* of H_{c2} at a fixed temperature, and in particular of its composition dependence. As a result of these early paramagnetic calculations it was concluded by MAKI and TSUNETO [MAK64[b]] that

$$H_{c20}/H^*_{c20} = (\sqrt{1 + 2\alpha^2} - 1)/\alpha^2 \qquad [15\text{-}5b](17\text{-}1)$$

and again by MAKI [MAK64[b]] that

$$H_{c20}/H^*_{c20} = 1/\sqrt{1 + \alpha^2} \qquad [15\text{-}6](17\text{-}2)$$

Although there is little numerical difference between these two results for α-values of interest (e.g., $\lesssim 10\%$ for $\alpha < 2$) the depression produced by (17-1) is the less severe. Nevertheless, as pointed out by both HANCOX [HAN66] and MAKI himself [MAK64[b]], when compared against the experimentally measured H_rs for a series of Ti–V alloys [BER63] even the (17-1)-calculated Pauli-limited H_{c2}s were uniformly too small (see Figure 15-3).

17.2.2 Further Developments of Mixed-State Theory

Further developments of the theory of paramagnetic limitation were aimed at improving the agreement between the experimental and calculated values of H_{c2}; in other words they were aimed at dealing with the accumulating body of evidence that the uncompensated effect of normal-state paramagnetism was overestimating the critical-field depression. One possible source of the discrepancy—the electron-phonon interaction—had already been identified by CLOGSTON [CLO62ᵃ] but subsequently forgotten, to be rediscovered, or at least reestablished in importance, by ORLANDO *et al.* [ORL79] in the late 1970s. Another mechanism for the relief of PPL—a strong-coupling correction to H_p—had been suggested long ago by HANCOX [HAN66], to be taken up again later by RAINER *et al.* [RAI73] and subsequently by ORLANDO and BEASLEY [ORL79].

In the meantime, attention was being exclusively focused on attempts to develop ways of increasing the mixed-state paramagnetism in order to decrease the Clogston-Chandrasekhar difference between the *n*-state and *s*-state free energies leading to a partial restoration of the original GLAG condition of zero difference. The mechanism selected by MAKI [MAK66] and WERTHAMER, HELFAND, and HOHENBERG [WER66] was a random spin-orbit scattering (SOS) of the quasiparticles. The resulting SOS-moderated PLP was characterized in Maki theory by the combined parameter $\beta_M^2 \equiv \alpha^2(3\Delta_{00}/\hbar)\tau_{so}$, (15-17), which contains both α (defined above) *and* an SOS relaxation time, τ_{so}; it was characterized in WHH theory separately by α (which was then common to both theories) and λ_{so}, the SOS frequency parameter, which is related to the τ_{so} of both WHH and Maki theory through $\lambda_{so} = 2\hbar/3\pi k_B T_c \tau_{so}$, (15-29). MAKI then dealt with the temperature dependences of H_{c2} and κ_1, while WHH theory considered only that of κ_1; both theories thus shared common ground in the temperature dependence of κ_1 represented in Maki theory by $\kappa_1(t, \beta_M)$ and in WHH theory by $\kappa_1(t, \alpha, \lambda_{so})$. Although α is calculable, λ_{so} is not, and must be treated as an adjustable parameter selected so as to enable theory to fit the experimental results (but see Section 15.5.4). In WHH theory λ_{so} must not be too large; indeed the WHH and Maki theories are valid only in the approximation that $\tau_{so} \gg \tau_{tr}$, the normal transport relaxation time (but see Section 15.8.2). But in numerous comparisons of theory and experiment, it has turned out that the data could be fitted only by combining a finite α with an infinite λ_{so}. Experimental tests of Maki-WHH theory using the individual alloys Ti-V(65 at.%) and Ti-Nb(44 at.%) were conducted by KIM and STRNAD [KIM66] and SHAPIRA and NEURINGER [SHA65], respectively, while the sequence: Ti-V(58 at.%), Ti-Nb(44 at.%), and Ti-Ta(52 at.%), was investigated by NEURINGER and SHAPIRA [NEU66].

In a subsequent study of Maki-WHH theory, BRAND [BRA72] selected for measurement two series of alloys: Ti-V(36, 70, and 82 at.%) and Ti-Nb(32, 55, 62, and 83 at.%). With regard to the Ti-Nb alloys, the rationale for choosing the particular composition sequence specified was that it provided for measurement samples in which the PPL was both strong

Table 17-1. Intercomparison of H_{p0}, H^{*n}_{c20}, H^{*s}_{c20} (kOe) and the Results of the WHH Fitting Procedure—After Brand [BRA72]

at.% Nb	32	55	62	83
H_{p0}	150	179	178	170
H^{*n}_{c20}	254	192	128	68
H^{*s}_{c20}	166	143	134	68
H^{WHH}_{c20}	155	143	123	68
α_n	2.40	1.52	1.01	0.564
α_s	1.56	1.13	1.06	0.565
λ_{so}	6	∞	5	∞
mean-free-path length (Å)	4.1	6.8	10	16

($H_{p0} < H^{*n}_{c20}$; 32, 55 at.% Nb) and weak ($H_{p0} > H^{*n,s}_{c20}$; 62, 83 at.% Nb). Using atomic constants and physical measurables, H_{p0}, $H^{*n,s}_{c20}$, and $\alpha_{n,s}$ were calculated and λ_{so} was determined by trial-and-error fitting of $H_{c2}(t)$, as derived from the WHH function ((15-28)), to the magnetically determined upper critical field temperature dependence, $H_u(t)$, as in Figure 15-9. Some further results of the experiments and associated calculations are listed in Table 17-1 and depicted in Figure 17-2, which demonstrate that: (i) the paramagnetic limitation parameter, α, whose variation with composition is dominated by that of ρ_n in the range considered, increases monotonically between Ti-Nb(83 at.%) and Ti-Nb(32 at.%) following a steady decrease in the electronic mean-free-path (cf. Table 5-2); (ii) the ratio $H^{*n}_{c20}/H^{*s}_{c20}$

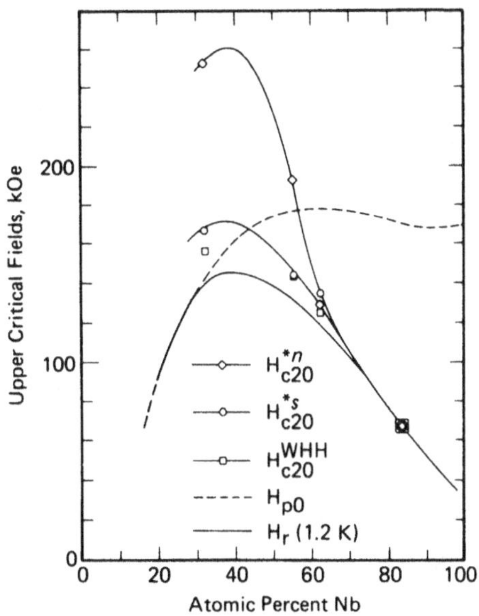

Figure 17-2. Calculated zero-K non-paramagnetically limited upper critical fields H^{*n}_{c20} and H^{*s}_{c20}, calculated zero-K Clogston limiting field H_{p0}, WHH-extrapolated zero-K resistive upper critical field H^{WHH}_{c20} (after BRAND [BRA72]), and as-measured 1.2-K resistive upper critical field, H_r (after BERLINCOURT and HAKE [BER63]), as functions of concentration in Ti-Nb—after BRAND [BRA72].

departs from unity as the alloys become dirtier; (iii) λ_{so} has no smooth variation, but with decreasing Nb content must be assigned the values ∞, 5, ∞, 6 in order that $H_{c2}^{\text{WHH}}(t)$ fits the experimental $H_u(t)$ curve. Thus, two of the alloys treated demanded that λ_{so} be infinite, which not only violated the restrictions placed on WHH theory, but also corresponded to an unphysically rapid spin-orbit scattering frequency.

A similar observation had been made earlier by NEURINGER and SHAPIRA [NEU66], who, in studies of Ti–Ta(52 at.%), noted that even the assignment of $\lambda_{so} = \infty$ was insufficient to raise $H_{c2}^{\text{WHH}}(t)$ up to the experimental $H_r(t)$ curve. This difficulty is resolvable if H_{p0} is able to be higher than predicted by the bare Clogston formula, (15-4), for then the paramagnetic limitation would be smaller and less spin-orbit scattering would be needed to match theory and experiment. The reintroduction of normal-state electron–phonon interaction [ORL79, ORL81], [BEA82] coupled with a relaxation of the $\tau_{so} \gg \tau_{tr}$ restriction [SCH81] now combine to accomplish this goal.

In the meantime, the experimental observation that the difference $H_r - H_{c2,\text{Maki-PPL}}$ [1] increased for alloys arranged in the sequence Ti–V, Ti–Nb, Ti–Ta, [NEU66] along with the atomic number (Z) of the solute (which increased in the sequence: 23(V), 41(Nb), 73(Ta)), coupled with the result from ABRIKOSOV and GOR'KOV [ABR62] that

$$\lambda_{so}/\lambda_{tr} = \tau_{tr}/\tau_{so} \sim (Ze^2/\hbar c)^4 \qquad (17\text{-}3)$$

gave rise to the idea that the addition of heavy-element solutes to a binary Ti–Nb base, already selected for its optimal superconducting properties, would still further increase H_{c2} by a spin-orbit-scattering reduction of the Pauli limitation. Uncritical belief in this principle has motivated several studies in which the following substitutions were made in Ti–Nb alloys:

Hf for Ti — HAWKSWORTH and LARBALESTIER [HAW80], WADA [WAD80]

Ta for Nb — SUENAGA and RALLS [SUE68, SUE69], HAWKSWORTH and LARBALESTIER [HAW80]

Zr for Ti, Ta for Nb — HORIUCHI *et al.* [HOR73]

Such deliberate attempts to increase H_{c2} through heavier-element substitutions for the basic components of Ti–Nb alloys have resulted in the development of some promising ternary (in particular, Ti–Nb–Ta) and quaternary (in particular, Ti–Zr–Nb–Ta) commercial-alloy superconductors.

[1] "Maki-PPL" indicates an H_{c2} calculated on the basis of the Pauli-limited Maki theory; H_r was found to be substantially higher than this [NEU66].

17.2.3 Influence of Electron–Phonon and Electron–Electron Interaction on the Paramagnetic Limit, H_{p0}

It was pointed out recently by ORLANDO *et al.* [ORL79] that the electron–phonon interaction parameter, λ, which comes into the McMillan equation for the transition temperature, (12-8), and whose presence has been shown to enhance (renormalize) the Fermi density-of-states, $n(E_F)$, and density-of-states-related quantities such as γ_v, (7-5), can by the same token increase the Clogston limiting field. At the same time, electron–electron interaction parameterized by α^e, which enhances the Pauli susceptibility according to $\chi_p^0/(1-\alpha^e)$, (7-3), to some extent counteracts the influence of λ. The result is that the Clogston paramagnetic limit, now represented by H_{p0}^{CL}, is modified (under the assumption of a second-order s/n transition) according to

$$H_{p0} \cong (1-\alpha^e)(1+\lambda+\lambda_s)H_{p0}^{CL} \tag{15-43}$$

where λ_s is a spin-fluctuation component that accompanies the presence of α^e-parameterized electron–electron interaction [ORL81], [BEA82]. Since α^e is generally less than λ (see Section 15.8.1, also [ORL81]) the effect of many-body interaction is to enhance H_{p0}^{CL}, thereby reducing the magnitude of the λ_{so} requirement (i.e., to draw τ_{so} closer to τ_{tr}). The right-hand side of (15-43) may also be furnished with a premultiplier in order to provide for the effect of a strong-coupling correction to the gap-energy [RAI73], [ORL79].

Semiempirical and empirical methods for determining electron–phonon coupling strengths have been considered in Sections 8.7.2 and 8.7.3. According to MCMILLAN's semiempirical approach [MCM68], λ can be calculated from the expression

$$\lambda = \frac{\mu^* \ln + 1.04}{(1-0.62\mu^*)\ln - 1.04} \tag{17-4}$$

already encountered in simplified form in Section 7.3.2, in which μ^*, the "Coulomb pseudopotential," is expressible in terms of the isotope-shift coefficient α_I and other quantities according to

$$\mu^* = (1-2\alpha_I)^{1/2}/\ln \cong 0.1 \tag{17-5}$$

where in both expressions $\ln \equiv \ln(\theta_D/1.45\,T_c)$.

MCMILLAN has thus been able to calculate for transition metals and many of their alloys a set of semiempirical values of λ which turned out to occupy a range of between about 0.5 (Ti–V alloys) and 0.9 (Zr–Nb alloys).

For Ti–Nb, values of θ_D and T_c are available from Figures 8-4 and 10-1. For Ti–Nb(40 at.%), for example, $\theta_D = 225$ K and $T_c = 8.7$ K, from which it can be concluded that $\lambda \cong 0.80$. Even assuming that α^e is as large as $\lambda/3$, the enhancement predicted by (15-43) still amounts to 32% of H_{p0}^{CL}.

17.2.4 Summary of Recent Advances in WHH Theory

As pointed out in Section 15.8.2, a recalculation of WHH theory by SCHOPOHL and SCHARNBERG [SCH81] removed the $\tau_{so} \gg \tau_{tr}$ restriction originally imposed by WHH. Thus, in recent years, two new results have been produced to reduce the severity of paramagnetic limitation: (i) the many-body correction which raises H_{p0} [ORL79, ORL81], [BEA82], and (ii) the scattering-time correction which permits more spin-orbit scattering to be introduced [SCH81].

17.3 Fundamental Determiners of the Upper Critical Field—Prospects for Raising H_{c2}

Equations, cited in preceding sections, that are primarily responsible for controlling the magnitude of the experimentally attainable upper critical field are

$$H_{c20}^{\min} = \frac{H_{c20}^{*} H_{p0}}{\sqrt{2H_{c20}^{*2} + H_{p0}^{2}}} \qquad (15\text{-}6)$$

$$H_{c20}^{*}\,(\text{dirty}) = 3.06 \times 10^{4} \rho_n \gamma_v T_c \qquad (\text{Oe}) \qquad (14\text{-}27)$$

$$H_{p0} \cong (1 - \alpha^{e})(1 + \lambda + \lambda_s) H_{p0}^{CL} \qquad (15\text{-}43)$$

$$H_{p0}^{CL} = \frac{3.5}{2\sqrt{2}} \left(\frac{k_B}{\mu_B}\right) T_c = 1.84 \times 10^{4} T_c \qquad (\text{Oe}) \qquad (15\text{-}3, 4)$$

$$\lambda_{so}/\lambda_{tr} \sim (Ze^2/\hbar c)^4 \qquad (17\text{-}3)$$

It is important to enquire into the ways in which the parameters of these equations can be manipulated to enable the various fields to be increased in the manner indicated by the following diagram:

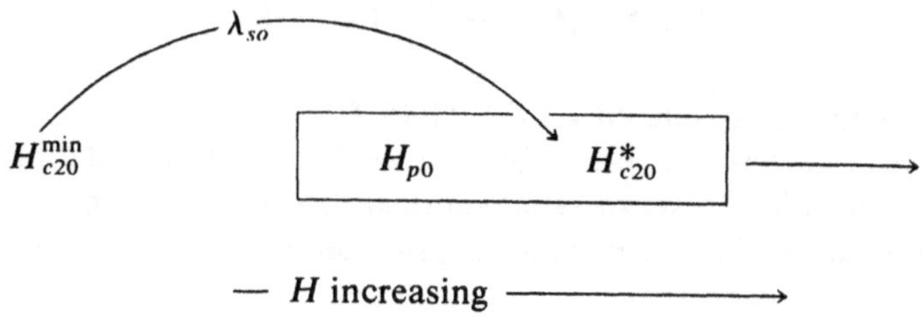

According to the diagram, our goal would be achieved by first increasing both H^*_{c20} and H_{p0}, then by introducing sufficient spin-orbit scattering to promote the experimental H_{c20} to as high a value as possible within the interval $H_{p0} - H^*_{c20}$. Prospective ways of accomplishing this are discussed below.

17.3.1 Prospects for Raising H^*_{c20}

According to (14-27), H^*_{c20} will increase in response to increases in one or all of the quantities ρ_n, γ, and T_c. The latter two are both $n(E_F)$-related and tend to vary in unison as functions of e/a ratio, (see e.g., Figure 10-8).

1. In Ti–TM alloys, γ generally rises to a maximum at $e/a \cong 4.5$ [HEI66], although for Ti–Nb in particular, the "maximum" is fairly flat throughout $4.5 < e/a < 4.9$ (Figure 8-4). In the interest of encouraging flux-pinning precipitate formation (see Section 21.13), the technical Ti–Nb alloy is usually made to contain about 34 at.% (~50 wt.%) Nb. At an e/a of 4.34, such an alloy is just on the threshold of the specific-heat plateau. Thus, Ti–Nb superconductors for use in very high fields, where flux-lattice softening considerations tend to be more important than flux-pinning potency, would benefit from a higher than usual Nb content.
2. Factors primarily responsible for T_c have already been discussed in Section 12.3.
3. Equation (14-27) also demonstrates that H^*_{c20} is directly proportional to ρ_n, the residual resistivity. This can be increased by, for example, the addition of strong-scattering centers. Ti–TM alloys having residual resistivities of typically 100 $\mu\Omega$ cm (near $e/a = 4.3$) with electronic mean-free-paths of about 5 Å [HAK67[b]] are already very dirty, but it is not inconceivable that further improvements to H^*_{c20} could be achieved through further increases in ρ_n. Normal scattering may be increased in two ways: (i) *intrinsically*, by introducing a high density of structural defects into the lattice—this will take place naturally during the cold deformation that is always administered as part of the J_c-optimization process, and also under high-energy particle radiation; (ii) *extrinsically*, through further addition of strong-scattering solutes. The strongest scatterers of conduction electrons in Ti-base alloys are the interstitial elements C, B, N, and O and the simple metals such as Al, Sn, and Si (see Figure 5-2). Unfortunately the interstitial elements are particularly rapid hardeners of their Ti-base hosts (Section 3.4) so that any benefits to H^*_{c20} are acquired at the risk of decreases in workability. Interstitial elements are always present in commercial Ti–Nb; it is therefore possible that the variations in H_{c2} noted among different samples of ostensibly the same alloy could well be due in part to variations in interstitial-impurity level. As a final note, (14-27) as it stands does not suggest the existence of a theoretical upper bound to ρ_n, but when very short mean-free-paths are being contemplated the possibilities of electron

localization and the breakdown of conventional alloy theory cannot be ignored [BEA82[a]].

17.3.2 Prospects for Raising H_{p0}^{CL}

Since Pauli paramagnetic limitation arises when the field-induced paramagnetic energy of the normal state, $\frac{1}{2}\chi_p^0 H_a^2$, becomes comparable to that of the superconducting state, $\frac{1}{2}\Delta_{00}^2 n(E_F)$, it would seem at first glance that the effect could be reduced by selecting an alloy with a low normal-state Pauli paramagnetic susceptibility. Such is not the case; in fact, since $n(E_F)$ enters into the expression for χ_p^0 as well as the right-hand side of (15-3a), PPL is *independent* of χ_p^0. Returning then to (14-3) and (14-4) above, it can be seen that the only way to increase H_{p0}^{CL} is to increase T_c. By so doing, H_{p0} and H_{c20}^* would increase together—a highly desirable result.

17.3.3 Prospects for Raising H_{p0}

For a given H_{p0}^{CL}, H_{p0} could in principle be enhanced by increasing the λs and reducing α^e. Unfortunately, apart from the fact that little independent control can be exerted over those parameters, since α^e is proportional to $n(E_F)$ by definition and λ is proportional to $n(E_F)$ via its "proportionality" to T_c/θ_D (as in Figure 8-7), both α^e and λ increase together.

1. According to Figure 8-7 and the accompanying discussion, λ is a monotonically increasing function of T_c/θ. Thus, any factor that raises T_c while keeping θ_D fixed or even reducing it will benefit λ. There is fortunately full compatibility between these two requirements. For example, Figure 8-1 suggests and Figure 10-8 directly demonstrates that there is an inverse scaling of T_c with θ_D. It follows that any solid-solution alloying that places T_c at a maximum on the T_c *versus* composition curve will also ensure that θ_D is at a minimum. Typical values of λ for Ti, Nb, and Ti–Mo(25 at.%) have been given in Section 15.8.1. For Ti–Nb(~35 at.%), with $T_c = 7.9$ K and $\theta_D \cong 222$ K, Figure 8-7 yields a λ-value of 0.8 (similar to that of Nb) and an electron–phonon enhancement factor of 1.8.

2. Electron-electron correlation (between opposite spins) and exchange (between parallel spins), jointly known as electron–electron interactions, enhance the spin paramagnetism of the electron gas according to

$$\chi_p = \chi_p^0/(1-\alpha^e) \tag{7-3}$$

where $\alpha^e \equiv 2n(E_F)\phi$ and ϕ is an electron–electron interaction potential. What little information that is available about α^e in transition elements has been assembled by GARDNER and colleagues (e.g., [GAR65, GAR65[a]]). It seems that for the paramagnetic transition elements, ϕ increases from 0.03 to 0.30 eV on proceeding from left to right along the rows of the periodic table, Section 7.3.1. To obtain the interaction parameter, α^e, for a given metal or alloy, values of ϕ within this range must be multiplied by the appropriate $n(E_F)$, which executes wide excursions as e/a increases along these rows. Calculations already performed in Section 15.8.1 suggest that the Stoner factor $(1-\alpha^e)$, which appears as a premultiplier to H_{p0}^{CL} in (15-43), will be significantly less than 1 for alloys with high $n(E_F)$s—precisely those which are being sought as candidates for high T_c and H_{c2}. There seems to be no escape from this disadvantage. To be sure, a large α^e will tend to be accompanied by strong spin fluctuations, which are reflected here in a nonzero value of λ_s, but the resulting correction to $(1+\lambda)$, although going in the right direction, is very small. α^e-data for transition elements and alloys are scarce; some available values are: $\alpha_{\text{Sc}}^e = 0.4$, $\alpha_{\text{Y}}^e = 0.1$ (Section 7.3.1), $\alpha_{\text{V}}^e \cong 0.45$, and $\alpha_{\text{Nb}}^e \cong 0.25$ [ORL81]. From this it can be guessed that for Ti-Nb(35 at.%) $\alpha^e \cong 0.3$, giving a Stoner factor of 0.7. The aggregate enhancement factor to be applied to H_{p0}^{CL} is, therefore, about 1.26.

Returning to the prognosis for increasing the Pauli limiting field, (15-43): since α^e scales with $n(E_F)$, and hence T_c, and λ is connected to T_c via Figure 8-7 (in particular by $\lambda \cong 13(T_c/\theta_D) + 0.35$), provided α^e is nonzero it follows that, although H_{p0}^{CL} appears to increase indefinitely with T_c, the modified quantity, H_{p0}, passes through a maximum[2] at some $T_c|_{H_{c2,\max}}$, a value which could conceivably be less than the maximally attainable T_c for a given class of alloy.

17.3.4 Prospects for Raising H_{c20} above $H_{c20}^{\min}$

As pointed out in Section 17.2.2 (and, of course, Chapter 15) and indicated in the diagram on p. 604, through the influence of spin-orbit scattering H_{c20} can be promoted from the minimum value $H_{c20}^{\min}$ prescribed by (15-6) to some value between H_{p0} and H_{c20}^* (the latter, in this case, being the higher)[3] through the inclusion in the alloy of high-Z scattering centers. The extent to which the spin-orbit scattering principle was applicable, according to the original WHH theory, was restricted by the scattering-relaxation requirement $\tau_{so} \gg \tau_{tr}$. This is no longer a limitation according to SCHOPOHL and SCHARNBERG [SCH81], whose results give free rein to the use of heavy-ion spin-orbit scattering for the elevation of H_{c20} in relation to H_{p0} and H_{c20}^*, now enhanced by the operation of the mechanisms outlined in the previous subsections.

[2]With λ as given above, and with $n(E_F) = AT_c + B$, $(1-\alpha^e)(1+\lambda)$ maximizes at $T_c|_{H_{c2,\max}} = 1/4\phi A - (B/2A + \theta_D/19)$, which, depending on ϕ, could conceivably limit H_{p0}.

[3]In Ti-Nb(32 at.%), for example, $H_{p0} = 150$ kOe and $H_{c20}^* = 254$ kOe, Table 17-1.

17.4 Influence of Metallurgical and Physical Variables on the Measured Upper Critical Field

17.4.1 Cold Deformation

The influence of cold deformation enters through its effect on ρ_n which, being already large in alloys within the concentration range of interest, is susceptible to only relatively small cold-work-induced change. For example, HAKE and LESLIE [HAK63] noted that the resistivity of Ti–Mo(16 at.%) increased by only about 11% in response to cold rolling of up to 90% reduction in thickness; the increase in ρ_n was expected to be accompanied by a corresponding small increase in H_r, the resistive upper critical field. With Ti–Nb(33 at.%) (heat treated and quenched), the change in critical field accompanying cold deformation of 97.2% was undetectable within an experimental error of ±0.5% [HEL71]. On the other hand, in a comparison of similarly-heat-treated $Ti_{64.7}$–$Nb_{33.3}$–$Ge_{2.0}$ samples, H_r was found to increase from 96.5 to 105.5 kOe as a result of the final cold work [HEL71]. It is possible only to speculate that this change may have been associated with the redistribution of a precipitated Ti_5Ge_3 phase.

17.4.2 Aging

As indicated in Table 17-2 the aging dependences of the upper critical fields of three classes of binary Ti–Nb alloy have been considered. In discussing the aging results, it should be borne in mind that in the concentration range below 30 at.% Nb the aged $\omega+\beta$-microstructure is permissible as a metastable phase, Figure 2-1, whilst both below and above that concentration limit, equilibrium α-phase precipitation has been observed in alloys with Nb contents of up to about 37 at.% [WES80, WES82, WES83] but not to any practical extent above about 42 at.% [NEA71]. This is the rationale guiding the grouping of Nb concentrations in Table 17-2.

(a) Low Concentration Ti–Nb Alloys. Although they are situated within the meta-equilibrium $\omega+\beta$ field, "low-concentration" Ti–Nb alloys when heavily deformed, tend to precipitate α-phase at dislocations and cell walls (Section 2.9 ff). X-ray diffractometry is a powerful tool for identifying the precipitation products of aging. Using it: (i) BAKER and SUTTON [BAK69] showed that in previously recrystallized 78% cold-swaged Ti–Nb(20 at.%), the deformation-induced α''-phase gave way to β- then α-phase during moderate-temperature (330°C) aging; (ii) VETRANO and BOOM [VET65] identified the development of α-phase during the aging at 400°C of previously recrystallized and 80% cold rolled Ti–Nb(22.5 at.%); (iii) RICKETTS *et al.* [RIC70] positively identified the formation of ω-phase during the 1-h

Table 17-2. Influence of Aging on the Upper Critical Field in Ti-Nb Alloys—Representative Studies

Nb Concentration at.%	Major Impurities or Additions	wt. ppm	Literature
Low Concentration Range			
20	O	1000	Baker and Sutton [BAK69]
22.5	O	<200	Vetrano and Boom [VET65]
26	O	440, 2400	Ricketts *et al.* [RIC70]
	Er	3000, 2900	
	Sc	3800, 4500	
Intermediate Concentration Range			
33	O	600	Heller [HEL71]
34	O	510	Rauch [RAU68]
35	O	3000	Witcomb and Dew-Hughes [WIT73]
	N	400	
High Concentration Range			
44	O	520, 1100, 2700	Rauch [RAU68]
45	O	2390	Reuter *et al.* [REU66]
67	O	340	Rauch [RAU68]

aging at temperatures near 400°C, and α-phase at temperatures near 500°C, of Ti-Nb(25 at.%) containing various small amounts of O, Er, and Sc, although in this case the metallurgical examination was conducted on a rolled foil, rather than the wire used for resistive critical field determination.

According to Figure 10-1, in all but the highest concentration Ti-Nb alloys (i.e., mean composition $\lesssim$55 at.% Nb), Nb enrichment of the matrix will increase its T_c. According to Figure 8-4, γ peaks at about 45 at.% Nb but decreases only slightly thereafter. It is to be expected, therefore, in agreement with most authors and the usual relationship $H^*_{c20} = 3.06 \times 10^4$ $\rho_n \gamma_v T_c$, (14-27), that in the low-concentration alloys H_r will tend to increase with aging as a consequence of the rejection of Nb to the matrix during precipitate ripening. For this reason it is interesting to note in the results of the measurements by RICKETTS *et al.* [RIC70] on Ti-Nb(26 at.%) an unexpected *inverse* proportionality between ρ_n and H_r, Figure 17-3. It is presumably the clearing of dislocations away from cell interiors (Section 2.11.2) that is responsible for the decrease in resistivity during optimal aging. This effect, which in the absence of other mechanisms would decrease

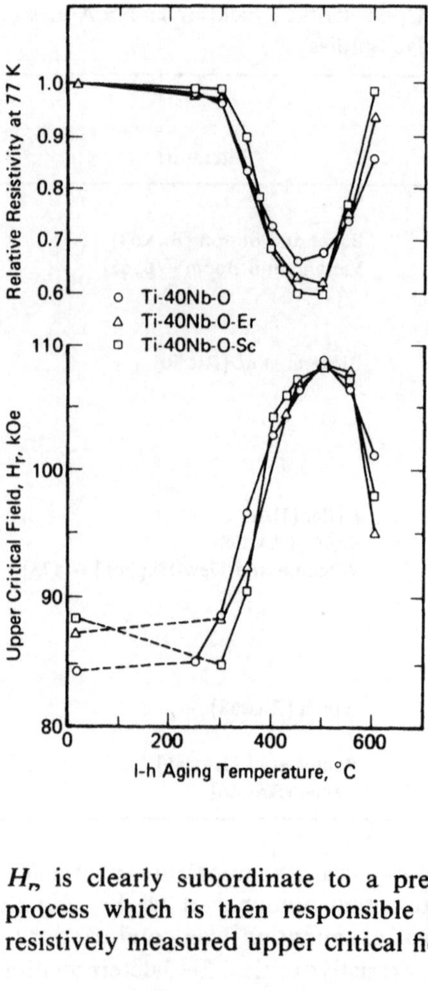

Figure 17-3. Variation of relative resistivity (measured at 77 K) and resistive upper critical field (H_r, kOe) with temperature of 1-h aging for a nominal Ti–40Nb alloy containing various small amounts of O, Er, and Sc, *viz*: 2400 ppm O (○); 1.67 vol.% Er_2O_3 plus 60 ppm O (△); 1.20 vol.% Sc_2O_3 plus 1140 ppm O (□) (H_r criterion: $\frac{1}{2}\mu V\,cm^{-1}$ at 1 mA in a 10-mil wire; i.e., $4 \times 10^{-7}\,\Omega$ cm)—after RICKETTS *et al.* [RIC69].

H_r is clearly subordinate to a precipitation-induced matrix-enrichment process which is then responsible for the net increase observed in the resistively measured upper critical field.

(b) Intermediate-Concentration Ti–Nb Alloys. As the Nb concentration increases, and H_{c2} approaches its peak value within the concentration range 35–39 at.% Nb [BER63], the response of H_r to aging, even though matrix enrichment may be taking place, is expected to be rather slight. Cold-worked-and-aged Ti–Nb(33.3, 34, and 35 at.%) alloys have been studied. RAUCH [RAU68] noted a quite erratic variation of H_r with 1-h aging temperature which he attributed to precipitation at moderate temperatures followed by O pickup at elevated temperatures. WITCOMB and DEW-HUGHES [WIT73], as a result of low-field magnetization studies of Ti-Nb(35 at.%), deduced that H_u was increasing with moderate-temperature aging time. HELLER [HEL71] studied the effect of aging at 300, 400, 500, and 600°C on the H_r of Ti–Nb(33.3 at.%) (and similar alloys doped with Ge). The small increases in the binary-alloy H_r that were noted after aging for short times ($\cong$1 h) at all of the temperatures mentioned were attributed to the usual solute-redistribution mechanism.

(c) High-Concentration Ti–Nb Alloys. In the high-Nb alloys with moderate levels of O, rather erratic dependences of H_r on the 1-h aging temperature of previously cold-worked alloys is seen in the results of RAUCH [RAU68]. The high-O (2700 ppm) alloy, however, exhibited a pronounced increase in H_r with 1-h aging temperature. REUTER [REU66] likewise noted an increase in H_r during the heat treatment of Ti–Nb(45 at.%) containing 2390 ppm O, attributing the effect to e/a-ratio changes associated with Ti_xO precipitation.

17.4.3 Influence of Temperature

(a) The Entire Temperature Range. The full temperature dependence of H_{c2} over the entire range from $t = 0$ to 1 is describable in terms of

$$H_{c2}(t) = \sqrt{2}\kappa_1(t, \beta_M)H_c(t) \qquad [14\text{-}10](17\text{-}6)$$

where $H_c/H_{c0} \cong (1 - t^2)$ ((8-7)) and the temperature dependence of $\kappa_1(t, \beta_M)$ according to Maki theory is given by (15-19) and has been depicted, for a series of integral values of β_M^2, in Figure 15-6(a). Involving as it does the unknown SOS relaxation time, (τ_{so}, (15-17) and (15-18)), β_M^2 is not independently calculable but must be treated as an empirical fitting parameter. Values of β_M^2, so obtained by BRAND [BRA72] for a series of Ti–Nb(32, 55, 62, and 83 at.%) alloys were, respectively: 0.27, 0, 0.06, and

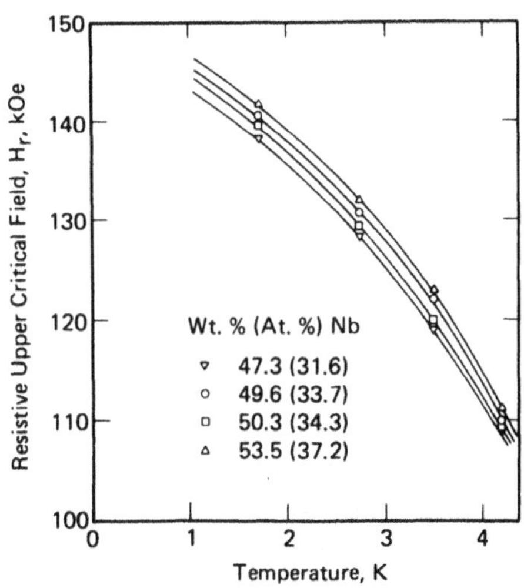

Figure 17-4. Temperature dependences of resistive upper critical fields for four optimized multifilamentary Cu/Ti–Nb conductors. H_r was the applied field corresponding to the "first appearance of resistance." The onset criterion was probably 10^{-12} Ω cm [HAW80[a]]; measuring current density J = 500–900 A cm^{-2}—after LARBALESTIER [LAR80].

0. Referring again to Figure 15-6(a), we note that for $0 < \beta_M^2 < 1$, $\kappa_1(t, \beta_M)$ changes by less than 20% throughout the *entire* temperature range, which for intermediate-concentration Ti–Nb alloys extends up to about 9 K. It follows that the dominant temperature dependence is again the $1 - t^2$ of the $H_c(t)$ factor; Figure 17-1 for a series of commercial Ti–Nb alloys has amply verified that statement.

If the $(1 - t^2)$ law were strictly obeyed, with $T_c = 9$ K a reduction of temperature of from 4.2 K to 1.2 K should increase the measured critical field by about 26%. Starting with an $H_{r,4.2\,K}$ of 110 kOe as in Figure 17-4, also for a series of technical alloys, this should amount to an increase of about 29 kOe; referring again to the figure, this prediction seems to be reasonably well borne out in practice.

(b) The Incremental Temperature Dependence Near T_c. Measurements taken from Figure 17-4 show that, whereas the average slope of $H_r(T)$ (in tesla K^{-1}) between 2 and 4 K is $(dH_r/dT)|_{2K}^{4K} = -1.22$ T K^{-1}, that for the more restricted temperature range near 4 K is the slightly greater, $(dH_r/dT)|_{3.5K}^{4K} = -1.68$ T K^{-1}. The resistive critical field temperature dependence obviously becomes steeper as T approaches T_c. In fact measurements of the alloy Ti–52Nb taken just below $T_c(= 9.0$ K) have given $(dH_r/dT)_{T_c} = -2.64$ T K^{-1} [HAw80].

The H-W nonparamagnetic theory of upper critical field temperature dependence which resulted in $-(dH_u/dT)_{T_c} = (0.68\,T_c)^{-1}H^*_{c20}$ ((14-29b)) in combination with (14-27) yields for the limiting measured upper critical field gradient near T_c

$$(dH_u/dT)_{T_c} = -4.50 \times 10^4 \rho_n \gamma_v \tag{17-7}$$

For Ti–Nb(34 at.%), with $\rho_n = 78 \times 10^{-6}\ \Omega$ cm (Table 5-2) and $\gamma_v = 1.05 \times 10^4$ erg cm^{-3} K^{-2} (Table 8-4), this equation yields

$$(dH_u/dT)_{T_c} = -3.69 \qquad (\text{T K}^{-1}) \tag{17-8}$$

in qualitative agreement with the directly measured result.

(c) Conclusion. The conclusion to be drawn from this is that if an increase in upper critical field of a few tesla is sufficiently important, as it may be in the development of very high-field magnets, advantage having been taken of the fundamental methods of H_{c2} enhancement discussed in Section 17.3 a reduction in the He-bath temperature of from 4.2 to 1.2 K can generally be relied on to provide an additional ~3 T.

17.5 Measurement of the Upper Critical Field

17.5.1 Transition Criteria

The upper critical field is usually measured magnetically (yielding H_u) or resistometrically (yielding H_r), the latter procedure being the most common. Since the transition from s to n occupies a finite magnetic field range (typically 0.5-0.7 T in optimized Ti-Nb alloys [LAR81]) the question of a fiduciary point becomes important. Different authors have regarded the critical field as corresponding to either the threshold, the mid-point, or the end, of the $s \rightarrow n$ transition. LARBALESTIER and HAWKSWORTH [LAR80], [HAW80, HAW81] recommended the use of the threshold value (i.e., the onset of resistance, or the *foot* of the $s \rightarrow n$ curve), WADA has used the midpoint of the $s \rightarrow n$ transition as his criterion [WAD81], while SUENAGA and RALLS, in their early studies [SUE69] of Ti-Nb-Ta alloys, chose the endpoint, H_{rn}, as the criterion for assigning a value to H_r.

The rationale offered by SUENAGA and RALLS for their selection of H_{rn} rather than H_{rs}, the field at the start of the transition, or even that at the mid-point, lay in their belief that whereas H_{rs} depended to some extent on the bulk pinning strength and consequently the metallurgical state of the sample, H_{rn} would tend to be more under the control of an intrinsic alloy property [SUE69]. These ideas are rooted in the fact that what is actually being measured is the ability of the material to support the chosen transport current density (typically 10 A cm^{-2}). At fields a little below H_{c2} such a current will remain stable or not depending on whether the relevant $\mathbf{J}_c \times \mathbf{B}(H_a)$ is less than or greater than the bulk pinning force; on the other hand, at H_{c2} itself this alternative is moot, since the Abrikosov lattice is on the point of vanishing. The "critical-current" nature of the measuring current had been noted much earlier by RALLS [RAL64] who drew attention to the fact that H_r-anisotropy had been observed in a cold-rolled strip. For reasons such as these, LARBALESTIER [LAR81] has recommended that in case the upper critical field is being estimated from a single potentiometric measurement the resulting H_r-value assigned to the sample should be accompanied by information as to: (i) the degree of cold work; (ii) the transition-point selected; (iii) the measuring current density. In measurements of bare superconductor, sensitivity of the measuring equipment should not be a problem; obviously at 10 A cm^{-2} the complete transition in an alloy such as Ti-Nb(34 at.%), with a residual resistivity of $78 \times 10^{-6}\ \Omega$ cm, will be accompanied by a voltage gradient along the sample of 0.8 mV cm^{-1}.

17.5.2 Measurement Current Density and Other Considerations

The pulsed-field difference between the H_{rs} and H_{rn} measured by SUENAGA and RALLS was about 0.7 T [SUE69]. Since in steady-field measurements performed on the same sample of commercial Ti–Nb wire the DC *onset* field was 0.25 T higher than H_{rs} (pulsed) and 0.45 T lower than H_{rn} (pulsed), it is clear that comparisons between the results of critical-field measurements performed using different techniques upon different samples of the same alloy must be made with considerable caution.

The question of measuring current density was first addressed by BERLINCOURT and HAKE [BER63] in connection with their classical series of critical-field measurements. They claimed that a J_c of less than 10 A cm^{-2} represented the best compromise between adequate detection sensitivity and the validity of the identification of the resulting H_r with the H_{c2} being sought. In support of that claim they offered the fact that their H_r was nearly independent of both the degree of cold-rolling to which the test alloys had been subjected and to the orientation of the applied field with respect to the rolling induced texture. The acceptability of the 10 A cm^{-2} criterion has endured rather well. In subsequent measurements of H_r in Ti–Nb and related alloys that value of current density has been widely favored: e.g., by CHANDRASEKHAR *et al.* [CHA63], EL BINDARI and LITVAK [ELB63], JONES *et al.* [JON64], and SALTER [SAL66], although more recently the tendency has been to employ even lower measuring-current densities such as 5 A cm^{-2} [HAW80], 1.3 A cm^{-2} [HEL71], and 1 A cm^{-2} [WAD81]. As early as 1964, RALLS [RAL64] compared three H_r *versus* composition curves for Ti–Nb prepared for data taken at current densities of 10^3, 10, and 1 A cm^{-2}, respectively, Figure 17-5. At the optimal composition (~45 at.% Nb) the resistive upper critical fields H_r(10 A cm^{-2}) and H_r(1 A cm^{-2}) (of which the latter was the higher) differed by only 1 kOe.

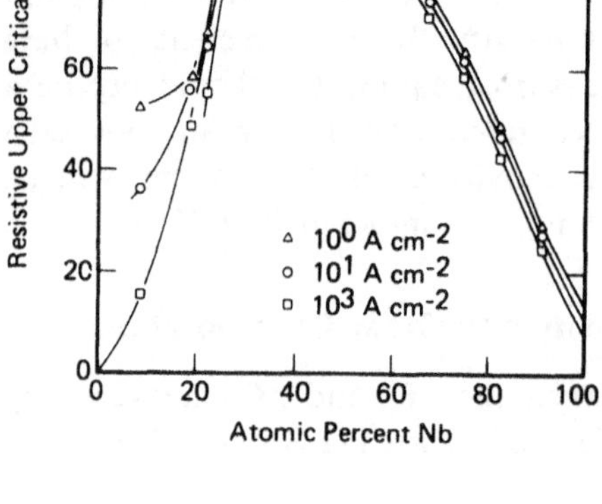

Figure 17-5. Influence of measuring current density, in association with a resistive-onset criterion of $\frac{1}{4}$ μV cm^{-1}, on the 4.2-K resistive upper critical fields of a series of bare heavily cold-worked 0.25 mm$^{\varnothing}$ Ti–Nb alloy wires—after RALLS [RAL64].

Obviously if a value of H_r corresponding to a vanishingly small test current density is desired, it is always possible to extrapolate a linear plot of J_c *versus* H_a down to the horizontal axis [TAD80].

17.6 Upper Critical Fields of Titanium–Niobium-Base Alloy Superconductors—Some Experimental Results

Systematic study of the upper critical field in Ti-alloy superconductors commenced with the work of BERLINCOURT and HAKE [BER63] which followed a comparable investigation by HULM and BLAUGHER [HUL61] of their superconducting transition temperatures. Numerous inquiries into the effects of metallurgical variables on H_{c2} were to follow and, as indicated in Section 17.2.2, in attempts to improve upon the H_{c2} of the basic Ti–Nb binary alloy, transition-metal substitutions were made on one hand for the Ti (using Zr or Hf) and on the other for the Nb (using V or Ta). In pursuing the same philosophy several quaternary alloys, in which substitutions were made for *both* of the primary constituents, were prepared and measured; the most important of these turned out to be Ti_{61}–Zr_6–Nb_{27}–Ta_6.

At the same time, the influence of simple-metal alloying on H_{c2} had not been neglected; the most comprehensive studies were conducted by ZWICKER *et al.* who in a series of publications [ZWI70], [HEL71, HEL71ª], [LOH71] reported on the effects of small additions of Al, Cu, Ga, Ge, Ag, In, and Sn. The influence of Si on the superconducting properties of Ti–Nb has also been considered, and in a series of recent studies INOUE *et al.* [INO80] investigated the upper critical fields of a family of Ti–Nb–Si alloys which had been prepared in amorphous form as continuous melt-spun ribbons.

Some of the most important results are mentioned in the following brief review. For a complete account the reader is directed to [COL83].

17.6.1 Upper Critical Field of Ti–Nb

The resistive upper critical field composition dependence of Ti–Nb as determined by RALLS [RAL64] at 4.2 K using three measuring-current densities, is depicted in Figure 17-5. These results, together with those of several more recent measurements also made at 4.2 K, and those of BERLINCOURT and HAKE [BER63] which were taken at 1.2 K, are assembled for comparison in Figure 17-6. As discussed already in Section 17.3 the fundamental basis for the general shape of the curves, which possess broad maxima within the composition range 34–42 at.% Nb, has to do with a competition between the composition dependences of ρ_n, γ_v, and T_c (see (14-27)). The rapid decrease in H_r with composition below about 36 at.% Nb

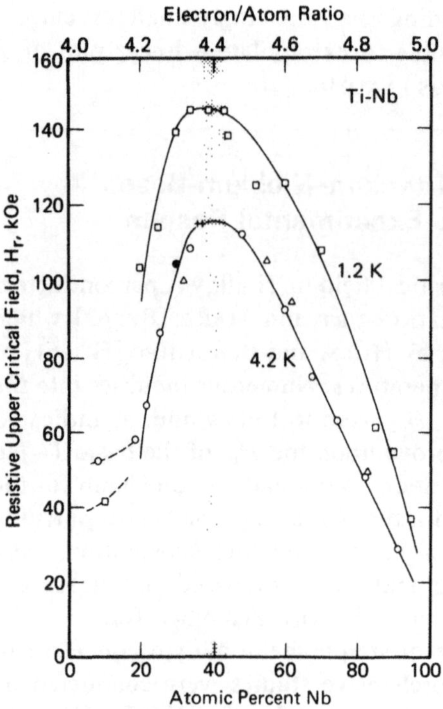

Figure 17-6. Resistive upper critical fields, H_r, as functions of Nb content in Ti–Nb alloys. Represented are: the early 1.2-K results of BERLINCOURT and HAKE [BER63] (□), the subsequent 4.2-K results of RALLS [RAL64] (○) ($\frac{1}{4}$ μV cm^{-1} or 10 A cm^{-2}), and the more recent results of BRAND [BRA72] (△) (from $J_c(H_a) \to 0$; $\sim 10^{-8}$ V cm^{-1} criterion), HAWKSWORTH and LARBALESTIER [HAW80] (++) (resistive onset at 5 A cm^{-2}), and WADA *et al.* [WAD80] (●) (resistive midpoint at 1 A cm^{-2}).

(i.e., 52 wt.% Nb) is an important reason for maintaining the Nb concentration above that level. Indicated by a vertical band in Figure 17-6 is the *practically* attainable composition limit for macroscopic α-phase precipitation in these alloys (see Section 1.4.3); only with a Nb concentration to the left of this band is it possible to develop, in the interests of J_c maximization, an adequately dense population of α-phase particles. However several authors such as SEGAL *et al.* [SEG80] and HILLMANN *et al.* [HIL79], following VETRANO and BOOM [VET65] who first proposed the idea, suggested that the best of both worlds could be achieved by choosing the initial mean alloy composition and the thermomechanical processing schedule in such a way that the precipitation of α-phase in the cell walls is accompanied by a rejection of Nb into the matrix and a movement of its H_r towards the peak of the H_r-composition curve.

17.6.2 Simple-Metal Additions to Ti–Nb

Although it is possible that simple metals, which when present in solid solution are strong scatterers of conduction electrons, should exert some influence on H_{c2} by way of the residual resistivity, ρ_n, their principal role turns out to be metallurgical rather than physical. As indicated in Section

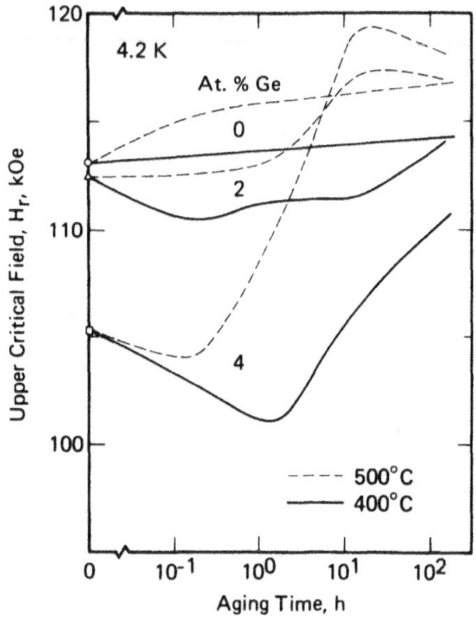

Figure 17-7. Resistive 4.2-K upper critical fields as functions of aging time at 400°C and 500°C of 1-mm$^{\varnothing}$ wires of previously cold-worked (97%) Ti_{63}-Nb_{33}-Ge_4 and Ti_{65}-Nb_{33}-Ge_2, compared with that of a Ti-Nb(33 at.%) control (criterion: 1 $\mu V\,cm^{-1}$ at 10 mA; i.e., $8 \times 10^{-7}\,\Omega\,cm$)—after HELLER [HEL71].

2.6 and with reference to Figure 2-4, so-called α-stabilizing additions (of which the simple metals are a particular class) encourage α-phase precipitation from both the $\omega+\beta$-phase and $\beta'+\beta$-phase of Ti-Nb. Studies of superconductivity in Ti-Nb-SM alloys have frequently included the administering of a matrix of aging heat treatments, the effects of which on H_{c2} have usually been interpreted in terms of this α-phase precipitation and the accompanying Nb-enrichment of the β-matrix just referred to. In the absence of heat treatment the presence of small percentages of simple metals in solid solution is usually detrimental to the critical fields of commercially interesting Ti-Nb alloys, such as Ti-Nb(33.3 at.%) [HEL71[a]]. However, during moderate-temperature aging it is possible for H_r to increase beyond that of the binary host due to the β-enrichment effect which, at sufficiently high solute concentrations, is associated not only with the α-phase referred to above but also with Ti_xSM_y intermetallic-compound precipitation. The critical-field properties of Ti_{65}-Nb_{33}-Ge_2 and Ti_{63}-Nb_{33}-Ge_4 as depicted in Figure 17-7 are examples of this mechanism in operation.

17.6.3 Substitutes for Ti in Ti–Nb Alloys

(a) Upper Critical Fields of Ti–Zr–Nb Alloys. One of the first systematic studies of H_r in concentrated Ti-Zr-Nb alloys was by RALLS and coworkers [RAL64] who investigated the binary sides and the Nb_x-$(Ti_{50}$-$Zr_{50})_{1-x}$ perpendicular bisector of the Nb-apex composition triangle. Other investigators of the variation of H_{c2} throughout the interior of the Ti-Zr-Nb

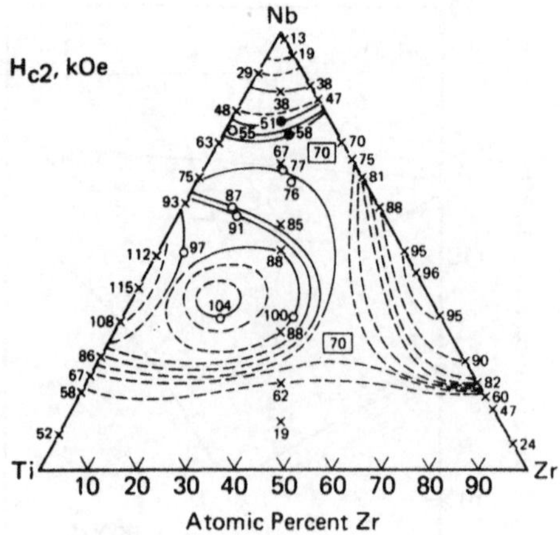

Figure 17-8. Ternary diagram (at.% linear) of the magnetically determined upper critical fields of ten Ti-Zr-Nb alloys as determined by DESORBO [DES67] plotted as measured except that two values which gave rise to a local maximum near the apex of the triangle (●) have, with careful justification, been reduced from 79.5 and 76.0 kOe to 58 and 51 kOe, respectively. The binary boundary-value data and ternary "axial" data, all denoted by ×, are due to RALLS [RAL64].

triangle were DESORBO *et al.* [DES67] (H_u, magnetic), LAZAREV *et al.* [LAZ68] (H_r), and ALEKSEEVSKII *et al.* [ALE68] (H_r). The results of the latter two groups have been reported by SAVITSKII *et al.* [SAV73, p. 313]. Some serious discrepancies mar the combined conclusions of DESORBO, LAZAREV, and ALEKSEEVSKII, but when these are resolved by a carefully reasoned readjustment of some of the data, a consistent picture emerges typified by Figure 17-8. A careful examination of the combined results reveals the following features: (i) a monotonic increase in H_{c2} for isoelectronic alloys (parallel to the Ti-Zr base) of compositions up to about Nb_{30}-$(Ti$-$Zr)_{70}$ followed by a flat valley lying between the Zr-Nb side and some high values of H_{c2} associated with the opposite side; (ii) a local maximum in H_{c2} centered about Ti_{40}-Zr_{20}-Nb_{40} which, however, by a slight rearrangement of the contours could equally well be regarded as being connected by a ridge to the maximum on the Ti-Nb edge.

(b) Upper Critical Fields of Ti-Hf-Nb Alloys. Although a few measurements were carried out during the 1960s [DES67], [BYC70[a]] the most extensive studies of the upper critical fields of Ti-Hf-Nb alloys were conducted in the late 1970s by: (i) HAWKSWORTH and LARBALESTIER, who measured seven binary Ti-Nb alloys and twenty-five ternary alloys containing 2.5, 5, and 10 at.% Hf [HAW80] and 15 and 20 at.% Hf [HAW81] at temperatures between 2 and 4.2 K; (ii) WADA *et al.* [WAD81], who investigated a family of eight Ti-Nb binary and thirty-four ternary alloys containing up to 15 at.% Hf. Very little improvement, if any, was won by the substitution of Hf for Ti: the 4.2-K results of HAWKSWORTH and LARBALESTIER showed a steady deterioration of H_r with added Hf, while at 2 K some slight but practically unimportant improvement in H_r (~0.2 T) seemed to accompany the substitution of 2.5 and 5 at.% Hf into some Ti-Nb alloys. The 4.2-K results of WADA *et al.*, which were for the most part in satisfactory agreement with the above, did exhibit a small maximum in H_r slightly offset from the Ti-Nb edge of the composition triangle; in particular, a maximal critical field for the entire series was obtained with Ti_{36}-Hf_3-Nb_{40}, whose 4.2-K H_r of 11.7 T represented at best a marginal increase over those achieved by the binary control alloys, *viz.*, 11.3 T by Ti-Nb(40 at.%) [WAD81] and Ti-Nb(44 at.%) [HAW80].

17.6.4 Substitutes for Nb in Ti–Nb Alloys

(a) Upper Critical Fields of Ti–Nb–V Alloys. The resistive upper critical fields of thirty-five ternary Ti-Nb-V alloys were measured by BELLIN, GATOS, and SADAGOPAN [BEL70]. The presence of V is not beneficial. In that the isoelectronic H_r *versus* composition curves all possess minima, it is always possible to find a binary alloy with an H_r higher than any given ternary.

(b) Upper Critical Fields of Ti–Nb–Ta Alloys. The resistive upper critical fields of a group of some twenty-seven Ti-Nb-Ta alloys were measured using pulsed field techniques at 4.2 K by SUENAGA and RALLS [SUE69]. The interesting and encouraging conclusion that the ternary system produced a maximum H_r(4.2 K) of 12.4 T in Ti_{64}-Nb_{30}-Ta_6, in agreement with the H_u(4.2 K) of 12.4 T in Ti_{60}-Nb_{36}-Ta_4 obtained by DESORBO [DES67] using magnetic techniques, was not substantiated by the more recent work

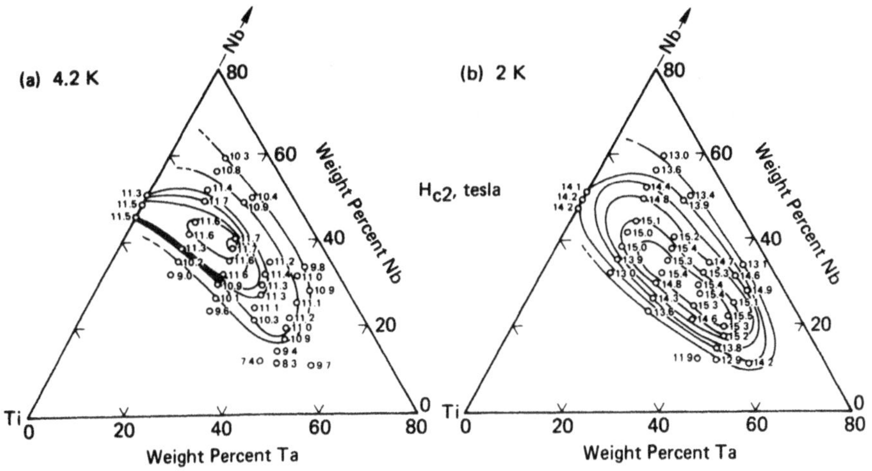

Figure 17-9. Resistive DC upper critical fields (criterion: resistive "onset" at $J = 5\ A\ cm^{-2}$) of bare cold-drawn Ti-Nb-Ta wires at (a) 4.2 K, (b) 2 K. The contours have been inserted, with considerable license, to guide the eye—data from LARBALESTIER [LAR80].

of HAWSKWORTH and LARBALESTIER [HAW80], [LAR80, LAR81]. Some of the results of the latter authors who studied forty-one Ti-Nb-Ta alloys, generally at 2 and 4.2 K [LAR81] but occasionally also at 1.7 and 3 K [HAW80], are depicted in Figure 17-9. At 4.2 K, the ternary alloys measured showed little improvement over the corresponding binaries. But since at 2 K, H_r rose to 15.2–15.4 T within the alloy group Ti-$Nb_{20\text{-}27.5}$-Ta_{15} [HAW80] and to 15.4–15.5 T for the alloy Ti_{65}-Nb_{25}-Ta_{10} [HAW80], [LAR81], it can be concluded that at this lower temperature the best ternary alloys have resistive upper critical fields some 1.15–1.25 T higher than those of Ti-Nb(36 and 37 at.%) [HAW80].

Detailed studies of Ti-Nb-Ta alloys within the framework of Maki-WHH theory were undertaken by SUENAGA and RALLS [SUE69], and again more recently by KHAN [KHA83].

(c) Upper Critical Fields of Ti–V–Ta Alloys. When calculated from their "normal-state properties" (see (14-27)) the H^*_{c20}s of Ti-V(34 at.%) and Ti-Nb(32 at.%) are about the same and equal to about 25 T (see [COL83, pp. 139 and 227]). The fact that V has a lower atomic number ($Z = 23$) than Nb ($Z = 41$) suggested to INOUE *et al.* [INO82] that, from a spin-orbit-scattering standpoint, the addition of Ta ($Z = 73$) would have a stronger influence on Ti-V than on Ti-Nb and would result in an exceptionally high upper critical field and, hopefully, a useful large high-field critical current density. Although J_c measurements have yet to be reported, the results so far are somewhat encouraging. In the Ti-V system, the maximum H^*_{c20} is obtained with Ti-V(40 at.%). Accordingly interest was focused on ternary systems containing a fixed 60 at.% of Ti. As far as the components of the GLAG expression for the upper critical field, (14-27), are concerned: ρ_n is the "mixture-rule" average of those of V-Ti(60 at.%) and Ta-Ti(60 at.%), hence decreases monotonically with the addition of Ta to Ti-V; T_c sags to a minimum between the V-Ti(60 at.%) and Ta-Ti(60 at.%) endpoints; no electronic specific heat results were reported, but H^*_{c20} computed from superconductive data (see (14-29))[4] decreased, mostly monotonically, between the 60-at.%-Ti endpoint values. On the other hand, and presumably on account of the spin-orbit-scattering effect, H_{c2}(2.0 K) rose to a substantial maximum for Ta concentrations within the range of 9–12 at.%. The highest H_{c2}(2.0 K) for the entire Ti-V-Ta system investigated (which included 4 alloys with 70 at.% Ti, 9 alloys with 60 at.% Ti and 4 alloys with 50 at.% Ti) was 14.3 T for Ti_{60}-V_{31}-Ta_9.

[4]Unfortunately, HAKE's clean-limit [HAK67[b]], rather than the correct dirty-limit, expression, was used.

Table 17-3. Systematics of Quaternary Alloy Critical Field Studies

Group IV			Group V			
Ti	Zr	Hf	V	Ta	Nb	References
×	×	×			×	*a, b*
×	×			×	×	*a, c*
×		×		×	×	*a, d*
×		×	×		×	*a*

[a]DeSorbo *et al.* [DES67].
[b]Rayevskii *et al.* [RAY69, RAY71].
[c]Horiuchi *et al.* [HOR73, HOR80].
[d]Hawksworth *et al.* [HAW80].

17.6.5 Substitutes for Both Ti and Nb in Ti-Nb Alloys

Of the six possible Ti-Nb-base quaternary alloy systems available by combining the groups IV and V transition elements, the four that have been investigated are indicated in Table 17-3. In the present discussion attention is focused on the relative merits of Hf and Ta substitutions on the upper critical field of Ti-Zr-Nb.

(a) Upper Critical Fields of Ti–Zr–Hf–Nb Alloys. The results of RAYEVSKII *et al.* [RAY69, RAY71] on three sequences of [Nb] = constant = 30, 50, and 70 at.% alloys, when displayed on pseudoternary composition triangles, showed that for the [Nb] = 50 at.% series, for example, H_r decreased upon the substitution of Zr for Ti (i.e., along the Ti-Nb/Zr-Nb edge) and then even more severely with the substitution of Hf (i.e., towards the Nb-Hf vertex). The deleterious influence of Hf substitution was confirmed by DESORBO [DES67], who showed that, commencing with Ti_{53}-Zr_4-Nb_{39}-Ta_4, the replacement of Zr by Hf[5] decreased H_u by about 1.1 T and a further increase of the Hf content to 20 at.% (plus a substitution of Zr for half of the remaining Ti, and removal of the Ta, leading to the alloy Ti_{20}-Zr_{20}-Hf_{20}-Nb_{40}) reduced it by another 1.6 T.

(b) Upper Critical Fields of Ti–Hf–Nb–Ta Alloys. From a series of measurements at 2, 3, and 4.2 K conducted by HAWKSWORTH and LARBALESTIER on a large matrix of compositions [HAW80], it is possible to derive useful information about the effect of Hf on the upper critical fields of Ti-Nb-Ta alloys. In substituting Hf for Ti at constant group-V element concentration (Nb_x-Ta_{10}), decreases in H_r were noted at 4.2 K and again

[5]Actually the new alloy was Ti_{50}-Hf_4-Nb_{42}-Ta_4.

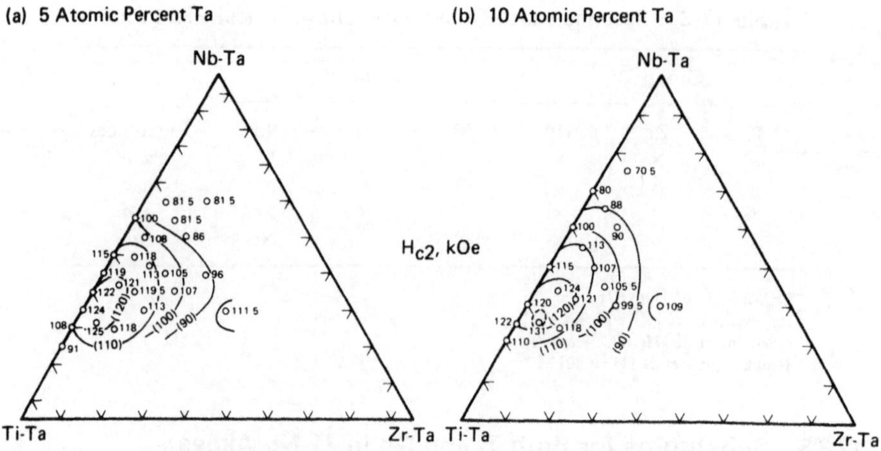

Figure 17-10. Pseudoternary diagrams (at.% linear) indicating the resistive upper critical fields (H_r, kOe, 4.2 K) of the quaternary alloys Ti-Zr-Nb-Ta. The full lines are in steps of 10 kOe. Two series of alloys are represented: (a) at.% Ta = constant = 5; (b) at.% Ta = constant = 10. Measurements were made on 0.25 mm$^{\varnothing}$ wire using a current of 5 mA (J = 10 A cm^{-2})—after HORIUCHI *et al.* [HOR73].

at 2 K. The deleterious influence of Hf, this time as a substitute for Zr in Ti-Zr-Nb-Ta, can again be appreciated by comparing the 4.2-K H_r of Ti_{60}-Zr_5-Nb_{25}-Ta_{10}, *viz.*, 13.1 T [HOR73], with that of Ti_{60}-Hf_5-Nb_{25}-Ta_{10}, *viz.*, 10.8 T [HAW80].

It may, therefore, be concluded on the basis of both of the above subsections that in seeking high-H_{c2} quaternary alloys the inclusion of Hf should be avoided. According to HORIUCHI *et al.*, the Ta-containing alloy just referred to had the highest upper critical field of any known alloy.

(c) Upper Critical Fields of Ti-Zr-Nb-Ta Alloys. The search for high-H_{c2} Ti-Zr-Nb-Ta alloys by HORIUCHI *et al.* culminated in the discovery of Ti_{60}-Zr_5-Nb_{25}-Ta_{10} and Kobe Steel Ltd.'s proprietary alloy "Cryozitt" (*viz.*, Ti_{61}-Zr_6-Nb_{27}-Ta_6), both of which may be regarded as having been derived from the Hitachi Ltd. alloy Ti_{60}-Zr_5-Nb_{35} by a partial replacement of Nb by Ta. Work leading up to the selection of an optimal composition, and its ultimate commercialization, consisted of a thorough exploration of the superconducting properties, particularly the upper critical fields of an extensive matrix of alloys. H_r measurements were made on alloy systems conforming to the following specifications: Ti-$Zr_{32.3}$-Ta_y ($0 < y < 12$ at.%) [HOR73], Ti-Zr_{30}-Nb_{30}-Ta_y ($0 < y < 15$ at.%) [HOR75],[6] $[(\text{Ti-Nb}_{30})_{95}\text{-Zr}_5]_{100-y}$-$Ta_y$ ($0 < y < 25$ at.%) [HOR74],[6] and two pseudoternary groups describable in terms of Ti-Zr-Nb-$Ta_{5,10}$ [HOR73, HOR74]. The

[6]Patent application.

results of the latter study as summarized in Figures 17-10 reveal, in the [Ta] = 10 at.% diagram, a maximum 4.2-K H_r of 131 kOe (13.1 T, the value referred to in subsection (b) above).

The upper critical field temperature dependences of Ti_{45}–$Zr_{7.5}$–Nb_{40}–$Ta_{7.5}$ in various metallurgical conditions have been measured and examined from the standpoint of Maki theory by KHAN [KHA83].

Chapter 18

Flux in Motion under the Influence of a Field Gradient

Chapter 16 has dealt with the critical state primarily as the *static* condition of a magnetized irreversible type-II superconductor. The critical state is, of course, attained in the first place by the *motion* of flux into the superconductor and the establishment of a flux density gradient whose magnitude at any place is controlled by the local pinning forces. When the applied field stops increasing or decreasing, the internal flux density gradient rests at its critical value, poised ready to move in response to any subsequent change, in either direction, of the external field. Numerous authors have inquired into the permanence of the critical state, as defined by the equality of a fluxoid driving force, F, to the aggregate or bulk pinning force, F_p. It seems that, although the critical state can for all practical purposes be regarded as permanent, it does undergo a slow logarithmic decay by a thermally activated process of flux creep. Similarly, if the sample is in its shielding critical state and the field increase is reinstated in the form of a small step, more flux will creep into the sample. In this chapter, *flux creep* is to be treated both as a thermally activated process and as an example of electromagnetic diffusion. If pinning is relatively weak, *flux flow* is able to take place. But since in the weak-pinning situation the critical flux-density gradient is necessarily small, flux flow usually occurs under the influence of a transport-current instigated Lorentz force.

This chapter deals primarily with the dynamic properties of *creep* (both thermally activated and electromagnetic), *flux flow*, and *flux jumping*, as they occur in response to a *magnetic* driving force—that is to say, the force exerted by a flux-density gradient. The influence of *transport current* on the mixed state is considered in Chapters 20ff and, in particular, in Chapter 23 where the motion of a uniform density of fluxoids under the influence of an applied-field instigated Lorentz force is considered.

18.1 Classes of Flux Motion

18.1.1 Flux Creep

Flux creep has been treated as the motion of groups of fluxoids, or vortices, referred to as "bunches," "bundles" [AND62], "fluxon bundles," or just "fluxons" [HEI74[a]]. The fluxons, which may consist of interacting groups of from a few to 10^3 fluxoids [BEA69], interact among themselves and with the pinning centers [HEI74]. A model of thermally activated creep was devised by ANDERSON [AND62] to explain two related experimental observations of KIM, HEMPSTEAD, and STRNAD [KIM62, KIM63[a]], *viz.*, (i) the temperature dependence of the pinning-force parameter $\alpha(T)$ (see Table 16-1), and (ii) an observed logarithmic decay of the critical state at constant temperature. ANDERSON's original model in which fluxoids in the presence of a flux-density gradient were thermally activated over pinning barriers of fixed height, has been criticized by CAMPBELL *et al.* [CAM68] and GRASSIE [GRA75, p. 94] on the grounds that it is the barrier height itself that responds most strongly to temperature, and by so doing controls the passage of flux. BEASLEY *et al.* [BEA69], in the most detailed study of flux creep since the work of KIM *et al.* [KIM62, KIM63[a]], did not take issue with this assessment of the situation but regarded the temperature dependence of the barrier height as a new feature, or generalization of the Anderson model, rather than as a breakdown of it. Underlying the model of flux creep as the thermally activated negotiation of pinning barriers is the idea of creep as a diffusional process. With this in mind, LUBELL and WIPF [LUB66] applied the usual equations of cylindrical heat diffusivity to a description of field change during the transition between successive critical states following a stepwise perturbation of the external field.

18.1.2 Flux Flow

Whereas flux creep specifically involves interaction between fluxoids and pinning centers, flux flow may be modelled as the free motion of vortices not subject to pinning forces. The flux lattice moves as an interacting system[1] across the material, driven by a flux-density gradient or Lorentz force, and retarded by the electric fields generated by the motion which thereby plays the role of a viscous drag [BAR65]. The velocity of flux flow, and consequently the magnitude of the viscosity coefficient η, has been measured by KROEGER [KRO66] on well-annealed (hence weakly pinned) Ti–Nb and Zr–Nb alloys. Although creep may take place in well-pinned material, and a sufficient condition for flow is the "absence" of pinning, in real materials

[1] According to BARDEEN and STEPHEN [BAR65], the supercurrent distribution moves as a rigid body, or single quantum state, through the metal.

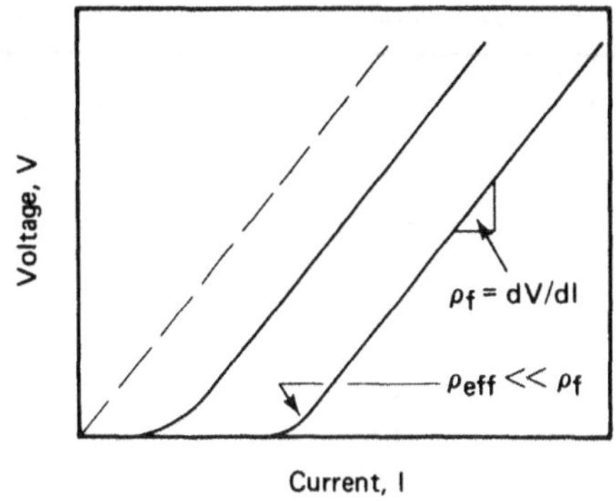

Figure 18-1. Schematic current-voltage characteristics for a type-II superconductor; pinning-site densities are imagined to increase from left to right in the diagram. Flux creep sets in (the critical current is exceeded) at current levels which increase with bulk pinning strength, F_p. During flux flow, the resistivity $\rho_f = dV/dI$, independent of F_p, is an intrinsic materials property—after KIM *et al.* [KIM65]; see also GANDOLFO [GAN67, p. 63].

a continuous transition between the states of creep and flow has been noted, particularly in the results of resistive measurement. This is depicted in Figure 18-1, which serves to illustrate two important points: (i) noticeable flux creep sets in at current levels (i.e., Lorentz-force levels), which increase as the lattice-defect density increases thereby emphasizing the connection with pinning centers; (ii) once flow has set in, the differential resistivity, $\rho_f = \Delta V/\Delta I$, is independent of defect density suggesting that flux flow, and its resistivity ρ_f, are intrinsic materials properties as required by BARDEEN–STEPHEN [BAR65] theory.

18.1.3 Flux Jumping

If pinning is strong, the application of an even stronger depinning force may result in the catastrophic dislodgment of the flux lattice—a flux jump. Our present understanding of this subject stems from two early observations: the first by KIM *et al.* [KIM62], who noted that in tube magnetization dH_a/dT must remain suitably small as the sample moves between critical states if a complete transition to the normal state is to be prevented; the second by ANDERSON and KIM [AND64], who, in an investigation of the thermal stability of the creep state in terms, essentially, of a competition between magnetic and thermal diffusivities, established a connection between flux motion and local heating. According to ANDERSON and KIM [AND64], the rate of local temperature rise was given by

$$\frac{dT}{dt} = -\left(\frac{K}{C^v}\right)\nabla^2 T + \left(\frac{1}{C^v}\right)\dot{Q}_v \tag{18-1}$$

where K is the thermal conductivity, C^v is the specific heat per unit volume, and where $\dot{Q}_v$, the power dissipation per unit volume, is related to the velocity of flux motion.

Clearly, adiabatic flux motion is subject to thermal runaway leading, under conditions described by WIPF and LUBELL [WIP65] and WIPF [WIP67],

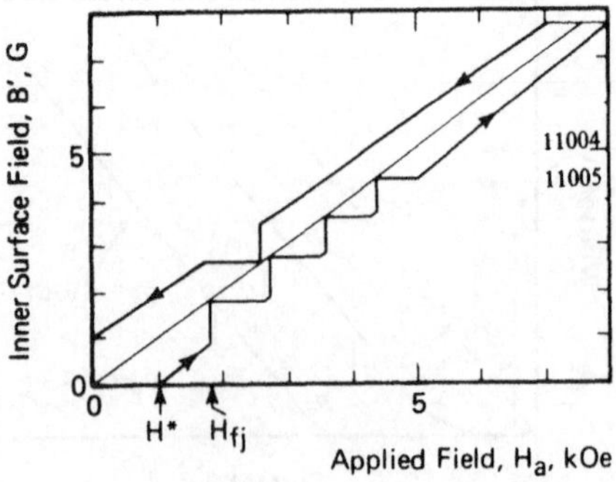

Figure 18-2. Flux jumping during the magnetization of a Ti–Nb tube (Westinghouse alloy HI 120, 75% cold worked; $1\frac{1}{4}$ in. length, $\frac{1}{4}$ in. O.D., 0.050 in. wall). The flux jump manifests itself as a temporary vanishing of the magnetization, i.e. of the flux-density gradient, $(B_0 - B')/w$, within the wall—see Figure 16-10(a). Results presented are due to HECHT and HARPER [HEC65].

to temperature increments sufficiently large to place the sample in the normal state. In tube-magnetization experiments, such as those described by HECHT and HARPER [HEC65], GANDOLFO [GAN67, GAN69], KROEGER [KRO66], and numerous others, the occurrence of a flux jump manifests itself as a temporary vanishing of the magnetization, i.e., of the field gradient, $(B_0 - B')/w$, within the tube wall. This has already been encountered in connection with Figure 16-11. Another excellent example is given in Figure 18-2 which represents a half-cycle of irreversible magnetization for a sample of 75%-cold-worked Ti–Nb alloy [HEC65]. Presented in this way flux jumping is the catastrophic breakdown of creep; thus, as a response to the application of a supercritical depinning force it represents an alternative to flux flow, a condition which itself must eventually be terminated by an abrupt transition to the normal state.

18.1.4 Summary

Creep, flow, and jumping may be treated either as magnetic or transport phenomena. This chapter focuses attention on their occurrence in response to a magnetic driving force; nevertheless it is instructive to draw attention, from time to time, to the transport-property aspects. Thus, a pair of schematic voltage-current characteristics are presented as part of Figure 18-3 to schematically illustrate the relationship between the three dynamical properties of the mixed state. The magnitudes of the creep and flow effects can be expressed in terms of the resistivities ρ_{cr} (or ρ_{eff} [GAN68]) and ρ_f, respectively. The former, a ratio of voltage gradient (E) to current density (J) during creep, is rather poorly defined but, when of measurable size, tends to lie within the range 10^{-13} Ω cm [KIM63[a]] to 10^{-11} Ω cm [GAN68]. The flux-flow resistivity on the other hand, as the slope of the E–J characteristic in the flow regime, is an accurately-determinable materials property. Since $\rho_f/\rho_n = H_a/H^*_{c20}$ ((14-32), see also Section 23.8) the magnitude of the flux-flow resistivity is less than, but comparable to, that of the normal-state residual resistivity which in hard superconducting alloys is typically

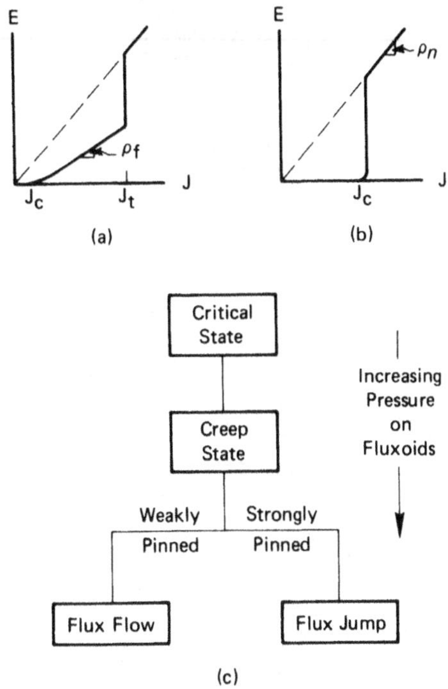

Figure 18-3. Schematic current-density, J, *versus* electric field, E, (i.e., current-voltage) characteristics of (a) weakly-pinned and (b) strongly-pinned bare superconducting wires. J_c and J_t are the critical- and take-off-current densities, respectively. (c) Interrelationships and distinctions between the three dynamical properties of the mixed state—*viz.*, flux creep, flux flow and flux jumping.

of the order of 5×10^{-5} Ω cm. Excellent examples of flux flow and associated instabilities are to be found in the work of GAUSTER *et al.* (see, for example, [GAU70]) to be considered in Chapter 23.

18.2 Physical Analogs of the Dynamic Mixed State

18.2.1 Mechanical and Thermal Analogs

In that it possesses long-range order, occasionally interrupted by point and line defects (dislocations), can be treated as an elastic continuum, may be partially immobilized by pinning centers but is otherwise able to creep, plastically flow, and eventually disintegrate catastrophically in response to an applied stress or a thermodynamic variable, the Abrikosov flux lattice can be usefully compared to a crystal lattice. Figure 18-4 serves to emphasize the parallelism (occasionally quantitative) that exists between these two classes of long-range ordered structure. TRÄUBLE, ESSMANN, and others used a Bitter-pattern decoration technique in association with electron microscropy to visualize the flux lattice, a structure whose morphologies and dislocation properties have been recently reviewed by HUEBENER [HUE79, pp. 75–82]. In the calculation of bulk pinning forces it has been useful, following LABUSCH [LAB67, LAB69, LAB69[a]] (see also Section 21.5),

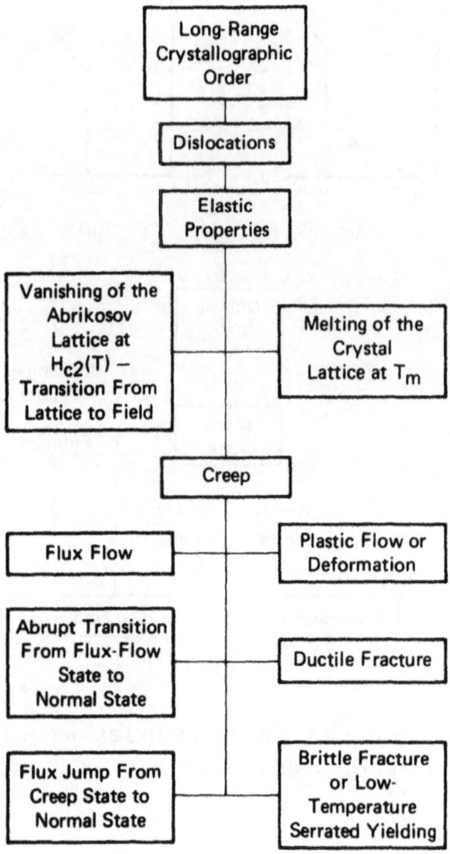

Figure 18-4. Metallurgical analog of the mixed state; common and parallel properties of flux lattices and crystal lattices.

to treat the flux lattice as an elastic continuum describable in terms of three elastic stiffness constants C_{11}, C_{44}, and C_{66}. In this model, as the temperature increases and the applied field approaches $H_{c2}(T)$, the flux lattice softens and loses its long-range order at the critical field in a sequence of processes comparable to the softening and melting of a crystal.

Figures 18-5(a) and (b), a pair of schematic stress-strain diagrams, which are to be compared with the E–J curves of Figures 18-3(a) and (b), respectively, describe the responses of ductile and brittle crystals to applied mechanical stresses. As shown, both solids have similar elastic moduli. But crystal A, which is supposed to be well annealed, soft, and free of strengthening solutes, plastically deforms at a low stress level, elongates plastically at constant true stress, and undergoes ductile failure at the point ×. The plastic deformation may be compared to flux flow, and the ductile fracture to the sudden transition to normal state at J_t, the "take-off point" in Figure 18-3(a). Crystal B on the other hand may be a brittle intermetallic compound or a highly solution-strengthened alloy. The yield strength for such a material

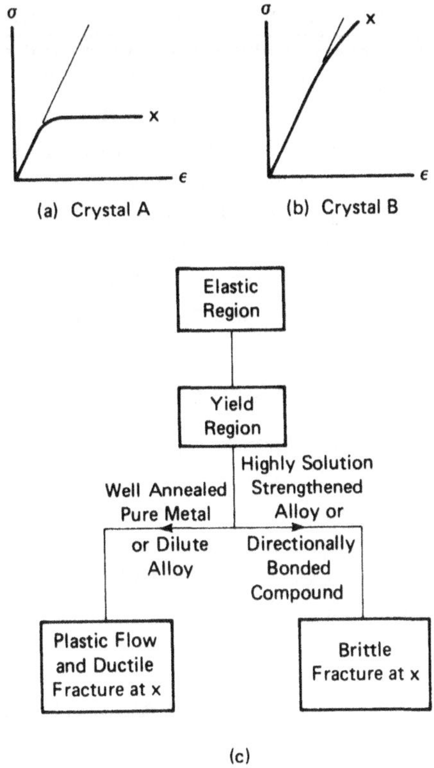

Figure 18-5. Schematic strain–stress characteristics of (a) ductile, and (b) strong and brittle metals, for use in conjunction with Figure 18-3 in a further comparison between the properties of flux lattices and crystal lattices.

is much higher than for Crystal A, but it may undergo brittle failure at lower values of strain. Brittle failure is equivalent to the flux jump of Figure 18-3(b). The microyielding which takes place near the elastic limits (towards the ends of the initially linear segments of the $\sigma(\varepsilon)$ curves), more pronounced in Crystal A than in Crystal B, may be associated with the flux creep that gives rise to the small exponentially curved initial segments ($E \cong 0$) of Figures 18-3(a) and (b). Some time dependence is, of course, implicit in the experiment referred to. Creep, being a thermally activated process, is usually investigated by studying the explicit time dependence of strain at constant temperature, preferably at constant stress. Thus, *flux creep* manifests itself as a logarithmic decay from the critical state, while *mechanical creep* occurs as a small but measurable time-dependent elongation at a stress level of about nine-tenths of the yield strength. Of course, if the need to prove the existence of a thermal-activation mechanism arises, temperature dependence measurements must be undertaken.

This heuristic comparison of the dynamic properties of flux and crystal lattices cannot be expected to offer complete mechanistic validity. Thus, whereas conventional brittle fracture is a localized phenomenon, the flux jump, although it may begin as a localized effect, quickly spreads throughout a large volume of lattice by virtue of the associated liberation of heat. Perhaps the mechanical effect more closely allied to flux jumping is low-temperature serrated yielding (Sections 3.9 and 3.10), which, taking place in a temperature regime in which the metal has a low specific heat (the lattice specific heat, $C_g \propto (T/\theta_D)^3$), also gives rise to a runaway heat pulse. This paragraph concludes with the suggestion that, in view of the catastrophic nature of flux-jump, just as the THOM theory of catastrophe [THO75[a]] is beginning to be applied to crack-propagation analysis [RAB79], it would be equally rewarding to investigate its applicability to the problem of flux-lattice depinning (but see [WIP78, p. 5]).

18.2.2 An Electrical Transport Analog

Being controlled by a localized thermally-activated fluxoid/pinning center interaction, flux creep is analogous to electrical transport by polaronic hopping. Flux flow, on the other hand, being responsive to the collective electrodynamics of the unpinned flux lattice is comparable to band conduction in a perfect crystal at low temperatures in the presence of a magnetic field—i.e., what might be termed "ideal residual magnetoconduction."

18.3 Electromagnetism of the Dynamic Mixed State

18.3.1 The Magnetic Driving Force

Expressions for the restraining force, F_p, necessary to immobilize a system of fluxoids equivalent to a field B with gradient dB/dr, as well as for the individual fluxoid restraining force, f_p, have been developed in Section 16.2. Using thermodynamic arguments, FRIEDEL, DE GENNES, and MATRICON [FRI63], have shown that the force per unit length acting in the r-direction on a single vortex in a nonuniform distribution of vortices is

$$f_\Phi = -\frac{\Phi_0}{4\pi}\frac{1}{\mu_{eq}}\frac{dB_z}{dr} \tag{16-10}$$

which is equivalent to a force per unit volume acting on the flux lattice of

$$F = -\frac{B_z}{4\pi}\frac{1}{\mu_{eq}}\frac{dB_z}{dr} \tag{16-7}$$

In the critical state, these forces are exactly balanced by the elementary and bulk pinning forces, respectively. Measurable flux motion occurs whenever $F > F_p$ or $f_\Phi > f_p$. Although creep is still detectable as f_Φ drops slightly below f_p, its rate decreases rapidly with f_ϕ [BEA69] (also see Section 18.5.2).

18.3.2 The Lorentz Driving Force

As pointed out in Section 16.2, the moving flux lattice experiences a force in the presence of transport current J_y given by

$$F = \frac{B_z J_y}{10} \tag{16-14}$$

This moving force gives rise to energy dissipation, the extent of which can be estimated by calculating the associated electromotive force (e.m.f.), which is, of course, in the same direction as the transport current.

18.3.3 Electromotive Force and Power Dissipation Associated with Flux Motion

As proposed by ANDERSON and KIM [AND64], who drew on the idea of electromagnetic induction as embodied in Faraday's law for the electric field induced in a conducting path enclosing magnetic flux changing at the rate $d\Phi/dt$, the e.m.f. associated with the flux motion is given (in practical c.g.s. units) by

$$\oint \mathbf{E} \cdot d\mathbf{l} = -1 \times 10^{-8} \frac{d\Phi}{dt} \quad \text{(volts)} \tag{18-2a}$$

or

$$\nabla \times \mathbf{E} = -1 \times 10^{-8} \frac{d\mathbf{B}}{dt} \tag{18-2b}$$

Flux traversing the width (x direction) of the conductor under either the magnetic or Lorentz driving forces referred to above would be expected to develop an e.m.f. in the direction of its length (or y direction). Suppose fluxoids, each of Φ_0 gauss cm^2, to be present at a density of n per cm^2 and moving at a velocity $v = dx/dt$; then the field E_y, or e.m.f. per unit length in the y-direction, is given by

$$E_y = -1 \times 10^{-8}(n\Phi_0\, dx)/dt \tag{18-3}$$

Thus, electric fields are present, either (i) across the diameter of a flux bundle creeping at velocity, v_{cr}, or (ii) along the length of a conductor in which a flux lattice of induction B is flowing with velocity v_f. They are given generally by

$$E_{i,y} = -1 \times 10^{-8} n\Phi_0 v_i \qquad (\text{v cm}^{-1}) \tag{18-4}$$

where i represents "cr" or "f," respectively. If the flowing flux is being steadily propelled by a Lorentz force, the power dissipated per unit volume is

$$\dot{Q}_v = E_{f,y} J_y$$

$$= 10^{-8} v_f B_z J_y \qquad (\text{W cm}^{-3}) \tag{18-5}$$

According to KROEGER [KRO66], the induction mechanism of ANDERSON and KIM [AND64] has been criticized by several authors (see [JON65], [LOR65], and [PEA65]) who pointed out that the induction approach just outlined requires the amount of flux enclosed by a loop consisting of sample and voltmeter to be increasing continuously—an impossible situation. Nevertheless, JOSEPHSON [JOS65], with the aid of a local-but-macroscopic argument, has reconfirmed the validity of (18-4) which we will, therefore, continue to use.

18.4 The Tube Magnetization Experiment in Flux Dynamics Studies

Variations of the basic tube magnetization experiment have found frequent use not only in the study of the static critical state, as discussed in Chapter 16, but also in studies of the dynamic phenomena of creep and flow. The material under test is usually machined to hollow cylindrical shape typically of dimensions: l = 15–30 mm, O. D. = 5–10 mm, wall thickness (w) = 0.4–3 mm. This is mounted coaxially within a superconducting solenoid which, depending on the design of the experiment, provides either a static or variable background field; an additional coaxial "tickler" coil may also be installed if a small field-step is to be applied. Instrumentation consists of magnetoresistive or Hall-effect sensors suitably mounted so as to monitor the applied external field, H_a, and the internal field which has usually been designated H' (Section 16.6.2); in addition, search coils or SQUID magnetometers may be employed to follow the ingress and egress of flux. Three general classes of tube- or cylinder-magnetization experiments, to be referred to herein as types A, B, and C, have been conducted. In *type-A* experiments, the applied field is slowly cycled and the H' *versus* H_a curves traced out, as in the original quasistatic thick-wall-tube investigations of KIM *et al.* [KIM62, KIM63], HECHT and HARPER [HEC65], and many others, who set out to study the magnetic properties of the critical state. In *type-B* experiments, typified by the work of GANDOLFO [GAN67], GANDOLFO *et al.* [GAN66, GAN68, GAN69], LUBELL and WIPF [LUB66], and BEASLEY *et al.* [BEA69], the field of a superconducting solenoid is employed to establish the critical state in a tubular or solid cylindrical sample and then set at some steady value by switching to the persistent mode; a small auxiliary solenoid usually of Cu is then used to apply an AC modulation field (some hundreds of Oe in amplitude) or an incremental field-step (some tens of Oe in height). The results of this type of experiment are analyzed in terms of flux jumping (e.g., [GAN69]) or magnetic diffusion (e.g., [LUB66], [GAN67, GAN68], [BEA69]), whichever is applicable. Finally, in *type-C* experiments, the total field applied to the outer surface of a thick-walled cylinder is increased at some uniform rate. In the work of KROEGER [KRO66] for example, dH_a/dt was varied from 1×10^4 to 4×10^5 Oe s^{-1}, sufficient to stimulate flux jumping in some samples and initiate flux flow in others. In the latter case, the velocity of viscous flow could be measured by timing

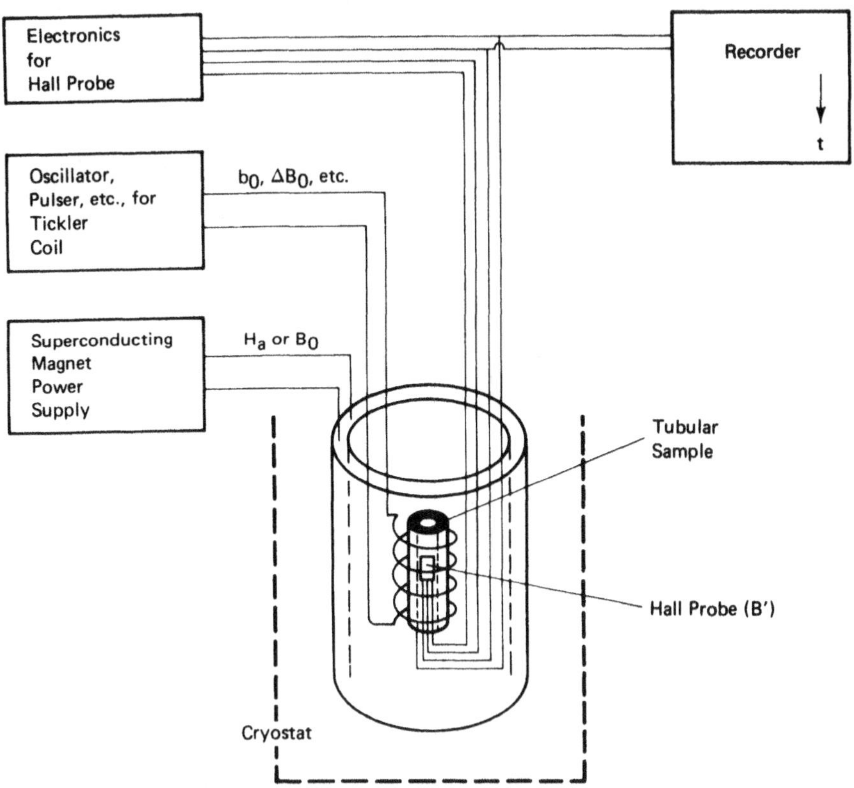

Figure 18-6. Typical experimental arrangement for tube-magnetization measurement—after GANDOLFO [GAN67, p. 26].

Table 18-1. Some Classic or Representative Magnetic Studies of Flux Creep, Flow, and Jumping

Phenomenon	Technique	Tube Magnetization Type	Property Measured or Discussed	Literature
Equilibrium magnetization	Tube magnetization	A	H' *versus* H_a magnetization loops	Kim *et al.* [KIM63]
Creep	Tube magnetization	A	Temperature dependence of magnetization	Kroeger [KRO66]
		B	Time dependence of critical state decay	Kim *et al.* [KIM62]
Creep	Analysis		—	Anderson [AND62]
Creep	Tube magnetization	A	$(H' - H_a)^{-1}$ *versus* $\langle B \rangle$ for various temperatures, etc.	Kim *et al.* [KIM63[a]]
	Potentiometry		Electrical resistance	
Creep	Analysis		—	Anderson and Kim [AND64]
Creep	Cylinder magnetization	B	Magnetic diffusivity	Lubell and Wipf [LUB66]
Creep	Tube magnetization	B	Magnetic diffusivity	Gandolfo [GAN67, GAN68]
Creep	Cylinder magnetization	B	Temperature and time dependence of flux motion	Beasley *et al.* [BEA69]
Flow	Potentiometry		Electrical resistivity	Kim *et al.* [KIM65]

Flow	Theory			Bardeen and Stephen [BAR65]
Flow	Tube magnetization	C	Flux-front transit time	Kroeger [KRO66]
Creep and flow	Potentiometry		Potential drop at constant J as function of H	Kim *et al.* [KIM64[a]]
Creep and flow	Review		Transport	Kim [KIM67]
Jumping	Cylinder magnetization	C	Jump initiation as function of dH_a/dt: heating effect of flux jump	Wipf and Lubell [WIP65]
Jumping	Tube magnetization	C	Pinning-center-dependence and dH_a/dt-dependence of jumping	Hecht and Harper [HEC65]
Jumping	Tube magnetization	C	The same as above	Kroeger [KRO66]
Jumping	Tube magnetization	B	Flux jumping in response to an AC field modulation	Gandolfo *et al.* [GAN66, GAN67, GAN69]
Jumping	Theory		Stability under isothermal and adiabatic conditions	Wipf [WIP67]
Jumping	Tube magnetization	C	Shielding and trapping stability in dynamically stabilized fabricated ("pancake"-stacked) cylinders	Haebel and Witzeling [HAE77]

the transit of the field as it progressed between the outer and inner surfaces of the test cylinder to which a pair of pickup coils were attached.

The experimental arrangements are generally as depicted in Figure 18-6; a number of classic and representative magnetic studies of creep, flow, and flux jumping are listed in Table 18-1.

18.5 Flux Creep

The idea of flux creep, based on an adaptation of conventional chemical-diffusion theory (e.g., [LEC59], [SHE63]) was introduced by ANDERSON [AND62] to explain the temperature dependence of the pinning parameter $\alpha(T)$ (see Table 16-1) and the time-dependence of critical-state decay, effects which had been first examined experimentally by KIM, HEMPSTEAD, and STRNAD [KIM62].

18.5.1 Experimental Observations—Temperature Dependence of the Critical State

A list of low-field critical state functions has been given in Table 16-1. The semiempirical relationship between local critical current density and local field preferred by KIM *et al.* [KIM62] was

$$J_c(r) \cong \alpha(T)/(b_0 + B(r)) \tag{16-41}$$

which, restated in terms of *mean* critical current densities and inductions, becomes

$$\langle J_c \rangle = \alpha(T)/(b_0 + \langle B \rangle) \tag{16-42}$$

The critical state tube magnetization experiments of KIM *et al.* [KIM62] were introduced in Section 16.6.1, where it was shown that since (16-42) was equivalent to

$$\pm\left(\frac{4\pi}{10}\right)\frac{\alpha(T)w}{H' - H_a} = b_0 + \langle B \rangle \tag{16-44}$$

where w was the wall thickness of the tube, the quantities $\alpha(T)$ and b_0, necessary for a determination of J_c, could be obtained from the slope and intercept of a plot of $(H' - H_a)^{-1}$ *versus* $\langle B \rangle$. A typical set of curves, as developed by KIM *et al.* [KIM62] from magnetization data obtained from sintered-Nb-powder tubes, is reproduced in Figure 18-7(a). These results, whose linearity substantiates the underlying assumption ((16-42)), provide

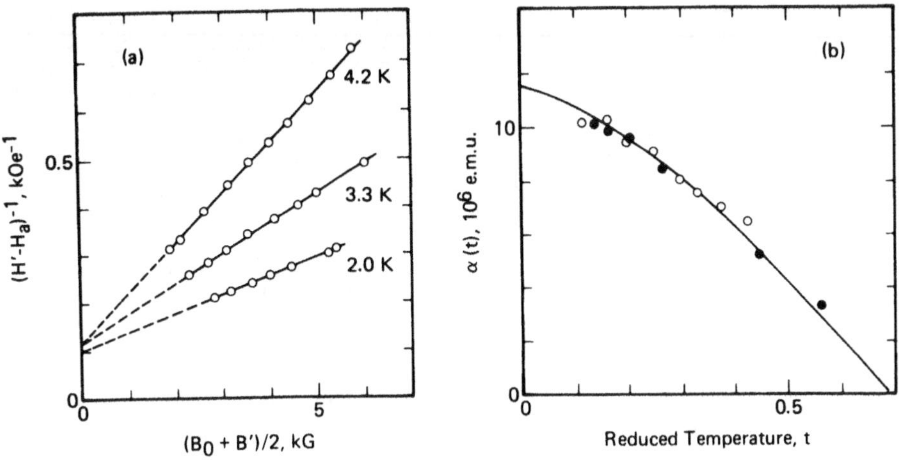

Figure 18-7. Results of a magnetization study of a sintered-Nb-powder tube. (a) Plot of $(H_a - H')^{-1}$ (proportional to $(B_0 - B')/w$, the field gradient in the tube wall) *versus* $\langle B\rangle \equiv (B_0 + B')/2$ (the mean field in the tube wall), the slope of which is a measure of the critical-state constant $\alpha(T)$—after KIM *et al.* [KIM62, KIM63[a]]. (b) Critical-state constant $\alpha(T)$ as a function of reduced temperature, $t = T/T_c$, for the sintered-Nb-powder tube (●) together with data for a Zr-75Nb tube (○). The full line represents the function $\alpha(T) = (H_c^2/8\pi)(p/d) - (28 \pm 9)(k_BT/d^4)$ (where H_c is the thermodynamic critical field), based on the Anderson theory of thermally-activated flux creep and fitted to the Zr-Nb data with $p \cong 7 \times 10^{-3}$ cm, and $d \cong 1 \times 10^{-5}$ cm—after ANDERSON [AND62].

three values of $\alpha(T)$ and three corresponding values of the temperature-dependent critical current density. A repetition of the tube-magnetization experiment over a wide temperature range enabled $\alpha(t)$ *versus* t to be plotted as in Figure 18-7(b), and provided sufficient data for testing the results of ANDERSON's flux-diffusion theory, to be considered below. Comparable studies of pressure-sintered Nb_3Sn tubes were conducted by GANDOLFO *et al.* [GAN66, GAN67].

18.5.2 Experimental Observations—Time-Dependence of Critical State Decay

The first experimental exploration of the durability of the critical state was also conducted by KIM, HEMPSTEAD, and STRNAD [KIM62]. Although the time-dependence of critical state decay has been subjected to more detailed study in recent years by BEASLEY *et al.* [BEA69], the essential features of the original observations have remained valid, namely, that (i) the trapped field gradient (or "persistent current") decreases logarithmically with time, and (ii) although the rate of decay is measurable near the critical point, the decay from subcritical states takes place immeasurably slowly.[2]

[2]According to KIM *et al.* [KIM62] at the 90% level of $J_{\theta C}$ seven years are required for a decay of one part in 10^4.

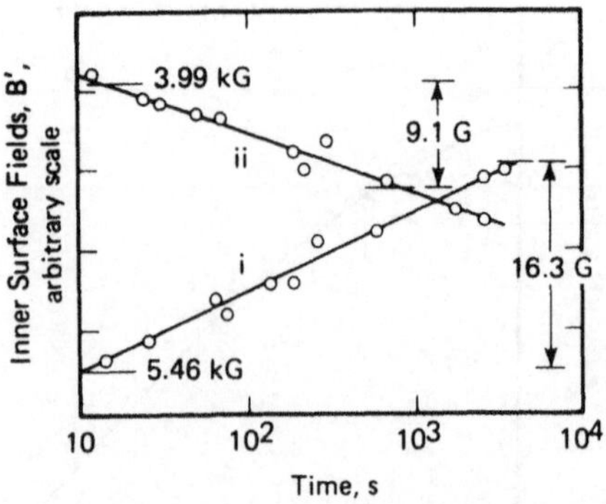

Figure 18-8. Field at the inner wall of a superconducting Zr-75Nb tube as function of time after the external field has been quickly raised (shielding state of the tube) and quickly lowered (trapping state), respectively, and held steady. The mechanism is discussed in Section 18.10 and illustrated in Figure 18-9. Curve (i) represents the decay of "shielding magnetization;" curve (ii) represents that of "trapped magnetization" (see Figure 16-7). A few representative data points indicate that the curves are experimental—after Kim *et al.* [Kim62].

In association with the type-A temperature-dependent magnetization experiments referred to above, Kim *et al.* [Kim62] also conducted type-B experiments during which a magnetoresistance probe, mounted along the axis of the cylinder, monitored the time-dependence of H' as the sample relaxed to a new critical state following a small readjustment of the external field to a new constant (one part in 10^4 for several h) value. Figure 18-8, a typical set of results representative of the relaxation of both the shielding and the trapping critical states (see Figure 16-7), shows for example a trapped field initially of 4000 G decaying at the rate of 4.5 G per decade of time.

Obviously, flux leakage by creep does not represent a serious technical problem. In the above example, after one year ($\sim 3 \times 10^7$ s) the field would have dropped by only about 30 G or 0.75%. In a practical sense, therefore, we may regard the critical state as permanent, an observation which augurs well for the durability of persistent currents in large superconducting solenoids, such as, for example, those intended for NMR body-imaging [Kau83], whose initial conditions in most cases will be well below critical.

18.6 The Thermal Activation Theory of Flux Creep

18.6.1 Development of the Theory

Recognizing that the decay of flux gradient in a magnetized superconductive cylinder was somewhat analogous to other time-relaxation processes, such as magnetic after-effect and plastic creep, Anderson [And62] constructed a model predicated upon the idea of a thermally activated depinning of flux bundles in response to a magnetic or electric force-field (hence of their "hopping" from one pinning site to the next), and in so doing was able to explain the temperature- and time-dependent effects observed by Kim *et al.* [Kim62]. If fluxoids are to self-diffuse among an array of pinning centers represented by energy barriers of height U_p, they would do so with a diffusivity proportional to $\exp(-U_p/k_BT)$. Here U_p is, of course, related to a temperature-dependent elementary pinning force, f_p, multiplied by some characteristic interaction distance, λ_L, the London penetration depth.[3] In the presence of a driving force such as a field gradient, this erstwhile isotropic random-walk motion becomes superimposed upon a drift in the direction of the force. The starting point in the analysis of thermally activated flux creep is always to take this into account by writing down an expression for a reduced diffusion activation energy in the form of $U_p(\equiv f_p\lambda_L)$ diminished by an energy term $U_L(\equiv f_L\lambda_L)$, which is formed from the product of a Lorentz force F (per unit volume), an "activation volume," V, and some characteristic interaction length which we have designated λ_L. The new hopping rate in the direction of the applied force is then

$$r = r_0 \exp[-(U_p - U_L)/k_B T] \tag{18-6}$$

where r_0 is some characteristic frequency of oscillation of the fluxoid in its potential well—about 10^{10} Hz. Numerous authors [BEA69], [WIL70], [DEW71], [TIN75] since ANDERSON [AND62] have presented different versions of the original analysis, but only BEASLEY [BEA69], who allowed the diffusivity to vary as a function of distance into the tube wall, has offered a more sophisticated approach. The following, a simplified version of the treatments of ANDERSON [AND62], WILLIAMS [WIL70], and TINKHAM [TIN75], preserves the essential physics.

The theory of ANDERSON [AND62] suggests that the fluxoids move not individually, but in clusters or "bundles," mutually interacting within a characteristic distance equal to the London penetration depth λ_L. According to the theory, the entire bundle is pinned down by some obstacle, the energy of interaction of which is reduced by the Lorentz force just referred to. In GANDOLFO's interpretation of ANDERSON's theory [GAN67], an *individual* fluxoid is held at the pinning site while tending to be dislodged by Lorentz forces acting on the *entire interacting bundle.* But with the introduction of the usual pinning force efficiency factor, p (see [AND62]), the distinction is moot.

With these considerations in mind, the individual pinning energy, U_p, based on the length l of a fluxoid of cross-sectional area $\pi\xi_0^2$ trapped at, say, a localized normal region is

$$U_p = p\frac{H_c^2}{8\pi}\pi\xi_0^2 l \tag{18-7}$$

and the elementary pinning force, again regarding λ_L as the characteristic

[3]Table 13-3 shows that for alloys of present interest, λ_L is greater than ξ_0 but only about 1/5th of λ_{GL}^d, a distance parameter also frequently used in this context (see Section 21.3.4).

distance, is

$$f_p = \frac{U_p}{\lambda_L} = p\frac{H_c^2}{8}\frac{\xi_0^2}{\lambda_L}l \tag{18-8}$$

For the dislodgement force, f_L, we consider the number of fluxoids (of number density, n, per unit area, and length l) that intersect an area $\pi\lambda_L^2$, and multiply by the Lorentz force (per unit volume) on a single fluxoid, *viz.*, $J\Phi_0/10$. The force, therefore, is given by

$$f_L = \left[\frac{J\Phi_0}{10}\right]\left[\frac{B}{\Phi_0}\right](\pi\lambda_L^2)l$$

$$\equiv \begin{matrix}\text{Lorentz}\\\text{Force}\end{matrix} \cdot n \cdot \begin{matrix}\text{Activation}\\\text{Volume}\end{matrix}$$

$$= \frac{JB}{10}\pi\lambda_L^2 l$$

which, with the aid of (16-42) and with $b_0 \ll B$, becomes

$$f_L = \alpha(T)\frac{\pi\lambda_L^2}{10}l \tag{18-9}$$

The magnitude of the net force on the bundle, f_n, is

$$f_n = f_p - f_L \tag{18-10a}$$

$$= p\frac{H_c^2}{8}\frac{\xi_0^2}{\lambda_L}l - \alpha(T)\frac{\pi\lambda_L^2}{10}l \tag{18-10b}$$

18.6.2 Temperature Dependence of the Critical Parameter, $\alpha_c(T)$

After defining the critical state as that in which, although f_n is still nonzero, flux creep is barely detectable, the following substitutions can be made:

$$f_{n,\text{critical}} = f_p - \alpha_c(T)\frac{\pi\lambda_L^2}{10}l \equiv f_c \tag{18-11}$$

leading to an expression for $\alpha_c(T)$ (the parameter used elsewhere without the subscript "c") of the following form:

$$\alpha_c(T) = 10p\frac{\xi_0^2}{\lambda_L^3}\frac{H_{c0}^2}{8\pi}(1-t^2)^2 - \frac{10}{\pi\lambda_L^2 l}f_c \tag{18-12}$$

in which (8-7) has been used to provide a temperature dependence for H_c.

Equation (18-12), as it stands, predicts a temperature dependence for α_c based on that of f_p without any need for the introduction of thermally activated flux creep. However, as pointed out by TINKHAM [TIN75, p. 178] at low values of t, while experiment is still calling for a temperature dependence (Figure 18-7(b)), $\alpha_c(T)$ in the form of (18-12) would freeze out unless the second term were itself strongly temperature dependent. That this will indeed by the case under thermally activated flux creep is demonstrated by the following argument: at the critical point, a combination of (18-6) with the definition $(f_p - f_L)_{\text{critical}} = f_{n,\text{critical}} = f_c$, yields for the "minimum-detectable" hopping frequency, r_c,

$$r_c = r_0 \exp(-f_c\lambda_L/k_BT) \tag{18-13a}$$

thereby permitting the substitution

$$f_c = k_BT\ln(r_0/r_c)/\lambda_L \tag{18-13b}$$

so that (18-12) becomes

$$\alpha_c(T) = \underbrace{10p\frac{\xi_0^2}{\lambda_L^3}\frac{H_{c0}^2}{8\pi}(1-t^2)^2}_{\text{Pinning Strength}} - \underbrace{\frac{10k_BT}{\pi\lambda_L^3 l}\ln\left(\frac{r_0}{r_c}\right)}_{\text{Flux Creep}} \tag{18-14}$$

which is the sum of a pinning term proportional to $(1-t^2)^2$ and a creep term proportional to t. Normalized to the zero-K critical pinning parameter, $\alpha_c(0)$, this obviously reduces to the form

$$\frac{\alpha_c(t)}{\alpha_c(0)} = (1-t^2)^2 - \gamma_c t \tag{18-15}$$

18.6.3 Commentary on Anderson's Theory

The Anderson model for thermally activated flux creep has come under severe criticism by several authors. As pointed out by TINKHAM [TIN75], at all except low temperatures (18-15) is dominated by the temperature dependence of the pinning strength, the $(1-t^2)^2 \cong (1-at^2)$ term, rather than by creep as gauged by $\gamma_c t$. This is in accord with the comments of

CAMPBELL *et al.* [CAM68] who noted, as a result of magnetic studies of J_c at temperatures of 4.2 and 2.1 K, that the effect of thermal diffusion was undetectable, and that the temperature dependence of J_c could be completely accounted for in terms of changes in pinning potential well depth. It seems that if further indirect evidence for the existence of thermally activated flux creep is to be sought, it will most likely be found in the isothermal temporal decay of the critical state, as initially studied by KIM *et al.* [KIM62] and subsequently by BEASLEY *et al.* [BEA69].

18.7 Current-Voltage Relationships in the Creep State

A useful current *versus* e.m.f. relationship in the creep state may be derived by starting with (18-10b) rewritten, with the aid of (18-8) and (16-42), as before, in the form

$$f_n\lambda_L = U_p - JB\frac{\pi\lambda_L^3 l}{10} \tag{18-16a}$$

$$= U_p\left(1 - \frac{J}{J_c}\right) \tag{18-16b}$$

in which the subscript c defines a state for which f_n is exactly zero. Next, with reference to (18-6), the net diffusion (creep) velocity may be written

$$v_{cr} = v_{cr0}\exp\left[-\frac{U_p}{k_BT}\left(1 - \frac{J}{J_c}\right)\right] \tag{18-17}$$

and then by simply applying (18-4), any motion of the flux lattice (induction B_z) is seen to be equivalent to an electric field normal to the direction of motion given by

$$E = -1\times 10^{-8}B_z v_{cr0}\exp\left[-\frac{U_p}{k_BT}\left(1 - \frac{J}{J_c}\right)\right] \tag{18-18}$$

which is consistent with the exponentially increasing threshold e.m.f.s noted experimentally (see Figures 18-1 and 18-3 for schematic illustration and the concluding remarks in Section 18-8 for further comment).

18.8 Time Dependence of the Critical State

The circuit e.m.f. due to changing flux is given by (18-2a). Consider for simplicity a tube of outer and inner radii R_0 and R_1 and wall thickness

w, whose magnetic state can be described by the Bean model (Section 16.4); then

$$2\pi R_0 E \cong -1 \times 10^{-8} \frac{d\Phi}{dt} \tag{18-19}$$

with

$$\Phi = \int_{R_1}^{R_0} 2\pi r B(r)\, dr \tag{18-20}$$

and $B(r)$, according to Section 16.4.2, given by

$$B(r) = B_0 - \left(\frac{dB_z}{dr}\right)(R_0 - r) \tag{16-18}$$

$$= B_0 + \mu_{\text{eq}}\left(\frac{4\pi}{10}\right) J(R_0 - r) \tag{18-21}$$

where the surface induction B_0 is to be held constant by a steady applied field. Integration of the time-dependent second term of (18-21) yields

$$\Phi = 2\pi\mu_{\text{eq}}\left(\frac{4\pi}{10}\right) J \frac{R_0^3}{2}\left(h^2 - \tfrac{2}{3}h^3\right) \tag{18-22}$$

(where $h = 1 - R_1/R_0 = w/R_0$; see Section 16.4.2) which upon differentiation becomes

$$\frac{d\Phi}{dt} = 2\pi R_0 \mu_{\text{eq}}\left(\frac{4\pi}{10}\right)\frac{dJ}{dt}\frac{w^2}{2} \tag{18-23}$$

A comparison of this with (18-19) and (18-18) in turn then makes it possible to write

$$\left[\mu_{\text{eq}}\left(\frac{4\pi}{10}\right)\frac{w^2}{2}\right]\frac{dJ}{dt} = -1 \times 10^8 E = B_z v_{\text{cr0}} \exp\left[-\frac{U_p}{k_B T}\left(1 - \frac{J}{J_c}\right)\right] \tag{18-24}$$

the solution of which, for long times, is of the form

$$J = J_c - J_c \frac{k_B T}{U_p} \ln t \tag{18-25}$$

where t is time. This relationship, which owes its origin to the hypothesis of thermally activated flux creep as embodied in the exponential form of (18-17), shows that the circulating current, initially representative of the critical state of a cylinder in constant external field, H_a, dies away logarithmically with time. The rate of decay is k_BT/U_p, a very important result which enables an elementary pinning force to be extracted from the results of flux-creep measurements.

BEASLEY *et al.* [BEA69] have made a detailed study of flux creep in solid magnetized cylinders. The measurements, which could be regarded as being both *indirect* investigations of the phenomenon, in that the logarithmic creep rate predicted by (18-25) was examined, and *direct* investigations, in the sense that a SQUID magnetometer was able to detect the passage across the sample surface of flux "increments"—presumably the "bundles" predicted by ANDERSON—led to the conclusion that the Anderson theory of thermally activated flux creep had been experimentally confirmed. On the other hand, CAMPBELL and EVETTS [CAM72, p. 100], after reviewing all the available results agree that, although thermal activation is a likely mechanism for creep, no direct evidence for it in the form of an Arrhenius-type plot has become available. In Section 18.11 flux creep is treated phenomenologically as electromagnetic diffusion. Finally, as pointed out by CAMPBELL [CAM82], the assignment of a mechanism to "flux creep" is confused by the experimental fact that what would be expected to be thermally activated creep is contiguous with the processes that, in four-terminal measurements, give rise to exponential current-voltage characteristics just below J_c in well-pinned materials (Section 23.6).

18.9 Relatively Recent Magnetic Studies of Flux Creep

18.9.1 Determination of Pinning Energy

In BEASLEY's hands the flux creep experiment was developed into a useful probe of elementary flux pinning mechanisms and was able to provide information on the size of the Anderson flux bundle. Experiments were performed on samples of extruded Pb–Tl alloy in which dislocations provided the principal flux-pinning sites.

A convenient starting point for the discussion is the net hopping energy $f_n\lambda_L$ (see (18-6), (18-10a), and (18-16a)) in the form

$$f_n\lambda_L = U_p - FV\lambda_L \qquad (18\text{-}26)$$

where F is the Lorentz force (per unit volume), V is the "activation volume," and λ_L is the characteristic interaction length referred to above. Critical current measurements, which are made under experimental conditions describable by

$$U_p = F_c V\lambda_L, \qquad F_c = J_c B/10 \tag{18-27}$$

naturally yield values for the quotient $U_p/V\lambda_L$. But as indicated above, the flux-creep measurement is unique in its ability to yield a value for U_p itself; consequently, by combining the two kinds of measurement, values for both U_p and $V\lambda_L$ can be obtained. The full expressions for the logarithmic creep rate, in terms of these parameters, are

$$\frac{d\Phi}{d\ln t} = \frac{\pi}{3} k_B T R_0^3 \frac{4\pi}{\gamma B} \frac{1}{V\lambda_L}(1 \pm \delta) \tag{18-28a}$$

$$\equiv \frac{\pi}{3} k_B T R_0^3 \left(\frac{4\pi}{10}\right) \frac{J_c}{U_p}(1 \pm \delta) \tag{18-28b}$$

with

$$\delta = \frac{1}{2}\left(\frac{4\pi}{10}\right) R_0 \left(\frac{1}{2} J_c \frac{\partial \ln U_p}{\partial B} - \frac{\partial J_c}{\partial B}\right).$$

and

$$\gamma \equiv \frac{\partial H(B)}{\partial B} \equiv \frac{1}{\mu_{eq}} \tag{16-2}$$

and where R_0 is radius of the cylinder.

The experimentally determined value of U_p was about 1 eV, much greater than the expected energy of interaction between a single fluxoid and a single dislocation, *viz.*, $\xi_0 \cong 10^{-2}$ eV [WEB63], [KRA67[a]]. Nevertheless, it is easy to justify the higher value in terms of an interaction energy $U_p \cong n_c^{1/2}\varepsilon_0$, where n_c is the actual number of contact points, in the following way: the cross section for intersection between a dislocation and a fluxoid is $\lambda_L \cdot$(fluxoid length); thus the number of contact points per unit volume, which is proportional to the fluxoid number density, B/Φ_0, and the dislocation density, n_w, is $\lambda_L(B/\Phi_0)n_w$. The number of contact points inside an activation volume, V, is accordingly $V\lambda_L(B/\Phi_0)n_w$. Taking $n_w \cong 10^{11}$ cm^{-2} along with the measured value of $V\lambda_L$, n_c was found to lie between 10^4 at $B = \frac{1}{2}H_{c2}$ and 10^3 at $B = H_{c2}$; thus $U_p \cong n_c^{1/2} 10^{-2} \cong 1$ eV, as measured. This conclusion justifies the surprisingly large barrier energy and at the same time confirms the $n_c^{1/2}$ law for interaction summation; a law which arose from the hypothesis that while n_c physical interactions actually take place most of them contribute nothing more than a uniform background energy, the fluctuations in their number and position (*viz.*, $n_c^{1/2}$) having to be relied on to provide the net pinning barrier [FIE69].

18.9.2 Evidence for Flux Clustering

ANDERSON [AND62] originally made the suggestion that flux lines spaced closer than λ_L the London penetration depth would assemble themselves into interacting clusters—internally ordered according to ABRIKOSOV's prescription. KIM *et al.* [KIM63[a]], using calibrated internally and externally mounted pickup coils, claimed to have detected the formation of such clusters during creep transit in tube magnetization experiments. Clusters (or "bundles") contained at least 20–50 flux quanta, corresponding to the lower detection limit of the equipment. The studies of BEASLEY *et al.* [BEA69] were comparable, but were favored by the superiority of the SQUID magnetometer over pickup coils in the detection of (moving) flux. It was concluded that during passage under creep conditions the flux assembled into clusters, typically of volume 10^{-10} cm^3, containing upwards of 10^3 fluxoids. The structure of such bundles or "fluxons" has been considered further by HEIDEN [HEI74[a]].

18.10 Flux Creep as Magnetic Diffusion

18.10.1 Atomic Diffusion

(a) Phenomenological Relationships. According to the standard theories of diffusion (see e.g., [LEC59], [LAZ60]) the phenomenological equations of atomic diffusion, in differential and vector form, respectively, are

$$J_A = -D_A \frac{dn_A}{dx}; \qquad \mathbf{J}_A = -D_A \nabla n_A \tag{18-29}$$

$$\frac{dJ_A}{dx} + \frac{dn_A}{dt} = 0; \qquad \nabla \cdot \mathbf{J}_A + \frac{dn_A}{dt} = 0 \tag{18-30}$$

$$\frac{dn_A}{dt} = D_A \frac{d^2 n_A}{dx^2}; \qquad \frac{dn_A}{dt} = \nabla \cdot [D_A \nabla n_A]$$

$$= D_A \nabla^2 n_A \tag{18-31}$$

Equation (18-29) is Fick's first law for a flux of atoms, J_A, across a plane normal to a concentration gradient, dn_A/dx. Equation (18-30) is a continuity statement, as is (18-31), Fick's second law—generally known as Poisson's equation—which embodies (18-29) and (18-30). The microscopic mechanism of diffusion is buried within D, the diffusivity. It is convenient to subdivide diffusion theory into two categories [LEC59]: (i) self-diffusion and (ii) chemical diffusion.

(b) Self-Diffusion. Self-diffusion is the accumulated result of the random walking of independent atoms among themselves. For the experiment to be interpretable, a starting cluster of atoms must be radioactively labelled. The process takes place quite independently of any concentration gradient (which, when present, plays the role of a kind of normalization factor) and yields a net flux across some selected surface only if it separates regions differing in concentration. Thus, in self-diffusion the concentration gradient itself does not provide a driving force. After expanding the concentration, n_A, about its value at an initial position, x_0, the atomic flux, J_A, can be expanded in terms of the moments, $\overline{X^n}$, of $P(x, t)$, the relative probability that an atom will have migrated a distance x from x_0 in time t. Thus,

$$J_A = (1/t)\left[n_A(x_0)\bar{X} - \frac{\partial n_A}{\partial x}\frac{\overline{X^2}}{2!} + \frac{\partial^2 n_A}{\partial x^2}\frac{\overline{X^3}}{3!} - \cdots\right] \tag{18-32}$$

which, since the odd moments vanish, reduces to

$$J_A \cong -(\overline{X^2}/2t)\frac{\partial n_A}{\partial x} \tag{18-33}$$

which is (18-29). Examples of self-diffusion are (i) tracer diffusion in an otherwise homogeneous metal, (ii) diffusion of phonons (heat), and (iii) solute diffusion in the limit of vanishing concentration gradient.

(c) Chemical Diffusion. Chemical diffusion presupposes the existence of some driving force which superimposes a drift on the otherwise random motion of the diffusing species. After performing an analysis comparable to that applied to self-diffusion, (18-32) is recovered; but because of the drift asymmetry none of the moments vanish. Next, assuming that $\bar{X}$ is approximately proportional to dn_A/dx, the first two terms of (18-32) again yield Fick's law which is thus seen to be more appropriate to a description of chemical diffusion than of self-diffusion. Examples of chemical diffusion (or "self-diffusion" in the presence of drift) are (i) diffusion of solute down a chemical potential gradient, and (ii) diffusion of flux vortices in the presence of a magnetic induction gradient or a Lorentz force.

Having recovered Fick's law from a phenomenological treatment of chemical diffusion it is difficult to recognize, especially in the context of solution theory, that J_A is not simply the response to a driving force represented explicitly by dn_A/dx. But if the $\bar{X}$ drift were induced by some other kind of force, only the leading nongradient term in (18-32) is required to guarantee flow. Thus, in that $dn_A/dx > 0$ is not a necessary condition for $J_A > 0$, Fick's law in this case is only conditionally true.

The point of this is that the driving force, if such exists, is primarily embodied in the diffusivity, D, a term which must be treated mechanistically.

(d) The Diffusivity, *D*. In self-diffusion, the diffusivity is given by the well-known Arrhenius equation

$$D = D_0 e^{-U/k_B T} \tag{18-34}$$

which is consistent with the Einstein equation, $\overline{X^2} = 2Dt$, for random walk or Brownian motion. In (18-34), U is the barrier activation energy, D_0 is the frequency factor, and D is variously known as the diffusivity, diffusion constant, or diffusion coefficient.

In generalized chemical diffusion an applied driving force, with which an energy ΔU may be associated, biases the process so that hopping takes place at rates of

$$r_\pm = r_0 \exp[-(U \overset{\swarrow\ \text{down}}{\underset{\nwarrow\ \text{up}}{\mp}} \Delta U)/k_B T] \tag{18-35}$$

down or up the energy gradient. From this point, the manner in which the calculation of $D = f(r_+ - r_-)$ proceeds depends upon the relative magnitudes of U, ΔU, and $k_B T$.

In the chemical diffusion of atoms, it is conventional [LEC59] to assume that $\Delta U \ll U$ and $\Delta U \ll k_B T$ such that

$$r_+ - r_- \cong r_0 \exp(-U/k_B T)(-2\Delta U/k_B T) \tag{18-36}$$

leading, in the manner outlined above, to a flow-equation of the form

$$J_A = D_0 \exp(-U/k_B T)\left[\frac{dn_A}{dx} + \left(\frac{n_A}{k_B T}\right)\left(\frac{\Delta U}{a}\right)\right] \tag{18-37}$$

where a is the atomic spacing. Equation (18-37) clearly reflects the influence of the superimposed energy gradient on the rate of diffusion. DARKEN'S form of this equation, in which the second term in square brackets is replaced by the logarithmic concentration dependence of the activity coefficient, is given in the standard texts [SHE63].

18.10.2 Magnetic Diffusion

In magnetic vortex diffusion on the other hand, following TINKHAM [TIN75, p. 176], one recognizes that $\Delta U \cong U$ and $U \gg k_B T$ permitting the term $(r_+ - r_-)$ to be treated in the following way:

$$r_+ - r_- = r_0 \exp(-U/k_BT)(e^{\Delta U/k_BT} - e^{-\Delta U/k_BT}) \qquad (18\text{-}38)$$

$$\cong r_0 \exp[-(U - \Delta U)/k_BT] \qquad (18\text{-}39)$$

which is, in effect, (18-6).

Next, rather than develop a full expression for the vortex-transport density (e.g., (18-37)) as in conventional diffusion theory, most workers simply rewrite (18-39) in the form of a creep-velocity relationship (as in (18-17)) and proceed from there. The rationale for so doing, as explained by TINKHAM [TIN75, p. 176], is that U is so great compared to k_BT that small uncertainties in it completely overwhelm any improvements in accuracy that might be gained by properly adjusting the premultipliers in (18-39) in such a way as to convert it from a jump-rate to a vortex-flux (J_ϕ) relationship (cf. (18-36) and (18-37)). For example, according to BEASLEY [BEA69], J_ϕ, the amount of magnetic flux flowing in, say, the $\hat{r}$ direction and crossing unit line in unit time, is adequately described by

$$J_\phi = \hat{r} B v_{\text{cr0}} \exp[-(U_p - U_L)/k_BT] \qquad (18\text{-}40)$$

Fick's second law (18-31) describes the temporal response of the medium to the accumulation of whatever is flowing. Thus, a particle flux leads to a change of concentration, phonon flux to a temperature change, and vortex flux to a change of induction (Table 18-2). With regard to the latter, if J_ϕ is given by (18-40), the growth-rate of induction is, according to BEASLEY [BEA69],

$$dB/dt = \nabla \cdot \hat{r} B v_{\text{cr0}} \exp[-(U_p - U_L)/k_BT] \qquad (18\text{-}41)$$

This, like (18-40), again avoids specific reference to a frequency factor (cf. (18-34)). Some of the essential elements of BEASLEY's work on the creep or diffusion of flux have been referred to above.

In experiments to be outlined below, and whose analyses were approached from phenomenological standpoints, LUBELL and WIPF [LUB66] and GANDOLFO [GAN67, GAN68], using type-B tube or cylinder magnetization data (Table 18-1), in association with conventional solutions of an electromagnetic equation having precisely the form of Fick's second law ((18-31)), succeeded in obtaining experimental values of a phenomenological electromagnetic diffusivity, D_{EM} and a related quantity, ρ_{eff}, the "effective creep resistivity."

Table 18-2. "First-Law" and "Second-Law" (Poisson's Equation) Formulations of Three Diffusion Phenomena

Formulations	Solute Diffusion	Heat Flow	Flux Creep
First-Law Formulation	$J_A = -D_A \frac{dn_A}{dx}$	$\dot{Q} = -K \frac{dT}{dx}$	$J_\phi = -D_{EM} \frac{dB}{dx}$
Flux	Atoms	Phonons	Fluxoids (vortices)
Units	J_A, (at.) $cm^{-2}\,s^{-1}$	$\dot{Q}$, erg $cm^{-2}\,s^{-1}$	J_ϕ, (gauss cm^2) $cm^{-1}\,s^{-1}$
	$\frac{dn_A}{dx}$, (at.) cm^{-4}	$\frac{dT}{dx}$, K cm^{-1}	$\frac{dB}{dx}$, gauss cm^{-1}
	D_A, $cm^2\,s^{-1}$	K, erg $cm^{-1}\,s^{-1}\,K^{-1}$	D_{EM}, $cm^2\,s^{-1}$
Comments		This equation is arranged such that K is a conductivity (cf. Ohm's law) rather than a diffusivity. Some dimensional conversion is associated with K since the quantity flowing is *energy* but the associated gradient is *temperature-gradient*.	dB/dx and J_ϕ are one linear dimension lower than their "solute-diffusion" counterparts, since the fluxoids themselves occupy one dimension.

Second-Law Formulation	$\frac{dn_A}{dt} = D_A \frac{d^2 n_A}{dx^2}$	$\frac{dT}{dt} = D_{TH} \frac{d^2 T}{dx^2}$	$\frac{dB}{dt} = D_{EM} \frac{d^2 B}{dx^2}$
Response to Flux	Concentration change	Temperature change	Field change
Units of D	$cm^2\, s^{-1}$	$D_{TH} = \frac{K}{C^v}$ $= \frac{erg\, cm^{-1}\, s^{-1}\, K^{-1}}{erg\, cm^{-3}\, K^{-1}}$ $= cm^2\, s^{-1}$	$cm^2\, s^{-1}$ (from above definition—see also Section 18.11)
Typical values in "operating" temperature range	10^{-10}	10^{0}	10^{3}

18.11 Phenomenological Investigation of Electromagnetic Diffusion

According to SMYTHE [SMY50, p. 390], Poisson's equation in the vector **B** can be derived from Maxwell's equation, provided that displacement current is neglected in the equation for $\nabla \times \mathbf{H}$—i.e., provided that the wavelength of the disturbance is large compared to the dimensions of the experiment. The classic result can be quickly achieved by three applications of Maxwell's equations, commencing with

$$\nabla \times \mathbf{B} = \mu_{\text{eq}}\left(\frac{4\pi}{10}\right)\mathbf{J} \tag{16-12}$$

$$= \mu_{\text{eq}}\left(\frac{4\pi}{10}\right)\frac{\mathbf{E}}{\rho_{\text{eff}}} \tag{18-42}$$

where **E** is an electric field and ρ_{eff} some effective resistivity of the medium. Taking the curl a second time

$$\nabla \times \nabla \times \mathbf{B} = \mu_{\text{eq}}\left(\frac{4\pi}{10\rho_{\text{eff}}}\right)\nabla \times \mathbf{E} \tag{18-43a}$$

$$\nabla\nabla \cdot \mathbf{B} - \nabla^2\mathbf{B} = -\mu_{\text{eq}}\left(\frac{4\pi}{\rho_{\text{eff}}}\right)10^{-9}\frac{\partial \mathbf{B}}{\partial t} \tag{18-43b}$$

which, since $\nabla \cdot \mathbf{B} = 0$, can be rewritten as

$$D_{EM}\nabla^2\mathbf{B} = \frac{\partial \mathbf{B}}{\partial t} \tag{18-44}$$

with

$$D_{EM} = \left(\frac{10^9}{4\pi\mu_{\text{eq}}}\right)\rho_{\text{eff}} \qquad (\text{cm}^2\,\text{s}^{-1}) \tag{18-45}$$

provided that ρ_{eff} is in Ω cm and $|\mu_{\text{eq}}| = 1$. This expression for D_{EM} was obtained by LUBELL and WIPF [LUB66] and GANDOLFO [GAN67, GAN68]. Before proceeding it is useful to check the dimensions of D_{EM}. Using the Cullwick four-dimensional [M], [L], [T], [Q] system for representing electrical and magnetic quantities, within which

$$[\rho] = [\text{R}][\text{L}] = [\text{M}][\text{L}]^3[\text{T}]^{-1}[\text{Q}]^{-2}$$

and

$$[\mu_{\text{eq}}] = [\text{B}]/[\text{H}] = \frac{[\text{M}][\text{T}]^{-1}[\text{Q}]^{-1}}{[\text{L}]^{-1}[\text{T}]^{-1}[\text{Q}]}$$

we find after substitution that

$$[D_{EM}] = [\mathrm{L}]^2[\mathrm{T}]^{-1} \qquad (\text{e.g., cm}^2\,\text{s}^{-1})$$

which, with reference to Table 18-2, is indeed the dimensions of a diffusivity.

Thus, under dynamic conditions the field **B** may be propagated *at absolute zero as an electromagnetic disturbance* that simulates diffusion with diffusivity D_{EM} characterized by an effective resistivity ρ_{eff}. At other temperatures this does not preclude additional contributions from thermally activated processes.

As SMYTHE [SMY50, p. 390] has pointed out, (18-44) for vector diffusion does not generally lead to scalar diffusion of the individual components. Exceptions are the three components of Cartesian geometry and the B_z of the cylindrical coordinate system of particular interest in tube magnetization.

It is now possible to apply several useful results from the standard texts to the problems of magnetic-field diffusion through solid and hollow cylinders, in response to an externally applied and maintained field-step of height ΔB_0 (the field being parallel to the axis of the cylinder—the z axis).

18.11.1 The Basic Equations

(a) Solid Cylinder of Radius *a*. With field at radius r [CAR59, p. 199],

$$\frac{B(r,t)}{\Delta B_0} = 1 - \frac{2}{a}\sum_{n=1}^{\infty}\frac{J_0(r\alpha_n)}{\alpha_n J_1(a\alpha_n)}e^{-D\alpha_n^2 t} \tag{18-46}$$

where the α_n are the positive roots of $J_0(a\alpha) = 0$ and $D \equiv D_{EM}$.

(b) Hollow Cylinder of Inner Radius *a* and Outer Radius *b* (Wall Field). With field within the wall at radius r [CAR59, p. 207],

$$\frac{B(r,t)}{\Delta B_0} = 1 - \pi\sum_{n=1}^{\infty}\frac{J_0(a\alpha_n)U_0(r\alpha_n)}{J_0(a\alpha_n) + J_0(b\alpha_n)}e^{-D\alpha_n^2 t} \tag{18-47}$$

where

$$U_0(r\alpha) = J_0(r\alpha)Y_0(b\alpha) - J_0(b\alpha)Y_0(r\alpha)$$

and the α_n are the positive roots of $U_0(a\alpha) = 0$.

(c) Hollow Cylinder of Inner Radius *a* and Outer Radius *b* (Core Field). With field in the interior at radius $r < a$ [JAE40],

$$\frac{B(t)}{\Delta B_0} = 1 - \frac{4}{a^2}\sum_{n=1}^{\infty}\frac{J_2(a\alpha_n)J_0(b\alpha_n)}{\alpha_n^2[J_0^2(b\alpha_n) - J_2^2(a\alpha_n)]}e^{-D\alpha_n^2 t} \tag{18-48}$$

where the α_n are the roots of $J_0(b\alpha)Y_2(a\alpha) - Y_0(b\alpha)J_2(a\alpha) = 0$.

In (18-46) through (18-48), J_i and Y_i are Bessel functions of the first and second kind, respectively, of order *i*. Equations (18-46) and (18-47) were derived by CARSLAW and JAEGER [CAR59, pp. 199, 207] from a Poisson equation in temperature; the latter equation was subsequently used by CRANK [CRA56, p. 76] to describe conventional "particle" diffusion in a hollow cylinder. Equation (18-48) was derived *ab initio* by JAEGER [JAE40], who started with Maxwell's equations and proceeded via the Laplace transformation method to derive an equation representative of magnetic shielding by a hollow cylinder.

In applying these results to the magnetization of superconducting cylinders, ΔB_0 is supposed to represent a step-wise increase of a previously increasing, but now steady, surface induction, B_0. The resulting $B(r, t)$ is superimposed on a preexisting critical induction gradient $(dB_z/dr)|_{\text{critical}}$ within the cylinder wall, as illustrated in Figure 18-9.

18.11.2 Application of the Electromagnetic Diffusion Equations to the Measurement of Creep Resistivity

(a) Solid Cylinder: Radius *a*. As pointed out by LUBELL and WIPF [LUB66], if only the field at the center of the cylinder ($r = 0$) is required as a function of reduced time $t' = D_{EM}t/a^2$, with $a\alpha_n \equiv \beta$, (18-46) simplifies [CAR59, p. 199] to

$$\frac{B(r=0,t')}{\Delta B_0} = 1 - 2\sum_{n=1}^{\infty} \frac{1}{\beta_n J_1(\beta_n)}\, e^{-\beta_n^2 t'} \tag{18-49a}$$

$$= 1 - C(t') \tag{18-49b}$$

where $C(t')$ has been tabulated by INGERSOLL *et al.* [ING48, p. 260]. LUBELL and WIPF [LUB66] investigated magnetic diffusion in a 13 mm O.D. cylinder of cold-drawn Zr–Nb(75%) that was solid except for a 2 mm$^{\varnothing}$ axial hole housing a Cu magnetoresistance probe whose purpose was to detect the growth of axial field in response to an externally applied field-step. The sample was raised into the shielding critical state in external fields of 15–35 kOe prior to the application of increasing-field pulses of rise-times of 2–40 Oe s^{-1} and amplitudes of 1–20 Oe. The procedure used was to measure the time taken for the internal field, $B(r = 0, t)$, to attain some selected relative value such as $\Delta B_0/2$ after which a value of $t' = D_{EM}t/a^2$ could be obtained from INGERSOLL's table [ING48, p. 260]. The diffusivity was then calculated, and a value of ρ_{eff} was derived from (18-45). The creep resistivity so obtained was found to be sensitive to dH_a/dt, the rise rate of the applied field-step, such that (using for example a step height, $\Delta B_0 = \Delta H_a$, of 3 Oe) $\rho_{\text{eff}} = 1.5 \times 10^{-10}\ (dH_a/dt)\ \Omega$ cm. In a continuation of the work

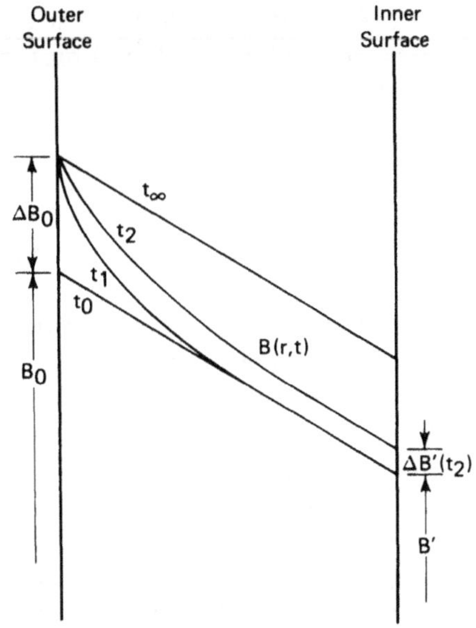

Figure 18-9. Temporal response of the induction within the wall of a magnetized superconducting cylinder to the instantaneous application of an increment ΔB_0 to a steady, but previously increasing, surface induction B_0; i.e., the transmission of a field-step through the wall of a cylinder—after GANDOLFO [GAN67, p. 58].

into the realm of low dH_a/dt with linearly increasing applied field increments, during which the response of the axial field probe was analyzed with the aid of another equation due to CARSLAW and JAEGER [CAR59, p. 328] (not listed above), LUBELL and WIPF [LUB66] found the creep resistivity to be represented by $\rho_{\text{eff}} = (1 \pm 0.2) \times 10^{-13} (dH_a/dt)\ \Omega$ cm.

(b) Tube: Inner Radius *a*, Outer Radius *b*, Internal Field. Creep resistivity in heavily cold-worked Ti–Nb(33 at.%) was measured by GANDOLFO *et al.* [GAN67, GAN68] by applying field-steps of 20–43 Oe (rise-time $\lesssim$1 ms) to a superconducting tube (32 mm length × 5.1 mm O.D. × 0.38 mm wall thickness) in a shielding critical state established by (previously increasing)

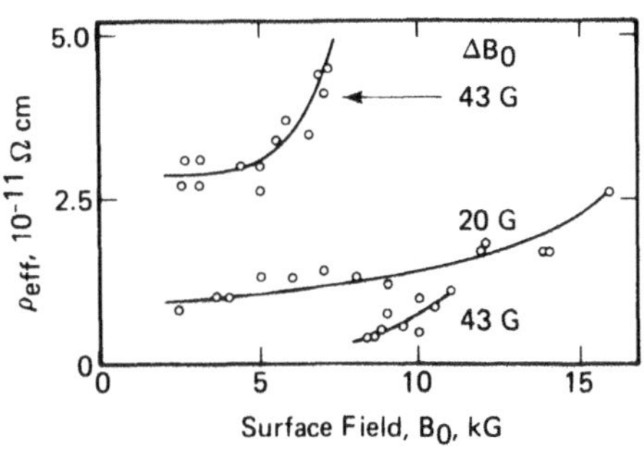

Figure 18-10. Results of a step-transmission type of tube-magnetization experiment. The effective resistivity, ρ_{eff}, may be obtained from the rate of growth of the internal field expressed in terms of the magnetic diffusivity, D_{EM}, (18-50b), with the aid of (18-45). It is implied that B_0 is the same as the applied field, H_a, used in previous figures. The tubular sample was machined from forged and cold-rolled Ti–Nb(33 at.%) rod. Upper two curves—as-machined; lower curve—aged 30 min/400°C. In all cases, $10^{-12}\ \Omega$ cm $< \rho_{\text{eff}} <$ $10^{-10}\ \Omega$ cm (and is therefore several orders of magnitude smaller than the flux-flow resistivity, ρ_f)—after GANDOLFO [GAN67, p. 60].

steady fields of up to 26 kOe. The data were analyzed with the aid of (18-48), which, for sufficiently long times, reduces to the first-term approximation [GAN67]

$$\frac{B(t)}{\Delta B_0} = 1 - \frac{4}{a^2\alpha_1^2}\left[\frac{J_2(a\alpha_1)J_0(b\alpha_1)}{J_0^2(b\alpha_1) - J_2^2(a\alpha_1)}\right] e^{-D\alpha_1^2 t} \tag{18-50a}$$

$$= 1 - A\, e^{-D\alpha_1^2 t} \tag{18-50b}$$

Thus, from the oscilloscope response to the growth of internal field, D_{EM} could be extracted and ρ_{eff} calculated. The magnitude of ρ_{eff}, which depended to some extent on those of the applied field-step and the DC background field, was found to be within the range of 10^{-12} to 10^{-10} Ω cm; Figure 18-10 is a typical set of results.

The inherent variability of ρ_{eff} may be understood with reference to Figures 18-1 and 18-3. Whereas the flow resistivity, ρ_f, defined as the slope $dV/dI = \rho_n H_a/H_{c20}^*$ ((14-32)), is a relatively large ($\sim\rho_n$) intrinsic property of the ideal material, the creep resistivity, rather poorly defined as the ratio E/J near $J = J_c$, is a function of the interaction of flux vortices with pinning centers, whose dependence on depinning force reflects the nonlinearity of the E *versus* J characteristic in the vicinity of J_c. The creep resistivity data are summarized in Table 18-3.

18.12 Magnetic Studies of Flux Flow

18.12.1 The Flow Viscosity Coefficient

When the pinning forces are weak, magnetic flux will flow across the superconductor under the influence of an appropriate driving force, at a velocity controlled by some viscosity coefficient, η, which (since $\nabla \times \mathbf{E} = -1 \times 10^{-8}\, \partial\mathbf{B}/\partial t$) must also be associated with some resistivity, say ρ_f, the flux-flow resistivity. As pointed out in Section 18.3.1, the force on a fluxoid in a field of gradient dB_z/dr is

$$\mathbf{f}_\phi = -\hat{r}\frac{\Phi_0}{4\pi}\frac{1}{\mu_{\text{eq}}}\frac{dB_z}{dr} \tag{16-10}$$

If this force is opposed only by viscous drag of coefficient η, such that a flux line moving with steady velocity $\mathbf{v}_f$ experiences a force

$$\mathbf{f}_v = -\eta\mathbf{v}_f \tag{18-51}$$

it follows that in steady-state equilibrium

$$\mathbf{f}_\phi + \mathbf{f}_v = 0 \tag{18-52}$$

Table 18-3. Typical Values of Creep and Flow Resistivity from Tube Magnetization Measurements

Phenomenon	Alloy	Condition	Formula	Chosen Field Gradient $G\,s^{-1}$	Field Range or Chosen Field kG	Creep or Flow Resistivity $\mu\Omega$ cm	Normal State Resistivity $\mu\Omega$ cm	Literature
Creep	Zr-75Nb	Cold drawn	$\rho_{\text{eff}} = 1.5 \times 10^{-10} \dfrac{dH_a}{dt}$	20	15 to 35	3×10^{-3}		Lubell and Wipf [Lub66]
			$\rho_{\text{eff}} = 1 \times 10^{-13} \dfrac{dH_a}{dt}$	1		1×10^{-7}		
Creep	Ti-Nb(33 at.%)	Cold rolled	$\rho_{\text{eff}} = D_{EM} \left[\dfrac{10^9}{4\pi\mu_{\text{eq}}}\right]^{-1}$		up to 26	2×10^{-5}	160	Gandolfo [Gan67]
Flow	Zr-Nb(75 at.%)	Annealed	$\rho_f = 10^{-9} \dfrac{B\Phi_0}{\eta}$		35	17	23	Kroeger [Kro66]
Flow	Ti-Nb(75 at.%)	Annealed	$\rho_f = 10^{-9} \dfrac{B\Phi_0}{\eta}$		35	15	28	Kroeger [Kro66]

in other words

$$\frac{\Phi_0}{4\pi}\frac{1}{\mu_{\rm eq}}\hat{r}\frac{dB_z}{dr} + \eta \mathbf{v}_f = 0 \tag{18-53}$$

This equation as it stands suggests that flow would take place in the presence of *any* field gradient. But such is not the case within the generally accepted critical state model, which requires that the gradient exceed some critical threshold ((16-9) and (16-11)) before flux motion commences and viscous forces are invoked. This additional force, defined by the elementary pinning force $\mathbf{f}_p$ (equal and opposite to the critical value of $\mathbf{f}_\phi$) when added to (18-52) gives:

$$\mathbf{f}_\phi + \mathbf{f}_v + \mathbf{f}_p = 0 \tag{18-54}$$

i.e.,

$$\eta \mathbf{v}_f = \mathbf{f}_p + \mathbf{f}_\phi \tag{18-55a}$$

$$= \frac{\Phi_0}{4\pi}\frac{1}{\mu_{\rm eq}}\hat{r}\left(\left.\frac{dB_z}{dr}\right|_{\rm critical} - \frac{dB_z}{dr}\right) \qquad (18\text{-}56)^4$$

with reference to (16-11) and (16-10), respectively. Equation (18-56) emphasizes that $\mathbf{v}_f$ is numerically greater than zero by an amount proportional to the extent that the flux-density gradient exceeds the critical value. In terms of magnitudes, then,

$$|\eta v_f| = |f_\phi| - |f_p| \tag{18-55b}$$

and, after replacing f_ϕ with its value from (16-10),

$$\left|\frac{1}{\mu_{\rm eq}}\frac{dB_z}{dr}\right| = \frac{4\pi}{\Phi_0}(|\eta v_f| + |f_p|) \tag{18-57}$$

which, as we shall see, is useful both in its present form and after the insertion of some boundary conditions.

18.12.2 Experimental Design for Viscosity Coefficient Measurement

The type-C experimental arrangement is suitable for viscosity coefficient measurement. In it, a pickup coil wound on the outer surface of the cylinder

[4] Recalling (16-13) and (16-17), this may be rewritten in current-density form as

$$\eta v_f = (\Phi_0/10)(J - J_c)$$

(see DEW-HUGHES [DEW71, DEW74]).

and another positioned inside are able to detect, during steady increase of the external field, H_a, the first penetration of flux into the cylinder followed by its arrival at the inner surface. Signals from these coils applied to an oscilloscope enable the transit time, and hence velocity, of the flux front to be measured. Other data needed are two special values of the *applied field*: (i) H_{c1}, its value corresponding to the first response of the external surface search coil, and (ii) H^{**}, its value at the moment that the inner detector coil signals the arrival of the flux front at the interior surface of the cylinder (wall thickness, w).

18.12.3 Analysis of the Flux-Flow Equations

In KROEGER's analysis of the flux-flow equations [KRO66] it was assumed (i) that the flux density at the advancing interface corresponded to an applied field, H_{c1}, and (ii) that the field gradient could be regarded as linear, in which case (18-57) (with magnitudes implied) became

$$\frac{H^{**} - H_{c1}}{w} = \frac{4\pi}{\Phi_0}(\eta v_f + f_p) \tag{18-58}$$

predicting that a plot of the measurables H^{**} *versus* v_f should also be linear with a slope of

$$m = 4\pi w\eta/\Phi_0 \tag{18-59a}$$

and an intercept of

$$H^{**}_{\text{int}} = 4\pi w f_p/\Phi_0 + H_{c1} \tag{18-59b}$$

As an interesting corollary, we note that $v_f = 0$ represents the critical-state condition that exists during the quasistationary type-A tube-magnetization experiment. It follows that (18-59b) should bear some resemblance to BEAN's quasistationary condition for critical state equilibrium, under which the field gradient within the wall was given simply by H^*/w (see Section 16.4.2). In the present dynamic or practical model, full penetration is gauged by the appearance at the "inner" wall of a field front of intensity H_{c1}; the corresponding critical field gradient in the BEAN sense would then be $(H^{**}_{\text{int}} - H_{c1})/w$. Equation (18-59b) as it stands yields the following expression for the elementary pinning force:

$$f_p = \frac{\Phi_0(H^{**}_{\text{int}} - H_{c1})}{4\pi w} \tag{18-60}$$

which is, of course, both formally and by the above argument closely related to the earlier expression $f_p = (\Phi_0/4\pi)(dH(B_z)/dr)|_{\text{critical}}$, ((16-11)).

Since

$$\frac{dB_z}{dt} = \frac{dB_z}{dr}\frac{dr}{dt} = \frac{dB_z}{dr} v_f \tag{18-61}$$

it follows from (18-57) (again, with magnitudes implied) that

$$\frac{dB_z}{dt} = \mu_{\text{eq}} \frac{4\pi}{\Phi_0}(\eta v_f + f_p) v_f \tag{18-62a}$$

which shows that dB_z/dt is independent of r (i.e., that the flux profile does not change shape as $B(r)$ increases with time) insofar as v_f is constant.

Equation (18-62a), either in its present form or after the substitution of m and H^{**}_{int} from (18-59a) and (18-59b) which yields

$$\frac{dH_a}{dt} = \frac{m}{w} v_f^2 + \frac{(H^{**}_{\text{int}} \cdot - H_{c1})}{w} v_f \tag{18-62b}$$

also shows that dH_a/dt is quadratic in v_f, verification of which would serve as additional confirmation of the theory.

18.12.4 Flux-Flow Viscosity in Weakly-Pinned Alloys—A Case Study of Annealed Ti–Nb and Zr–Nb Alloys

Samples of Ti–Nb(75 at.%) and Zr–Nb(75 at.%) conforming to the specifications of Table 18-4 have been studied by KROEGER [KRO66] using both type-A and type-C tube magnetization techniques. The results of the type-A experiments, presented in Figure 18-11, yielded values of the critical state full penetration threshold applied field H^*; they also emphasized the

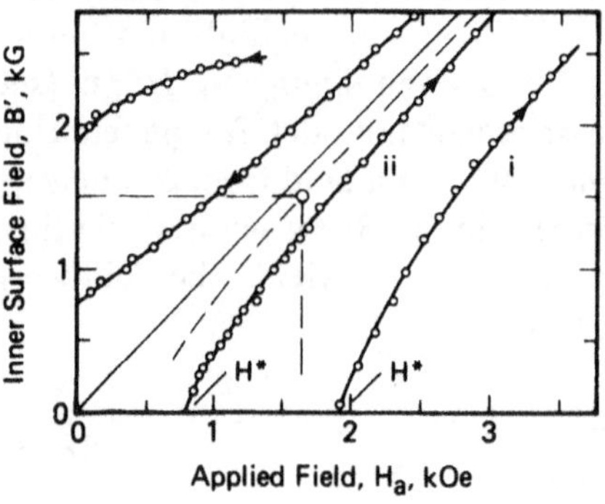

Figure 18-11. Results of a type-A tube-magnetization study of two annealed-and-quenched alloys (see Table 18-4), *viz.*, (i) Zr–Nb(75 at.%) and (ii) Ti–Nb(75 at.%). Indicated are the full-penetration-threshold fields H^*, and an assumed equilibrium $B' - H_a$ curve (dashed line) to aid in the estimation of a typical value for $\mu_{\text{eq}0} = B_0/H_a$. According to the figure, $\mu_{\text{eq}0}$ for sample (ii) is approximately 0.9. Magnetization data due to KROEGER [KRO66, p. 44].

Table 18-4. Analysis of the Flux-Flow-Velocity Experiment of KROEGER [KRO66]

Alloy[a]	Sample Dimensions mm			Lower Critical Field	Applied Field at Type-A Penetration Threshold	From Figure 18-11 and the Evaluation of Eqs. (18-59a), (18-59b), and (18-60)[b]				Deduced Critical Current Densities of Annealed Alloys
						Intercept	Slope	Viscosity Coefficient	Elementary Pinning Force	
	l	O.D.	I.D.	H_{c1} Oe	(Figure 18-11)	H^{**}_{int} Oe	m Oe cm^{-1} s	$\eta = \frac{\Phi_0 m}{4\pi w}$ dyn cm^{-2} s	$f_p = \frac{\Phi_0(H^{**}_{\text{int}} - H_{c1})}{4\pi w}$ dyn cm^{-1}	$J_c = \frac{10 f_p}{\Phi_0} = \frac{10(H^{**}_{\text{int}} - H_{c1})}{4\pi w}$ A cm^{-2}
Zr–Nb(75 at.%)	27	9.6	3.2	640 ± 20	~ 1900	1790	8.3	4.3×10^{-7}	3.0×10^{-5}	1.43×10^{3}
Ti–Nb(75 at.%)	27	9.6	3.2	570 ± 20	~800	840	9.2	4.7×10^{-7}	0.7×10^{-5}	0.33×10^{3}

[a] Condition in both cases: encapsulated under argon, annealed 2 h/1250°C and ice-water quenched.
[b] $\Phi_0 = 2.07 \times 10^{-7}$ gauss cm^2.

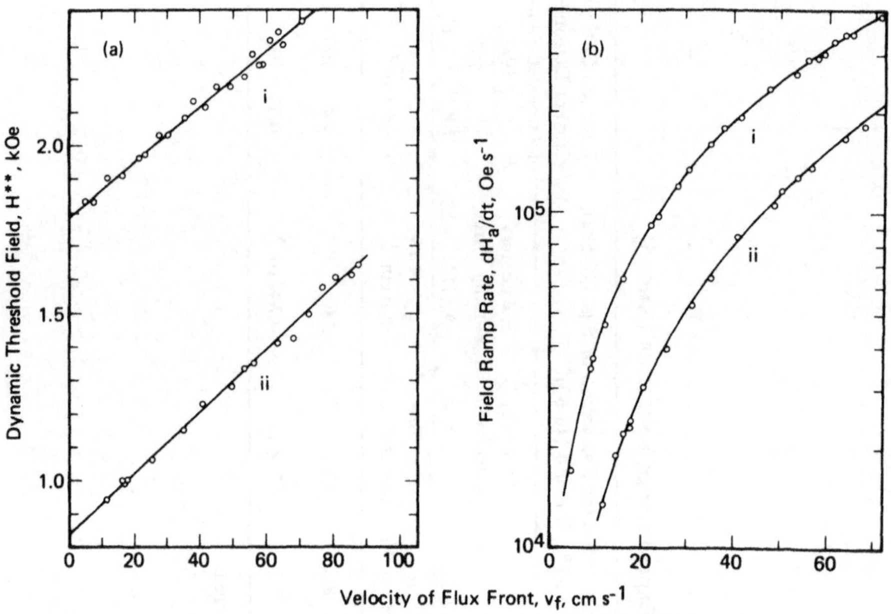

Figure 18-12. Results of a type-C tube-magnetization study of flux flow in the two samples referred to in Figure 18-11. (a) Plot of the dynamic threshold applied field, H^{**}, *versus* flux front velocity, v_f, the linearity of which serves to verify (18-58). (b) Plot of the rate of increase of the applied field, dH_a/dt, *versus* flux front velocity v_f; the points are experimental data; the curve has been calculated using (18-62b)—after KROEGER [KRO66, pp. 41, 42].

expected correlation between the degree of annealing and the approach to magnetic reversibility, and suggested, with reference to an imaginary equilibrium B' *versus* H_a curve shown as a dashed line in the figure, that 0.9 would be a reasonable value for μ_{eq0} within the field range considered. The results of the type-C dynamic magnetization experiment in terms of the applied field H^{**} *versus* the transit-timed v_f are given in Figure 18-12(a); as expected the intercept H^{**}_{int} ($v_f = 0$) is very close to the quasistatically measured H^* (Figure 18-11). Finally, values of η and f_p were obtained from the slopes and intercepts, respectively (see (18-59)), of the demonstrably linear H^{**} *versus* v_f plot. Figure 18-12(b) compares the measured dH_a/dt *versus* v_f data with the predictions of (18-62b) as an additional confirmation of the model.

18.12.5 Conclusion—Relationship Between Pinning Strength and the Dynamics of Flux Motion

Demonstrated below is the satisfactory agreement that has been obtained between the values of η just calculated from the results of direct flow retardation measurement and those obtained from the results of other types of flux-flow investigation. It is a sufficient for the time being, and with reference to (18-61), to draw attention to the following points: (i) Whereas in weakly pinned material flux flow can be expected under the influence of a magnetic induction gradient, if the pinning is sufficiently strong the large gradients that result from a sufficiently high applied-field ramp rate can lead to the runaway condition known as flux jumping. (ii) In the present alloys flux flow takes place for ramp rates of up to at least 4×10^5 Oe s^{-1}. (iii) The stronger the pinning the easier it is to preserve the critical state condition $v_f = 0$ in the presence of a time-varying applied field. (iv) In the more strongly pinned of the two alloys considered in the tube magnetization experiments (the Zr–Nb(75 at.%), see Figure 18-11) the critical state is maintained for applied-field ramp rates of up to about 10^4 Oe s^{-1}. (v) Finally, in order to provide a semiquantitative reference by means of which the pinning strengths of the alloys considered here can be compared with those generally encountered in hard superconductivity, estimates of critical current densities have been made in the manner outlined below for inclusion in the final column of Table 18-4.

18.13 Flow Resistivities and Critical Current Densities of Annealed Ti–Nb(75 at.%) and Zr–Nb(75 at.%) Alloys

18.13.1 Flow Resistivity

Flux-flow resistivity, as a transport property, was first studied and discussed in a series of papers by KIM *et al.* [KIM65, KIM67, KIM69]. According to Section 18.3.3, the viscous motion of flux is accompanied by the dissipation of energy. This may be regarded as being associated with a flow resistivity, an expression for which is easily obtained by taking (18-56) in its current-density form (see footnote 4), *viz.*,

$$\eta v_f = (\Phi_0/10)(J - J_c) \qquad (18\text{-}63)$$

and inserting it into (18-5) to yield, in c.g.s.-practical units,

$$\dot{Q}_v = 10^{-9} \frac{B_z \Phi_0}{\eta} J^2 \left(1 - \frac{J_c}{J}\right) \quad (\mathrm{W\,cm^{-3}}) \qquad (18\text{-}64)$$

Equation (18-64) shows that, provided $J > J_c$, energy is dissipated in a flux-flow resistivity given by

$$\rho_f = 10^{-9} \frac{B_z \Phi_0}{\eta} \quad (\mathrm{ohm}) \qquad (18\text{-}65)$$

That this is a differential resistivity can be seen, following KIM *et al.* [KIM65], by taking the differential forms of (18-4) and (18-63), giving, respectively,

$$\Delta E = 10^{-8} B_z \Delta v_f \qquad [18\text{-}4](18\text{-}66)$$

and

$$\Delta J = \frac{10\eta}{\Phi_0} \Delta v_f \qquad [18\text{-}63](18\text{-}67)$$

from which it follows that

$$\rho_f = \frac{\Delta E}{\Delta J} = 10^{-9} \frac{B_z \Phi_0}{\eta}$$

Values of ρ_f have been determined from the results of KROEGER'S [KRO66] type-C tube-magnetization study of flow viscosity. They are listed in Table 18-3 where they may be compared with the corresponding normal-state residual resistivities (ρ_n) and the creep-resistivity (ρ_{eff}) results of LUBELL and WIPF [LUB66] and GANDOLFO *et al.* [GAN67, GAN68]. The flow resistivity is seen to be several orders of magnitude larger than ρ_{eff}, and comparable in magnitude to ρ_n. The field dependence of ρ_f is treated as a transport property in Section 23.8.

Creep is a property of the heavily pinned cold-worked alloys while flow in this context is confined to the precipitate-free annealed samples.[5] As will be shown below, exposure of a *precipitated* Zr–Nb alloy to the same magnetic conditions that induced flow in the annealed samples may result in flux jumping.

18.13.2 Critical Current Density

A combination of (16-10) and (16-13) shows that the critical Lorentz force on an individual fluxoid is $f_{\phi c} = \Phi_0 J_c/10$. After inserting the condition $f_v = 0$ (i.e., $\eta v_f = 0$) for the critical state, (18-54) then yields

$$\frac{\Phi_0 J_c}{10} = -f_p \qquad \text{for the elementary pinning force}$$

$$= -\frac{\Phi_0 (H_{\text{int}}^{**} - H_{c1})}{4\pi w} \qquad \text{from (18-60)} \qquad (18\text{-}68)$$

[5] As discussed in Chapters 1 and 2 (see, in particular, Figures 1-7 and 2-7), alloys such as Ti–Nb(33 at.%) and Zr–Nb(75 at.%) in addition to possessing cold-worked deformation structures (Section 2.8) may also contain precipitates after suitable heat treatment. The Nb-rich alloy Ti-Nb(75 at.%) is single-phase bcc; so also is Zr–Nb(75 at.%) after quenching from 1250°C.

These in turn lead to the following pair of expressions for J_c:

$$|J_c| = \begin{cases} \dfrac{10 f_p}{\Phi_0} & (18\text{-}69) \\[2ex] \dfrac{10(H_{\text{int}}^{**} - H_{c1})}{4\pi w} & (18\text{-}70) \end{cases}$$

either one of which, when evaluated from the results of KROEGER's experiments on annealed alloys [KRO66], yields the expectedly low values of critical current density listed in Table 18-4.

18.13.3 Corollary

In an examination of the implications of the above relationships it is interesting to note that, whereas (18-69) expresses J_c in terms of the elementary pinning force f_p, implicit in the mean-gradient format of (18-70) is the idea of a bulk pinning force, F_p, that balances the bulk critical Lorentz force $J_c B/10$ as in (16-15). With this in mind, (18-69) and (18-70) combine to yield

$$F_p = -\frac{B}{10} J_c = \frac{B}{\Phi_0} f_p = n f_p \tag{18-71}$$

where n, as before, is the number of fluxoids per unit area. This leads to the interesting conclusion that the extension of the elementary pinning force expression (16-11), (generally attributed to FRIEDEL *et al.* [FRI63]) to bulk materials by way of the assumptions used here—*viz.*, (i) that dB_z/dr = constant = $(B_0 - b')/w$ (as in the Bean model) and (ii) that $n\Phi_0 = B_z$ (equivalent to stating that all fluxoids in the field are individually pinned) is tantamount to adopting $F_p = n f_p$ as a pinning-force summation rule. This is the simplest of all such rules; pinning-force summation in general is discussed in Chapter 21.

18.14 Magnetic Studies of Flux Jumping

18.14.1 The Use of Tube and Cylinder Magnetization Techniques

Tube and cylinder magnetization experiments have frequently been used in studies of the critical state. Section 16.6 has dealt with the properties of conventional $4\pi\langle M\rangle$ *versus* H_a and B' *versus* H_a hysteresis loops and with experiments designed to measure the flux-penetration profiles.

However, the hysteresis loops are not always as smooth as those depicted in Figures 16-6 and 16-7. From the outset, flux jumping has been a frequently observed phenomenon, an example of which, within the context of tube magnetization, has been given in Figure 18-2. With reference to that figure, as H_a, applied parallel to the axis of a Ti–Nb tube (31.8 mm length, 6.4 mm O.D., and 1.3 mm wall thickness), is very slowly increased from zero, the interior remains shielded until H_a reaches H^*, at which point flux has just penetrated the wall. B' and H_a then increase in unison until H_a attains the value H_{fj}, whereat part or all of the tube suddenly becomes normal and "transparent" to the applied field. As soon as the tube recovers from this "quench" or "flux-jump" it again shields the bore from the applied field until the quench repeats itself. If the effect is represented as a succession of critical states in a field-penetration diagram as in Figure 16-11(a), it becomes apparent that it should be approximately periodic—a kind of fixed-frequency relaxation oscillation. Indeed, as numerous authors have discovered (see SWARTZ and BEAN [SWA68] for references), the spacing between the flux jumps has turned out to be approximately constant at a given temperature and sweep-rate of the applied field. Upon variation of these parameters it has been noted, however, that the spacing between flux jumps is maximum at temperatures intermediate between zero and T_c, and that at constant temperature the spacing decreases with increasing sweep rate.

At the time of its discovery flux jumping was correctly identified as the result of thermal avalanching and subsequently WIPF and LUBELL [WIP65], with the aid of a suitably placed carbon resistance thermometer, were able to directly measure the heights of the temperature pulses.

A premature transition to the normal state is occasionally evident in the results of magnetization studies. The failure of the material to attain its short-sample critical current density is generally known as "degradation."[6] As will be demonstrated below, the kind of degradation exhibited here is geometry- or size-related and, in the case of a coil, is not a reflection of any form of nonuniformity in the wire from which it is wound. Before leaving this topic it is useful to make two related points: (i) a monolithic tube is equivalent to a fully dense coil of bare superconducting wire; (ii) for reasons to be considered in detail below, although flux jumping is no longer a problem of practical importance, its study in superconductive tubes has directly led, in a manner also to be dealt with below, to an understanding of the mechanisms of intrinsic superconductor instability.

A few of the essential characteristics of tube or cylinder instability can be understood with reference to Figure 18-13, a set of three critical state

[6]This term should not be construed as "deterioration" since no permanent changes in materials properties are supposed to be taking place.

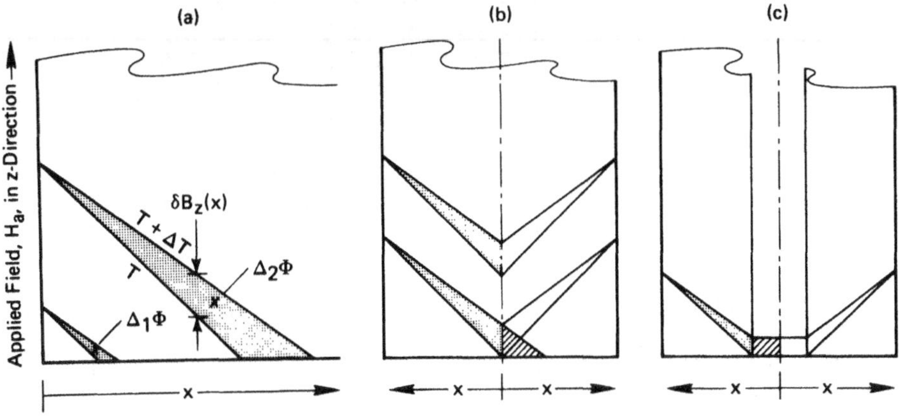

Figure 18-13. Influence of a temperature increment, ΔT, on the critical flux-density gradients. Represented are: (a) a one-sided slab, (b) a solid cylinder or two-sided slab, and (c) a hollow cylinder or tube—after WIPF [WIP67].

diagrams devised by WIPF [WIP67]. Each figure represents the effect of a small temperature increase on the critical flux-density gradient. A temperature increment ΔT results in a reduction ΔJ_c in the critical current density, a corresponding decrease in dB_z/dx, and an ingress of flux $\Delta\Phi = \int \delta B_z(x)\, dx$ per unit thickness, proportional to the cross-hatched areas in Figure 18-13. The applied field and surface induction are assumed to be constant and equal, and ΔT is supposed to be the same in each of the three diagrams. Figure 18-13(a), which represents a one-sided plate, shows that the flux increment $\Delta\Phi$, and with it the danger of instability, increases with H_a. Figure 18-13(b) shows that once the flux has penetrated to the center of a *solid* cylinder (or two-sided plate) $\Delta\Phi/\Delta T$ becomes constant, which suggests, in agreement with observation [WIP65, WIP67], that if flux jumping did not take place up to the point of full penetration it is unlikely to occur subsequently. Apart from $\Delta\Phi$ considerations, it is also likely that the symmetry of the two-sided configuration depicted in Figure 18-13(b) would discourage flux instability once central penetration has been achieved. Finally Figure 18-13(c) indicates that in a *hollow* cylinder (tube) the danger of flux jumping becomes enhanced at the moment the field reaches the inner wall, since any further increase in the field will be accompanied by a sudden demand for the extra flux needed to establish a uniform field in the bore, a requirement which is absent with the solid cylinder.

Using cylinder or tube magnetization techniques, measurements have been made by WIPF and LUBELL [WIP65], KROEGER [KRO66], and GANDOLFO *et al.* [GAN67, GAN69] of: (i) the influence on H_{fj}, the field at which the *first flux jump* takes place, of the applied-field ramp-rate dH_a/dt and the metallurgical condition of the sample; (ii) the influence of the applied

Table 18-5. Flux Jumping in Tube Magnetization—Three Case Studies

		Specimen Type and Dimensions, mm						Applied Fields		
Alloy	Condition	Type	Length	O.D.	I.D.	Wall Thickness	Experiment Type[a]	Ambient Field, H_a kOe	Field (or Auxiliary Field) Ramp Rate dH_a/dt (or dh_a/dt) Oe s^{-1}	Purpose
Zr–75Nb	Cold drawn	Cylinder	35–50	3–18			C	≤12	10^1–10^5	dh_a/dt-dependence of H_{fj} and temperature rise during flux jump.[b]
Zr–Nb (75 at.%)	Precipitated (30 min/800°C)	Tube	27	9.6	3.2		C	≤4	(2–40) × 10^4	dh_a/dt-dependence of H_{fj}, and metallurgical conditions for flux jump within the framework of comparative flux-jump/flux-flow studies.[c]
Zr–Nb (67 at.%)	As cast	Tube	32	16	8					
Ti–Nb (33 at.%)	Forged at ~300°C and cold rolled	Tube	38	5.1		0.381	B	10 and 20	(0.26–3.5) × 10^5	Flux jumping induced by an alternating auxiliary field h_0, sin ωt. Hence, $dh_a/dt = \omega h_0 \cos \omega t_{fj}$.[d]

[a] Refer to Table 18-1.
[b] Wipf and Lubell [WIP65].
[c] Kroeger [KRO66].
[d] Gandolfo *et al.* [GAN69].

field strength and the rate of change dh/dt of a small modulation field on the *spacing between* flux jumps. In their classic paper, WIPF and LUBELL [WIP65] also investigated, both indirectly and directly, the surface and internal temperature changes associated with adiabatic flux jumping. The experimental details of the three studies to be discussed are summarized in Table 18-5.

The sweep-rate dependent H_{fj} had already been studied in Nb_3Sn cylinders by CORSAN [COR64], who noted that at the low sweep rates employed (33–330 Oe s^{-1}) H_{fj} decreased with increasing dH_a/dt. Experiments by MORTON [MOR65[a]] with a thick-walled Zr–Nb tube of 8.5 mm$^{\varnothing}$ gave a similar result, although a reduction of the diameter to 3.55 mm removed the sweep-rate dependence. According to CORSAN [COR64], the interval between flux jumps, Δh, was related to the sweep rate of the external field by

$$dH_a/dt = a \exp(-b\Delta h) \tag{18-72}$$

where a and b are constants. Subsequently WIPF [WIP67] concluded, on the basis of calculation, that

$$H_{fj} \propto c - \ln(dH_a/dt) \tag{18-73}$$

where c is a constant.

18.14.2 The Experiments of Wipf and Lubell

WIPF and LUBELL [WIP65] adopted the position that in adiabatic flux jumping the stored energy of the field H_{fj} is converted into heat, causing the sample temperature to rise locally to T_f, a temperature such that, according to SWARTZ and BEAN [SWA68],

$$H_{fj}^2(T_i) = H_{c2}^2(T_f) = 8\pi \int_{T_i}^{T_f} C^v(T)\, dT \tag{18-74}$$

where C^v is the specific heat per unit volume. With an experimentally estimated $C^v \cong 46.5\,T^4$ erg cm^{-3} K^{-1} for Zr-75Nb in the superconductive state [WIP65] (but cf. the Gorter-Casimir result, (8-9)), temperatures of $T_i = 4.2$ K and $T_f = 10.1$ K are consistent via (18-74) with $H_{fj} = 4.9$ kOe, a field that, after allowance has been made for the influence of induced normal currents [WIP65], is closely related to H_{c2} (at 10.1 K). Experiment showed that in all but the smallest diameter cylinders, and provided that dH_a/dt was greater than about 2-3 $\times 10^2$ Oe s^{-1}, the measured H_{fj} was in good agreement with the calculated value. At low field-sweep rates ($\lesssim$100 Oe s^{-1}), higher values of H_{fj} were encountered (Figure 18-14(a)).

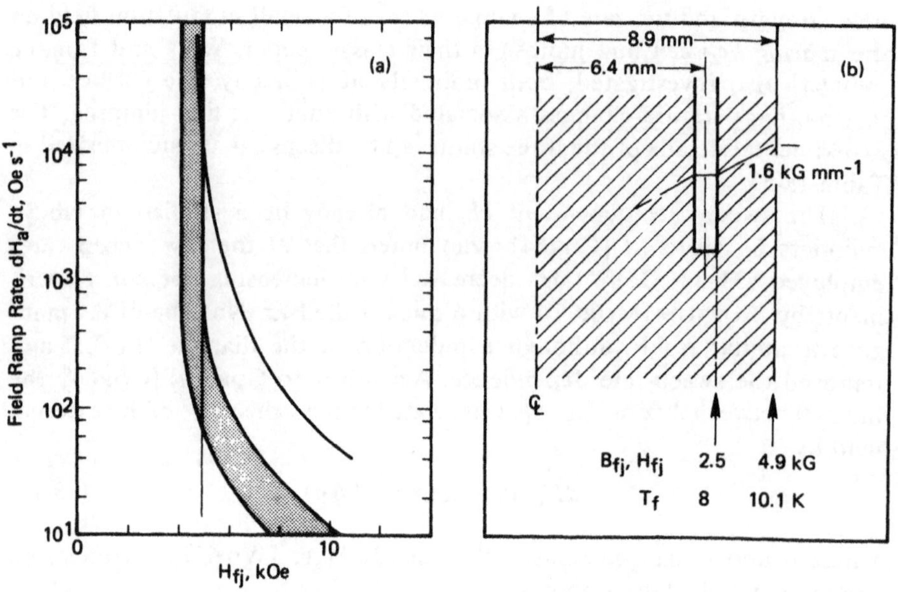

Figure 18-14. (a) Results of a type-C cylinder magnetization study of flux jumping in cold-drawn Zr–Nb(75 at.%) ($T_c \cong 11$ K). The flux-jump threshold field, H_{fj}, is plotted against the rate of increase of the applied axial field, dH_a/dt. Included in the shaded band are a multitude of data points for cylinders of diameters from 12 to 18 mm. The isolated line refers to a 3 mm$^{\varnothing}$ cylinder. The vertical line is positioned at H_{c2} (10.1 K). (b) Cross section of the cylindrical sample indicating the internal-field gradient and the position of the carbon-resistance thermometer used for measuring the local temperature rise during flux jumping—after WIPF and LUBELL [WIP65].

With the aid of a submerged carbon-resistance thermometer (Figure 18-14(b)), WIPF and LUBELL [WIP65] were able to make a direct determination of the local temperature rise during flux jumping. After adiabatic transition in a thermally insulated sample a temperature of 8 K was recorded, which is again in agreement with (18-74) provided that the flux density at the thermometer position was inserted into the formula, a correction having been made for the gradient in the specimen.

Taken together these results indicate that flux jumping is initiated at the surface under the control of the surface temperature and field, but that once flux movement is under way the local temperature rises are governed, through (18-74), by the local flux density, which in the shielding critical state decreases with depth below the surface.

18.14.3 The Experiments of Kroeger

In a study of flux jumping in precipitated and as-cast Zr–Nb alloys, which was juxtaposed against the investigation of flux-flow viscosity in the

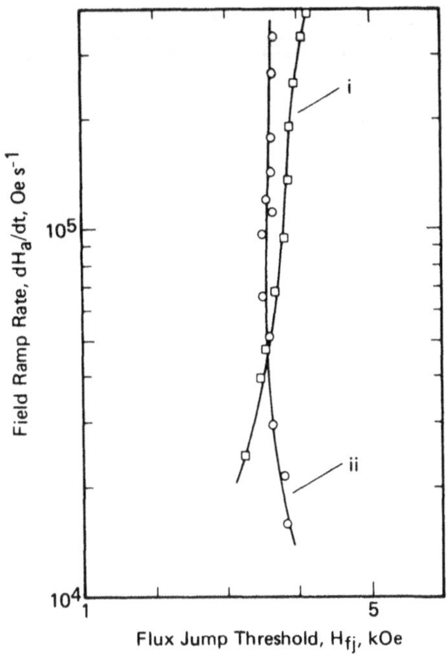

Figure 18-15. Results of type-C tube magnetization studies of two samples of Zr–Nb alloy (see Table 18-5). Curves are: (i) as-cast Zr–Nb(67 at.%), and (ii) precipitation-heat-treated (30 min/800°C) Zr–Nb(75 at.%). The flux-jump threshold, H_{fj}, is plotted against the rate of increase of the applied axial field, dH_a/dt. The curves are experimental; a few representative data points are included—after KROEGER [KRO66, p. 35].

weakly-pinned annealed Zr–Nb and Ti–Nb alloys treated above, KROEGER [KRO66] measured H_{fj} as a function of applied-field ramp-rate, dH_a/dt, within the range 2–40 × 10^4 Oe s^{-1}. The results, which are summarized in Figure 18-15, indicate that whereas in the precipitated sample H_{fj} tended to be independent of dH_a/dt over the range concerned (in agreement with the results depicted in Figure 18-14(a)), the same was not true for the as-cast material. KROEGER [KRO66] suggested that in the more weakly pinned as-cast, but not heat treated, alloy the flux was not completely restrained by pinning forces and could move, viscously to some extent, further into the sample before the flux jump took place. The power dissipation associated with the flow component of flux motion resulted in premature flux jumping at fields below H_{fj} (see Section 18.18 below) unless dH_a/dt was made sufficiently large to guarantee adiabatic (i.e., WIPF-LUBELL [WIP65]) conditions.

18.14.4 The Experiments of Gandolfo

If an applied field is increased, at a steady rate, sufficiently high to guarantee adiabatic flux jumping, a plot of the induction nearest to the bore of the cylinder, B', *versus* the applied field, H_a, increases in a step-wise manner as in Figure 18-2. Since, following a quench, the critical magnetization rebuilds itself (or "starts over") and increases until the next quench occurs, as in Figure 16-11(a), it is to be anticipated that the effect should occur at regular intervals of the applied field. A major increase in the applied field, however, should be accompanied by a decrease in the interval between such jumps, as the shielding and trapping curves converge at high fields (Figure 18-2).

If H_a is cycled about some steady mean value, a repeatable sequence of flux jumps with intervening relaxation to the superconducting state may take place. In general, a cycle of events such as that depicted by the inset

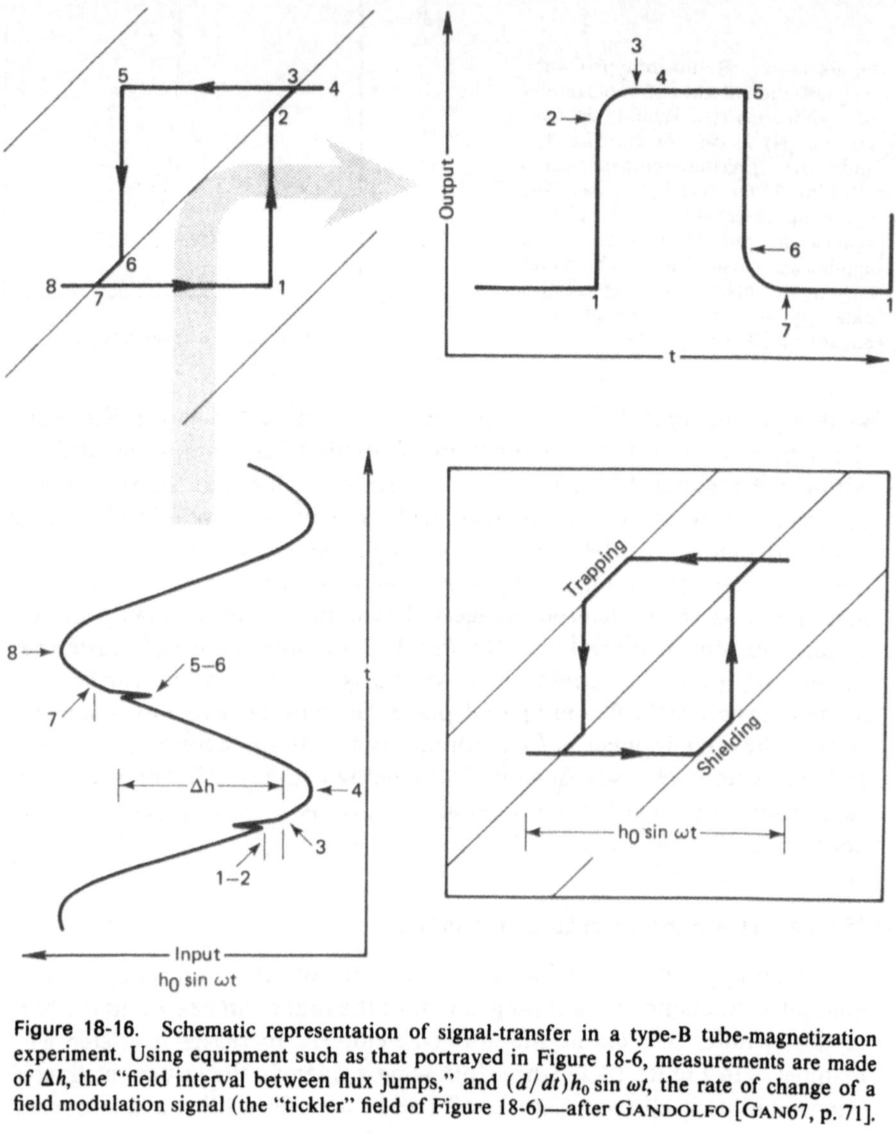

Figure 18-16. Schematic representation of signal-transfer in a type-B tube-magnetization experiment. Using equipment such as that portrayed in Figure 18-6, measurements are made of Δh, the "field interval between flux jumps," and $(d/dt)h_0 \sin \omega t$, the rate of change of a field modulation signal (the "tickler" field of Figure 18-6)—after GANDOLFO [GAN67, p. 71].

in Figure 18-16 is anticipated. GANDOLFO *et al.* [GAN67, GAN69] investigated this so-called "steady-state flux jumping" by applying a sinusoidal field modulation, $h_0 \sin \omega t$, of amplitude $h_0 = 390\text{–}690$ Oe and frequency $\omega/2\pi = 20\text{–}100$ Hz, to steady bias fields of 10 and 20 kOe. The resulting flux-jump loop is shown in Figure 18-16, in which it may be noticed that the applied field was not able to complete its excursion to the shielding or trapping lines before flux jumping took place—possibly as a consequence of incomplete thermal relaxation during the time interval permitted by the field sweep rate. Under the conditions of the experiment, the flux-jump loop served as a transfer function between the sinusoidal "input" field, $h_0 \sin \omega t$ and the square-wave "output" field as detected by means of a Hall-effect probe located in the bore of the tube.

With reference to Figure 18-16 the following sequence of events occurs:

1 to 2	The flux jump takes place and the tube wall becomes magnetically transparent.
2 to 3	The wall, before recovery, transmits a portion of the sinusoidal signal.
3 to 4 } 4 to 5 }	The tube has recovered and shields the bore from the changing signal.
5 to 6	A second flux jump marks the beginning of the second half-cycle of magnetic events.
6 to 7	A flux-transparent region again transmits a portion of the sinusoidal signal.

The measurable quantity of interest in these experiments was Δh, the input-field difference between the field at point "3" (corresponding to the threshold of the flat top of the output trace) and that at the base of the spike "5–6". Δh was the quantity regarded by GANDOLFO [GAN67] as the field interval between flux jumps and the "difference between the fields at the sample surface and that deep within the sample at the time the flux jump occurs," (see also Figure 16-10(a)).

The results, summarized in Figure 18-17, (i) demonstrate the reproducibility of the flux-jump effect, and (ii) in the sense that Δh may be regarded as a "local H_{ij}," are in general agreement with Figures 18-14 and 18-15. The decrease of magnetization with increasing applied field, which permits convergence of the shielding and trapping curves in Figure 18-2, is to some extent responsible for the separation of the Δh(20 kOe) and Δh(10 kOe) curves of Figure 18-17.

Flux jumping represents a catastrophic mixed state instability. This discussion of the phenomenon concludes with the note that, although flux jumping is no longer a problem in contemporary conductors or magnets fabricated from them, a knowledge of its fundamental causes is essential to an understanding of superconductor stability in general.

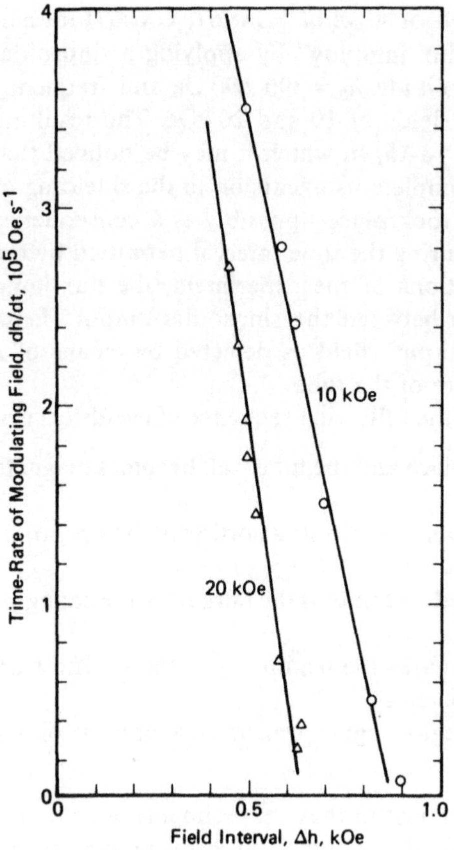

Figure 18-17. Results of a type-B tube magnetization study of flux jumping in a sample of Ti–Nb(33 at.%) (as-machined from a 300°C-forged and cold-rolled ingot). Field interval between flux jumps, Δh, *versus* the rate of change of a superimposed applied field, dh_a/dt; cf. Figures 18-14 and 18-15—after GANDOLFO [GAN67, p. 73].

18.15 Magnetic Instability in Tube Magnetization

18.15.1 Incomplete and Full Flux Jumping—Historical Background and Present Status

Quantitative study of magnetic stability under field-sweep conditions commenced with the experimental investigations by WIPF and LUBELL [WIP65] of adiabatic flux-jumping in Zr–75Nb cylinders. Publication of the results of these experiments was quickly followed by the work of WIPF [WIP67] and SWARTZ and BEAN [SWA68], who independently concluded that the response of a critical-state superconductor to a magnetic (or thermal—see (18-74)) disturbance may take the form of either: (i) a shift to a "new" critical state taking place under conditions of "full stability" [WIP67]; (ii) a nonpropagating local movement of the flux lattice—the so-called "limited instability" or "incomplete flux jump" [SWA68]—setting in at fields greater than some H_{fi}; or (iii) a local displacement of the flux lattice leading via thermal avalanching to a quenching of the superconductive state—the "runaway instability" [WIP67], "full flux jump" [SWA68], or now simply "flux-jump"—taking place adiabatically at fields above H_{fj} where stored energy converted into heat is sufficient to drive the superconductor into the normal state. The regime of full magnetic stability—stability against even limited flux jumping—has been considered by both WIPF [WIP67] and SWARTZ and BEAN [SWA68]; the regime of "limited stability," $H_{fi} < H_a < H_{fj}$, has been treated only by WIPF [WIP67], while the conditions of flux jumping, $H_a \gtrless H_{fj}$, were initially dealt with by WIPF and LUBELL [WIP65] and again by SWARTZ and BEAN [SWA68].

The research to be discussed in the final sections of this chapter was conducted at a time when flux jumping was thought to be the only important obstacle to the achievement of short-sample critical current densities in large coils, and was, therefore, the principal concern of the early magnet designers and builders. The terms "limited instability" and "incomplete flux jump," along with numerous others that have appeared from time to time in the lexicon of superconductor stability [WIP78], are now obsolete, having been replaced by more apt descriptions of the physical phenomena concerned. In the treatments of Sections 18-16 and 18-17 it is noted that the energy of the flux jump is expended in heating up the superconductor, elevating its temperature either to some "safe" value less than T_c or to a value above T_c, in the case of a runaway flux jump or quench. Clearly, under the adiabatic limited-flux-jump conditions the superconductor is enthalpy stabilized by its intrinsic heat capacity. It is then but a short conceptual step to imagine the attachment of extra heat-absorbing material to the superconductor, or to introduce into the problem and its solution the enthalpy associated with a surrounding liquid-He film. An up-to-date description of these and all other aspects of the stability of bare and clad (composite) He-cooled superconductors is given in Chapter 25.

18.15.2 Intrinsic Stability Considerations

The movement of flux, or an increment of heat or critical current density, will initiate a feedback cycle [WIP70] of the kind exemplified by Figure 18-18 in which "input" and "output" are represented by temperature increments. Suppose the temperature to increase by a small amount ΔT, then $J_c \cong \alpha(T)/(b_0 + B)$, (16-42), will decrease through the temperature dependence of the pinning parameter, $\alpha(T)$; thus, the gradient of the induction $dB_z/dr = -\mu_{eq}(4\pi/10)J_c$, (16-17), will decrease, calling for an increase of flux $\Delta\Phi$ through the sample surface; the flux movement dissipates heat[7] and raises the temperature by $\Delta T'$. Obviously, the terms of stability

[7]"Generated" by the E_y of (18-3) in association with J_c.

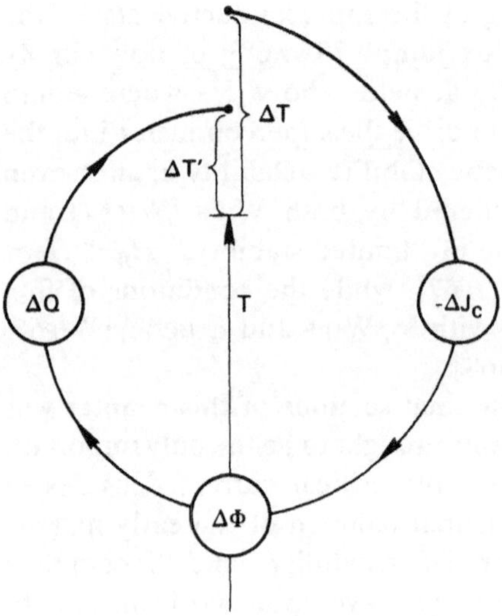

Figure 18-18. Stability cycle intrinsic to the superconductor, indicating the interrelationship between $J_c(T)$, the ingress of flux, $\Delta\Phi$, and heat liberated ΔQ. Stability is assured provided $\Delta T' < \Delta T$, the initiating temperature increment.

are dictated by the conditions under which the curve of Figure 18-18 spirals inwards (as shown) or outwards.

The starting point of any quantitative analysis is (i) a statement of the static and dynamic conditions of equilibrium or (ii) a description of the temporal response of the medium to magnetic and thermal disturbances as in Table 18-6. In dealing with magnetic stability, several interrelated factors must be taken into consideration. The dominant materials properties are D_{TH} and D_{EM} the thermal and electromagnetic diffusivities, respectively, as defined in (T-9) and (T-10) of Table 18-6. These quantities, of course, control the rates of propagation of the thermal and magnetic disturbances, their relative magnitudes determining whether the process is to be classified as "adiabatic" or "isothermal" [WIP67]. The full thermal transport relationship, including the source or sink term $(1/C^v)\partial Q_v/\partial t$ is (T-7). As pointed out by WIPF [WIP67], the applicable form of this equation depends on the time rate of the applied magnetic disturbance. According to LUBELL and WIPF [LUB66], $D_{EM} \propto dH_a/dt$; consequently, the high rates of propagation that accompany *rapid changes of applied field* permit (T-7) to be approximated by $\partial T/\partial t = (1/C^v)\,\partial Q_v/\partial t$; conversely, for *small values of dH_a/dt* heat is conducted away as fast as it is produced, and the nonadiabatic simple diffusional form of (T-7) is to be used. With this in mind, and with further reference to Table 18-6, WIPF'S [WIP67] calculation of the highest shielding field for full critical-state stability, H_{fi}, will be outlined.

Table 18-6. Investigation of the Thermomagnetic Stability of the Critical State

A. Force Balance

(T-1)	$\mathbf{F}_{\text{crit}} + \mathbf{F}_p = 0$	Critical state
(T-2)	$\mathbf{F} + \mathbf{F}_p = \eta \mathbf{v}_f$	Flux-flow state
(T-3)	$\mathbf{F}_{\text{crit}} + \Delta\mathbf{F} + \mathbf{F}_p + \Delta\mathbf{F}_p < 0$	Full stability of the critical state
(T-4)	$\Delta\mathbf{F} + \Delta\mathbf{F}_p < 0$	
(T-5)	$\mathbf{F}_{\text{crit}} + \Delta\mathbf{F} + \mathbf{F}_p + \Delta\mathbf{F}_p > 0$	Limited or runaway instability
(T-6)	$\Delta\mathbf{F} + \Delta\mathbf{F}_p > 0$	

B. Thermal and Magnetic Diffusion

Temperature

Phenomenological Equation of Thermal Transport (in terms of *volume* specific heat, C^v, and heat per unit volume, Q_v)

(T-7) $$\frac{\partial T}{\partial t} = D_{TH}\frac{\partial^2 T}{\partial x^2} + \frac{1}{C^v}\frac{\partial Q_v}{\partial t}$$

(T-9) $$D_{TH} = K/C^v$$

Influence of ΔT on the Pinning Force ($\beta \equiv \delta B$)

(T-11) $$\frac{\partial T}{\partial t} = \frac{1}{C^v}\frac{\partial Q_v}{\partial t}$$

(T-13) $$\Delta Q_v = J_c \Delta\Phi = -\frac{1}{\mu_{\text{eq}}}\left(\frac{10}{4\pi}\right)\frac{\partial B}{\partial x}\int_x \beta\, dx$$

(T-15) $$\Delta F_p = \frac{\partial F_p}{\partial B}\beta + \frac{\partial F_p}{\partial T}\Delta T$$

Field

Phenomenological Equation for the Diffusion of Magnetic Flux

$$\frac{\partial B}{\partial t} = D_{EM}\frac{\partial^2 B}{\partial x^2}$$ (T-8)

$$D_{EM} = 10^9 \rho_{\text{eff}}/4\pi\mu_{\text{eq}}$$ (T-10)

Influence of δB on the Lorentz Force ($\beta \equiv \delta B$)

$$F = -\frac{1}{\mu_{\text{eq}}}\left(\frac{1}{4\pi}\right)\frac{\partial B}{\partial x}B$$ (T-12)

$$\Delta F = -\frac{1}{\mu_{\text{eq}}}\left(\frac{1}{4\pi}\right)\left(\frac{\partial \beta}{\partial x}B + \frac{\partial B}{\partial x}\beta\right)$$ (T-14)

C. Coupling of Thermal and Magnetic Effects—Wipf's [WIP67] Stability Criterion

(T-16) $$\Delta F_p + \Delta F = 0$$

(T-17) $$-(4\pi\mu_{\text{eq}})\frac{\partial F_p}{\partial B}\beta + \frac{10}{C^v}\frac{\partial B}{\partial x}\frac{\partial F_p}{\partial T}\int_x \beta\, dx + \frac{\partial \beta}{\partial x}B + \frac{\partial B}{\partial x}\beta = 0$$

18.16 Upper Shielding Limit of Full Critical State Stability, H_{fi}—The Lower Bound of the Incomplete Flux-Jump Regime

As indicated in Figure 18-18, the stability loop includes the effects of induction (B_z or Φ) and temperature. With reference to Figure 18-13, we note that the ingress of flux at fixed H_a produces two effects: (i) it reduces dB_z/dx, and consequently the "Lorentz force," $F = (-1/\mu_{eq}4\pi)B_z(dB_z/dx)$ ((16-7)), by an amount ΔF; (ii) the corresponding temperature increase reduces the pinning strength by an as yet unspecified amount, ΔF_p. Table 18-6(b), which is a quantitative statement of these observations, shows quite generally (albeit with the proviso that $\Delta T = (1/C^v)\Delta Q_v$) how F is influenced by the change of induction and F_p by the corresponding change of temperature. If the superconductor is to be fully stable $|\Delta F| < |\Delta F_p|$, while the *limit* of full stability is defined by $|\Delta F| = |\Delta F_p|$ leading to WIPF's stability criterion [WIP67] in the form of (T-17). To find the limiting field, (16-34) and (16-41) are first combined to form the critical-state relationship $-F_p = F_{\text{critical}} = \alpha(T)B/(b_0 + B)$, which leads, following the path clearly delineated in WIPF's original article [WIP67], to the following expression for H_{fi}, the upper shielding field for full critical state stability:

$$H_{fi} = [\pi^3 C^v \alpha(T)/(-\partial\alpha/\partial T)]^{1/2} \tag{18-75a}$$

Using a comparable argument, SWARTZ and BEAN [SWA68] obtained the result

$$H_{fi} = [\pi^3 C^v J_c/(-\partial J_c/\partial T)]^{1/2} \tag{18-75b}$$

while SUTTON [SUT73], by means of very straightforward alternative approach, obtained a similar expression (see also TINKHAM [TIN75, p. 184]) for a field that they regarded as the threshold for full flux jumping (i.e., the field referred to by WIPF [WIP67], and in the following discussion, as H_{fj}).

The explicit temperature dependence of H_{fi} may be calculated by taking for the specific heat the form $C^v = \beta_s T^3 \equiv \beta_s T_c^3 t^3$ (see (8-9)) and assuming either

$$\alpha(T) = \alpha_0(1 - t^2)^2 \qquad \text{WIPF [WIP67]} \tag{18-76}$$

or

$$J_c(T) = J_{c0}(1 - t) \qquad \text{SWARTZ and BEAN [SWA68]} \tag{18-77}$$

Substitution of these forms into (18-75a) and (18-75b), respectively, yields

$$H_{fi} = (T_c^2/2)[\pi^3 \beta_s t^2(1 - t^2)]^{1/2} \qquad \text{WIPF [WIP67]} \tag{18-78a}$$

and

$$H_{fi} = T_c^2[\pi^3\beta_s t^3(1-t)]^{1/2} \qquad \text{Swartz and Bean [Swa68]} \tag{18-78b}$$

defining functions that maximize at $t = 0.71$ and $t = 0.75$, respectively.

It turns out as a result of measurements by Sukharevskii *et al.* [Suk68] that the total specific heat of superconducting Ti–Nb(25 at.%) is rather well described in the interval 4–6 K by a quadratic temperature dependence of the form: $C = \beta_s T^2$ with $\beta_s = 0.63 \times 10^3$ erg g^{-1} $K^{-2} = 3.45 \times 10^3$ erg cm^{-3} K^{-2}.[8] Under this condition, (18-78a) and (18-78b) convert to

$$H_{fi} = (T_c/2)[\pi^3\beta_s t(1-t^2)]^{1/2} \qquad \text{after Wipf} \tag{18-79a}$$

and

$$H_{fi} = T_c[\pi^3\beta_s t^2(1-t)]^{1/2} \qquad \text{after Swartz and Bean} \tag{18-79b}$$

defining functions which peak at $t = 0.58$ and 0.67, respectively. Taking $T_c \cong 6.5$ K (the measured [Suk68] transition temperature of Ti–Nb(25 at.%)) the corresponding values of H_{fi} according to (18-79a) and (18-79b) are 659 and 818 Oe, respectively.

18.17 Upper Bound of the Incomplete Flux-Jump Regime, H_{fj}—The Threshold of Runaway Instability

In magnetization studies of Zr–75Nb cylinders, in which the time rate of the applied field was varied from about 10 to 10^5 Oe s^{-1}, Wipf and Lubell [Wip65] noted that although at low values of dH_a/dt the sample was immune from flux jumping and thereby existed in the stable regime discussed above, for $dH_a/dt \gtrsim 10$ kOe s^{-1}, an adiabatic flux-jump condition was approached in which the local movement of flux triggered a thermal avalanche followed by a complete quenching of the superconductivity. Wipf and Lubell [Wip65], and subsequently Swartz and Bean [Swa68], suggested that the energy previously stored in the field, if applied adiabatically to the superconductor, would raise its temperature above the critical point in the manner described by (18-74). The final temperature, T_f, attained by this process is, of course, such that $H_{fj}(T_i) = H_{c2}(T_f)$, which in turn, according to those authors, can be represented by $H_{c2} = H_{c20}(1 - T_f/T_c)$ close to

[8] A density of 5.5 g cm^{-3} was assumed in this case; see Table 8-4.

the transition temperature. Thus, if $C^v(T) = \beta_s T^2$ (as it was for superconductive Ti–Nb(25 at.%) at temperatures of about 4–6 K) T_f, and consequently H_{fj}, are determinable through

$$\frac{t_f^3 - t_i^3}{(t_f - 1)^2} = \frac{3H_{c20}^2}{8\pi\beta_s T_c^3} \tag{18-80}$$

in which the *t*s are reduced temperatures.

Bearing in mind the presence of temperature and field gradients within the sample during flux jumping (Section 18.15.2), it is probably just about as accurate, and certainly very much simpler, to assume that T_f is itself the transition temperature. Then a straightforward integration of $H_{fj}^2 = 8\pi\int_{T_i}^{T_c} C^v(T)\,dT$, which is (18-74) with T_f replaced by T_c, directly leads to

$$H_{fj} = T_c^{3/2}[(8\pi\beta_s/3)(1 - t_i^3)]^{1/2} \tag{18-81}$$

Again with reference to Ti–Nb(25 at.%), insertion of the values of T_c and β_s employed above, accompanied by the assumption that the ambient temperature prior to adiabatic flux jump is 4.2 K, leads to $H_{fj} = 2.4$ kOe.

More recently, in studies of flux-jump instability at 4.3 K in 6.7 mm$^\varnothing$ rods of Ti–52Nb as part of an investigation of an improved stabilization technique (see Section 25.9.2), SUTTON [SUT73] found the threshold fields, within which thermal disturbances of amplitudes of 0.2 and 0.3 K were able to initiate flux jumps, to be 5.8 and 5.3 kOe, respectively. If we take $T_c = 8.5$ K (the measured value) and $t_i \cong 0$, an application of (18-81) in conjunction with the same value of β_s assumed for Ti–Nb(25 at.%) leads to $H_{fj} = 4.2$ kOe, which must, under the circumstances, be regarded as being in quite good agreement with experiment. NOEL *et al.* [NOE71], [BOY71] studied flux jumping in solid cylinders of Ti–60Nb in the presence of a slowly increasing external field, and, without presupposing a particular functional form for $J_c(B, T)$, have developed equations that permit the determination of the internal magnetic induction distribution, B_z, after a flux jump as well as the critical current density.

It may be concluded that the analyses presented above provide useful semiquantitative descriptions of the physics of critical-state stability. It is of considerable interest to study the difference between H_{fi} and H_{fj}. As an aid to this inquiry, Figure 18-19 compares these fields as functions of *t*, in this case defined as the reduced *ambient* temperature.

18.18 The Concept of "Limited Instability"

According to Figure 18-19 there exists a range of fields, defined by $H_{fi} < H_a < H_{fj}$, within which the superconductor is sufficiently stable against small disturbances (temperature excursions, for example) that the

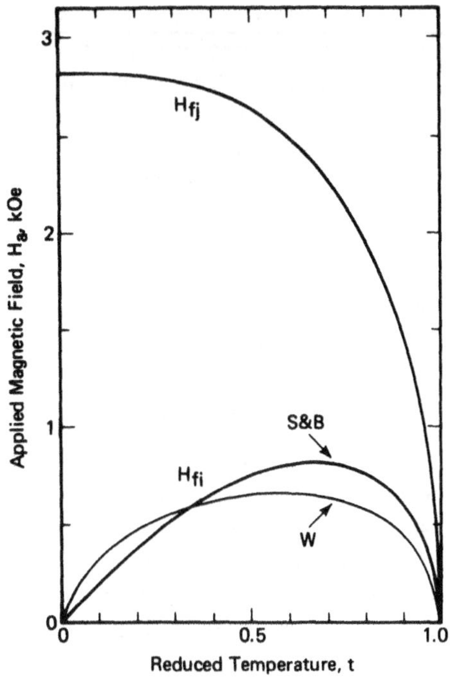

Figure 18-19. Calculated temperature dependences of the two flux-jump fields, H_{fi} and H_{fj}, for Ti–Nb(25 at.%) ($\beta_s = 3.45 \times 10^3$ erg cm^{-3} K^{-2}, $T_c = 6.5$ K). H_{fi}, the highest shielding field for full critical state stability (or "onset of limited instability") has been discussed by WIPF [WIP67] ((18-79a), curve "W"), SWARTZ and BEAN [SWA68] ((18-79b), curve "S&B"), and also by GAUSTER *et al.* [GAU69]. (According to a recent review by WIPF, the concept of "limited instability" is no longer in favor [WIP78, Appendix].) H_{fj}, the threshold field for runaway instability, has been calculated from (18-81).

flux motion that does occur is unable to trigger a full flux jump. The existence of a regime of "limited instability" between the fields H_{fi} and H_{fj} was discovered experimentally by WIPF [WIP80]. Other evidence is to be found in the work of: (i) WISCHMEYER, who noted that bursts of 100 or so flux quanta, increasing in number as dH_a/dt increased, emerged from localized regions of the specimen surface at fields above H_{fi} but yet preceded flux jumping [WIS64, WIS65]; and (ii) BOROVIK *et al.* [BOR66], who have detected flux bursts of between 5×10^4 and 2×10^7 quanta—some one to three orders of magnitude smaller than would have been expected from a full flux jump.

Both WIPF [WIP67] and SWARTZ and BEAN [SWA68] recognized limited flux jumping as a consequence of the retardation of flux motion by viscous drag, $\eta_B \mathbf{v}_f = \mathbf{F}_p + \mathbf{F}$ (in terms of bulk forces, but see ((18-55a))), an effect which favors the diffusion of heat away from the region of the moving bundle. It was supposed that once H_{fi} was exceeded, fluxoids would be able to be accelerated away from their pinning centers by the applied disturbance, but that if their movement was constrained to a slow rate compared to thermal diffusion they would come to rest before initiating an avalanche [WIP70]. Approaching the problem from this standpoint, WIPF [WIP67] developed an expression for the maximum velocity of flux motion consistent with recovery from the immediate effect of the disturbance.

In the years following WIPF's original investigations, the theory of superconductor stability has steadily developed in sophistication. As indicated in Section 18.15.1, flux-jump instability is no longer a problem in modern composite conductors, which, however, still must be designed to cope with thermal disturbances, usually mechanical in origin. An integrated approach to all aspects of superconductor stability is presented in Chapter 25.

18.19 Insights into Superconductor Stabilization Derived from Flux-Jump Studies

18.19.1 The Stability Cycle

Central to any analysis of the stability of a bare superconductor is its cyclical response to a small thermal disturbance, as in Figure 18-18. The following simplified analysis, which recovers *en route* some of the results presented above, emphasizes the essential physics of the stability problem, with particular reference to adiabatic stability.

Consider a two-sided slab of unit width and thickness, w, Figure 18-20. At the initial temperature T the induction B_z (relative to that at the surface) at distance x from the center line is given by the usual formula

$$B_z/x = -\mu_{\text{eq}}(4\pi/10)J_{yc} \qquad [16\text{-}17](18\text{-}82)$$

Suppose the temperature of the slab to be raised by an amount ΔT in time

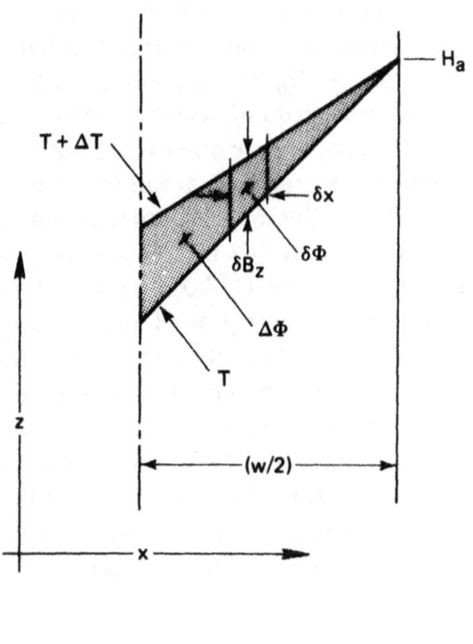

Figure 18-20. Change of critical flux density gradient in a two-sided slab of unit thickness and half-width $w/2$ under the influence of a temperature increment, ΔT, in response to which an amount of flux $\delta\Phi$ enters the elementary volume δB_z. δx and flux $\Delta\Phi$ enters the entire volume between the critical isotherms.

Δt bringing about a proportionate reduction $\Delta J_c = (dJ_c/dT)\Delta T$ in the critical current density. The induction at x will then undergo an increase δB_z given by

$$\delta B_z/x = -\mu_{\text{eq}}(4\pi/10)\Delta J_{yc} \tag{18-83}$$

resulting in an ingress of flux into volume δx at x given by

$$\delta\Phi = \delta B_z \cdot \delta x = -\mu_{\text{eq}}(4\pi/10)\Delta J_{yc} \cdot x \cdot \delta x \tag{18-84}$$

The total flux that rushes into each half of the sample, and which, of course, must flow through each elementary volume in time Δt in order to establish the new critical state, is, therefore,

$$\Delta\Phi = \int_0^{w/2} \delta\Phi \tag{18-85a}$$

$$= \mu_{\text{eq}}(4\pi/10)(w^2/8)(-dJ_c/dT)\,\Delta T \tag{18-85b}$$

The incoming flux is responsible for an electric field in the y direction (normal to the surface of the diagram) given by

$$|E_y| = 10^{-8}\Delta\Phi/\Delta t \qquad [18\text{-}3](18\text{-}86)$$

and consequently for a heat dissipation per unit volume of

$$\Delta Q_v = E_y J_{yc}\Delta t = 10^{-8} J_{yc}\Delta\Phi \tag{18-87}$$

The corresponding temperature increase is $\Delta T' = \Delta Q_v/C^v$, where C^v is the volume specific heat of the material. Thus,

$$\Delta T' = 10^{-8} J_{yc}\mu_{\text{eq}}(4\pi/10)(w^2/8C^v)(-dJ_c/dT)\,\Delta T \tag{18-88}$$

It follows, with reference to Figure 18-18, that stability is assured provided that $\Delta T' < \Delta T$, i.e., provided that

$$\frac{\Delta T}{\Delta T'} = \left(\frac{10^9}{4\pi\mu_{\text{eq}}}\right)\left(\frac{8C^v}{w^2}\right)\frac{1}{J_c}\left(-\frac{dJ_c}{dT}\right)^{-1} > 1 \tag{18-89}$$

In that it is assumed that the heat generated by the moving flux is confined to the superconductor, (18-89) is the condition for *adiabatic* flux-jump stability.

18.19.2 Stability and Degradation

Degradation is defined as the failure of a material to support its expected flux-density gradient or critical current density. In tube magnetization experiments it has been frequently noted that the first flux-jump may occur before the field has penetrated to the internal wall. Similarly, early solenoids tightly wound from unstabilized conductor closely resembled a solid cylinder of the same material and underwent quenching at current densities well below those exhibited by the component wires during short-sample testing. Although, as explained in connection with Figure 18-13(c), the presence of the bore space contributed to the instability, premature quenching can also be traced to a size-effect degradation of critical-state properties.

The factor w that appears in (18-89) serves to meter, or limit, the amount of flux that is set in motion by a given thermal disturbance. Its occurrence in the denominator demonstrates why solenoids wound from monofilamentary wire would be less stable than the wire itself, which in turn would be less stable than the same cross section of superconductor subdivided into separate filaments. This, of course, suggests that the degradation encountered in the early studies was a property of wire and coil geometry rather than a consequence of longitudinal nonuniformity of the wire or the presence of other such manufacturing defects.

For the reason embodied in (18-89), commercial superconductors are subdivided into a multitude of fine filaments each of diameter w sufficiently small to prevent flux jumping. Flux jump and all other aspects of superconductor stability are discussed in detail in Chapter 25.

Chapter 19

Magnetization and Critical Current Density

19.1 Principles of Magnetic Critical Current Density Measurement

This chapter, which concludes a series of discussions of the magnetic-field-induced mixed state of type-II superconductors, considers the application of magnetic measurements to critical current density determination, and in so doing makes a transition into the next group of chapters (in Volume 2) which deal with the current-carrying mixed state itself—i.e., with flux pinning and flux motion under the influence of a transport current.

In the usual four-terminal technique of critical current density measurement severe experimental difficulties are encountered if the conductors under test demand that large currents be supplied to them and transported through heavy leads into and out of the cryostat. Since magnetic hysteresis and the ability of a conductor to transport current in a transverse magnetic field are both consequences of the operation of a common basic mechanism—the pinning of an Abrikosov flux lattice—magnetization measurements on irreversible superconductors can be made to yield values of the critical current density. Since such measurements are usually best carried out on relatively large-diameter samples (i.e., large in comparison with the usual laboratory magnet wire), they are particularly useful for the measurement of critical current density in conductors of high current carrying capacities, such as those which would need to be supplied with thousands of amperes in conventional four-terminal resistometric tests. Since almost any magnetization measurement is adaptable to J_c determination, the technique finally selected from a large field of candidates should represent a satisfactory compromise between convenience and accuracy.

Table 19-1. Magnetic Techniques for Critical Current Density Determination

Procedure	Principle	Literature
(a) Tube magnetization	$(H_a - H')/w = \pm(4\pi/10)J_c$	Kim *et al.* [KIM62] Kim *et al.* [KIM63] Kim *et al.* [KIM63[a]] Gandolfo *et al.* [GAN66] Gandolfo *et al.* [GAN67] Gandolfo *et al.* [GAN69]
(b) Saturation magnetization reversal	$4\pi[\langle M(H)\rangle_+ - \langle M(H)\rangle_-] = \mu_{eq}\frac{2}{3}(4\pi/10)R_0J_c$	Fietz *et al.* [FIE64] Fietz *et al.* [FIE67] Fietz *et al.* [FIE69] Agapov *et al.* [AGA75] Kroeger *et al.* [KRO75]
(c) Harmonic analysis	$V_3 = -(4h_0^2 fNR_0/J_c) \times 10^{-8}$ (volts)	Bean [BEA64] Ullmaier [ULL66] Kroeger *et al.* [KRO73] Kroeger *et al.* [KRO75]
(d) Static field profile analysis	$dB_z(r)/dr = \pm\mu_{eq}(4\pi/10)J_{\theta c}$	Coffey [COF67] Voigt [VOI68] Sikora [SIK68] Weber *et al.* [WEB73] Frankel *et al.* [FRA77]
(e) Dynamic field profile analysis (field-modulation technique)	Parametric determinations of $B(x)$ and x followed by $dB(x)/dx = \pm\mu_{eq}(4\pi/10)J_c(x)$	Campbell [CAM69] Rollins *et al.* [ROL74] Kroeger *et al.* [KRO75]
(f) Torque magnetometry	$J_c = 30\pi\lvert\tau_{max}\rvert/4VR_0H_a$	Mathur *et al.* [MAT73] Deis *et al.* [DEI73]
(g) Vibrating sample magnetometry	$\langle M\rangle_+ \propto J_c$ (Bean) $J_c = 15(I_+ - I_-)/\mu_{eq}R_0V$ (Fietz *et al.*)	Mathur *et al.* [MAT72] Mathur *et al.* [MAT74] Deis *et al.* [DEI74]

Table 19-1 presents seven classes of magnetic measurement that have been used by the authors listed for the evaluation of critical current densities. The relative merits of techniques (b), (c), and (e) have been investigated by KROEGER *et al.* [KRO75], who used them to measure the critical current field dependences of samples of Nb (annealed), Ti–Nb(90 at.%) (annealed, also cold worked), and Ti–Nb(25 at.%) (annealed and aged), and compared the results with those of conventional resistometry.

The principles underlying the seven basic procedures listed in Table 19-1 are outlined below. For detailed discussions of the individual methods the cited literature should be consulted, and for a critical discussion of methods (b), (c), and (e), the article by KROEGER *et al.* [KRO75] is recommended. The techniques to be described are all applicable to the "critical state" regime wherein the magnetization, $4\pi M$, is commensurate with the applied field, H_a. As will be pointed out in Chapter 20 (see in particular Figure 20-3), this implies that the strength of the applied field is in the "low" to "intermediate" range—typically of the order of 10 kOe for technical hard superconductors. In such fields, in which J_c is relatively high, large conductors will pass extremely high critical currents (frequently in the range of tens of thousands of amperes), for the measurement of which contactless methods would be eminently suitable.

In examining the descriptions of the "saturation magnetization reversal" method of Section 19.3 and its dynamic counterparts referred to in Section 19.6, a relationship between the requirements for high J_c and the presence of magnetic hysteresis will be recognized. AC hysteretic loss, together with eddy-current loss, which must be taken into consideration in composite conductors (see Section 19.9), are the subjects of an entire chapter in Volume 2. It will also be noted that the possibility of surface pinning (or the existence of a surface barrier to the ingress and egress of flux) is generally ignored in the following discussions which usually treat the superconductor's surface in the reversible approximation, referred to in, for example, Section 16.7.1 and Section 20.6.1. The surface barrier *is* mentioned briefly in Section 19.4, but for a full account of the influence of field-dependent surface pinning on magnetic hysteresis (i.e., AC loss) within the critical state regime the work of CLEM [CLE79] is recommended.

19.2 Static Tube Magnetization

In the tube magnetization experiment designed by KIM *et al.* [KIM62, KIM63, KIM63[a]], the field $B' = \mu_{eq0} H'$, measured at the internal surface of a hollow cylinder of wall thickness w using, for example, a magnetoresistance probe, is traced as a function of the longitudinal applied field H_a, as depicted in Figure 16-7. The mean induction gradient, taken as $(H_a - H')/w$, then related to the critical circulating supercurrent, $J_{\theta c}$, by means of the Maxwell equation

$$\frac{H_a - H'}{w} = -\left(\frac{4\pi}{10}\right) J_{\theta c} \qquad [16\text{-}43](19\text{-}1)$$

The assumption is that BEAN's linear approximation is adequate (Section 16.5), and that pinning is isotropic such that $J_{\theta c}$ is equal to the critical transport current in the transverse-field configuration. GANDOLFO [GAN66, GAN69, GAN67] has applied the tube-magnetization technique to samples of Ti-Nb(33 at.%) (plus 0.1 at.% O) in tubular form (32 mm length × 5.1 mm O.D. × 0.38 mm wall). After measuring H' *versus* H_a in the first quadrant (see Figure 18-2), $J_{\theta c}$ was determined from (19-1) and plotted against $\langle B\rangle = (B_0 + B')/2$, the mean field in the sample wall (all under the practical assumption that B and H were numerically interchangeable), as in Figure 19-1. The data were found to fit $J_c = \alpha(4.2\text{ K})/(b_0 + \langle B\rangle)$, (16-42), with

$$\left.\begin{aligned} \alpha(4.2\text{ K}) &= 0.47 \times 10^6 \text{ kOe A cm}^{-2} \\ b_0 &= 11.5 \text{ kOe} \end{aligned}\right\} \text{"as received" —curve i, Figure 19-1}$$

$$\left.\begin{aligned} \alpha(4.2\text{ K}) &= 1.10 \times 10^6 \text{ kOe A cm}^{-2} \\ b_0 &= 12.0 \text{ kOe} \end{aligned}\right\} \text{above plus 30 min/400°C —curve ii, Figure 19-1}$$

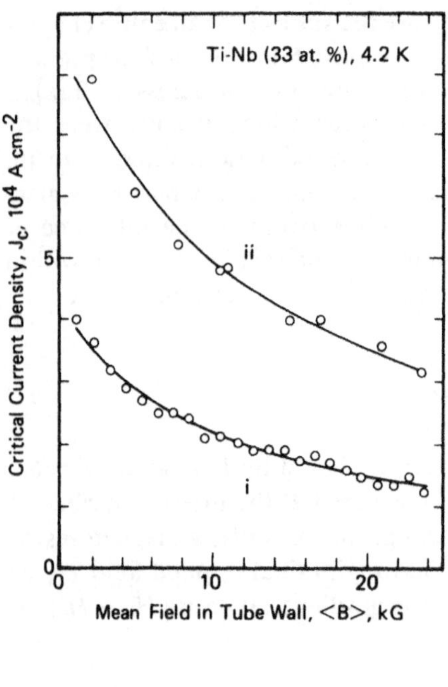

Figure 19-1. Results of type-A tube magnetization measurements of critical current density in Ti-Nb(33 at.%). Sample conditions: (i) cast, 300°C-forged and cold-rolled; (ii) above plus aged 30 min/400°C. J_c is seen to be well fitted in each case to $\alpha/(b_0 + \langle B\rangle)$ with values of α and b_0 as listed in the text—after GANDOLFO [GAN67, p. 43].

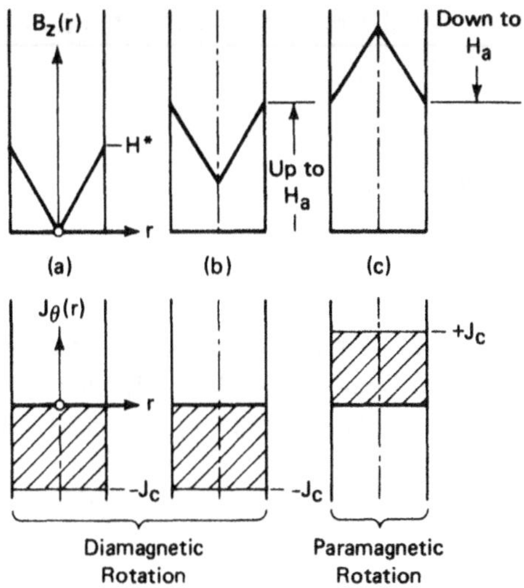

Figure 19-2. Critical induction and current profiles (Bean model) corresponding to the slow increase of a longitudinal field through H^* to some arbitrary value H_a, followed by a return to H_a from a field above H_{c2}.

19.3 Saturation-Magnetization Reversal

The saturation-magnetization reversal method is based on (16-18), (16-19), (16-20), and the fact that, under the Bean approximation (*viz.*, $J_c = 0$, or constant), the mean magnetization $4\pi\langle M_z(r)\rangle$ of a cylinder depends on the magnitude of the applied longitudinal field, H_a, only insofar as it controls the degree of "filling" of the test cylinder with critical current (Figure 16-3). Once a cylinder of radius R_0 is "full" at $H_a = H^*$ (Figure 16-3(c)) $4\pi\langle M_z\rangle$ can no longer increase—i.e., it "saturates" at the value $\mu_{eq}(4\pi/10)(R_0/3)J_{\theta c}$ given by (16-26). Figures 19-2(a) and (b) show the effect on $J_c(r)$ and $B_z(r)$ of increasing H_a beyond H^*. Similarly Figure 19-2(c) represents the effect of reducing H_a from elevated fields. Since $4\pi M_z(r)$ is linear in r with $|\text{slope}| = \mu_{eq}(4\pi/10)J_c$, (16-20), the magnetizations represented in Figures 19-2(b) and 19-2(c) differ only in sign. It follows that, for the field increasing to H_a,

$$4\pi\langle M\rangle_+ = \mu_{eq}(4\pi/10)(J_cR_0/3) \qquad [16\text{-}26](19\text{-}2)^1$$

[1]This simple relationship which follows the Bean model has assumed that $B_0 = H_a$ (i.e., that $\mu_{eq0} = 1$, see Section 16.7), otherwise (19-2) would need to be written as

$$4\pi\langle M\rangle_+ = 4\pi M_{eq} + \mu_{eq}(4\pi/10)(J_cR_0/3)$$

in order to take into account the equilibrium moment, $4\pi M_{eq}$, that would be present in response to the applied field H_a in the absence of pinning ($J_c = 0$)—but see the Fietz model to follow.

and for the field decreasing from above H_{c2} to H_a

$$-4\pi\langle M\rangle_- = \mu_{\rm eq}(4\pi/10)(J_cR_0/3) \tag{19-3}$$

from which

$$4\pi[\langle M\rangle_+ - \langle M\rangle_-] = \mu_{\rm eq}\tfrac{2}{3}(4\pi/10)R_0J_c \tag{19-4}$$

If we now retain this convenient relationship while admitting a critical current density field dependence (i.e., adopt what might be termed a "semi-Bean" approach), (19-4) becomes

$$4\pi[\langle M(H)\rangle_+ - \langle M(H)\rangle_-] = \mu_{\rm eq}\tfrac{2}{3}(4\pi/10)R_0J_c(H) \tag{19-5}$$

enabling $J_c(H)$ to be determined as a result of a series of measurements in which the magnetization is cycled up to, and down to, respectively, a sequence of values of H_a. Equation (19-5) is in fact none other than the relationship derived by FIETZ *et al.* [FIE64, FIE67, FIE69] and others [AGA75] after taking the first approximation to the exact expression for the magnetization difference and assuming a generalized field dependence for J_c.

It is useful briefly to outline the analysis offered by FIETZ *et al.* It proceeds along much the same lines as before excepting that J_c, instead of being regarded as constant, is expanded as a Taylor series about some chosen value of r, say R. The (now) nonlinear $J_c(r)$ profiles (e.g., Figure 16-5) are expressed, for H_a increasing and H_a decreasing, respectively, in the forms

$$J_+(r) = J_+(R) + \sum_{n=1} \frac{1}{n!}\frac{\partial^n J_+(r)}{\partial r^n}\bigg|_R (r-R)^n \tag{19-6a}$$

$$J_-(r) = J_-(R) + \sum_{n=1} \frac{1}{n!}\frac{\partial^n J_-(r)}{\partial r^n}\bigg|_R (r-R)^n \tag{19-6b}$$

Since $dB_z/dr = -\mu_{\rm eq}(4\pi/10)J_c(r)$, (16-13), the induction profiles corresponding to (19-6) are obtained by integration, yielding

$$B_+(r) = B_0 + \mu_{\rm eq}(4\pi/10)\left[J_+(r-R) + J'_+\frac{(r-R)}{2!} + \cdots\right] \tag{19-7a}$$

$$B_-(r) = B_0 + \mu_{\rm eq}(4\pi/10)\left[J_-(r-R) + J'_-\frac{(r-R)}{2!} + \cdots\right] \tag{19-7b}$$

where the $B(r)$ satisfy the $(r = R_0)$ boundary condition

$$B_0 = H_a + 4\pi M_{eq} \tag{19-8}$$

(which is identical to the $B_0 = \mu_{eq0} H_a$ postulate of Section 16.7) where M_{eq} is the reversible magnetization (independent of r in the reversible material) in equilibrium with the applied field, H_a, and where J_+^n represents the nth derivative of $J_+(r)$ evaluated at $r = R$. The mean magnetizations $4\pi\langle M\rangle_+$ and $4\pi\langle M\rangle_-$ are then obtained by integrating the $B(r)$s across the cylinder; i.e., by recalling (19-8) while performing as before (see Section 16.6) the integration

$$4\pi\langle M_z\rangle = \frac{1}{\pi R_0^2}\int_0^{R_0} 2\pi r\,dr[B_z(r) - H_a] \tag{16-37}$$

The results are

$$4\pi\langle M\rangle_+ = 4\pi M_{eq} + 2\mu_{eq}(4\pi/10)\left[\frac{J_+R}{3!} - \frac{J_+'R^2}{4!} + \frac{J_+''R^3}{5!} - \cdots\right] \tag{19-9a}$$

$$4\pi\langle M\rangle_- = 4\pi M_{eq} + 2\mu_{eq}(4\pi/10)\left[\frac{J_-R}{3!} - \frac{J_-'R^2}{4!} + \frac{J_-''R^3}{5!} - \cdots\right] \tag{19-9b}$$

where all the derivatives of J are evaluated at the surface. Recalling that $J_+(B) = -J_-(B)$, and in addition (according to FIETZ and WEBB [FIE69]) that $J_+^n = (-1)^n J_-^n = (-1)^{n+1} J_+^n(r = R_0)$, it follows that: (i) the corresponding terms of the two series are equal in magnitude; (ii) whereas series (19-9a) alternates as shown, the terms in series (19-9b) are all negative. It follows by substraction and addition, respectively, that

$$4\pi[\langle M\rangle_+ - \langle M\rangle_-] = 4\mu_{eq}(4\pi/10)\left[\frac{J_+R}{3!} + \frac{J_+''R^3}{5!} + \frac{J_+'''R^5}{7!} + \cdots\right] \tag{19-10a}$$

$$4\pi[\langle M\rangle_+ + \langle M\rangle_-] = 2(4\pi M_{eq}) - 4\mu_{eq}(4\pi/10)\left[\frac{J_+'R^2}{4!} + \frac{J_+'''R^4}{6!} + \cdots\right] \tag{19-10b}$$

where the current is evaluated at the surface and where no assumptions have been made about the functional form of $J(B)$. If the series converges rapidly, i.e., if the flux profiles are nearly linear, (19-10a) and (19-10b) reduce to

$$4\pi[\langle M\rangle_+ - \langle M\rangle_-] \cong \mu_{eq\frac{2}{3}}(4\pi/10)J_cR_0 \qquad [19\text{-}4](19\text{-}11)$$

$$4\pi[\langle M\rangle_+ + \langle M\rangle_-] \cong 2(4\pi M_{eq}) \qquad (19\text{-}12)$$

Equation (19-11) is just the result derived above ((19-4)) using the Bean approximation, while (19-12) provides a means of determining a value for the equilibrium magnetization, M_{eq}, required to satisfy the relationship $B_0 = \mu_{eq0}H_a$, which is referred to above and is needed in converting between applied field and surface-boundary induction (see Section 16.7).

To perform the experiment, the magnetization of a virgin sample is measured with field increasing from zero to $H_a > H^*$. The value so obtained, *viz.*, $4\pi\langle M\rangle_+$, is then associated with that obtained by reducing the field to the same final value from a starting field above H_{c2}. FIETZ and WEBB successfully applied their hysteretic magnetization technique to a study of flux-pinning in Nb–Ti and Nb–Ta alloys [FIE69]. AGAPOV *et al.* reanalyzed the method and applied it to measurements of critical current density in the Soviet alloy NT-50 either bare or after incorporation into monofilamentary and multifilamentary Cu-, brass-, and Cu–Ni-matrix composites [AGA75].

19.4 Harmonic Analysis

If a transformer whose core is an irreversible type-II superconductor in the mixed state is supplied with a small sinusoidal current, the harmonic content of the output signal is a function of the degree of nonlinearity inherent in the minor hysteresis loop described by the core magnetization; this, in turn, is a function of the flux pinning and consequently of the critical transport current density. As pointed out in Section 16.4.2, BEAN [BEA64] was the first to explore this effect, and demonstrated its utility by measuring the critical current density of an "artificial" type-II superconductor. The experimental arrangement typically consists (e.g., [KRO73, KRO75]) of a cylindrical sample (length : diameter = 5–10 : 1) tightly wound with a 300- to 400-turn Cu pickup coil placed inside a Cu solenoid (the primary) capable of supplying a maximum AC field amplitude (h_0) of about 600 Oe to the sample, which in turn is located along the axis of a 72 kOe superconducting solenoid that supplies the steady background field, H_a. According to BEAN's original analysis [BEA64], provided that $h_0 \ll H_a$, and with the assumption that J_c is independent of B wherever B exists within the specimen (Bean model), the third harmonic of the output signal is

$$V_3 = -(4h_0^2 fNR_0/J_c) \times 10^{-8} \qquad \text{(volts)} \qquad [16\text{-}32](19\text{-}13)$$

where the signal is $h = h_0 \sin 2\pi ft$ (Oe), N is the number of turns in the secondary winding of radius R_0 (cm), and J_c is in A cm^{-2}.

The convenience of the method is a result of several simplifying assumptions, *viz.*, the constancy of $J_c(r)$ and the disregard of surface effects. Thus, although excellent agreement with the results of four-terminal measurement were initially obtained [BEA64], the method in its original simplicity failed to yield satisfactory results when applied to a sample of Nb-Zr [ULL66]. The possibility of encountering serious discrepancies, coupled with the difficulty of estimating the conditions under which the underlying assumptions of the model might be justified, led to the design of improved variations of the harmonic analysis method. The first improvement was to take into account the existence of a surface barrier to the ingress and egress of flux, an effect which has been explained by the existence of surface pinning (see for example [CLE79]) and referred to occasionally as "surface superconductivity." According to ULLMAIER and GAUSTER [ULL66[a]] (see Section 22.5), a variation, ΔH_s, of the applied longitudinal magnetic field within certain limits is able to be accommodated by the generation of surface supercurrents which thereby shield the interior of the sample against change of induction. LEBLANC and colleagues [LEB66], [BEL68] considered the problem in detail and demonstrated that these supercurrents may have densities some two orders of magnitude higher than those of the bulk and can be regarded as flowing in an indefinitely thin layer of the surface. In accord with this picture, LOVE [LOV66] simply reduced the h_0 of (19-13) by a ΔH_s of appropriate magnitude and went on to apply the Bean equation. In the technique devised by ULLMAIER [ULL66] for AC critical current density determination, an output voltage waveform generated in response to a triangular-wave signal input, upon comparison with a calculated trace based on BEAN's critical-state model for field penetration, yielded values for ΔH_s, the surface critical current, and the bulk critical current density. In a further discussion of the technique, KROEGER *et al.* [KRO73] removed a serious restriction of the initial model, *viz.*, the requirement that $h_0 \ll H_a$; the flux density was assumed to vary linearly with depth into the specimen as in BEAN's original picture, but full penetration of the modulating field to the core of the sample was permitted. Although this additional complication prevented the development of expressions for ΔH_s and J_c in useful analytical form, a computer program was written that enabled these quantities to be obtained using data extracted from plots of V_3 (the third-harmonic voltage output) *versus* h_0, further analysis of which could provide information on the applicability of the model itself. KROEGER *et al.* used their technique to measure J_c in a sample of Nb-Zr(40%), and compared the results with those obtained using the original Bean model and LOVE's modification of it [KRO73]. They also measured several samples of Ti-Nb alloy for comparison with the results of three other J_c-measurement techniques, including resistometry [KRO75].

As initially conceived, the method of KROEGER *et al.* [KRO73] was intended to accommodate a relatively large h_0. But in practice it was found that agreement between calculation and experiment deteriorated for modulation-field depths greater than $0.3R_0$. With regard to ΔH_s, although both diamagnetic (opposing flux *entry*) and paramagnetic (opposing flux *exit*) shielding components were considered, the measured quantity was the sum of $\Delta H_{s,en}$ and $\Delta H_{s,ex}$, and the method was insensitive to surface-field "rectification." KROEGER *et al.* generalized their analysis to take into account nonlinearity in $B(r)$ but found the result to be impractically complicated.

19.5 Static Field Profile Analysis

By now it must be recognized that any field-gradient determination is automatically a measurement of critical current density averaged over the appropriate interval of field and space. The tube magnetization measurement has already been cited as a primitive example of this, and one that yields a J_c averaged over the entire sample cross section. Saturation magnetization reversal as presented above yields, in principle, a bulk surface critical current density $J_+(R_0)$, (19-10a), but in practice yields a bulk average value, (19-4) and (19-11). Harmonic analysis samples J_c within a depth that depends on the field modulation amplitude h_0.

As originally presented by COFFEY [COF67], the method of static field profile determination is capable of providing, with the aid of miniature Hall effect probes, a profile of the magnetic induction and consequently, following $dB_z(r)/dr|_{\text{crit}} = \pm\mu_{\text{eq}}(4\pi/10)J_{\theta c}(r)$, (16-18), a microscopic picture of the distribution of critical current density across the diameter of the sample. A typical pair of results—a set of field profiles and a J_c field dependence at a selected radial position—has been presented in Figures 16-11(a) and (b). Unfortunately, although simple but informative, this is not a nondestructive measuring technique. For a nondestructive probe of the internal distribution of flux-pinning sites and critical current density, the dynamic procedures referred to below and elsewhere are recommended.

19.6 Dynamic Field Profile Analysis

Two modulation techniques are available for the measurement of induction profiles, both of which are, of course, capable of yielding the radial dependences of critical current density.

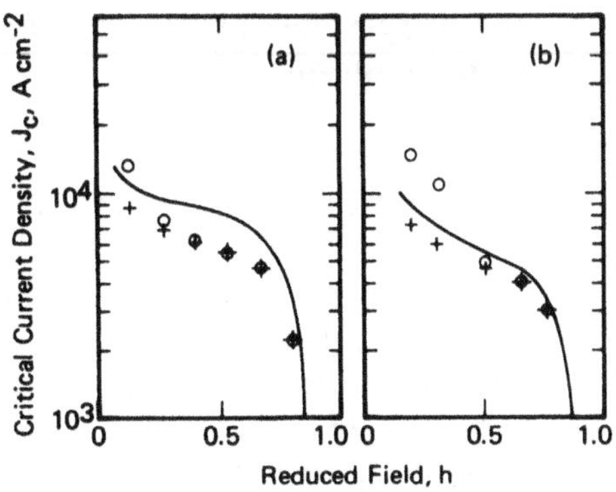

Figure 19-3. Results of a pair of dynamic-field-profile measurements of J_c *versus* reduced applied field, $h = H_a/H_{c2}$ in (a) Ti-Nb(95.5 at.%) and (b) Ti-Nb(98.5 at.%) alloys as conducted by ROLLINS *et al.* [ROL74] (+, cycled to positive field; O, cycled to negative field) compared with the results of saturation-magnetization-reversal studies of the same two samples by FIETZ and WEBB [FIE67, FIE69] (——). The metallurgical condition of the samples was cast, cold-swaged 91.2%, and chemically polished—after ROLLINS *et al.* [ROL74].

1. CAMPBELL's method, Figure 16-12(b), as discussed in [CAM69], made possible construction of a field profile point by point, one for each selected value of the modulation-field amplitude. In a subsequent application of the technique in a study of flux pinning, CAMPBELL [CAM78] measured the reversible penetration of an AC ripple signal into superconducting Pb–Tl alloys. The reversible penetration depth (which he referred to as λ') was compared with the theoretical predictions of a model based on a revision of the YAMAFUJI–IRIE [IRI67], [YAM67] and LABUSCH [LAB69[a]] flux-pinning principles. Irreversible surface pinning (referred to above, and discussed in more detail in Chapter 22) was eliminated by diffusing a layer of Tl into the surface of the specimen.

2. The method of ROLLINS *et al.* [ROL74] yields the complete profile in a single sweep of the modulation field either (i) to "negative field"—the profile labelled $B_0 + b_0$ in Figure 16-15, or (ii) to "positive field"—the profile $B_0 - b_0$.[2] Using their method, the authors developed sets of J_c *versus* H_a data for two Ti-doped Nb alloys from the slopes of the corresponding field profiles (averaged over the linear regions of depth $\lesssim 25\ \mu$m, Figure 16-16). They showed that the results compared very favorably with those obtained on the same alloys by FIETZ and WEBB [FIE67, FIE69] using the saturation-magnetization reversal method, Figure 19-3.

KROEGER *et al.* [KRO75] compared the Campbell-method-determined J_c field-dependences ($\lesssim 60$ kOe) of samples of Nb and several Ti–Nb alloys with those obtained using saturation-magnetization reversal, harmonic analysis, and conventional resistometry. With regard to the flux-profile approach, they concluded that on account of difficulties associated with the interpretation of flux-gradient data, it would be generally desirable, in

[2]In "*cycling to positive field*", H_a is increased to $3H_{c2}$ (sample in the normal state, well beyond H_{c3}) and then reduced to its desired final value, H_{final}. In "*cycling to negative field*," H_a is increased in the negative direction to $|H_a| = 3H_{c2}$, reduced to zero, and then increased in the positive direction to $H_a = H_{\text{final}}$. Quite different profiles were obtained at H_{final} in the two cases.

any study of flux pinning, to bring several independent techniques to bear on the problem.

19.7 Torque Magnetometry

Although not relevant to the chapter as a whole, which is otherwise devoted to the contactless magnetic determination of critical current density, reference is briefly made here to an interesting torsion-balance measurement technique. In an experiment devised by EGGENDORFER [EGG73], the sample in the form of a disc is suspended from a torsion balance in a magnetic field directed along the axis of suspension. Suitably attached leads permit current to flow radially across the sample. This Corbino-disc arrangement, in which the torque is proportional to the Lorentz force per unit volume, provides an excellent means of directly measuring pinning forces and for studying flux flow in the absence of field gradients and edge effects.

MATHUR *et al.* [MAT73] applied torque magnetometry to the measurement of critical current density in multifilamentary composite superconductors. Their sample, a 1-in. length of superconducting wire, was vertically suspended, in the transverse field of a "12 in." electromagnet, from the movement of a Cahn Type-RG electronic microbalance rotated 90° from its usual weighing position and thereby adapted for use as a torque magnetometer.[3] In a typical experiment, as the magnet is rotated, the torque reaction developed by the balance quickly increases up to some saturation value, τ_{max}, (Figure 19-4). To maximize the acquisition of data, a full 180° of rotation is completed. The magnet is then turned back to the starting position during which the torque reverses until it saturates at a value $-\tau_{max}$.

WIPF (see [MAT73]) has shown that for a cylinder of radius R_0 (cm) and volume V (cm^3) the maximum torque τ_{max} (dyn cm), associated with an induction B (gauss), is related to the bulk pinning force F_p (dyn cm^{-3}) by

$$\tau_{max} = 4VR_0F_p/3\pi \tag{19-14}$$

assuming a linear superposition of elementary pinning forces, $F_p = N_p f_p$ (see Section 20.2). Taking $F_p = -BJ_c/10$ as before ((16-14)) it follows, according to MATHUR *et al.* [MAT73], that

$$J_c = 30\pi|\tau_{max}|/4VR_0H_a \qquad (\text{A cm}^{-2}) \tag{19-15}$$

with the assumption that $\mu_{eq0} = 1$ (see Section 16.7). In tests of the procedure,

[3]The Leeds and Northrup wall galvanometer has often found a similar application [MCG69, p. 180].

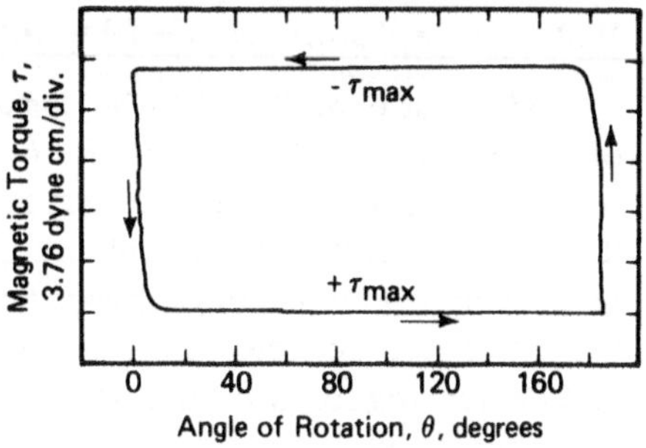

Figure 19-4. Magnetic torque as a function of the angle of rotation in a fixed field of 8.45 kOe for a sample consisting of a 1-in. length of Cu/Ti–Nb composite wire. Strand specifications: diameter, 0.285 mm; Cu/SC ratio, 1:1; number of filaments, 400; filament diameter, 10 μm; twist pitch, 3.4 mm—after DEIS *et al.* [DEI73].

magnetically determined critical current densities were compared with data taken by conventional resistometry. It was found that provided the superconductor was fully penetrated by the magnetic field (a prerequisite for the validity of (19-14)), reasonably good agreement between the results of the two approaches was obtained, Figure 19-5. Torque magnetometry may, of course, be extended into the high-field range if a split superconducting magnet is available.

19.8 Vibrating-Sample Magnetometry

MATHUR *et al.* have devised two interesting applications of the vibrating-sample magnetometer to the measurement of critical current density in multifilamentary composite superconductors [MAT72, MAT74], [DEI74].

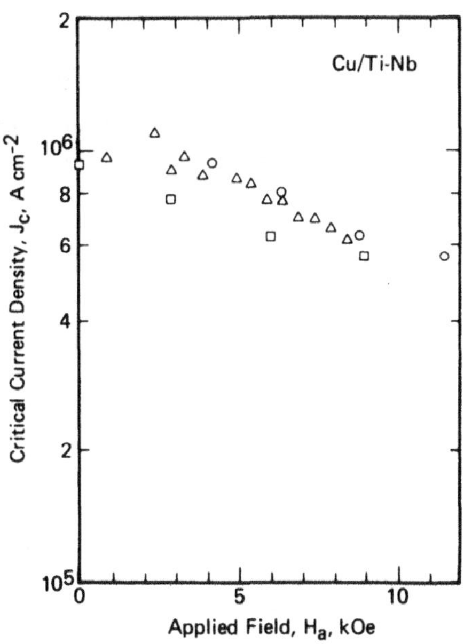

Figure 19-5. Critical current density *versus* applied magnetic field for the Cu/Ti–Nb composite referred to in the previous figure. The results of torque magnetometry (△) are compared with the results of conventional four-terminal resistometry (hair-pin samples, 10^{-10} Ω cm criterion) (□; and ○, after the self-field has been incorporated into the external field)—after DEIS *et al.* [DEI73] and MATHUR *et al.* [MAT73].

19.8.1 Adaptation of the Magnetization-Reversal Technique

The first experiment to be considered is a straightforward application of the method of FIETZ *et al.* [FIE64, FIE67, FIE69]. The magnetic moment of a sample, in the form of a small coil mounted with its axis parallel to that of a superconducting solenoid, is measured as a function of applied field using a Foner vibrating-sample magnetometer. According to (19-11), the difference between the mean saturation magnetizations in increasing and decreasing longitudinal fields, for a cylindrical superconducting strand of radius R_0 is

$$\langle M\rangle_+ - \langle M\rangle_- = \mu_{eq} J_c R_0/15 \qquad (19\text{-}16)$$

Thus, if the sample to be measured is of length l (cm) and contains n_f filaments each of volume $\pi R_0^2 l$ (cm^3), the critical current density is directly proportional to the measured difference of total magnetic moment, $I_+ - I_-$ (gauss cm^3), according to

$$J_c = 15(I_+ - I_-)/\mu_{eq}\pi R_0^3 n_f l \qquad (\text{A cm}^{-2}) \qquad (19\text{-}17)$$

Setting $\mu_{eq} = 1$ (and $\mu_{eq0} = 1$), which are good approximations in high-κ_{GL} materials away from H_{c1} (see e.g., Sections 16.7 and 20.5), MATHUR *et al.* [MAT74] employed this expression and the associated technique, with considerable success, for the measurement of critical current density in a study of pinning force relationships in multifilamentary Cu/Ti–Nb and other superconductive composites.

19.8.2 Measurement of Critical Current Density Anisotropy

(a) Principle of the Measurement. Pinning-force and critical-current-density anisotropies are discussed in Chapter 21. Interest in these problems arises through the possibility of a conductor's being exposed to a longitudinal component of the applied field. As a consequence of the metallurgically induced pinning-force anisotropy inherently associated with the wire's cold-drawn microstructure (see Section 2.8), the circumferential critical current, $J_{\theta c}$, associated with a longitudinal applied field will generally be different from the longitudinal transport critical current density, J_c. MATHUR *et al.* [MAT72], again using a Foner magnetometer, have studied the ratios of these critical supercurrents as developed by suitably oriented applied fields in a series of commercial multifilamentary Cu/Ti–Nb composites. The basic measurement, which essentially consisted of comparing the "saturated" transverse and longitudinal filament magnetizations,[4] relied on for

[4]The magnetizations were "saturated" in the sense of Section 19.3. In transverse and longitudinal applied fields, they will be here designated M_{0t} and M_{0l}, respectively.

its analysis the proportionality, first recognized by BEAN [BEA64], between the magnetization and the critical current density (as expressed in terms of (16-26), for example, see also Section 19.3). According to the authors,

$$\frac{J_{\theta c}(H)}{J_c(H)} = 1.29\,\frac{M_{0l}(H)}{M_{0t}(H)} \tag{19-18}$$

where the 1.29 was a geometrical coefficient related to filament shape and orientation in the magnetic field [MAT72], and in which field-dependences of the J_cs and M_0s are included. Anisotropy ratios as high as 3.4:1 were encountered, suggesting that the usual longitudinal critical current density (J_c) is a poor approximation to the value that should be used in calculations involving critical currents induced by axial magnetic fields. By the same token, caution should be taken in interpreting critical current density data derived from tube-magnetization studies.

(b) Magnetization Studies of Twisted Multifilamentary Superconducting Composites. When a composite conductor is exposed to a time-varying transverse magnetic field, contributions to the total magnetization (M_t) from both the hysteresis (irreversibility) of the superconducting component (M_{0t}) and the eddy currents which circulate along the filaments and across the matrix between them ($\alpha_m M_{0t}$) must be considered. Thus, as will be explained in Section 26.3,

$$M_t = M_{0t}(1 + \alpha_m) \cong M_{0t}(1 + \alpha_{m0}\dot{H}_a) \tag{19-19}$$

in which

$$\alpha_{m0} = \frac{4}{3 \times 10^8}\left(\frac{1}{w\lambda J_c}\right)\left(\frac{l_p^2}{\rho_\perp}\right) \tag{19-20}$$

where w is the filament diameter, λ is superconductor/entire conductor volume fraction (the filling factor), $4l_p$ is the twist pitch, and $\rho_\perp$ is the transverse resistivity of the conductor. According to the above approximation, M_t increases linearly with $\dot{H}_a$ at a rate proportional to α_{m0}. The latter, which is a measure of the strength of the induced circulating currents, can be reduced either by (i) reducing the twist pitch, the effect of which is depicted in Figure 19-6(a), or (ii) cutting an untwisted composite into shorter and shorter pieces, as illustrated in Figure 19-6(b). In either case, the transverse magnetization of the superconducting component, M_{0t}, for use in (19-18), can be obtained by extrapolating the results to zero $\dot{H}_a$, a task that is most accurately accomplished when the coefficient α_{m0} is small to begin with. In the longitudinal orientation, the moment of the composite is dominated by contributions from the circumferential supercurrents.

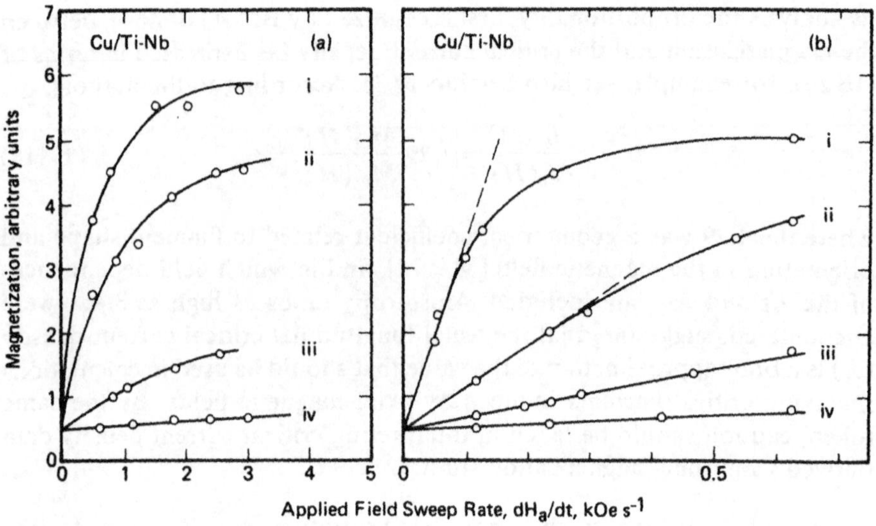

Figure 19-6. Two magnetization studies of interfilamentary "coupling" in multifilamentary Cu/Ti–Nb composites. Magnetization at 10 kOe *versus* the applied field sweep rate, dH_a/dt. (a) Various twist pitches, *viz.*, (i) ∞, (ii) 51 mm, (iii) 25 mm, and (iv) 13 mm—after CRITCHLOW and ZEITLIN [CRI70]. (b) Several pieces of untwisted composite, of lengths (i) $L = 30$ cm, (ii) $L/2$, (iii) $L/4$, and (iv) $L/8$—after MATHUR *et al.* [MAT72]. *Composite specifications*: (a) strand diameter, 0.20 mm; Cu/SC ratio, 3:1, number of filaments, 121; filament diameter, 9 μm; (b) strand diameter, 0.51 mm; Cu/SC ratio, 1:1; number of filaments, 261; filament diameter, 20 μm.

Transverse magnetization in twisted multifilamentary superconducting composites has been considered in detail by WILSON and WALTERS in their classic 1970 paper [WIL70[a]] and in an earlier report published in 1969. The purpose of twist can be traced back to the elimination of flux jumping by the subdivision of the superconductor into a number of filaments sufficient to transport the desired current, each of a diameter d (cm) less than about $1500/J_c$ (A cm^{-2}),[5] see Section 25.7, accompanied by the recognition that these filaments will act cooperatively under a changing applied magnetic field unless the whole composite is given an axial twist. WILSON *et al.* measured the magnetization by integrating the output of a "superconducting-core" transformer, exposed to the DC background field of a superconducting magnet, as in the experiments of FIETZ [FIE64, FIE67]. They went on to plot M_t *versus* $\dot{H}_a$, and obtained M_{0t} by extrapolating to zero $\dot{H}_a$. CRITCHLOW and ZEITLIN [CRI70] performed a similar series of experiments on commercial and experimental twisted multifilamentary conductors, and clearly verified that twisting reduced the total AC magnetization. Plots of

[5]According to WILSON *et al.* [WIL70[a]], typical Ti–Nb composite filament diameters guaranteeing intrinsic stability in fields of 10, 30, 50, and 70 kOe appear to be: 30, 50, 70, and 100 μm, respectively, but see Chapter 25.

M_t *versus* $\dot{H}_a$, typical examples of which are given in Figure 19-6(a), were comparable to those obtained by WILSON *et al.* and shared with them the following characteristics: (i) at infinite and large twist pitches, M_t rose very rapidly with $\dot{H}_a$; (ii) as the twist pitch was reduced, M_t became smaller, $M_t(\dot{H}_a)$ tended to become linear, and the curves could be conveniently back-extrapolated to the common intercept, $M_{0t}(H)$, the (hysteretic) magnetization of the superconducting component of the composite at the particular value of the DC bias field being applied.

In the experiments of MATHUR *et al.* [MAT72], the magnetic field was supplied by a superconducting solenoid mounted with its axis in the vertical direction. The samples for circumferential critical current density, $J_{\theta c}$, measurement consisted of $\frac{1}{2}$-in.-long *bundles* of wire, while the longitudinal critical current was developed in small $\frac{1}{2}$-in.-long, $\frac{3}{32}$-in. diameter *solenoids* (wound with about 12 in. of wire) mounted with their axes parallel to that of the field magnet. Magnetization was measured using a vibrating-sample magnetometer. The multifilamentary wires examined were untwisted; thus interfilamentary coupling effects were studied not with respect to twist pitch as in Figure 19-6(a) but as a function of sample length. Figure 19-6(b) is a representative plot of the transverse magnetization, M_t, *versus* $\dot{H}_a$ (0–700 Oe s^{-1}) in a background field of 30 kOe. The samples under measurement were all of the same mass, but were subdivided into up to eight separate pieces. Figure 19-6(b) is in accord with Figure 19-6(a) and demonstrates the influence of interfilamentary coupling (matrix eddy currents) on the total magnetization. $M_{0t}(H)$ was obtained by extrapolation, as before, and combined with $M_{0l}(H)$ in order to determine, with the aid of (19-18), the ratios of the circumferential ($J_{\theta c}$) to the longitudinal (J_c) critical current densities.

References

A

[ABB61] ABBISS, C. P., HUZAN, E., and JONES, G. O., Thermal Expansion of Aluminum at Low Temperatures, *Proc. VIIth Int. Conf. on Low Temperature Physics* (Aug.-Sept. 1960) ed. by G. M. Graham *et al.*, University of Toronto Press, Toronto, 1961, pp. 688-689.

[ABR52] ABRIKOSOV, A. A., Influence of Size on the Critical Field of Superconductors of the Second Group (in Russian), *Dokl. Akad. Nauk. SSSR* **86**, 489-492 (1952).

[ABR57] ABRIKOSOV, A. A., On the Magnetic Properties of Superconductors of the Second Group, *Sov. Phys. JETP* **5**, 1174-1182 (1957), [Transl. of *Zh. Eksp. Teor. Fiz.* **32**, 1442-1452 (1957)].

[ABR62] ABRIKOSOV, A. A., and GOR'KOV, L. P., Spin-Orbit Interaction and the Knight Shift in Superconductors, *Sov. Phys. JETP* **15**, 752-757 (1962), [Transl. of *Zh. Eksp. Teor. Fiz.* **42**, 1088-1096 (1962)].

[ACK82] ACKERMAN, D. A., and ANDERSON, A. C., Dilatometry at Low Temperatures, *Rev. Sci. Instrum.* **53**, 1657-1660 (1982).

[ADE52] ADENSTEDT, H. K., PEQUIGNOT, J. R., and RAYMER, J. M., The Titanium-Vanadium System, *Trans. ASM* **44**, 990-1003 (1952).

[AGA74] AGARWAL, K. L., Low Temperature Calorimetry of Transition Metal Alloys, *Ph.D. Thesis*, 1974, University of New Orleans, Louisiana.

[AGA75] AGAPOV, Yu. P., DROZHENKO, V. G., DUDIN, V. I., and MIKHAILOV, V. N., The Relationship Between the Magnetization and Critical Current Density in Hard (Type-II) Superconductors, *Sov. J. Low Temp. Phys.* **1**, No. 4, 212-215 (1975) [Transl. of *Fiz. Nizk. Temp.* **1**, 428-434 (1975)].

[ALA67] ALAIS, P., and SIMON, Y., Skin Effect in a Type-II Superconductor, *Phys. Rev.* **158**, 426-432 (1967).

[ALB72] ALBERT, H., and PFEIFFER, I., Über die Temperaturabhängigkeit des Elastizitätsmoduls von Niob-Titan-Legierungen, *Z. Metallkd.* **63**, 126-131 (1972).

[ALB76] ALBERT, H., and PFEIFFER, I., Temperaturabhängigkeit der Festigkeitseigenschaften des Hochfeldsupraleiters NbTi50, *Z. Metallkd.* **67**, 356-360 (1976).

[ALE67] ALEKSEEVSKII, N. E., DUBROVIN, A. V., MIKHAILOV, N. N., SOKOLOV, V. I., and FEDOTOV, L. N., Principal Properties of Superconducting Wire Made from an Alloy of the 65 BT Type in Samples and Solenoids, *Sov. Phys. Dokl.* **11**, 993-996 (1967), [Transl. of *Dokl. Akad. Nauk. SSSR* **171**, 566-569 (1966)].

[ALE67[a]] ALEKSEEVSKII, N. E., IVANOV, O. S., RAYEVSKII, I. I. and STEPANOV, N. V., Constitution Diagram of the System Niobium-Titanium-Zirconium and Superconducting Properties of the Alloys, *Phys. Met. Metallogr.* **23**, No. 1, 28-35 (1967), [Transl. of *Fiz. Met. Metalloved.* **23**, 28-36 (1967)].

[ALE68] ALEKSEEVSKII, N. E., IVANOV, O. S., RAYEVSKII, I. I., and STEPANOV, N. V., Critical Magnetic Fields of Niobium-Titanium-Zirconium Alloys, *Sov. Phys. Dokl.* **12**, 898-900 (1968), [Transl. of *Dokl. Akad. Nauk. SSSR* **176**, 305-307 (1967)].

[ALL75] ALLEN, P. B., and DYNES, R. C., Transition Temperature of Strong-Coupled Superconductors Reanalyzed, *Phys. Rev. B*: **12**, 905-922 (1975).

[ALL82] ALLEN, P. B., personal communication.

[ALT67] ALTMANN, S. L., and BRADLEY, C. J., The Band Structure of Hexagonal Close-Packed Metals, II Sc, Ti, Y and Zr, *Proc. Phys. Soc., London* **92**, 764-775 (1967).

[AME54] AMES, S. L., and MCQUILLAN, A. D., The Resistivity-Temperature-Concentration Relationships in the System Niobium-Titanium, *Acta Metall.* **2**, 831-836 (1954).

[AND61] ANDRES, K., The Measurement of Thermal Expansion at Low Temperatures, *Cryogenics* **2**, 93-97 (1961).

[AND61[a]] ANDERSON, P. W., Localized Magnetic States in Metals, *Phys. Rev.* **124**, 41-53 (1961).

[AND62] ANDERSON, P. W., Theory of Flux Creep in Hard Superconductors, *Phys. Rev. Lett.* **9**, 309-311 (1962).

[AND62[a]] ANDRES, K., OLSEN, J. L., and ROHRER, H., Mechanical Effects at the Superconducting Transition, *IBM J. Res. Dev.* **6**, 84-88 (1962).

[AND63] ANDERSON, O. L., A Simplified Method for Calculating the Debye Temperature from Elastic Constants, *J. Phys. Chem. Solids* **24**, 909-917 (1963).

[AND63[a]] ANDRES, K., The Thermal Expansion of Some Metals, *Low Temperature Physics—LT8, Proc. VIIIth Int. Conf. on Low Temperature Physics*, (Sept. 1962), ed. by R. O. Davies, Butterworths, London, 1963, pp. 397-398.

[AND64] ANDERSON, P. W., and KIM, Y. B., Hard Superconductivity: Theory of the Motion of Abrikosov Flux Lines, *Rev. Mod. Phys.* **36**, 39-43 (1964).

[AND64[a]] ANDRES, K., Thermische Ausdehnung von Metallen bei tiefen Temperaturen, *Phys. Kondens. Mater.* **2**, 294-333 (1964).

[ANG32] ANGUS, W. R., Ionic Diamagnetic Susceptibilities, *Proc. Roy. Soc. London, Ser. A*: **136**, 569-578 (1932).

[ANJ79] ANJANEYULU, Y., and ROLLINS, R. W., Magnetic Flux Pinning in Proton-Irradiated Thick Nb Samples, *J. Appl. Phys.* **50**, 410-417 (1979).

[ARN74] ARNDT, R., and EBELING, R., Einfluss von Gefügeparametern auf die Stromtragfähigkeit von Niob-Titan-Supraleitern, *Z. Metallkd.* **65**, 364-373 (1974).

[ASH65] ASHCROFT, N. W., and WILKINS, J. W., Low Temperature Electronic Specific Heat of Simple Metals, *Phys. Lett.* **14**, 285-287 (1965); also personal communication (N. W. Ashcroft).

[ATO66] ATOMICS INTERNATIONAL, Investigation of Supercurrent Instabilities in Type II Superconductors, *Quart. Progr. Report No. 1, NASA-CR-76060*, AI-66-87, N66-29908, Atomics International, Canoga Park, California, May 1966.

[ATO67] ATOMICS INTERNATIONAL, Investigation of Supercurrent Instabilities in Type II superconductors, *Final Report, NASA-CR-86651*, AI-67-83, N67-31510, Atomic International, Canoga Park, California, May 1967.

B

[BAC68] BACHMANN, D., HEDRICH, D., KALSCH, E., RÖSCHEL, E., RAUB, E., and ZWICKER, U., Über den Einfluss von Zusätzen auf Verarbeitbarkeit, Gefuge,

Elektrische and Supraleitende Eigenschaften der Legierungen TiNb33 and TiNb40, *Z. Metallkd.* **59**, 426–432 (1968).

[BAG58] BAGARIATSKII, Yu. A., and NOSOVA, G. I., A More Accurate Determination of Atomic Coordinates of the Metastable ω-Phase in Ti–Cr Alloys, *Soviet Phys. Crystallogr.* **3**, 15–26 (1958), [Transl. of *Kristallografiya* **3**, 17–28 (1958)].

[BAG59] BAGARIATSKII, Yu, A., NOSOVA, G. I., and TAGUNOVA, T. V., Factors in the Formation of Metastable Phases in Titanium-Base Alloys, *Sov. Phys. Dokl.* **3**, 1014–1018 (1959), [Transl. of *Dokl. Akad. Nauk. SSSR* **122**, 593–596 (1958)].

[BAK67] BAKER, C., and TAYLOR, M. T., Some Problems of Relating the Critical Current Density to Dislocation Distribution in Worked Superconducting Alloys, *Philos. Mag.* **16**, 1129–1132 (1967).

[BAK69] BAKER, C., and SUTTON, J., Correlation of Superconducting and Metallurgical Properties of a Ti-20 at.% Nb Alloy, *Philos. Mag.* **19**, 1223–1255 (1969).

[BAK70] BAKER, C., Effect of Heat Treatment and Nitrogen Addition on the Critical Current Density of a Worked Niobium 44 wt.% Titanium Superconducting Alloy, *J. Mater. Sci.* **5**, 40–52 (1970).

[BAK71] BAKER, C., The Shape-Memory Effect in a Titanium-35 wt.% Niobium Alloy, *Met. Sci. J.* **5**, 92–100 (1971).

[BAL72] BALCERZAK, A. T., and SASS, S. L., The Formation of the ω Phase in Ti–Nb Alloys, *Metall. Trans.* **3**, 1601–1605 (1972).

[BAN82] BANERJEE, I., YANG, Q. S., FALCO, C. M., and SCHULLER, I. K., Superconductivity of Nb/Cu Superlattices, *Solid State Commun.* **41**, 805–808 (1982).

[BAN82[a]] BANERJEE, I., personal communication.

[BAR55] BARRON, T. H. K., On the Thermal Expansion of Solids at Low Temperatures, *Philos. Mag.* **46**, 720–734 (1955).

[BAR57] BARDEEN, J., COOPER, L. N., and SCHRIEFFER, J. R., Theory of Superconductivity, *Phys. Rev.* **108**, 1175–1204 (1957).

[BAR57[a]] BARRON, T. H. K., Grüneisen Parameters for the Equation of State of Solids, *Ann. Phys.* (*N.Y*). **1**, 77–90 (1957).

[BAR57[b]] BARRON, T. H. K., and MORRISON, J. A., On the Specific Heat of Solids at Low Temperatures, *Can. J. Phys.* **35**, 799–810 (1957).

[BAR59] BARDEEN, J., RICKAYZEN, G., and TEWORD, L., Theory of the Thermal Conductivity of Superconductors, *Phys. Rev.* **113**, 982–992 (1959).

[BAR61] BARRON, T. H. K., Thermodynamic and Vibrational Properties of Simple Solids at Low Temperatures, *Proc. VIIIth Int. Conf. Low Temperature Physics*, (1960), ed. by G. M. Graham *et al.*, North-Holland, Amsterdam, 1961, pp. 655–669.

[BAR64] BARRON, T. H. K., and LEADBETTER, A. J., The Thermal Properties of Alkali Halide Crystals IV. Analysis of Thermal Expansion Measurements, *Proc. Roy. Soc. London, Ser. A* **279**, 62–81 (1964).

[BAR65] BARDEEN, J., and STEPHEN, M. J., Theory of the Motion of Vortices in Superconductors, *Phys. Rev. A*: **140**, 1197–1207 (1965).

[BAR65[a]] BARON, V. V., BYCHKOVA, M. I., and SAVITSKII, E. M., Superconducting Properties of Niobium-Titanium Alloys and Influence of Alloying Additions on Critical Density of Current, in *Soviet Superconductor Research*, Nauka Publishing House, Moscow, USSR, 1965 [Transl. No. JPRS-36575, TT 66-33006, July 1966] pp. 39–47.

[BAR66] BARNES, L. J., and HAKE, R. R., Specific Heat and Magnetization of a Pauli-Paramagnetic Superconductor, *Ann. Acad. Sci. Fenn.* **A-VI**, 78–84 (1966).

[BAR67] BARNES, L. J., and HAKE, R. R., Calorimetric Evidence for Pauli-Paramagnetic Superconductivity, *Phys. Rev.* **153**, 435–437 (1967).

[BAR67[a]] BARRON, T. H. K., and MUNN, R. W., Analysis of the Thermal Expansion of Anisotropic Solids: Application to Zinc, *Philos. Mag.* **15**, 85–103 (1967).

[BAR69] BARBEE, T. W., Jr., Effects of Dislocation Distribution on the Specific Heat of Deformed Type II Superconductors, *Phys. Status Solidi* **31**, 535-544 (1969).

[BAR70] BARON, V. V., and BYCHKOVA, M. I., Individual Effects of Plastic Deformation and the Decomposition of the Solid Solution on the Critical Current of Superconducting Alloys of the Nb-Ti and Nb-Zr Systems, in *Physics and Metallurgy of Superconductors*, ed. by E. M. Savitskii and V. V. Baron, Consultants Bureau, NY (Plenum Publishing Co. 1970) pp. 48-52; (Proc. of the Second and Third Conf. on Metallurgy, Physical Chemistry and Metal Physics of Superconductors, Moscow, USSR, May 1965 and May 1966); first published in *Metalloved., Fiz-Khim., Metallofiz. Sverkhprovodnikov*, Izdatelstvo Nauka, Moscow, USSR, 1967, pp. 44-48.

[BAR70[a]] BARANOV, I. A., SHMULEVICH, R. S., SYTNIKOV, V. A., KARASIK, V. R., and VASIL'EV, N. G., Effect of Microinhomogeneities on the Transformation Temperature of Superconducting Alloys, in *Physics and Metallurgy of Superconductors*, ed. by E. M. Savitskii and V. V. Baron, Consultants Bureau, NY (Plenum Publishing Co. 1970) pp. 94-97; (Proc. of the Second and Third Conf. on Metallurgy, Physical Chemistry and Metal Physics of Superconductors, Moscow, USSR, May 1965 and May 1966); first published in *Metalloved., Fiz.-Khim., Metallofiz. Sverkhprovodnikov*, Izdatelstvo Nauka, Moscow, USSR, 1967, pp. 82-86.

[BAR70[b]] BARIŠIĆ, S., LABBE, J., and FRIEDEL, J., Tight Binding and Transition-Metal Superconductivity, *Phys. Rev. Lett.* **25**, 919-922 (1970).

[BAR78] BAR-SAGI, J., Microscopic Investigation of the Proximity Effect in a Finite Geometry, *Phys. Rev. B*: **18**, 3105-3111 (1978).

[BAR80] BARRON, T. H. K., COLLINS, J. G., and WHITE, G. K., Thermal Expansion of Solids at Low Temperatures, *Adv. Phys.* **29**, 609-730 (1980).

[BAS57] BASINSKI, Z. S., The Instability of Plastic Flow of Metals at Very Low Temperatures, *Proc. Roy. Soc. London, Ser. A*: **240**, 229-242 (1957).

[BAT64] BATT, R. H., Apparatus for Heat Capacity Measurements in the Range 0.24 to 20°K; Applications to Some Transition-Metal Superconductors and Ferromagnetic-Superconductors, *Ph.D. Thesis*, 1964, University of California, Lawrence Radiation Laboratory, Berkeley, California.

[BAT73] BATTERMAN, B. W., MARACCI, G., MERLINI, A., and PACE, S., Diffuse Mössbauer Scattering Applied to Dynamics of Phase Transformations, *Phys. Rev. Lett.* **31**, 227-230 (1973).

[BEA62] BEAN, C. P., Magnetization of Hard Superconductors, *Phys. Rev. Lett.* **8**, 250-253 (1962).

[BEA64] BEAN, C. P., Magnetization of High-Field Superconductors, *Rev. Mod. Phys.* **36**, 31-39 (1964)

[BEA69] BEASLEY, M. R., LABUSCH, R., and WEBB, W. W., Flux Creep in Type-II Superconducturs, *Phys. Rev.* **181**, 682-700 (1969).

[BEA82] BEASLEY, M. R., New Perspectives on the Physics of High-Field Superconductors, *Advances in Cryogenic Engineering* (*Materials*), Vol. 28, ed. by R. P. Reed and A. F. Clark, Plenum Press, New York, 1982, pp. 345-360.

[BEA82[a]] BEASLEY, M. R., personal communication.

[BEL68] BELANGER, B. C., The Behavior of Electric Currents and Magnetic Fields in Nonideal Type II Superconductors: Longitudinal Case, *Ph.D. Thesis*, 1968, University of Southern California, Los Angeles, California.

[BEL69] BELLIN, P. H., SADAGOPAN, V., and GATOS, H. C., Ternary Superconducting Alloys of the Titanium-Niobium-Vanadium System: Transition Temperature Variation, *J. Appl. Phys.* **40**, 3982-3984 (1969).

[BEL70] BELLIN, P. H., GATOS, H. C., and SADAGOPAN, V., Critical Field in the Ti-Nb-V System, *J. Appl. Phys.* **41**, 2057-2059 (1970).

[BEN72] BENNEMANN, K. H., and GARLAND, J. W., Theory for Superconductivity in d-Band Metals, in *Superconductivity in d- and f-Band Metals* (Proc. AIP Conf., Rochester, NY, Oct. 1971), ed. by D. H. Douglass, AIP Conf. Proc. No. 4, Amer. Inst. Phys., New York, 1972, pp. 103–137.

[BER51] BERMAN, R., The Thermal Conductivities of Some Dielectric Solids at Low Temperatures, *Proc. Roy. Soc. London Ser. A*: **208**, 90–108 (1951).

[BER52] BERMAN, R., and MACDONALD, D. K. C., The Thermal and Electrical Conductivity of Copper at Low Temperatures, *Proc. Roy. Soc. London, Ser. A*: **211**, 122–128 (1952).

[BER58] BERGER, L. W., WILLIAMS, D. N., and JAFFEE, R. I., Mechanical Properties and Heat Treatment of Titanium-Niobium Alloys, *Trans. ASM* **50**, 384-393 (1958).

[BER62] BERLINCOURT, T. G., and HAKE, R. R., Upper Critical Fields of Transition Metal Alloy Superconductors, *Phys. Rev. Lett.* **9**, 293-295 (1962).

[BER63] BERLINCOURT, T. G., and HAKE, R. R., Superconductivity at High Magnetic Fields, *Phys. Rev.* **131**, 140–157 (1963).

[BER63[a]] BERLINCOURT, T. G., Pulsed Magnetic Field Studies of Superconducting Transition Metal Alloys at High and Low Current Densities, in *Low Temperature Physics—LT8* (Proc. 8th Int. Conf., London, UK, Sept. 1962), ed. by R. O. Davies, Butterworths, London, 1963, pp. 338–341.

[BES79] BEST, K. J., GENEVEY, D., HILLMANN, H., KREMPASKY, L., POLAK, M., and TURCK, B., Anisotropy of the Critical Current in Solid Solution Superconductor NbTi, *IEEE Trans. Magn.* **MAG-15**, 395-397 (1979).

[BET68] BETTERTON, J. O., Jr., and SCARBROUGH, J. O., Low-Temperature Specific Heats of Zr-Ti, Zr-Hf and Zr-Sc Alloys, *Phys. Rev.* **168**, 715–725 (1968).

[BEV76] BEVOLO, A. J., GIBSON, E. D., VERHOEVEN, J. D., and FINNEMORE, D. K., Specific Heat of Composite Superconductors in the Proximity-Effect Regime, *Phys. Rev. B*: **14**, 114–117 (1976).

[BIS78] BISCHOF, J., and RYSKA, A., Investigation of Thermal Conductivity of Superconducting Materials Within the Temperature Range from 4.2 to 300 K, *J. Phys. (Paris), Colloque, Suppl. 8* **39**, C6-675 to 676 (1978).

[BLA57] BLACKMAN, M., On the Thermal Expansion of Solids, *Proc. Phys. Soc., London, Ser. B*: **70**, 827-832 (1957).

[BLA65] BLAUGHER, R. D., The Temperature Dependence of the Lower Critical Field for Some Non-Ideal Type II Superconductors, *Phys. Lett.* **14**, 181–182 (1965).

[BLA67] BLANDIN, A. P., Theory of the Hume-Rothery Rules, in *Phase Stability in Metals and Alloys*, ed. by P. S. Rudman, J. Stringer, and R. I. Jaffee, McGraw-Hill, New York, 1967, pp. 115–124.

[BLA68] BLACKBURN, M. J., and WILLIAMS, J. C., Phase Transformations in Ti–Mo and Ti–V Alloys, *Trans. TMS-AIME* **242**, 2461-2469 (1968).

[BLA70] BLACKBURN, M. J., Some Aspects of Phase Transformations in Titanium Alloys, in *The Science, Technology, and Application of Titanium*, ed. by R. I. Jaffee and N. E. Promisel, Pergamon Press, London, 1970, pp. 633–643.

[BOG58] BOGOLIUBOV, N. N., Concerning a New Method in the Theory of Superconductivity. III, *Sov. Phys. JETP* **7**, 41–46 (1958), [Transl. of *Zh. Eksp. Teor. Fiz.* **34**, 73–79 (1958)].

[BOR66] BOROVIK, E. S., FOGEL', N. Ya., and LITVINENKO, Yu. A., Magnetic Flux Jumps in Hard Superconductors in Pulsed Magnetic Fields, *Sov. Phys. JETP* **22**, 307–312 (1966), [Transl. of *Zh. Eksp. Teor. Fiz.* **49**, 438–446 (1965)].

[BOY71] BOYER, L., FOURNET, G., MAILFERT, A., and NOEL, J. L., Études de Sauts de Flux Limités dans les Supraconducteurs de Seconde Espèce, *Rev. Phys. Appl.* **6**, 501–504 (1971).

[BOY74] BOYER, R. R., TAGGART, R., and POLONIS, D. H., Effects of Thermal and Mechanical Processes on the β-III Titanium Alloy, *Metallography* **7**, 241–251 (1974).

[BRA67] BRAMMER, W. G., Jr., and RHODES, C. G., Determination of Omega Phase Morphology in Ti-35% Nb by Transmission Electron Microscopy, *Philos. Mag.* **16**, 477–486 (1967).

[BRA75] BRAND, R. A., Pinning in the Paramagnetic Limit, in *Low Temperature Physics—LT14*, (Proc. 14th Int. Conf. Otaniemi, Finland, Aug. 1975), ed. by M. Krusius and M. Vuorio, Elsevier North-Holland, New York, 1975, pp. 485–488.

[BRE67] BREWER, L., Viewpoints of Stability of Metallic Structures, in *Phase Stability in Metals and Alloys*, ed. by P. S. Rudman, J. Stringer, and R. I. Jaffee, McGraw-Hill, New York, 1967, pp. 39–61.

[BRE73] BRECHNA, H., *Superconducting Magnet Systems*, Springer-Verlag, New York, 1973.

[BRO53] BROWN, A., ZEMANSKY, M. W., and BOORSE, H. A., The Superconducting and Normal Heat Capacities of Niobium, *Phys. Rev.* **92**, 52–58 (1953).

[BRO55] BROTZEN, F. R., HARMON, E. L., Jr., and TROIANO, A. R., Decomposition of Beta Titanium, *Trans. TMS-AIME* **203**, 413–417 (1955).

[BRO64] BROWN, A. R. G., CLARK, D., EASTABROOK, J., and JEPSON, K. S., The Titanium-Niobium System, *Nature* **201**, 914–915 (1964).

[BRO65] BROWN, A. R. G., JEPSON, K. S., and HEAVENS, J., High-Speed Quenching in Vacuum, *J. Inst. Met.* **93**, 542–544 (1965).

[BUC61] BUCHER, E., and MÜLLER, J., Supraleitung in hexagonalen Ti-V- und Ti-Nb-Legierungen, *Helv. Phys. Acta.* **34**, 410–413 (1961).

[BUC64] BUCHER, E., HEINIGER, F., MUHEIM, J., and MÜLLER, J., Superconductivity and Electronic Properties of Transition Metal Alloys, *Rev. Mod. Phys.* **36**, 146–149 (1964).

[BUC65] BUCHER, E., HEINIGER, F., and MÜLLER, J., Anomalies in the Superconducting Transition of HCP Titanium Alloys, in *Low Temperature Physics-LT9* (Proc. 9th. Int. Conf., Columbus, OH, Sept. 1964), ed. by J. G. Daunt *et al.*, Plenum Press, New York, 1965, pp. 482–486.

[BUT77] BUTLER, W. H., Electron-phonon Coupling in the Transition Metals: Electronic Aspects, *Phys. Rev. B*: **15**, 5267–5282 (1977).

[BUY70] BUYNOV, N. N., VOZILKIN, V. A., and RAKIN, V. G., Structural Study of the Superconductive Alloy 65 BT, *Phys. Met. Metallogr.* **29**, No. 5, 115–119 (1970), [Transl. of *Fiz. Met. Metalloved.* **29**, 1005–1009 (1970)].

[BYC70] BYCHKOVA, M. I., BARON, V. V., and SAVITSKII, E. M., Effect of Heat Treatment on the Superconducting Properties of Nb–Ti Alloys, in *Physics and Metallurgy of Superconductors*, ed. by E. M. Savitskii and V. V. Baron, Consultants Bureau, New York (Plenum Publishing Corp., 1970) pp. 53–60; (*Proc. of the Second and Third Conf. on Metallurgy, Physical Chemistry and Metal Physics of Superconductors*, Moscow, USSR, May 1965 and May 1966); first published in *Metalloved, Fiz.-Khim., Metallofiz. Sverkhprovodnikov*, Izdatelstvo Nauka, Moscow, USSR, 1967, pp. 64–68.

[BYC70[a]] BYCHKOVA, M. I., BARRON, V. V., and SAVITSKII, E. M., Three-Component Alloys Based on the Nb–Ti System, in *Physics and Metallurgy of Superconductors*, ed. by E. M. Savitskii and V. V. Baron, Consultants Bureau, NY (Plenum Publishing Corp. 1970) pp. 90–93 (Proc. of the Second and Third Conf. on Metallurgy, Physical Chemistry and Metal Physics of Superconductors, Moscow, USSR, May 1965 and May 1966); first published in *Metalloved, Fiz.-Khim., Metallofiz. Sverkhprovodnikov*, Izdatelstvo Nauka, Moscow, USSR, 1967, pp. 79–82.

C

[CAH61] CAHN, J. W., On Spinodal Decomposition, *Acta. Metall.* **9**, 795–801 (1961).

[CAL61] CALVERLEY, A., MENDELSSOHN, K., and ROWELL, P. M., Some Thermal and Magnetic Properties of Tantalum, Niobium and Vanadium at Helium Temperatures, *Cryogenics* **1**, 26–33 (1961).

[CAM64] CAMPBELL, A. M., EVETTS, J. E., and DEW-HUGHES, D., The Behaviour of Type II Superconductors, *Philos. Mag.* **10**, 333–338 (1964).

[CAM68] CAMPBELL, A. M., EVETTS, J. E., and DEW-HUGHES, D., Pinning of Flux Vortices in Type II Superconductors, *Philos. Mag.* **18**, 313–343 (1968).

[CAM69] CAMPBELL, A. M., The Response of Pinned Flux Vortices to Low-Frequency Fields, *J. Phys. C* **2**, 1492–1501 (1969).

[CAM72] CAMPBELL, A. M., and EVETTS, J. E., *Critical Currents in Superconductors*, Taylor and Francis Ltd., London; Barnes and Noble Books, New York, 1972.

[CAM78] CAMPBELL, A. M., Pinning Parameters in the PbTl System, *Philos. Mag., Ser. B*: **37**, 169–174 (1978).

[CAM82] CAMPBELL, A. M., personal communication.

[CAP63] CAPE, J. A., Superconductivity and Localized Magnetic States in Ti–Mn Alloys, *Phys. Rev.* **132**, 1486–1492 (1963).

[CAP66] CAPE, J. A., Pauli-Spin and Spin-Orbit Effects in the Magnetization of a Superconducting Ti–Mo Alloy, *Phys. Rev.* **148**, 257–263 (1966).

[CAP82] CAPPELLETTI, R. L., personal communication.

[CAR47] CARSLAW, H. S., and JAEGER, J. C., *Conduction of Heat in Solids*, Clarendon Press, Oxford, 1947.

[CAR59] CARSLAW, H. S., and JAEGER, J. C., *Conduction of Heat in Solids, 2nd edition*, Clarendon Press, Oxford, 1959.

[CAR66] CAROLI, C., CYROT, M., and DE GENNES, P. G., The Magnetic Behavior of Dirty Superconductors, *Solid State Commun.* **4**, 17–19 (1966).

[CAT66] CATTERALL, J. A., High-Field Superconductivity and Its Applications, *Metall. Rev.* **11**, 25–46 (1966).

[CHA58] CHANDRASEKHAR, B. S., and HULM, J. K., The Electrical Resistivity and Superconductivity of Some Uranium Alloys and Compounds, *J. Phys. Chem. Solids* **7**, 259–267 (1958).

[CHA62] CHANDRASEKHAR, B. S., A Note on the Maximum Critical Field of High-Field Superconductors, *Appl. Phys. Lett.* **1**, 7–8 (1962).

[CHA63] CHANDRASEKHAR, B. S., HULM, J. K., and JONES, C. K., Temperature Dependence of the Upper Critical Field in Some Niobium Solid Solution Alloys, *Phys. Lett.*, **5**, 18–20 (1963).

[CHA68] CHANG, C. T.-M., Macroscopic Configuration of Magnetic Fields and Electric Currents in Type II Superconductors, *Ph.D. Thesis*, 1968, University of Southern California, Los Angeles, California.

[CHA70] CHARLESWORTH, J. P., and MADSEN, P. E., Effect of Heat Treatment on the Superconducting Critical Current of Cold-Worked Titanium-45 at.% Niobium: Part I, *Report No. AERE-R-6534*, Oct., 1970, Atomic Energy Research Establishment, Harwell, UK.

[CHA72] CHANDRASEKARAN, V., TAGGART, R., and POLONIS, D. H., Fracture Modes in a Binary Titanium Alloy, *Metallography* **5**, 235–250 (1972).

[CHA72[a]] CHANDRASEKARAN, V., TAGGART, R., and POLONIS, D. H., Phase Separation Processes in the Beta Phase of Ti–Mo Binary Alloys, *Metallography* **5**, 393–398 (1972).

[CHA73] CHANDRASEKARAN, V., COTTON, W., NARASIMHAN, S., TAGGART, R., and POLONIS, D. H., Study of Phase Transformation and Superconductivity, *Annual Progr. Report No. RLO-2225-T-13-19*, Oct. 1973, University of Washington, Seattle, Washington.

[CHA74] CHANDRASEKARAN, V., TAGGART, R., and POLONIS, D. H., The Influence of Constitution and Microstructure on the Temperature Coefficient of Resistivity in Ti-Base Alloys, *J. Mater. Sci.* **9**, 961-968 (1974).

[CHA74[a]] CHALLIS, L. J., Kapitza Resistance and Acoustic Transmission Across Boundaries at High Frequencies, *J. Phys. C: Solid State Phys.* **7**, 481-495 (1974).

[CHA78] CHANDRASEKARAN, V., TAGGART, R., and POLONIS, D. H., An Electron Microscopic Study of the Aged Omega Phase in Ti-Cr Alloys, *Metallography* **11**, 183-198 (1978).

[CHE60] CHENG, C. H., WEI, C. T., and BECK, P. A., Low Temperature Specific Heat of Body-Centered Cubic Alloys of 3d Transition Elements, *Phys. Rev.* **120**, 426-436 (1960).

[CHE62] CHENG, C. H., GUPTA, K. P., VAN REUTH, E. C., and BECK, P. A., Low-Temperature Specific Heat of Body-Centered Cubic Ti-V Alloys, *Phys. Rev.* **126**, 2030-2033 (1962).

[CHE72] CHEN, A.-B., WEISZ, G., and SHER, A., Temperature Dependence of the Electron States and dc Electrical Resistivity of Disordered Binary Alloys, *Phys. Rev.* **5**, 2897-2924 (1972).

[CHI73] CHILDS, G. E., ERICKS, L. J., and POWELL, R. L., Thermal Conductivity of Solids at Room Temperature and Below, *NBS Monograph No. 131*, National Bureau of Standards, Boulder, Colorado, 1973.

[CHR65] CHRISTIAN, J. W., Phase Transformations, in *Physical Metallurgy*, ed. by R. W. Cahn, North-Holland, Amsterdam, 1965.

[CHU67] CHUNG, D. G., and BUESSEM, W. R., The Elastic Anisotropy of Crystals, *J. Appl. Phys.* **38**, 2010-2012 (1967).

[CLA68] CLARK, A. F., Low Temperature Thermal Expansion of Some Metallic Alloys, *Cryogenics* **8**, 282-289 (1968).

[CLA68[a]] CLARKE, J., The Proximity Effect Between Superconducting and Normal Thin Films in Zero Field, *J. Phys. (Paris)* **29**, C2 3-16 (1968).

[CLA69] CLARKE, J., Supercurrents in Lead-Copper-Lead Sandwiches, *Proc. Roy. Soc. London, Ser. A*: **308**, 447-471 (1969).

[CLA70] CLARK, A. F., and KROPSCHOT, R. H., Low Temperature Specific Heat and Thermal Expansion of Alloys, *Bull. Inst. Int. Froid, Annexe 1970-2* (Proc. Comm. 1, Conf. on Cryophysics and Cryoengineering, Tokyo, Japan, 1970), pp. 249-254.

[CLA73] CLAPP, P. C., A Localized Soft Mode Theory for Martensitic Transformations, *Phys. Status Solidi B*: **57**, 561-569 (1973).

[CLA81] CLARK, A. F., FUJII, G., and RANNEY, M. A., The Thermal Expansion of Several Materials for Superconducting Magnets, *IEEE Trans. Magn.* **MAG-17**, 2316-2319 (1981).

[CLE79] CLEM, J. R., Theory of AC Losses in Type-II Superconductors with a Field-Dependent Surface Barrier, *J. Appl. Phys.* **50**, 3518-3530 (1979).

[CLO62] CLOGSTON, A. M., GOSSARD, A. C., JACCARINO, V., and YAFET, Y., Orbital Paramagnetism and the Knight Shift of d-Band Superconductors, *Phys. Rev. Lett.* **9**, 262-266 (1962).

[CLO62[a]] CLOGSTON, A. M., Upper Limit for the Critical Field in Hard Superconductors, *Phys. Rev. Lett.* **9**, 266-267 (1962).

[CLO64] CLOGSTON, A. M., JACCARINO, V., and YAFET, Y., Interpretation of Knight Shifts and Susceptibilities of Transition Metals: Platinum, *Phys. Rev. A*: **134**, 650-660 (1964).

[COF67] COFFEY, H. T., Distribution of Magnetic Fields and Currents in Type II Superconductors, *Cryogenics* **7**, 73-77 (1967).

[COL63] COLLINS, J. G., On the Calculation of the Temperature Variation of the Coefficient of Thermal Expansion for Materials of Cubic Structure, *Philos. Mag.* **8**, 323-332 (1963).

[COL64] COLLINS, J. G., and WHITE, G. K., Thermal Expansion of Solids, *Prog. Low Temp. Phys.* **4**, 450-479 (1964).

[COL66] COLLING, D. A., RALLS, K. M., and WULFF, J., High-Field Superconductivity of Tantalum-Titanium Alloys, *Trans. TMS-AIME* **236**, 1218-1223 (1966).

[COL69] COLLINGS, E. W., and HO, J. C., Enhancement of Superconducting Transition Temperatures in Martensitic Ti-Mo Alloys, *Phys. Lett. A* **29**, 306-307 (1969).

[COL70] COLLINGS, E. W., and HO, J. C., Magnetic Susceptibility and Low Temperature Specific Heat of High-Purity Titanium, *Phys. Rev.* **2**, 235-244 (1970).

[COL70a] COLLINGS, E. W., and HO, J. C., Physical Properties of Titanium Alloys, in *The Science, Technology and Application of Titanium*, ed. by R. I. Jaffee and N. E. Promisel, Pergamon Press, London, 1970, pp. 331-347.

[COL70b] COLLINGS, E. W., and HO, J. C., Influence of Microstructure on the Superconductivity of a Dilute Ti-Mo Alloy, *Phys. Rev.* **1**, 4289-4294 (1970).

[COL70c] COLLINGS, E. W., Local-Moment Paramagnetism in Liquid Tin-Manganese Alloys and the Occurrence of a Residual Paramagnetism, *Solid State Commun.* **8**, 381-385 (1970).

[COL70d] COLLINGS, E. W., and HO, J. C., Experimental Determination of the Orbital Contributions to the Magnetic Susceptibilities of hcp and bcc Titanium, *Phys. Lett. A* **31**, 193-194 (1970).

[COL71] COLLINGS, E. W., and HO, J. C., Density of States of Transition Metal Binary Alloys in the Electron-to-Atom Ratio Range 4.0 to 6.0, *Proc. 3rd Materials Research Symposium, Electronic Density of States*, Nat. Bur. Stand. (U.S.) Spec. Publ. 323, Dec. 1971, pp. 587-596.

[COL71a] COLLINGS, E. W., ENDERBY, J. E., and HO, J. C., Influence of Generalized Order-Disorder on the Electron States in Five Classes of Binary Alloy System, *3rd Materials Research Symposium, Electronic Density of States*, Nat. Bur. Stand. (U.S.) Spec. Publ. 323, Dec. 1971, pp. 483-492.

[COL71b] COLLINGS, E. W., and HO, J. C., Empirical Determination of the Electronic Specific Heat Enhancement Factor in Transition Metal Alloy Superconductors, *Phys. Status Solidi B*: **43**, K123-K127 (1971).

[COL71c] COLLINGS, E. W., Paramagnetic Effects Associated with Mn Additions to Liquid Al-Sn Alloys, *J. Phys. (Paris)* **32**, C1-516 to 518 (1971).

[COL71d] COLLINGS, E. W., and HO, J. C., Magnetic Susceptibility and Low Temperature Specific Heat of Ti, Zr, and Hf, *Phys. Rev.* **4**, 349-356 (1971).

[COL71e] COLLINGS, E. W., and GEHLEN, P. C., Some Observations on the Temperature Dependences of Magnetic Susceptibility of Ti and Ti-Al Alloys, *J. Phys. F: Metal Phys.* **1**, 908-918 (1971).

[COL71f] COLLINGS, E. W., BOYD, J. D., and HO, J. C., Enhancement of the Superconducting Transition Temperature in Deformation-Induced Structures, in *Low Temperature Physics—LT12* (Proc. 12th Int. Conf., Kyoto, Japan, Sept., 1970) ed. by E. Kanda, Academic Press of Japan, Tokyo, 1971, p. 316.

[COL72] COLLINGS, E. W., HO, J. C., and JAFFEE, R. I., Superconducting Transition Temperature, Lattice Instability, and Electron-to-Atom Ratio of Transition-Metal Binary Solid Solutions, *Phys. Rev.* **5**, 4435-4449 (1972).

[COL72a] COLLINGS, E. W., Experimental Studies of Density-of-States-Related Properties in Ti-Mo Alloys, *Proceedings of the Michigan State University Summer School on Alloys*, ed. by W. M. Hartmann *et al.*, 1972, pp. 236-269 (unpublished).

[COL72[b]] COLLINGS, E. W., and SMITH, R. D., The Magnetic Susceptibility of Niobium, *J. Less-Common Metals* **27**, 389-401 (1972).

[COL72[c]] COLLINGS, E. W., and HO, J. C., The Inverse Scaling of the Low Temperature Specific Heat Parameters γ and Θ_D in Transition-Metal Binary Alloys and its Dependence on Phase Stability, *Phys. Lett. A* **40**, 337-338 (1972).

[COL72[d]] COLES, B. R., Magnetism in Solid Solution Alloys, *Proceedings of the Michigan State University Summer School on Alloys*, Aug.-Sept. 1972, ed. by W. M. Hartmann *et al.* pp. 142-195 (unpublished).

[COL73] COLLINGS, E. W., and GEGEL, H. L., A Physical Basis for Solid Solution Strengthening and Phase Stability in Alloys of Titanium, *Scr. Metall.* **7**, 437-443 (1973).

[COL73[a]] COLLINGS, E. W., HO, J. C., and JAFFEE, R. I., Physics of Titanium Alloys—II: Fermi Density-of-States Properties, and Phase Stability of Ti-Al and Ti-Mo, in *Titanium Science and Technology*, ed. by R. I. Jaffee and H. M. Burte, Plenum Press, New York, 1973, pp. 831-842.

[COL73[b]] COLLINGS, E. W., ENDERBY, J. E., GEGEL, H. L., and HO, J. C., Some Relationships Between the Electronic and Mechanical Properties of Ti Alloys Discussed from the Standpoint of Fundamental Alloy Theory, in *Titanium Science and Technology*, ed. by R. I. Jaffee and H. M. Burte, Plenum Press, New York, 1973, pp. 801-814.

[COL73[c]] COLLVER, M. M., and HAMMOND, R. H., Superconductivity in "Amorphous" Transition-Metal Alloy Films, *Phys. Rev. Lett.* **30**, 92-95 (1973).

[COL74] COLLINGS, E. W., Anomalous Electrical Resistivity, bcc Phase Stability, and Superconductivity in Titanium-Vanadium Alloys, *Phys. Rev.* **9**, 3989-3999 (1974).

[COL75] COLLINGS, E. W., HO., J. C., and UPTON, P. E., Low-Temperature Calorimetric Studies of Superconductivity and Microstructure in Titanium-Vanadium Alloys, *J. Less-Common Metals* **42**, 285-301 (1975).

[COL75[a]] COLLINGS, E. W., and HO., J. C., Superconducting Transition Temperature in Martensitic Titanium-Base Transition Metal Binary Alloys, *J. Less-Common Metals* **41**, 157-164 (1957).

[COL75[b]] COLLINGS, E. W., Magnetic Studies of Omega-Phase Precipitation and Aging in Titanium-Vanadium Alloys, *J. Less-Common Metals* **39**, 63-90 (1975).

[COL75[c]] COLLINGS, E. W., and GEGEL, H. L., (eds.) *Physics of Solid Solution Strengthening*, Plenum Press, New York, 1975.

[COL75[d]] COLLINGS, E. W., and GEGEL, H. L., Physical Principles of Solid Solution Strengthening in Alloys, in *Physics of Solid Solution Strengthening*, ed. by E. W. Collings and H. L. Gegel, Plenum Press, New York, 1975, pp. 147-182.

[COL76] COLLINGS, E. W., and SMITH, R. D., Magnetic Susceptibility of Ti-Nb Alloys, *J. Less-Common Metals* **48**, 187-198 (1976).

[COL76[a]] COLLINGS, E. W., and HO, J. C., Solute-Induced Lattice Stability as it Relates to Superconductivity in Titanium-Molybdenum Alloys, *Solid State Commun.* **18**, 1493-1495 (1976).

[COL76[b]] COLLINGS, E. W., and HO, J. C., Calorimetric Studies of Superconductive Proximity Effects in Titanium-Vanadium Alloys, in *Low Temperature Physics—LT14*, (Proc. 14th Int. Conf., Aug., 1975) ed. by M. Krusius and M. Vuorio, American Elsevier, New York, 1975, pp. 273-276.

[COL77] COLLINGS, E. W., JELINEK, F. J., HO, J. C., and MATHUR, M. P., Magnetic and Thermal Properties of Stainless Steels and Inconels at Cryogenic Temperatures, *Advances in Cryogenic Engineering* (*Materials*), Vol. 22, ed. by K. D. Timmerhaus, R. P. Reed and A. F. Clark, Plenum Press, New York, 1977, pp. 159-173.

[COL78] COLLINGS, E. W., Anomalous Electrical Resistivity and Magnetic Susceptibility Temperature Dependences in Ti-V Alloys Exhibiting Reversible Soft-Phonon-

Induced Structural Inhomogeneous, in *Electrical Transport and Optical Properties of Inhomogeneous Media*, ed. by J. C. Garland and D. B. Tanner, AIP Conf. Proc. No. 40, Amer. Inst. Phys., New York, 1978, pp. 410–415.

[COL78a] COLLINGS, E. W., and WHITE, J. J. III, Deformation- and Solute-Induced Microstructural Effects in the Superconductivity of Transition-Metal Alloys, in *Transition Metals 1977*, ed. by M. J. G. Lee, J. M. Perz, and E. Fawcett, The Institute of Physics, Conf. Ser. No. 39, 1978, pp. 645–649.

[COL78b] COLLINGS, E. W., and CAPPELLETTI, R. L., Diffuse Cluster Superparamagnetism in Inconel X750, *Advances in Cryogenic Engineering* (*Materials*), Vol. 24, ed. by K. D. Timmerhaus, R. P. Reed, and A. F. Clark, Plenum Press, New York, 1978, pp. 240–248.

[COL79] COLLINGS, E. W., Magnetic Studies of Phase Equilibria in Ti–Al(30–57 at.%) Alloys, *Metall. Trans.* **10A**, 463–474 (1979).

[COL79a] COLLINGS, E. W., and KING, H. W., Austenitic Stainless Steel as a Magnetic Transition Metal Alloy, in *The Metal Science of Stainless Steels*, ed. by E. W. Collings and H. W. King, The Metallurgical Society AIME, Warrendale, Pennsylvania, 1979, pp. 1–21.

[COL79b] COLLINGS, E. W., and LEDBETTER, H. M., Sound Velocity Anomalies Near the Spin Glass Transition in an Austenitic Stainless Steel Alloy, *Phys. Lett. A* **72**, 53–56 (1979).

[COL79c] COLLINGS, E. W., and HART, S. C., Low Temperature Magnetic Susceptibility and Magnetization Studies of Some Commercial Austenitic Stainless Steels, *Cryogenics* **19**, 521–530 (1979).

[COL80] COLLINGS, E. W., The Magnetic Character of Austenitic Stainless Steel, *Advances in Cryogenic Engineering* (*Materials*), Vol. 26, ed. by A. F. Clark and R. P. Reed, Plenum Press, New York, 1980, pp. 37–40.

[COL82] COLLINGS, E. W., Response of Superconducting Transition Temperature and Other Physical Properties to Phase Transformation, Precipitation, and Aging in Titanium-Base Transition-Metal-Binary Alloys, in *Titanium and Titanium Alloys, Scientific and Technological Aspects*, ed. by J. C. Williams and A. F. Belov, Plenum Press, New York, 1982, pp. 1407–1419.

[COL82a] COLLINGS, E. W., Magnetic Investigations of Electronic Bonding and α-through-γ Phase Equilibria in the Titanium Aluminium System, in *Titanium and Titanium Alloys, Scientific and Technological Aspects*, ed. by J. C. Williams and A. F. Belov, Plenum Press, New York, 1982, pp. 1391–1402.

[COL83] COLLINGS, E. W., *A Sourcebook of Titanium Alloy Superconductivity*, Plenum Press, New York, 1983.

[COL84] COLLINGS, E. W., Previously unpublished research.

[COM67] COMEY, K. R., Jr. The Superconductivity of a Titanium–Niobium Alloy, *M.S. Thesis*, 1967, Massachusetts Institute of Technology, Cambridge, Massachusetts.

[CON75] CONRAD, H., DE MEESTER, B., DÖNER, M., and OKASAKI, K., Strengthening of Alpha Titanium by the Interstitial Solutes C, N, and O, in *Physics of Solid Solution Strengthening*, ed. by E. W. Collings and H. L. Gegel, Plenum Press, New York, 1975, pp. 1–45.

[CON81] CONRAD, H., Effect of Interstitial Solutes on the Strength and Ductility of Titanium, *Prog. Mater. Sci.* **26**, 123–403 (1981).

[COO61] COOPER, L. N., Superconductivity in the Neighborhood of Metallic Contacts, *Phys. Rev. Lett.* **6**, 689–690 (1961).

[COO73] COOK, H. E., On the Nature of the Omega Transformation, *Acta. Metall.* **21**, 1445–1449 (1973).

[COO74] COOK, H. E., A Theory of the Omega Transformation, *Acta. Metall.* **22**, 239–247 (1974).

[Coo75] Cook, H. E., On First-Order Structural Phase Transitions-I. General Considerations of Pre-Transition and Nucleation Phenomena, *Acta. Metall.* **23**, 1027–1039 (1975).

[Coo75a] Cook, H. E., On First-Order Structural Phase Transitions-II. The Omega Transformation, *Acta. Metall.* **23**, 1041–1054 (1975).

[Cor56] Corak, W. S., Goodman, B. B., Satterthwaite, C. B., and Wexler, A., Atomic Heats of Normal and Superconducting Vanadium, *Phys. Rev.* **102**, 656–661 (1956).

[Cor64] Corsan, J. M., The Effect of Rate of Change of Magnetic Field on Flux-Jumping in Nb_3Sn Cylinders, *Phys. Lett.* **12**, 85–86 (1964).

[Cor67] Cornish, D. N., The Effect of Mechanical Loading on Composite Conductors, *Proc. 2nd Int. Conf. on Magnet Technology*, Oxford, England, July 1967, pp. 507–512.

[Cor75] Cordier, T. E., and McDonald, W. K., Production and Fabrication of 2500-lb Nb–Ti Ingots to Rod, *IEEE Trans. Magn.* **MAG-11**, 280–282 (1975).

[Cou67] Courtney, T. H., and Wulff, J., Quarternary Solid Solution Superconductors. *Phys. Lett A* **25**, 477–479 (1967).

[Cou69] Courtney, T. H., and Wulff, J., Omega Phase Formation in Superconducting Ti Alloys, *Mater. Sci. Eng.* **4**, 93–97 (1969).

[Cow68] Cowan, J. A., Pawlowicz, A. T., and White, G. K., Thermal Expansion of Polycrystalline Titanium and Zirconium, *Cryogenics* **8**, 155–157 (1968).

[Cra56] Crank, J., *The Mathematics of Diffusion*, Clarendon Press, Oxford, 1956.

[Cri70] Critchlow, P. R., and Zeitlin, B., Magnetization Measurements on Twisted Superconducting Composite Conductors, *J. Appl. Phys.* **41**, 4860–4865 (1970).

[Cur79] Curtis, C. W., and McDonald, W. K., Suitability of NbTi Containing up to 2 wt.% Ta for Use in Fabrication of Superconductors, *IEEE Trans. Magn.* **MAG-15**, 768–771 (1979); see also *Final Report to ERDA on Contr. No. RL-E-77-0189.*

[Cus58] Cusack, N., *The Electrical and Magnetic Properties of Solids*, Longmans, New York, 1958.

D

[Dan69] Danner, S., and Dummer, G., Kalorimetrische Bestimmung der Übergangstemperatur zur Supraleitung von verdünnten Titan–Rhodium-Legierungen, *Z. Phys.* **222**, 243–252 (1969).

[Dau55] Daunt, J. G., The Electronic Specific Heats in Metals, *Prog. Low Temp. Phys.* **1**, 202–223 (1955).

[Dav75] Davis, L. A., Hardness/Strength Ratio of Metallic Glasses, *Scri. Metall.* **9**, 431–435 (1975).

[Dav79] Davis, R., Flower, H. M., and West, D. R. F., Martensitic Transformations in Ti–Mo Alloys, *J. Mater. Sci.* **14**, 712–722 (1979).

[Daw70] Dawson, C. W., and Sass, S. L., The As-Quenched Form of the Omega Phase in Zr–Nb Alloys, *Metall. Trans.* **1**, 2225–2233 (1970).

[Dec69] Decker, D. L., and Lacquer, H. L., Magnetization Studies on Superconducting Vanadium-Gallium, *J. Appl. Phys.* **40**, 2817–2822 (1969).

[Def70] de Fontaine, D., Mechanical Instabilities in the b.c.c. Lattice and the Beta to Omega Phase Transformation, *Acta. Metall.* **18**, 275–279 (1970).

[Def71] de Fontaine, D., Paton, N. E., and Williams, J. C., The Omega Phase Transformation in Titanium Alloys as an Example of Displacement Controlled Reactions, *Acta. Metall.* **19**, 1153–1162 (1971).

[DEG63] DE GENNES, P. G., and GUYON, E., Superconductivity in "Normal" Metals, *Phys. Lett.* **3**, 168-169 (1963).

[DEG64] DE GENNES, P. G., and MATRICON, J., Collective Modes of Vortex Lines in Superconductors of the Second Kind, *Rev. Mod. Phys.* **36**, 45-49 (1964).

[DEG64[a]] DE GENNES, P. G., Boundary Effects in Superconductors, *Rev. Mod. Phys.* **36**, 225-237 (1964).

[DEG74] DEGTYAREVA, V. F., KARIMOV, Yu. S., and RABINKIN, A. G., Superconductivity and Magnetic Susceptibility of the α and ω Modifications of Titanium and Zirconium, *Sov. Phys. Solid State* **15**, 2293-2294 (1974), [Transl. of *Fiz. Tverd. Tela (Leningrad)* **15**, 3436-3438 (1973)].

[DEI73] DEIS, D., MATHUR, M. P., ASHKIN, M., and JONES, C. K., Investigation of Flux Pinning and Critical Current Densities in Type 2 Superconductors, *Tech. Rept. July, 1972-June, 1973, ONR Contr. No. N00014-73-0026*, Westinghouse Res. Lab., 1973.

[DEI74] DEIS, D. W., MATHUR, M. P., ASHKIN, M., and JONES, C. K., Investigation of Flux Pinning and Critical Current Densities in Type 2 Superconductors, *Tech. Rept. July, 1973-June 1974, ONR Contr. N00014-73-C-0026*, Westinghouse Res. Lab., 1974.

[DEJ65] DE JONG, L. N. J., and BLAISSE, B. S., A Simple Model for the Magnetic Behavior of Superconductors with an Anchored Filament, *Physica (The Hague)* **31**, 317-325 (1965).

[DEL53] DELAZARO, D. J., and ROSTOKER, W., The Influence of Oxygen Contents on Transformations in a Titanium Alloy Containing 11 Per Cent Molybdenum, *Acta. Metall.* **1**, 676-677 (1953).

[DEL74] DELAEY, L., KRISHNAN, R. V., TAS, H., and WARLIMONT, H., Thermoelasticity, Pseudoelasticity and the Memory Effects Associated with Martensitic Transformations, *J. Mater. Sci.* **9**, 1521-1535 (1974).

[DEL80] DELANNOY, M., and PERRIN, G., On the Determination of the Volume Dependence of Grüneisen Parameters in Cubic and Non-Cubic Solids, *J. Phys. Solids* **41**, 11-16 (1980).

[DEN63] DENBIGH, J. S., and LOMER, W. M., Induced Orbital Paramagnetism in Metallic Chromium, *Proc. Phys. Soc., London* **82**, 156-158 (1963).

[DES63] DESORBO, W., Size Factor and Superconducting Properties of Some Transition Metal Solutions, *Phys. Rev.* **130**, 2177-2187 (1963).

[DES65] DESORBO, W., Solute Size and Valence Effect in Some Superconducting Alloys of Transition Elements, *Phys. Rev. A* **140**, 914-919 (1965).

[DES67] DESORBO, W., LAWRENCE, P. E., and HEALY, W. A., Probable Upper Limit of H_{c2} For Some Higher-Order Ductile Transition-Metal Superconducting Alloys, *J. Appl. Phys.* **38**, 903-904 (1967).

[DEU69] DEUTSCHER, G., and DE GENNES, P. G., Proximity Effects, in *Superconductivity*, ed. by R. D. Parks, Marcel Dekker Inc., New York, 1969, pp. 1005-1033.

[DEW71] DEW-HUGHES, D., The Metallurgical Enhancement of Type II Superconductors, *Rep. Prog. Phys.* **34**, 821-873 (1971).

[DEW74] DEW-HUGHES, D., Practical Superconductive Materials, in *Superconducting Machines and Devices*, ed. by S. Foner and B. B. Schwartz, Plenum Press, New York, 1973, pp. 87-137.

[DEW79] DEW-HUGHES, D., Physical Metallurgy of A15 Compounds, in *Treatise on Materials Science and Technology, Volume 14: Metallurgy of Superconducting Materials*, ed. by T. Luhman and D. Dew-Hughes, Academic Press, New York, 1979, pp. 137-169.

[Doi66] Doi, T., Mitani, M., and Umezawa, T., Superconducting Properties of Nb-Zr-Ti Ternary Alloys (Studies of Hard Superconductor, I), *Nippon Kinzoku Gakkaishi* **30**, 133-138 (1966).

[Doi66[a]] Doi, T., Ishida, H., and Umezawa, T., Study of Nb-Zr-Ti Phase Diagram (Studies of Hard Superconductor II), *Nippon Kinzoku Gakkaishi* **30**, 139-145 (1966).

[Doi66[b]] Doi, T., Ishida, H., and Umezawa, T., Effect of Cold Working and Heat Treatment on the Critical Current Characteristics of Nb-25 at.% Zr and Nb-40 at.% Zr-10 at.% Ti Alloys (2) (Studies of Hard Superconductor, IV), *Nippon Kinzoku Gakkaishi* **30**, 220-225 (1966).

[Doi68] Doi, T., Mitani, M., Kobayashi, M., and Ohara, H., Superconducting Alloys and Apparatus for Generating Superconducting Magnetic Field, *US. Patent 3,408,604* (Oct. 1968).

[Doi68[a]] Doi, T., Ishida, F., Kawabe, U., and Kitada, M., Critical Current of Superconducting Nb(Cb)-Zr-Ti Alloys in High Magnetic Field, *Trans. TMS-AIME* **242**, 1793-1800 (1968).

[Dom78] Domb, E. R., and Johnson, W. L., The Upper and Lower Critical Fields of High-κ Amorphous Superconductors, *J. Low Temp. Phys.* **33**, 29-40 (1978).

[Dub68] Dubeck, L., and Setty, K. S. L., Thermal Conductivity of Niobium-Titanium, *Phys. Lett.* **27A**, 334-335 (1968).

[Dum68] Dummer, G., and Oftedal, E., Supraleitung und Spezifische Wärme von Titan-Rhodium-Legierungen, *Z. Phys.* **208**, 238-248 (1968).

[Duw53] Duwez, P., The Martensitic Transformation Temperature in Titanium Binary Alloys, *Trans. ASM* **45**, 934-940 (1953).

[Dyn72] Dynes, R. C., McMillan's Equation and the T_c of Superconductors, *Solid State Commun.* **10**, 615-618 (1972).

[Dyn76] Dynes, R. C., and Allen, P. B., Maximum T_c—Optimistic Evidence, in *Superconductivity in d- and f-Band Metals* (Proc. 2nd Rochester Conference, 1976) ed. by D. H. Douglass, Plenum Press, New York, 1976, pp. 15-28.

E

[Eas71] Easton, D. S., Scarbrough, J. O., and Betterton, J. O., Jr., Effect of Heat Treatment and Impurities on the Current Density of Transition Metal Superconductors, *Report No. ORNL-4703, N72-11656*, June, 1971, Oak Ridge National Lab., Oak Ridge, Tennessee.

[Eas75] Easton, D. S., and Koch, C. C., Tensile Properties of Superconducting Composite Conductors and Nb-Ti Alloys at 4.2K, in *Shape-Memory Effect in Alloys*, ed. by J. Perkins, Plenum Press, New York, 1975, pp. 431-444.

[Eas76] Easton, D. S., and Koch, C. C., Magnet Development, and Stress Effects on Superconductors, *Report No. ORNL-5189* (*1976*); *Magnetic Fusion Energy Materials Technology Program Annual Progress Report for Period Ending June 30, 1976*, ed. by S. Peterson, Oak Ridge National Laboratory, Oak Ridge, Tennessee.

[Eas76[a]] Easton, D. S., Kroeger, D. M., and Moazed, A., Thermomechanical Heat Generation in Copper and a Nb-Ti Superconducting Composite, *Appl. Phys. Lett.* **29**, 382-384 (1976).

[Eas77] Easton, D. S., and Koch, C. C., Mechanical Properties of Superconducting Nb-Ti Composites, in *Advances in Cryogenic Engineering* (*Materials*), Vol. 22, ed. by K. D. Timmerhaus, R. P. Reed, and A. F. Clark, Plenum Press, New York, (1977) pp. 453-462.

[ECK76] ECKERT, C., and HANDSTEIN, A., Experimental Investigation of the Flux Density Profile in Hard Superconductors, *Phys. Status. Solidi A*: **37**, 171–178 (1976).

[EFI70] EFIMOV, Yu. V., BARON, V. V., and SAVITSKII, E. M., Superconducting and Mechanical Properties of Three-Component Alloys Based on the V–Ti System, in *Physics and Metallurgy of Superconductors*, ed. by E. M. Savitskii and V. V. Baron, Consultants Bureau, New York (Plenum Publishing Corp. 1970) ppp. 98–101; (Proc. of the Second and Third Conf. on Metallurgy, Physical Chemistry and Metal Physics of Superconductors, Moscow, USSR, May 1965 and May 1966); first published in *Metalloved, Fiz.-Khim., Metallofiz. Sverkhprovodnikov*, Izdatelstvo Nauka, Moscow, USSR, 1967, pp. 86–89.

[EGG73] EGGENDORFER, G., Direct Mechanical Measurement of Pinning Forces on Type-II Superconductors, *J. Low Temp. Phys.* **10**, 715–718 (1973).

[EIL67] EILENBERGER, G., Determination of $\kappa_1(T)$ and $\kappa_2(T)$ for Type-II Superconductors with Arbitrary Impurity Concentration, *Phys. Rev.* **153**, 584–598 (1967).

[EKI77] EKIN, J. W., Mechanisms for Critical-Current Degradation in NbTi and Nb_3Sn Multifilamentary Wires, *IEEE Trans. Magn.* **MAG-13**, 127–130 (1977).

[EKI78] EKIN, J. W., Fatigue and Stress Effects in NbTi and Nb_3Sn Multifilamentary Superconductors, in *Advances in Cryogenic Engineering* (*Materials*), Vol. 24, ed. by K. D. Timmerhaus, R. P. Reed, and A. F. Clark, Plenum Press, New York, 1978, pp. 306–316.

[ELB63] EL BINDARI, A., and LITVAK, M. M., The Upper Critical Field of Nb–Zr and Nb–Ti Alloys, *Report No. BSD-TDR-63-62, AMP-101, N63-14201*, Jan., 1963, Avco-Everett Res. Lab., Everett, Massachusetts.

[ELB63[a]] EL BINDARI, A., and LITVAK, M. M., Critical Current and Field in Non-Ideal Superconductors, *Report No. BSD-TDR-63-148, Res. Report 162, N63-22202*, July, 1963, Avco-Everett Res. Lab., Everett, Massachusetts.

[ELB64] EL BINDARI, A., and LITVAK, M. M., Critical Current and Field in Non-Ideal Superconductors, *Rev. Mod. Phys.* **36**, 98–102 (1964).

[ELB67] EL BINDARI, A., BERNERT, R. E., and HOPPIE, L. O., Performance of Stabilized Nb–40%Ti in a High Current Density 50-kG Superconducting Solenoid, *Report No. 270, NASA N67, 32997*, May, 1967, Avco-Everett Res. Lab., Everett, Massachusetts.

[ELB68] EL BINDARI, A., BERNERT, R. E., and HOPPIE, L. O., Performance of a Cu-Clad Nb–40%Ti in a High-Current-Density 50-kG Superconducting Solenoid, in *Advances in Cryogenic Engineering*, Vol. 13, ed. by K. D. Timmerhaus, Plenum Press, New York, 1968, pp. 30–34.

[ELL65] ELLIOT, R. P., *Constitution of Binary Alloys*, McGraw-Hill Book Company, New York, 1965.

[ELR82] ELROD, S. A., MILLER, J. R., and DRESNER, L., The Specific Heat of NbTi from 0 to 7T Between 4.2 and 20K, in *Advances in Cryogenic Engineering* (*Materials*), Vol. 28 ed. by R. P. Reed and A. F. Clark, Plenum Press, New York, 1982, pp. 601–610.

[EMB66] EMBURY, J. D., KEH, A. S., and FISHER, R. M., Substructural Strengthening in Materials Subject to Large Plastic Strains, *Trans. TMS-AIME* **236**, 1252–1260 (1966).

[ERD68] ERDMANN, J. C., and JAHODA, J. A., Affected Volume and Temperature Rise During Discontinuous Slip at Low Temperatures, *J. Appl. Phys.* **39**, 2793–2797 (1968).

[EVA66] EVANS, D. J., and WINSTANLEY, C. J., Measurement of the Coefficient of Thermal Expansion of Solids and Its Precise Variation with Temperature, *J. Sci. Instrum.* **43**, 772–773 (1966).

[EVA73] EVANS, D., An Hypothesis Concerning the Training Phenomenon Observed in Superconducting Magnets, *Report No. RL-73-092*, Aug., 1973, Rutherford High Energy Lab., Chilton, England.

F

[FAH70] FAHMY, A. A., and RAGAI, A. N., Thermal-Expansion Behavior of Two-Phase Solids, *J. Appl. Phys.* **41**, 5108–5111 (1970).

[FAH74] FAHMY, A. A., and RAGAI-ELLOZY, A. N., A Discrete Element Method for the Calculation of the Thermal Expansion Coefficients of Unidirectional Fiber Composites, in *Thermal Expansion—1973* (AIP Conf. Proc. No. 17) ed. by R. E. Taylor and G. L. Denman, Amer. Inst. Phys., New York, 1974, pp. 231–240.

[FAU82] FAULKNER, J. S., The Modern Theory of Alloys, *Prog. Mater. Sci.* **27**, 1–187 (1982).

[FED63] FEDOTOV, S. G., and BELOUSOV, O. K., The Elastic Properties of Alloys of Titanium with Molybdenum, Vanadium, and Niobium, *Sov. Phys. Dokl.* **8**, 496–498 (1963), [Transl. of *Dokl. Akad. Nauk SSSR* **150**, 77–80 (1963)].

[FED64] FEDOTOV, S. G., and BELOUSOV, O. K., Elastic Constants of the System Titanium-Niobium, *Phys. Met. Metallogr.* **17**, No. 5, 83–86 (1964), [Transl. of *Fiz. Met. Metalloved.* **17**, 732–736 (1964)].

[FED66] FEDOTOV., S. G., Dependence of the Elastic Properties of Titanium Alloys on Their Composition and Structure, in *Titanium and Its Alloys*, ed. by I. I. Kornilov, Akademiya Nauk SSSR (1963); Transl. by Israel Program for Scientific Translations Ltd., IPST. Cat. No. 1454, 1966, pp. 199–215.

[FED73] FEDOTOV, S. G., Peculiarities of Changes in Elastic Properties of Titanium Martensite, in *Titanium Science and Technology*, ed. by R. I. Jaffe and H. M. Burte, Plenum Press, New York, 1973, pp. 871–880.

[FEH77] FEHER, A., REIFFERS, M., PETROVIČ, P., and JÁNOŠ, Š., The Electrical Resistivity of Nb–Ti and Nb–Ti–Zr Alloys at Low Temperatures, *Phys. Status Solidi A*: **39**, K67–K69 (1977).

[FER66] FERREIRA DA SILVA, J., VAN DUYKEREN, N. W. J., and DOKOUPIL, Z., Low Temperature Specific Heats of Unannealed Niobium Wires in Magnetic Fields, *Phys. Lett.* **20**, 448–449 (1966).

[FER66[a]] FERREIRA DA SILVA, J., VAN DUYKEREN, N. W. J., and DOKOUPIL, Z., Low Temperature Specific Heats of Unannealed Niobium Wires in Magnetic Fields, *Physica* **32**, 1253–1273 (1966).

[FER70] FERRIER, M., Description Physique du Comportement des Supraconducteurs, *Ann. Mines*, 11–33 (Jan. 1970).

[FET69] FETTER, A. L., and HOHENBERG, P. C., Theory of Type-II Superconductors, in *Superconductivity*, ed. by R. D. Parks, Marcel Dekker Inc., New York, 1969, pp. 817–923.

[FEV73] FEVRIER, A., and MORIZE, D., The Effect of Magnetic Field on the Thermal Conductivity and Electrical Resistivity of Different Materials, *Cryogenics* **13**, 603–606 (1973).

[FEY55] FEYNMAN, R. P., Application of Quantum Mechanics to Liquid Helium, *Prog. Low Temp. Phys.* **1**, 17–53 (1955).

[FIC71] FICKETT, F. R., Magnetoresistance of Very Pure Polycrystalline Aluminum, *Phys. Rev. B*: **3**, 1941–1952 (1971).

[FIC72] FICKETT, F. R., A Preliminary Investigation of the Behavior of High Purity Copper in High Magnetic Fields, *Annual Report, Project No. 186*, Int. Copper Res. Assn. 1972.

[FIE64] FIETZ, W. A., BEASLEY, M. R., SILCOX, J., and WEBB, W. W., Magnetization of Superconducting Nb-25% Zr Wire, *Phys. Rev. A*: **136**, 335-345 (1964).

[FIE67] FIETZ, W. A., Reversible and Irreversible Magnetic Properties of Some Type-II Alloy Superconductors, *Ph.D. Thesis*, 1967, Cornell University, Ithaca, New York.

[FIE67[a]] FIETZ, W. A., and WEBB, W. W., Magnetic Properties of Some Type-II Alloy Superconductors Near the Upper Critical Field, *Phys. Rev.* **161**, 423-433 (1967).

[FIE69] FIETZ, W. A., and WEBB, W. W., Hysteresis in Superconducting Alloys—Temperature and Field Dependence of Dislocation Pinning in Niobium Alloys, *Phys. Rev.* **178**, 657-667 (1969); Erratum in *Phys. Rev.* **185**, 862 (1969).

[FIN69] FINK, H. J., and JOINER, W. C. H., Surface Nucleation and Boundary Conditions in Superconductors, *Phys. Rev. Lett.* **23**, 120-123 (1969).

[FIS67] FISHER, R. M., and LALLY, J. S., Microplasticity Detected by an Acoustic Technique, *Can. J. Phys.* **45**, 1147-1159 (1967).

[FIS70] FISHER, E. S., and DEVER, D., Relation of the C′ Elastic Modulus to Stability of b.c.c. Transition Metals, *Acta. Metall.* **18**, 265-269 (1970).

[FIS70[a]] FISHER, E. S., and DEVER, D., The Single Crystal Elastic Moduli of Beta-Titanium and Titanium-Chromium Alloys, in *The Science, Technology, and Application of Titanium*, ed. by R. I. Jaffee and N. E. Promisel, Pergamon Press, London, 1970, pp. 373-381.

[FIS75] FISHER, E. S., A Review of Solute Effects on the Elastic Moduli of BCC Transition Metals, in *Physics of Solid Solution Strengthening*, ed. by E. W. Collings and H. L. Gegel, Plenum Press, New York, 1975, pp. 199-225.

[FIS75[a]] FISHER, E. S., WESTLAKE, D. G., and OCKERS, S. T., Effects of Hydrogen and Oxygen on the Elastic Moduli of Vanadium, Niobium, and Tantalum Single Crystals, *Phys. Status Solidi A*: **28**, 591-602 (1975).

[FIS77] FISHER, E. S., and KIM, S. H., Critical Current Changes and Fatigue Damage After 4.2 K Strain Cycling of Superconducting Composites, *IEEE Trans. Magn.* **MAG-13**, 112-115 (1977).

[FLA78] FLACHBART, K., FEHER, A., JÁNOŠ, Š., MÁLEK, Z., and RYSKA, A., Thermal Conductivity of Nb-Ti Alloy in the Low-Temperature Range, *Phys. Status Solidi B*: **85**, 545-551 (1978).

[FLE74] FLEMING, M. C., *Solidification Processing*, McGraw-Hill Book Company, New York 1974.

[FLO82] FLOWER, H. M., DAVIS, R., and WEST, D. R. F., Martensite Formation and Decomposition in Alloys of Titanium Containing β-Stabilizing Elements, in *Titanium and Titanium Alloys, Scientific and Technological Aspects*, ed. by J. C. Williams and A. F. Belov, Plenum Press, New York, 1982, pp. 1703-1715.

[FOU70] FOURNET, C., and MAILFERT, A., Behavior of Type II Superconductors in Time Dependent Magnetic Fields, *Elektrotech. Čas.* **21**, 303-313 (1970).

[FOW73] FOWLKES, C. W., ANGERHOFER, P. E., NEWTON, R. N., and CLARK, A. F., Characterization of a Superconducting Coil Composite, *Interim Report* (prepared for Annapolis Laboratory, Nav. Ship Res. and Dev. Center by Cryogenic Div., N.B.S.), National Bureau of Standards, Boulder, Colorado, 1973.

[FRA61] FRASER, D. B., and HOLLIS HALLETT, A. C., The Coefficient of Linear Expansion and Grüneisen γ of Cu, Ag, Au, Fe, Ni, and Al from 4°K to 300°K, in *Proc. VIIth Int. Conf. Low Temperature Physics*, (Toronto, 1960), ed. by G. M. Graham *et al.*, North-Holland, Amsterdam, 1961, pp. 689-692.

[FRA77] FRANKEL, D. J., and GARWIN, E. L., Field Replication and Flux Shielding in Annular Superconductors, *IEEE Trans. Magn.* **MAG-13**, 205-208 (1977).

[FRE79] FREDERKING, T. H. K., personal communication.

[FRI56] FRIEDEL, J., On Some Electrical and Magnetic Properties of Metallic Solid Solutions, *Can. J. Phys.* **34**, 1190-1211 (1956).

[FRI63] FRIEDEL, J., DE GENNES, P. G., and MATRICON, J., Nature of the Driving Force in Flux Creep Phenomena, *Appl. Phys. Lett.* **2**, 119–121 (1963).

[FRI64] FRIEDEL, J., *Dislocations,* Pergamon Press, New York, 1964 (Transl. of *Les Dislocations,* Gauthier-Villars, Paris, 1956).

[FUL67] FULDE, P., and MOORMANN, W., Thermodynamical Properties of Superconducting Contacts, *Phys. kondens. Mater.* **6**, 403–415 (1967).

[FUL74] FULLER, E. R., GRANATO, A. V., HOLDER, J., and NAIMON, E. R., Ultrasonic Properties of Solids, in *Methods of Experimental Physics, Volume 11, Solid State Physics,* ed. by R. V. Coleman, Academic Press, New York, 1974, pp. 371–441.

G

[GAN66] GANDOLFO, D. A., and HARPER, C. M., Study of Properties of High-Field Superconductors at Elevated Temperatures, *Final Report No. NASA-CR-80154, N67-12913,* Oct., 1966, Radio Corporation of America, Camden, New Jersey.

[GAN67] GANDOLFO, D. A., Motion of Quantized Flux Lines in Hard Superconductors in the Presence of Time-Varying Magnetic Fields, *Ph.D. Thesis,* 1967, Temple University, Philadelphia, Pennsylvania.

[GAN68] GANDOLFO, D. A., DUBECK, L. W., and ROTHWARF, F., Effective Resistivity of a Hard Superconductor During Transmission of a Magnetic Field Step, *Solid State Commun.* **6**, 799–803 (1968).

[GAN69] GANDOLFO, D. A., DUBECK, L., and ROTHWARF, F., Steady-State Flux Jumping in Superconducting Niobium Titanium Tubes in Superimposed A.C. and D.C. Magnetic Fields, *J. Appl. Phys.* **40**, 2066–2070 (1969).

[GAR65] GARDNER, W. E., and PENFOLD, J., The Magnetic Susceptibility of Scandium, *Philos. Mag.* **11**, 549–559 (1965).

[GAR65[a]] GARDNER, W. E., PENFOLD, J., and TAYLOR, M. A., The Magnetic Susceptibility and Electrical Resistivity of Yttrium, *Proc. Phys. Soc., London,* **85**, 963–966 (1965).

[GAR68] GARLAND, J. W., BENNEMANN, K. H., and MUELLER, F. M., Effect of Lattice Disorder on the Superconducting Transition Temperature, *Phys. Rev. Lett.* **21**, 1315–1319 (1968).

[GAR72] GARLAND, J. W., and BENNEMANN, K. H., Theory for the Pressure Dependence of T_c for Narrow-Band Superconductors, in *Superconductivity in d- and f-Band Metals,* (Proc. AIP Conf., Rochester, New York, Oct. 1971) AIP Conf., Proc. No. 4, ed. by D. H. Douglass, Amer. Inst. Phys., New York, 1972, pp. 255–292.

[GAU69] GAUSTER, W. F., HORTON, J. L., KROEGER, D. M., LEICHSENRING, G. F., LUBELL, M. S., and MCDOWELL, J. D., Magnetic Flux Flow and Superconductor Stabilization, *Report No. ORNL-TM-2534,* April, 1969, Oak Ridge National Lab, Oak Ridge, Tennessee.

[GAU70] GAUSTER, W. F., IHARA, S., and LUBELL, M. S., Magnetic Flux Flow and Superconductor Stabilization, *Report No. ORNL-TM-2980,* May, 1970, Oak Ridge National Lab., Oak Ridge, Tennessee.

[GEA54] GEARY, P. J., Flexure Devices, *B.S.I.R.A. Research Report M. 18,* British Scientific Instrument Research Association, 1954.

[GEG73] GEGEL, H. L., HO, J. C., and COLLINGS, E. W., An Electronic Approach to Solid Solution Strengthening in Titanium Alloys, *Inst. Met. (London) Monogr. Rep. Ser.* **1**, PAP 116, 544–548 (1973).

[GEG73[a]] GEGEL, H. L., and HOCH, M., Thermodynamics of α-Stabilized Ti–X–Y Systems, in *Titanium Science and Technology,* ed. by R. I. Jaffee and H. M. Burte, Plenum Press, New York, 1973, pp. 923–933.

[GEI66] GEILIKMAN, B. T., and KRESIN, V. Z., Critical Temperature for Ordinary and Anomalous Superconductors, *Sov. Phys. Solid State* **7**, 2659-2663 (1966), [Transl. of *Fiz. Tverd. Tela* **7**, 3294-3301 (1965)].

[GIL76] GILMORE, C. M., FREEDMAN, M., and IMAM, M. A., The Relationship of Axial Strain Induced by Cyclic Torsion to Metal Stability and Fatigue Failure in Ti-6Al-4V, *Eng. Fract. Mech.* **8**, 9-15 (1976).

[GIL77] GILABERT, A., Effet de Proximité Entre un Métal Normal et un Supraconducteur, *Ann. Phys. (France)* **2**, 203-252 (1977).

[GIN50] GINZBURG, V. L., and LANDAU, L. D., On the Theory of Superconductivity, *Zh. Eksp. Teor. Fiz.* **20**, 1064-1082 (1950).

[GIT75] GITTINS, A., Fracture in Hot Working of Steel, *J. Aust. Inst. Met.* **20**, 184-200 (1975).

[GIT75[a]] GITTUS, J., *Creep, Viscoelasticity, and Creep Fracture in Solids*, John Wiley & Sons, New York, 1975.

[GOE67] GOEDEMOED, S. H., KES, P. H., JACOBS, F. T. A., and DEKLERK, D., The Irreversible Behavior of Superconducting Niobium Wires, *Physica (The Hague)* **35**, 273-289 (1967).

[GOL59] GOLDENSTEIN, A. W., METCALFE, A. G., and ROSTOKER, W., The Effect of Stress on the Eutectoid Decomposition of Titanium-Chromium Alloys, *Trans. ASM* **51**, 1036-1052 (1959).

[GOL69] GOLIK, V. R., PETRENKO, N. S., and POPOV, V. P., Coefficients of Thermal Expansion of the Niobium-Zirconium-Titanium Alloy and Organic Glass Between 10 and 300 K, *Ukr. Fiz. Zh.* **14**, No. 2, 294-296 (1969) [English transl. in *Ukrainian Phys. J.*].

[GOO61] GOODMAN, B. B., Simple Model for the Magnetic Behavior of Superconductors of Negative Surface Energy, *Phys. Rev. Lett.* **6**, 597-599 (1961).

[GOO62] GOODMAN, B. B., The Magnetic Behavior of Superconductors of Negative Surface Energy, *IBM J. Res. Dev.* **6**, 63-67 (1962).

[GOO64] GOODMAN, B. B., Type II or London Superconductors, *Rev. Mod. Phys.* **36**, 12-19 (1964).

[GOO66] GOODMAN, B. B., Type II Superconductors, *Rep. Prog. Phys.* **29**, 445-483 (1966).

[GOP66] GOPAL, E. S. R., *Specific Heats at Low Temperatures*, Plenum Press, New York, 1966.

[GOR58] GOR'KOV, L. P., On the Energy Spectrum of Superconductors, *Sov. Phys. JETP* **34**, 505-508 (1958), [Transl. of *Zh. Eksp. Teor. Fiz.* **23**, 402-407 (1957)].

[GOR59] GOR'KOV., L. P., Microscopic Derivation of the Ginzburg-Landau Equations in the Theory of Superconductivity, *Sov. Phys. JETP* **36**, 1364-1367 (1959), [Transl. of *Zh. Eksp. Teor. Fiz.* **36**, 1918-1923 (1959)].

[GOR60] GOR'KOV, L. P., The Critical Supercooling Field in Superconductivity Theory, *Sov. Phys. JETP* **10**, 593-599 (1960), [Transl. of *Zh. Eksp. Teor. Fiz.* **37**, 833-843 (1959)].

[GOR60[a]] GOR'KOV, L. P., Theory of Superconducting Alloys in a Strong Magnetic Field Near the Critical Temperature, *Sov. Phys. JETP* **10**, 998-1007 (1960), [Transl. of *Zh. Eksp. Teor. Fiz.* **37**, 1407-1416 (1959)].

[GOR64] GORTER, C. J., Some Remarks on Superconductivity of the Second Kind, *Rev. Mod. Phys.* **36**, 27-31 (1964).

[GOR69] GORINA, N. B., GRUZNOV, YU. A., IL'ICHEV, A. I., LOSINA, T. N., MATORIN, V. I., MIKHAILOV, N. N., PROKOSHIN, A. F., SOKOLOV, V. I., and FEDOTOV, L. N., Physico-Technological Research on Superconducting Alloy 65 BT, in *Precision Alloys* ed. by N. N. Potapov, Joint Publications Research Service, Washington, DC, 1969, pp. 18-30, [Transl. of *Pretsizionnyye Splavy* Moscow, USSR, Aug. 1968].

[GRA75] GRASSIE, A. D. C., *The Superconducting State*, Sussex University Press, Sussex, 1975.

[GRU26] GRÜNEISEN, E., Zustand des festen Körpers, in *Handbuch der Physik, Band X, Thermische Eigenschaften der Stoffe*, Springer, Heidelberg, 1926, pp. 1-59.

[GUB80] GUBSER, D. U., FRANCAVILLA, T. L., WOLF, S. A., and LEIBOWITZ, J. R., (eds.) *Inhomogeneous Superconductors—1979*, AIP Conf. Proc. No. 58, Amer. Inst. Phys., New York, 1980.

[GUE69] GUERTLER, W., GUERTLER, M., and ANASTASIADIAS, E., *A Compendium of Constitutional Ternary Diagrams of Metallic Systems*, WADC Tech. Rep. 58-615, Parts I, II, and III, Israel Program for Scientific Translations, 1969.

[GUL71] GULLBERG, R. B., TAGGART, R., and POLONIS, D. H., On the Decomposition of the Beta Phase in Titanium Alloys, *J. Mater. Sci.* **6**, 384-389 (1971).

[GUZ66] GUZEI, L. S., SOKOLOVSKAYA, E. M., and GRIGOR'EV, A. T., Phase Diagram of the Niobium-Titanium System, *Vestn. Mosk. Univ., Khim.* **21**, No. 5, 79-82 (1966).

[GYG64] GYGAX, S., OLSEN, J. L., and KROPSCHOT, R. H., Four Critical Fields in Superconducting Indium-Lead Alloys, *Phys. Lett.* **8**, 228-230 (1964).

H

[HAC71] HACKWORTH, J. V., HOCH, M., and GEGEL, H. L., Thermodynamics of Titanium Alloys: 1. Development of a Triple Knudsen Cell and its Use to Study the Activity of Copper in the Ti-Cu System, *Metall. Trans.* **2**, 1799-1805 (1971).

[HAE77] HAEBEL, E. U., and WITZELING, W., Differences in Shielding Versus Trapping Behaviour of Dynamically Stabilized NbTi Cylinders, *Cryogenics* **17**, 119-120 (1977).

[HAH70] HAHN, T. A., Thermal Expansion of Copper from 20 to 800 K—Standard Reference Material 736, *J. Appl. Phys.* **41**, 5096-5101 (1970).

[HAK61] HAKE, R. R., LESLIE, D. H., and BERLINCOURT, T. G., Electrical Resistivity, Hall Effect and Superconductivity of some b.c.c. Titanium-Molybdenum Alloys, *J. Phys. Chem. Solids* **20**, 177-186 (1961).

[HAK61[a]] HAKE, R. R., Specific Heats of Some Cubic Superconducting Titanium-Molybdenum Alloys Between 1.1 and 4.3 Degrees K, *Phys. Rev.* **123**, 1986-1994 (1961).

[HAK62] HAKE, R. R., LESLIE, D. H., and BERLINCOURT, T. G., Low-Temperature Resistivity Minima and Negative Magnetoresistivities in Some Dilute Superconducting Ti Alloys, *Phys. Rev.* **127**, 170-179 (1962).

[HAK63] HAKE, R. R., and LESLIE, D. H., High-Field Superconducting Properties of Ti-Mo Alloys, *J. Appl. Phys.* **34**, 270-276 (1963).

[HAK63[a]] HAKE, R. R., LESLIE, D. H., and RHODES, C. G., Giant Anisotropy in the High-Field Critical Currents of Cold-Rolled Transition Metal Alloy Superconductors, in *Low Temperature Physics—LT8*, (Proc. 8th Int. Conf. London, UK, 1962), ed. by R. O. Davies, Butterworths, London, 1963, pp. 342-344.

[HAK64] HAKE, R. R., and CAPE, J. A., Calorimetric Investigation of Localized Magnetic Moments and Superconductivity in Some Alloys of Titanium with Manganese and Cobalt, *Phys. Rev. A*: **135**, 1151-1160 (1964).

[HAK65] HAKE, R. R., Mixed-State Paramagnetism in High-Field Type-II Superconductors, *Phys. Rev. Lett.* **15**, 865-868 (1965).

[HAK67] HAKE, R. R., Magnetization and Resistive Behavior of Pauli-Paramagnetic Superconductors, in *Low Temperature Physics and Chemistry*, (Proc. 10th Int. Conf. Moscow, USSR, Aug.-Sept. 1966), Vol. IIA, *Superconductivity*, ed. by N. V. Zavaritskii *et al*, Viniti, Moscow, USSR, 1967, pp. 480-489.

[HAK67[a]] HAKE, R. R., Upper-Critical-Field Limits for Bulk Type-II Superconductors, *Appl. Phys. Lett.* **10**, 189–192 (1967); Erratum in *Appl. Phys. Lett.* **15**, 107 (1969).

[HAK67[b]] HAKE, R. R., Paramagnetic Superconductivity in Extreme Type-II Superconductors, *Phys. Rev.* **158**, 356–376 (1967).

[HAK69] HAKE, R. R., Thermodynamics of Type-I and Type-II Superconductors, *J. Appl. Phys.* **40**, 5148–5169 (1969).

[HAK69[a]] HAKE, R. R., Evidence for Fluctuation Superconductivity in Bulk Type-II Superconductors, *Phys. Rev. Lett.* **23**, 1105–1108 (1969).

[HAK70] HAKE, R. R., Apparent High-Magnetic-Field Fluctuation Superconductivity in the 2–3 T_c Range, *Phys. Lett.* **32A**, 143–144 (1970).

[HAK71] HAKE, R. R., High-Field Fluctuation Superconductivity in a Bulk Extreme-Type-II Superconductor, *Physica* (*Utrecht*) **55**, 311–316 (1971).

[HAK74] HAKE, R. R., Evidence for Magnetic-Field-Induced Reduction of the Fluctuation Dimensionality in Bulk Type-II Superconductors Just Above the Upper Critical Field, H_{c2}, in *Low Temperature Physics—LT13, Vol. 3: Superconductivity*, (Proc. 13th. Int. Conf. Low Temperature Physics, Boulder, Colorado, 1972) ed. by K. D. Timmerhaus *et al.*, Plenum Press, New York 1974, pp. 638–641.

[HAK75] HAKE, R. R., MONTGOMERY, A. G., and LUE, J. W., Low-Temperature Anomalies in the Normal-State Resistivity of Some Disordered Superconducting Transition-Metal Alloys, in *Low Temperature Physics—LT14* (Proc. 14th Int. Conf. Low Temperature Physics, Otaniemi, Finland, 1975), ed. by M. Krusius *et al.*, American Elsevier, New York, 1975, pp. 122–125.

[HAM69] HAMMOND, C., and KELLY, P. M., The Crystallography of Titanium Alloy Martensites, *Acta Metall.* **17**, 869–882 (1969).

[HAM70] HAMMOND, C., and KELLY, P. M., Martensitic Transformations in Titanium Alloys, in *The Science, Technology and Application of Titanium*, ed. by R. I. Jaffee and N. E. Promisel, Pergamon Press, London, 1970, pp. 659–676.

[HAM72] HAMPSHIRE, R. G., and TAYLOR, M. T., Critical Supercurrents and the Pinning of Vortices in Commercial Nb-60at.%Ti, *J. Phys. F: Metal. Phys.* **2**, 89–106 (1972).

[HAM74] HAMPSHIRE, R. G., The Critical Current Density of Nb-60at.%Ti and Nb-25at.%Zr Superconductors in Small Magnetic Fields, *J. Phys. D: Appl. Phys.* **7**, 1847–1863 (1974).

[HAN51] HANSEN, M., KAMEN, E. L., KESSLER, H. D., and MCPHERSON, D. J., Systems Titanium-Molybdenum and Titanium-Columbium, *Trans. TMS-AIME* **191**, 881–888 (1951).

[HAN54] HANKE, H., *Prüfung Metallischer Werkstoffe*, VEB Verlag Technik, Berlin, 1954.

[HAN58] HANSEN, M., and ANDERKO, K., *Constitution of Binary Alloys*, Second Edition, McGraw-Hill, New York, 1958.

[HAN66] HANCOX, R., The Paramagnetic Limit in Type-II Superconductors, *Report No. CLM-R 60*, 1966, UK Atomic Energy Authority, Abingdon, UK.

[HAR63] HARDEN, J. L., and ARP, V., The Lower Critical Field in the Ginzburg–Landau Theory of Superconductivity, *Cryogenics* **3**, 105–108 (1963).

[HAT68] HATT, B. A., and RIVLIN, V. G., Phase Transformations in Superconducting Ti–Nb Alloys, *J. Phys. D: Appl. Phys.* **1**, 1145–1149 (1968).

[HAU64] HAUSER, J. J., THEUERER, H. C., and WERTHAMER, N. R., Superconductivity in Cu and Pt by Means of Superimposed Films with Lead, *Phys. Rev. A*: **136**, 637–641 (1964).

[HAU66] HAUSER, J. J., THEUERER, H. C., and WERTHAMER, N. R., Proximity Effects Between Superconducting and Magnetic Films, *Phys. Rev.* **142**, 118–126 (1966).

[HAW80] HAWKSWORTH, D. G., and LARBALESTIER, D. C., Enhanced Values of B_{c2} in Nb–Ti Ternary and Quaternary Alloys, in *Advances in Cryogenic Engineering*

(*Materials*), Vol. 26, ed. by A. F. Clark and R. P. Reed, Plenum Press, New York. 1980, pp. 479–486.

[HAW80[a]] HAWKSWORTH, D. G., and LARBALESTIER, D. C., The High Field J_c and Scaling Behavior in Nb–Ti and Alloyed Nb–Ti Superconductors, in *Proc. of the 8th Symposium on Engineering Problems of Fusion Research, Pt. I*, San Francisco, CA, 13–16 Nov. 1979, IEEE, 1980, pp. 249–254.

[HAW81] HAWKSWORTH, D. G., and LARBALESTIER, D. C., Further Investigations of the Upper Critical Field and the High Field Critical Current Density in Nb–Ti and its Alloys, *IEEE Trans. Magn.* **MAG-17**, 49–52 (1981).

[HAY65] HAYDEN, H. W., MOFFATT, W. G., and WULFF, J., *The Structure and Properties of Materials, Volume 3—Mechanical Behavior*, John Wiley & Sons, New York, 1965.

[HEA46] HEARMON, R. F. S., The Elastic Constants of Anisotropic Materials, *Rev. Mod. Phys.* **18**, 409–436 (1946).

[HEC65] HECHT, R., and HARPER, C. M., Study of Properties of High-Field Superconductors at Elevated Temperatures, *Report No. NASA-CR-67157, N65-34464*, July, 1965, Defense Electronic Products, Radio Corp. of America, Camden, New Jersey.

[HEI64] HEINIGER, F., and MÜLLER, J., Bulk Superconductivity in Dilute Hexagonal Titanium Alloys, *Phys. Rev.* **134**, 1407–1409 (1964).

[HEI66] HEINIGER, F., BUCHER, E., and MÜLLER, J., Low Temperature Specific Heat of Transition Metals and Alloys, *Phys. Kondens. Mater.* **5**, 243–284 (1966).

[HEI74] HEIM, J. R., Superconducting Coil Training and Instabilities Due to the Bauschinger Effect, *Fermi National Accelerator Lab. Tech. Memo. TM-334*, Fermi National Acceleratory Laboratory, Batavia, Illinois, 1974.

[HEI74[a]] HEIDEN, C., On the Nature of Flux Transport Noise, in *Low Temperature Physics—LT-13, Volume 3: Superconductivity*, (Proc. 13th Int. Conf. Low Temperature Physics, Boulder, Colorado, 1972) ed. by K. D. Timmerhaus *et al.*, Plenum Press, New York, 1974, pp. 75–78.

[HEL64] HELFAND, E., and WERTHAMER, N. R., Temperature and Purity Dependence of the Superconducting Critical Field, H_{c2}, *Phys. Rev. Lett.* **13**, 686–690 (1964).

[HEL66] HELFAND, E., and WERTHAMER, N. R., Temperature and Purity Dependence of the Superconducting Critical Field, H_{c2}, II, *Phys. Rev.* **147**, 288–294 (1966).

[HEL71] HELLER, W., Über den Einfluss von Germaniumzusätzen auf supraleitende Eigenschaften und Umwandlungsvorgänge in technischen Titan–Niob-Legierungen, *D. Eng. Dissertation*, 1971, Universität Erlangen-Nürnberg, Erlangen, West Germany.

[HEL71[a]] HELLER, W., LÖHBERG, R., and BABISKIN, J., Einfluss von Zusätzen dritter Elemente auf das kritische Magnetfeld von Titan–Niob-Legierungen, *J. Less-Common Metals* **24**, 265–276 (1971).

[HEL72] HELLER, W., LÖHBERG, R., and ZWICKER, U., Einfluss der Verankerungskräfte für Flussfäden auf die kritische Stromdichte in Magnetfeldern bis 10 Tesla bei supraleitenden Titan–Niob-Legierungen, *Z. Metallkd.* **63**, 735–739 (1972).

[HIC68] HICKMAN, B. S., Precipitation of the Omega Phase in Titanium-Vanadium Alloys, *J. Inst. Met.* **96**, 330–337 (1968).

[HIC69] HICKMAN, B. S., The Formation of Omega Phase in Titanium and Zirconium Alloys: A Review, *J. Mater. Sci.* **4**, 554–563 (1969).

[HIC69[a]] HICKMAN, B. S., Omega Phase Precipitation in Alloys of Titanium with Transition Metals, *Trans. TMS-AIME* **245**, 1329–1335 (1969).

[HIL61] HILLERT, M., A Solid-Solution Model for Inhomogeneous Systems, *Acta. Metall.* **9**, 525–535 (1961).

[HIL62] HILSCH, P., Zum Verhalten von Supraleitern im Kontakt mit Normalleitern, *Z. Phys.* **167**, 511–524 (1962).

[HIL67] HILL, R., *The Mathematical Theory of Plasticity*, Oxford University Press, Oxford, 1967.

[HIL73] HILLMANN, H., Entwicklung harter Supraleiter, vorzugsweise am Beispiel Nb-Ti, Teil I, *Metall (Berlin)* **27**, 797-805 (1973).

[HIL73a] HILLMANN, H., Werkstoffe für supraleitende Wechselfeldmagnete mit Ummagnetisierungszeiten der Grössenordnung Sekunde, *Forschungsbericht T 73-03*, Vacuumschmelze GmbH, Hanau, Germany, April 1973.

[HIL74] HILLMANN, H., Ausscheidungen und Flussverankerung bei Hochfeldsupraleitern aus Niobtitan, *Siemens Forsch. u. Entwickl. Bericht* **3**, 197-204 (1974).

[HIL76] HILLMANN, H., Werkstoffe für dynamisch beanspruchbare supraleitende Magnete, *Forschungsbericht T 76-13*, Vacuumschmelze GmbH, Hanau, Germany, June 1976.

[HIL79] HILLMANN, H., BEST, K. J., HOEFLICH, H., RUDOLPH, J., PFEIFFER, I., and WEBER, H., Hochfeldsupraleiter aus NbTi mit Stromfähigkeiten über 3000 A fur die Anwendung in Magnetsystem mit Wechselfeldkomponenten, *Forschungsbericht NT 1022*, Vacuumschmelze GmbH, Hanau, Germany, April 1979.

[HIL79a] HILLMANN, H., personal communication.

[HIR62] HIRSHFELD, A. T., LEUPOLD, H. A., and BOORSE, H. A., Superconducting and Normal Specific Heats of Niobium, *Phys. Rev.* **127**, 1501-1507 (1962).

[HO69] HO, J. C., GEHLEN, P. C., COLLINGS, E. W., and JAFFEE, R. I., Phase Stability and Solution Strengthening of Solid Solution Phase Titanium Alloys, *Tech. Rep. AFML-TR-70-1*, Sept., 1969, Battelle Memorial Institute, Columbus, Ohio.

[HO69a] HO, J. C., and COLLINGS, E. W., The Influence of ω-Phase on the Superconductivity in Ti-Mo Alloys, *Phys. Lett. A*: **29**, 206-207 (1969).

[HO71] HO, J. C., and COLLINGS, E. W., Enhancement of the Superconducting Transition Temperatures of Ti-Mo(5, 7 at.%) Alloys By Mechanical Deformation, *J. Appl. Phys.* **42**, 5144-5150 (1971).

[HO71a] HO, J. C., BOYD, J. D., and COLLINGS, E. W., Bulk Superconductivity in a Non-Homogeneous Ti-Mo(10 at.%) Alloy, in *Low Temperature Physics—LT12* (Proc. 12th Int. Conf., Kyoto, Japan, Sept. 1970), ed. by E. Kanda, Academic Press of Japan, Tokyo, 1971, p. 366.

[HO72] HO, J. C., and COLLINGS, E. W., Anomalous Electrical Resistivity in Titanium-Molybdenum Alloys, *Phys. Rev. B*: **6**, 3727-3738 (1972).

[HO73] HO, J. C., and COLLINGS, E. W., Physics of Titanium Alloys—I: Alloying and Microstructural Effects in the Superconductivity of Ti-Mo, in *Titanium Science and Technology*, ed. by R. I. Jaffee and H. M. Burte, Plenum Press, New York, 1973, pp. 815-830.

[HO74] HO, J. C., and COLLINGS, E. W., Calorimetric Studies of Superconductive Proximity Effects in a Two-Phase Ti-Fe(7.5 at.%) Alloy, in *Low Temperature Physics—LT13, Volume 3: Superconductivity* (Proc. 13th Int. Conf., Boulder, Colorado, 1972), ed. by K. D. Timmerhaus *et al.*, Plenum Press, New York, 1974, pp. 403-407.

[HOC71] HOCH, M., and VISWANATHAN, R., Thermodynamics of Titanium Alloys: III, The Ti-Mo System, *Metall. Trans.* **2**, 2765-2767 (1971).

[HOC81] HOCHSTUHL, P., NbTi65—Strukturelle Instabilitäten und Training, *KfK. Report No. 3141*, July, 1981, Kernforschungszentrum Karlsruhe, Karlsruhe, Germany.

[HOC82] HOCHSTUHL, P., and OBST, B., Beta-Phase Instability and Martensitic Transformation in Ti-22a/oNb Alloy, *J. Phys. (Paris), Suppl. to No. 12*, **43**, pp. C4-133 to C4-138 (1982).

[HOP71] HOPFIELD, J. J., On the Systematics of High T_c in Transition Metal Materials, *Physica (Utrecht)* **55**, 41-49 (1971).

[HOR73] HORIUCHI, T., MONJU, Y., and NAGAI, N., Superconducting Transition Temperatures and Resistive Critical Fields of Superconducting Ti-Nb-Zr-Ta Alloys, *Nippon Kinzoku Gakkaishi* **37**, 882-887 (1973).

[HOR73[a]] HORIUCHI, T., MONJU, Y., TATARA, I., and NAGAI, N., Phase Transformations of Ti-30Nb-30Zr-7Ta Superconducting Alloy, *Nippon Kinzoku Gakkaishi* **37**, 1057-1064 (1973).

[HOR74] HORIUCHI, T., MONJU, Y., and NAGAI, N., Supraleitende Legierung, *German Patent No. 2,350,199* (July 1974).

[HOR75] HORIUCHI, T., MONJU, Y., TATARA, I., and NAGAI, N., Verfahren zur Behandlung einer supraleitenden Legierung, *German Patent 2,347,400* (April 1975).

[HOR80] HORIUCHI, T., MATSUMOTO, K., and MONJU, Y., Effect of Ta and Zr Additions to Ti-Nb Alloys on Superconducting Properties, in *Proceedings of the 8th Symposium on Engineering Problems of Fusion Research, Pt. I*, San Francisco, CA, 13-16 Nov., 1979, IEEE, 1980, pp. 274-277.

[HOR80[a]] HORIUCHI, T., personal communication.

[HOU70] HOUGHTON, R. W., and SARACHIK, M. P., Anomalous Electrical Resistivity and the Existence of Giant Magnetic Moments in Ni-Cu Alloys, *Phys. Rev. Lett.* **25**, 238-239 (1970).

[HSI80] HSIANG, T. Y., and FINNEMORE, D. K., Magnetic Field Dependence of J_c in SNS Junctions, *Solid State Commun.* **33**, 847-848 (1980).

[HUE79] HUEBENER, R. P., *Magnetic Flux Structures in Superconductors*, Springer-Verlag, New York, 1979.

[HUL61] HULM, J. K., and BLAUGHER, R. D., Superconducting Solid Solution Alloys of the Transition Elements, *Phys. Rev.* **123**, 1569-1580 (1961).

[HUL64] HULBERT, J. A., The Temperature Dependence of the Paramagnetically Limited Critical Field in a Superconductor of the Second Kind, *Phys. Lett.* **8**, 34-35 (1964).

[HUL72] HULM, J. K., and BLAUGHER, R. D., Transition-Metal Superconductors—Experimental Survey, in *Superconductivity in d- and f-Band Metals* ed. by D. H. Douglass, *AIP Conf. Proc. No. 4*, Amer. Inst. Phys., New York, 1972, pp. 1-16.

I

[IKE74] IKEBE, M., KOBAYASHI, N., and MUTO, Y., Phonon Thermal Conductivity in Superconducting Ta and Ta-Nb Alloys below 1K, *J. Phys. Soc. Jpn.* **37**, 278 (1974).

[IKE74[a]] IKEBE, M., KOBAYASHI, N., and MUTO, Y., Lattice Thermal Conductivity of Superconducting Nb-Mo and Nb-Ta Alloys Below 1K, *Phys. Lett. A*: **47**, 277-278 (1974).

[IKE77] IKEBE, M., NAKAGAWA, S., HIRAGA, K., and MUTO, Y., Anomalous Phonon Thermal Resistivity in Superconducting $Ti_{55}Nb_{45}$ Alloys, *Solid State Commun.* **23**, 189-192 (1977).

[ILI80] ILIYEV, L. B., POPOV, V. P., and PERKAKOV, V. A., Low-Temperature Thermal Expansion of Polycrystalline Titanium, *Phys. Met. Metallogr.* **47**, No. 3, 189-191 (1980), [Transl. of *Fiz. Met. Metalloved* **47**, 668-670 (1979)].

[IMG61] IMGRAM, A. G., WILLIAMS, D. N., WOOD, R. A., OGDEN, H. R., and JAFFEE, R. I., Metallurgical and Mechanical Characteristics of High-Purity Titanium-Base Alloys, *WADC Technical Report 59-595, Part II*, March, 1961.

[IMP65] IMPERIAL METAL INDUSTRIES (KYNOCH) LTD., Superconducting Alloys, *Belgium Patent No. 659,033* (July 1965).

[ING48] INGERSOLL, L. R., ZOBEL, O. J., and INGERSOLL, A. C., *Heat Conduction, with Engineering and Geological Applications*, McGraw-Hill Book Company, New York, 1948.

[ING69] INGLESFIELD, J. E., Perturbation Theory and Alloying Behavior I. Formalism, *J. Phys. C: Solid State Phys.* **2**, 1285-1292 (1969).

[INH73] INHABER, H., Models of Magnetic Hysteresis in Type II Superconductors, *Cryogenics* **13**, 261-271 (1973).

[INO80] INOUE, A., KIMURA, H. M., MASUMOTO, T., SURYANARAYANA, C., and HOSHI, A., Superconductivity of Ductile Ti-Nb-Si Amorphous Alloys, *J. Appl. Phys.* **51**, 5475-5482 (1980).

[INO82] INOUE, K., WADA, H., KURODA, T., and TACHIKAWA, K., Superconducting Properties of V-Ti-Ta Ternary Alloys, *Proceedings of the International Cryogenic Materials Conference, Kobe, Japan, 1982*, ed. by K. Tachikawa and A. F. Clark, Butterworths, London, pp. 57-60.

[IRI67] IRIE, F., and YAMAFUJI, K., Theory of Flux Motion in Non-Ideal Type II Superconductors, *J. Phys. Soc. Japan* **23**, 255-268 (1967).

[ISH68] ISHIDA, F., DOI, T., and UMEZAWA, T., The Effect of Precipitation on the Critical Current Density of Nb-5 at.% Zr-60 at.% Ti Alloy in Magnetic Field, *Nippon Kinzoku Gakkaishi* **32**, 893-897 (1968).

[IWA77] IWASA, Y., LEUPOLD, M. J., and WILLIAMS, J. E. C., Stabilization of Large Superconducting Magnets: Experimental Models, *IEEE Trans. Magn.* **MAG-13**, 20-23 (1977).

J

[JAC67] JACOBS, A. E., Calculation of the Superconducting Transition Temperature of Superimposed Films in the Diffusion Approximation, *Phys. Rev.* **162**, 375-382 (1967).

[JAC68] JACOBS, A. E., and GINSBERG, D. M., Experimental Test of the Theory of the Proximity Effect in Superconductors, *Phys. Rev.* **175**, 569-575 (1968).

[JAC69] JACKSON, J., Transient Heat Transfer and Stability of Superconducting Composites, *Cryogenics* **9**, 103-105 (1969).

[JAC72] JACOBS, S. F., BERTHOLD, J. W., III, and OSMUNDSEN, J., Ultraprecise Measurement of Thermal Expansion Coefficients—Recent Progress, in *Thermal Expansion—1971*, AIP Conf. Proc. No. 3, ed. by M. G. Graham and H. E. Hagy, Amer. Inst. Phys., New York, 1972, pp. 1-12.

[JAE40] JAEGER, J. C., Magnetic Screening by Hollow Circular Cylinders, *Philos. Mag.* **29**, 18-31 (1940).

[JAE62] JAEGER, J. C., *Elastic Fracture and Flow*, second edition, Methuen and Co., New York, 1962.

[JAF70] JAFFEE, R. I., and PROMISEL, N. E., (eds.) *The Science, Technology, and Application of Titanium* (Proc. First Int. Conf. on Titanium, London, England), Pergamon Press, London, 1970.

[JAF73] JAFFEE, R. I., and BURTE, H. M., (eds.) *Titanium Science and Technology* (Proc. Second Int. Conf. on Titanium, Boston, U.S.A.), Plenum Press, New York, 1973.

[JAY63] JAYARAMAN, A., KLEMENT, W., and KENNEDY, G. C., Solid-Solution Transitions in Titanium and Zirconium at High Pressures, *Phys. Rev.* **131**, 644-649 (1963).

[JEL74] JELINEK, F. J., Thermal Expansion Data, in *Materials Research for Superconducting Machinery*, ed. by A. F. Clark, R. P. Reed, and E. C. van Reuth, *Part II, AD-A004 586*, 1974, pp. 57-70.

[JEL75] JELINEK, F. J., Thermal Expansion Data, in *Materials Research for Superconducting Machinery*, ed. by A. F. Clark, R. P. Reed, and E. C. van Reuth, *Part IV, AD-A019 230*, 1975, pp. 24-65.

[JEN65] JENSEN, M. A., MATTHIAS, B. T., and ANDRES, K., Electron Density and Electronic Properties in Noble-Metal Transition Elements, *Science* **150**, 1448-1450 (1965).

[JEP70] JEPSON, K. S., BROWN, A. R. G., and GRAY, J. A., The Effect of Cooling Rate on the Beta Transformation in Titanium-Niobium and Titanium-Aluminum Alloys, in *The Science, Technology, and Application of Titanium*, ed. by R. I. Jaffee and N. E. Promisel, Pergamon Press, London, 1970, pp. 677-690.

[JOH75] JOHNSON, W. L., POON, S. J., and DUWEZ, P., Amorphous Superconducting Lanthanum-Gold Alloys Obtained by Liquid Quenching, *Phys. Rev. B*: **11**, 150-154 (1975).

[JOH79] JOHNSON, W. L., Superconductivity and Electronic Properties of Amorphous Transition Metal Alloys, *J. Appl. Phys.* **50**, 1557-1563 (1979).

[JOI64] JOINER, W. C. H., and BLAUGHER, R. D., Magnetic and Resistive Transitions of Some Mo-Re Alloys, *Rev. Mod. Phys.* **36**, 67-69 (1964).

[JON64] JONES, C. K., HULM, J. K., and CHANDRASEKHAR, B. S., Upper Critical Field of Solid Solution Alloys of the Transition Elements, *Rev. Mod. Phys.* **36**, 74-76 (1964).

[JON65] JONES, R. G., RHODERICK, E. H., and ROSE-INNES, A. C., On the Origin of Resistance in the Mixed State of Type II Superconductors, *Phys. Lett.* **15**, 214-215 (1965).

[JOS65] JOSEPHSON, B. D., Potential Differences in the Mixed State of Type II Superconductors, *Phys. Lett.* **16**, 242-243 (1965).

K

[KAD71] KADYKOVA, G. N., Phase Transformations in Superconducting Niobium-Titanium Alloys, *Russ. Metall.* **4**, 160-162 (1971), [Transl. of *Izvest. Akad. Nauk SSSR, Met.* **4**, 226-230 (1971)].

[KAR72] KARASIK, V. R., VASIL'EV, N. G., and VYSOTSKIĬ, V. S., Surface Superconductivity of the Ti-22 at.% Nb Alloy, *Sov. Phys. JETP* **35**, 945-949 (1972), [Transl. of *Zh. Eksp. Teor. Fiz.* **62**, 1818-1826 (1972)].

[KAT75] KATOH, S., TSUBAKIHARA, H., OKADA, T., and SUITA, T., Influence of Precipitates to Superconducting Properties of Nb-Ti Alloy in Connection with Hardness, *J. Nucl. Sci. Technol.* **12**, 193 (1975).

[KAT79] KATAHARA, K. W., MANGHNANI, M. H., and FISHER, E. S., Pressure Derivatives of the Elastic Moduli of BCC Ti-V-Cr, Nb-Mo, and Ta-W Alloys, *J. Phys. F: Metal Phys.* **9**, 773-790 (1979).

[KAT79[a]] KATAHARA, K. W., NIMALENDRAN, M., MANGHNANI, M. H., and FISHER, E. S., Elastic Moduli of Paramagnetic Chromium and Ti-V-Cr Alloys, *J. Phys. F: Metal Phys.* **9**, 2167-2176 (1979).

[KAU70] KAUFMAN, L., and BERNSTEIN, H., *Computer Calculations of Phase Diagrams*, Academic Press, New York, 1970.

[KAU73] KAUFMAN, L., and NESOR, H., Phase Stability and Equilibria as Affected by the Physical Properties and Electronic Structure of Titanium Alloys, in *Titanium Science and Technology*, ed. by R. I. Jaffee and H. M. Burte, Plenum Press, New York, 1973, pp. 773-800.

[KAU83] KAUFMAN, L., CROOKS, L. E., DAVIS, P., and MARGULIS, A. R., Advances in Imaging Technology: Nuclear Magnetic Resonance, *IEEE Trans. Magn.* **MAG-19**, 516-521 (1983).

[KEA74] KEATING, D. T., and LAPLACA, S. J., Neutron Diffraction Determination of the Number and Displacement of the Atoms in the Diffuse ω-Phase of $Zr_{0.8}Nb_{0.2}$, *J. Phys. Chem. Solids* **35**, 879-891 (1974).

[KER69] KERNER, E. H., The Elastic and Thermo-elastic Properties of Composite Media, *Proc. Phys. Soc., London, Sect. B*: **69**, 808-813 (1956).

[KHA83] KHAN, H. R., Microstructure Related Superconducting and Mechanical Properties of Some Nb-Ti Alloys with Additions of Ta and Zr, *IEEE Trans. Magn.* **MAG-19**, 745-748 (1983).

[KIM62] KIM, Y. B., HEMPSTEAD, C. F., and STRNAD, A. R., Critical Persistent Currents in Hard Superconductors, *Phys. Rev. Lett.* **9**, 306-309 (1962).

[KIM63] KIM, Y. B., HEMPSTEAD, C. F., and STRNAD, A. R., Magnetization and Critical Supercurrents, *Phys. Rev.* **129**, 528-535 (1963).

[KIM63[a]] KIM, Y. B., HEMPSTEAD, C. F., and STRNAD, A. R., Flux Creep in Hard Superconductors, *Phys. Rev.* **131**, 2486-2495 (1963).

[KIM64] KIM, Y. B., Superconducting Magnets and Hard Superconductivity, *Physics Today* **17**, 21-30 (Sept. 1964).

[KIM64[a]] KIM, Y. B., HEMPSTEAD, C. F., and STRNAD, A. R., Resistive States of Hard Superconductors, *Rev. Mod. Phys.* **36**, 43-45 (1964).

[KIM65] KIM, Y. B., HEMPSTEAD, C. F., and STRNAD, A. R., Flux-Flow Resistance in Type-II Superconductors, *Phys. Rev. A*: **139**, 1163-1173 (1965).

[KIM66] KIM, Y. B., and STRNAD, A. R., Temperature Dependence of Upper Critical Fields of Type II Superconductors, in *Quantum Fluids*, ed. by D. F. Brewer, North-Holland, Amsterdam, and John Wiley & Sons, New York, 1966, pp. 68-73.

[KIM67] KIM, Y. B., Transport Phenomena in Type-II Superconductors, in *Low Temperature Physics and Chemistry* (Proc. 10th Int. Conf. Moscow, USSR, Aug-Sept. 1966), Vol. IIA, *Superconductivity*, ed. by N. V. Zavaritskii *et al.* Viniti, Moscow, USSR, 1967, pp. 43-58.

[KIM69] KIM, Y. B., and STEPHEN, M. J., Flux Flow and Irreversible Effects, in *Superconductivity*, ed. by R. D. Parks, Marcel Dekker, New York, 1969, pp. 1107-1162.

[KIM80] KIMURA, H., and IZUMI, O. (eds), *Titanium '80: Science and Technology* (Proc. Fourth Int. Conf. on Titanium, Kyoto, Japan), The Metallurgical Society of AIME, Warrendale, Pennsylvania, 1980.

[KIN64] KINSEL, T., LYNTON, E. A., and SERIN, B., Magnetic and Thermal Properties of Second-Kind Superconductors. I. Magnetization Curves, *Rev. Mod. Phys.* **36**, 105-107 (1964).

[KIR70] KIRKPATRICK, S., VELICKÝ, B., and EHRENREICH, H., Paramagnetic Ni-Cu Alloys: Electronic Density of States in the Coherent-Potential Approximation, *Phys. Rev. B*: **1**, 3250-3263 (1970).

[KIT49] KITTEL, C., Interpretation of the Thermal Conductivity of Glasses, *Phys. Rev.* **75**, 972-974 (1949).

[KIT68] KITTEL, C., *Introduction to Solid State Physics*, 3rd Edition, John Wiley & Sons, New York, 1968.

[KIT69] KITADA, M., and DOI, T., Precipitation Phenomena of Nb-60Ti-5Zr Superconducting Alloy, *Nippon Kinzoku Gakkaishi* **33**, 1115-1121 (1969).

[KIT70] KITADA, M., and DOI, T., Relation Between the Anisotropy of Applied Magnetic Field-Critical Current Density Characteristics and Microstructures for Superconducting Nb-40Zr-10Ti Alloy Sheets, *Nippon Kinzoku Gakkaishi* **34**, 5-10 (1970).

[KIT70[a]] KITADA, M., and DOI, T., Discontinuous Precipitation of Solution Treated Nb-40Zr-10Ti Superconducting Alloy, *Nippon Kinzoku Gakkaishi* **34**, 361-365 (1970).

[KIT70[b]] KITADA, M., and DOI, T., Discontinuous Precipitation of Cold Worked Nb-40Zr-10Ti Superconducting Alloy, *Nippon Kinzoku Gakkaishi* **34**, 365-369 (1970).

[KIT70[c]] KITADA, M., and DOI, T., Precipitation and Superconducting Properties of Nb-40Zr-10Ti Alloy, *Nippon Kinzoku Gakkaishi* **34**, 369-374 (1970).

[KIT70[d]] KITADA, M., and DOI, T., Effects of V, Mo or Ta Additions on Superconducting Properties of Nb-40Zr-10Ti Alloy, *Nippon Kinzoku Gakkaishi* **34**, 1082-1086 (1970).

[KIT72] KITADA, M., Relation Between the Applied Magnetic Field *vs.* Critical Current Density Characteristics and the Microstructures for Superconducting Nb-Ti-Zr Sheets, *Nippon Kinzoku Gakkaishi* **36**, 1064-1069 (1972).

[KIT73] KITADA, M., Superconductivity of Nb-Ti-Ge Alloys, *Nippon Kinzoku Gakkaishi* **37**, 104-109 (1973).

[KLA75] KLAFFKY, R. W., MOHAN, N. S., and DAMON, D. H., Lattice Thermal Conductivity of Deformed and Annealed Aluminum Alloys in the Temperature Range 1.3-60K, *Phys. Rev. B*: **11**, 1297-1307 (1975).

[KLE56] KLEMENS, P. G., Thermal Conductivity of Solids at Low Temperatures, in *Encyclopedia of Physics*, Vol. 14, ed. by S. Flügge, Springer, Berlin, 1956, pp. 198-281.

[KLE58] KLEMENS, P. G., Thermal Conductivity and Lattice Vibrational Modes, in *Solid State Physics*, Vol. 7, ed. by F. Seitz and D. Turnbull, Academic Press, New York, 1958, pp. 1-96.

[KLE64] KLEMENS, P. G., and TEWORDT, L., Reduction of the Lattice Thermal Conductivity of Superconductors Due to Point Defects, *Rev. Mod. Phys.* **36**, 118-120 (1964).

[KLE64[a]] KLEINER, W. H., ROTH, L. M., and AUTLER, S. H., Bulk Solution of Ginzburg-Landau Equations for Type II Superconductors: Upper Critical Field Region, *Phys. Rev. A*: **133**, 1226-1227 (1964).

[KOB78] KOBE STEEL LTD., High Critical Current Density High Tensile Strength Superconductor, *Technical Data, Kobe Steel Ltd., Asada Research Lab.* (Oct. 1978).

[KOB80] KOBE STEEL LTD., High Critical Current Density High Tensile Strength Superconductor, *Technical Data, Kobe Steel Ltd., Asada Research Lab.* (May 1980).

[KOC77] KOCH, C. C., and EASTON, D. S., A Review of Mechanical Behavior and Stress Effects in Hard Superconductors, *Cryogenics* **17**, 391-413 (1977).

[KOS69] KOS, J. F., and LAMARCHE, J. L. G., Thermal Expansion of the Noble Metals Below 15°K, *Can. J. Phys.* **47**, 2509-2518 (1969).

[KOU70] KOUL, M. K., and BREEDIS, J. F., Phase Transformations in Beta Isomorphous Titanium Alloys, *Acta Metall.* **18**, 579-588 (1970).

[KOU70[a]] KOUL, M. K., and BREEDIS, J. F., Reply to Comments on "Phase Transformations in Beta Isomorphous Titanium Alloys," *Scri. Metall.* **4**, 877-880 (1970).

[KOU70[b]] KOUL, M. K., and BREEDIS, J. F., Strengthening of Titanium Alloys by Shock Deformation, in *The Science, Technology, and Application of Titanium* ed. by R. I. Jaffee and N. E. Promisel, Pergamon Press, London, 1970, pp. 817-828.

[KRA67] KRAMER, D., and RHODES, C. G., Omega Phase Precipitation and Superconducting Critical Transport Currents in Titanium-22at.% Niobium (Columbium), *Trans. TMS-AIME* **239**, 1612-1615 (1967).

[KRA67[a]] KRAMER, E. J., and BAUER, C. L., First Order Dislocation-Magnetic Fluxoid Interactions, *Philos. Mag.* **15**, 1189-1199 (1967).

[KRE75] KRESIN, V. Z., and PARKHOMENKO, V. P., Thermodynamic Properties of Strong-Coupling Superconductors, *Sov. Phys. Solid State* **16**, 2180-2184 (1975), [Transl. of *Fiz. Tverd. Tela* **16**, 3363-3370 (1974)].

[KRI54] KRIESSMAN, C. J., and CALLEN, H. B., The Magnetic Susceptibility of the Transition Elements, *Phys. Rev.* **94**, 837-844 (1954).

[KRI79] KRISHNAN, R. S., SRINIVASAN, R., and DEVANARAYANAN, S., *Thermal Expansion of Crystals*, Pergamon Press, New York, 1979.

[KRO66] KROEGER, D. M., Flux Motion in High-Field Type-II Superconducting Cylinders, *Ph.D. Thesis*, 1966, Vanderbilt University, Nashville, Tennessee.

[KRO73] KROEGER, D. M., KOCH, C. C., and COGHLAN, W. A., Measurement of Critical Bulk and Surface Current Densities in Type II Superconductors in the Mixed State, *J. Appl. Phys.* **44**, 2391-2396 (1973).

[KRO75] KROEGER, D. M., KOCH, C. C., and CHARLESWORTH, J. P., A Comparison of Methods for Measuring Flux Gradients in Type II Superconductors, *J. Low Temp. Phys.* **19**, 493–512 (1975).

[KRO77] KROEGER, D. M., EASTON, D. S., and MOAZED, A., Stress-Induced Heating in Commercial Conductors and its Possible Influence on Magnetic Performance, *IEEE Trans. Magn.* **MAG-13**, 120–123 (1977).

[KRO77[a]] KROEGER, F. R., and SWENSEN, C. A., Absolute Linear Thermal-Expansion Measurements on Copper and Aluminum from 5 to 320K, *J. Appl. Phys.* **48**, 853–864 (1977).

[KUB56] KUBO, R., and OBATA, Y., Note on the Paramagnetic Susceptibility and the Gyromagnetic Ratio in Metals, *J. Phys. Soc. Jpn* **11**, 547–550 (1956).

[KUH78] KUHN, H. A., and DIETER, G. E., Workability in Bulk Forming Processes, *Proceedings of the 4th Int. Conf. on Fracture*, Waterloo, Canada, June, 1977, Pt. 1, pp. 307–324.

[KUN70] KUNAKOV, Ya. N., KACHUR, Ye. V., and PAKHOMOV, V. Ya., Effect of Various Factors on the Superconducting Properties of Nb-Ti Alloys, in *Physics and Metallurgy of Superconductors*, ed. by E. M. Savitskii and V. V. Baron, Consultants Bureau, New York (Plenum Publishing Corp., 1970), pp. 64–68; (Proc. of the Second and Third Conf. on Metallurgy, Physical Chemistry and Metal Physics of Superconductors, Moscow, USSR, May 1965 and May 1966); first published in *Metalloved, Fiz.-Khim., Metallofiz. Sverkhprovodnikov*, Izdatelstvo Nauka, Moscow, USSR, 1967, pp. 59–63.

L

[LAB67] LABUSCH, R., Elastische Konstanten des Flussfadengitters in Supraleitern zweiter Art, *Phys. Status Solidi* **19**, 715–719 (1967).

[LAB69] LABUSCH, R., Elastic Constants of the Fluxoid Lattice Near the Upper Critical Field, *Phys. Status Solidi* **32**, 439–441 (1969).

[LAB69[a]] LABUSCH, R., Calculations of the Critical Field Gradient in Type-II Superconductors, *Cryst. Lattice Defects* **1**, 1–16 (1969).

[LAI70] LAIBOWITZ, R. B., SADAGOPAN, V., and SEIDEN, P. E., Strong-Coupling Parameters for Superconductors, *Phys. Lett. A*: **31**, 133–135 (1970).

[LAN70] LANCHESTER, P. C., The Low Temperature Magnetic Thermal Expansion Coefficients of $NiCl_2.6H_2O$, *J. Phys. C: Solid State Phys.* **3**, 2411–2419 (1970).

[LAR80] LARBALESTIER, D. C., Nb-Ti Alloy Superconductors—Present Status and Potential for Improvement, in *Advances in Cryogenic Engineering* (*Materials*), Vol. 26, ed. by A. F. Clark and R. P. Reed, Plenum Press, New York, 1980, pp. 10–36.

[LAR81] LARBALESTIER, D. C., Niobium-Titanium Superconducting Alloys, in *Superconductor Materials Science, Fabrication and Applications*, (Proc. NATO Advanced Study Institute on the Science and Technology of Superconducting Materials, Sintra, Portugal, 1980), ed. by S. Foner and B. B. Schwartz, Plenum Press (NATO ASI Ser. B: Physics, Vol. 68), New York, 1981, pp. 133–199.

[LAR82] LARSEN, D. C., (ed.) *Thermal Expansion* 7 (Proc. Chicago Conf., Nov., 1979), Plenum Press, New York, 1982.

[LAR84] LARBALESTIER, D. C., and WEST, A. W., New Perspectives on Flux Pinning in Niobium-Titanium Composite Superconductors, *Acta. Metall.* **32**, 1871–1881 (1984).

[LAZ60] LAZARUS, D., Diffusion in Metals, in *Solid State Physics*, Vol. 10, ed. by F. Seitz and D. Turnbull, Academic Press, New York, 1960, pp. 71–99.

[LAZ68] LAZAREV, B. G., OVCHARENKO, O. N., MATSAKOVA, A. A., and VOLOTSKAYA, V. G., Superconducting Properties of Niobium-Base Alloys, *Sov. Phys. JETP* **27**, 549–552 (1968), [Transl. of *Zh. Eksp. Teor. Fiz.* **54**, 1031–1036 (1968)].

[LAZ72] LAZAREV, B. G., LAZAREVA, L. S., and GORIDOV, S. I., Critical Currents in Superconductive Wires of Deformed Alloys Based on Niobium, *Sov. Phys. Dokl.* **17**, 265–267 (1972), [Transl. of *Dokl. Akad. Nauk, SSSR* **203**, 329–331 (1972)].

[LAZ73] LAZAREV, B. G., OVCHARENKO, O. N., and MARTYNOV, I. S., Direct Observation of Magnetic Microstructure in Deformed Niobium-Based Superconducting Alloys, *Sov. Phys. Dokl.* **17**, 810–812 (1973), [Transl. of *Dokl. Akad. Nauk, SSSR* **205**, 1343–1345 (1972)].

[LAZ73[a]] LAZAREV, B. G., and GORIDOV, S. I., Critical Current Value for a Superconductor Niobium-Alloy Wire as a Function of its Copper Coating Thickness, *Sov. Phys. Dokl.* **17**, 902–903 (1973), [Transl. of *Dokl. Akad. Nauk, SSSR* **206**, 85–86 (1972)].

[LEB66] LEBLANC, M. A. R., and MATTES, H. G., Degradation of Low Frequency Alternating Currents in Non-Ideal Type II Superconductors, *Solid State Commun.* **4**, 267–270 (1966).

[LEC59] LE CLAIRE, A. D., Random Walks and Drift in Chemical Diffusion, in *Colloque sur la Diffusion à l'État Solide* (Symposium held in Saclay, France, July, 1958) North-Holland, Amsterdam, 1959, pp. 1–11.

[LEC72] LECHEVET, J., NEIGHBOR, J. E., and SHIFFMAN, C. A., Thermodynamics of the Proximity Effect: Specific-Heat Jumps in Lamellar Lead–Tin Eutectic Alloys, *Phys. Rev. B*: **5**, 861–864 (1972).

[LED77] LEDBETTER, H. M., and READ, D. T., Orthorhombic Elastic Constants of an NbTi/Cu Composite Superconductor, *J. Appl. Phys.* **48**, 1874–1879 (1977).

[LED78] LEDBETTER, H. M., Elastic Constants at Low Temperatures: Recent Measurements on Technological Materials at NBS, in *Advances in Cryogenic Engineering* (*Materials*), Vol. 24, ed. by K. D. Timmerhaus, R. P. Reed, and A. F. Clark, Plenum Press, New York, 1978, pp. 103–119.

[LED79] LEDBETTER, H. M., and COLLINGS, E. W., Low Temperature Magnetically-Induced Elastic Constant Anomalies in Three Manganese Stainless Steels, in *The Metal Science of Stainless Steels*, ed. by E. W. Collings and H. W. King, Metallurgical Society AIME, Warrendale, Pennsylvania, 1979, pp. 22–40.

[LED80] LEDBETTER, H. M., Sound Velocities and Elastic-Constant Averaging for Polycrystalline Copper, *J. Phys. D: Appl. Phys.* **13**, 1879–1884 (1980).

[LED81] LEDBETTER, H. M., MOULDER, J. C., and AUSTIN, M. W., Young's Modulus of a Copper-Stabilized Niobium-Titanium Superconductive Wire, *Wire* (*Germany*) **31**, 27–29 (1981).

[LED82] LEDBETTER, H. M., personal communication, see also [LED78].

[LEU64] LEUPOLD, H. A., and BOORSE, H. A., Superconducting and Normal Specific Heats of a Single Crystal of Niobium, *Phys. Rev. A* **134**, 1322–1328 (1964).

[LEV68] LEVI, L., *Applied Optics*, John Wiley & Sons, New York, 1968, pp. 107–114.

[LIE81] LIEDTKE, I., Untersuchungen des elektrischen Widerstandes in Abhängigkeit von Temperatur und Konzentration im System Niob–Titan, *Diplomarbeit*, 1981, Universität Karlsruhe, Germany.

[LIV63] LIVINGSTON, J. D., and SCHADLER, H. W., The Effect of Metallurgical Variables on Superconducting Properties, *Prog. Mater. Sci.* **12**, 183–287 (1963).

[LOH71] LÖHBERG, R., Über den Einfluss von Kupferzusätzen auf die supraleitenden Eigenschaften und Phasenumwandlungen von technischen Titan-Niob-Legierungen, *D.Eng. Dissertation*, 1971, Universität Erlangen-Nürnberg, Erlangen, Germany.

[LOH73] LÖHBERG, R., HELLER, W., and ZWICKER, U., Influence of Pinning Forces of Titanium–Niobium Base Alloys on the Critical Current up to 10 Tesla, in *Titanium*

Science and Technology, ed. by R. I. Jaffee and H. M. Burte, Plenum Press, New York, 1973, pp. 859–870.

[LON63] LONDON, H., Alternating Current Losses in Superconductors of the Second Kind, *Phys. Lett.* **6**, 162–165 (1963).

[LOR65] LORCH, H. O., On the Movement of Flux Lines in a Type II Superconductor, *Phys. Lett.* **17**, 196–197 (1965).

[LOV66] LOVE, G. R., and PICKLESHEIMER, M. L., The Kinetics of Beta-Phase Decomposition in Niobium(Columbium)-Zirconium, *Trans. TMS-AIME* **236**, 430–435 (1966).

[LOV66a] LOVE, G. R., Near-Surface Effects in Superconducting Niobium-Zirconium Alloys, *J. Appl. Phys.* **37**, 3361–3366 (1966).

[LUB66] LUBELL, M. S., and WIPF, S. L., Magnetic Diffusivity in a Type II Superconductor, *J. Appl. Phys.* **37**, 1012–1014 (1966).

[LUB71] LUBELL, M. S., and KERNOHAN, R. H., Comparison of Calculated and Measured Lower Critical Field for Some Nb–Ti Alloys, *J. Phys. Chem. Solids* **32**, 1531–1539 (1971); first published as Oak Ridge National Lab. *Report No. ORNL-TM-3206* and NASA George C. Marshall Space Flight Center *Report No. NASA-CR-117493, N71-21595* (Oct. 1970).

[LUH68] LUHMAN, T. S., TAGGART, R., and POLONIS, D. H., A Resistance Anomaly in Beta Stabilized Ti–Cr Alloys, *Scri. Metall.* **2**, 169–172 (1968).

[LUH69] LUHMAN, T. S., TAGGART, R., and POLONIS, D. H., Correlation of Superconducting Properties with the Beta to Omega Phase Transformation in Ti–Cr Alloys, *Scri. Metall.* **3**, 777–783 (1969).

[LUH70] LUHMAN, T. S., Superconductivity and Constitution of Titanium Base Transition Metal Alloys, *Ph.D. Thesis*, 1970, University of Washington, Seattle, Washington.

[LUH70a] LUHMAN, T. S., TAGGART, R., and POLONIS, D. H., The Effects of Step Quenching and Aging on the Superconducting Transition in Beta Stabilized Ti–Cr Alloys, *Scri. Metall.* **4**, 611–615 (1970).

[LUH71] LUHMAN, T. S., TAGGART, R., and POLONIS, D. H., The Effect of Omega Phase Reversion on the Superconducting Transition in Titanium-Base Alloys, *Scri. Metall.* **5**, 81–86 (1971).

[LUH79] LUHMAN, T., and DEW-HUGHES, D., (eds.) *Treatise on Materials Science and Technology, Volume 14: Metallurgy of Superconducting Materials*, Academic Press, New York, 1979.

[LUK64] LUKE, C. A., TAGGART, R., and POLONIS, D. H., The Metastable Constitution of Quenched Titanium and Zirconium-Base Binary Alloys, *Trans. ASM* **57**, 142–149 (1964).

[LYN64] LYNTON, E. A., *Superconductivity*, Methuen Monograph, Methuen and Company, London, John Wiley & Sons, New York, 2nd edition, 1964.

M

[MAK64] MAKI, K., The Magnetic Properties of Superconducting Alloys, I, *Physics* **1**, 21–30 (1964).

[MAK64a] MAKI, K., The Magnetic Properties of Superconducting Alloys, II, *Physics* **1**, 127–143 (1964).

[MAK64b] MAKI, K., and TSUNETO, T., Pauli Paramagnetism and Superconducting State, *Prog. Theor. Phys.* **31**, 945–956 (1964).

[MAK65] MAKI, K., Specific Heat of the Mixed State in Superconducting Alloys, *Phys. Rev. A*: **139**, 702–705 (1965).

[MAK65a] MAKI, K., and TSUZUKI, T., Magnetic Properties of Intrinsic London Superconductors, *Phys. Rev. A*: **139**, 868–877 (1965).

[MAK66] MAKI, K., Effect of Pauli Paramagnetism on Magnetic Properties of High-Field Superconductors, *Phys. Rev.* **148**, 362–369 (1966).

[MAR64] MARSH, D. M., Plastic Flow in Glass, *Proc. Roy. Soc. London Ser. A*: **279**, 420–435 (1964).

[MAR64[a]] MARSH, D. M., Plastic Flow and Fracture of Glass, *Proc. Roy. Soc. London Ser. A*: **282**, 33–44 (1964).

[MAR73] MARTIN, D. L., BRADLEY, L. L. T., CAZEMIER, W. J., and SNOWDON, R. L., Automatic Calorimetry in the 3–30 K Range: the Specific Heat of Copper, *Rev. Sci. Instrum.* **44**, 675–684 (1973).

[MAR78] MARGOLIN, H., HAZAVEH, F., and YAGUCHI, H., The Grain Boundary Contribution to the Bauschinger Effect, *Scri. Metall.* **12**, 1141–1145 (1978).

[MAS80] MASUMOTO, T., INOUE, A., SAKAI, S., KIMURA, H., and HOSHI, A., Superconductivity of Ductile Nb-Based Amorphous Alloys, *Trans. Jpn. Inst. Met.* **21**, 115–122 (1980).

[MAT57] MATTHIAS, B. T., Superconductivity in the Periodic System, in *Progress in Low Temperature Physics*, Volume II, ed. by C. J. Gorter, North-Holland, Amsterdam, 1957, pp. 139–150.

[MAT59] MATTHIAS, B. T., COMPTON, V. B., SUHL, H., and CORENZWIT, E., Ferromagnetic Solutes in Superconductors, *Phys. Rev.* **115**, 1597–1598 (1959).

[MAT70] MATTHIAS, B. T., Superconductivity and the Periodic System, *Am. Sci.* **58**, 80–83 (1970).

[MAT72] MATHUR, M. P., WALKER, M. S., DEIS, D. W., and JONES, C. K., Anisotropy of the Critical Current Density in the NbTi Filaments of Round Composite Superconductors, *J. Appl. Phys.* **43**, 3831–3833 (1972).

[MAT73] MATHUR, M. P., DEIS, D. W., and ASHKIN, M., Critical Currents of Superconducting Conductors Measured by a New Technique, *J. Appl. Phys.* **44**, 5116–5119 (1973).

[MAT74] MATHUR, M. P., ASHKIN, M., and DEIS, D. W., Scaling Laws for Flux Pinning in Nb and NbTi Multifilamentary Conductors, *J. Appl. Phys.* **45**, 3627–3633 (1974).

[MAY61] MAYKUTH, D. J., HOLDEN, F. C., WILLIAMS, D. N., OGDEN, H. R., and JAFFEE, R. I., The Effects of Alloying Elements in Titanium: Volume B, Physical and Chemical Properties, Deformation and Transformation Characteristics, *DMIC Report 136B*, May 29, 1961, Battelle Memorial Institute, Columbus, Ohio.

[MCC71] MCCABE, K. K., and SASS, S. L., The Initial Stages of the Omega Phase Transformation in Ti–V Alloys, *Philos. Mag.* **23**, 957–970 (1971).

[MCG69] MCGUIRE, T. R., and FLANDERS, P. J., Direct Current Magnetic Measurements, in *Magnetism and Metallurgy*, ed. by A. Berkowitz and E. Kneller, Academic Press, New York, pp. 123–188.

[MCI73] MCINTURFF, A. D., and CHASE, G. G., Effect of Metallurgical History on Critical Current Density in NbTi Alloys, *J. Appl. Phys.* **44**, 2378–2384 (1973).

[MCM68] MCMILLAN, W. L., Transition Temperature of Strong-Coupled Superconductors, *Phys. Rev.* **167**, 331–344 (1968).

[MCM68[a]] MCMILLAN, W. L., Tunneling Model of the Superconducting Proximity Effect, *Phys. Rev.* **175**, 537–542 (1968).

[MEN35] MENDELSSOHN, K., A Discussion on Superconductivity and Other Low Temperature Phenomena, *Proc. Roy. Soc. London, Ser. A*: **152**, 34–41 (1935).

[MEN36] MENDELSSOHN, K., and MOORE, J. R., Supraconductive Tantalum, *Philos. Mag.* **21**, 532–543 (1936).

[MEN50] MENDELSSOHN, K., and OLSEN, J. L., Heat Transport in Superconductors, *Proc. Phys. Soc., London, Sect. A* **63**, 2–13 (1950).

[MEN52] MENDELSSOHN, K., and ROSENBERG, H. M., The Thermal Conductivity of Metals at Low Temperatures. II: The Transition Elements, *Proc. Phys. Soc., London, Sect. A* **65**, 388-391 (1952).

[MEN61] MENDELSSOHN, K., and ROSENBERG, H. M., The Thermal Conductivity of Metals at Low Temperatures, in *Solid State Physics*, Vol. 12, ed. by F. Seitz and D. Turnbull, Academic Press, New York, 1961, pp. 223-273.

[MEN64] MENDELSSOHN, K., On Different Types of Superconductivity, *Rev. Mod. Phys.* **36**, 50-51 (1964).

[MEN71] MENDIRATTA, M. G., LUTJERING, G., and WEISSMANN, S., Strength Increase in Ti 35Wt Pct Nb Through Step-Aging, *Metall. Trans.* **2**, 2599-2605 (1971).

[MET74] METALS AND CERAMICS INFORMATION CENTER, *Handbook on Materials for Superconducting Machinery*, Nov. 1974 and Supplements, Battelle-Columbus Laboratories, Columbus, Ohio.

[MEY76] MEYER, G. E., COLLINGS, E. W., FIORENTINO, R. J., JELINEK, F. J., and CARMICHAEL, D. C., Experimental Evaluation of Hydrostatic Extrusion for the Fabrication of Multifilament Superconducting Wire, *Progress Report to Energy Research and Development Administration*, Sept. 1976, Battelle-Columbus Laboratories, Columbus, Ohio.

[MIN66] MINNIGERODE, G., Zur Kontaktwirkung in supraleitenden Doppelschichten aus Cu/Pb, *Z. Phys.* **192**, 379-408 (1966).

[MIS75] MISSELL, F. P., OLIVEIRA, N. F., Jr., and SHAPIRA, Y., Absence of the Influence of Pauli Spin Paramagnetism on H_{c2} for a Ti-11 at.% Nb Alloy, *Phys. Status Solidi B*: **67**, K11-K13 (1975).

[MIT68] MITSUBISHI ELECTRIC CORPORATION, Alliages Superconducteurs, *French Patent 1,512,769* (Feb. 1968).

[MOL65] MOLCHANOVA, E. K., *Phase Diagrams of Titanium Alloys*, [Transl. of *Atlas Diagram Sostoyaniya Titanovyk Splavov*], Israel Program for Scientific Translations, 1965.

[MON69] MONTGOMERY, D. B., The Magnetic and Mechanical Aspects of Resistive and Superconducting Systems, in *Solenoid Magnet Design*, Wiley Interscience, New York, 1969, p. 155.

[MON78] MONJU, Y., and TATARA, I., Stress Effects on the Critical Current of Ti-Nb-Zr-Ta Multifilamentary Superconductors, *Cryogenic Engineering (Japan)* **13**, 169-174 (1978).

[MOO66] MOORMANN, W., Zur Theorie der Übergangstemperaturen von supraleitenden Doppelschichten (Proximity-Effekt), *Z. Phys.* **197**, 136-158 (1966).

[MOO68] MOON, J. R., A Possible Source of Instability in Niobium Alloy Superconductors, *Philos. Mag.* **18**, 229-236 (1968).

[MOO73] MOOIJ, J. H., Electrical Conduction in Disordered Alloys, *Phys. Status Solidi A*: **17**, 521-530 (1973).

[MOO80] MOORE, J. P., personal communication.

[MOR59] MOREL, P., Calcul Semi-Empirique de la Temperature de Transition des Superconducteurs et de l'Effet de la Pression sur la Transition, *J. Phys. Chem. Solids* **10**, 277-285 (1959).

[MOR63] MORIN, F. J., and MAITA, J. P., Specific Heats of Transition Metal Superconductors, *Phys. Rev.* **129**, 1115-1120 (1963).

[MOR65] MORRISH, A. H., *Physical Principles of Magnetism*, John Wiley & Sons, New York, 1965.

[MOR65[a]] MORTON, N., Some Effects of Sample Geometry Upon the Occurrence of Flux Jumps in Hard Superconductors, *Phys. Lett.* **19**, 457-459 (1965).

[MOR73] MORTON, N., JAMES, B. W., WOSTENHOLM, G. H., SANDERSON, R. J., and BLACK, M. A., The Thermal Conductivity of Niobium-Zirconium Alloys at Low Temperatures, *Cryogenics* **13**, 665–670 (1973).

[MOR74] MORTON, N., JAMES, B. W., WOSTENHOLM, G. H., and NICHOLLS, R. J., The Low-Temperature Thermal Conductivity of Niobium-Zirconium Alloys, *J. Phys. F: Metal Phys.* **4**, L94–L97 (1974).

[MOR75] MORTON, N., JAMES, B. W., WOSTENHOLM, G. H., and NUTTALL, S., The Thermal and Electrical Conductivities of Niobium-65% Titanium Alloys, *J. Phys. F: Metal Phys.* **5**, 2098–2104 (1975).

[MOR77] MORTON, N., JAMES, B. W., WOSTENHOLM, G. H., TAYLOR, I. A., and TAYLOR, G. M., The Lattice Thermal Conductivity of Superconducting Vanadium–Titanium Alloys, *Cryogenics* **17**, 447–450 (1977).

[MOS70] MOSKALENKO, B. A., and PUPSOVA, B. H., Influence of Alloying on the Plastic Deformation of Alpha-Titanium at Low Temperatures (in Russian), *Report from the Physico-Technical Institute for Low Temperatures of the Ukrainian Academy of Sciences*, Kharkov, 1970—Physics of Condensed Matter.

[MOS73] MOSS, S. C., KEATING, D. T., and AXE, J. D., Neutron Study of the Beta-to-Omega Instability in $Zr_{0.80}Nb_{0.20}$, in *Phase Transitions, 1973*, ed. by L. E. Cross, Pergamon Press, New York, 1973, pp. 179–188.

[MOS80] MOSKALENKO, V. A., STARTSEV, V. I., and KOVALEVA, V. N., Low Temperature Peculiarities of Plastic Deformation in Titanium and its Alloys, in *Titanium '80: Science and Technology*, ed. by H. Kimura and O. Izumi, The Metallurgical Society of AIME, Warrendale, Pennsylvania, 1980, pp. 821–830.

[MOT58] MOTT, N. F., and JONES, H., *The Theory of Properties of Metals and Alloys*, Dover Publications reprint, New York, 1958.

[MUH59] MÜHLSCHLEGEL, B., Die thermodynamischen Funktionen des Supraleiters, *Z. Phys.* **155**, 313–327 (1959).

[MUL76] MULAY, L. N., and BOUDREAUX, E. A., (eds.), *Theory and Applications of Molecular Diamagnetism*, Wiley Interscience, New York, 1976.

[MUL76[a]] MULAY, L. N., and BOUDREAUX, E. A., (eds.), *Theory and Applications of Molecular Paramagnetism*, Wiley Interscience, New York, 1976.

[MUL82] MULAY, L. N., personal communication.

[MUL82[a]] MULAY, L. N., and MULAY, I. L., Magnetometry: Instrumentation and Analytical Applications Including Catalysis, Bioscience, Geoscience, and Amorphous Materials, *Anal. Chem. Rev.* **54**, 216R–227R (1982).

[MYR] MYRON, H. W., FREEMAN, A. J., and MOSS, S. C., Electronically Induced Lattice Instabilities in bcc Zr, *Solid State Commun.* **17**, 1467–1470 (1975).

N

[NAK59] NAKAGAWA, Y., Magnetic Susceptibility of Liquid Alloys of Copper with Cobalt, Iron, Manganese, and Chromium, *J. Phys. Soc. Jpn* **14**, 1372–1379 (1959).

[NAK74] NAKAYAMA, Y., KOIKE, Y., and TOYODA, T., A.C. Losses of Nb-Ti Superconducting Wire with Hollow Cylindrical Shape in Longitudinal D.C. Fields, *Proc. 5th Int. Cryo. Eng. Conf.*, Kyoto, May, 1974, ed. by K. Mendelssohn, IPC Science and Technology Press, UK, 1974, pp. 129–131.

[NAK74[a]] NAKAMURA, H. H., and LARSEN, D. C., Thermal Expansion Behavior of Boron/Epoxy and Graphite/Epoxy Advanced Composite Materials, in *Thermal Expansion—1973*, AIP Conf. Proc. No. 17, ed. by R. E. Taylor and G. L. Denman, Amer. Inst. Phys., New York, 1974, pp. 117–128.

[NAM72] NAMJOSHI, K. V., MITRA, S. S., and VETELINO, J. F., Comparison of Various Lattice Dynamical Models for the Calculation of Thermal Expansion of Solids, in *Thermal Expansion—1971*, AIP Conf. Proc. No. 3, ed. by M. G. Graham and H. E. Hagy, Amer. Inst. Phys., New York, 1972, pp. 237-243.

[NAR66] NARLIKAR, A. V., and DEW-HUGHES, D., Superconductivity in Deformed Niobium Alloys, *J. Mater. Sci.* **1**, 317-335 (1966).

[NAR70] NARAYANAN, G. H., and ARCHBOLD, T. F., Comments on "Phase Transformations in Beta Isomorphous Titanium Alloys," *Scri. Metall.* **4**, 873-876 (1970).

[NAR71] NARAYANAN, G. H., LUHMAN, T. S., ARCHBOLD, T. F., TAGGART, F., and POLONIS, D. H., A Phase Separation Reaction in a Binary Titanium-Chromium Alloy, *Metallography* **4**, 343-358 (1971).

[NEA71] NEAL, D. F., BARBER, A. C., WOOLCOCK, A., and GIDLEY, J. A. F., Structure and Superconducting Properties of Nb44% Ti Wire, *Acta Metall.* **19**, 143-149 (1971).

[NED66] NEDYUKHA, M. N., and CHERNYY, V. G., Elastic Properties of Niobium Alloys with Tantalum, Vanadium, and Titanium, *Phys. Met. Metallogr.* **22**, No. 1, 115-117 (1966), [Transl. of *Fiz. Metal. Metalloved* **22**, 114-116 (1966)].

[NEU66] NEURINGER, L. J., and SHAPIRA, Y., Effect of Spin-Orbit Scattering on the Upper Critical Field of High-Field Superconductors, *Phys. Rev. Lett.* **17**, 81-84 (1966).

[NIS68] NISHIMURA, T., and ZWICKER, U., Supraleitung und Phasenumwandlung bei Titanlegierungen, III: Zusätze von Eisen und Niob, *Z. Metallkd.* **59**, 69-73 (1968).

[NOE71] NOEL, J. L., MAILFERT, A., and FOURNET, G., Étude Expérimentale des Sautes de Flux Limités dans les Supraconducteurs Impurs de Seconde Espèce: Répartition de l'Induction Magnétique avant et après Saut de Flux, *J. Phys. (Paris)* **32**, 191-204 (1971).

[NOM80] NOMURA, H., SINCLAIR, M. N. L., and IWASA, Y., Acoustic Emission in a Composite Copper NbTi Conductor, *Cryogenics* **20**, 283-289 (1980).

O

[OBS78] OBST, B., and BRANDT, E. H., Phase Transition from the Triangular to the Square Flux Line Lattice in Type II Superconductors with Small κ, *Phys. Lett. A*: **64**, 460-462 (1978).

[OBS80] OBST, B., PATTANAYAK, D., and HOCHSTUHL, P., Structural Effects in the Superconductor NbTi65, *J. Low Temp. Phys.* **41**, 595-609 (1980).

[OHM75] OHMER, M. C., WOLLAN, J. J., and LAWSON, L. O., Experimental Hysteretic Loss for a Series of Superconducting Filamentary NbTi Wires and a Field Dependent Critical State Model, *Tech. Report AFML-TR-75-10*, Air Force Materials Laboratory, July 1975.

[ONN11] ONNES, H. K., Electrical Resistance of Pure Metals at Liquid Helium Temperatures, *Electrician* **67**, 657 (1911).

[ONS49] ONSAGER, L., Statistical Hydrodynamics, *Nuovo Cimento* **6**, (Suppl. 2) 279-287 (1949).

[ORL79] ORLANDO, T. P., MCNIFF, E. J., Jr., FONER, S., and BEASLEY, M. R., Critical Fields, Pauli Paramagnetic Limiting and Material Parameters of Nb_3Sn and V_3Si, *Phys. Rev. B*: **9**, 4545-4561 (1979).

[ORL81] ORLANDO, T. P., and BEASLEY, M. R., Pauli Limiting and the Possibility of Spin Fluctuations in the A15 Superconductors, *Phys. Rev. Lett.* **46**, 1598-1600 (1981).

[OSA80] OSAMURA, K., MATSUBARA, E., MIYATANI, T., MURAKAMI, Y., HORIUCHI, T., and MONJU, Y., Effect of Cold Working on Precipitation Behaviour in Superconducting Ti-Nb Alloys, *Philos. Mag. A*: **42**, 575-589 (1980).

[OTT70] OTTE, H. M., Mechanism of the Martensitic Transformation in Titanium and its Alloys, in *The Science, Technology and Application of Titanium*, ed. by R. I. Jaffee and N. E. Promisel, Pergamon Press, London, 1970, pp. 645–657.

P

[PAR69] PARKS, R. D., (ed.) *Superconductivity*, Marcel Dekker, New York, 1969.

[PAR78] PARR, R. A., and JOHNSTON, M. H., Growth Parameters for Aligned Microstructures in Directionally Solidified Aluminum-Bismuth Monotectic, *Metall. Trans.* **9A**, 1825–1828 (1978).

[PAS78] PASZTOR, G., and SCHMIDT, C., Dynamic Stress Effects in Technical Superconductors and the "Training" Problem of Superconducting Magnets, *J. Appl. Phys.* **49**, 886–899 (1978).

[PAT74] PATON, N. E., and WILLIAMS, J. C., Deformation and Solid Solution Strengthening of bcc Titanium–Vanadium Alloys, *Annual Report to ONR on Contract No. N00014-71-C-0363*, Jan. 1974.

[PAT81] PATTANAYAK, D., OBST, B., and WOLFSTIEG, U., Röntgenographische Phasenbestimmung an Niob-Titan-Supraleitern, *Z. Metallkd.* **72**, 481–486 (1981).

[PAU56] PAULING, L., The Electronic Structures of Metals and Alloys, in *Theory of Alloy Phases*, American Society for Metals, 1956, pp. 220–242.

[PAU67] PAULING, L., *The Chemical Bond*, Cornell University Press, Ithaca, New York, 1967, Chap. 11.

[PEA65] PEARL, J., On the Nature of Resistance in the Mixed State of Type II Superconductors, *Phys. Lett.* **17**, 12–14 (1965).

[PEI55] PEIERLS, R. E., *Quantum Theory of Solids*, Clarendon Press, Oxford, 1955.

[PER70] PERIERA, F. N. D. D., BARNES, C. H., and GRAHAM, G. M., Sensitive Dilatometer for Low Temperatures and the Thermal Expansion of Copper below 10 K, *J. Appl. Phys.* **41**, 5050–5054 (1970).

[PER72] PERIERA, F. N. D. D., and GRAHAM, G. M., Thermal Expansion of the N.B.S. Standard Copper Below 30 K, in *Thermal Expansion—1971*, AIP Conf. Proc. No. 3, ed. by M. G. Graham and H. E. Hagy, Amer. Inst. Phys., New York, 1972, pp. 65–71.

[PER75] PERKINS, J., (ed.) *Shape-Memory Effect in Alloys*, Plenum Press, New York, 1975.

[PER75[a]] PERKINS, J., EDWARDS, G. R., SUCH, C. R., JOHNSON, J. M., and ALLEN, R. R., Thermomechanical Characteristics of Alloys Exhibiting Martensitic Thermoelasticity, in *Shape-Memory Effect in Alloys*, ed. by J. Perkins, Plenum Press, New York, 1975, pp. 273–299.

[PET72] PETTIFOR, D. G., Theory of the Crystal Structure of Transition Metals at Absolute Zero, in *Metallurgical Chemistry*, ed. by O. Kubaschewski, National Physical Laboratory, HMSO, 1972, pp. 191–199.

[PET79] PETTIFOR, D. G., Theory of the Heats of Formation of Transition-Metal Alloys, *Phys. Rev. Lett.* **42**, 846–850 (1979).

[PFE68] PFEIFFER, I., and HILLMANN, H., Der Einfluss der Struktur auf die Supraleitungseigenschaften von NbTi 50 and NbTi 65, *Acta Metall.* **16**, 1429–1439 (1968).

[PIE65] PIEHLER, H. R., Plastic Deformation and Failure of Silver-Steel Filamentary Composites, *Trans. TMS-AIME* **233**, 12–16 (1965).

[PIN55] PINES, D., Electron Interaction in Metals, in *Solid State Physics*, Vol. 1, ed. by F. Seitz and D. Turnbull, Academic Press, New York, 1955, pp. 367–450.

[PIP55] PIPPARD, A. B., Trapped Flux in Superconductors, *Trans. Roy. Soc.* (*London*) **248A**, 97–129 (1955).

[POL69] POLONIS, D. H., A Study of Phase Transformations and Superconductivity, *Progr. Report No. 10 and Report No. RLO-1375-18*, Oct. 1969, University of Washington, Seattle, Washington.

[POL70] POLONIS, D. H., A Study of Phase Transformations and Superconductivity, *Annual Progr. Report No. RLO-2225-T-13-6*, Oct. 1970, University of Washington, Seattle, Washington.

[POL71] POLONIS, D. H., Study of Phase Transformations and Superconductivity, *Annual Progr. Report No. RLO-2225-T-13-9, N72-24786*, Oct. 1971, University of Washington, Seattle, Washington.

[POO83] POON, S. J., Structural Dependence of the Superconducting Transition Temperature in Liquid-Quenched Alloys Based on the Group IV B Elements, *Solid State Commun.* **47**, 431-434 (1983).

[POO83[a]] POON, S. J., HASANIAN, S. K., and WONG, K. M., Analysis of Upper Critical Field Curves in Amorphous Superconductors, *Phys. Lett. A*: **93**, 495-498 (1983).

[POT81] POTT, R., GÜNTHERODT, G., and BACH, H., Thermal Expansion of La_3S_4 at Low Temperatures, *Solid State Commun.* **39**, 1293-1296 (1981).

[PRE74] PREKUL, A. F., RASSOKHIN, V. A., and VOLKENSHTEIN, N. V., Effect of Spin Fluctuations on the Superconducting and Normal Properties of Alloys of Ti Containing V, Nb or Ta, *Sov. Phys. JETP* **40**, 1134-1136 (1974), [Transl. of *Zh. Eksp. Teor. Fiz.* **67**, 2286-2292 (1974)].

[PRE76] PREKUL, A. F., SHCHERBAKOV, A. S., and VOLKENSHTEIN, N. V., Resistivity and Anomalous Superconducting Transition in $Ti_{1-x}Fe_x$ Alloys ($0 < x \leq 0.2$), *Sov. J. Low Temp. Phys.* **2**, 684-686 (1976), [Transl. of *Fiz. Nizk Temp.* **2**, 1399-1404 (1976)].

[PRO68] PROKOSHKIN, D. A., VASIL'YEVA, Ye. V., and POPOV, N. N., Dislocation Structure of Niobium-Titanium Alloys, *Phys. Met. Metallogr.* **25**, No. 5, 198-200 (1968), [Transl. of *Fiz. Met. Metalloved.* **25**, 953-955 (1968)].

[PUD69] PUDALOV, V. M., and KHAIKIN, M. S., Dilatometer with a Sensitivity of 10^{-4} Angstrom, *Cryogenics* **9**, 128-130 (1969).

R

[RAB79] RABINOVITCH, A., and BAHAT, D., Catastrophe Theory: A Technique for Crack Propagation Analysis, *J. Appl. Phys.* **50**, 231-234 (1979).

[RAD66] RADEBAUGH, R., and KEESOM, P. H., Low-Temperature Thermodynamic Properties of Vanadium, I: Superconducting and Normal States, *Phys. Rev.* **149**, 209-212 (1966).

[RAI73] RAINER, D., BERGMANN, G., and ECKHARDT, U., Strong-Coupling Calculation of the Upper Critical Field of Amorphous Lead, *Phys. Rev. B*: **8**, 5324-5325 (1973).

[RAL64] RALLS, K. M., Effect of Metallurgical Structure on the Superconductivity of Nb-Zr and Nb-Ti Alloys, *Ph.D. Thesis*, 1964, Massachusetts Institute of Technology, Cambridge, Massachusetts.

[RAL65] RALLS, K. M., ROSE, R. M., and WULFF, J., Limiting Magnetic Fields for High-Field Superconductivity, *J. Appl. Phys.* **36**, 1295-1298 (1965).

[RAS72] RASSMANN, G., and ILLGEN, L., Zum Zusammenhang zwischen Gefüge und kritischer Stromdichte bei supraleitenden binären Titan-Niob-Legierungen, *Neue Hütte* **17**, 321-328 (1972).

[RAS72[a]] RASSMANN, G., and ILLGEN, L., Zum Zusammenhang zwischen Gefüge und kritischer Stromdichte bei supraleitenden Titan-Niob-Legierungen mit Zusätzen von α-Stabilisatoren, *Neue Hütte* **17**, 547-553 (1972).

[RAS72[b]] RASSMANN, G., and ILLGEN, L. Zum Zusammenhang zwischen Gefüge und kritischer Stromdichte bei supraleitenden Titan-Niob-Legierungen mit Zusätzen von neutralen Elementen, *Neue Hütte* **17**, 718–723 (1972).

[RAS73] RASSMANN, G., and ILLGEN, L., Zum Zusammenhang zwischen Gefüge und kritischer Stromdichte bei supraleitenden Titan-Niob-Legierungen mit Zusätzen von β-Stabilisatoren, *Neue Hütte* **18**, 33–40 (1973).

[RAU68] RAUCH, G. C., Aging Phenomena and Superconductivity in Niobium-Titanium Alloys, *Ph.D. Thesis*, 1968, Massachusetts Institute of Technology, Cambridge, Massachusetts.

[RAU68[a]] RAUCH, G. C., COURTNEY, T. H., and WULFF, J., Aging in Nb(Cb)-Ti-O Superconductors, with Appendix, *Trans. TMS-AIME* **242**, 2263–2270 (1968).

[RAY69] RAYEVSKII, I. I., STEPANOV, N. V., SKRYABINA, M. A., DUBROVIN, A. V., ALEKSEEVSKII, N. Ye., and IVANOV, O. S., Phase Structure and Superconductive Properties of Alloys on the System Niobium–Titanium–Zirconium–Hafnium, *Phys. Met. Metallogr.* **27**, No. 2, 42–55 (1969), [Transl. of *Fiz. Met. Metalloved.* **27**, 235–249 (1969)].

[RAY71] RAYEVSKII, I. I., STEPANOV, N. V., DUBROVIN, A. V., IVANOV, O. S., and ALEKSEEVSKII, N. Ye., Superconductive Properties of Niobium–Titanium–Zirconium–Hafnium Alloys, *Phys. Met. Metallogr.* **31**, No. 1, 71–76 (1971), [Transl. of *Fiz. Met. Metalloved.* **31**, 72–78 (1971)].

[REA78] READ, D. T., Metallurgical Effects in Niobium–Titanium Alloys, *Cryogenics* **18**, 579–584 (1978).

[REE67] REED, R. P., and MIKESELL, R. P., Low Temperature (295 to 4 K) Mechanical Properties of Selected Copper Alloys, *J. Mater.* **2**, 370–391 (1967).

[REE77] REED, R. P., MIKESELL, R. P., and CLARK, A. F., Low Temperature Tensile Behavior of Copper-Stabilized Niobium–Titanium Superconducting Wire in *Advances in Cryogenic Engineering* (*Materials*), Vol. 22, ed. by K. D. Timmerhaus, R. P. Reed, and A. F. Clark, Plenum Press, New York, 1977, pp. 463–471.

[REI67] REINBACH, R., Verwendung einer Niob-Tantal-Titan-Legierung zur Herstellung von Supraleitern, *German Patent 1,237,786* (Mar. 1967).

[REI73] REID, C. N., ROUTBORT, J. L., and MAYNARD, R. A., Elastic Constants of Ti–40at.% Nb at 298 K, *J. Appl. Phys.* **44**, 1398–1399 (1973).

[REU66] REUTER, F. W., RALLS, K. M., and WULFF, J., Microstructure and Superconductivity of a 44.7 At. Pct. Niobium (Columbium)–54.3 At. Pct. Titanium Alloy Containing Oxygen, *Trans. TMS-AIME* **236**, 1143–1151 (1966).

[RIC65] RICKAYZEN, G., *Theory of Superconductivity*, John Wiley & Sons, New York, 1965.

[RIC69] RICKETTS, R. L., Correlation of Structure and Superconducting Properties in Niobium-Titanium Quaternary Alloys, *Ph.D. Thesis*, 1969, Massachusetts Institute of Technology, Cambridge, Massachusetts.

[RIC70] RICKETTS, R. L., COURTNEY, T. H., SHEPARD, L. A., and WULFF, J., Correlation of Structure and Superconducting Properties in Nb(Cb)–Ti Quaternary Alloys, *Metall. Trans.* **1**, 1537–1544 (1970).

[RIZ74] RIZZUTO, C., Formation of Localized Moments in Metals: Experimental Bulk Properties, *Rep. Prog. Phys.* **37**, 147–229 (1974).

[ROD70] RODRIGUEZ-GONZALEZ, F. A., Application of Neutron Activation Analysis to the Study of Interstitial Solid Solution of Oxygen in Niobium–Titanium Alloys, *Ph.D. Thesis*, 1970, University of Texas, Austin, Texas.

[ROL71] ROLINSKI, E. J., HOCH, M., and OBLINGER, C. F., Determination of Thermodynamic Interaction Parameters in Solid V–Ti Alloys Using the Mass Spectrometer, *Metall. Trans.* **2**, 2613–2618 (1971).

[ROL74] ROLLINS, R. W., KUEPFER, H., and GEY, W., Magnetic Field Profiles in Type-II Superconductors with Pinning Using a New AC Technique, *J. Appl. Phys.* **45**, 5392–5398 (1974).

[ROL77] ROLLINS, R. W., and ANJANEYULU, Y., Position-Dependent Flux Pinning by Proton-Irradiation Damage in Thick Niobium Samples, *J. Appl. Phys.* **48**, 1296–1300 (1977).

[ROR65] RORER, D. C., ONN, D. G., and MEYER, N., Thermodynamic Properties of Molybdenum in Its Superconducting and Normal State, *Phys. Rev. A*: **138**, 1661–1668 (1965).

[ROS63] ROSENBERG, H. M., *Low Temperature Solid State Physics*, Clarendon Press, Oxford, 1963.

[ROS64] ROSENBLUM, B., and CARDONA, M., Observation of Surface Superconductivity Above the Thermodynamic Critical Field in Type I Superconductors, *Phys. Lett.* **9**, 220–221 (1964).

[RUD64] RUDMAN, P. S., An X-Ray Diffuse-Scattering Study on the Nb–Ti B.C.C. Solution, *Acta Metall.* **12**, 1381–1388 (1964).

[RUD69] RUDY, E., Ternary Phase Equilibria in Transition Metal–Boron–Carbon–Silicon Systems. Part V. Compendium of Phase Diagram Data, *Technical Report AFML-TR-65-2, Part V*, Aerojet-General Corporation, May 1969.

S

[SAI63] SAINT-JAMES, D., and DE GENNES, P. G., Onset of Superconductivity in Decreasing Fields, *Phys. Lett.* **7**, 306–308 (1963).

[SAI69] SAINT-JAMES, D., THOMAS, E. J., and SARMA, G., *Type II Superconductivity*, Pergamon Press, London, 1969.

[SAL64] SALTER, L. C., Jr., Investigation of Current Degradation Phenomenon in Superconducting Solenoids, *First Quart. Prog. Report, Contr. No. NAS 8-5356 to NASA*, Atomics International, Canoga Park, California, Oct. 1964.

[SAL66] SALTER, L. C., Jr., Investigation of Current Degradation Phenomenon in Superconducting Solenoids, *Summary Report, Contr. No. NAS 8-5356 to NASA, N66-23800*, Atomics International, Canoga Park, California, Jan. 1966.

[SAL79] SALMON, D. R., *Low Temperature Data Handbook, Titanium and Titanium Alloys, NPL Report QU53* (*N 80 23448*), National Physical Laboratory, May 1979.

[SAS69] SASS, S. L., The ω Phase in a Zr-25at.%Ti Alloy, *Acta. Metall.* **17**, 813–820 (1969).

[SAS72] SASS, S. L., The Structure and Decomposition of Zr and Ti b.c.c. Solid Solutions, *J. Less-Common Metals* **28**, 157–173 (1972).

[SAV71] SAVITSKII, E. M., BARON, V. V., EFIMOV, Yu. V., BYCHKOVA, M. I., and KOZLOVA, N. D., Use of Experimental Planning to Study Superconductor Systems, *Dokl. Phys. Chem.* **196**, 176–179 (1971), [Transl. of *Dokl. Akad. Nauk SSSR* (*Phys. Chem. Sect.*) **196**, No. 5, 1145–1148 (1971)].

[SAV71[a]] SAVITSKII, E. M., EFIMOV, Yu. V., and KOZLOVA, N. D., Structure and Properties of V–Ta–Ti Alloys, *Russ. Metall.* **No. 4** 149–153 (1971), [Abridged Transl. of *Izv. Akad. Nauk SSSR*, (*Met.*) **No. 4**, 214–219 (1971)].

[SAV73] SAVITSKII, E. M., BARON, V. V., EFIMOV, Yu. V., BYCHKOVA, M. I., and MYZENKOVA, L. F., *Superconducting Materials*, Plenum Press, New York, 1973.

[SCA69] SCALAPINO, D. J., The Electron–Phonon Interaction and Strong-Coupling Superconductors, in *Superconductivity*, ed. by R. D. Parks, Marcel Dekker, New York, 1969, pp. 449–558.

[SCA80] SCANLAN, R. M., personal communication.

[SCH63] SCHOFIELD, B. H., Acoustic Emission from Metals—Its Detection, Characteristics, and Source, *Trans. of Symposium on Physics and Nondestructive Testing*, San Antonio, Texas, 1963, pp. 63-91.

[SCH70] SCHILLER, A., RAYROUX, J. M., and HORVATH, I., Steady-State Stability of Composite Superconductors with a Copper-Solder Stabilizing Substrate, *Proc. 3rd Int. Cryo. Eng. Conf.*, Berlin, May, 1970, Iliffe Science and Technology Publ. Ltd., UK, 1970, pp. 341-345.

[SCH75] SCHMIDT, C., Thermal Boundary Resistance (Kapitza Resistance) at Interfaces Between Copper and Type-II Superconductors, *J. Appl. Phys.* **46**, 1372-1375 (1975).

[SCH76] SCHMIDT, C., Investigation of the Training Problem of Superconducting Magnets, *Appl. Phys. Lett.* **28**, 463-465 (1976).

[SCH77] SCHMIDT, C., Effect of Dynamic Stress on Commercial Superconductors: A Test Facility, *Rev. Sci. Instrum.* **48**, 597-601 (1977).

[SCH77[a]] SCHMIDT, C., Superconductors Under Dynamic Mechanical Stress, *IEEE Trans. Magn.* **MAG-13**, 116-119 (1977).

[SCH79] SCHMIDT, C., Simple Method to Measure the Thermal Conductivity of Technical Superconductors, e.g., NbTi, *Rev. Sci. Instrum.* **50**, 454-457 (1979).

[SCH81] SCHOPOHL, N., and SCHARNBERG, K., Upper Critical Fields in the Presence of Electron-Spin and Spin-Orbit Effects, *Physica (Utrecht)* **107B**, 293-294 (1981).

[SEG79] SEGAL, H. R., HEMACHALAM, K., DE WINTER, T. A., and STEKLY, Z. J. J., Development of NbTi Conductor for High Field Applications, *IEEE Trans. Magn.* **MAG-15**, 807-809 (1979).

[SEG80] SEGAL, H. R., HRYCAJ, T. M., STEKLY, Z. J. J., DE WINTER, T. A., and HEMACHALAM, K., NbTi-Based Conductors for Use in 12 Tesla Toroidal Field Coils, in *Proc. of the 8th Symposium on Engineering Problems of Fusion Research, Pt. 1*, San Francisco, California, 13-16 Nov., 1979, IEEE, 1980, pp. 255-259.

[SEG81] SEGAL, H. R., DE WINTER, T. A., STEKLY, Z. J. J., and HEMACHALAM, K., The Use of NbTiTa as a High Field Superconducting Alloy, *IEEE Trans. Magn.* **MAG-17**, 53-56 (1981).

[SEI79] SEIBT, E., Investigations of a Steel Reinforced NbTi Superconducting Flat Cable for Toroidal Magnets, *IEEE Trans. Magn.* **MAG-15**, 804-806 (1979).

[SEL56] SELWOOD, P. W., *Magnetochemistry*, Wiley Interscience, New York, 1956.

[SHA62] SHAPOVAL, E. A., The Upper Critical Field of Superconducting Alloys, *Sov. Phys. JETP* **14**, 628-632, (1962), [Transl. of *Zh. Eksp. Teor. Fiz.* **41**, 877-885 (1961)].

[SHA64] SHAPIRO, J. M., TAYLOR, D. R., and GRAHAM, G. M., A Sensitive Dilatometer for Use at Low Temperatures, *Can. J. Phys.* **42**, 835-847 (1964).

[SHA65] SHAPIRA, Y., and NEURINGER, L. J., Upper Critical Fields of Nb-Ti Alloys—Evidence for the Influence of Pauli Paramagnetism, *Phys. Rev. A*: **140**, 1638-1644 (1965).

[SHC73] SHCHETKIN, I. S., and KHARCHENKO, T. N., Superconductivity and Electron Structure of a Solid Solution of Titanium in Niobium, *Sov. Phys. JETP* **37**, 491-493 (1973), [Transl. of *Zh. Eksp. Teor. Fiz.* **64**, 964-969 (1973)].

[SHE58] SHEARD, F. W., Calculation of the Thermal Expansion of Solids from the Third-Order Elastic Constants, *Philos. Mag.* **3**, 1381-1390 (1958).

[SHE63] SHEWMON, P. G., *Diffusion in Solids*, McGraw-Hill Book Company, New York, 1963.

[SHI64] SHIFFMAN, C. A., COCHRAN, J. F., GARBER, M., and PEARSALL, G. W., Specific Heat Measurements and Proximity Effect in Tin-Lead Eutectic Alloys, *Rev. Mod. Phys.* **36**, 127-130 (1964).

[SHI77] SHIBATA, M., and ONO, K., On the Minimization of Strain Energy in the Martensitic Transformation of Titanium, *Acta Metall.* **25**, 35-42 (1977).

[SHM70] SHMIDT, V. V., Third Critical Field of Superconductors with a Film on the Surface, in *Physics and Metallurgy of Superconductors*, ed. by E. M. Savitskii and V. V. Baron, Consultants Bureau, New York (Plenum Publishing Corp., 1970) pp. 105-113, (Proc. of the Second and Third Conf. on Metallurgy, Physical Chemistry and Metal Physics of Superconductors, Moscow, USSR, May 1965 and May 1966); first published in *Metalloved., Fiz.-Khim., Metallofiz. Sverkhprovodnikov*, Izdatelstvo Nauka, Moscow, USSR, 1967.

[SHU37] SHUBNIKOV, L. W., KOTKEVICH, W. I., SHEPELEV, J. D., and RIABININ, J. N., *Zh. Eksp. Teor. Fiz.* **7**, 221 (1937).

[SHU69] SHUNK, F. A., *Constitution of Binary Alloys, Second Supplement*, McGraw-Hill Book Company, New York, 1969.

[SIK68] SIKORA, A., MAKIEJ, B., and TROJNAR, E., Distribution of the Magnetic Induction in the Type II Current-Carrying Superconductor, *Phys. Lett. A*: **27**, 175-176 (1968).

[SIK82] SIKKA, S. K., VOHRA, Y. R., and CHIDAMBARAM, R., Omega Phase in Materials, *Prog. Mater. Sci.* **27**, 245-310 (1982).

[SIL58] SILCOCK, J. M., An X-ray Examination of the ω Phase in TiV, TiMo, and TiCr Alloys, *Acta Metall.* **6**, 481-493 (1958).

[SIL63] SILCOX, J., and ROLLINS, R. W., Hysteresis in Hard Superconductors, *App. Phys. Lett.* **2**, 231-233 (1963).

[SIM78] SIMPSON, M. A., and SMITH, T. F., Thermal Expansion in the Superconducting State, *J. Low Temp. Phys.* **32**, 57-65 (1978).

[SIN68] SINHA, A. K., Low Temperature Specific Heat of B.C.C. Titanium Molybdenum Alloys, *J. Phys. Chem. Solids* **29**, 749-754 (1968).

[SIN75] SINHA, S. K., and HARMON, B. N., Electronically Driven Lattice Instabilities, *Phys. Rev. Lett.* **35**, 1515-1518 (1975).

[SIN76] SINHA, S. K., and HARMON, B. N., Phonon Anomalies in d-Band Metals and Their Relationship to Superconductivity, in *Superconductivity in d- and f-Band Metals*, (Proc. 2nd Rochester Conference, 1976) ed. by D. H. Douglass, Plenum Press, New York, 1976, pp. 269-296.

[SMI72] SMITH, T. F., Determination of Pressure Dependence of T_c for d- and f-Band Superconductors, in *Superconductivity in d- and f-Band Metals*, (AIP Conf. Proc. No. 4) ed. by D. H. Douglass, Amer. Inst. Phys., New York, 1972, pp. 293-319.

[SMI72[a]] SMITH, T. F., Pressure Dependence of the Superconducting Transition Temperature for Vanadium, *J. Phys. F: Metal Phys.* **2**, 946-956 (1972).

[SMI75] SMITH, T. F., and SHELTON, R. N., Pressure Dependence of T_c for the Transition Metals: bcc Solid Solutions in the System Zr-Nb-Mo-Re, *J. Phys. F: Metal Phys.* **5**, 911-922 (1975).

[SMI76] SMITH, H. G., WAKABAYASHI, N., and MOSTOLLER, M., Phonon Anomalies in Transition Metals, Alloys and Compounds, in *Superconductivity in d- and f-Band Metals*, (Proc. 2nd Rochester Conference, 1976) ed. by D. H. Douglass, Plenum Press, 1976, pp. 223-249.

[SMI76[a]] SMITH, T. F., and FINLAYSON, T. R., Thermal Expansion of BCC Solid Solution Alloys in the System Zr-Nb-Mo-Re, *J. Phys. F: Metal Phys.* **6**, 709-723 (1976).

[SMI77] SMITH, T. F., and WHITE, G. K., Grüneisen Parameters, Electron-Phonon Enhancement and Superconductivity for Pd-H Alloys, *J. Phys. F: Metal Phys.* **1**, 1029-1038 (1977).

[SMY50] SMYTHE, W. R., *Static and Dynamic Electricity*, McGraw-Hill Book Company, New York, 1950.

[SOE69] SOENO, K., and KURODA, T., Kinetics of Beta-Phase Decomposition and the Precipitation of Alpha-Zirconium in Nb-40at.%Zr-10at.%Ti Superconducting Alloy, *Nippon Kinzoku Gakkaishi* **33**, 791-795 (1969).

[SOL76] SOLDATOV, V. P., ABRAIMOV, V. V., and STARTSEV, V. I., Softening of Superconductors with an Anomalous Thermal Conductivity, *Sov. J. Low Temp. Phys.* **2**, 606-609 (1976).

[SOU73] SOULIN, R. J., Jr., SCHOOLEY, J. F., and EVANS, G. A., Jr., Simple Instrumentation for the Inductive Detection of Superconductivity, *Rev. Sci. Instrum.* **44**, 1537-1538 (1973).

[SPA77] SPARKS, L. L., Magnetothermal Conductivity of Selected Pure Metals and Alloys, in *Advances in Cryogenic Engineering (Materials)*, Vol. 22, ed. by K. D. Timmerhaus, R. P. Reed and A. F. Clark, Plenum Press, New York, 1977, pp. 119-127.

[SPA78] SPARKS, L. L., Magnetic Field Effect on Thermal Conductivity of Selected Metals, in *Advances in Cryogenic Engineering (Materials)*, Vol. 24, ed. by K. D. Timmerhaus, R. P. Reed and A. F. Clark, Plenum Press, New York, 1978, pp. 224-231.

[STE78] STEKLY, Z. J. J., SEGAL, H. R., HEMACHALAM, K., DE WINTER, T. A., and COLLING, D. A., Development of NbTi Conductors for 10T-14T Operation, *Final Report, Dept. of Energy Contr. No EG-77-C-02-4140*, Sept., 1978, Magnetic Corp. of America.

[STE81] STEINMAN, S. G., and FISHER, E. S., Elastic Properties of Transition Metals, in *Treatise on Materials Science and Technology*, Vol. 21, ed. by F. Y. Fradin, Academic Press, New York, 1981, pp. 223-295.

[STO78] STOCKS, G. M., TEMMERMAN, W. M., and GYORFFY, B. L., Complete Solution of the Korringa-Kohn-Rostoker Coherent-Potential Approximation Equations: Cu-Ni Alloys, *Phys. Rev. Lett.* **41**, 339-341 (1978).

[STR64] STRONGIN, M., PASKIN, A., SCHWEITZER, D. G., KAMMERER, O. F., and CRAIG, P. P., Surface Superconductivity in Type I and Type II Superconductors, *Phys. Rev. Lett.* **12**, 442-444 (1964).

[STR68] STRONGIN, M., KAMMERER, O. F., CROW, J. E., PARKS, R. D., DOUGLASS, D. H., Jr., and JENSEN, M. A., Enhanced Superconductivity in Layered Metallic Films, *Phys. Rev. Lett.* **21**, 1320-1323 (1968).

[SUD68] SUDEREVA, S. V., BUYNOV, N. N., and RAKIN, V. G., Electron Microscopic and X-Ray Diffraction Analysis of the Quenched Alloy Ti-25at.%Nb, *Phys. Met. Metallogr.* **26**, No. 5, 14-20 (1968), [Transl. of *Fiz. Met. Metalloved.* **26**, 781-788 (1968)].

[SUD70] SUDAREVA, S. V., On the Relations Between the Real Structure of Alloys and Model Representations of Hard Superconductors, *Phys. Met. Metallogr.* **30**, No. 5, 19-29 (1970), [Transl. of *Fiz. Met. Metalloved.* **30**, 915-923 (1970)].

[SUE68] SUENAGA, M., OLSON, J., and RALLS, K., High-Field Superconductivity, *Report No. UCRL-18043*, Mar. 1968, Lawrence Radiation Lab., University of California, Berkeley, California, pp. 141-144.

[SUE69] SUENAGA, M., and RALLS, K. M., Some Superconducting Properties of Ti-Nb-Ta Ternary Alloys, *J. Appl. Phys.* **40**, 4457-4463 (1969).

[SUK68] SUKHAREVSKII, B. Ya., and ALAPINA, A. V., Some Features of the Temperature Dependence of the Specific Heat of a Niobium-Titanium Alloy at the Transition to the Superconducting State, *Sov. Phys. JETP* **27**, 897-899 (1968), [Transl. of *Zh. Eksp. Teor. Fiz.* **54**, 1675-1679 (1968)].

[SUK69] SUKHAREVSKII, B. Ya., ALAPINA, A. V., DUSHECHKIN, Yu. A., and SHCHETKIN, I. S., Investigation of the Influence of the Concentration of Solid Solutions of the Nb-Ti System and of Lattice Defects on Singularities of the Superconducting Transitions, *Collected Papers of Soviet-Japanese Conference on Low-Temperature Physics*, Novosibirsk, 1969.

[SUK71] SUKHAREVSKII, B. Ya., SHCHETKIN, I. S., and FAL'KO, I. I., Investigation of the Superconducting State of Solid Solutions of the Niobium-Titanium System,

Sov. Phys. JETP **33**, 152–155 (1971), [Transl. of *Zh. Eksp. Teor. Fiz.* **60**, 277–282 (1971)].

[SUM80] SUMITOMO ELECTRIC INDUSTRIES, LTD., Nb–Ti Superconducting Wire—High Current Rectangular Wire, *Company Data*, June 1980.

[SUN75] SUN, C. T., and GRAY, W. H., Theoretical and Experimental Determination of Mechanical Properties of Superconducting Coil Composites, *Proc. of the 6th Symposium on the Engineering Problems of Fusion Research*, San Diego, California, 1975, pp. 261–265.

[SUT73] SUTTON, J., New Method of Improving the Stability of Superconductors Against Flux Jumping, *J. Appl. Phys.* **44**, 461–471 (1973).

[SUZ75] SUZUKI, T., and WUTTIG, M., Analogy Between Spinodal Decomposition and Martensitic Transformation, *Acta Metall.* **23**, 1069–1076 (1975).

[SWA68] SWARTZ, P. A., and BEAN, C. P., A Model for Magnetic Instabilities in Hard Superconductors: The Adiabatic Critical State, *J. Appl. Phys.* **39**, 4991–4998 (1968).

[SWA73] SWANSON, M. L., and QUENNEVILLE, A. F., The Effect of Compressional Plastic Deformation on the Superconducting Transition Temperatures of Tin Alloys, *Scri. Metall.* **7**, 1011–1017 (1973).

[SWA81] SWARTZENDRUBER, L. J., BENNETT, L. H., IVES, L. K., and SHULL, R. D., The Titanium–Aluminum Phase Diagram: The α/α_2 Phase Boundary, *Mater. Sci. Eng.* **51**, P1–P9 (1981).

[SWE62] SWENSON, C. A., The Temperature and Pressure Dependence of Critical Field Curves, *IBM J. Res. Dev.* **6**, 82–83 (1962).

T

[TAD80] TADA, E., ANDO, T., OKA, K., and SHIMAMOTO, S., Superconducting Characteristics of Nb–Ti Alloy in High Magnetic Field, *Report No. JAERI-M8785*, Tokai Research Establishment, JAERI, Feb. 1980.

[TAM67] TAMAKI, S., and TAKEUCHI, S., Localized Impurity States in Liquid Metals: Paramagnetic Susceptibilities of Manganese in Liquid Metals, *J. Phys. Soc. Jpn* **22**, 1042–1045 (1967).

[TAN62] TANIGUCHI, S., TEBBLE, R. S., and WILLIAMS, D. E. G., The Magnetic Susceptibilities of Some Transition Metal Alloys and the Corresponding Density of States Curves, *Proc. Roy. Soc. London, Ser. A*: **265**, 502–518 (1962).

[TAY68] TAYLOR, M. T., WOOLCOCK, A., and BARBER, A. C., Strengthening Superconducting Composite Conductors for Large Magnet Construction, *Cryogenics* **8**, 317–319 (1968).

[TED79] TEDROW, P. M., and MESERVEY, R., Experimental Test of the Theory of High-Field Superconductivity, *Phys. Rev. Lett.* **43**, 384–387 (1979).

[TEW63] TEWORDT, L., Theory of the Intrinsic Electronic Thermal Conductivity of Superconductors, *Phys. Rev.* **129**, 657–663 (1963).

[THO63] THORSEN, A. C., and JOSEPH, A. S., de Haas–van Alphen Effect in Zirconium, *Phys. Rev.* **131**, 2078–2081 (1963).

[THO75] THOMPSON, S. J., and FLEWITT, P. E. J., The Defect Structure and Superconducting Transition of Cold-Worked Niobium, *J. Less-Common Metals* **40**, 269–283 (1975).

[THO75[a]] THOM, R., *Structural Stability and Morphogenesis*, Benjamin, Reading, Massachusetts, 1975.

[TIM56] TIMOSHENKO, S., *Strength of Materials, Part II: Advanced Theory and Problems*, Third Edition, Robert E. Kreiger Publishing Co., Huntington, New York, 1956.

[TIN75] TINKHAM, M., *Introduction to Superconductivity*, McGraw-Hill Book Company, New York, 1975.

[TIT64] TITTMANN, B., HAMILTON, D., and JAYARAMAN, A., Superconducting Behavior of Pressure-Induced Zirconium Polymorph, *J. Appl. Phys.* **35**, 732–733 (1964).

[TOU67] TOULOUKIAN, Y. S., (ed.) *Thermophysical Properties of High Temperature Solid Materials, Vol. 2, Part 1: Nonferrous Binary Alloys*, The MacMillan Co., New York, 1967.

[TOU75] TOULOUKIAN, Y. S., KIRBY, R. K., TAYLOR, R. E., and DESAI, P. D., Thermal Expansion; Metallic Elements and Alloys, in *Thermophysical Properties of Matter*, Vol. 12, Plenum Press, New York, 1975.

[TOU77] TOULOUKIAN, Y. S., KIRBY, R. K., TAYLOR, R. E., and LEE, T. Y. R., Thermal Expansion; Nonmetallic Solids, in *Thermophysical Properties of Matter*, Vol. 13, Plenum Press, New York, 1977.

[TRA66] TRÄUBLE, H., and ESSMANN, U., High-Resolution Method for the Investigation of the Magnetic Structures of Superconductors, *Phys. Status Solidi* **18**, 813–828 (1966).

[TRA67] TRÄUBLE, H., and ESSMANN, U., Observation of the Magnetic Structure of Type II Superconductors, *Phys. Status Solidi* **20**, 95–111 (1967).

[TUM70] TUMMALA, R. R., and FRIEDBERG, A. L., Thermal Expansion of Composite Materials, *J. Appl. Phys.* **44**, 5104–5107 (1970).

[TWE64] TWEEDALE, J. G., *The Mechanical Properties of Metals, Assessment and Significance*, American Elsevier Publishing Co., New York, 1964.

[TYS75] TYSON, W. R., Solution Hardening by Interstitials in Close-Packed Metals, in *Physics of Solid Solution Strengthening*, ed. by E. W. Collings and H. L. Gegel, Plenum Press, New York, 1975, pp. 47–77.

U

[UIT77] VAN UITERT, L. G., O'BRYAN, H. M., LINES, M. E., GUGGENHEIM, H. J., and ZYDZIK, G., Thermal Expansion—An Empirical Correlation, *Mat. Res. Bull.* **12**, 261–268 (1977).

[ULL66] ULLMAIER, H. A., AC Measurements on Hard Superconductors, *Phys. Status Solidi* **17**, 631–643 (1966).

[ULL66[a]] ULLMAIER, H. A., and GAUSTER, W. F., Some Surface Current Phenomena in Type-II Superconductors, *J. Appl. Phys.* **37**, 4519–4521 (1966).

[ULY64] UL'YANOV, R. A., and KOVTUN, S. F., Effect of Alloying on the Volume Resistivity of Titanium, *Phys. Met. Metallogr.* **17**, No. 4, 24–30 (1964), [Transl. of *Fiz. Metal. Metalloved.* **17**, 505–511 (1964)].

[UML67] UMLAUF, E., Wärmeleitung und kalorische Effekte in supraleitendem Niobium, *Z. Phys.* **206**, 415–430 (1967).

[UPT72] UPTON, P. E., Low Temperature Specific Heats of Titanium–Vanadium Alloys, *M.S. Thesis*, 1972, Wichita State University, Wichita, Kansas.

V

[VAN64] VAN DER HOEVEN, B. J. C., Jr., and KEESOM, P. H., Specific Heat of Niobium Between 0.4 and 4.2 K, *Phys. Rev. A*: **134**, 1320–1321 (1964).

[VAN74] VAN DEGRIFT, C. T., A Sensitive Displacement Transducer Using an Extremely Reentrant 84 MHz Cavity Oscillator, *Rev. Sci. Instrum.* **45**, 1171–1172 (1974).

[VAN80] VAN SCIVER, S. W., and HARTWIG, K. T., Boundary Resistance in Aluminum Stabilized Conductors, in *Advances in Cryogenic Engineering* (*Materials*), Vol. 26, ed. by A. F. Clark and R. P. Reed, Plenum Press, New York, 1980, pp. 499–504.

[VAR56] VARLEY, J. H. O., The Thermal Expansion of Pure Metals and the Possibility of Negative Coefficients of Volume Expansion, *Proc. Roy. Soc. London, Ser. A*: **237**, 413–421 (1956).

[VAR76] VARMA, C. M., and DYNES, R. C., Empirical Relations in Transition Metal Superconductivity, in *Superconductivity in d- and f-Band Metals* (Proc. 2nd Rochester Conference, 1976), ed. by D. H. Douglass, Plenum Press, New York, 1976, pp. 507–533.

[VER76] VERVIN, B. I., PARKHOMENKO, T. A., PUSTALOV, V. V., and STARTSEV, V. I., Low-Temperature Plasticity of Superconducting Materials of Niobium–Titanium Alloys, *Sov. Phys. Dokl.* **21**, No. 5, 276–277 (1976), [Transl. of *Dokl. Akad. Nauk SSSR*, **228**, 586–589 (1976)].

[VER80] VERKIN, B. I., personal communication.

[VET65] VETRANO, J. B., and BOOM, R. W., High Critical Current Superconducting Titanium–Niobium Alloy, *J. Appl. Phys.* **36**, 1179–1180 (1965).

[VET68] VETRANO, J. B., GUTHRIE, G. L., KISSINGER, H. E., BRIMHALL, J. L., and MASTEL, B., Superconductivity Critical Current Densities in Ti–V Alloys, *J. Appl. Phys.* **39**, 2524–2528 (1968).

[VET70] VETELINO, J. F., NAMJOSHI, K. V., and MITRA, S. S., Mode-Grüneisen Parameters and Thermal-Expansion Coefficient of NaCl, CsCl, and Zinc-Blende-Type Crystals, *J. Appl. Phys.* **41**, 5141–5144 (1970).

[VIJ68] VAN VIJFEIJKEN, A. G., On the Theory of Vortices in Type-II Superconductors, *Phillips Research Report, Suppl. No. 8*, 1–114 (1968).

[VOH79] VOHRA, Y. K., SIKKA, S. K., and CHIDAMBARAM, R., Electronic Structure of Omega Phase of Titanium and Zirconium, *J. Phys. F: Metal Phys.* **9**, 1771–1782 (1979).

[VOI68] VOIGT, H., The Spatial Distribution of Magnetic Fields and the Critical Current Density in Superconducting Lead-Bismuth Alloys, *Z. Phys.* **213**, 119–151 (1968).

W

[WAD80] WADA, H., TACHIKAWA, K., and ROSE, R. M., A Metallurgical Study on Superconducting Ti–Nb Binary and Ternary Alloys, in *Titanium '80: Science and Technology*, ed. by H. Kimura and O. Izumi, The Metallurgical Society of AIME, Warrendale, Pennsylvania, 1980, pp. 745–753.

[WAD81] WADA, H., TACHIKAWA, K., and KATO, T., Superconducting Properties of Ti–Nb–Hf Alloys, *IEEE Trans. Magn.* **MAG-17**, 61–64 (1981).

[WAN78] WANG, R. Y. P., and CAPPELLETTI, R. L., Specific Heat Measurements in Superconducting and Normal State Ta–Nb–V Ternary Solid Solutions, *J. Low Temp. Phys.* **32**, 481–494 (1978).

[WEB63] WEBB, W. W., Dislocations in Superconductors, *Phys. Rev. Lett.* **11**, 191–193 (1963).

[WEB73] WEBER, H. W., and RIEGLER, R., Measurement of the Flux Density Distribution in Type-II Superconductors with Several Micro-Field-Probes, *Solid State Commun.* **12**, 121–124 (1973).

[WER63] WERTHAMER, N. R., Theory of the Superconducting Transition Temperature and Energy Gap Function of Superposed Metal Films, *Phys. Rev.* **132**, 2440–2445 (1963).

[WER66] WERTHAMER, N. R., HELFAND, E., and HOHENBERG, P. C., Temperature and Purity Dependence of the Superconducting Critical Field, H_{c2}, III: Electron Spin and Spin-Orbit Effects, *Phys. Rev.* **147**, 295-302 (1966).

[WER69] WERTHAMER, N. R., The Ginzburg-Landau Equations and Their Extensions, in *Superconductivity*, ed. by R. D. Parks, Marcel Dekker, New York, 1969, pp. 321-369.

[WES67] WESTINGHOUSE ELECTRIC CORPORATION, Superconductive Alloys, *British Patent 1,089,786* (Nov. 1967).

[WES75] WESTON, W. F., Low-Temperature Elastic Constants of a Superconducting Coil Composite, *J. Appl. Phys.* **46**, 4458-4465 (1975).

[WES79] WESTERHOLT, K., BACH, H., WENDEMUTH, R., and METHFESSEL, S., High Critical Magnetic Field Superconductor La_3S_4, *Solid State Commun.* **31**, 961-965 (1979).

[WES80] WEST, A. W., and LARBALESTIER, D. C., Transmission Electron Microscopy of Commercial Filamentary Nb-Ti Superconducting Composites, in *Advances in Cryogenic Engineering (Materials)*, Vol. 26, ed. by A. F. Clark and R. P. Reed, Plenum Press, New York, 1980, pp. 471-478.

[WES81] WEST, A. W., and LARBALESTIER, D. C., Microstructure Superconducting Property Relationships in a Fermilab Nb-46.5w/oTi Filamentary Superconducting Composite, *IEEE Trans. Magn.* **MAG-17**, 65-68 (1981).

[WES82] WEST, A. W., and LARBALESTIER, D. C., Alpha-Titanium Precipitation in Niobium-Titanium, in *Advances in Cryogenic Engineering (Materials)*, Vol. 28, ed. by R. P. Reed and A. F. Clark, Plenum Press, New York, 1982, pp. 337-344.

[WES83] WEST, A. W., and LARBALESTIER, D. C., α-Ti Precipitates in High Current Density Multifilamentary Niobium Titanium Composites, *IEEE Trans. Magn.* **MAG-19**, 548-551 (1983).

[WES84] WEST, A. W., and LARBALESTIER, D. C., Microstructural Changes Produced in a Multifilamentary Nb-Ti Composite by Cold Work and Heat Treatment, *Metall. Trans.* **15A**, 843-852 (1984).

[WHI61] WHITE, G. K., Thermal Expansion of Solids at Low Temperatures, *Proc. VIIth Int. Conf. Low Temp. Phys.*, Toronto, 1960, ed. by G. M. Graham *et al.*, North-Holland, Amsterdam, 1961, pp. 685-688.

[WHI61[a]] WHITE, G. K., Measurement of Thermal Expansion at Low Temperatures, *Cryogenics* **1**, 151-158 (1961).

[WHI61[b]] WHITE, G. K., Thermal Expansion at Low Temperatures II. Electronic Component in Metals, *Philos. Mag.* **6**, 815-818 (1961).

[WHI62] WHITE, G. K., Thermal Expansion of Vanadium, Niobium, and Tantalum at Low Temperatures, *Cryogenics* **2**, 292-296 (1962).

[WHI62[a]] WHITE, G. K., Thermal Expansion at Low Temperatures—V. Dilute Alloys of Manganese in Copper, *J. Phys. Chem. Solids* **23**, 169-171 (1962).

[WHI62[b]] WHITE, G. K., Thermal Expansion at Low Temperatures IV. Normal and Superconducting Lead, *Philos. Mag.* **7**, 271-278 (1962).

[WHI63] WHITE, G. K., Thermal Expansion of Solids at Low Temperature, in *Low Temperature Physics—LT8* (Proc. 8th Int. Conf. London, UK, 1962) ed. by R. O. Davies, Butterworths, 1963, pp. 394-396.

[WHI72] WHITE, G. K., Expansion Coefficient of Coppers at 283 K and Low Temperatures, in *Thermal Expansion—1971* (AIP Conf. Proc. No. 3), ed. by M. G. Graham and H. E. Hagy, Amer. Inst. Phys., New York, 1972, pp. 59-64.

[WHI72[a]] WHITE, G. K., and COLLINS, J. G., Thermal Expansion of Copper, Silver, and Gold at Low Temperatures, *J. Low Temp. Phys.* **7**, 43-75 (1972).

[WHI73] WHITE, G. K., and COLLINS, J. G., Thermal Expansion of Alkali Halides at Low Temperatures II. Sodium, Rubidium, and Caesium Halides, *Proc. Roy. Soc. London Ser A*: **333**, 237–259 (1973).

[WHI74] WHITE, G. K., Reference Materials at Low Temperatures, in *Thermal Expansion—1973* (AIP Conf. Proc. No. 17) ed. by R. E. Taylor and G. L. Denman, Amer. Inst. Phys., New York, 1974, pp. 1–7.

[WHI74[a]] WHITE, G. K., COLLINS, J. G., SHEARD, F. W., and SMITH, T. F., Thermal Expansion of Solids at Low Temperature, in *Thermal Expansion—1973* (AIP Conf. Proc. No. 17), ed. by R. E. Taylor and G. L. Denman, Amer. Inst. Phys., New York, 1974, pp. 278–279.

[WHI74[b]] WHITE, J. J. III, Estimation of Confidence Intervals for the Nonlinear Parameters of a Least-Squares Fit. Application to Cholesteryl Nonanoate Data, *Appl. Phys. (Berlin)* **5**, 57–61 (1974).

[WHI76] WHITE, J. J. III, and COLLINGS, E. W., Analysis of Calorimetrically Observed Superconducting Transition-Temperature Enhancement in Ti–Mo(5 at.%)-Based Alloys, in *Magnetism and Magnetic Materials—1976* (AIP Conf. Proc. No. 34), ed. by J. J. Becker and G. H. Lander, Amer. Inst. Phys., New York, 1976, pp. 75–77.

[WHI83] WHITE, G. K., personal communication.

[WIL69] WILLIAMS, J. C., and BLACKBURN, M. J., The Influence of Misfit on the Morphology and Stability of the Omega Phase in Titanium-Transition Metal Alloys, *Trans. TMS-AIME* **245**, 2352–2355 (1969).

[WIL70] WILLIAMS, J. E. C., *Superconductivity and its Applications*, Pion Ltd., London, 1970.

[WIL70[a]] WILSON, M. N., WALTERS, C. R., LEWIN, J. D., SMITH, P. F., and SPURWAY, A. H. Experimental and Theoretical Studies of Filamentary Superconducting Composites, *J. Phys. D: Appl. Phys.* **3**, 1517–1583 (1970).

[WIL71] WILLIAMS, J. C., HICKMAN, B. S., and LESLIE, D. H., The Effect of Ternary Additions on the Decomposition of Metastable β-Phase Titanium Alloys, *Metall. Trans.* **2**, 477–484 (1971).

[WIL73] WILLIAMS, J. C., Kinetics and Phase Transformations: A Critical Review, in *Titanium Science and Technology*, ed. by R. I. Jaffee and H. M. Burte, Plenum Press, New York, 1973, pp. 1433–1494.

[WIL75] WILLBRAND, J., and SCHLUMP, W., Einfluss von Ausscheidungsdichte und Teilchengrösse auf die Stromtragfähigkeit von NbTi-Supraleitern, *Z. Metallkd.* **66**, 714–719 (1975).

[WIL75[a]] WILLBRAND, J., ARNDT, R., EBELING, R., and MOHS, R., Optimierung supraleitender NbTi Legierungen, *Forschungsbericht T* 75-35, Krupp Forschungsinstitut, Essen, Germany, Nov. 1975.

[WIL78] WILLIAMS, J. C., Precipitation in Titanium-Base Alloys, in *Precipitation Processes in Solids*, ed. by K. C. Russell and H. I. Aaronson, The Metallurgical Society AIME, Warrendale, Pennsylvania, 1978, pp. 191–221.

[WIL79] WILLIAMS, I. S., GOPAL, E. S. R., and STREET, R., The Specific Heat of Strained and Annealed Chromium, *J. Phys. F: Metal Phys.* **9**, 431–445 (1979).

[WIL81] WILLIAMS, I. S., GOPAL, S. R., and STREET, R., A Note on the Effect of Strain Upon the Heat Capacity of Gadolinium Near the Curie Temperature T_c = 293 K, *Phys. Status Solidi A*: **67**, 83–88 (1981).

[WIL82] WILLIAMS, J. C., Phase Transformations in Ti Alloys—A Review of Recent Developments, in *Titanium and Titanium Alloys, Scientific and Technological Aspects*, ed. by J. C. Williams and A. F. Belov, Plenum Press, New York, 1982, pp. 1477–1498.

[WIL82ª] WILLIAMS, J. C., and BELOV, A. F., (eds), *Titanium and Titanium Alloys, Scientific and Technological Aspects* (Proc. Third Int. Conf. on Titanium, Moscow, USSR), Plenum Press, New York, 1982.

[WIP65] WIPF, S. L., and LUBELL, M. S., Flux Jumping in Nb-25%Zr Under Nearly Adiabatic Conditions, *Phys. Lett.* **16**, 103-105 (1965).

[WIP67] WIPF, S. L., Magnetic Instabilities in Type-II Superconductors, *Phys. Rev.* **161**, 404-416 (1967).

[WIP70] WIPF, S. L., Flux Jumps and Stabilization, *Elektrotech. Čas.* **21**, 314-322 (1970), Proc. EUROPHYS Applied Superconductivity Colloquium, Smolenice, 1969.

[WIP78] WIPF, S. L., Stability and Degradation of Superconducting Current-Carrying Devices, *Technical Report No. LA-7275*, Dec., 1978, Los Alamos Scientific Laboratory, Los Alamos, New Mexico.

[WIP80] WIPF, S. L., personal communication.

[WIS64] WISCHMEYER, C. R., and KIM, Y. B., Flux Migration and Instabilities in Hard Superconductors, *Bull. Am. Phys. Soc.* **9**, 439 (1964).

[WIS65] WISCHMEYER, C. R., Spatial Variations of Flux Motion in Hard Superconductors, *Phys. Lett.* **19**, 543-545 (1965).

[WIT73] WITCOMB, M. J., and DEW-HUGHES, D., Superconductivity of Heat-Treated Nb-65at.%Ti Alloy, *J. Mater. Sci.* **8**, 1383-1400 (1973).

Y

[YAM67] YAMAFUJI, K., and IRIE, F., On the Concept of Pinning Force in Type II Superconductors, *Phys. Lett. A*: **25**, 387-388 (1967).

[YAS64] YASUKŌCHI, K., OGASAWARA, T., USUI, N., and USHIO, S., Magnetic Behavior in Superconducting Nb-Zr Wire, *J. Phys. Soc. Jpn* **19**, 137-138 (1964).

[YAS64ª] YASUKŌCHI, K., OGASAWARA, T., USUI, N., and USHIO, S., Magnetic Behavior and Effect of Transport Current on it in Superconducting Nb-Zr Wire, *J. Phys. Soc. Jpn* **19**, 1649-1661 (1964).

[YAS66] YASUKŌCHI, K., OGASAWARA, T., USUI, N., KOBAYASHI, H., and USHIO, S., Effect of External Current on the Magnetization of Non-Ideal Type II Superconductors, *J. Phys. Soc. Jpn* **21**, 89-99 (1966).

[YAT72] YATES, B., *Thermal Expansion*, Plenum Press, New York, 1972.

[YOS56] YOSHIDA, S., and TSUYA, Y., The Temperature Dependence of the Electrical Resistivity of the β-Phase Titanium-Molybdenum Alloys, *J. Phys. Soc. Jpn* **11**, 1206-1207 (1956).

[YOS58] YOSIDA, K., Paramagnetic Susceptibility in Superconductors, *Phys. Rev.* **110**, 769-770 (1958).

[YOS82] YOSHIDA, K., TAKAHASHI, Y., TADA, E., SHIMADA, M., TOKUCHI, A., TADA, N., and SHIMAMOTO, S., Mechanical Properties of the Japanese LCT Coil Conductor, in *Advances in Cryogenic Engineering* (*Materials*), Vol. 28, ed. by R. P. Reed and A. F. Clark, Plenum Press, New York, 1982, pp. 781-790.

Z

[ZAI82] ZAITLIN, M. P., Heat Capacity and Density of States of a Normal-Metal-Superconductor Bilayer, *Phys. Rev.* **25**, 5729-5736 (1982).

[ZEN48] ZENER, C., *Elasticity and Anelasticity of Metals*, University of Chicago Press, Chicago, 1948.

[ZHE81] ZHENG, J. Q., KETTERSON, J. B., FALCO, C. M., and SCHULLER, I. K., Superconducting and Transport Properties of NbTi Layered Metals, *Physica* (*Utrecht*) **108B**, 945–946 (1981).

[ZUB79] ZUBECK, R. B., BARBEE, T. W., Jr., GEBALLE, T. H., and CHILTON, F., Effects of Plastic Deformation on the Superconducting Specific-Heat Transition of Niobium, *J. Appl. Phys.* **50**, 6423–6436 (1979).

[ZWI65] ZWICKER, U., Einflusse von Platinmetallen auf technologische Eigenschaften und Leitfähigkeit von Titan-Niob-Legierungen, *Z. Metallkd.* **56**, 222–228 (1965).

[ZWI68] ZWICKER, U., MEIER, T., and RÖSCHEL, E., Verarbeitbarkeit, Phasengleichgewichte, Elektrische und Supraleitfähigkeit im System Niob-Titan-Zirkonium, *J. Less-Common Metals* **14**, 253–268 (1968).

[ZWI70] ZWICKER, U., LÖHBERG, R., and HELLER, W., Metallkundliche Probleme und Supraleitung bei Legierungen auf Basis Titan-Niob, die als Werkstoffe für die Herstellung von supraleitenden Magneten dienen können, *Z. Metallkd.* **61**, 836–847 (1970).

[ZWI74] ZWICKER, U., *Titan und Titanlegierungen*, Springer-Verlag, Heidelberg, 1974.

Symbols and Abbreviations

Location of Greek Letters

α	A
β	B
Γ, γ	C
Δ, δ	D
ϵ	E
η	G
Θ, θ	H
κ	K
λ	L
μ	M
ν, ξ	N
π	P
ρ	R
σ	S
τ	T
Φ, ϕ	U
χ	X
Ψ, ψ	Y
Ω, ω	Z

A, a

A	ampère.
A	displacement amplitude of an oscillating particle.
A	Zener's elastic anisotropy ratio; $A = C_{44}/C'$.
Å	angstrom unit; 1 Å = 10^{-10} m.
AC	alternating current, alternating.
AR	reduction in area (as in, for example, 75% AR).
ARR	area reduction ratio, A_i/A_f; cf. percent area reduction, $100(A_i - A_f)/A_i$.
a	lattice parameter.
at.%	atomic percent; "A − B(n at.%)" indicates n at.% of element B dissolved in A; "A − B_n" indicates the same.

α

α	abbreviation of $\alpha(T)$.
α	coefficient of linear thermal expansion.
α	hexagonal-close-packed (hcp) crystal structure in Ti-base alloys.
α	Maki–WHH paramagnetic limitation parameter; $\alpha = \sqrt{2}\, H^*_{c20}/H_{p0}$.
α_c	critical value of the above, separating regions of first-order and second-order superconducting transitions.
α_c	coefficient of linear thermal expansion of a composite.
α_e	coefficient of the electronic component of linear thermal expansion.
α_g	coefficient of the lattice component of linear thermal expansion.
α_I	isotope-shift coefficient.
α_m	ratio of eddy-current to hysteretic loss in a composite conductor.
α_{m0}	coefficient of the relationship $\alpha_m = \alpha_{m0}\dot{H}_a$.
$\alpha_{n,s}$	Maki-WHH α calculated from normal-state and superconducting-state properties, respectively.
α_2	an ordered hexagonal phase in Ti-base alloys.
α'	solute-lean, hcp, martensite.
α''	solute-rich, orthorhombic, martensite.
α^m	generalized martensite (includes α' and α'' indiscriminately).
α^e	electron–electron interaction parameter.
α^T	a temperature coefficient of electrical resistivity; $\alpha^T = (1/\rho)(d\rho/dT)_H$.
$\alpha(T)$	pinning-force parameter connecting the bulk pinning force to a function of the applied field; abbr. α.
$\alpha_c(T)$	critical value of the above; applied in (super)-critical-state hopping theory.
$\alpha^2(\omega)$	average electron–phonon interaction in T_c-theory.

B, b

B	magnetic induction.
$\langle B \rangle$	mean value of B within the sample (wall).
B_0	magnetic induction in thermal equilibrium with H_a in a reversible superconductor; $B_0 = \mu_{eq0} H_a$.
B_0	magnetic induction just below the surface of a reversible or irreversible (pinned) superconductor.
B_{c2}	magnetic induction corresponding to H_{c2}.
B'	magnetic induction at the interior wall of a tube; related to H', the field in the bore, by $B' = \mu_{eq0} H'$.
B^*	value of B_0 corresponding to H^*.
$B_z(r)$	magnetic induction in the axial direction of a cylinder (as function of r); abbr. $B(r)$.
$B(x)$	B as a function of distance into a slab.
BCS	Bardeen, Cooper, and Schrieffer (theory of superconductivity).
b_0	amplitude of an oscillating field superimposed on B_0.
b_0	constant (with the dimensions of B) in the Kim and other critical-state relationships; corresponding field strength, h_0.
bcc	body-centered cubic.
bct	body-centered tetragonal.
bracket notation	( ) lattice plane;
	{ } generalized lattice plane;
	[ ] crystallographic direction;

⟨ ⟩	generalized crystallographic direction;
[X]	concentration of X; dimension of X.

β

β	body-centered-cubic (bcc) crystal structure in Ti-base alloys.
β	angle between the applied magnetic field and the surface of a sample.
β	lattice specific heat coefficient.
$\beta_{n,s}$	same as the above, but with particular reference to the normal and superconducting states, respectively.
β_A	Abrikosov's flux-lattice number; $\beta_A = 1.16$ for a triangular lattice.
β_M	Maki combined paramagnetic limitation and spin-orbit-scattering parameter; $\beta_M^2 = 1.11\alpha^2/\lambda_{so}$.
β_v	coefficient of volume expansion.
β_e	electronic component of the above.
β'	a β-phase leaner in solute than the average.
β''	a β-phase richer in solute than the average.
β'''	a third β-phase in quaternary-alloy phase decomposition.

C, c

C	heat capacity, molar (usually) or generic, either at constant pressure or constant volume; also used with subscripts: *n* (normal state); *s* (superconducting state); *r* (rounded); *u* (unrounded); *m* (magnetic).
C_e	electronic heat capacity, molar, generic, or otherwise; also used with superscript-*v* and extra subscripts: *s* (superconducting); *n* (normal); *r* (rounded); *m* (mixed state).
C_g	lattice specific heat.
C_k	heat capacity associated with the *k*th vibrational mode.
C_m	Curie constant in magnetic susceptibility.
C_m	dimensionless model specific-heat function such as those due to Gorter and Casimir or BCS (as tabulated by Mühlschlegel).
C_M	the product $n(E_F)\langle I^2\rangle$ in McMillan's theory; see also the η of Hopfield's theory.
C_0	numerical constant in coupling-strength or T_c theory.
C_p	molar heat capacity at constant pressure.
C_v	molar heat capacity at constant volume.
C^g	heat capacity per unit mass.
C^v	heat capacity per unit volume.
C_e^v	electronic heat capacity per unit volume.
C'	elastic shear stiffness modulus; shear constant; $C' = (C_{11} - C_{12})/2$.
C_{ij}	elastic stiffness moduli (elastic constants) of either a crystal lattice or a fluxoid lattice in the Voigt notation.
CPA	coherent potential approximation.
CSIRO	Commonwealth Scientific and Industrial Research Organization (Australia).
c	concentration in general; solute concentration.
c_A	concentration of A; also written [A].
c	velocity of light.
cw	cold work.

Γ, γ

Γ	half-width at half maximum of a Gaussian distribution; $(\Gamma^2/\ln 2) = 2\sigma^2$, where σ is the standard deviation.
Γ_s	half-width at half maximum of an asymmetrical distribution.
γ	abbr. for "intermetallic compound".
γ	Euler's constant; $\gamma = 1.78$.
γ	electronic specific heat coefficient (generic or molar).
$\gamma_{n,s}$	same as above but with particular reference to normal and superconducting states, respectively.
γ^0	"bare" electronic specific heat coefficient (ignoring electron–phonon interaction).
γ_v	electronic specific heat coefficient per unit volume.
γ_G	Grüneisen "constant" or parameter.
γ_e	electronic component of γ_G.
γ_m	magnetic component of γ_G.
γ_0	Grüneisen constant at zero K.
γ_∞	Grüneisen constant in the high-temperature limit.

D, d

D	abbr. for "deformed."
D	solute or tracer diffusion coefficient (or diffusivity).
D_A	the above, emphasizing the diffusivity of atoms.
D_0	prefactor or frequency factor in the diffusivity expression.
D_e	single-electron diffusivity in proximity-effect theory.
$D_{n,s}$	thickness of normal and superconducting layers, respectively, in proximity-effect theory.
D_t	threshold layer thickness in the above theory.
$D(t)$	deviation function in coupling-strength theory.
D_{EM}	electromagnetic diffusivity.
D_{TH}	thermal diffusivity.
D_{BCS}	BCS deviation function.
DC	direct current, direct, steady.
d	diameter of the indentor impression in a hardness measurement.
d	dimension of unit cell.
d	distance between atomic planes.
d	lattice spacing of a fluxoid lattice.
d	subband or deformation-cell diameter.
d	width of a lamellar precipitate.

Δ, δ

Δ	separation of energy bands.
Δ_0	BCS half gap energy.
Δ_{00}	the above at zero K; $\Delta_{00} = 3.53\, k_B T_c$.
Δ'_{00}	half gap energy for arbitrary coupling strength; $\Delta'_{00} = \eta_{H_c} \Delta_{00}$.
Δ_{GL}	Ginzburg–Landau gap function.
Δ_{eff}	effective gap energy for a proximity-effect couple.
ΔB_0	increment of surface induction; surface field step.

ΔC	specific heat jump (usually at T_c or T_s).
ΔC_{sn}	specific heat jump at T_{sn} for a proximity-effect bilayer.
$\Delta\Phi$	amount of flux associated with a major flux movement.
ΔH_s	"surface barrier" field such that $\Delta H_s = \Delta H_{en} + \Delta H_{ex}$.
ΔH_{en}	barrier to flux *en*try.
ΔH_{ex}	barrier to flux *ex*it.
ΔT	increment of temperature.
ΔT_K	temperature drop across an interface; see also R_K.
ΔU	energy associated with some driving force in diffusion theory.
Δg	increment of specific Gibb's free energy.
Δh	field interval between flux jumps; a "local H_{fj}."
$\Delta\theta$	angular deflection.
$\Delta\rho$	increment of electrical resistivity.
δ	electron-scattering strength parameter; $0 < \delta < 1$.
δ	elongation (percent) at fracture.
δ_0	relative gap energy; $\delta_0 = 2\Delta'_{00}/k_B T$.

E, e

E	Young's modulus.
E	electric field strength.
E	energy of a band electron.
E_F	energy of the above at the Fermi level; abbr. for "Fermi level."
EDAX	energy dispersive analysis of x-rays.
e	electronic charge in e.s.u.
eV	electron-volt.
e/a	electron/atom ratio.
e.m.f.	electromotive force.

ε

ε	percent deformation; strain.

F, f

F	"finesse" of an optical resonator; $F = \pi\sqrt{\rho}/(1-\rho)$.
F	force per unit volume of an assembly of pinned fluxoids.
F	Helmholtz free energy, see also F^H.
F_c	critical (or "breakaway") value of F; $F_c + F_p = 0$.
F_D	Debye vibrational free energy.
F_p	appropriately summed bulk pinning force.
$F(\omega)$	phonon density of states.
$F(T_c)$	distribution of transition temperatures.
F^H	Helmholtz free energy; used when confusion with F, the specific force on a flux lattice, is possible.
f	frequency.
f	thermal conductivity ratio; $f = K^s_{es}/K^n_{es}$.
f	volume fraction, mole fraction, etc., frequently with subscripts; $0 < f < 1$.

f_i	volume fraction of the ith component of a mixture.
f_p	individual fluxoid pinning force; elementary pinning force.
f_Φ	individual fluxoid driving force; the net force on the individual fluxoid is $f_\Phi + f_p$.
f_L	driving force on a flux bundle; in Anderson's theory of flux creep the net force on the bundle is $f_L + f_p$.
f_n	net force on a flux bundle in hopping (creep) theory; $f_n = f_L + f_p$.
f_c	critical (or perhaps "supercritical") value of f_n; $f_c = f_{n,\mathrm{crit}}$.
f_R	fractional change in electrical resistance in phase-transformation diagnostics.
f_v	viscous force (drag) on a moving (flowing) fluxoid; $f_v = -\eta v_f$.
f^H	Helmholtz free energy per unit volume.
fcc	face-centered cubic.
fct	face-centered tetragonal.

G, g

G	elastic shear modulus.
G	gauss.
$\mathbf{G}$	in phonon scattering (umklapp process), a reciprocal lattice vector.
G	Gibbs free energy.
$G_{n,s}$	the above in the normal and superconducting states, respectively.
$G(0)$	either of the above in zero magnetic field.
$G(H_a)$	either of the above in an applied field, H_a.
G_L	longitudinal elastic shear modulus.
G.P.	Guinier-Preston (zone).
G-L	Ginzburg-Landau.
GLAG	Ginzburg-Landau-Abrikosov-Gor'kov.
g	thermal conductivity ratio; $g = K^s_{eg}/K^n_{eg}$.
g	Gibbs free energy per unit volume; free energy density; specific free energy.
$g(c)$	concentration dependence of the above.
$g_{m,n,s}$	Gibbs free energy density for the mixed, normal, and general superconducting states, respectively.
$g(0)$	the above in zero magnetic field.
$g(H_a)$	the above in an applied field H_a.

η

η	individual fluxoid flow-viscosity coefficient; fluxoid drag $= \eta v_f$.
η_B	*bulk* flux-flow-viscosity coefficient; bulk drag $= \eta_B v_f$.
η	parameter involving atomic-like properties in Hopfield's theory for superconductive interaction; see also the C_M of McMillan's theory.
η_0	parameter analogous to the above due to Bennemann and Garland.
η_{H_c}	strong-coupling modification to Δ_{00}; $\Delta'_{00} = \eta_{H_c}\Delta_{00}$.

H, h

The Critical Magnetic Field Representations

H_c	thermodynamic critical field.
H_{c1}	lower critical field.

H_{c2}	upper critical field.
H_{c3}	"third" or surface-sheath critical field.
H_r	resistively measured upper critical field.
H_u	magnetically or otherwise measured upper critical field.
H^*_{c2}	upper critical field in the absence of paramagnetic limitation; Goodman–Maki–Eilenberger upper critical field.
H^{*n}_{c2}	the above calculated from "normal-state" parameters (if a distinction need be drawn).
H^{*s}_{c2}	the above calculated from "superconducting-state" parameters.
H_p	paramagnetic limit of the upper critical field, Clogston-derived or otherwise.
H_p^{CL}	the above, Clogston-derived when a distinction need be made.
$H_{()0}$	all of the above critical fields at zero K.
H_{c20}^{WHH}	WHH-extrapolated zero-K resistive (measured) upper critical field.

H	magnetic field in general.
H^*	value of an external longitudinally applied magnetic field whose induction in a pinned superconductor just "reaches" the opposite, or shielded, wall.
H^{**}	value of a dynamic applied field under the above conditions.
H'	field within the bore of a hollow superconducting cylinder in a longitudinal external magnetic field; $B' = \mu_{eq0}H'$.
H_a	applied magnetic field strength.
H_m	amplitude of an applied AC magnetic field.
H_s	see ΔH_s.
Hy	henry.
Hz	hertz (cycles per second).
H_V	Vickers hardness number.
H_{fi}	low-field limit of the field-range of so-called "limited instability" defined by: $H_{fi} < H_a < H_{fj}$; it is also the upper shielding field for full critical state stability; instability commences above H_{fi}.
H_{fj}	threshold field for runaway instability, i.e., a flux jump.
H^*_{c1}	position of the first (rounded) maximum in the magnetization-*versus*-field-strength curve for an irreversible superconductor; $H^*_{c1} \to H_{c1}$ as $F_p \to 0$.
HT	heat treated.
H–W	Hohenberg and Werthamer's theory of the upper critical field.

Reduced Magnetic Field Representations

h	reduced magnetic field; $h = H/H_{c2}$.
h_c	reduced thermodynamic critical field; $h_c = H_c/H_{c2}$; appears as $h_c(t)$.
h^*	reduced upper critical field defined by $h^* = H_{c2}/(-dH_{c2}/dt)_{t=1}$; an equivalent definition is $h^* = 0.68H_{c2}/H^*_{c20}$.
$\bar{h}$	reduced upper critical field defined by $(\pi^2/4)\bar{h} = h^*$.
h_{c2}	reduced upper critical field defined by $h_{c2} = H^*_{c2}/H^*_{c20}$.

See Table 15-3 for further discussion of reduced and asterisk notations, respectively.

---

h	heat transfer coefficient.
h	thermal conductivity ratio; $h = K^s_{ge}/K^n_{ge}$.
h	hour(s).
h	Planck's constant.
$\hbar$	$h/2\pi$.

h_K	Kapitza thermal conductance.
h_0	amplitude of an oscillating magnetic field modulation; dh/dt is the time-rate of the modulation.
h_0	field strength corresponding to b_0 (especially Kim theory).
hcp	hexagonal close packed.

θ

θ	Debye temperature.
$\theta_{D,\infty}$	high-temperature limit of the above.
θ_W	Weiss temperature in magnetism.

I

I	electric current.
I	magnetic moment.
I.D.	inside diameter or dimension.
$\langle I^2 \rangle$	average over the Fermi surface of the squared electron–phonon matrix elements.
IBQ	ice-brine quenched; quenched into a slurry of ice in salt brine.
Intl	interstitial element such as B, C, N, or O.

J

J	electric current density.
J	joule.
J_A	flux of atoms across a plane normal to a concentration gradient dn_A/dx.
J_c	critical current density.
J_t	take-off current density.
J_i	Bessel function of the first kind of order i (cf. Y_i).
J_θ	current density in the circumferential direction in a cylindrical conductor.
$J_{\theta c}$	critical value of the above.
$J_c(\parallel)$	J_c measured with the applied magnetic field normal to the current direction but *parallel* to the flat face of a ribbon conductor.
$J_c(\perp)$	the above but with the applied field *perpendicular* to the flat face.
$J_+(r)$	$J_{\theta c}(r)$ current profile for a superconducting cylinder in the shielding critical state—H_a increasing; abbr. J_+.
$J_-(r)$	the above for the trapping critical state—H_a decreasing; abbr. J_-.

K, k

K	bulk modulus.
K_S	adiabatic bulk modulus.
K_T	isothermal bulk modulus.
K	thermal conductivity (generic).
$K(T)$	the above, emphasizing temperature dependence.
K_{sub}	thermal conductivity component or contribution according to: sub = e, electronic thermal conductivity; g, phonon thermal conductivity

eg, electron-lattice contribution (intrinsic or "ideal" electron scattering);
es, electron-impurity contribution;
ge, phonon-electron contribution;
gd, phonon-defect contribution;
gi, phonon-phonon contribution (intrinsic or "ideal" phonon scattering);
u, umklapp-scattering contribution.

$\mathbf{K}_i$ phonon wavevector.
k abbr. for 10^3.
k as a subscript in lattice-dynamical theory; a property of the kth vibrational mode.
$\mathbf{k}$ electronic wavevector.
k occasionally as abbr. for $4\pi/10$.
k decay constant for superconductive interaction in proximity-effect theory.
k_n the above in the "normal" component of a bilayer.
k_s the above in the superconducting component of a bilayer such that $k_s^{-1} \cong \xi_{GL}$.
k_B Boltzmann's constant.
ksi 10^3 psi.

κ

κ_{GL} Ginzburg-Landau parameter (valid near $t = 1$); also used with subscript-1 to emphasize its $t = 1$ value; abbr. κ.
κ_{GL}^c the above in the clean limit.
κ_{GL}^d the above in the dirty limit.
κ_{GL1} see κ_{GL}.
$\kappa_1(t)$ Maki's first "Ginzburg-Landau" parameter as function of t.
κ_{10} special value of the above at zero K.
κ_{11} the above at $t = 1$.
$\kappa_2(t)$ Maki's second parameter as function of t.
κ_{20} special value of the above at zero K.
κ_{21} the above at $t = 1$.
$\kappa_3(t)$ Maki's third parameter as function of t.
κ_{30} special value of the above at zero K.
κ_{31} the above at $t = 1$.
$\kappa_2^*(t)$ defined as $\kappa_2(t, \beta_M^2)$ with $\beta_M = 0$.
κ_i^* see Table 15-3 for further discussion of asterisk notation.

L, l

L load imposed on the indentor in a hardness measurement.
L Lorenz ratio (number or constant).
L repeat distance for the constituents of an *in situ* laminate.
LCT Large Coil Task (Japan).
l abbr. for "liquid."
l length of an object; may appear as a superscript.
l length of a pinned fluxoid.
l electronic mean-free-path length; also specialized to s-wave scattering mean-free-path when a distinction is needed, as in Eilenberger's theory.

l_{tr}	the above, emphasizing the inclusion of both s-wave and p-wave scattering.
l_{so}	mean-free-path for spin-orbit scattering.
l_g	phonon mean-free-path length.
l_{ge}	the above, referring to phonon-electron scattering.
l_i	characteristic distance between impurity centers.
l_p	quarter twist pitch of a composite conductor.
ln	abbr. for $\ln(\theta_D/1.45T_c)$ in McMillan's theory.

λ

λ	a Lamé elastic constant (see also μ).
λ	superconductor/entire-conductor volume ratio for a composite superconductor.
λ	wavelength of light.
λ	electron-phonon coupling constant or interaction parameter.
λ_s	linear correction term to the above to allow for the effect of spin fluctuations.
λ_t	total interaction parameter, emphasizing the inclusion of both electron-phonon and electron-electron effects.
λ_d	dominant phonon wavelength in thermal-conductivity theory.
λ_L	characteristic distance of interaction between a fluxoid and a pinning site.
λ_L	London penetration depth.
λ_{L0}	the above at zero K.
λ_{GL}	Ginzburg-Landau penetration depth.
λ_{GL}^c	the above in the clean limit.
λ_{GL}^d	the above in the dirty limit.
λ	s-electron scattering-frequency parameter in H-W theory.
λ_{so}	spin-orbit scattering-frequency parameter.
λ_{tr}	*s*- and *p*-electron (anisotropic) scattering-frequency parameter as in Eilenberger's theory.

M, m

M	atomic mass (in T_c theory).
M	longitudinal elastic modulus.
M	mean atomic weight, molar weight.
M	magnetization (i.e. magnetic moment, I, per unit volume); $4\pi M$ is also referred to as the magnetization.
$M_{m,n,s}$	the above for the mixed, normal, and general superconducting states, respectively.
M_s	temperature at the start of a martensitic $\beta \rightarrow \alpha'$ or $\beta \rightarrow \alpha''$ transformation; martensite-start transus.
M_{eq}	magnetization of a reversible (ideal, unpinned) superconductor in an applied magnetic field, H_a; $B_0 = \mu_{eq0}H_a = H_a + 4\pi M_{eq}$.
$M_z(r)$	magnetization as function of r of a superconducting cylinder in the axial (z) direction.
$\langle M_z(r)\rangle$	the above averaged over r; abbr. $\langle M_z\rangle$.
$\langle M\rangle_{sat}$	maximum or saturation value of the above.
$\langle M\rangle_+$	abbr. of the above in the case of an applied field *increasing* to some $H_a > H^*$.
$\langle M\rangle_-$	abbr. of the above in the case of an applied field *decreasing* from $H_a > H_{c2}$.
$M_{0t}(H)$	hysteretic magnetization of a composite conductor in the transverse direction; it is field dependent to the extent that J_c is field dependent in the field-range concerned.

$M_{0l}(H)$ the above, but in the longitudinal direction.
$M_t(H)$ total (hysteretic + eddy-current) transverse magnetization of a composite conductor.
$M_l(H)$ the above, but in the longitudinal direction.
MFTF Mirror Fusion Test Facility (USA).
m abbr. for 10^{-3}.
m abbr. for mixed state.
m electronic mass.
m^* effective electronic mass (of a conduction electron in a crystal).
m meter.
min minute.

μ

μ abbr. for 10^{-6}.
μ a Lamé elastic constant (see also λ).
μ magnetic permeability; $\mu = B/H$.
μ^* Coulomb pseudopotential for electron-electron interaction.
μ_B Bohr magneton.
μ_{eq} mixed-state differential magnetic permeability; $\mu_{eq} = dB/dH(B)$.
μ_{eq0} static equilibrium permeability of a reversible superconductor; $\mu_{eq0} = B_0/H_a$.

N, n

Density-of States Representations

$N(E)$ number of states per unit energy range centered about E (density-of-states) per mole for one direction of spin (half the total value).
$N(E_F)$ the above with respect to E_F (the Fermi density-of-states).
$n(E)$ density-of-states per atom for one direction of spin.
$n_v(E)$ $n(E)$ per unit volume.
$n(E_F)$ Fermi density-of-states per atom for one direction of spin. This is the $N(0)$ of BCS and McMillan.
$n_v(E_F)$ $n(E_F)$ per unit volume.
$n^d(E_F)$ $n(E_F)$ for d electrons.
$n^s(E_F)$ $n(E_F)$ for s electrons.
$n^{\chi}(E_F)$ $n(E_F)$ calculated from the Pauli paramagnetic susceptibility when a distinction is needed.
$n^{\gamma}(E_F)$ $n(E_F)$ calculated from the electronic specific heat coefficient when a distinction is needed.

N newton.
N metal purity code; e.g. $5N = 99.999\%$ of the nominal species.
$\mathcal{N}$ Avogadro's number.
N_p effective pinning-site density.
N_{eff} an effective electron/atom ratio.
n abbr. for 10^{-9}.
n abbr. for $n(E_F)$.
n abbr. for $n_v(E)$.
n subscript indicating normal-state property.

n	number of fluxoids per unit area; $n = B/\Phi_0$.
n	number of occupied states per atom in an electron band.
n_A	concentration of diffusing atoms; corresponding concentration gradient, dn_A/dx.
n_e	number density of conduction electrons.
n_s	number density of superelectrons.
n_W	dislocation density, especially within deformation-cell walls.

ν, ξ

ν	characteristic phonon frequency (in BCS theory).
ν	Poisson's ratio.
ν_E	Einstein lattice-vibrational frequency.
ξ	a coherence length.
ξ_0	BCS-calculated coherence length.
$\xi_{n,s}$	coherence length in "normal" and superconducting layers, respectively, under proximity-effect conditions.
ξ_{GL}	Ginzburg–Landau coherence length.
ξ_{GL}^c	the above in the clean limit.
ξ_{GL}^d	the above in the dirty limit.
$\xi_{()0}^{()}$	the three above at zero K.
$\xi(l)$	general electromagnetic coherence length.
ξ_{0P}	Pippard electromagnetic coherence length.

O

Oe	oersted.
O.D.	outside diameter or dimension.
OPW	orthogonalized plane wave.
OFHC	oxygen-free high-conductivity (grade of unalloyed Cu).

P, p

PPL	Pauli paramagnetic limitation.
p	pinning-force "efficiency factor."
p_F	Fermi momentum; $p_F = \sqrt{2mE_F}$.
p_{eff}	effective magneton number.
ppm	parts per million (usually by wt.); "wt.ppm" refers specifically to ppm by weight.
ppt	abbr. for precipitate.
psi	pounds per square inch.

Q

Q	abbr. for "quenched".
Q	heat energy.
Q_h	magnetic hysteretic energy dissipation.
Q_{hs}	the above per unit surface area.
$\dot{Q}$	heat power per unit area; heat flux; the so-called "cooling" of stability theory.
$\dot{Q}_v$	heat power dissipated per unit volume.

R, r

R	electrical resistance.
R_K	Kapitza thermal resistance.
R_0	outside radius of a tube; radius of a cylinder.
R_1	inside radius of a tube.
RF	radio frequency.
RT	room temperature.
$R(H, T)$	electrical resistance in a magnetic field H at temperature T.
RRR	residual resistance ratio.
r	a distance in a radial direction.
$\hat{r}$	unit vector in the r (or radial) direction.
$\langle r \rangle$	time-averaged displacement from rest, R_0, of a particle oscillating in an asymmetric potential well.
r	fluxoid-hopping rate.
r_0	rate-constant of the above.
r_c	critical or minimum-detectable fluxoid-hopping rate.

ρ

ρ	electrical resistivity (generic); $\rho = \sigma^{-1}$.
ρ	reflectances of the surfaces of an optical resonator.
ρ_d	mass density.
ρ_D	specific dislocation resistivity.
ρ_{eff}	effective resistivity due to flux creep.
ρ_{cr}	same as above.
ρ_f	flux-flow resistivity.
ρ_G	Gor'kov's dirtiness parameter.
ρ_n	residual (low-temperature normal-state) electrical resistivity.
ρ_i	"ideal" component of electrical resistivity due to phonon scattering.
ρ_s	"impurity" component of electrical resistivity due to impurity (e.g., solute) scattering.
$\rho_\perp$	transverse resistivity of a composite superconductor.

S

S	entropy; also with subscripts e and s for "electronic" and "superconducting," respectively.
S	signal from a pickup coil.
S	area of the Fermi surface.
S_F	area of the free-electron Fermi sphere.
S_M	slope of the curve of magnetization, $4\pi M$, *versus* applied magnetic field, H_a.
S_{ij}	elastic compliance moduli in Voigt notation.
S, S_{eq}	model vibrational stiffness constants.
SC	abbr. for "superconductor."
SM	"simple metal," such as Al, Sn, etc; or a metalloid.
SAD	selected-area diffraction; a special technique of transmission electron microscopy.
SEM	scanning electron microscopy.
SOS	spin-orbit scattering.
STQ	solution treated and quenched.

STEM	scanning transmission electron microscopy.
s	asymmetry parameter in a T_c-distribution function.
s	dimensionless stiffness-anharmonicity parameter.
s	second(s).
s	spin (quantum number).
s	subscript indicating superconducting-state property.

σ

σ	"fitting" prefactor to D_{BCS} in superconductor thermal expansion.
σ	standard deviation.
σ	electrical conductivity; $\sigma = \rho^{-1}$.
σ_s	solute or "impurity" contribution to the electrical conductivity.
σ	stress.
σ_B	the above at fracture; in some cases this may be identified with the ultimate tensile stress, Y_{Ult} or UTS.
$\sigma_{0.01}$	the "proportional limit;" the stress at 0.01% plastic strain.
$\sigma_{0.1,0.2}$	the "0.1%-," "0.2%-offset yield strengths," respectively; i.e., the stresses at 0.1% and 0.2%, respectively, plastic strain.

T, t

T	temperature in general.
T_c	superconducting transition temperature.
T_s	the above in a magnetic field, H_a
T_{c0}	the modal value in a T_c distribution.
T_i	initial temperature of a flux-jumping sample.
T_f	final temperature of the above.
T_{cn}	transition temperature of a normal metal as a consequence of its being in proximity to a superconductor.
T_{cs}	transition temperature of an isolated superconductor within the context of proximity-effect theory.
T_{sn}	transition temperature of a bilayer.
TB	tight binding (approximation).
TM	transition metal.
TEM	transmission electron microscopy.
T-T-T	"time, temperature, and transformation;" a type of diagram representing the kinetics of phase transformation.
TESPE	Toroidal energy-storage experiment (European).
t	reduced temperature in general.
t	reduced temperature; $t = T/T_c$.
t_c	reduced transition temperature in proximity-effect theory, $t_c = T_{sn}/T_{cs}$.
t_s	reduced transition temperature in a magnetic field; $t_s = T_s/T_c$.
t	thickness of a strip, ribbon, or layer; may appear as a superscript.
t	time.

τ

τ	any characteristic relaxation time.

τ	electron transport scattering relaxation time (corresponding to mean-free-path l).
τ_{tr}	the above, but corresponding to mean-free-path l_{tr}.
τ_s	electron-spin relaxation time.
τ_{so}	spin-orbit-scattering relaxation time.
τ	torque experienced by a suspended object in response to an applied magnetic field rotating in the horizontal plane (torque magnetometry).
τ_{max}	maximum value of the above.

U

U	activation energy for fluxoid diffusion.
U	internal energy.
U_0	the above at zero K.
U_D	Debye lattice-vibrational internal energy.
U_L	fluxoid/pinning-site de-pinning energy.
U_p	pinning potential; flux-pinning barrier height.
UTS	ultimate tensile stress; same as Y_{Ult}.

Φ, ϕ

Φ	magnetic flux entering a superconductor or threading a coil.
Φ_0	flux quantum; $\Phi_0 = hc/2e$.
$\varnothing$	diameter of a small wire or object; frequently used as a superscript.
ϕ	electron–electron interaction potential.

V, v

V	voltage.
V	volume, molar or otherwise, of a sample.
V	electron-(phonon)-electron pairing potential.
V_n	the above for the "normal" component of a proximity-effect bilayer.
V_s	the above for the superconducting component.
V_{app}	apparent pairing potential; $V_{app} = V/(1 + \lambda)$.
v_g	acoustic wave velocity.
v_F	velocity of conduction electrons at the Fermi surface; Fermi velocity.
v_f	velocity of flux flow.
v_{cr}	velocity of flux creep.
v_{cr0}	"initial value" of the above.
$v_{L,T,Torsion}$	velocity of longitudinal, transverse, and torsional, respectively, elastic waves.

W, w

W	abbr. for watt.
W	thermal resistivity.
W_e	electronic component of thermal resistivity.
W_g	phonon component of thermal resistivity.
W_{sub}	thermal resistivity components; see K_{sub} for definitions.

WQ water quenched.
WHH Werthamer, Helfand, and Hohenberg (theory of the upper critical field).
w diameter of a superconducting filament in a composite.
w thickness of a superconducting slab or cylinder wall.
wt.% weight percent; "A-B(n wt.%) indicates n wt.% of element B dissolved in A; "A-nB" indicates the same.

X, x

$\overline{X^n}$ nth moment of the distribution $P(x, t)$, the probability that an atom will diffuse a distance x from x_0 in time t.
x diffusion distance.
x_0 reference position of diffusing particle.

χ

χ magnetic susceptibility (generic) or in the units emu (cm^3) per mole.
χ^g magnetic susceptibility in the units emu (cm^3) per gram.
χ^v the same, in the units emu (cm^3) per cm^3—hence dimensionless.
χ_{sub} magnetic susceptibility component or contribution according to:
sub = i, ion-core diamagnetism;
L, Landau diamagnetism of the conduction electrons;
orb, orbital paramagnetism;
p, Pauli paramagnetism of the conduction electrons including the effects of electron-electron interactions;
so, spin-orbit component.
χ^{super} magnetic susceptibility component according to:
super = d, d electrons;
s, s electrons;
0, "bare" noninteracting electrons;
g, gram susceptibility;
v, volume susceptibility.
χ_G Gor'kov's function of ρ_G, a dirtiness parameter.

Y

Y yield strength (see $\sigma_{0.1}$ and $\sigma_{0.2}$).
Y_i Bessel function of the second kind of order i (cf. J_i).
Y_{Ult} ultimate tensile strength (stress), sometimes identical to σ_B.

Ψ, ψ

Ψ digamma function.
Ψ superconducting order parameter of Ginzburg-Landau theory such that $|\Psi|^2 = n_s$.
ψ reduction in area (%) at position of fracture.

Z

Z	atomic number.
$\langle Z \rangle$	mean atomic number of an alloy.

Ω, ω

Ω_{ij}	thermodynamic interaction parameter between elements i and j.
ω	angular frequency ($=2\pi f$) of an AC field, voltage, etc.
ω	ω-phase.
ω_l	angular frequency of a linear harmonic oscillator.
ω_0	maximum phonon frequency (high-frequency cutoff) expressed as a temperature.
$\langle \omega \rangle$	average phonon frequency expressed as a temperature.
$\langle \omega^2 \rangle$	mean-square of the phonon "frequencies."
$\omega_{\log}$	defined as $\omega_{\log} = 2\langle \omega \rangle - \langle \omega^2 \rangle^{1/2}$.

Index of Plotted and Tabulated Data

Subject Index

B

C

D

M

N

O

P

Q

R

S

GPSR Compliance

The European Union's (EU) General Product Safety Regulation (GPSR) is a set of rules that requires consumer products to be safe and our obligations to ensure this.

If you have any concerns about our products, you can contact us on ProductSafety@springernature.com

In case Publisher is established outside the EU, the EU authorized representative is:

Springer Nature Customer Service Center GmbH
Europaplatz 3
69115 Heidelberg, Germany

Batch number: 10712232

Printed by Printforce, the Netherlands